T0122338

Advances in Intelligent Systems and Computing

Volume 308

About this Series

The series "Advances in Intelligent Systems and Computing" contains publications on theory, applications, and design methods of Intelligent Systems and Intelligent Computing. Virtually all disciplines such as engineering, natural sciences, computer and information science, ICT, economics, business, e-commerce, environment, healthcare, life science are covered. The list of topics spans all the areas of modern intelligent systems and computing.

The publications within "Advances in Intelligent Systems and Computing" are primarily textbooks and proceedings of important conferences, symposia and congresses. They cover significant recent developments in the field, both of a foundational and applicable character. An important characteristic feature of the series is the short publication time and world-wide distribution. This permits a rapid and broad dissemination of research results.

More information about this series at http://www.springer.com/series/11156

Lakhmi C. Jain · Srikanta Patnaik
Nikhil Ichalkaranje
Editors

Intelligent Computing, Communication and Devices

Proceedings of ICCD 2014, Volume 1

Editors
Lakhmi C. Jain
Faculty of Education, Science, Technology and Mathematics
University of Canberra
Canberra, ACT
Australia

and

University of South Australia
Mawson Lakes, SA
Australia

Srikanta Patnaik
Department of Computer Science and Engineering
SOA University
Bhubaneswar, Odisha
India

Nikhil Ichalkaranje
Department of Premier and Cabinet
Office of the Chief Information Officer
Adelaide, SA
Australia

ISSN 2194-5357 ISSN 2194-5365 (electronic)
ISBN 978-81-322-2011-4 ISBN 978-81-322-2012-1 (eBook)
DOI 10.1007/978-81-322-2012-1

Library of Congress Control Number: 2014944718

Springer New Delhi Heidelberg New York Dordrecht London

Printed on acid-free paper

Springer is part of Springer Science+Business Media (www.springer.com)

Preface

The Organizing Committee is delighted to present the high-quality papers presented in the first International Conference on Intelligent Computing, Communication and Devices (ICCD 2014) organized by SOA University during 18–19, April 2014. The title was chosen as this converges three upcoming technologies for the next decade. In recent times, "Intelligence" is the buzzword for any discipline and many scholars are working in these areas.

In simple terms "intelligence" is the ability to think and learn. Looking back to its origin and development, reports say that since 1956 when artificial intelligence was formally found, it has enjoyed tremendous success over the past 60 years. During the 1960s, the subject was dominated by traditional artificial intelligence following the principle of physical symbolic system hypothesis to get great success, particularly in knowledge engineering. During the 1980s, Japan proposed the fifth generation computer system (FGCS), which is knowledge information processing forming the main part of applied artificial intelligence. During the next two decades, key technologies for the FGCS was developed such as VLSI architecture, parallel processing, logic programming, knowledge base system, applied artificial intelligence and pattern processing, etc. The last decade observed the achievements of intelligence in mainstream computer science and at the core of some systems such as Communication, Devices, Embedded Systems, Natural Language Processor, and many more.

ICCD 2014 covers all dimensions of intelligent sciences in its three tracks, namely Intelligent Computing, Intelligent Communication, and Intelligent Devices. Intelligent Computing track covers areas such as Intelligent and Distributed Computing, Intelligent Grid and Cloud Computing, Internet of Things, Soft Computing and Engineering Applications, Data Mining and Knowledge Discovery, Semantic and Web Technology, Hybrid Systems, Agent Computing, Bioinformatics, and Recommendation Systems.

At the same time, Intelligent Communication covers communication and network technologies, including mobile broadband and all optical networks that are the key to groundbreaking inventions of intelligent communication technologies. This covers Communication Hardware, Software and Networked Intelligence,

Mobile Technologies, Machine-to-Machine Communication Networks, Speech and Natural Language Processing, Routing Techniques and Network Analytics, Wireless Ad Hoc and Sensor Networks, Communications and Information Security, Signal, Image and Video Processing, Network Management, and Traffic Engineering.

The Intelligent Device is any equipment, instrument, or machine that has its own computing capability. As computing technology becomes more advanced and less expensive, it can be built into an increasing number of devices of all kinds. The Intelligent Device covers areas such as Embedded Systems, RFID, RF MEMS, VLSI Design and Electronic Devices, Analog and Mixed-Signal IC Design and Testing, MEMS and Microsystems, Solar Cells and Photonics, Nanodevices, Single Electron and Spintronics Devices, Space Electronics, and Intelligent Robotics.

The "Call for Paper" for this conference was announced in the first week of January 2014 and due to shortage of time we have to keep a very tight deadline for paper submission, i.e., 15 March 2014. But to our surprise, we received 324 papers, which were considered for review and editing. Of these 324 papers, 163 papers were accepted for presentation and publication whereas 147 papers were registered, which are covered in this proceeding.

I am sure the participants would have shared a good amount of knowledge during the two days of this conference. I wish all success in their academic endeavors.

Srikanta Patnaik

Contents

ICCD-2014 Conference Committee

Chief Patron
Prof. Manojranjan Nayak
President, Siksha 'O' Anusandhan University, India

Patron
Prof. R.P. Mohanty
Vice Chancellor, Siksha 'O' Anusandhan University, India

General Chair
Prof. Chitta Ranjan Das, Penn State University, USA

Program Chair
Prof. Srikanta Patnaik, Siksha 'O' Anusandhan University, India

Program Co-Chair
Prof. Kwang Baek Kim, Silla University, South Korea

Finance Chair
Prof. Manas Kumar Mallick, Siksha 'O' Anusandhan University, India

Convener
Dr. Alok Kumar Jagadev, Siksha 'O' Anusandhan University, India

Co-Convener
Dr. Ajit Kumar Nayak, Siksha 'O' Anusandhan University, India

International Advisory Committee
Prof. Florin Popentiu Vlădicescu, UNESCO Chair in Information and Communication Engineering, City University, London
Prof. Ishwar Sethi, Oakland University, USA
Prof. Reza Langari, Texas A&M University, USA

Prof. Rabi Mahapatra, Texas A&M University, USA
Prof. Kazumi Nakamatsu, University of Hyogo, Japan
Prof. Ma Maode, Nanyang Technological University, Singapore
Prof. Bruno Apolloni, Università degli Studi di Milano, Italy
Prof. Yadavalli Sarma, University of Pretoria, South Africa
Prof. DeSouza, Guilherme N., West University of Missouri-Columbia, Missouri
Prof. Zbigniew Michalewicz, School of Computer Science, Australia
Prof. Akshya Kumar Swain, The University of Auckland, New Zealand
Prof. Zhihua Cui, Taiyuan University of Science and Technology, China
Prof. Rajib Mall, IIT Kharagpur, India
Prof. Ashish Ghosh, ISI Kolkata, India
Prof. P.K. Dash, Siksha 'O' Anusandhan University, India
Prof. R.K. Mishra, Siksha 'O' Anusandhan University, India
Prof. P.K. Nanda, Siksha 'O' Anusandhan University, India
Prof. G.C. Bose, Siksha 'O' Anusandhan University, India
Prof. R.K. Hota, Siksha 'O' Anusandhan University, India

Program Committee
Dr. Xiaolong Li, Indiana State University, USA
Dr. Yeon Mo Yang, Kumoh University, Korea
Dr. Sugam Sharma, Iowa State University, USA
Dr. Arturo de la Escalera Hueso, Intelligent Systems Lab, Spain
Dr. Debiao He, Wuhan University, China
Dr. Nadia Nouali-Taboudjemat, Research Centre on Scientific and Technical Information, Algeria
Prof. Doo Heon Song, Yong-in SongDam College, South Korea
Prof. Baojiang Zhong, Soochow University, China
Prof. Nitaigour Mahalik, California Sate University, Fresno, CA
Prof. Guangzhi Qu, Oakland University, USA
Prof. Peng-Yeng Yin, National Chi Nan University, Taiwan
Dr. Yaser I. Jararweh, Jordan University of Science and Technology, Jordan
Dr. Jayanthi Ranjan, Information Management and Systems, Ghaziabad

Organizing Committee

Organizing Chair
Dr. Debahuti Mishra

Organizing Co-Chair
Badrinarayan Sahu
Sarada Prasanna Pati

Organizing Committee
Dr. Renu Sharma
Dr. Shazia Hasan
Trilok Nath Pandey
Sharmistha Kar
Sashikala Mishra
Saumendra Kumar Mohanty
Debasish Samal
Jyoti Mohanty
Shruti Mishra
Prabhat Kumar Sahoo
Priyabrata Pattnaik
Aneesh Wunnava
Sarbeswar Hota
Debabrata Singh
Kaberi Das

Publicity Chair
Dr. B.K. Pattanayak

Publicity Co-Chair
Bibhu Prasad Mohanty
Dr. D.B. Ramesh

Publicity Committee
Chinmaya Kumar Swain
Shrabanee Swagatika
Pandab Pradhan
Sandeep Kumar Satapathy
Subrat Kumar Nayak
Kulamala Vinod Kumar
Madhuri Rao
Meera Nayak
Susmita Panda
Jeevan Jyoti Mahakud

Technical Chair
Prof. Niva Das

Technical Co-Chair
Dr. Guru Prasad Mishra
B.M. Acharya

Technical Committee
Dr. Sukanta Sabut
Dr. Satyanarayan Bhuyan
Dr. Benudhar Sahu
Dr. Mihir Narayan Mohanty
Gyana Ranjan Patra
Smita Prava Mishra
Manoranjan Parhi
Minakhi Rout
Nibedan Panda
Ambika Prasad Mishra
Satya Ranjan Das
Sarita Mahapatra
Barnali Sahoo
Bandana Mahapatra
Alaka Nanda Tripathy

About the Editors

Dr. Nikhil Ichalkaranje is currently serving as Senior ICT Strategist, Office of the Chief Information Officer (CIO), Department of Premier and Cabinet, the Government of South Australia. He holds Doctor of Philosophy in Computer Systems Engineering from University of South Australia (March 2006). He served in various capacities such as Assistant Director, Cyber Security Programs and Security Strategy, Senior Technology Advisor, Department of Broadband, Communications and Digital Economy (DBCDE), Government of Australia. He has produced useful and meaningful research outcomes from diverse methodologies, perspectives and concepts to solve real-life problems, and support policy makers in forming evidence-based policy. He has a strong academic record of research publications in the computer science and telecommunication industry including journals, conference papers, and edited international books from renowned publishers.

Prof. Lakhmi C. Jain is with the Faculty of Education, Science, Technology, and Mathematics at the University of Canberra, Australia and University of South Australia, Australia. He is a Fellow of the Institution of Engineers, Australia. Professor Jain founded the KES International, a professional community for providing opportunities for publications, knowledge exchange, cooperation, and teaming. Involving around 5,000 researchers drawn from universities and companies worldwide, KES facilitates international cooperation and generates synergy in teaching and research. KES regularly provides networking opportunities for the professional community through one of the largest conferences of its kind in the area of KES. His interests focus on artificial intelligence paradigms and their applications in complex systems, security, e-education, e-healthcare, unmanned air vehicles, and intelligent agents.

Prof. Srikanta Patnaik is presently serving as Professor of Computer Science and Engineering, SOA University, Bhubaneswar, India. He holds Doctor of Philosophy in Engineering from Jadavpur University, India. He has published more than 80 research papers and articles in international journals and magazines of repute. He has supervised 10 research scholars for their PhDs. He has completed various funded projects as Principal Investigator from various funding agencies of India. He is Editor-in-Chief of two international journals, namely *International Journal of Information and Communication Technology* and *International Journal of Computational Vision and Robotics*, published from Inderscience Publishing House, England, and also Series Editor of Springer Book Series on *Modeling and Optimization in Science and Technology* (MOST).

Neutrosophic Logic Applied to Decision Making

Henrik Madsen, Grigore Albeanu, Bernard Burtschy and Florin Popentiu-Vladicescu

Abstract Decision making addresses the usage of various methods to select "the best", in some way, alternative strategy (from many available) when a problem is given for solving. The authors propose the usage of neutrosophic way of thinking, called also Smarandache's logic, to select a model by experts when degrees of trustability, ultrastability (falsehood), and indeterminacy are used to decide. The procedures deal with multi-attribute neutrosophic decision making and a case study on e-learning software objects is presented.

Keywords Neutrosophic sets · Decision making · e-learning

H. Madsen (✉)
Department of Informatics and Mathematical Modelling, Technical University of Denmark, Lyngby, Denmark
e-mail: hm@imm.dtu.dk

G. Albeanu
Department of Mathematics and Computer Science, Spiru Haret University, 13, Ion Ghica Str., 030045 Bucharest, Romania
e-mail: g.albeanu.mi@spiruharet.ro

B. Burtschy
Telecom ParisTech, Informatique et Réseaux, 46, rue Barrault, 75634 Paris Cedex 13, France
e-mail: bernard.burtschy@telecom-paristech.fr

F. Popentiu-Vladicescu
"UNESCO" Department, University of Oradea, 1, University Str., Oradea, Romania
e-mail: popentiu@imm.dtu.dk

L.C. Jain et al. (eds.), *Intelligent Computing, Communication and Devices*, Advances in Intelligent Systems and Computing 308,
DOI 10.1007/978-81-322-2012-1_1

1 Introduction

Neutrosophic models have been proposed by Smarandache [13], in order to measure simultaneously the truth, indeterminacy, and falsity.

This paper revisits the subject of digital learning objects (DLOs) ranking discussed by Albeanu and Popentiu [1] and Albeanu and Duda [3] providing an extension to include the indeterminacy component. Therefore, specific neutrosophic models are obtained. The reported results address the software reliability engineering subject following the modules described by Popentiu [11, 12] and Albeanu and Popentiu [2].

Linguistic variables used by many applications are considered under neutrosophy in order to capture more aspects related on their vague or incomplete specification. Such extension was proposed by Smarandache [13, 14].

Let the sets *T, I,* and *F* be the neutrosophic components and represent the membership/truth value, indeterminacy value, and nonmembership/falsehood value for a given set *A* of universe *U*, or a proposition *p*. The universe is a classical set of entities (objects, processes, tutors, trainers, learners, etc.). The set *A* (respectively, the proposition *p*) is described by *T*(*A*), *I*(*A*), and *F*(*A*) (respectively, *T*(*p*), *I*(*p*), and *F*(*p*)) where $T(\cdot)$, $I(\cdot)$, and $F(\cdot)$ are numbers, intervals, or unions of sets/intervals describing the range of membership (respectively truth) value, the range of indeterminacy (hesitation degree), and the nonmembership (respectively, falsehood) value associated to *A* (respectively, *p*).

Let *U* be a set composed by items x_i, $i = 1, 2, \ldots, n$, $n > 0$. A neutrosophic set *A* is defined as $\{(T(x_i), I(x_i), F(x_i)) \mid i = 1, 2, \ldots, n\}$. There are many ways to interpret the given components. If x_i is the *i*th learning unit and the research objective is to establish a subjective degree of some characteristic of x_i, by conducting *M* interviews, then $T(x_i)$ can be the proportion of subjects indicating x_i as "high quality," $F(x_i)$ can be defined as the proportion of subjects who labeled the item as "poor quality," and $I(x_i)$ will give the proportion of subjects that are not able to provide their appreciation. If $I(x_i) = 0$ and $F(x_i) = 1 - T(x_i)$ then the model of Zadeh [15] is obtained, $T(x_i)$ being viewed as the degree of membership to the set of "high quality" objects. If $T(x_i) + F(x_i) \leq 1$ then the model of Atanassov [8] is obtained. The neutrosophic model extends the classic interpretation to support $T(x_i) + F(x_i) + I(x_i) \geq 1$, that means paraconsistency: one subject will declare the item as "high quality" according to some criteria, as "poor quality" for other criteria, and "not decided" about some criteria. In this case, the sum of the components is greater than unity.

More possible interpretations, including the study of paradoxes, can be found in [8, 14] to mention only a few references.

2 Neutrosophic Decision Making Strategies

Following [3], two types of analysis are considered. For the first one, the instructor has to select the most appropriate learning object to be used during training/teaching activities. The second case addresses the existence of p users/experts $\{e_1, e_2, \ldots, e_p\}$ evaluating every learning object.

A nonempty set of criteria/attributes $\{c_1, c_2, \ldots, c_n\}$ taking into consideration weights indicating the importance (priority) of every criterion is used to decide about the "best" item. Both trainer/teacher and users/experts use linguistic variables or positive degrees/scales to describe the level of adequacy for every learning object related to every criterion.

The neutrosophic components are associated to every learning object related on some criterion. In [3], both membership and nonmembership levels were considered, their sum being less than or equal one, and the quantity $1 - T(\cdot) - F(\cdot)$ was called the hesitancy degree. This paper replaces the hesitancy degree with the indeterminacy $I(\cdot)$ in such a way that $T(\cdot) + F(\cdot) + I(\cdot) \geq 1$.

Let us consider the set of learning objects be $U = \{u_1, u_2, \ldots, u_m\}$. The matrix of performance is obtained and should be used to choose the "awarded" learning object. In this manner a three dimensional array of values is obtained: $\mathrm{OCE} = \{(T_{ijk}, I_{ijk}, F_{ijk}); i = 1, 2, \ldots, m; j = 1, 2, \ldots, n; k = 1, 2, \ldots, p\}$ describing the degree of acceptance, the degree of nonacceptance (rejection), and the degree of indeterminacy by every user/expert k, of the learning object i, related to the criterion j. As in [4], every user/expert e_k considers a normalized weight w_{jk} (a positive real number) for every criterion c_j according to his/her appreciation of the criterion importance, or utility. Therefore, a matrix $W = (w_{jk};\ j = 1, 2, \ldots, n;\ k = 1, 2, \ldots, p)$ is built, with $\sum_{j=1}^{n} w_{jk} = 1$. If the criterion importance is defined as constant (and applied by all experts) W is a one dimensional array. For $p = 1$, the single expert model is obtained. This is useful when a teacher/trainer has to select one learning object from a set of "similar" learning objects. However, when $p > 1$ the multi-expert case will be obtained. Such a model can be useful to assess the degree of acceptance of the learning object by students, or by members of the quality assessment team.

Let $\alpha_{ik} = \min\{w_{jk}T_{ijk} | j = 1, 2, \ldots, n\}$, $\beta_{ik} = \min\{w_{jk}I_{ijk} | j = 1, 2, \ldots, n\}$, $\gamma_{ik} = \max\{w_{jk}F_{ijk} | j = 1, 2, \ldots, n\}$ be the evaluation of the object i by the expert k. This is one possible reduction method based on min/max operators due to the single value approach for the degrees of acceptance, indeterminacy, and rejection. Other models, for generalized cases considering sets of values, or union of sets/intervals can be inspired by [14].

A decision can be made using a linear combination based on the preference levels associated to the degrees of acceptance (θ), indeterminacy (λ), and rejection (μ), where $\theta + \lambda + \mu = 1$. In this case, the available data can be reduced to $\{\omega_{ik} | i = 1, 2, \ldots, m; k = 1, 2, \ldots, p\}$, where $\omega_{ik} = \theta\alpha_{ik} + \lambda\beta_{ik} + \mu\gamma_{ik}$. The acceptance degree as a preferred criterion is a good choice ($\theta = 1$, $\lambda = 0$, and $\mu = 0$).

Any user/expert will rank the objects in decreasing order of the associated value: $\omega_{i_1k} \geq \omega_{i_2k} \geq \cdots \geq \omega_{i_nk}$, the object with index i_1 being the favorite (best alternative). If more experts are used to analyze the objects overall quality in order to establish a hierarchy then two approaches are possible: (a) a *group decision* making; (b) a *black box* super ranking starting from existing rankings.

Group decision making is based on consensus. As a natural fact, users/experts are resistant to option changing [9], and the model should consider membership and nonmembership degrees, and the indeterminacy degree:

$$f_k(x, y, z; \alpha_{ik}, \beta_{ik}, \gamma_{ik}) = 1 - \frac{1}{e^{(x-\alpha_{ik})^2+(y-\beta_{ik})^2+(y-\gamma_{ik})^2}},$$

where (*x*, *y*, *z*) describes the average neutrosophic performance (a *centroid*) when consider all users/experts and learning objects.

In the second case, a set of positive weights r_j associated to the ranking positions *j* (*j* = 1, 2, …, *m*) can be considered and, for every object, the associated score is computed:

$$s_i = \sum_{k=1}^{p} r_j \omega_{i_jk}, \quad i=1,2,\ldots,m,$$

where i_j gives the index of object ranked at *j*th by the *k*th expert/user. Using this strategy, the ranking process can be successfully solved.

An alternative approach is to consider that every expert/user *k* is able to provide a preference degree of the object *i* over object *j* (pairwise comparison), denoted by q_{ij}^k, computed as a neutrosophic similarity index. Many approaches for similarity evaluation can be proposed. Distance-based similarity analysis can use

$$d_{ij}^k = \mathrm{dist}(\langle \alpha_{ik}, \beta_{ik}, \gamma_{ik} \rangle, \langle \alpha_{jk}, \beta_{jk}, \gamma_{jk} \rangle),$$

where dist is any distance function.

When apply the aggregation method described above, a table having the row *i*, dedicated to the object o_i can be structured as: $(u_k, \langle \alpha_{ik}, \beta_{ik}, \gamma_{ik} \rangle)$, $k = 1, 2, \ldots, p$. In order to identify the similarity of the learning objects under neutrosophic evaluation by users/experts, a matrix of distances can be computed based on the neutrosophic extensions/variants of distances described by Hung et al. [7].

3 Complex Metrics for Digital Learning Object Evaluation

According to [1], DLOs are "powerful units for building learning, education, or training materials based on ICT recent developments." There are many types of digital resources to be used in e-Learning content development [10], and for every category a specific set of criteria can be used by experts (mainly e-education experts) to recommend some optimum configuration. As identified by [1], some common criteria for every type of DLO include: *content quality*, *standards compliance*, *learning goal alignment*, *accessibility* and *interaction usability*, and *reusability*. Also, specific aspects concerning the DLO media type are necessary to be considered. More aspects on multimedia quality are covered by Brotherton et al. [5], and Morales et al. [10].

A ranking of DLOs may have the following types of relevance: *algorithmic* (query-object matching), *topical* (real world—object approximation), *pertinence, cognitive or personal* (information object—information need/perceived), and *situational* (object-generator relation). Many ranking metrics were proposed in literature (see [1]). One metric is dedicated to Course-Similarity Topical (CST) Relevance Ranking: Two courses are considered similar if they have a predefined percentage of learning objects in common.

$$\mathrm{CST}(o,c) = \sum_{i=1}^{\mathrm{NC}} \mathrm{SR}(c,c_i)p(o,c_i),$$

where NC is the total number of courses, n is the total number of DLOs, o represents the learning object to be ranked, c is the course where it will be inserted or used, c_i is ith course available in the system, $p(o, c) = 1$ if and only if o appear in c, and

$$\mathrm{SR}(c_1,c_2) = \sum_{i=1}^{n} p(o_i,c_1)p(o_i,c_2).$$

A special metric considering object learning metadata fields under user/expert evaluation is Basic Personal (BP) Relevance Ranking.

Let m be the total number of objects under ranking, o represents the learning object to be ranked, f represents a field in the metadata standard and v is a value that the f field could take, and val(o, f) represents the value of the field f in the object o.

Let f_i be the ith field considered for the calculation of the metric and NF the total number of fields. The frequencies for each metadata field are calculated by:

$$\mathrm{freq}(u,f,v) = \frac{1}{m} \sum_{i=1;\ o_i \text{ used by } u}^{m} \mathrm{count}(o_i,f,v),$$

where count(o, f, v) $= 1$ if and only if val(o, f) $= v$, otherwise is equal zero. Finally, the BP metric is given by:

$$\mathrm{BP}(o,u) = \sum_{i=1;\, f_i \text{ appears in } o}^{\mathrm{NF}} \mathrm{freq}(u, f_i, \mathrm{val}(o, f_i)).$$

Similar to the calculation of the BP metric, the n objects contained in the course can be "averaged" to create a set of relative frequencies for different fields of the learning object metadata record.

4 Practical Experience

A course on software reliability engineering designed and updated using the learning object methodology was reported in [2].

This approach permits the replacement of old assets with new assets developed recently and keeping the course up-to-date. Being a specialized topic, only a small group of learners attends to this course.

Different versions of DLOs are evaluated by learners to select that version which is "best one" to be included in the next release of the course. The described methodology was applied considering data collected from 25 learners [12].

The following types of assets were evaluated: Text, Audio, Video, Webpage, Quiz, Project, and Essay. Based on neutrosophic models, a questionnaire was used during an interview on quality of available assets, depending on the content quality, standard compliance, accessibility, and interaction usability.

The data analysis provides both quality indicators and interesting experience working with neutrosophic models.

The obtained results are interesting to motivate future investigation and development of new metrics and algorithms for neutrosophic data analysis.

5 Conclusion

In this paper, the usage of neutrosophic sets was proposed in order to evaluate the quality of learning objects based on the multi-criteria approach. Single expert/learner and multi-experts/multi-learners cases were discussed and experimental results on learning objects covering subjects from reliability engineering field were reported.

Acknowledgments During this research, the authors were supported by their departments according to the institutional research strategy and associated research programs.

References

1. Albeanu, G., Popentiu-Vladicescu, F.: Recent soft computing approaches in digital learning object evaluation. In: 8th International Scientific Conference eLearning and Software for Education, pp. 16–21 (2012)
2. Albeanu, G., Popentiu-Vladicescu, F.: On designing learning objects for a software reliability engineering course. In: 7th International Scientific Conference eLearning and Software for Education, pp. 105–110
3. Albeanu, G., Duda, I.G.: Intuitionistic fuzzy approaches for quality evaluation of learning objects. In: 17th ISSAT International Conference on Reliability and Quality in Design, International Society of Science and Applied Technologies, pp. 258–262 (2011)
4. Atanassov, K.T.: Intuitionistic fuzzy sets. Physica-Verlag, Heidelberg (1999)
5. Brotherton, M.D., Huynh-Thu Q., Hands D.S., Brunnström K.: Subjective multimedia quality assessment. IEICE Trans. Fundam. **E89**(A 11) 2920–2932 (2006)
6. Hung, W.-L., Yang, M.-S.: On similarity measures between intuitionistic fuzzy sets. Int. J. Intell. Syst. **23**, 364–383 (2008)
7. Kandasamy, V.W.B., Smarandache, F., Ilanthenral, K.: Social fuzzy matrices for socials scientists. http://arxiv.org/ftp/arxiv/papers/0707/0707.4637.pdf (2007)
8. Mich, L., Fedrizzi, M., Gaio L.: Approximate reasoning in the modeling of consensus in group decisions. In: Klement, E.P., Slany., W. (eds.) Fuzzy logic in artificial intelligence, pp. 91–102 (Springer, Heidelberg 1993)
9. Morales, M.E.M., Gómez, A.D.A., García, P.F.J., Therón S.R.: Supporting the quality of learning objects through their ranking visualization, iJET—4 (Special Issue 1) "SIIE'2008", pp. 24–29 (2009)
10. Popentiu-Vladicescu, F.: Software reliability engineering. Course book of series of advanced mechatronics systems (Debrecen 2012)
11. Popentiu-Vladicescu, F.: Home page of course 02445 software reliability. http://www2.imm.dtu.dk/~popen/Software_Reliability.html (2013)
12. Smarandache, F.: A unified field in logics: neutrosophic logic. American Research Press, Rehoboth (1995)
13. Smarandache, F.: Neutrosophy, neutrosophic logic, neutrosophic set, neutrosophic probability and statistics. http://www.gallup.unm.edu/~smarandache/neutrosophy.htm
14. Zadeh, L.A.: Fuzzy sets. Inf. Control **8**(3), 338–353 (1965)

Transforming e-Government to Smart Government: A South Australian Perspective

Akhilesh Harsh and Nikhil Ichalkaranje

Abstract Over the last few years, the concept of e-government has enabled governments to serve the public using the Internet. It also allowed governments to capture, process and report on data efficiently and improve on their decision making. However, the advances in smart technologies, better informed and connected citizens, and global connected economies have created opportunities, forcing governments to rethink their role in today's society. The governments are beginning to take the concept of e-government to a new level by realising the power of data they hold to improve their services, to enable an integrated, seamless service experience, to engage with citizens, codevelop policies and implement solutions for well-being of the community and transforming themselves into 'smart government'. The emergence of social media, mobile apps, big data analytics and mashup technologies is empowering citizens to connect with government in new way. This paper discusses the steps taken by South Australian (SA) Government to transform itself into a modern, smart government through its initiatives such as open data and modern public service. The views expressed in this paper are observation of the authors, and not of the government of South Australia.

Keywords Open government · Smart government · Open data · Big data · South Australia · South Australian Government

A. Harsh (✉) · N. Ichalkaranje
Office of the Chief Information Officer, Government of South Australia,
Adelaide, Australia
e-mail: aharsh@alumni.carnegiemellon.edu

N. Ichalkaranje
e-mail: nikhil.Ichalkaranje@sa.gov.au

L.C. Jain et al. (eds.), *Intelligent Computing, Communication and Devices*,
Advances in Intelligent Systems and Computing 308,
DOI 10.1007/978-81-322-2012-1_2

1 Introduction

'Finding smarter ways of operating is a challenge for any large organisation and South Australia's public sector is no exception' [9]. South Australia (SA) is one of the few places in the world which fits with the sentiment of big enough to be good but small enough to be great due to its size, geographic location, climate, widespread availability of goods and services, low personal risk, an effective infrastructure and most importantly its people. This creates right nurturing condition for innovation. True to above sentiment, recently the South Australian Government continued its commitment to create a modern and open public sector through its Modern Public Service Policy. This policy sets a transformational agenda for the future of SA government through five perspectives: responsive, open, productive, collaborative and innovative government [4], each having its own initiatives for government agencies to lead. Open government and open data are among key initiatives which aim to provide government with social and economic benefits. To achieve this change, government is applying the right combination of technology, leadership, an engaged workforce and a culture of creativity and innovation.

2 Open Data: Social and Economic Benefits

Tim Berners-Lee known as the 'father of the Internet' endorsed the notion of sharing and distribution of data to empower the general public in a form 'that allows for the direct manipulation for various analysis, mapping, visualisation or other initiatives' [5]. There are significant developments underway by UK, US and Australian governments to release and publish datasets to the general public by dedicated online data portals. These governments have realised the social and economic benefits this may have on their policies and economies. The emergence of open technologies such as XML, JSON and GeoJSON has fuelled this trend of 'open, smart and digital government' which empowers general public to gain insights, to coproduce services and to some extent know their environment (smart cities initiatives).

Governments have been leveraging the benefit provided by the development of mobile apps by entrepreneurs, created from the data supplied by governments. Customers are willing to pay for useful apps that are service oriented and have a positive impact on their day-to-day lives. This, in turn, provides social benefits to the community and saves government resources in investing in major service industries. This coproduction trend will shape the future of government services in health, education, transport, etc. In South Australia, an 'open data declaration' was signed that commits the government to proactively release data and is seen as one of the key election issues. There are wide-ranging political impacts as this can be viewed as governments trying to open up, and allowing the public to raise questions that would not have been possible in the past. The public becomes better informed about whether the government is performing and conforming to highest ethical standards.

Currently, governments across the world are struggling with a framework to measure and understand the social and economic impacts of open data. 'The choice between either giving access to data inexpensively and widely, or restricting access and managing data as a source of revenue is widely discussed amongst the international government communities' [6]. 'The direct impact of Open Data on the EU27 economy alone was estimated at €32 Billion in 2010, with an estimated annual growth rate of 7 %' [1].

Several countries, and their jurisdictions, have matured considerably in their capacity to publish datasets and have developed guidelines and procedures to assist government agencies in publishing. Figure 1 shows the level of maturity of open data in various jurisdictions around Australia, and around the world. Currently, UK open data portal, data.gov.uk, has 17,837 datasets that are either linked or stored. The Australian data portal, data.gov.au, has 3,362 datasets. Several other jurisdictions within Australia have their own open data portals; for example, the South Australian Government's open data portal data.sa.gov.au has 233 datasets and is constantly being enhanced to accommodate more datasets with diverse range of formats including geospatial data as shown in Fig. 2.

3 Developing Smart Government Through Open Data

Smart government is 'the implementation of a set of business processes and underlying information technology capabilities that enable information to flow seamlessly across government agencies and programs to become intuitive in providing high quality citizen services across all government programs and activity domains' [8]. Open data are instrumental in the transformation from e-government to smart government.

Open data portals usually provide access to data in two ways: stored or linked. Some datasets are stored on the data portal, whereas some datasets are linked to various government agency sites. Normally, it is up to the government agencies to identify which datasets should be published, as they are the custodians of the data. The challenge for open data Web portals such as South Australia's data.sa.gov.au is to ensure that the growth in the numbers of dataset continues, along with the real-time linked datasets. Another challenge is to empower citizens to be able to combine datasets seamlessly to create unique insights.

Around the world, governments are faced with a cultural challenge to implement their open data and open government initiatives. The closed culture within government, which is caused by a general fear of the disclosure of government failures and any resultant democratic impact, is the biggest challenge for transforming into an open government. To overcome this barrier, successful governments are taking measured actions in the following areas:

- Community Engagement and Coproduction
- Financial Investment

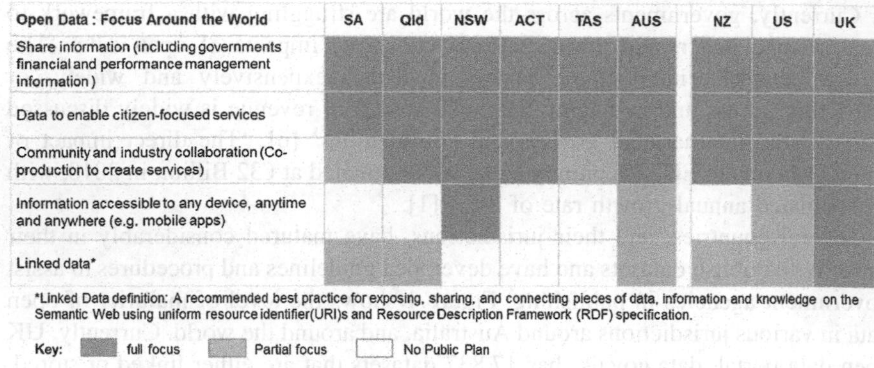

Fig. 1 Open data focus around the world

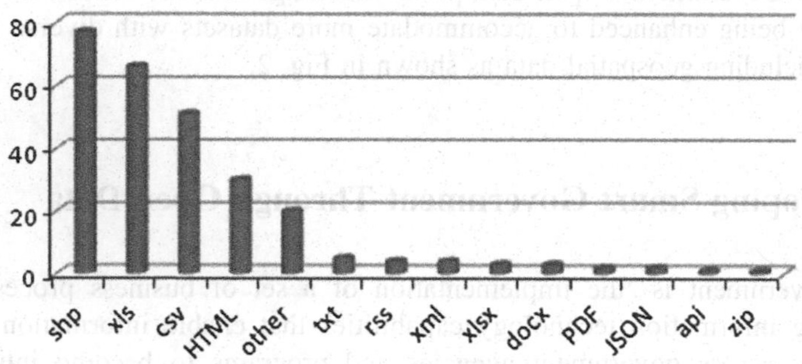

Fig. 2 Number of formats published on data.sa.gov portal

- Automation
- Collaboration
- Governance

3.1 Community Engagement and Coproduction

Community engagement is one of the most powerful ways of improving and transforming governments' services. It allows government agencies to identify the needs of the people and provide their services accordingly. It is where customers drive the solutions, not the government. It also motivates government agencies to allow customers to develop innovative solutions. One of the marketing methods that has been highly successful not only for customers is events such as Unleashed and GovHack, where over a single weekend, an event provides an 'opportunity for web and application developers, open data and visualisation gurus, user experience folk, accessibility peeps, augmented reality-ists and mobile masters to create new mashups, data visualisations and apps.' [10] using licensed open data, released by the governments.

At times, government agencies can be very protective and secretive about their data due to various reasons such as privacy, confidentiality, intellectual property, etc. The graph below shows the growth in number of datasets on the South Australian Government Open Data Portal [2]. Initially, there was a slow growth in the number of datasets as South Australian Government agencies were not sure of the advantages of publishing their data. However, just prior to the 'Unleashed' event held from 11 July 2013 to 13 July 2013, there was a sharp increase in the number of datasets published. This is because the South Australian Government agencies realised the potential of getting something created that could assist in providing Governments' services. This also allows customers to provide feedback of what their needs and requirements are Fig. 3.

3.2 Financial Investment

Government agencies around the world have been facing the challenge of finding new ways to fund the open data exercise. It is quite easy for government agencies to publish datasets as once-off. However, smart governments would provide endless supply of real-time data that can be used in a variety of government services such as e-transport, e-health or e-education. To satisfy this, government agencies require resources to maintain, and further develop strategies to ensure that they meet the changing demands of users. Currently, in South Australia, government agencies are funding the publishing of datasets without any additional funding and are investigating various funding models to sustain this, whereas the Canadian Government recently announced funding to support an initiative that will

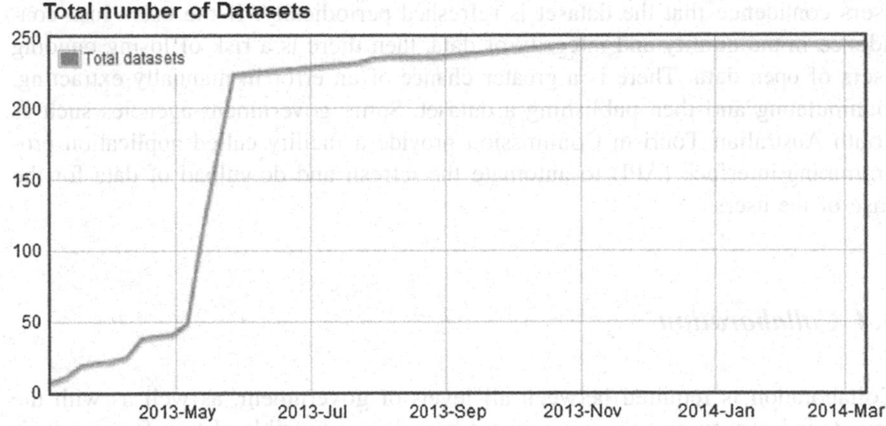

Fig. 3 Total number of datasets

assist all levels of Government to publish datasets. 'The Open Data Institute will work with governments, academic institutions and the private sector to solve challenges facing Open Government efforts and realise the full potential of Open Data. These partners will work on development of common standards, the integration of data from different levels of government and the commercialisation of data, allowing Canadians to derive greater economic benefit from datasets that are made available by all levels of government' [3].

Apart from conducting further research, individual government agencies that produce and supply data require adequate resources to maintain the infrastructure and produce newer datasets. There are a range of costs associated with maintaining and supplying data including hardware, software licensing, etc. To address this issue of ongoing costs to deliver and maintain datasets, government agencies must foresee and plan. Every business case for a new system should incorporate strategies to automate the release of data, as well as strategies to ensure that controls are in place to ensure the validity and integrity of data is maintained.

3.3 Automation

Automation is crucial in identifying and maintaining datasets, to make the process more efficient in updating data, and hence improving the data quality. Automation is the only method to provide real-time data that can be replied upon. Appropriate automated mechanisms can be implemented in data repositories and data registers that can identify and validate datasets which can be published. For example, if a dataset qualifies against a range of set requirements such as licensing, copyrights, third party, privacy, intellectual property, etc., then these datasets can be automated to be published. This reduces the timeframe in identifying and publishing datasets.

Automation can also assist in ensuring that data are up to date and provides users confidence that the dataset is refreshed periodically. If the users lose confidence in the quality and integrity of data, then there is a risk of losing ongoing users of open data. There is a greater chance of an error in manually extracting, manipulating and then publishing a dataset. Some government agencies such as South Australian Tourism Commission provide a facility called application programming interface (API) to automate the refresh and download of data for the ease of the users.

3.4 Collaboration

Collaboration is required between all levels of government, as well as with the private industry to give the government best chance possible of transforming their services. For users to mashup data or develop a product that provides integrated

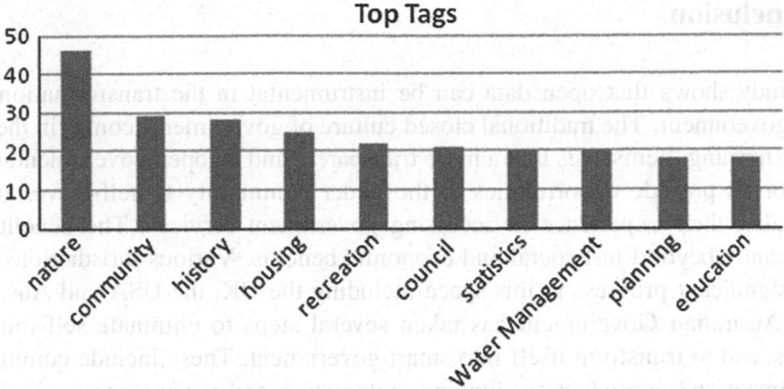

Fig. 4 Top tags with each datasets

service, it is important that appropriate data are available from various domains. Currently, as shown in Fig. 4, the South Australian Government data portal has variety of datasets, with 'nature' being the top tag. 'Tags' are keywords or terms that are associated with each dataset. Each dataset can have one or more 'tags'. However, there are hardly any datasets related to 'health', which reduces the possibility of conducting any 'health'-related analysis, and develop any meaningful services using these data. To transform from e-government to smart government, agencies must collaborate to provide a seamless integrated service to its citizens across all its domains.

3.5 Governance

It is extremely important to ensure that appropriate governance models are in place to ensure that specific areas or individuals are accountable for certain aspects of publishing data. For ongoing publishing, it is important to identify and allocate the roles which include data authority, data owner, data custodian, data steward, data publisher, etc. Appropriate governance practices can help government agencies collaborate and assist each other. As open data portals contain datasets from various agencies including education, health, spatial and community data, it is extremely important to have a cohesive and a unified governance model. 'Agencies and inter-agency groups must review and, where appropriate, revise existing policies and procedures to strengthen their data management and release practices to ensure consistency' [7].

4 Conclusion

This study shows that open data can be instrumental in the transformation to a smart government. The traditional closed culture of governments comes in the way of transforming themselves into a more transparent and an open government. Open data portals provide opportunities to the wider community to self-serve, and to personalise their experience in accessing government services. The benefits are significantly beyond just social and economic benefits. Various jurisdictions have made significant progress in this space including the UK, the USA and Australia. South Australian Government has taken several steps to eliminate self-imposed barriers, and to transform itself into smart government. These include community engagement and coproduction, funding, automation and governance.

References

1. Capgemini Worldwide: The open data economy: unlocking economic value by opening government and public data|resource (online). Available at: http://www.capgemini.com/resources/the-open-data-economy-unlocking-economic-value-by-opening-government-and-public-data (2013). Accessed 15 Mar 2014
2. Data.sa.gov.au.: Welcome—data.sa.gov.au (online). Available at: http://data.sa.gov.au (2014). Accessed 12 Feb 2014
3. Digitaljournal.com.: Federal budget provides funding to create open data institute (online). Available at: http://www.digitaljournal.com/pr/1732697 (2014). Accessed 12 Feb 2014
4. Dpc.sa.gov.au.: Office for Public Sector Renewal|dpc.sa.gov.au. (online). Available at: http://dpc.sa.gov.au/renewal (2014) Accessed 8 Apr 2014
5. Gurstein's Community Informatics: open data: empowering the empowered or effective data use for everyone? (online). Available at: http://gurstein.wordpress.com/2010/09/02/open-data-empowering-the-empowered-or-effective-data-use-for-everyone/ (2014). Accessed: 12 Feb 2014
6. Open data: an international comparison of strategies (online). Available at: http://www.epractice.eu/files/European%20Journal%20epractice%20Volume%2012_1.pdf (2014) (Accessed: 8 Apr 2014)
7. Project-open-data.github.io.: Project open data (online). Available at: http://project-open-data.github.io/policy-memo/ (2014). Accessed 12 Feb 2014
8. Rubel, T.: Smart government: creating more effective information and services (online). Available at: http://www.govdelivery.com/pdfs/IDC_govt_insights_Thom_Rubel.pdf (2014). (Accessed: 15 Mar 2014)
9. Stronger.sa.gov.au.: Building a stronger South Australia|Government of South Australia (online). Available at: http://stronger.sa.gov.au/ (2014) Accessed 6 Apr 2014
10. Unleashedadelaide.dptiapps.com.au.: Unleashed Adelaide|South Australian node of GovHack (online). Available at: http://Unleashedadelaide.dptiapps.com.au (2014). Accessed 12 Feb 2014

An Online Content Syndication Tool with Built-in Search and Social Sharing

Zijie Tang and Kun Ma

Abstract Information Syndication is a convenient way to access information in the Internet age. There are many websites and applications providing information subscription service. However, there are so many drawbacks of current methods, such as content from various media is very complex, so that these services cannot meet the requirements of users. We aim to meet the needs of users in all aspects, such as high retrieval speed, filtering out advertisements, and soft-text, automatically push the information people are interested into their smart devices. First, we design a RSS source listener to grab RSS feeds in time. Second, we have developed the user interface to provide subscription service.

Keywords Information syndication · RSS · NoSQL · Listener

1 Introduction

RSS is the abbreviation Really Simple Syndication, An RSS feed (often called channel) includes full or summarized text, and its metadata with the compatible XML file format [1]. Users can get information by the RSS reader or aggregator instead of browsing heterogeneous websites. Currently, there are several ways to obtain information through the Internet, such as a search engine, surfing the Internet, and social network sites [2].

Z. Tang · K. Ma (✉)
School of Information Science and Engineering, University of Jinan, Jinan, China
e-mail: ise_mak@ujn.edu.cn

L.C. Jain et al. (eds.), *Intelligent Computing, Communication and Devices*,
Advances in Intelligent Systems and Computing 308,
DOI 10.1007/978-81-322-2012-1_3

2 Architecture

2.1 RSS Source Listener

RSS source listener is the most basic and important part of this system, it is responsible for monitoring all of the RSS sources which from spider or user to add, then capturing all this sources' changes and store them in the storage. We design a RSS spider just like Googlebot to make RSS source gathered. The RSS spider will search "News RSS," "Technology RSS" or other similar keywords through search engine and analyze the results page for RSS source. After we get massive RSS sources, we then classify their contents. Finally, we put them into RSS source storage.

The next step is to get RSS Feeds from a RSS Source. We propose an optimized listening algorithm to help us complete this task efficiently. Initially, our listeners read the latest information of the RSS source in a short time interval. With the increase of reading times, it will generate the success rate of updates that is the frequency of updating of each RSS Source. For low frequencies of updated source, we enlarge the interval time to read, so that we can reduce the number of concurrent and the pressure of the server (Fig. 1).

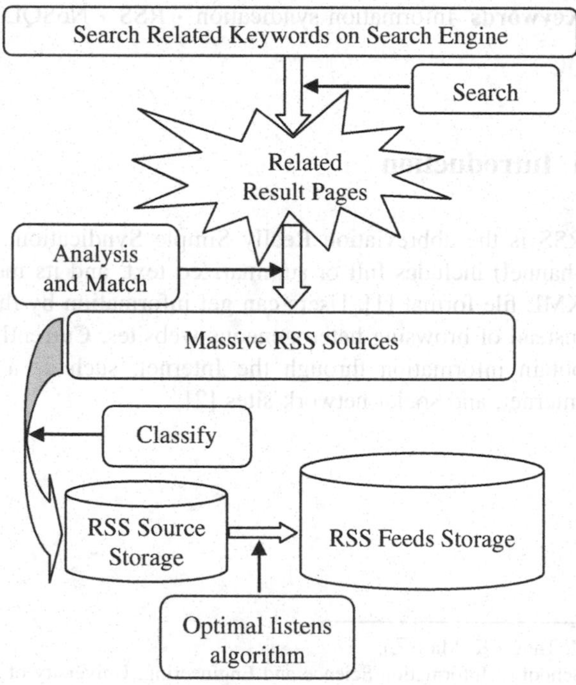

Fig. 1 RSS listener workflow

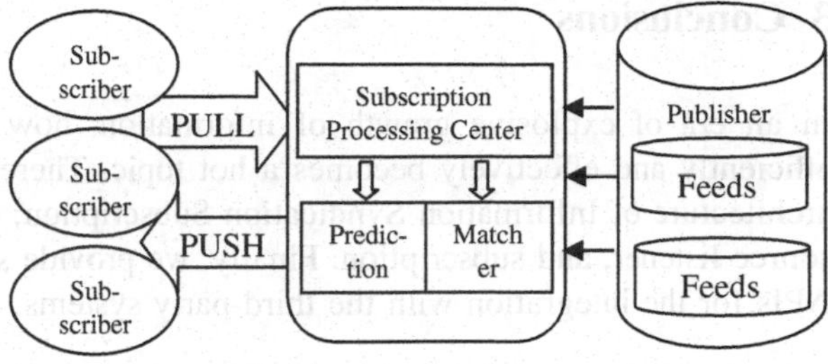

Fig. 2 Architecture of subscription

2.2 *Subscription*

First, we abstract a concept model of subscription from our process. The RSS feeds that are stored in the storage act as a publisher, and the subscriber can pull the information on their own initiative can also receive the information from the system push (Fig. 2).

Thus, the subscription processing center is the focus of our design, which is represented by two important components—Router and Matcher. Prediction is responsible for searching massive database efficiently. Matcher is responsible for matching precision.

Another feature is the combination of push and pull [3]. When a user subscribes to a source or topic, users can search for and read to pull the content they are interested, compared with the traditional search engines that return a large number of results based on keywords. We have done much work on the analysis of natural language to better understand the needs of users. On the other hand, our system picked the hottest and latest information push to user's smart device.

2.3 *RESTful Web Service APIs*

We design the core business to implement the different clients (desktop client, mobile client, and Web user interface) using RESTful Service API. This allows for easy integration with existing third-party system. The communication uses the form in API with the JavaScript object notation (JSON) [4] format to communicate with each other. Table 1 summarizes its provided functionalities.

Table 1 RSSReader service resources

Resources	URL	Parameters	Description
List of RSS sources	GET:/ rss_sources/	Secret_key=?&rss_name=?	Find the RSS sources by name
List of RSS feeds	GET:/ rss_feeds/	Secret_key=?&keywords=?	Find the RSS feeds by name
List of RSS sources	POST:/ rss_feeds/	Secret_key=?&keywords=?	Post a RSS source

3 Conclusions

In an era of explosive growth of information, how to access the information efficiently and effectively becomes a hot topic. Therefore, we have proposed an architecture of Information Syndication Subscription, which is composed of RSS source listener, and subscription. Finally, we provide some RESTful Web service APIs for the integration with the third-party systems.

Acknowledgment This work was supported by the Doctoral Fund of University of Jinan (XBS1237), the Student Research Training of University of Jinan (SRT13310), and the Teaching Research Project of University of Jinan (J1344).

References

1. Board, Advisory RSS: RSS 2.0 Specification. (2007)
2. Luo, W., Qi, X., Hengartner, U.: Facecloak: an architecture for user privacy on social networking sites. In International Conference on Computational Science and Engineering CSE'09, vol. 3, IEEE (2009)
3. Kendall, J.E., Kendall, K.E.: Information delivery systems: an exploration of web pull and push technologies. Commun. AIS **1**(4es), 1 (1999)
4. Crockford, D.: The application/json media type for javascript object notation (json) (2006)

Customizing EPCglobal to Fit Local ONS Requirements

Shiju Sathyadevan, C.A. Akhila and M.K. Jinesh

Abstract Internet of Things (IoT) is the new network of physical object that has the ability to automatically transfer data over a network. This paper proposes an architecture to extend the current object identification scheme in order to custom build the same to uniquely identify each object associated with an RFID tag. The electronic product code (EPC) is a very popularly used identification scheme to identify objects and is stored in the RFID tags. The work described in this paper is based on the EPCglobal framework. Our research study focused primarily on extending the EPCglobal framework, thereby defining a customized identification scheme for each object in an IoT platform. A lookup called object naming service (ONS) is used to locate information about these objects in the EPC network. Object name service makes use of Internet's existing domain name system (DNS) for looking up information about an EPC.

Keywords Radio-frequency identification (RFID) · Electronic product code (EPC) · Object name service (ONS) · Name authority pointer (NapTr) · EPC tag data standard

1 Introduction

The Internet of Things (IoT) is defined as a collection of uniquely identifiable physical objects that has the ability to automatically transfer data such as information about the objects or real-time sensor data into the Internet [1, 2]. Radio-

---

S. Sathyadevan (✉) · C.A. Akhila · M.K. Jinesh
Amrita Center for Cyber Security, Amrita Vishwa Vidyapeetham, Amritapuri, Kollam 690525, India
e-mail: shiju.s@am.amrita.edu

C.A. Akhila
e-mail: akhila.ca@gmail.com

M.K. Jinesh
e-mail: jinesh@am.amrita.edu

L.C. Jain et al. (eds.), *Intelligent Computing, Communication and Devices*, Advances in Intelligent Systems and Computing 308,
DOI 10.1007/978-81-322-2012-1_4

frequency identification (RFID) [3] is considered as the endow of IoT. RFID is the method of communication in the IoT platform, although it also may contain other sensor technologies or wireless technologies. RFID is not a new technology. It is a technology that uses radio wave communication to exchange data between a reader device and a small electronic tag embedded in an object for the purposes of identification. The tags serve like barcodes or magnetic stripes on modern credit cards. They contain a unique identification about the object carrying the tag. The unique identification serial number contained in each RFID tag is called electronic product code (EPC).

EPCglobal Inc is a subsidiary of the global not-for-profit standards organization GS1, and it supports the global adoption of the electronic product code (EPC) and related industry-driven standards. The electronic product code can universally identify every physical object as it is possible to assign a unique identifier to each one of them. EPCs are not designed solely for use with RFID data carriers, but normally, it is used along with the RFID.

EPCglobal network [4] architecture uses existing Internet standards and infrastructure, and object name service (ONS) uses the Internet's existing domain name system (DNS) for looking up information about an electronic product code [5]. The following section describes how to identify each object uniquely in the IoT platform using the ONS described by EPCglobal.

2 EPC Framework

The electronic product code is an object identification scheme that could uniquely identify objects associated with an RFID tag [6]. This unique number is used to identify objects and get more information about that object through the appropriate lookups using Internet-based technologies. For that reason, the information about an object is not definitely stored on the RFID tag, but instead supplied by distributed servers on the Internet. Pure identity EPC URI is the name given to the EPC representation. The RFID tag stores the EPC in hexadecimal format. There are several standard defined by EPCglobal to translate the tag value into pure URI [7]. Object name service (ONS) is an idea put forward by the EPCglobal to transform the URI encodings into URLs. For that purpose, ONS defines a standard for the translation of EPC into domain names and then the traditional DNS [8, 9] is employed to extract the URL corresponding to the given tag. A particular type of DNS record called naming authority pointer (NAPTR) [5, 10] is used by EPC. A major disadvantage is the missing authorization and authentication connected to ONS queries [11, 14–16].

3 Related Work

The method for identifying objects using EPC along with the RFID technology is clearly defined in [5, 7]. Almost all applications that use the electronic product code rely upon RFID tags as a data carrier [12]. The EPC takes the form of uniform resource identifier (URI). Due to the memory limitation of RFID tags, EPC is not stored in RFID as its pure URI form; instead, it is encoded in a binary/hex format, which is called the "EPC binary encoding." The structure of the EPC URI syntax and binary format, also the encoding and decoding rules to allow conversion between these representations, is defined in the EPCglobal tag data standard defines [7]. The GS1 system of identification defines seven identification keys that are currently supported by the EPC identifiers. There is a well-known relationship between the EPC and GS1 key. EPC scheme sgtin for trade item corresponds to GTIN GS1 key [7]. EPC schemes that correspond to GS1 keys are sgtin, sscc, sgln, grai, giai, gdti, gsrn, usdod, gid, adi, and cpi. The following is an example of a pure identity EPC URI, for sgtin EPC scheme urn: epc:id:sgtin:00037000.06542.773346595.

Each EPC scheme provides a namespace of the identifiers that is used to identify a particular type of physical objects. The "id" namespace is used for the EPC that is encoded on the RFID tags and its service may look up using ONS. This "id" namespace is further divided into sub-namespaces based on the different schemes of physical objects including SGTIN, SSCC, etc. URI is encoded as binary/hex in RFID tags.

Let us see how the hex value on RFID tag is encoded into EPC URI. For example, hex value on RFID tag is: 30700048440663802E185523.

EPC Binary:

0011000001110000000000000100100001000100000001100110001110000000 00101110000110000101010100100011 (Table 1).

Each sub-namespaces defined by the tag data standard have different structure depending on what they identify, how they are used, and how they are assigned. The EPC tag data translation engine defines the encoding and decoding rules for EPC [7]. A description of how the encoding of sgtin-96 bit scheme works is given here. The header 00110000 (8 bits) states that the tag is encoded using sgtin-96 bit scheme. Each sub-namespace defined by the EPC has its own header value and it is unique. The filter value of three bits is additional data used for fast filtering. The three-bit partition field defines how the subsequent fields are partitioned to obtain the valid data for each company prefix and item references. For sgtin-96 bit scheme, the company prefix may vary from 40 to 20 bits (12–6 decimal digits) and the item reference number may vary from 24 to 4 bits (1–7 decimal digits). The value 100 indicates that company prefix number is encoded using 27 bits (8 decimal digits); and the item reference number is encoded using 17 bits (5 decimal digits). Finally, the serial number value is the last 38 bits (12 decimal digits). The Table 2 shows how to divide the binary tag value of sgtin-96 bits scheme in order to get the corresponding URI [7].

Table 1 EPC binary encoding to EPC URI

Header	Filter	Partition	Company prefix	Item_References	Serial Number
001100 00	011	100	000000000 001001000 010001000	0000110011000111	00000000101110 000110000101010 100100011

urn:epc:id:sgtin-96:3:00037000.06542.773346595

Table 2 The formation of sgtin-96

Header	Filter	Partition	Company prefix	Item references	Serial number
8	3	3	20–40	24–4	32

The ASCII notation of sgtin-96 bit scheme as follows urn: epc: id: sgtin: Company Prefix.Item References.Serial Number. There is an engine called tag data translation engine by Fosstrak [13] (open source RFID software platform that implements the GS1 EPC network specifications) that translates the hex encoded on the RFID tags into EPC URI.

The information about the physical object is stored in the Web and needs to get back that URL. To retrieve the URL for a particular object, the URI must convert into a domain name in order to query the DNS [8, 9]. This is the function of object name service (ONS). Since ONS contains pointer to services, a simple A record or IP address is insufficient. Therefore, ONS is implemented using DNS with NAPTR record. The NAPTR takes the form as follows [5, 10] (Table 3).

To do so, EPCglobal has reserved the sub-domain onsepc.com for ONS resolution. The procedure to construct the domain name is as follows: (1) remove urn:epc from URI; (2) remove serial number; (3)invert order of the fields; (4) replace ":" with "."; (5) append sub-domain onsepc.com. In the above example, the result of translating the EPC urn: epc: id: sgtin. 089123. 097124. 906943274877 into an ONS query is therefore encoded as 097124. 08912. sgtin.id.onsepc.com [5].

Table 3 NAPTR record

Order	Pref	Flag	Service	Regexp	Replacement
0	0	U	EPC + EPCIS	!^.*$!http://example.com/autoid/ cgi-bin/epcis.php!	.(a period)

The client uses the following procedure to retrieve the correct URL from the NAPTR record that corresponds to the given EPC. The procedure is as follows: (1) select those records with the service field names as the desired service. (2) Among that result set, select the record having lowest preference value. (3) Then, extract the service URL from the record by extracting the substring between the initial! ^.*$! and the final ! character. The extracted URL is given to the Web server and the Webpage that dovetail with the URL is displayed.

4 Proposed System and Implementation

When a company requires a unique code for their product, the current practice is that they have to register with GS1 that provides them with a unique EPC code. In the current system, only products (e.g., assets, logistic, shipping, etc.) are given unique codes, which are mainly used for identification, capturing, and sharing. However, the model proposed in this paper can be extended to provide unique codes to each, and every entity in IoT could be devices, products, vehicles, humans, animals, etc., by assigning unique header values to each of these categories. We have deployed the identification system currently employed by EPCglobal but customized it in order that it could be used locally and can be configured to include any new category. The addition of a new category will only demand the user to provide the TDT engine with the newly defined formation table and partition table to fit the new scheme. The current identification scheme by EPC Global is customized by modifying the RFID tag data and defining a scheme for conversion from RFID hex/binary to EPC.

Here, we propose a system that allows setting customizable value for RFID tag and also define the convention for translating the customizable hex/binary value into EPC URI. We extend the tag data translation engine (converts the hex value into URI) by adding the new scheme along with the existing schemes defined by the EPCglobal. Each scheme defined by the EPCglobal has its own header value to identify the type of product; for sgtin scheme (for trade item), the defined header value is 00110000[1]. The header values are unique and independent. There are some unused header values that are reserved for future use. Here, we use that header values to define the 96-bit encoding scheme, which can be used locally. The header for new custom tag value is 00111101 and is mainly defined to identify devices. Table 4 and Table 5 shows the division of binary data in order to convert them into EPC URI.

Then the URI is passed to the ONS resolver to convert it into domain name. The conversion is as follows.

1. Remove urn:epc from URI
2. Remove serial number

Table 4 The formation of customized standard (formation table)

Header	Filter	Partition	Category	Unique_id	Serial number
8 bits	3 bits	3 bits	20–40 bits	26–46 bits	16 bits

Table 5 Partition table for customized standard (partition table)

Partition value	Category		Unique_id	
	Bits	Max digit	Bits	Max digit
0	40	13	26	10
1	37	12	29	11
2	34	11	32	12
3	30	10	36	13
4	27	9	39	14
5	24	8	42	15
6	20	7	46	16

Table 6 Sample "named.config" setting file

```
zone "onsepc.com" {
        type master;
        file "/etc/bind/db.onsepc.com";
    }
```

3. Invert order of the fields
4. Replace ":" with "."
5. Append sub-domain onsepc.com.

The domain name is used to query the NAPTR record in the DNS server. For the implementation, the BIND [8] open source software on a Linux platform was used. Configuration was based on the BIND installation manual. There were no special settings used, and the server is accessible via requests at ports TCP/UDP 53 as per the default settings for DNS requests. Settings for the ONS server are stored in the "named.conf" (Table 6) file in the installation directory of BIND. An ONS server configured as a master server for the domain ".onsepc.com." Sub-domains in the DNS context are also called "zones," so the files containing the appropriate information on each server are called "zone files." NapTr is added under the sub-zone (Table 7). Using Java code, all the URLs are retrieved by setting the bind in the Linux as a primary DNS server. All the retrieved URLs are stored in a Web server (Fig. 1).

Table 7 Sample zone file

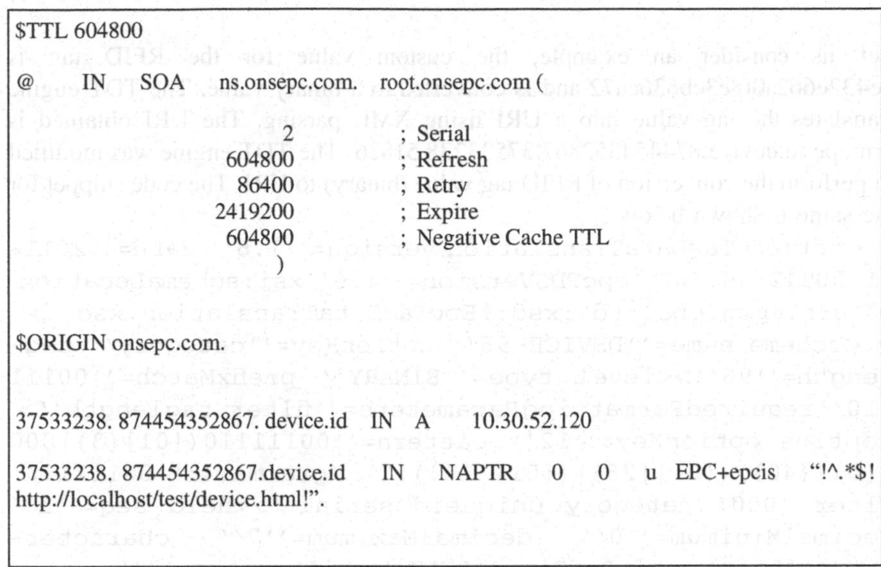

```
$TTL 604800
@       IN      SOA     ns.onsepc.com.   root.onsepc.com (
                        2               ; Serial
                   604800               ; Refresh
                    86400               ; Retry
                  2419200               ; Expire
                   604800               ; Negative Cache TTL
                        )

$ORIGIN onsepc.com.

37533238. 874454352867. device.id   IN   A   10.30.52.120

37533238. 874454352867.device.id   IN   NAPTR   0   0   u   EPC+epcis   "!^.*$!
http://localhost/test/device.html!".
```

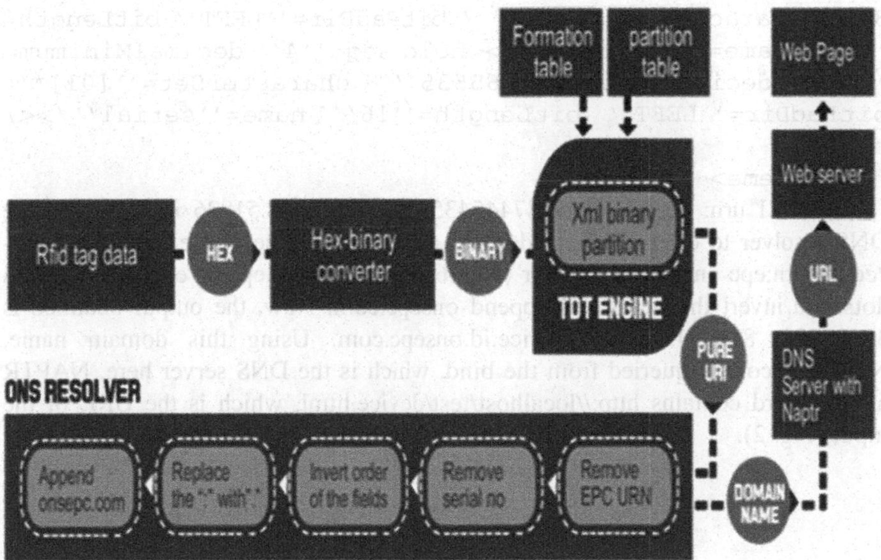

Fig. 1 The architecture diagram of proposed system

4.1 Sample Implementation

Let us consider an example, the custom value for the RFID tag is 3e432e662a0f8e3cb636ca72 and is converted to a binary value. The TDT engine translates the tag value into a URI using XML parsing. The URI obtained is urn:epc:id:device:874454352867.37533238.51826. The TDT engine was modified to perform the conversion of RFID tag value (binary) to URI. The code snippet for the same is shown below

```
<tdt:epcTagDataTranslation version=``1.6'' date=``2011-
01-20T12:20:00Z''epcTDSVersion=``1.6''xsi:schemaLocation
=``urn:epcglobal:tdt:xsd:1EpcTagDataTranslation.xsd''>
<scheme name=``DEVICE-96'' optionKey=``category'' tag-
Length=``96''><level type=``BINARY'' prefixMatch=``00111
110''requiredFormattingParameters=``filter,taglength''>
<option optionKey=``12'' pattern=``00111110([01]{3})000
([01]{40})([01]{26})([01]{16})'' grammar=```00111110'
filter '000' category uniqueid serial''><field seq=``1''
decimalMinimum=``0'' decimalMaximum=``7'' character-
Set=``[01]*'' bitPadDir=``LEFT'' bitLength=``3'' name=
``filter''/><field seq=``2'' decimalMinimum=``0'' decimal-
Maximum=``999999999999'' characterSet=``[01]*'' bitPad-
Dir=``LEFT'' bitLength=``40'' name=`` category ''/><field
seq=``3'' decimalMinimum=``0'' decimalMaximum=``67108
863'' characterSet=``[01]*'' bitPadDir=``LEFT'' bitLength=
``26'' name=`uniqueid''/><field seq=``4'' decimalMinimum=
``0'' decimalMaximum=``65535'' characterSet=``[01]*''
bitPadDir=``LEFT'' bitLength=``16'' name=``serial''/></
option>
</scheme>
```

The URI urn: epc:id:device:874454352867.37533238.51826 is passed to the ONS resolver to convert it into domain name. As per the above-mentioned procedure urn:epc and serial number (51826) is removed. Replace every colon with dots and invert the fields and append onsepc.com. Now, the output obtained is 37533238. 874454352867. device.id.onsepc.com. Using this domain name, NAPTR record is queried from the bind, which is the DNS server here. NAPTR name record contains http://localhost/test/device.html, which is the URL of the input (Fig. 2).

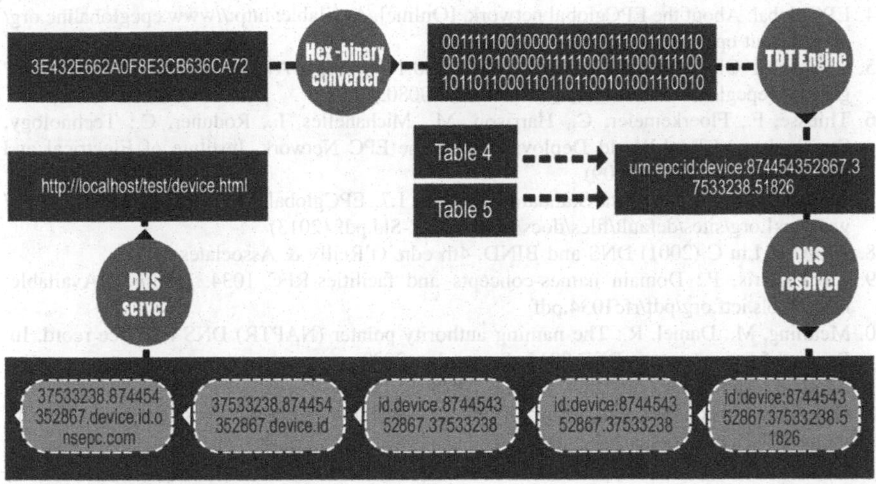

Fig. 2 Flow diagram using sample date describes the sample implementation using customized data

5 Conclusion

Electronic product code (EPC) is a unique number that identified a physical object, designed to store on RFID tags. EPC is designed as a flexible framework. This allows existing naming structures to be incorporated into the EPC system that can support many existing coding schemes. This paper defines a new model to issue customized values for RFID tag data by extending the EPCglobal framework. Also the work in this paper shows how to extend the tag data translation engine, an open source tool, for converting the hex on RFID tags based on the modified scheme definition into EPC URI, along with the existing schemes defined by the EPC.

ONS is a name lookup for EPC tags. ONS does not hold the actual EPC data. ONS can return a list of network accessible service endpoints that pertain to the EPC in question. Since ONS contains pointer to services, simple A record is insufficient, and therefore, Naptr record is used.

References

1. Aggarwal, C., Ashish, N., Sheth, A.: The Internet of Things: A Survey From The Data-Centric Perspective, book chapter in "managing and mining sensor data", Springer, Berlin (2013)
2. Atzori, L., Iera, A., Morabito, G.: The Internet of Things: A Survey. Elsevier, Amsterdam (2010)
3. Domdouzis, K., Kumar, B., Anumba, C.: Radio-Frequency Identification (RFID) Applications: A Brief Introduction. Elsevier, Amsterdam (2007)

4. EPCglobal: About the EPCglobal network. [Online]. Available: http://www.epcglobalinc.org/about/about epc network.html
5. EPCglobal Object Name Service (ONS): 1.0.1 [Online]. Available: http://www.gs1.org/gsmp/kc/epcglobal/ons/ons_1_0_1-standard-20080529.pdf
6. Thiesse, F., Floerkemeier, C., Harrison, M., Michahelles, F., Roduner, C.: Technology, Standards, and Real-World Deployments of the EPC Network. Institute of Electrical and Electronics Engineers (2009)
7. EPCglobal: EPC Tag Data Standards Version 1.7. EPCglobal. [Online]. Available: http://www.gs1.org/sites/default/files/docs/tds/tds_1_7-Std.pdf (2013)
8. Albitz P, Liu C (2001) DNS and BIND, 4th edn. O'Reilly & Associates
9. Mockapetris, P.: Domain names-concepts and facilities-RFC 1034. [Online]. Available: http://tools.ietf.org/pdf/rfc1034.pdf
10. Mealling, M., Daniel, R.: The naming authority pointer (NAPTR) DNS resource reord. In: Request for comments—RFC 2915, September 2000
11. Landt, J.: The history of RFID. J IEEE **24**, 8–11 (2005)
12. GSI Identification Keys (ID Keys) [Online]. Available : http://www.gs1india.org/upload/menu/GS1_Identification__key.pdf
13. Fosstrak Open Source RFID Platform [Online]. Available: https://code.google.com/p/fosstrak/
14. Fabian, B., Günther, O., Spiekermann, S.: Security analysis of the object name service. In: Proceedings of 1st IEEE Workshop on Security, Privacy and Trust in Pervasive and Ubiquitous Computing (SecPerU 2005)
15. Fabian, B., Günther, O.: Distributed ONS and its impact on privacy. In: Proceedings of IEEE International Conference on Communications (IEEE ICC2007), Glasgow, (2007)
16. Garcia-Alfaro, J., Barbeau1, M., Kranakis, E.: Evaluation of anonymized ONS queries. Nov 2009

Distributed CloudIMS: Future-Generation Network with Internet of Thing Based on Distributed Cloud Computing

Hamid Allouch and Mostafa Belkasmi

Abstract The next-generation network, cloud computing, and Internet of thing are a challenging and promising paradigm shift in IT world technology. Diminishing the cost for users for provisioning anywhere connecting at anytime from anywhere network, CloudIMS consists of interconnecting heterogeneous access technology and to respond to a major challenge for serving the increase in demand and scalable network access to share pool of configurable resource of enabling a convenient cloud computing. This paper mainly focused on common approach to integrate the IP multimedia subsystem (IMS), the Internet of thing, and cloud computing under the name of CloudIMS architecture which makes multimedia service easy to deploy on a cloud platform. We present the state of art of the different elements of CloudIMS. Moreover, we examine the layers designed for CloudIMS based on next-generation network access for mobile communication devices between different types of technologies (3GPP and non-3GPP), such as global system for mobile communication (GSM), wireless network, worldwide interoperability for microwave access (WiMAX), Universal Mobile Telecommunications System (UMTS) and long-term evolution (LTE). Finally, we present an architecture of CloudIMS according to our point of view, followed by a discussion of a use case for the future networks.

Keywords IP multimedia subsystem · Cloud computing · Virtualization · M2M · Internet of things · Interworking network

H. Allouch (✉) · M. Belkasmi
ENSIAS, Mohammed V University at Souissi, Rabat, Morocco
e-mail: hamid.allouch@um5s.net.ma

M. Belkasmi
e-mail: m.belkasmi@um5s.net.ma

L.C. Jain et al. (eds.), *Intelligent Computing, Communication and Devices*,
Advances in Intelligent Systems and Computing 308,
DOI 10.1007/978-81-322-2012-1_5

1 Introduction

Since its first launch in last 50 years, it has made the transition from closed to open networks, the Internet is evolving in capacity, size, availability, and technology continuously and is becoming the world's largest communication system recognized. From the first public mobile services and personal communications technologies, the development of future networks seems to have identifiable trajectories. It is difficult to predict the technological change and popularity of services. The mobile and Internet services increased dramatically since then, achieving near market penetration of cent per cent in some areas, and the growth continues in emerging global markets.

Mobile broadband services have extreme growth potential, as society shifts from a computer/Internet basis to a wireless/Internet basis. Voice telephony is undergoing extreme changes as well; the question is not whether voice communication over the Internet will supplant the old telephony, but how soon.

Since we believe in IMS as future next-generation network, we propose the architecture and vision strategy to improve the quality of services and fast data delivery for IMS based on effort in many areas, storage as cloud, automation as in Internet of thing, and network as IMS.

This paper comes for turning up some challenges of IMS architecture, storage, and processing system; hence, the fundamental contribution and the raisons of Internet evolution to a massive network with billions of computers include the following aspects:

- Easy user deployment of multiple services in simultaneous sessions with securing the architecture.
- Provide the quality of service (QoS) required for enjoying, rather than suffering, real-time multimedia sessions by designing.
- Respond to the evolution of cellular network and the network of convergence where users can access virtually the network everywhere (any location, with any movement), at any time (at night, or any time of day), and any way (computer laptop, PDA, Mobile call, etc.).

To ensure continuity of service or whatever the location, speed, and time, our vision is to first give a solution that operators or providers can exploit to maximize use of existing bandwidth to a top-quality service and second response to have a network that can provide the interconnection of these three operators with their different technologies.

This paper presents the current trends in interconnecting IMS, cloud computing, and IoT research propelled by applications and the need for convergence in several interdisciplinary technologies. Specifically, in Sect. 2, we present the overall survey and evolution of telecommunication, cloud computing, and Internet of thing and the technologies that will achieve it followed by some common definitions in the area along with some trends and taxonomy. We discuss our motivation of the architecture and the need of convergence with a new approach in defining our architecture CloudIMS distributed network vision in Sect. 3. A use case study of multimedia on the platform is given, and we conclude with discussions on open challenges and future trends.

2 Evolution of Network and Application

2.1 Evolution of Telecommunication (NGN)

Based on SIP and IP protocols, the IMS standard defines a generic architecture as in Fig. 1, for offering voice over IP (VoIP) and multimedia services [1, 2].

IMS is divided into three functional layers: the connectivity or transport layer; the control session layer; and service layer which contains various types of application servers such as telephony application server (TAS) that maintains the SIP call state. TAS is comprised of service logic which provides simple call processing services such as digit analysis and routing, call setup, forwarding, and waiting. Also the open services access AS which provides the flexibility to enable application developer partner who is located outside of core IMS domain to develop new applications. and we can integrate it via signaling SIP. IMS provides integrated services to its end of users and a platform for application providers to host their content on its servers [3, 4]. The core network consists of the following elements:

A database Home Subscriber Server (*HSS*): HSS is the main database used by the IMS; it contains and stores user profiles, stores all subscriber, service-related data, and the users of a domain. It provides the location and authentication information based on requests from the I- or S-CSCF, or the AS. HSS database is the same as the HLR in the existing mobile network.

Application Server (*AS*): Application servers provide application services including IP telephony, multimedia applications, voice call, and video conferencing applications (e.g., presence, PTT, instant messaging, supplementary services, and conferencing).

Proxy Call Session Control Function (*P-CSCF*): P-CSCF is first point and the gateway to UEs to the IMS network. PCSCF is a SIP-enabled proxy server, and all user requests, signaling, and control information pass through it.

The principal functions performed by the P-CSCF are forwarding a received SIP register request from the UE to another entry point using the home domain name.

Interrogating Call Session Control Function (*I-CSCF*): I-CSCF acts as the point of contact, for user connections and sessions, regardless of whether a user belongs to the same network, or a roaming user from another network. There may be multiple I-CSCFs within an operator's network.

Serving Call Session Control Function (*S-CSCF*): S-CSCF is the most important element of IMS core. Most of its functions are related to registration, session,

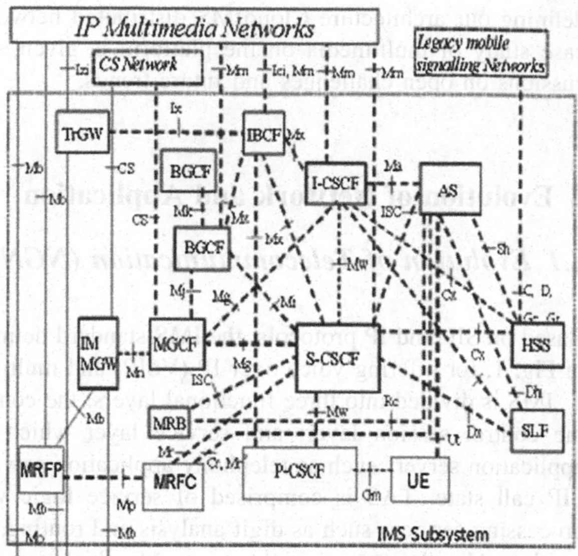

Fig. 1 Standard IMS architecture from [1]

and application, and it forwards the SIP request or response to the UE and generation of CDRs. The serving CSCF (S CSCF) performs the session control services for the UE. It maintains a session state as needed by the network operator for support of the services.

Media Gateway control Fonction (*MGCF*): MGCF connects the media plan of PSTN/PLMN to IMS media plan and provides interworking between IMS and PLMN/PSTN.

2.2 Evolution of Cloud Computing

The demand on system resources grows exponentially; see Fig. 2; all providers want to reduce the cost and simplify the management of their business. Regarding following huge application by Internet, they can be implemented all in the cloud. We can imagine that cloud computing is the giant computer that is equivalent to or more than 32 hosts/cluster and has more than 2048 processor cores, plus 32TB of RAM, more than 3 million IOPs, superiors, or equal to 1,280 virtual machines and more than 16PB of storage .

Cloud computing is an evolution of several computing paradigms, such as Internet delivery, pay-per-use-on-demand utility computing, virtualization, grid computing, distributed computing, storage elasticity, content outsourcing, and Web 2.0 [5]. The infrastructure referred to as a "cloud" enables on-demand provisioning of services across the world [6].

Fig. 2 Most application evolution of demand for resource

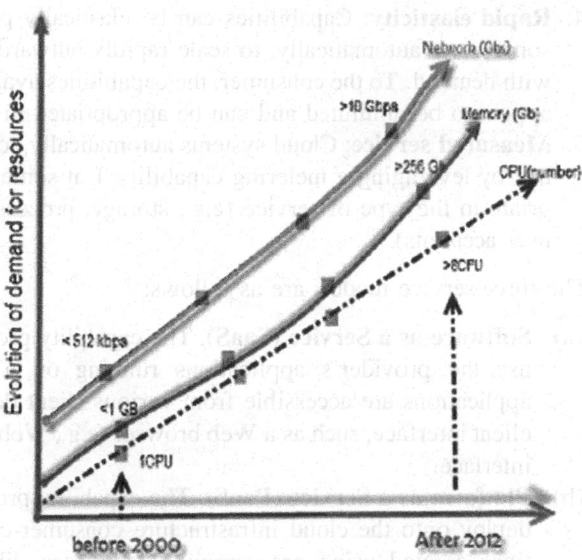

There are numerous definitions of cloud computing [3, 6]. According to the US National Institute of Standards and Technology (NIST) Definition of Cloud Computing, NIST SP 800-145 [6]:

> Cloud computing is a model for enabling ubiquitous, convenient, on-demand network access to a shared pool of configurable computing resources (e.g., networks, servers, storage, applications, and services) that can be rapidly provisioned and released with minimal management effort or service provider interaction. This cloud model is composed of **five essential characteristics**, **three service models**, and **four deployment models**.

The five Essential Characteristics are as follows (see Fig. 3):

1. **On-demand self-service**: A consumer can unilaterally provision computing capabilities, such as server time and network storage, as needed automatically without requiring human interaction with each service provider.
2. **Broad network access**: Capabilities are available over the network and accessed through standard mechanisms that promote use by heterogeneous thin or thick client platforms (e.g., mobile phones, tablets, laptops, and workstations).
3. **Resource pooling**: The provider's computing resources are pooled to serve multiple consumers using a multi-tenant model, with different physical and virtual resources dynamically assigned and reassigned according to consumer demand. There is a sense of location independence in that the customer generally has no control or knowledge over the exact location of the provided resources but may be able to specify location at a higher level of abstraction (e.g., country, state, or datacenter). Examples of resources include storage, processing, memory, and network bandwidth.

4. **Rapid elasticity**: Capabilities can be elastically provisioned and released, in some cases automatically, to scale rapidly outward and inward commensurate with demand. To the consumer, the capabilities available for provisioning often appear to be unlimited and can be appropriated in any quantity at any time.
5. **Measured service**: Cloud systems automatically control and optimize resource use by leveraging a metering capability 1 at some level of abstraction appropriate to the type of service (e.g., storage, processing, bandwidth, and active user accounts).

The three service models are as follows:

(a) **Software as a Service (SaaS)**. The capability provided to the consumer is to use the provider's applications running on a cloud infrastructure. The applications are accessible from various client devices through either a thin client interface, such as a Web browser (e.g., Web-based email), or a program interface.

(b) **Platform as a Service (PaaS)**. The capability provided to the consumer is to deploy onto the cloud infrastructure consumer-created or acquired applications created using programming languages, libraries, services, and tools supported by the provider.

(c) **Infrastructure as a Service (IaaS)**. The capability provided to the consumer is to provision processing, storage, networks, and other fundamental computing resources where the consumer is able to deploy and run arbitrary software, which can include operating systems and applications.

The Four Deployment Models are (Fig. 4) **as follows**:
Private cloud: The cloud infrastructure is provisioned for exclusive use by a single organization comprising multiple consumers. The main characteristics are as follows:

- Operated solely for an organization, typically within the firewall
- Low total cost of ownership
- Greater control over security, compliance, QoS
- Easier integration and support existing applications

Community cloud: The cloud infrastructure is provisioned for exclusive use by a specific community of consumers from organizations that have shared concerns (e.g., mission, security requirements, policy, and compliance considerations). It may be owned, managed, and operated by one or more organizations in the community, a third party, or some combination of them, and it may exist on or off premises; the main characteristics are like private cloud.
Public cloud: The cloud infrastructure is provisioned for open use by the general public. It may be owned, managed, and operated by a business, academic, or government organization, or some combination of them. It exists on the premises of the cloud provider; the main characteristics are as follows:

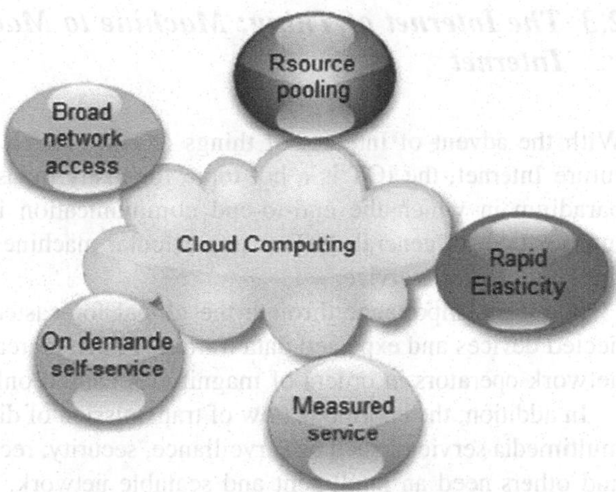

Fig. 3 A vision of a cloud computing essential characteristics

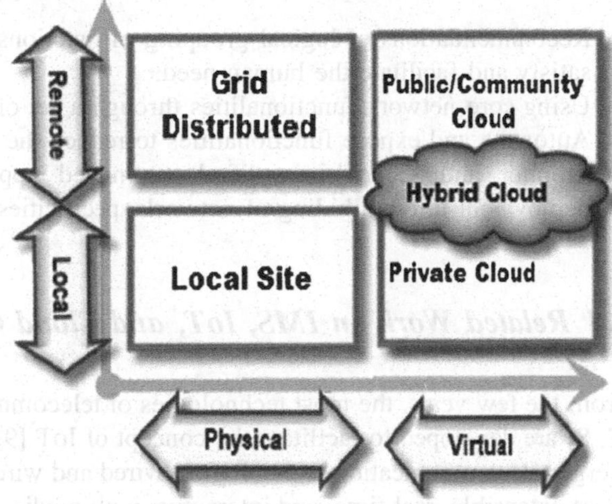

Fig. 4 Cloud computing deployment models

- Accessible over the Internet for general consumption
- Low acquisition costs and less administrative burden
- On-demand capacity and limited offerings

Hybrid cloud: The cloud infrastructure is a composition of two or more distinct cloud infrastructures (private, community, or public) that remain unique entities, but are bound together by standardized or proprietary technology that enables data and application portability (e.g., cloud bursting for load balancing between clouds); the main characteristics are as follows:

- Composition of 2, 3, or more interoperable clouds, enabling data and application portability
- Focus to deliver the best of both clouds.

2.3 The Internet of Thing: Machine to Machine the Future Internet

With the advent of Internet of things (IOT) referred as one typical view on the future Internet, the IOT is a hot topic for many industry and academia. IOT is a paradigm in which the end-to-end communication is executed without human intervention. In general, IOT or in particular machine to machine (M2M) is not a direct subscriber service.

IOT gain importance through the global forecasted numbers in terms of connected devices and expected data traffic. A huge increase in data traffic is faced by network operators in orders of magnitudes caused only by M2M.

In addition, the enormous flow of transmission of different applications and rich multimedia services, such as surveillance, security, recognition, health monitoring, and others need an intelligent and scalable network. The digital video holds the key in the future-generation network.

The functions and applications of Internet of thing Fig. 5 are:

- Recommendations of logical grouping of functions configured or triggered to satisfy and facilitate the human need.
- Using core network functionalities through a set of open interfaces.
- Automate and expose functionalities to reduce the cost
- Monitor and security simplified, optimized application development and deployment through hiding of network specificities from applications.

2.4 Related Work on IMS, IoT, and Cloud Computing

From the few years, the most technologies of telecommunication and applications [7, 8] are developed to facilitate the concept of IoT [9]. The concept of IMS is to merge telecommunication technologies, wired and wireless networks, and provide more extensible, real-time, and interactive multimedia services for next-generation networks under the all-IP environment [1, 3, 10].

The work in [3] presents the distributed and secured IMS architecture. The authors present a security classification that is critical to know in IMS, and they proposed secured and distributed architecture of IMS.

IoT research [11] presents the current trends in IoT research propelled by applications and the need for convergence in several interdisciplinary technologies. Specifically, they present the overall IoT vision and the technologies that will achieve by some common definitions in the area along with some trends and taxonomy of IoT. They discuss several application domains in IoT with a new approach in defining cloud-centric IoT vision. A case study of data analysis on the Aneka/Azure cloud platform is given with discussions on open challenges and future trends.

Nimbits [12] is a platform as a service (PaaS); it is an open source data logging cloud server built on cloud computing architecture that provides connectivity

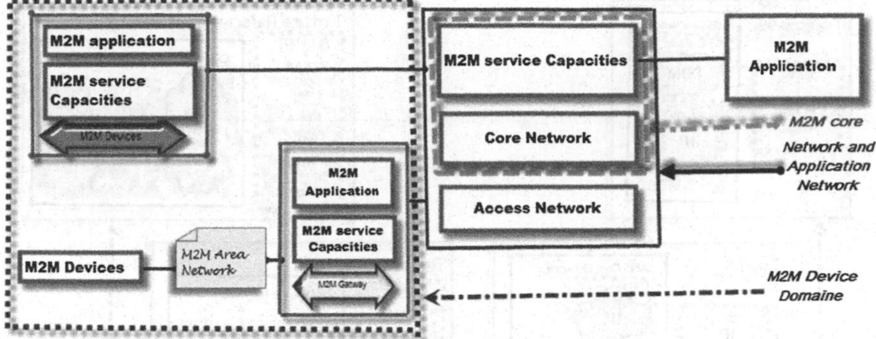

Fig. 5 ETSI M2M architecture

between the Internet of things using data points. Users can develop and use software and hardware solutions that seamlessly connect to the cloud and each other. Also users can record and share sensor data on the cloud freely. With Nimbits, users can create data points on the cloud and feed changing numeric, text-based, or xml values into them.

3 The CloudIMS Framework Architecture: Our Proposed Future Network

3.1 Our Vision of CloudIMS Platform

The evolution of telecommunication in simple concept is shown in Fig. 6. The all-IP architecture was planned promptly after R99 (the forerunner of IMS). Due to the architecture being too complex, the development work was divided into R5 (Release5) in 2000. Before R5, the IMS was not expected and included. R6 was completed in 2005, and the IMS was formally brought into the 3GPP standard. Further, IMS-related functions tend toward stability in R6 and were released in 2005. The follow-up R7 (Release 7) also adopted the concept of fixed mobile convergence. In the future, more access technologies and service frameworks will be integrated into 3GPP specification.

3.2 Architecture of CloudIMS Platform

We think and believe that our proposed architecture of CloudIMS will play an important role in future network Internet. The concept of IMS controls and enables an increased IMS service, and combines cloud and IoT; we proposed a common

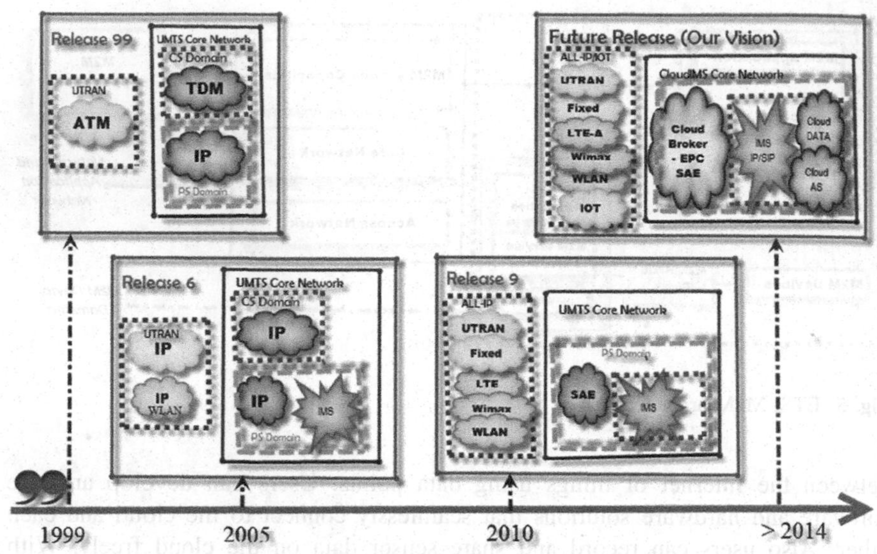

Fig. 6 The evolution of telecommunication (NGN)

framework for future Internet as illustrated in the Fig. 7 of CloudIMS network platform. The CloudIMS is divided into four layers:

i. **Cloud Networks**: where we can make application servers of different things and HSS database of IMS.
ii. **IMS Core Networks**: The control network is composed of different element of controls p/i/s-cscsf.
iii. **Cloud Broker engine/EPC Networks**: control the different radio access network and can connect the different technologies network.
iv. **Access network**: connect to the access network (GSM, UMTS, and WLAN), sensor network (WSN), and so on.

In this architecture of CloudIMS, the next-generation network for multimedia wired and wireless environments consists of heterogeneous access technologies with cloud control radio engine, data cloud, the datacenter of HSS, and application server cloud (ASC). The motivations of our framework are responding in the most section of availability, mobility, security, and scalability. We have the maximum total workload used for system. CloudIMS sizing is limited by the size of all max components.

The centralized architecture has a limitation of capacity of resource, because of single server and limitation of bandwidth; we can see this limitation in performance of centralized by of SIP signaling toward a network.

The architecture proposed with different elements is described in Fig. 7; the significance of different element is described in first section; here, we give the challenge of distributed CloudIMS, and we can see different type of interconnection element between UE and core CloudIMS.

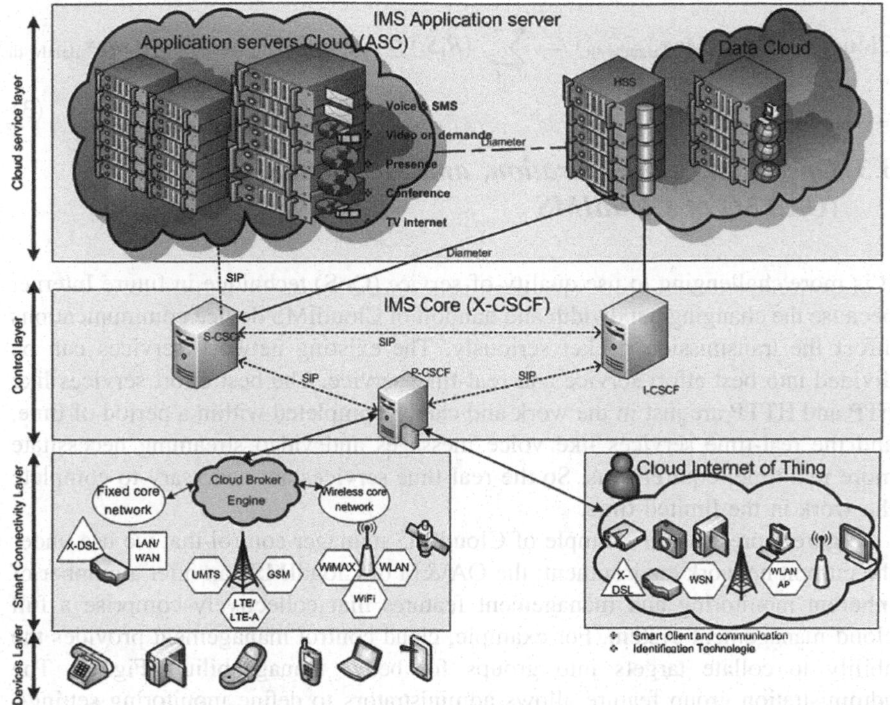

Fig. 7 Proposed architecture CloudIMS

We can illustrate the limitation of centralized, multiple technologies, so the distributed core CloudIMS proposed network architecture by Eqs. (1), (2), and (3):

$$\text{CloudIMS} = \sum_{i} N_{\text{vitualisations}} + \sum_{j} N_{\text{Automation}} + \sum_{m} N_{\text{IoT}} + \sum_{l} N_{\text{CoreIMS}} \tag{1}$$

where $N_{\text{vitualisations}}$ is the number of resource node cloud (database, application server, etc.). $N_{\text{Automation}}$ is the number of resource triggers, decision, or transfers for decision, N_{IoT} is the number of smart sensor and automation node to execute the decisions, and N_{CoreIMS} is the number of resource network IMS (x-CSCF).

The limitation of specific resource capacity centralized can be explained in the following equation:

$$\text{CloudIMS}_{\text{Total}_{\text{Size}}}\left(\text{Max}_{\text{capacity}}\right) = \sum_{i=1}^{N} (R_i S_i) \leq \text{CloudIMS}_{\text{Components}_{\text{Size}}}\left(\text{Max}_{\text{capacity}}\right) \tag{2}$$

$$\text{CloudIMS}_{\text{Total}_{\text{Size}}}(\text{Max}_{\text{Bandwith}}) = \sum_{i=1}^{N=\max} (R_i S_i) \leq \text{CloudIMS}_{\text{Components}_{\text{Size}}}(\text{Max}_{\text{Bandwith}}) \quad (3)$$

3.3 Operation, Administration, and Maintenance (OA&M) of CloudIMS

It is more challenging to use quality-of-service (QoS) technique in future Internet because the changing bandwidth and handoff of CloudIMS device communications affect the transmission packet seriously. The existing network services can be divided into best effort service and real-time service. The best effort services like FTP and HTTP are just in the work and can be completed within a period of time, and the real-time services like voice messages and video streaming necessitate more real-time requirements. So the real-time services are necessary to complete the work in the limited time.

Figure 8 presents an example of CloudIMS manager control that we use under the current network environment; the OA\&M of CloudIMS can offer a number of inherent monitoring and management features that collectively comprise a full cloud management system. For example, cloud control management provides the ability to collate targets into groups for better manageability (Fig. 8). The administration group feature allows administrators to define monitoring settings, compliance standards, and cloudIMS policies through templates and also organize each target in multiple hierarchies, such as line of business and lifecycle status.

3.4 Discussion and Motivation of Proposed CloudIMS Features and Use Case Study Characteristics

There are multiple challenges that can be foreseen in CloudIMS, the next generation of telecommunications network.

A first set of challenges lies in the need to deploy greater capacity storage in a low cost-effective manner: The new video formats with definitions much greater than HD (2K, 4K, and 8K) have been developed for digital cinema, and these are starting to appear on high-end devices. No convenient physical medium exists for the transport and delivery of films in these formats; therefore, CloudIMS networks will be an essential delivery channel. Legal and illegal file sharers are increasingly exchanging HD video content online. New camera and display technologies may enable a greater range of colors to be captured and displayed colors which the human eye can already see, but which are not offered due to the restricted capabilities of current multimedia systems. Given these developments, operators can expect an ever-increasing flow of multimedia content and applications such as Internet TV, catch-up TV over their networks, streaming already

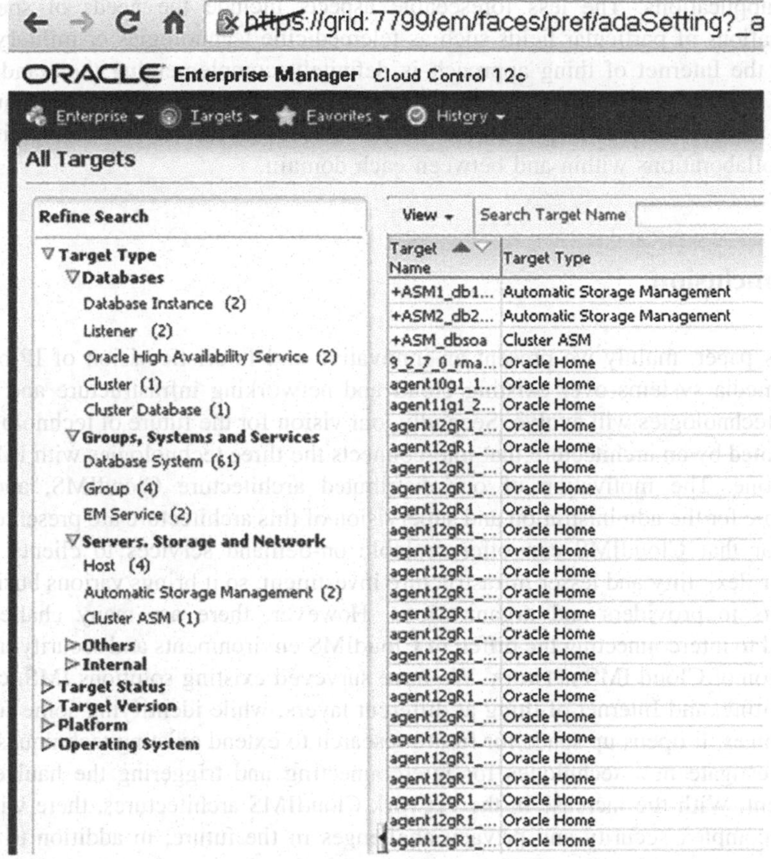

Fig. 8 Cloud control to monitor the instance and platform of cloudIMS and other networks

limited network resources, and ultimately requiring more investment in capacity. So cloud computing in CloudIMS can respond to this storage and capacity with minimum cost.

A second set of challenges concerns the number of users and services: The trend is that the users of the next generation of telecommunication network will not only be human but also machines, and different smart scenario like a smart meter to a fridge, moving by a camera or a car. This will increase the number of services and users Internet from a few billions to hundreds of billions, with a significant impact on the traffic, the addressing, the data storage, etc. So the IMS is the standard, the common umbrella, with SIP signalization and responds to the increased users and extensible services especially for the device of 5G.

A third set of challenges is for the network to cater with a plurality of natures of traffic: from the remote monitoring of a static food-vending machines, leading to send some bits every now and then, up to fast-moving multimedia real-time applications. The less foreseeable aspects include the needs of specific applications of particular fields such as telemedicine technologies or military.

So the Internet of thing approach is definitely complementing the standardization aspects rather than competing with them, and both domains will continue to define new systems and applications in the foreseeable future, with competitions and collaborations within and between each domain.

4 Conclusion

In this paper, mainly we present an innovative and futuristic vision of IP-based multimedia systems over existing cloud and networking infrastructure and how these technologies will evolve. Secondly, our vision for the future of technology is illustrated by an architecture that interconnects the three technologies with IMS as backbone. The motivation of our distributed architecture CloudIMS, and an example for the administration and supervision of this architecture are presented. It is clear that CloudIMS can offer scalable on-demand services to clients with greater flexibility and lesser infrastructure investment, so it brings various business benefits to providers and organizations. However, there are many challenges related to interconnecting the different CloudIMS environments and security in our adoption of Cloud IMS network. We have surveyed existing solutions IMS, cloud computing, and Internet of thing at different layers, while identifying some future challenges. It opens up space for future research to extend existing techniques and to investigate new techniques for interconnecting and triggering the haul environment. With the increase in the network CloudIMS architectures, there will be more complex security and privacy challenges in the future; in addition to this, building trust in the CloudIMS is an interesting future area of research. With the increasing and extensible usage of the CloudIMS services, it is possible to collect sufficient evidence from the CloudIMS providers on the level of trust on each of their services. This can help the service providers, infrastructure providers, and the end users to better choose the right services from the ever-growing CloudIMS vendors.

References

1. GPP, GSM, and ETSI: TS 23.228 IP Multimedia Subsystem (IMS); Stage 2 (Version 12.3.0, Release 12; 2013-12-17). link: http://www.3gpp.org/DynaReport/23228.htm
2. Sultan, A., Mulligan, U.: NGN standardization as a strength In: Bertin, E. et al. (eds.) Telecommunication Services Evolution, LNCS 7768, pp. 77–89. Springer, Berlin (2013)
3. Allouch, H., Belkasmi, M.: Design of transparent distributed IMS network: security challenges risk and signaling analysis. J. IJNGN **4**(4), 1–18 (2012). doi:10.5121/ijngn.2012.4401

4. Allouch, H., Belkasmi, M.: Risk analytic approach and cost analysis for interworking on the new secured IMS architecture. J. Inform. Secur. Res. **3**(3), 130–147 (2012)
5. Pallis, G.: Cloud computing: the new frontier of internet computing. IEEE Internet Comput. **14**, 70–73 (2010)
6. Mell, P., Grance, T.: The NIST definition of cloud computing. NIST Special Publication 800-145, p. 7 (2011)
7. Broll, G., Rukzio, E., Paolucci, M., Wagner, M., Schmidt, A., Hussmann, H.: Pervasive service interaction with the internet of things. IEEE Internet Comput. **13**(6), 74–81 (2009)
8. Hong, S., Kim, D., Ha, M., Bae, S., Park, S.J., Jung, W., Kim, J.: SNAIL: an IP-based wireless sensor network approach to the internet of things. IEEE Wirel. Commun. **17**(6), 34–42 (2010)
9. Atzori, L., Iera, A., Morabito, G.: The internet of things: a survey. Comput. Netw. **54**, 2787–2805 (2010)
10. Chang, K.-D., Chen, C.-Y., Chen, J.-L., Chao, H.-C.: Challenges to next generation services in IP multimedia subsystem. J. Inform. Process. Syst. **6**(2), 129–146 (2010)
11. Gubbi, J., et al.: Internet of Things (IoT): A Vision, Architectural Elements, and Future Directions. Elsevier, Amsterdam (2013)
12. Nimbits: http://www.nimbits.com/. Retrieved by 20 July 2012

Genetic Algorithm-Based Query Expansion for Improved Information Retrieval

Pragati Bhatnagar and Narendra Pareek

Abstract This paper is focused toward query expansion, which is an important technique for improving retrieval efficiency of an information retrieval system. In particular, the paper proposes an evolutionary approach for improving efficiency of pseudo-relevance feedback-based query expansion (PRFBQE). In this method, the candidate terms for query expansion are selected from an initially retrieved list of documents, ranked on the basis of co-occurrence measure of the terms with the query terms. Top *n* selected terms create a term pool. From this term pool, genetic algorithm (GA) is used to select a thematically rich combination of terms, which provides the terms for expanding the query. We call this method as genetic algorithm-based query expansion (GABQE). The experiments were performed on standard CISI dataset. The results are quite motivating, and one can clearly observe the difference in the result when GA is not used and when GA is used. The paper uses GA for improving pseudo-relevance feedback (PRF)-based query expansion, but at the same time, it can also be generalized and tested for other types of query expansions, where terms may be selected in a different way, but a good combination of expansion terms can be obtained using GA.

Keywords Information retrieval · Query expansion · Genetic algorithm

1 Introduction

Query expansion has been widely investigated as a method for improving the performance of information retrieval system. Though a lot of work has been done in this area, obtaining a proper expansion of query is still an unsolved problem. Different researchers are coming up with different techniques of query expansion.

P. Bhatnagar (✉) · N. Pareek
Department of Computer Science, M.L. Sukhadia University, Udaipur, Rajasthan, India
e-mail: pragatibhat@gmail.com

L.C. Jain et al. (eds.), *Intelligent Computing, Communication and Devices*,
Advances in Intelligent Systems and Computing 308,
DOI 10.1007/978-81-322-2012-1_6

A popular type of query expansion is **pseudo-relevance feedback-based query expansion (PRFBQE)**. The most important concern in query expansion is the source for the expansion terms and criteria for selecting and ranking the expansion terms. Xu and Croft [1, 2] provided an efficient co-occurrence-based measure for ranking the query expansion terms. However, Cao et al. [3] even question the basic notion of goodness of a term. They argue that a goodness criterion, which is based on the frequency of terms in PRF-based documents or their distribution in corpus, is itself not appropriate. The authors then propose to integrate a term classification process to predict the usefulness of expansion terms. Some work has been done for using genetic algorithm (GA) for information retrieval and query expansion. Most of the work has been done to tune the weights of query terms or matching functions. Pathak et al. [4] have used GA for improving the efficiency of matching function of an information retrieval system. Horng [5] used GA to tune the weight of retrieved query terms. The experiment has been done on Chinese data collection. Araujo [6] have used GA for query expansion based on stemming and morphological variations. Cecchini [7] has used GA along with the notion of thematic context to improve query expansion. The proposed techniques place emphasis on searching for novel material that is related to the search context.

The above papers are somewhat related to our work; however, our work is different in the sense that it has been used to achieve an improved ranking of query expansion terms obtained using PRFBQE. In PRFBQE, terms are ranked independently of each other. We observed that this leads to a serious problem, as the terms have dependence over one another. A proper combination of these terms, which is cohesive, can improve the result dramatically. Thus, given n candidate terms, we are interested in selecting a suitable subset of the terms that optimize precision/recall of the query. This makes the problem as an optimization problem, and since the search space is very large, we cannot have a polynomial time algorithm for solving the problem. Thus, it is appropriate to use GA for finding a cohesive selection of expansion terms that optimize the performance of query. Based on these notions, we present the idea of genetic algorithm-based query expansion (GABQE). After explaining our approach, we provide an algorithm for performing GABQE. We performed experiments on standard CISI dataset (benchmark dataset for information retrieval). The results are quite motivating, and one can clearly observe the difference in the result when GA is not used and when GA is used.

2 Proposed Approach

We have tried to improve the performance of PRFBQE by using GA. We call it **GAQBE**. This helps us to provide thematically rich collection of expansion terms. GABQE approach is divided two parts: construction of term pool and selection of expansion terms from the term pool. In order to present our approach, in Sect. 2.1, we discuss about the construction of term pool, in Sect. 2.2, we discuss the GA-based approach for selection of expansion terms.

2.1 Construction of Term Pool

In order to construct the term pool, we first retrieve top n documents for the query using a matching function. In our problem, a query is selected and its Okapi measure is used as a matching function. The Okapi measure is given by following equation:

$$\text{Okapi}(Q, D_i) = \sum_{T \in Q} w \frac{(k_1 + 1)tf}{K + tf} \times \frac{(k_3 + 1)qtf}{k_3 + qtf} \tag{1}$$

Q is the query that contains words T.
k_1, b, k_3 are constant parameters (k_1 = 1.2, $b = 0.75$, $k_3 = 7.0$)

$$K \text{ is } k_1(1 - b) + \left(b.\frac{dl}{avdl}\right)$$

tf is term frequency of term in document D_i
qtf is term frequency in query Q

$$w \text{ is } \log \frac{(N - n + 0.5)}{(n + 0.5)}$$

N is number of documents, n is number of documents containing the term.
dl and $avdl$ are document length and average document length.
All documents and sorted on the basis of Okapi measure. All the unique terms of top N documents are selected and are ranked on the basis of their co-occurrence with query terms. Top m terms co-occurring with original query terms are selected as candidate terms for expansion. For our experiments, we have used well-known Jaccard coefficient as a co-occurrence measure, which is given as:

$$\text{Jaccard\_co}(t_i, t_j) = \frac{d_{ij}}{d_i + d_j - d_{ij}} \tag{2}$$

where t_i and t_j are the terms for which co-occurrence is to be calculated and d_i and d_j are the number of documents in which terms occur, respectively, and d_{ij} is the number of documents in which t_i and t_j co-occur.

We can apply this coefficient to measure the similarity between the query terms and terms in the documents. Incorporating inverse document frequency and applying normalization, we define degree of co-occurrence of a candidate term with a query term as follows:

$$\text{co\_degree}(c, t_i) = \log_{10}(\text{co}(c, t_i) + 1) * (idf(c)/\log_{10}(D)) \tag{3}$$

$$idf(c) = \log_{10}(N/N_c) \tag{4}$$

where

N	number of documents in the corpus
D	number of top ranked documents used
c	candidate term listed for query expansion
t_j	*j*th term of the document
Nc	number of documents in the corpus that contain c
Nco(c, t_j)c	number of documents in the corpus that contain c

Above formula can be used for finding similarity of a term c with individual query term. To obtain a value measuring how good c is for whole query Q, we need to combine its degrees of co-occurrence with all individual original query terms t_1, t_2, t_3.... So, we use

$$\text{Suitability for } Q = f(c, Q) = \prod_{t_i \text{ in } Q} (\delta + \text{co\_degree}(c, t_i))^{idf(t_i)} \tag{5}$$

Above equation provides a suitability score for ranking the terms co-occurring with entire query. The terms of the document are ranked on the basis of similarity value obtained and top m terms form a term pool.

2.2 Genetic Algorithm for Selecting Expansion Terms

We have discussed the approach for developing the term pool. The term pool contains the good candidate terms that may be suitable for query expansion. Now, we have to select an optimal combination of a subset of these terms, which are cohesive among themselves and are better suited for query expansion. In order to apply GA, we require a proper fitness function. Moreover, the performance of GA is very much dependent on proper representation of chromosome, proper selection, and tuning of crossover and mutation operators.

2.2.1 Representation of Chromosome

We have used a chromosome representation where each gene represents a specific candidate term. One particular combination of expansion terms represents a chromosome. Considering number of terms as 10, chromosomes are represented in following way

t_1	t_2	t_3	t_4	t_5	t_6	t_7	t_8	t_9	t_{10}

where each t_i represents a term index.

2.2.2 Fitness Function

Fitness function is based on the suitability or goodness of the query in retrieving the relevant documents, which is measured by recall or precision. We have used recall of the retrieved result as a fitness function. Recall is given by:

$$\text{Recall} = \frac{|R_a|}{|R|} \tag{6}$$

R Set of relevant documents retrieved
R Set of all relevant documents

Selection, crossover, and mutation are the GA operators that are applied to the above chromosomes. Standard single-point crossover was used here. The algorithm for GAQBE is as follows

2.2.3 Algorithm for GABQE

Final algorithm for GABQE is presented in Table 1.

3 Experiments and Results

We tested the algorithm on CISI dataset. This dataset provides a benchmark for testing efficiency of an information retrieval system. CISI data consist of 1,460 abstracts from information retrieval papers and 112 queries. In order to perform PRFBQE, two important parameters need to be set: n (number of top documents to be retrieved), m (number of expansion terms). After extensive experiment on corpus, the values were set as $n = 10$ and $m = 10$.

For setting GA parameters, chromosome length was set to 10, as number of expansion terms was 10. Other parameters were fixed after extensive experimentation. Population size and final number of generations were 40 and 50, respectively. Crossover and mutation rates were kept as: 0.7 and 0.03.

The results were evaluated and compared for: without query expansion, PRFBQE and GABQE. The improvement in the result can be observed from recall precision curve as shown in Fig. 1. The effect of GA was observed for all the queries. Figure 1a shows 10-point recall precision for query number 28. It can be observed that GABQE is showing improvement over standard PRFBQE. Figure 1b gives average recall precision curve for all the queries. Again, it is observed that GABQE improves the results on average.

The effect of GABQE can be observed from generationwise average fitness curve. Fitness is measured by recall (Eq. 6). Figure 2a shows such graph for query number 28. Similar graph is presented for generationwise average of average fitness for all the queries in Fig. 2b. As it can be observed, average recall is

Table 1 Algorithm developed for selecting expansion terms using GABQE

```
Algorithm:   GABQE
Input : Document corpus D
        Query collection Q
Process:
          Select the  query q from Q
          Preprocess the documents in document collection D
          Calculate similarity measure of each document  d in D w.r.t
          query q using equation 1
          Sort documents in D according to their similarity measure with
          q
          Retrieve top n documents giving document collection R
          Find all unique terms of top n retrieved documents giving term
          collection T
          Find candidate expansion terms giving term collection C
              (a)Calculate co-occurrence  between each query term qi and
                 each term ti in T using jaccard similarity (equation 2)
              (b)Calculate similarity of entire query Q with each term ti
                 in T using equation 3
              (c)Calculate the suitability score of each term ti using
                 equation 5
              (d) Sort the terms in T on the basis of suitability score
              (e )Retrieve  top  m  terms  of  T  giving  candidate  expansion
                 Collection  C
          Perform following to select expansion terms by applying  GA
              Generate initial population randomly from the term pool.
              Repeat
                  Form new population using selection, crossover, mutation
                  operation(in   pair of 2)
                  Expand the original query by adding terms of the individual
                  population  member.
                  Retrieve the initial set of documents using tentatively
                  expanded      query
                  Calculate the fitness of the expanded query using recall based
                  measure  (equation 6)
              Until the population converges or for maximum number of generation
          Return the terms obtained in final generation of GA as final set of
          expansion terms.
Output : Set of expansion terms
```

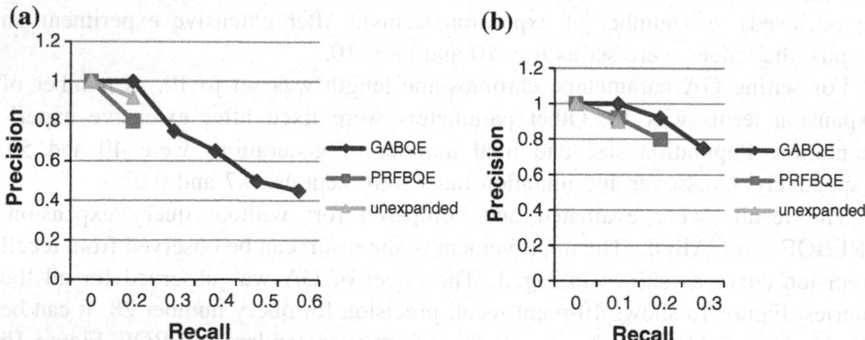

Fig. 1 **a** Recall precision graph for query 28. **b** Average recall precision graph

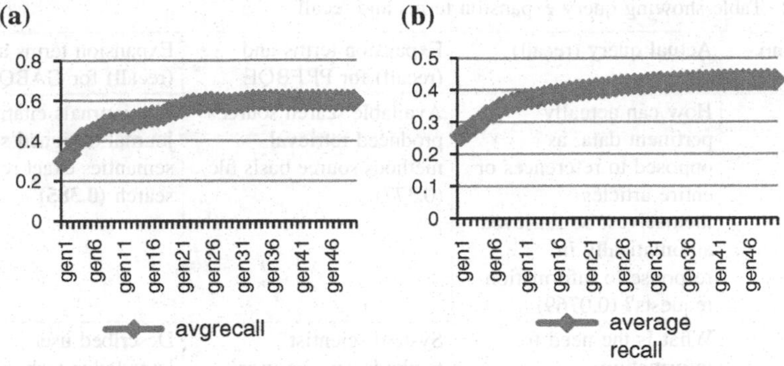

Fig. 2 **a** Generationwise average recall for query no. 28. **b** Generationwise average recall for all queries

increasing and slowly reaches to convergence. This shows that GA is able to improve the fitness (recall); hence, efficiency of information retrieval is increased.

In order to analyze the result we observed the expansion terms obtained without GA and with GA, we observed and analyzed expansion terms for individual query. For almost all the queries, GA-based expansion is providing better terms. Table 2 shows original query, query expansion terms obtained with PRFBQE, and query expansion terms obtained after applying GABQE. Due to space limitation, the result is presented for query number: 2, 11, 12, 23, and 28. The term in bracket indicates recall obtained. Highest recall is given in bold. Last row indicates average recall of all queries. For all the queries listed below as well as on average, our approach performs better. We observe that in case of expanding the query simply with PRFBQE, without applying GA, recall is increasing in some queries, while it is decreasing marginally in some cases. When applying GABQE, recall remains same or is increasing in almost all queries. However, for some queries, performance is detoriating as the term pool itself does not contain good candidate expansion terms. In general, we can see that GA effects query expansion positively; however, the actual effect may vary from query to query.

It can be observed that, in query 2, terms obtained from GABQE: 'search, semantic, retrieval citation, and file' are more focused and useful for expanding the query. Similarly, in query 28, 'asca' and 'sdi' are more related to query and are more focused terms. We observed that these terms have many meanings, but in this context, 'asca' is a terminology related to scientific classification taxonomy, whereas 'sdi.' comes from 'sdi biomed,' providing high-quality laboratory testing products used with laboratory chemical analyzers. Such observations can be made for other queries. So, we can say that application of GABQE helps in expanding the query in such a manner that it provides a better selection of thematically rich expansion terms and hence improves retrieval efficiency.

Table 2 Table showing query expansion terms and recall

Query no.	Actual query (recall)	Expansion terms and (recall) for PRFBQE	Expansion terms and (recall) for GABQE
2	How can actually pertinent data, as opposed to references or entire articles themselves, be retrieved automatically in response to information requests? (0.0769)	Available search sources produced retrieval methods source basis file (0.377)	File journals citation journal designed source semantics exact retrieval search **(0.385)**
11	What is the need for information consolidation, evaluation, and retrieval in scientific research? (0.1890)	System scientist methods science user document documents literature discussed analysis (0.1957)	Described user knowledge technical national subject scientists new analysis available **(0.2598)**
12	Give methods for high-speed publication, printing, and distribution of scientific journals. (0.4615)	Year citations papers work including articles able total abstracting primary (0.0769)	Journal citation multiple abstracting past coverage references highly source national **(0.6514)**
23	Amount of use of books in libraries. Relation to need for automated information systems. (0.2917)	Circulation journal designed people even available copies american articles stack (0.375)	Large, useful, material classification circulation, requirements longer universities available **(0.3958)**
28	Computerized information systems in fields related to chemistry. (0.2833)	Title considerably similarities estimating synthesis mathematics english citation cited alternative (0.2333)	Easily included asca chemical title sdi, estimating, alternative file compounds **(0.6)**
All queries	0.15	0.224	**0.326**

4 Conclusion

This paper suggests used of GABQE in order to improve retrieval efficiency of an information retrieval system. The experiments have been done on standard CISI collection. The comparison of the result has been done on the basis of recall. The results were compared for unexpanded query, PRFBQE and GABQE. It was observed that GA is providing a more cohesive and better selection of expansion terms. The improvement of the result can be observed from the graph. Further, we have also analyzed the result by observing the better expansion terms obtained by using our approach.

References

1. Xu, J.: Solving the Word Mismatch problem through Text analysis. Ph.D. Thesis, vol. 11, University of Massachusetts, Department of Computer Science, Amherst, USA (1997)
2. Xu, J., Croft, W.B.: Improving the effectiveness of information retrieval with local context analysis. ACM Trans. Inf. Syst. **18**(1), 79–112 (2000)
3. Cao, G., Nie, J.Y., Gao, J.F., Robertson, S.: Selecting good expansion terms for pseudo relevance feedback. In: 31st Annual International ACM SIGIR Conference on Research and Development in Information Retrieval, pp. 243–250 (2008)
4. Pathak, P., Gordon, M., Fan, W.: Effective information retrieval using genetic algorithm based matching functions adaption. In: Proceedings 33rd Hawai International Conference on Science (HICS), Hawaii, USA (2000)
5. Horng, J., Yeh, C.: Applying genetic algorithms to query optimization in document retrieval. Inf. Process. Manage. **36**, 737–759 (2000)
6. Araujo, L., Aguera J.P.: Improving query expansion with stemming terms: a new genetic algorithm approach. In: 8th European Conference on Evolutionary Computation in Combinatorial Explosion, pp. 182–193, Springer-Verlag Berlin, Heidelberg (2008)
7. Cecchini, R.L., Lorenzetti, C.M., Maguitman, A.G., Brignole, N.B.: Using genetic algorithms to evolve a population of topical queries. Inf. Process. Manage. **44**, 1863–1878 (2008)

Genetic Algorithm-Based Adaptive PID Controller

Shradhanand Verma and Rajani K. Mudi

Abstract Conventional PID controllers (CPID) usually fail to provide satisfactory performance for integrating and nonlinear systems due to large overshoots and oscillation. Nonlinear and adaptive PID controllers (APID) are being developed toward achieving the desired control performance for such systems. In this study, we make an attempt to develop a genetic algorithm-based adaptive PID controller (GA-APID) in order to attain adequate servo as well as regulatory performance. While designing our GA-APID, first we formulate the structure of the APID controller followed by its optimal parameter estimation for a given system using genetic algorithm. Performances of GA-APID for nonlinear and integrating systems are compared with those of CPID and APID reported in the leading literature. From detailed performance analysis, GA-APID is found to provide significantly improved performance over others.

Keywords Genetic algorithm · PID controller · Adaptive control

1 Introduction

PID controllers in its different forms are mainly used in process industries due to their simple structures and ease of implementation [1]. The parallel form of the conventional PID controller (CPID) with three adjustable gain parameters is extensively studied [2, 3]. Such controllers can provide reasonably acceptable performances for first-order and second-order linear self-regulating processes.

S. Verma (✉) · R.K. Mudi
Department of Instrumentation and Electronics Engineering, Jadavpur University, Sec-3, Block—LB/8, Salt-Lake, Kolkata 700098, India
e-mail: shradhanandei@gmail.com

R.K. Mudi
e-mail: rkmudi@yahoo.com

L.C. Jain et al. (eds.), *Intelligent Computing, Communication and Devices*, Advances in Intelligent Systems and Computing 308,
DOI 10.1007/978-81-322-2012-1_7

However, their performances for nonlinear and integrating systems are usually not satisfactory due to associated intolerably large overshoot [2, 4, 5]. Several attempts have been made to overcome this limitation [6–11] by developing adaptive PID controllers (APID) through nonlinear parameterization. However, most of the APID parameters are selected based on trial and error, or sometimes through heuristics [8–11]. Therefore, their performances may not be optimal. Keeping in mind this point, and the excellent optimization power of genetic algorithms (GA) [12], in this study, we attempt to develop GA-based adaptive PID (GA-APID) controller toward achieving optimum performance. Here, GA-APID design involves two steps—first we define the structure of the adaptive PID; then, GA is used to find its best set of parameters with respect to a given closed-loop performance index or objective function. Performances of the developed GA-APID for nonlinear and integrating systems with dead time are compared with other PID controllers. Considerably improved performance with respect to a large number of indices justifies the effectiveness of the developed GA-APID.

2 The Proposed PID Controller

2.1 The Conventional PID Controller

The discrete form of a conventional PID controller (CPID) can be expressed as

$$u^c(k) = K_p\left[e(k) + \frac{\Delta t}{T_i}\sum_{i=0}^{k} \mathrm{e}(i) + \frac{T_d}{\Delta t}\Delta \mathrm{e}(k)\right]$$
$$\text{or} \quad u^c(k) = K_p e(k) + K_i \sum_{i=0}^{k} e(i) + K_d \Delta e(k). \tag{1}$$

In (1), $e(k) = r - y(k)$ is the process error, where r is the set point and y is process output, K_p is the proportional gain, $K_i = K_p\left(\frac{\Delta t}{T_i}\right)$ is the integral gain, $K_d = K_p\left(\frac{T_d}{\Delta t}\right)$ is the derivative gain, T_i is the integral time, T_d is the derivative time, and Δt is the sampling period. Proper selection of the three tuning parameters—K_p, T_i, and T_d—is a critical task to attain the desired closed-loop performance. Through decades, various methods have been developed for the tuning of PID parameters [3]. Among them, Ziegler-Nichols (ZN) continuous cycling method [13] is most widely used by practicing engineers for the initial settings of PID parameters [3]. In this study also, we have used ZN continuous cycling method [4] for the initial settings of the PID parameters, i.e., $K_p = 0.6K_u$, $T_i = 0.5t_u$, and $T_d = \frac{t_u}{8}$, where K_u is the ultimate gain and t_u is the ultimate period.

2.2 The Adaptive PID Controller

We have already mentioned that CPID usually fail to provide acceptable performance for nonlinear and integrating systems. In order to overcome such limitations, a number of attempts have been made in [5, 8–11] for online adjustments of various gain parameters of CPID, thereby making it an APID, so that an overall improved performance is achieved. In this study, we will concentrate on the APID presented in [8], where the three gain constants, i.e., $K_P, K_i,$ and K_d, are continuously modified by an online updating factor alpha (α) with the following simple heuristic relations:

$$K_p^m(k) = K_p(1 + k_1|\alpha(k)|) \tag{2}$$

$$K_i^m(k) = K_i(k_2 + k_3\alpha(k)) \tag{3}$$

$$K_d^m(k) = K_d(1 + k_4|\alpha(k)|). \tag{4}$$

Here,

$$\alpha(k) = e_N(k) \times \Delta e_N(k), \tag{5}$$

where $e_N(k) = \frac{e(k)}{|r|}$; and $\Delta e_N(k) = e_N(k) - e_N(k-1)$.

Now, the APID can be redefined as

$$u^m(k) = K_p^m(k)e(k) + K_i^m(k)\sum_{i=0}^{k} \mathrm{e}(i) + K_d^m(k)\Delta \mathrm{e}(k). \tag{6}$$

In Eq. (6), $K_p^m(k), K_i^m(k)$ and $K_d^m(k)$ are the modified proportional, integral, and derivative gains, respectively, at kth instant and $u^m(k)$ is the corresponding control action. $k_1, k_2, k_3,$ and k_4 are the four additional positive constants. The objective behind such online gain adjustments as described by relations (2)–(4) is that when the process is moving toward the set point, control action will be less aggressive to avoid possible large overshoots and/or undershoots, and when the process is moving away from the set point, control action will be more aggressive to make a rapid convergence of the system. Following this gain-adaptive technique, in [8], a significantly improved performance of APID is found for high-order and nonlinear systems both in set point and load disturbance responses. However, out of the *seven* parameters of [8], i.e., $K_P, K_i, K_d, k_1, k_2, k_3,$ and k_4,, the first *three* constants, i.e., $K_P, K_i,$ and K_d, are selected based on ZN ultimate cycle rule, whereas the remaining *four* constants, i.e., $k_1, k_2, k_3,$ and k_4, are chosen by trial. Therefore, there is further scope to achieve improved performance if we can find the most appropriate settings of these parameters. Keeping in mind this objective and the GA as a powerful optimization tool, we are motivated to develop the proposed GA-APID. In the present work, *all* the seven parameters (i.e., $K_P, K_i, K_d, k_1, k_2, k_3,$ and k_4) are selected through optimization using binary coded GA. In the next section, we will describe the optimization process.

2.3 GA-Optimized APID

2.3.1 Objective Function of the Genetic Algorithm

In the present optimization problem, *integral absolute error* (*IAE*) is defined as the objective function or fitness function. The *IAE* is calculated as $IAE = \int_0^\infty |e(t)|dt$.

2.3.2 Different Operations Used in GA

Encoding—Here, the population size is 10. We use 4 bits for each of the 7 variables (i.e., $K_P, K_i, K_d, k_1, k_2, k_3$, and k_4) in a particular solution. So each chromosome has 28 bits. Here, we consider the following ranges of variables: K_P, K_i, K_d are ± 20 % of their respective CPID, $k_1[05]$, $k_2[01]$, $k_3[05]$, and $k_4[030]$.

Decoding—We convert encoded binary value of each variable to decimal value. Then, we bring this decimal value in its defined range to get the real value of the optimization variable. The following linear mapping rule is used for this purpose:

$X_i = X_i^L + \frac{X_i^U - X_i^L}{2^n - 1} \times$ (decimal value of the *i*th variable in the binary string). where $X_i^L \le X_i \le X_i^U$, X_i = real value of the *i*th optimization variable, X_i^U = upper *limitof i*th variable, X_i^L = lower limit of *i*th variable, and n = no. of bits used for each variable (here it is 4 bits). The decoded values thus obtained are used to simulate the closed-loop system response in MATLAB. This process is repeated for all the 10 solutions. From each closed-loop response, we calculate the value of the objective function, *IAE*.

Selection—We sort the values of *IAE* in ascending order and select the 50 % fittest roots (best solutions) from the top for the next stage, i.e., crossover.

Crossover—50 % chromosomes will go for crossover. These chromosomes are known as parent. The crossover operator produces two children for each parent pair. So here, we will get 50 % children after crossover. Therefore, after crossover, total population will again be 100 % = 50 % (parents) + 50 % (children).

Mutation—To avoid local optima, we mutate the strings. The mutation probability (percentage of bits in a population mutated in each iteration) is generally kept low for steady convergence (here, it is ≈1.78 %).

Convergence check—To ensure the convergence, we draw a curve between performance index (*IAE*) and number of iteration. Initially, the value of the objective function *IAE* goes on decreasing rapidly, but after some iteration, its value remains almost constant, which indicates the optimization is complete.

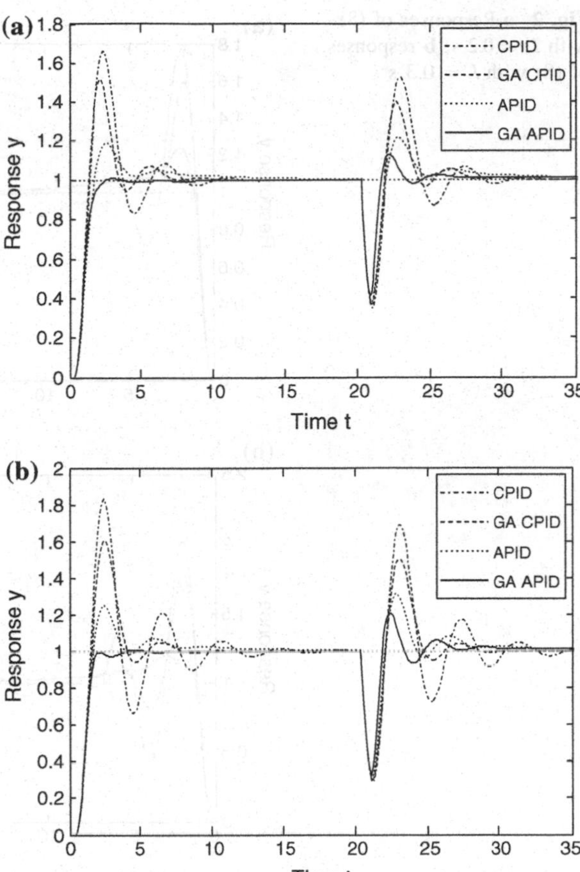

Fig. 1 **a** Responses of (7) with $L = 0.3$ s; **b** responses of (7) with $L = 0.4$ s

3 Results

For simulation study, we consider the following nonlinear (7) and integrating (8) systems with dead time (L):

$$\frac{d^2y}{dt^2} + \frac{dy}{dt} + 0.2y^2 = u(t - L) \tag{7}$$

$$G_p(s) = \frac{e^{-Ls}}{s(s+1)} \tag{8}$$

Performance of our GA-APID is compared with conventional PID (CPID), GA-optimized conventional PID (GA-CPID), and the APID of [8]. For detailed comparison, in addition to the response characteristics, several performance indices, such as percentage overshoot ($\%OS$), rise time (t_r), settling time (t_s), integral

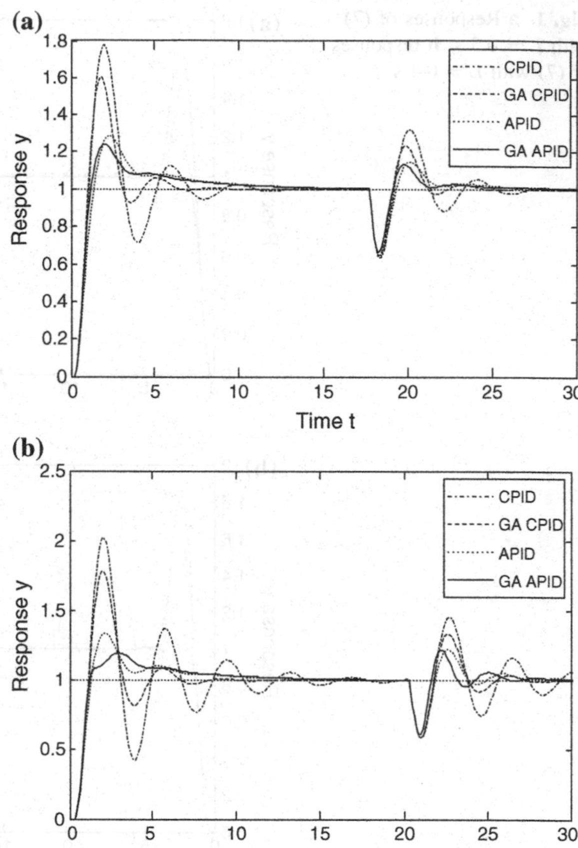

Fig. 2 **a** Responses of (8) with $L = 0.2$ s; **b** responses of (8) with $L = 0.3$ s

absolute error (*IAE*), and integral time absolute error (*ITAE*), are calculated for each setting. Closed-loop response curves (Figs. 1 and 2) for different PID controllers are presented as follows: CPID (– - –), GA-CPID (– –), APID (- - -), and GA-APID (—).

Table 1 and Fig. 1 present the performance comparison of different controllers for the nonlinear process of (7) with $L = 0.3$ and 0.4 s. Figure 1a and 1b shows remarkably improved performance of our proposed GA-APID during both set point change and load disturbance applied at $t = 20$ s, and this fact is clearly established from the various indices of Table 1. Responses of the integrating system of (8) with two different values of dead time ($L = 0.2$ and 0.3 s) are shown in Fig. 2. Table 2 provides the detailed performance comparison. In this case also, we find similar performance of GA-APID, though not to the same extent as that of the previous example.

From the above performance analysis, we observe that like CPID, GA-CPID also fails to provide acceptable performance due to excessively large overshoot,

Table 1 Performance analysis of (7)

L(s)	Controllers	%OS	t_r(s)	t_s(s)	IAE	ITAE
0.3	CPID	66.10	1.4	9.3	4.12	46.60
	GA-CPID	51.60	1.4	6.8	3.11	32.76
	APID	19.18	1.7	10.6	2.95	34.88
	GA-APID	0.97	2.6	2.1	1.93	20.10
0.4	CPID	83.11	1.5	13.3	5.78	72.40
	GA-CPID	60.77	1.6	7.4	3.81	41.39
	APID	25.15	1.7	10.8	3.27	39.45
	GA-APID	0.97	4.0	3.3	2.31	28.01

Table 2 Performance analysis of (8)

L(s)	Controllers	%OS	t_r(s)	t_s(s)	IAE	ITAE
0.2	CPID	77.50	1.1	10.2	3.45	27.19
	GA-CPID	60.21	1.1	6.5	2.22	15.28
	APID	28.75	1.4	11.0	2.46	19.44
	GA-APID	24.19	1.2	10.6	2.14	16.09
0.3	CPID	102.20	1.2	17.1	5.67	58.74
	GA-CPID	78.23	1.2	7.8	3.14	25.72
	APID	33.80	1.3	11.0	2.69	24.26
	GA-APID	19.54	1.2	11.8	2.42	21.27

possibly due to their linear control law, whereas our GA-optimized adaptive PID (GA-APID) can significantly improve the performance over APID, which justifies the usefulness of GA for further enhancement of APID [8].

4 Conclusion

In this work, we studied the performance of (GA-APID) controller for nonlinear and integrating systems with dead time under both set point change and load disturbance. Simulation results revealed that GA-APID is capable of providing remarkably improved servo as well as regulatory performance compared to even GA-CPID, and significantly overall improved performance in comparison with the recently reported APID.

References

1. Shinsky, F.G.: Process Control Systems—Application, Design, and Tuning. McGraw-Hill, New York (1998)
2. Astrom, K.J., Hang, C.C., Person, P., Ho, W.K.: Towards intelligent PID control. Automatica **28**(1), 1–9 (1992)
3. Ang, K.H., Chong, G.C.Y., Li, Y.: PID control system analysis, design, and technology. IEEE Trans. Control Sys. Technology **13**(4), 559–576 (2005)
4. Dey, C., Mudi, R.K., Simhachalam, D.: An auto-tuning PID controller for integrating plus dead-time processes. Adv. Mater. Res. **403–408**, 4934–4943 (2012)
5. Dey, C., Mudi, R.K., Simhachalam, D.: A simple nonlinear PD controller for integrating processes. ISA Trans. **53**(1), 162–172 (2014)
6. Seborg, D.E., Edgar, T.F.: Adaptive control strategies for process control: a survey. AICHE J. **32**(6), 881–913 (1986)
7. Kristiansson, B., Lennartson, B.: Robust and optimal tuning of PI and PID controllers. IEE Proc. Control Theory Appl. **149**(1), 17–25 (2002)
8. Dey, C., Mudi, R.K.: An improved auto-tuning scheme for PID controllers. ISA Trans. **48**(4), 396–409 (2009)
9. Mudi, R.K., Dey, C.: Performance improvement of PI controllers through dynamic set-point weighting. ISA Trans. **50**, 220–230 (2011)
10. Dey, C., Mudi, R.K., Lee, T.T.: Dynamic set-point weighted PID controller. Control Intell. Syst. **37**(4), 212–219 (2009)
11. Mudi, R.K., Dey, C., Lee, T.T.: An improved auto-tuning scheme for PI controllers. ISA Trans. **47**, 45–52 (2008)
12. Goldberg, D.E.: Genetic Algorithm in Search Optimization and Machine Learning. Addison-Wesley Publishing Co., Inc., MA (1989)
13. Ziegler, J.G., Nichols, N.B.: Optimum settings for automatic controllers. ASME Trans. **64**, 759–768 (1942)

Text Mining for Phishing E-mail Detection

Masoumeh Zareapoor and K.R. Seeja

Abstract Phishing e-mails are threats to online banking transactions as it mislead the customer to disclose their valuable information which results in monetary losses. Common approach is to extract some specific features from phishing e-mails in a semiautomatic way by using small scripts which is a very tedious process. This paper proposes text mining for extracting distinguishing features from a collection of e-mails consists of both phishing and legitimate for better detection of phishing attack. Proposed method first convert the e-mails to a vector representation and then feature selection techniques are used for selecting best features for classification. The proposed method is evaluated by using a data set collected from the HamCorpus of SpamAssasssin project (legitimate e-mail) and the publicly available Phishing-Corpus (phishing e-mail) and found that text mining-based phishing detection is simple, fast, and more accurate than the state-of-the-art methods.

Keywords Text mining · Phishing · Classification · Feature selection

1 Introduction

In recent years, e-mails have become a common and important medium of communication for most internet users. Phishing e-mails are a type of semantic attacks that are created by malicious people to mimic real e-banking e-mails. Generally, the attackers send large number of fake e-mails pretending to be from legitimate and well-known financial organizations such as banks or online banking service providers such as PayPal. In the content of e-mails, the attacker insists the victims to enter or update

M. Zareapoor · K.R. Seeja (✉)
Department of Computer Science, Jamia Hamdard University, New Delhi, India
e-mail: seeja@jamiahamdard.ac.in

M. Zareapoor
e-mail: mzarea@jamiahamdard.ac.in

L.C. Jain et al. (eds.), *Intelligent Computing, Communication and Devices*,
Advances in Intelligent Systems and Computing 308,
DOI 10.1007/978-81-322-2012-1_8

their personal and sensitive information for avoiding of fraud or more security. Most of the phishing e-mails have high visual similarities to that of the financial organization that they mimic. Some of these e-mails look exactly like the real ones and incautious internet users easily fall into this kind of scam. Victims of e-banking phishing e-mail expose their bank account number, password, credit card number, and other important information needed for financial transaction to the attacker. The attacker then misuses this information to make transactions from the victims account. This issue not only affects normal users of the internet, but also causes a big problem for companies and organizations those are misused by the attackers.

Phishing attacks can be classified into two main categories [1]—deceptive phishing and malware phishing. Malware phishing refers to malicious software that can be installed on victim's machines. Malware can be introduced as an e-mail attachment, as a downloadable file from a web site, or by exploiting known security vulnerabilities. The main focus of this research was deceptive phishing that is commonly done through e-mails.

2 Related Work

Numerous techniques have been developed to overcome the phishing attack problem. They include black listing and white listing [1], network and content based filtering [2], firewalls [1, 3], client side tool bars [1–3], server side filters [2, 3], and user awareness [4]. Each of these techniques has merits as well as demerits. Among this, the most popular technique is feature-based filtering. Server side filters and classifiers extract features from the e-mail and then train a classifier to classify the e-mail as phishing or legitimate. SpamAssassin [5] is a widely used rule-based host level filter. Attackers try to find out the rules and thus bypass these filters by appropriately constructing the e-mail. PILFER [6] is a phishing e-mail classifier that is trained using 10 features extracted from e-mails, but it shows high misclassification rate. Abu-Nimeh et al. [7] extracted 43 features from a collection of ham and phishing e-mails and used a collection of classifiers to evaluate their performance. They found that the random forest classifier outperforms the others with these features. Many other researchers such as Miyamoto et al. [8], Fette et al. [6], Toolan et al. [9], and Basnet and Sung [10] also performed similar studies on machine learning techniques with different set of features extracted by using different feature selection techniques. In all these research works, the feature identification is performed manually and extracted by using scripts. It is found that most of the research in this area is enhancement of the feature set. The focus of the proposed work is automatic extraction of features by using text mining techniques.

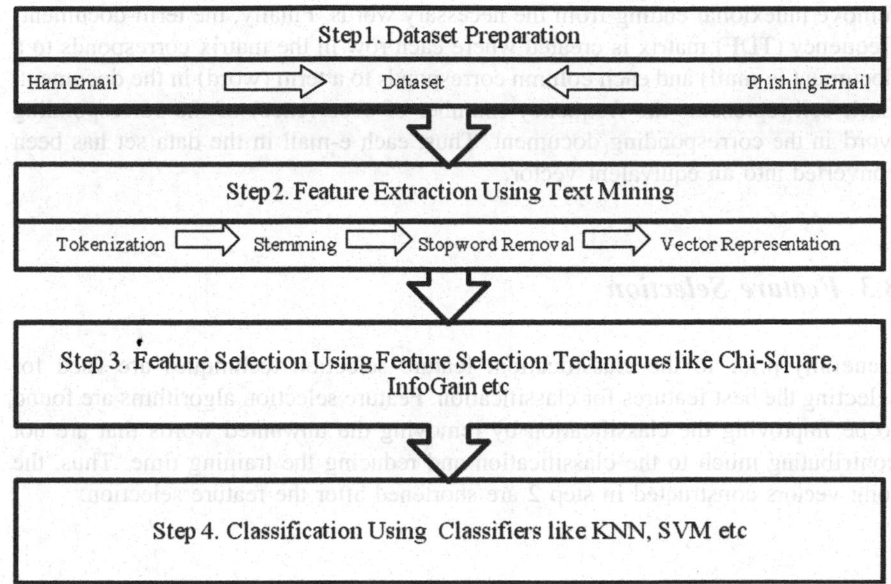

Fig. 1 Proposed methodology

3 Proposed Methodology

The proposed methodology is outlined in Fig. 1.

3.1 Data set Preparation

Data set is prepared by collecting a group of e-mails from the publicly available corpus of legitimate and phishing e-mails. Then the e-mails are labeled as legitimate and phishing correspondingly.

3.2 Feature Extraction Using Text Mining

Text mining techniques were used in this research for the automatic extraction of the word-based features from the data set prepared in step 1. Tokenization is performed to separate words from the e-mail by using white space (space, tab, newline) as the delimiter. Then the words that do not have any significant importance in building the classifier are removed. This is called stop word removal and stop words are words such as a, the, and that. Then stemming is performed to remove inflexional ending from the necessary words. Finally, the term-document-frequency (TDF) matrix is created where each row in the matrix corresponds to a document (e-mail) and each column corresponds to a term (word) in the document. Each cell represents the frequency (number of occurrence) of the corresponding word in the corresponding document. Thus, each e-mail in the data set has been converted into an equivalent vector.

3.3 Feature Selection

Generally prior to the classification, feature selection techniques are used for selecting the best features for classification. Feature selection algorithms are found to be improving the classification by removing the unwanted words that are not contributing much to the classification and reducing the training time. Thus, the long vectors constructed in step 2 are shortened after the feature selection.

3.4 Classification

Classification models are built by using the best features selected in step 3 for classifying the data set into phishing and legitimate.

4 Implementation

The two publicly available data sets selected for testing the proposed methodology are the Ham Corpus from the SpamAssassin project [5] and the publicly available Phishing Corpus [11]. Then the e-mails collected from the SpamAssassin ham corpora are labeled as legitimate and those collected from the PhishingCorpus as phishing. The size of the data set is shown in Table 1.

These e-mails are then converted into TDF matrix by following the procedure in Sect. 3.2. The matrix has 1,067 columns corresponds to features and 1,900 rows corresponds to different e-mails. Three different feature selection techniques namely chi-square [12], InfoGain [13], and GainRatio [14] were used for selecting the best 15 features. Then five different well-known classifiers namely Naïve Bayes [12], Random Forest [10, 12, 15, 16], Support Vector Machine [15, 17],

Table 1 Data set

Number of phishing e-mails	Number of legitimate e-mails	Total number of e-mails
500	1,400	1,900

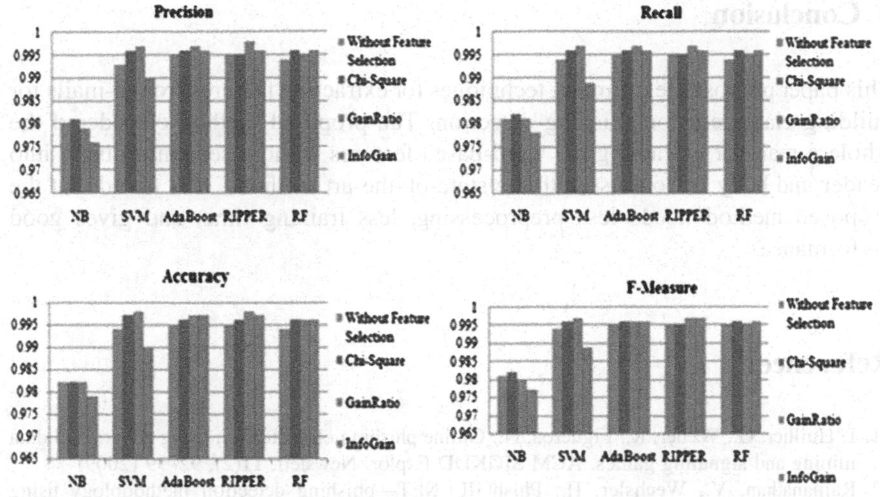

Fig. 2 Performance of classifiers

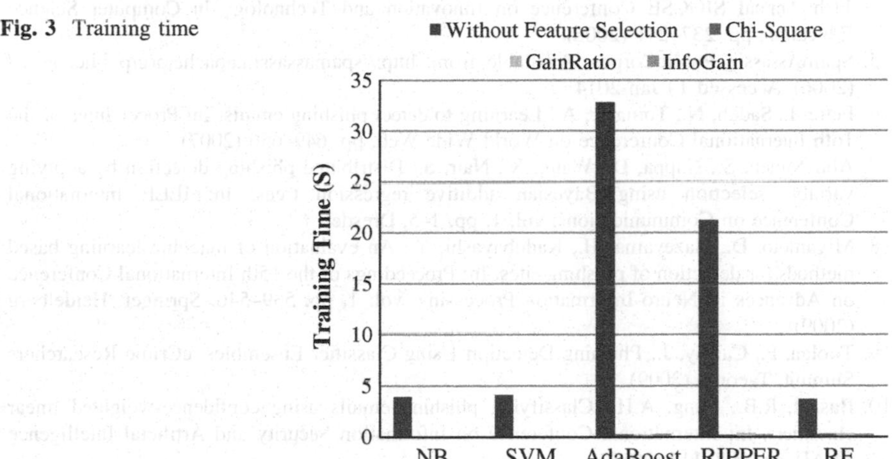

Fig. 3 Training time

Ripper [16], and AdaBoost [2, 3] were used to evaluate the effect of these feature selection techniques in classification. Four relevant performance metrics—Accuracy, Precision, Recall, and F-measure [8, 15] were used for evaluation. Figure 2 shows the complete comparison of the selected feature selection techniques and classifiers. The performance of the classifiers is also evaluated in terms of training time and is shown in Fig. 3.

5 Conclusion

This paper proposes text mining techniques for extracting features from e-mails for building classifiers for phishing detection. The proposed method considered the whole e-mail for extracting the word-based features without separating them into header and body in contrast with the state-of–the-art methods. It is found that the proposed method needs less preprocessing, less training time, and gives good performance.

References

1. L'Huillier, G., Weber, R., Figueroa, N.: Online phishing classification using adversarial data mining and signaling games. ACM SIGKDD Explor. Newslett. **11**(2), 92–99 (2009)
2. Ramanathan, V., Wechsler, H.: PhishGILLNET—phishing detection methodology using probabilistic latent semantic analysis, AdaBoost and co-training. J. Inf. Secur. **2012**, 1–22 (2012)
3. Ramanathan, V., Wechsler. H.: Phishing detection and impersonated entity discovery using conditional random field and latent dirichlet allocation. J. Comput. Secur. **34**, 123–139
4. Robila, S.A., Ragucci, J.W.: Don't be a phish: steps in user education. In: Proceedings of the 11th Annual SIGCSE Conference on Innovation and Technology in Computer Science Education, pp. 237–241 (2006)
5. SpamAssassin PublicCorpus: Available from: http://spamassassin.apache.org/publiccorpus/ (2006). Accessed 14 Jan 2014
6. Fette, I., Sadeh, N., Tomasic, A.: Learning to detect phishing emails. In: Proceedings of the 16th International Conference on World Wide Web, pp. 649–656 (2007)
7. Abu-Nimeh, S., Nappa, D., Wang, X., Nair, S.: Distributed phishing detection by applying variable selection using Bayesian additive regression trees. In: IEEE International Conference on Communications, vol. 1, pp. 1–5, Dresden
8. Miyamoto, D., Hazeyama, H., Kadobayashi, Y.: An evaluation of machine learning based methods for detection of phishing sites. In: Proceedings of the 15th International Conference on Advances in Neuro-Information Processing, vol. 1, pp. 539–546. Springer, Heidelberg (2009)
9. Toolan, F., Carthy, J.: Phishing Detection Using Classifier Ensembles. eCrime Researchers Summit, Tacoma (2009)
10. Basnet, R.B., Sung, A.H.: Classifying phishing emails using confidence-weighted linear classifiers. In: International Conference on Information Security and Artificial Intelligence (ISAI), pp. 108–112 (2010)
11. PhishingCorpus: Available from: http://monkey.org/wjose/wiki/doku.php (2006). Accessed 14 Jan 2014
12. Roglia, E., Cancelliere, R., Meo, R.: Classification of chestnuts with experiments on feature selection and noise. Universit'a di Torino, Dipartimento di Informatica corso Svizzera, Italy
13. L'Huillier, G., Hevia, A., Weber, R., Rıos, S.: Latent semantic analysis and keyword extraction for phishing classification. Department of Computer Science, University of Chile
14. Karegowda, A.G.: Comparative study of attribute selection using gain ratio and correlation based feature selection. Int. J. Inf. Technol. Knowl. Manage. **2**, 271–277

15. Abu-Nimeh, S., Nappa, D., Wang, X., Nair, S.: A comparison of machine learning techniques for phishing detection. In: Proceedings of the eCrime Researchers Summit, vol. 1, pp. 60–69, Pittsburgh
16. DNSBL: Spam database lookup. Available from: http://www.dnsbl.info/
17. Huang, H., Qian, L., Wang, Y.: A SVM-based technique to detect phishing URLs. Inf. Technol. J. **11**, 921–925 (2012)

A Multi-objective Cat Swarm Optimization Algorithm for Workflow Scheduling in Cloud Computing Environment

Saurabh Bilgaiyan, Santwana Sagnika and Madhabananda Das

Abstract As the world is progressing towards faster and more efficient computing techniques, cloud computing has emerged as an efficient and cheaper solution to such increasing and demanding requirements. Cloud computing is a computing model which facilitates not only the end-users but also organizational and other enterprise users with high availability of resources on demand basis. This involves the use of scientific workflows that require large amount of data processing, which can be costly and time-consuming if not properly scheduled in cloud environment. Various scheduling strategies have been developed, which include swarm-based optimization approaches as well. Due to the presence of multiple and conflicting requirements of users, multi-objective optimization techniques have become popular for workflow scheduling. This paper deals with cat swarm-based multi-objective optimization approach to schedule workflows in a cloud computing environment. The objectives considered are minimization of cost, makespan and CPU idle time. Proposed technique gives improved performance, compared with multi-objective particle swarm optimization (MOPSO) technique.

Keywords Cloud computing · Workflow scheduling · Multi-objective cat swarm optimization (MOCSO) · Cost minimization · Makespan · CPU idle time

S. Bilgaiyan (✉) · S. Sagnika · M. Das
School of Computer Engineering, KIIT University, Bhubaneswar, Odisha, India
e-mail: saurabhbilgaiyan01@gmail.com

S. Sagnika
e-mail: santu.hmm@gmail.com

M. Das
e-mail: mndas_prof@kiit.ac.in

L.C. Jain et al. (eds.), *Intelligent Computing, Communication and Devices*,
Advances in Intelligent Systems and Computing 308,
DOI 10.1007/978-81-322-2012-1_9

1 Introduction

In the last two decades, online computing services have become more popular than traditional offline computing services. Cloud computing has emerged as a dominant processing environment which has made it uncomplicated for demanding users to access their required services from among a wide variety of available and configurable computing resources. It shifts computation and information from client machines to over the networks where cloud service providers are connected [1–3].

Scheduling of tasks on resources in a network-based computing environment, such as cloud computing, is always a challenging task. It is the process of allocating limited number of resources among a set of tasks that require the services of these resources. The main aim of scheduling is to minimize the cost and time of task completion while maximizing the quality of services (QoS). Task scheduling on cloud is termed as a NP-complete problem because of dynamic nature of cloud environment [4, 5].

Most real-life scheduling applications require scheduling which satisfies multiple objectives, which are generally contradictory. Hence, the requirement arises to perform multi-objective scheduling that can satisfy such multiple objectives all together. For such problems, there is no solitary solution, but generally a group of unique solutions that achieve trade-off between the objectives can be found out. These set of solutions are known as non-dominated solutions and are represented graphically by Pareto front. Multi-objective workflow scheduling consists of a set of user-defined conflicting requirements [4, 6].

This paper aims to achieve optimization, considering the objectives as minimization of computation cost, makespan and CPU idle time. The authors propose an MOCSO technique to achieve scheduling. The proposed technique has been implemented, and the results have been compared with multi-objective particle swarm optimization (MOPSO) technique. This method MOCSO achieved optimal results in lesser number of iterations than MOPSO. Good convergence is achieved as discussed in the results.

2 Related Work

Scientific workflows represent a set of computational tasks having dependencies between them. Different applications are modelled as workflows for computation [7]. A major challenge is the allocation of these tasks in a manner to reduce the execution time and cost. The size of relocating data and the associated overhead both lead to a huge amount of data processing. So, various techniques have been discussed under this section for scheduling workflows in cloud computing systems [8].

ACO, BCO, GA and PSO have been applied to solve scheduling strategies in cloud systems that are market oriented. ACO was found to show the best performance among all [9].

A multi-objective PSO model has been designed to find an optimal solution, minimize task execution cost, transfer time and task execution time. It finds best trade-off and also effectively utilizes cloud resources and improves QoS [10].

A heuristic algorithm for static workflow scheduling comprising energy consumption, makespan, reliability and economic cost is defined that performs better than bicriteria heuristic algorithms [11].

A PSO algorithm is proposed for batch processing workflow scheduling. In this proposed technique, generational distance (GD) and execution time are taken as performance measures where GD represents the proximity between obtained solution and actual Pareto solution. In this proposed work, a substitute deep memory with particle swarm optimization (DMPSO) is implemented using dynamic grouping and scheduling optimization (DGSO) and standard PSO. The experimental results show that DGSO gives better searching and average GD while the execution time of the algorithm increases with the number of iterations [12, 13].

An artificial bee colony optimization algorithm is proposed for workflow scheduling in cloud computing environment. The proposed algorithm optimizes the computation time and server utilization. The experiment is done using CloudSim tool, and results are compared to existing GA. The proposed technique shows a better performance over the GA [14].

An ant colony optimization algorithm (ACO) is introduced for task scheduling in a cloud computing environment. The basic ACO is enhanced for minimizing the execution time and makespan of the all scheduled tasks. The experiment is carried out using CloudSim toolkit, and results show the better performance of extended ACO over existing simple ACO [15].

3 Workflow Design

The authors have characterized a general workflow as a directed acyclic graph (DAG). The DAG is denoted by $G = (W, D)$. W represents a set of tasks in the workflow, where $W = \{W_1, W_2, W_3, \ldots W_n\}$. D represents the dependencies among these tasks. These tasks need to be mapped on a set of available resources $S = \{S_1, S_2, S_3, \ldots S_n\}$, all geographically distributed throughout the world. Each resource has its own storage/memory, designated as $M = \{M_1, M_2, M_3, \ldots M_n\}$.

The execution time and cost of every task on all resources are known as per a hypothetically assumed pricing policy, within range defined by some policies of Gogrid and Amazon Web Services. The cost of transferring data between tasks is also known.

Figure 1 shows a sample workflow model having 14 tasks along with their interdependencies that the authors have used for experimenting. The tasks are represented by $W_1, W_2, \ldots, W_{14}$, and their dependencies are shown as $d_{i,j}$, which means a dependency on task W_j on W_i. In Fig. 2, the available 4 resources are depicted in the form of $S_1, \ldots, S_4$ with the per unit data transfer cost between 2 any two resources S_i, and S_j is denoted by $tc_{Si,Sj}$. Size of transferred data is assumed to be constant.

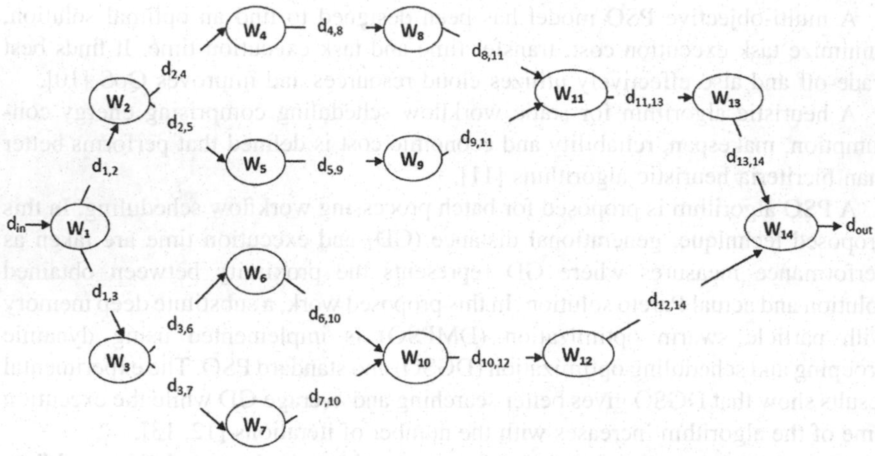

Fig. 1 A sample experimental workflow

Fig. 2 A sample resource distribution architecture with local storages

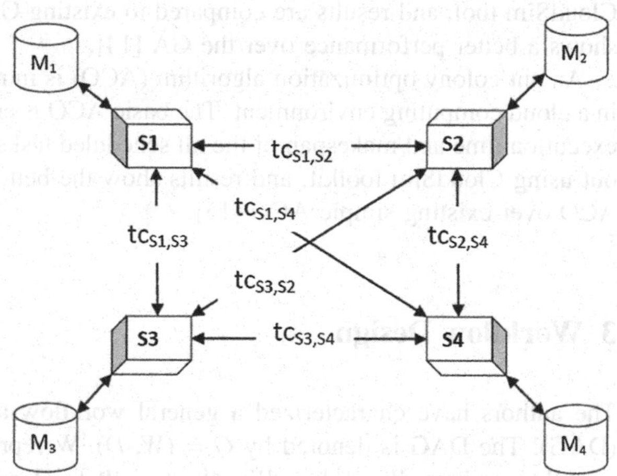

4 Multi-objective Scheduling Model

The authors represent the current multi-objective problem in the following mathematical format.

$$T\_\text{cost}(S_x) = \sum_y (\text{ec}_{yx} + \sum_z \text{tc}_{yz}), \quad \forall\ \text{Sch}(W_y) = S_x \text{ and } \text{Sch}(W_z) \neq S_x \qquad (1)$$

where $T\_cost(S_x)$ is the total cost of executing tasks assigned to resource S_x and transmitting data to other tasks dependent on those tasks. ec_{yx} is the cost of executing task W_y on S_x. tc_{yz} represents transmission cost between tasks W_y and

W_z, which are on different resources. Sch is a specific scheduling map, and Sch(W_y) is the resource where W_y is being executed.

$$\text{Max\_cost(Sch)} = \text{Max}(T\_\text{cost}(S_x)), \quad \forall\, S_x \tag{2}$$

where Max_cost(Sch) signifies that the tasks are fairly distributed over resources.

$$T\_\text{time}(S_x) = \sum_y \text{tt}_{yx} \tag{3}$$

where T_time(S_x) signifies total time of executing all tasks running on S_x. Here, *tt* gives the total running time of task W_y on S_x.

$$\text{Makespan(Sch)} = \max(T\_\text{time}(S_x)), \quad \forall\, S_x \tag{4}$$

where Makespan(Sch) indicates the total time between start and finish of total schedule.

$$T\_\text{idle(Sch)} = \sum_x (\text{Makespan(Sch)} - T_{\text{time}(Sx)}) \tag{5}$$

where T_idle(Sch) indicates the amount of time the resources remain idle, found out by summation of the idle time of each resource S_x till all tasks are completed.

$$\text{Minimize(Max\_cost(Sch), Makespan(Sch), } T\_\text{idle(Sch))}, \quad \forall\, \text{Sch} \tag{6}$$

Equation 6 is the fitness function for this problem.

5 General CSO Algorithm

The common behaviour of cats in real world has inspired the development of a new swarm-based optimization technique known as cat swarm optimization (CSO), as proposed in 2007 by Chu and Tsai. The activity of cats in general includes spending maximum time resting but alert, with slow and calculated methods (*Seeking mode)* else while chasing targets, they move with high velocity, converging towards the target (*Tracing mode*) [16–18]. The CSO optimization technique makes use of this behaviour of cats to search complex solution spaces for optimal solutions, using an initial population of cats that are randomly divided into seeking and tracing modes, as per a defined mixture ratio (MR). The cats move closer to solutions by updating best results in the memory. This process continues iteratively by redistributing cats into either mode each time, till all cats achieve the best solution. A universal CSO algorithm can be described as follows [19–21].

CSO algorithm
1. Generate N cats over required number of dimensions D
2. Allocate random velocities to all cats
3. Randomly distribute cats to seeking and tracing modes as per defined MR
4. Calculate fitness of all cats and memorize the non-dominated cats
5. For each cat, if cat is in seeking mode, perform seeking mode operations, else perform tracing mode operations on it and move it to its new position
6. If termination condition is not satisfied, goto step 3, else stop

6 Proposed Algorithm

The authors put forward an approach that utilizes this CSO technique to address the multi-objective scheduling problem as described in Sect. 4. The different tasks of a workflow are represented by dimensions, and each cat denotes a schedule between the set of tasks and available resources, and all cats follow the steps of CSO till the best schedule is achieved or termination criteria is reached. Here, best schedule refers to a set of non-dominated optimal schedules that achieve balance between the required objectives, namely computation cost, makespan and CPU idle time. These various solutions form a graph which is known as Pareto front.

6.1 Seeking Mode

Seeking mode represents the resting condition of cats, wherein they remain alert and look around to their surroundings. This mode uses certain parameters that determine a cat's behaviour. Those are the following:

- Seeking memory pool (SMP)—The number of replicas for each cat that are to be made. One among them will replace the original cat later.
- Count of dimension to change (CDC)—The number of allotments in each copy that will be modified. Each copy will be evaluated, and one of the best will be selected for replacement.

When a cat is in seeking mode, it performs the following steps.

Seeking mode steps
1. Generate replicas of the cat as per SMP
2. Randomly change dimensions of each replica as per CDC
3. Assess fitness values of all replicas
4. Find the best non-dominated replicas
5. Substitute the cat with a randomly selected non-dominated replica

6.2 Tracing Mode

Tracing mode depicts movement of cats towards targets with high velocity, while spending a high amount of energy. In tracing mode, a cat performs the following steps.

Tracing mode steps	
1. Find the new velocity $V_{i\_d}^{t+1}$ for cat i by using the formula $V_{i\_d}^{t+1} = w^* V_{i\_d}^{t} + c^* r^* (X_{best\_d} - X_{i\_d}^{t})$ where $V_{i\_d}^{t}$ is the velocity at tth iteration, $X_{i\_d}^{t}$ is the position in tth iteration, X_{best_d} is the current global best position in dth dimension, c represents a constant and r is a random number between 0 and 1	(7)
2. Change cats to a new position to the next best position $X_{i\_d}^{t+1}$ by adding the new velocity as per $X_{i\_d}^{t+1} = X_{i\_d}^{t} + V_{i\_d}^{t+1}$	(8)
3. Limit the updated position of cat within desired range	
4. Calculate fitness of all cats	
5. Update the set of solutions with non-dominated cats	

Figure 3 represents the complete algorithm flow.

7 Input Data for Experiment

The authors have assumed a hypothetical workflow having 14 tasks to be distributed over 4 resources that are located in different countries across the world, as illustrated previously. The cost of execution and communication is presumed on the basis of pricing policies followed by some well-known service providers, i.e. Amazon Web Services, Mosso, Gogrid, etc. The execution time for each task for each resource is presumed as suitable for experimenting. The experiments have been performed, and resultant graphs were generated using MATLAB as a tool. Following are the input data tables, where Table 1 represents the execution time for tasks on resources, Table 2 gives communication costs between different resources and Table 3 denotes execution time of each task on each resource.

8 Results and Discussion

The authors have performed various experiments using the proposed algorithm on MATLAB tools and have compared the results with an existing MOPSO algorithm for the same problem. The population size was preset at 50 while the number of iterations has been varied from 100 to 300. The following results have been obtained.

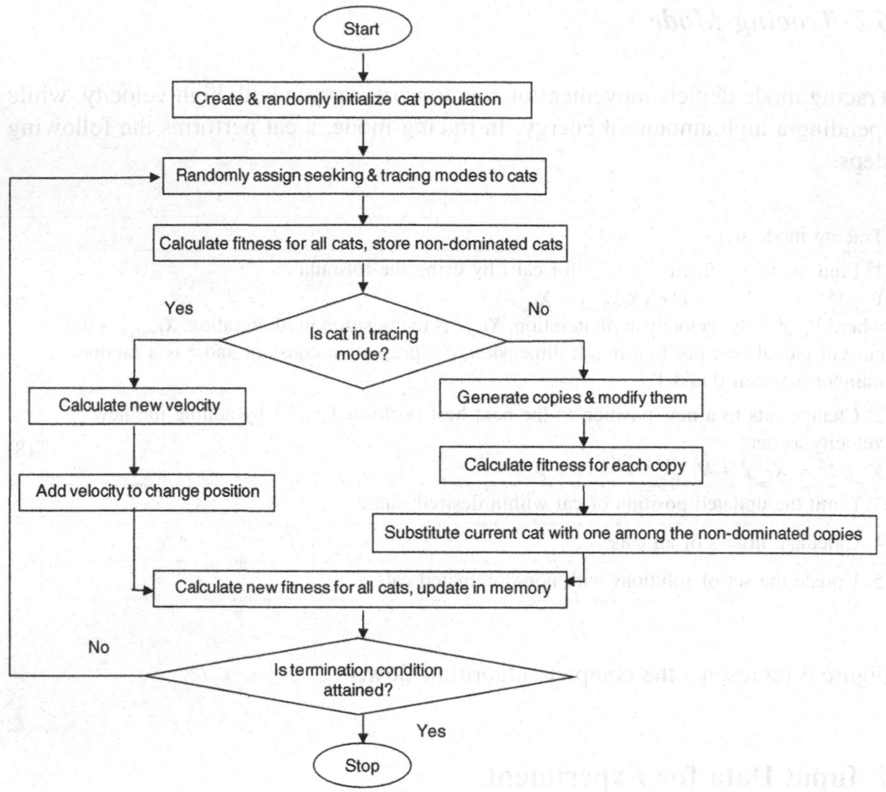

Fig. 3 Basic steps of MOCSO

8.1 Output Graphs Generated by MATLAB

See Figs. 4, 5, and 6.

8.2 Analysis of Resultant Graphs

Figures 4, 5, and 6 show pairs of graphs where in each figure, (a) represents Pareto front obtained by MOPSO and (b) represents Pareto front obtained by MOCSO for a particular number of iterations. By comparing the graph pairs, it can be observed that MOCSO approach generally provides more number of solutions than MOPSO for a fixed number of iterations, which are also closer to and better distributed over the Pareto front. On analysing the results for different iterations, it can be seen that MOCSO achieves good convergence and faster attainment of a convex Pareto front as compared with MOPSO. The rationale behind better performance of

Table 1 Execution cost matrix (in cents)

	S1	S2	S3	S4
T1	1.56	1.59	1.59	1.72
T2	1.83	2.01	2.24	2.14
T3	1.65	1.68	1.71	1.82
T4	2.21	2.00	2.33	2.48
T5	2.11	2.03	1.98	1.87
T6	1.79	2.19	1.63	1.50
T7	1.50	1.57	1.88	1.72
T8	2.50	2.03	2.16	2.42
T9	1.57	1.93	1.82	1.75
T10	2.19	2.03	2.25	2.36
T11	1.50	1.83	1.76	2.23
T12	2.31	2.98	2.50	2.27
T13	1.52	1.93	1.61	1.74
T14	2.39	2.14	2.04	1.96

Table 2 Communication cost matrix (in cents/MB)

	S1	S2	S3	S4
S1	0	0.12	0.17	0.07
S2	0.12	0	0.20	0.11
S3	0.17	0.20	0	0.13
S4	0.07	0.11	0.13	0

MOCSO can be attributed to the fact that it exhibits intelligent updating of positions rather than the random updating mechanism followed by MOPSO, which saves energy and hence increases speed and efficiency of reaching at best solutions.

9 Conclusion and Future Scope

This paper has presented a new and more efficient approach to use a swarm-based technique to solve the multi-objective scheduling problem in a cloud computing environment. This technique has proved to be faster and highly convergent over existing MOPSO technique, which is already among the dominant optimization methods. This is owing to the smart mechanism of position updating in MOCSO that reduces unnecessary energy expenditure and moves closer towards the

Table 3 Execution time matrix (in hours)

	S1	S2	S3	S4
T1	0.51	0.44	0.31	0.44
T2	0.93	0.84	0.46	0.84
T3	0.75	0.66	0.52	0.66
T4	1.20	1.15	1.11	1.15
T5	1.08	0.92	0.78	0.92
T6	0.63	0.59	0.52	0.59
T7	0.32	0.24	0.10	0.24
T8	0.91	0.78	0.62	0.78
T9	0.74	0.66	0.91	0.66
T10	0.42	0.36	0.23	0.36
T11	0.55	0.50	0.40	0.50
T12	0.88	0.72	0.53	0.72
T13	0.61	0.54	0.49	0.54
T14	1.16	0.98	0.81	0.98

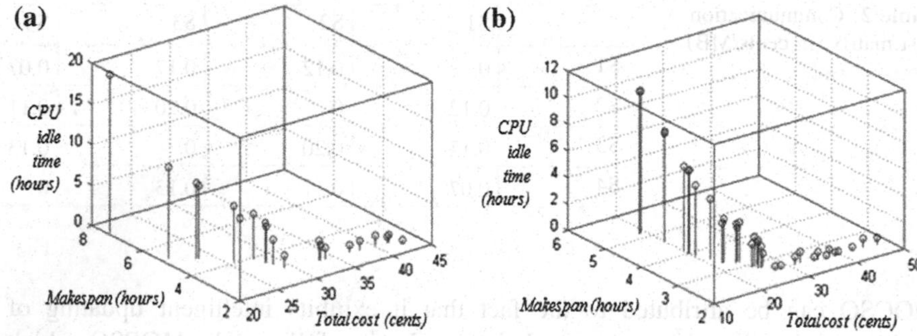

Fig. 4 Optimal results for 100 iterations **a** for MOPSO **b** for MOCSO

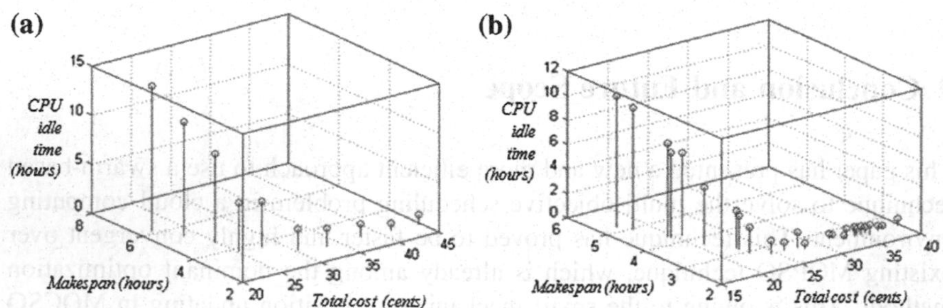

Fig. 5 Optimal results for 200 iterations **a** for MOPSO **b** for MOCSO

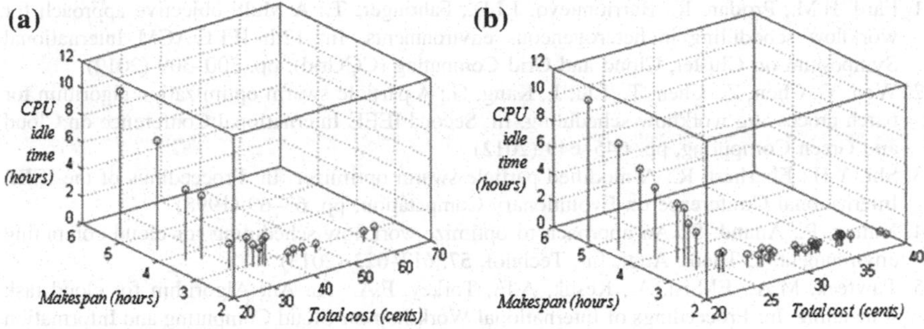

Fig. 6 Optimal results for 300 iterations **a** for MOPSO **b** for MOCSO

solution in each iteration. Future scope in this field can consist of improving the running time of this algorithm and involving more number of real-time conflicting objectives, which the authors believe can be effectively handled by this technique. Finding competent solutions to the workflow scheduling problem on cloud can increase the QoS and extend the reach of cloud computing over even more business and enterprise areas.

References

1. Buyya, R., Yeo, C.S., Venugopal, S., Broberg, J., Brandic, I.: Cloud computing and emerging IT platforms: vision, hype, and reality for delivering computing as the 5th utility. Future Gener. Comput. Syst. **25**, 599–616 (2009)
2. Dikaiakos, M.D., Pallis, G., Katsaros, D., Mehra, P., Vakali, A.: Distributed internet computing for IT and scientific research. IEEE Internet Comput. **13**, 10–13 (2009)
3. Chaisiri, S., Lee, B.S., Niyato, D.: Optimization of resource provisioning cost in cloud computing. IEEE Trans. Serv. Comput. **5**, 164–177 (2012)
4. Fard, H.M., Prodan, R., Fahringer, T.: A truthful dynamic workflow scheduling mechanism for commercial multicloud environments. IEEE Trans. Parallel Distrib. Syst. **24**, 1203–1212 (2013)
5. Jangra, A., Saini, T.: Scheduling optimization in cloud computing. Int. J. Adv. Res. Comput. Sci. Softw. Eng. **3**, 62–65 (2013)
6. Wang, X.J., Zhang, C.Y., Gao, L., Li, P.G.: A survey and future trend of study on multi-objective scheduling. In: Fourth IEEE International Conference on Natural Computation, pp. 382–391 (2008)
7. Gil, Y., Deelman, E., Ellisman, M., Fahringer, T., Fox, G., Gannon, D., Goble, C., Livny, M., Moreau, L., Myers, J.: Examining the challenges of scientific workflows. IEEE Compu. Soc. **40**, 24–32 (2007)
8. Szabo, C., Sheng, Q.Z., Kroeger, T., Zhang, Y., Yu, J.: Science in the cloud: allocation and execution of data-intensive scientific workflows. J. Grid Comput. (2013)
9. Singh, L., Singh, S.: A survey of workflow scheduling algorithms and research issues. Int. J. Comput. Appl. **0975–8887**(74), 21–28 (2013)
10. Ramezani, F., Lu, J., Hussain, F.: Task Scheduling Optimization in Cloud Computing Applying Multi-objective Particle Swarm Optimization. Service-Oriented Computing. Lecture Notes in Computer Science 8274, pp. 237–251. Springer, Berlin (2013)

11. Fard, H.M., Prodan, R., Barrionuevo, J.J.D., Fahringer, T.: A Multi-objective approach for workflow scheduling in heterogeneous environments. In: 12th IEEE/ACM International Symposium on Cluster, Cloud and Grid Computing (CCGrid), pp. 300–309 (2012)
12. Wen, Y., Chen, Z., Chen, T., Liu, J., Kang, G.: A particle swarm optimization algorithm for batch processing workflow scheduling. In: Second IEEE International Conference on Cloud and Green Computing, pp. 645–649 (2012)
13. Shi, Y.H., Eberhart, R.: A modified particle swarm optimizer. In: Proceedings of the IEEE International Conference on Evolutionary Computation, pp. 63–69 (1998)
14. Kumar, P., Anand, S.: An approach to optimize workflow scheduling for cloud computing environment. J. Theor. Appl. Inf. Technol. **57**, 617–623 (2013)
15. Tawfeek, M.A., El-Sisi, A., Keshk, A.E., Torkey, F.A.: An Ant Algorithm for cloud task scheduling. In: Proceedings of International Workshop on Cloud Computing and Information Security (CCIS), pp. 169–172 (2013)
16. Shojaee, R., Faragardi, H.R., Alaee, S., Yazdani, N.: A new cat swarm optimization based algorithm for reliability-oriented task allocation in distributed systems. In: 6th IEEE International Symposium on Telecommunications, pp. 861–866 (2012)
17. Sharafi, Y., Khanesar, M.A., Teshnehlab, M.: Discrete binary cat swarm optimization algorithm. In: 3rd IEEE International Conference on Computer, Control & Communication (IC4), pp. 1–6 (2013)
18. Chu, S.C., Tsai, P.W.: Computational intelligence based on the behavior of cats. Int. J. Innov. Comput. Inf. Control **3**, 163–173 (2007)
19. Pradhan, P.M. Panda, G.: Solving multiobjective problems using cat swarm optimization. Expert Syst. Appl. **39**, 2956–2964 (2011)
20. Santosa, B., Ningrum, M.K.: Cat Swarm optimization for clustering. In: IEEE International Conference on Soft Computing and Pattern Recognition, pp. 54–59 (2009)
21. Tsai, P.W., Pan, J.S., Chen, S.M., Liao, B.Y., Hao, S.P.: Parallel cat swarm optimization. In: Proceedings of the Seventh International Conference on Machine Learning and Cybernetics, Kunming, pp. 3328–3333 (2008)

A Case Study of User Behavior in Information Retrieval

Venkata Udaya Sameer and Rakesh Chandra Balabantaray

Abstract Information retrieval has taken an important turn when the researchers started using user behavior to improve their ranking algorithms. With the advent of user behavior in information retrieval, the interactive information retrieval is beginning to make its mark. In this paper, we discuss how user behavior is being used in information retrieval. We survey various strategies used for incorporating user behavior into information retrieval. The key fact is that taking absolute feedback about whether the retrieved documents are relevant or not is very difficult, and if we can take the implicit feedback from the user in the form of user behavior, we can arrive at a better learning function for the algorithm. We, in this paper, would like to provide a case study of various approaches in using user behavior for information retrieval.

Keywords User behavior · Information retrieval · Click-through data · Search · Relevance · Precision · Learning · SVM · Rank · Score

1 Introduction

There is a huge need to satisfy the end user when it comes to the relevancy of the documents/data being retrieved. In IR terms, the precision has to be improved for the end user at the first/top few results. In recent evolutions of IR systems, user behavior has been taken into account to provide a better information retrieval

V.U. Sameer (✉) · R.C. Balabantaray
Department of Computer Science and Engineering, International Institute of Information Technology, Bhubaneswar 751003, India
e-mail: sachmeer4u@gmail.com

R.C. Balabantaray
e-mail: rakeshbray@gmail.com

L.C. Jain et al. (eds.), *Intelligent Computing, Communication and Devices*,
Advances in Intelligent Systems and Computing 308,
DOI 10.1007/978-81-322-2012-1_10

system. When user behavior is taken into account, the objective is to please the user, i.e., to make the user of the system get the documents he is looking for. That thin line between giving the best relevant documents and retrieving the information the user might be looking for is the primary cause in involving the user behavior into information retrieval. The process of involving user behavior in information retrieval makes use of the psychology of the users, classic information retrieval methods, and the field of human–computer interaction (HCI).

Though each user is different in using the information retrieval system, the collective information that can be collected of all users can be very useful. In traditional information retrieval systems, the same result is expected for a query for any user, but using the user behavior, the information retrieval function is personalized. That is the reason many terms have been coined to describe it, such as personalized information retrieval (PIR) and interactive information retrieval (IIR).

2 User Behavior in Information Retrieval

The ranking module is one of the most important modules of a search engine. It can either make or break its success. For a given query, there may be hundreds or thousands of relative documents, but only a few of them are to be shown to the user at a time. Hence, it is very important to fetch the most relevant documents and display them to the user. It is invariably the top 10–20 documents that decide the success of the search engine and so each one of the entries is important.

It is very difficult to get the judgment of relevance done by external assessors. Though there are other means to take explicit feedback from the users such as conducting surveys, they are very costly. So there is a need to take implicit feedback from users into account. One of the advantages of using implicit feedback is that essentially, the feedback is freely available and it captures the natural use of the search engine. Similarly, a search engine could use implicit feedback to adapt to a specific document collection.

Though implicit feedback provides free and vast information about the user's behavior, it is also noisy and has a few biases. Joachims and Radlinski [1] discuss various biases involved in implicit feedback. There exists a rank bias also known as presentation bias, where the pages displayed at the top of the results are most looked at. So even an excellent result at a lower position might get unnoticed compared to that higher up the order. They also show that it is not safe to consider implicit feedback as absolute feedback but to be considered relative. In relative feedback, we can infer mainly from the links that the user did not click on. Suppose there is a page at position 3 which the user clicked on, without clicking the pages at positions 1 and 2. So it means that the user deliberately missed top 2 pages. The general insight here is that we must evaluate user actions in comparison with the alternatives that the user observed before making a decision. This naturally leads to feedback in the form of pairwise relative preferences like "A is better than B." In contrast, if we took a click to be an absolute statement like "A is good," we would face the difficult task of explicitly correcting for the biases. Joachims and Radlinski [1] also stressed upon the use of pairwise preferences. In a controlled user study, they found that pairwise relative preferences extracted from clicks are quite accurate. About 80 % of the pairwise preferences agree with expert labelled data. This is particularly promising, since two human judges only agree with each other about 86 % of the time. It means that the preferences extracted from clicks are not much less accurate than manually labeled data. However, we can collect them at essentially no cost and in much larger quantities. Furthermore, the preferences from clicks directly reflect the actual users' preferences, instead of the judges' guesses at the users' preferences.

Radlinski and Joachims [2] present an approach to learning retrieval functions by analyzing which links the users click on in the presented ranking. This leads to a problem of learning with preference examples such as "for query *q*, document *d*a should be ranked higher than document *d*b." In this formulation, a support vector machine (SVM) algorithm is presented that leads to a convex program and that can be extended to nonlinear ranking functions. Experiments show that the method can successfully learn a highly effective retrieval function for a meta-search engine.

Radlinski and Joachims in [2] address the task of learning rankings of documents from search engine logs of user behavior. In this paper, instead of relying on the passively collected click-through data, they show that an active exploration strategy can provide data that lead to much faster learning. To see the limitations of passively collected data, consider the typical interactions of search engine users. Users very rarely evaluate results beyond the first page, so the data obtained are strongly biased toward documents already ranked highly. Highly relevant results that are not initially ranked highly may never be observed and evaluated, usually leading to the learned ranking never converging to an optimal ranking. To avoid this presentation effect, they propose that the ranking presented to users be optimized to obtain useful data, rather than strictly in terms of estimated document relevance. For example, one possibility would be to intentionally present unevaluated results in the top few positions, aiming to collect more feedback on them. However, such an ad hoc approach is unlikely to be useful in the long run and would hurt user satisfaction. They instead introduce principled modifications that can be made to the rankings presented. These changes, which do not substantially reduce the quality of the ranking shown to users, produce much more informative training data and quickly lead to higher-quality rankings being shown to users.

Agichtein et al. in [3] consider two complementary approaches to ranking with implicit feedback: (1) treating implicit feedback as independent evidence for ranking results and (2) integrating implicit feedback features directly into the ranking algorithm. Zareh Bidoki et al. in [4] suggest A3CRank which is an adaptive ranking technique based on connectivity, content, and click-through data. Their method tries to aggregate ranking algorithms such as BM25, PageRank, and

TF-IDF. They have used reinforcement learning to incorporate user behavior and find a measure of user satisfaction for each ranking algorithm. Furthermore, OWA, an aggregation operator, is used for merging the results of various ranking algorithms. A3CRank adapts itself to user needs and makes use of user clicks to aggregate the results of ranking algorithms.

Song et al. in [5] discuss the problem of relevance judgments and about learning to rank. They generalize pairwise preference judgments to relative judgments and formulate the problem of relative judgments in a formal way and then propose a new strategy called select-the-best-ones to solve the problem. Radlinski et al. in [6] have proposed a taxonomy of duplication that shows how Web results can be redundant in terms of information content (or utility) to users in different ways as exact duplicates, as navigational alternatives, and as equally useful content. Users behave differently when presented with different types of duplicates, suggesting that the utility of the classes of duplicates is different and that they should be treated differently when evaluating the quality of search results. Moreover, if two results are not duplicated, the effect of presentation bias is much smaller than average, hence suggesting that if one could control for duplication, clicks would become a much stronger relevance signal.

Jung et al. in [7] address three issues related to using click data as implicit relevance feedback: (1) How click data beyond the search results' page might be more reliable than just the clicks from the search results' page; (2) whether we can further subselect from this click data to get even more reliable relevance feedback; and (3) how the reliability of click data for relevance feedback changes when the goal becomes finding one document for the user that completely meets their information needs. And they show that to achieve the maximal precision of the feedback data, the "last visited document" of each search session is the more reliable source of implicit relevance feedback data.

3 Case Study

We conducted an experiment to know the importance of implementing user behavior in the ranking process. We have taken Google, Bing, and Yahoo as search engines which do take user behavior, user location, etc, into account and other search engines such as ixquick, DuckDuckgo, and gibiru which do not take information about the user and compare the both.

Table 1 shows the different queries we used for the experiment. In reality, not all the users form the queries correctly. Many times, the queries are ambiguous or there may be more than one match to the search term the user has entered. In this case, the information about the user helps a lot in the ranking process. The below queries are fired by a computer science student studying M. Tech.

Table 1 Queries used for the survey

S. No.	Intended query	Entered query
1	Support vector machine	SVM
2	IIIT Bhubaneswar	IIIT
3	Arvind Kejriwal	Arvind
4	Anna Hazare	Anna
5	Post offices in Bhubaneswar	Post offices
6	A.R. Rehman	Rehman
7	Aam Aadmi Party	Aam Admi
8	A.P.J. Abdul Kalam	Kalam
9	GATE exam in India	GATE
10	Leaky bucket algorithm in communication networks	Leaky bucket

3.1 Scoring

The user who participated in the survey gives the scores. For the top 10 results displayed by both the search engines, the user gives a score. The score would tell how many results in top 10 are relevant for that user. If for a query *q*, the search engine has shown 10 pages, and out of that, the user feels only 8 are relevant, then the score for that query '*q*' against that search engine will be 8.

Taking the top 10 results given by the search engine into account, Table 2 shows the judgement table where the numbers in the cells indicate the relevance score given by the participant in the survey for the results shown by the search engines.

The bar graph 1 shows the comparative study for both the search methodologies being used. The average precision of the method when information about the user is taken into account is 6.8 (average for Bing, Yahoo, and Google), and when it is not taken into account, it is 2.3 (average for the rest). Clearly, incorporating information about user behavior is advantageous.

4 Approaches to Incorporating Implicit Feedback

An approach to incorporating implicit feedback is by learning from pairwise preferences. For a query, a user may prefer document *d*b over *d*a where score (*d*a) > score (*d*b). If we wish to take the absolute feedback, we need to have it in

Table 2 Relevance scores

Query/Search engine	Ixquick	DuckDuckGo	Gibiru	Bing	Yahoo	Google
Query 1	2	1	3	8	3	9
Query 2	2	6	8	10	9	9
Query 3	2	2	2	6	6	8
Query 4	0	1	0	3	2	6
Query 5	0	0	0	5	2	10
Query 6	2	0	0	4	2	5
Query 7	3	9	7	9	9	7
Query 8	2	3	4	7	5	10
Query 9	2	4	0	8	5	10
Query 10	1	1	3	10	8	10
Average score	1.6	2.7	2.7	7	5.1	8.4

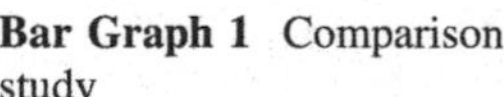

Bar Graph 1 Comparison study

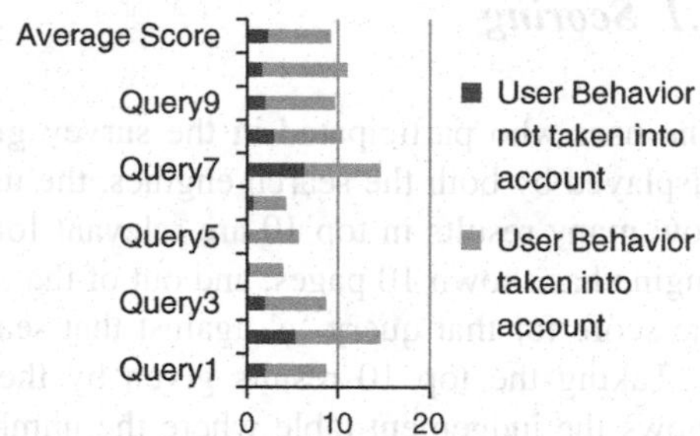

the form *d*a is relevant or *d*a is not relevant. But when we use relative preferences, the equation is always *d*i is better than *d*j. To use these pairwise preferences, translate the problem to binary classification where a positive example would be (*q*, *u*, *d*b, *d*a) and negative example would be (*q*, *u*, *d*a, *d*b) where *q* is the query, u is the user, and *d*a and *d*b are the documents with score (*d*a) > score (*d*b) and user chose to click on *d*b over *d*a. Also, while using pairwise preferences, a utility function can be calculated of the form *f* (*q*, *u*, *d*b) > *f* (*q*, *u*, *d*a) and RankingSVM approach can be used to train the system.

Implicit feedback can be incorporated into the ranking module in two ways. One way is to use the user behavior as independent evidence and re-rank the results obtained by the ranking module using the implicit feedback. The other way is to give place to the implicit feedback while calculating the score itself. The former approach uses machine learning concepts such as Learning to Rank, RankNet, and AdaNet, and the user behavior can be implemented as features which then could be used to tune the ranking process. For a given query q, implicit score is computed for each result d and the score is merged with original rank given by the algorithm.

Zareh Bidoki et al. [4] used a different approach. Though they also used machine learning approaches to adapt to the user needs, they have used operator weighted averaging (OWA) operator to merge results obtained by different ranking algorithms such as BM25, Page Rank and TF-IDF and used reinforcement learning to incorporate user behaviour. Then, use this factor to find weight of each algorithm. Compute the OWA vector for aggregation, and finally, compute the final weight of resulting OWA vector. This approach has its merits. It is scalable to add new algorithm, and also, it is adaptive to user preferences.

One more strategy is to generate pairwise preferences and also to include the entire click information, i.e., by not just noting the click information in the first results' page but also following the user to know the chain of links that he clicks from the first result. The dwell time that is nothing but the time the user spends on a page could decide the relevance of the page. We can keep a threshold for the dwell time and take the list of pages the user found interesting in his chain of links. There are many other features when it comes to the click-through data. They are time on page, cumulative time, time on domain, time on short URL, title overlap, summary overlap, query URL overlap, query length, click frequency, etc.

Calculating the query difficulty level would also contribute to incorporating user behavior in ranking. For queries which are difficult and ambiguous, the information about user would help the retrieval process. If a user enters "IR" as the query, it would be ambiguous for the system whether to retrieve "Information Retrieval" or "Infra-Red". If the system knows that the user is a computer science student, then probability is more for "Information Retrieval" to match the need of the user.

The below are the user behavior features that can be used.

- Browsing features
  - Position: Position of the URL in current ranking.
  - ClickFrequency: Number of clicks for this query, URL pair.
  - ClickProbability: Probability of a click for this query and URL
  - ClickDeviation: Deviation from expected click probability.
  - IsNextClicked: 1 if clicked on next position, 0 otherwise
  - IsPreviousClicked: 1 if clicked on previous position, 0 otherwise
  - IsClickAbove: 1 if there is a click above, 0 otherwise
  - IsClickBelow: 1 if there is a click below, 0 otherwise
- Domain Features
  - TimeOnPage: Page dwell time
  - CumulativeTimeOnPage: Cumulative time for all subsequent pages after search.

- TimeOnDomain: Cumulative dwell time for this domain.
- TimeOnShortUrl: Cumulative time on URL prefix.
- IsFollowedLink: 1 if followed link to result, 0 otherwise.
- IsExactUrlMatch: 0 if aggressive normalization used, 1 otherwise.
- IsRedirected: 1 is initial URL same as final URL, 0 otherwise.
- ClicksFromSearch: Number of hops to reach page from query.

• Query-Text features

- TitleOverlap: Words shared between query and title.
- SummaryOverlap: Words shared between query and snippet.
- QueryURLOverlap: Words shared between query and URL.
- QueryDomainOverlap: Words shared in the domain.
- QueryLength: Number of tokens in the query.
- QueryNextOverlap: Match with next query.

5 Conclusion

In this paper, we tried to survey about personalized information retrieval systems. We discussed the credibility of implicit feedback and the way how it is different from absolute feedback. We also focussed on different approaches to use the implicit feedback in ranking systems through machine learning approaches or by using it as independent evidence. We believe that a lot of information is now available with the rise in use of search engines and that information can be effectively used to improve the retrieval process. As we already stated, we could train the system for a particular document collection as in the case of intranet, or we can train the system according to the classes of users as is the case with Internet and large intranets. Judging the query difficulty level is also a key factor to use implicit feedback. We would like to conclude that the future search systems would be much powerful with the advent of implicit feedback being incorporated.

References

1. Joachims, T., Radlinski, F.: Search engines that learn from implicit feedback. IEEE Comput. **40**(8), 34–40. ISSN 0018-9162
2. Radlinski, F., Joachims, T.: Active exploration for learning rankings from clickthrough data. In: KDD'07. ACM 9781595936097/07/0008
3. Agichtein, E., Brill, E., Dumais, S.: Improving web search ranking by incorporating user behavior information. In: SIGIR'06, Aug 6–11 2006. ACM 1-59593-369-7/06/0008
4. Zareh Bidoki, A.M., Ghodsnia, P., Yazdani, N., Oroumchian, F.: A3CRank: an adaptive ranking method based on connectivity, content and click-through data. Inf. Process. Manage. **46**, 159–169 (2010)

5. Song, R., Guo, Q., Zhang, R., Xin, G., Wenb,J.-R., Yu, Y., Hon, H.-W.: Select-the-best-ones: a new way to judge relative relevance. Inf. Process. Manage. **47**, 37–52
6. Radlinski, F., Bennett, P.N., Yilmaz, E.: Detecting duplicate web documents using clickthrough data. In: WSDM'11. ACM 978-1-4503-0493-1/11/02
7. Jung, S., Herlocker, J.L., Webster, J.: Click data as implicit relevance feedback in web search. Inf. Process. Manage. **43**, 791–807 (2007)

Fuzzy Mapping of Robotic Arm Based on LabVIEW

Prasad Kulkarni and S.L. Patil

Abstract Robotic arm control is difficult task when it comes to find out mathematical model, forward, or inverse kinematics of robotic arm, but if a person has expert knowledge about the controlling of arm and he knows where and how the arm should be moved then one can put fuzzy table lookup approach. This paper deals with simple mapping of robotic arm using this same approach with the help of LabVIEW software and arduino board.

Keywords Fuzzy lookup table · LabVIEW · Arduino board

1 Introduction

The difficult part in robotic arm is to find joint angles given the *x*, *y*, *z* coordinate positions which consist of complex mathematical equations which is commonly an inverse kinematics problem, for trajectory planning for robotic arm inverse kinematics play an important role as we are required to find right set of angles for which arm moves through its specified trajectory [1]. A fuzzy table lookup approach was used by Wang and Mendel for truck backer-upper control which uses human knowledge for controlling of truck [2]. The same heuristic approach was used by Martinez and Bowels for positioning of robotic arm [3]. This approach use persons experience and knowledge for finding out the right set of joint angles of robotic arm.

P. Kulkarni (✉) · S.L. Patil
Department of Instrumentation and control, College of Engineering, Pune, India
e-mail: kulkarni.prasad40@gmail.com

S.L. Patil
e-mail: slp.instru@coep.ac.in

L.C. Jain et al. (eds.), *Intelligent Computing, Communication and Devices*,
Advances in Intelligent Systems and Computing 308,
DOI 10.1007/978-81-322-2012-1_11

This paper demonstrates the same idea of fuzzy mapping on robotic arm using LabVIEW software and arduino board.

2 Implementation of System

The system consists of robotic arm which has 3 DOF (degree of freedom) with attached 4 servo motors, controlling of these servos is through arduino UNO board, which is 8-bit Atmel AVR microcontroller, compatible with LabVIEW. LabVIEW is a system design platform and development environment for graphical programming language. This software allows user to program graphically, and it targets various types of embedded hardware boards such as FPGA, DSP, and Arduino [4]. System runs in open loop and there is no position sensors attached; signals are given through microcontroller to servo motors, as arduino board is compatible with LabVIEW and it has some dedicated blocks present in LabVIEW which are specific to the arduino. All the programming is done using LabVIEW.

3 Designing Fuzzy Systems

This approach uses human knowledge where expert knows what to do, but cannot express, so this knowledge can be expressed in terms of IF-THEN rule [2, 5]. For example, IF something happens THEN what action should be taken will be decided by human knowledge, so human knowledge will acts as black box and inputs will produce outputs depending upon human knowledge; this gives input–output data pairs, and this will form fuzzy systems.

4 A Table LookUp Scheme

Consider set of input and output data pair as given below:

$$(x_1^{(1)}, x_2^{(1)}; y^{(1)}), \left(x_1^{(2)}, x_2^{(2)}; y^{(2)}\right), \left(x_1^{(3)}, x_2^{(3)}; y^{(3)}\right), \ldots$$

where x_1 and x_2 are inputs and y is the output, and $x_1^{(i)}$, $x_2{}^{(i)}$, $y^{(i)}$ are data pairs. The approach is creating fuzzy rules using input, output pair such that $y = f(x_1, x_2)$.

For the inputs x_1, x_2 if there is *m*, *p* linguistic terms then total rules can be obtained are

$$N = m \times p \tag{1}$$

where *N* is no of rules, *m* and *p* are no of linguistic terms for x_1 and x_2, respectively [5].

Same can be applied up to n inputs. In this approach, five steps are involved [2]:

Step 1 Dividing input and output spaces in to fuzzy region:

Here, x_1, x_2, y are the inputs and output, respectively, having some set of input–output data pair and they lie in domain intervals $[x_1^+, x_1^-]$,$[x_2^+, x_2^-]$ and $[y^+, y^-]$. (The domain interval means their maximum–minimum range.). Then divide these inputs and output into regions, e.g., Small(SM), Big(BG), and Medium(ME) called linguistic terms and requires some membership functions to define them and these can be triangular, trapezoidal, and Gaussian or user defined depending upon the application and choice of expert (Table 1).

Step 2 Generating fuzzy rules from data pairs:

In this step, we assign membership values or degree to input and output variables forming rules. For example, $x_1^{(1)}$ assigned 0.4° in SM region, $x_2^{(2)}$ assigned 0.6° in ME region producing 0.8 membership value for output $y^{(3)}$ in the BG region. From this, we can obtain one rule.

IF x_1 is SM and x_2 is ME THEN y is BG.

Here, "and" will produce output when inputs are simultaneously occurring.

Step 3 Assigning a Degree to each rule:

When we have large no of data pair then each pair of data generates one rule, and chances of conflicting rules also increases. Conflicting rule means having same IF part but different THEN part. In order to select a rule out of many conflicting rules, we find degree of rule (D) defined as

$$D(Rule) = \mu(x_1) \times \mu(x_2) \times \mu(y) \qquad (2)$$

where $\mu()$gives membership value for input and output.

It is possible to assign degree of one to each data pair but it will increase no of rules so expert knowledge will tell which rule to assign higher degree.

Step 4 Create the fuzzy rule base

This step forms fuzzy rule table depends upon the number of membership function for each input and output. This will generate number of rules given by Eq. (1). This table will tell which rules are active. Instead of using whole rule base, it is possible use few rules that will satisfy our requirement.

Step 5 Mapping based on fuzzy rule base

This step involves which defuzzification method to use after generating the rule base. Depending upon the "and" and "or" operations rules are evaluated by taking minimum, maximum of degree of input variables, most commonly used defuzzification method is center of gravity (COG) is also used in this paper for mapping.

Table 1 Fuzzy rule base

y				
x	Center	Right	Left m	Between
Center	✔			
Left m			✔	
Right m		✔		
Right				✔
Left				✔

5 Fuzzy System Design for Robotic Arm

This section explains fuzzy mapping of robotic arm and its implementation using LabVIEW software. Here, fuzzy mapping have been applied for two sections:

1. Base of robotic arm, i.e., servo 1.
2. Motors that helps in proper gripping of objects of different lengths, i.e., servos 3, 2.

6 Mapping for Base of Robotic Arm

Mapping of base has been explained by referring Fig. 1, in this figure two half semicircle are plotted, which shows arm has base rotation of 0–180° and region between outer and inner half semicircles is workspace for arm, where arm can move for gripping of object. Here, *x* axis is 46 cm and *y* axis is 23 cm which is maximum reach of arm. Now, mapping in terms of user will move the pointer over this half semicircle at its desired angle so depending upon the *x, y* co-ordinates

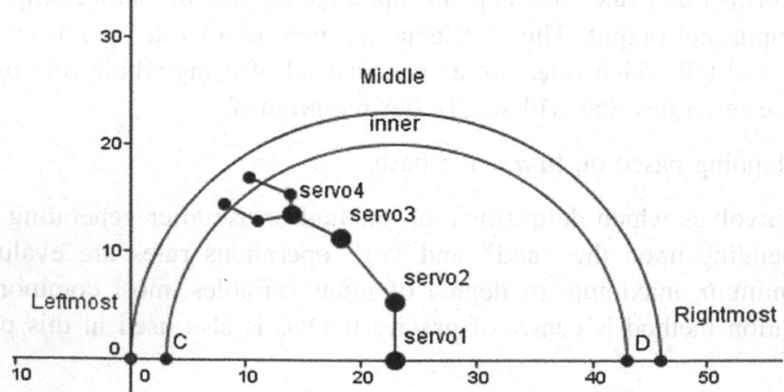

Fig. 1 Mapping of base

angle will generate. For example, $(x, y) = (0, 0)$ left side gives 0°, $(x, y) = (46, 0)$ right side gives 180°, and center gives 90° for $(x, y) = (23, 23)$, like this all angles have been mapped by experimentation. Considering the designing of fuzzy x, y coordinates are inputs and θ_1 is output, i.e., servo1. For x, y axis, 5, 4 linguistic terms have been created, respectively, so maximum possible rules are 20 but after no of trials rules have been reduced to five by considering appropriate membership functions and their ranges.

As pointer moves over the semicircle x, y values increases θ_1 start increasing from 0° and as we move further x decreases and y still increases and θ_1 reaches 180°. No of rules can be reduced by eliminating common rules and by adjusting membership functions. Choice of membership functions also depends upon the application and user (Fig. 2).

7 Mapping for Gripping of Object

In this, mapping is done on servo 3, 2, i.e., θ_3 and θ_2 mapping in terms of height of object will be given as user input through LabVIEW and required angles will be obtained for gripping of object, so θ_3, θ_2 are function of height. Idea behind this is if a person has knowledge about where the object should be hold. For example, if object is small, then arm should be in inclined position and for objects having longer length angles should be such that it should hold in middle. So up to 10 cm length of object, angles have been found out with no of trials and this data is mapped. Around each centimeter of object length, angles have been found out by deciding (Fig. 3) where to hold object. So, eleven rules have been created. $\theta_2 \in [0, 180]$, $\theta_3 \in [60, 180]$, and $h \in [0, 10]$.

This data pair is not fixed we can have different angles for the same height for, e.g., height of 2 cm also obtained by setting $\theta_2 = 174$ and $\theta_3 = 87°$. But problem with this is that even if it reaches 2 cm, it does not help in proper gripping of object. Above, data pairs vary from person to person and its choice (Table 2).

8 Kinematics of Robotic Arm

Mapping of second section is compared with kinematics of robotic arm. Kinematics consists of forward and inverse kinematics (Fig. 4). Forward kinematics is a method for determining the orientation and position of end-effector given the joint angles and link lengths [6]. In the figure shown above $L_0 = 7$ cm, $L_1 = 10$ cm, $L_3 = 7$ cm are link lengths and joint1 $0 \leq \theta_1 \leq 180$, joint 2.0 $\leq \theta_2 \leq 180$, joint 3 $60 \leq \theta_3 \leq 180$ are joint angles, for 60° θ_3 moves down which is minimum angle limit, whereas θ_2 minimum angle limit for down is 180°. Here, base is assumed fixed at x_0, $y_0 = (0, 0)$, so joint 2 x_1, y_1 position $= (0, L_1)$. Now, joint 3 position is

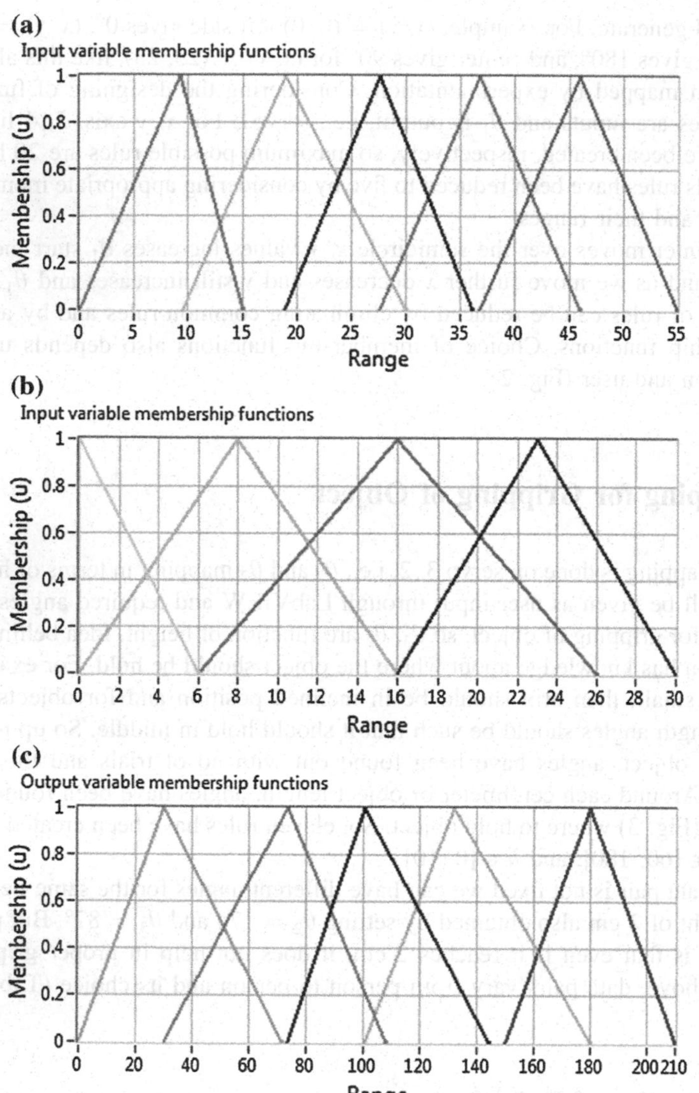

Fig. 2 Membership functions for **a** x axis **b** y axis **c** servo1, i.e., θ_1

$x_2 = L_1 \cdot \text{Cos}\,(\theta_2)$ and $y_2 = L_1 \cdot \text{Sin}\,(\theta_2)$, and joint 4 position is $x_3 = L_2.\text{Cos}\,(\theta_3)$ and $y_3 = L_2.\text{Sin}\,(\theta_3)$.

End–effector position is

$$X = x_1 + x_2 + x_3 = L_1 \cdot \text{Cos}(\theta_2) + L_2 \cdot \text{Cos}(\theta_3) \tag{3}$$

$$Y = y_1 + y_2 + y_3 = L_0 + L_1 \cdot \text{Sin}(\theta_2) + L_2 \cdot \text{Sin}(\theta_3) \tag{4}$$

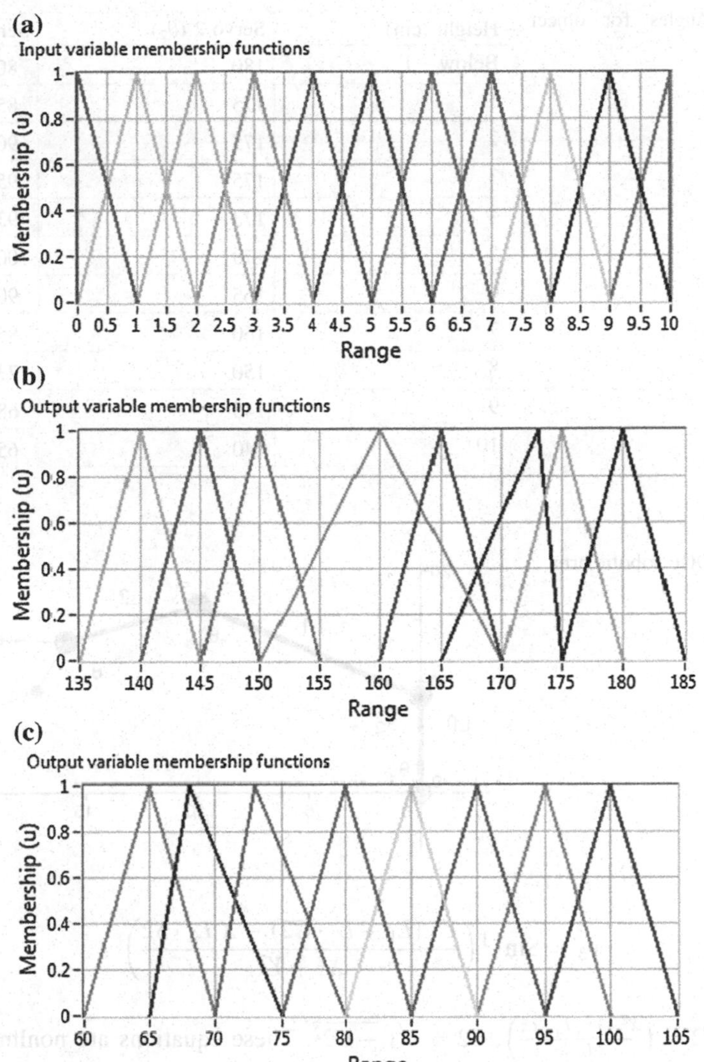

Fig. 3 Fuzzy membership functions for **a** height of object **b** servo 2, i.e., θ_2 **c** servo 3, i.e., θ_3

These two equations gives forward kinematics and helps in finding out *X, Y* position given the joint angles, but if we are interested in joint angles given the *X, Y* position then we use inverse kinematics. For above, arm inverse kinematic equations are given below.

$$\theta_2 = \mathrm{Cos}^{-1}\left(\frac{X^2 + Y^2 - L_1^2 - L_2^2}{2 \cdot L_1 \cdot L_2}\right) \tag{5}$$

Table 2 Angles for object gripping

Height (cm)	Servo 2 (θ_2)	Servo 3 (θ_3)
Below −1	180	80
1	175	85
2	175	90
3	175	95
4	173	93
5	173	100
6	165	90
7	160	85
8	150	73
9	145	68
10	140	65

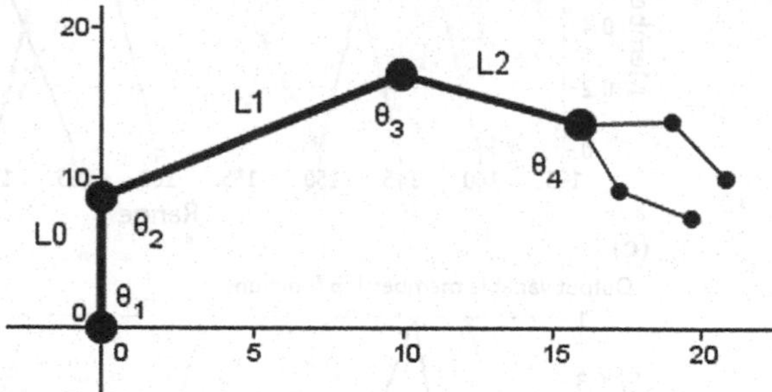

Fig. 4 3 DOF robotic arm

$$\theta_3 = \mathrm{Sin}^{-1}\left(\frac{Y \cdot (L_1 + L_2 \cdot C2) - X.L_2 \cdot S2}{X^2 + Y^2}\right) \tag{6}$$

where, $C2 = \left(\frac{X^2+Y^2-L_1^2-L_2^2}{2 \cdot L_1 \cdot L_2}\right)$, $S2 = \sqrt{1 - C2^2}$. These equations are nonlinear and provide multiple solutions which are might be out of workspace of robotic arm. Equations (5) and (6) are of two planer robotic arms, only addition is L_0 link attached to the base.

Considering θ_2 and θ_3 reference positions (0, 0) for which we obtained X, Y positions = (17, 7). Now, by setting different angle positions, we obtained different X, Y positions (Table 3).

These are values when θ_3 is parallel to the surface and θ_2 is varying and vice versa. From these, we observe that as θ_2 increase or decreases X reduces that is its holding ability also reduces. So for proper gripping, one needs to find right set of pairs for θ_2 and θ_3. Also one needs to consider constraints on joint angles, i.e., maximum and minimum limit of each joint angles of robotic arm.

Table 3 *x, y* positions by kinematics

X	Y	θ_2	θ_3
17	7	0	0
16.8	5.2	−10	0
16.3	3.5	−20	0
15.6	2	−30	0
14.6	0.5	−40	0
16.8	8.7	10	0
16.3	10.4	20	0
16.9	7.6	0	5
16.8	8.2	0	10
16.5	9.3	0	20
16.7	5.1	0	−15
16.06	3.5	0	−30

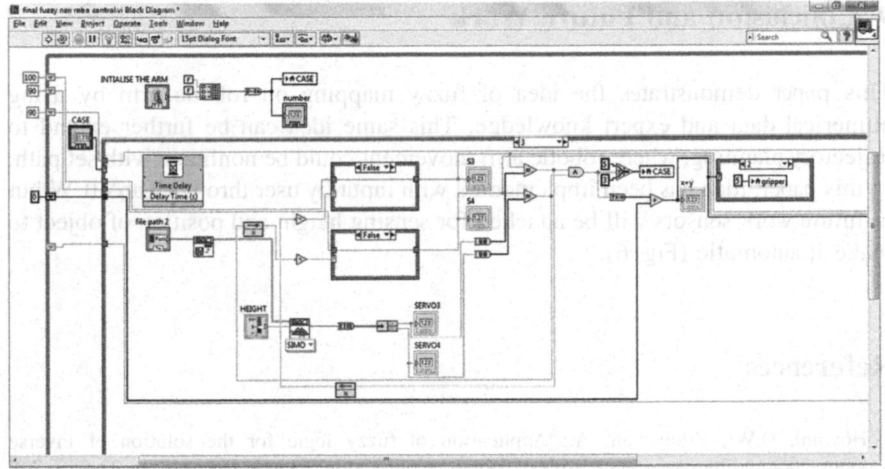

Fig. 5 Block diagram

Snap shot of block diagram 6 and front panel 6 is shown below, on front panel as we move pointer over this red semicircle we obtain arm movement in 0–180° (Fig. 5).

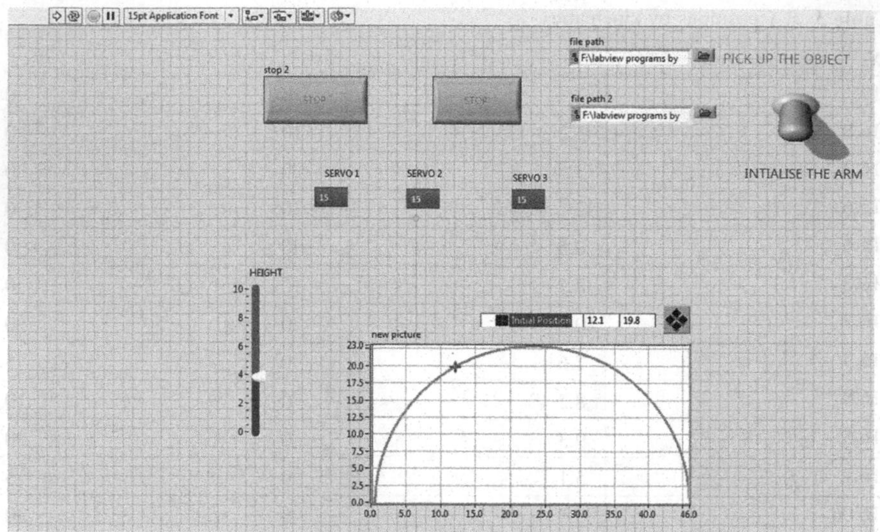

Fig. 6 Front panel

9 Conclusion and Future Work

This paper demonstrates the idea of fuzzy mapping on robotic arm by using numerical data and expert knowledge. This same idea can be further extend to trajectory planning, where robotic arm movement could be nonlinear with set path. In this paper, idea has been implemented with input by user through LabVIEW but in future work sensors will be attached for sensing height and position of object to make it automatic (Fig. 6).

References

1. Howard, D.W., Zilouchian, A.: Application of fuzzy logic for the solution of inverse kinematics and hierarchical controls of robotic manipulator. J. Intell. Robot. Syst. **23**, 217–247 (1998)
2. Wang, L.X., Mendel, J.M.: Generating fuzzy rules from learning by examples. Syst., Man cybern. **22**, 1414–1427 (1992)
3. Martinez, J., Bowles, J., Mills, P.: A fuzzy logic positioning system for an articulated robot arm. In: Proceedings of the 5th IEEE International Conference on Fuzzy Systems, vol. 1, pp. 251–257 (1996)
4. Elliott, C., Vijaykumar, V., Zink, W., Hansen, R.: A programming environment for laboratory automation and measurement. J. Assoc. Lab. Autom. **12**, 17–24 (2007)
5. Wang, L.X.: A course in fuzzy systems and control, Prentice-Hall International, Inc
6. Spong, M.W., Vidysagar, M.: Robot modelling and control, 1st edn., Wiley, Inc

An Improvement on NSGA-II for Finding Fast the Better Optimal Solution in Multi-objective Optimization Problem

Praloy Shankar De and B.S.P. Mishra

Abstract Nowadays, the most real-life problems are multi-objective or many objective in nature, which needs to be optimized to give a promising solution to the user. But the problem comes when we have to select the most significant solution from the large solution space. As a result, by applying genetic algorithm, we get a front which contains number of optimal solutions named as Pareto-optimal front. To select the most significant solution from a number of optimal solutions is a very difficult task as all solutions are nondominated to each other. The decision maker has to select a single solution. In this paper, we have shown an improvement on NSGA-II in order to select quickly the most significant and acceptable solution by the decision maker.

Keywords Multi-objective optimization · Pareto-optimal front · Genetic algorithm · NSGA-II · Decision maker

1 Introduction

Optimization means the process by which we can get the feasible solutions for solving any objective-based problem. The feasible solutions are corresponding to extreme values of one or more objectives [1]. It can be done for maximizing or minimizing the objectives. We can take a solution which is maximum or minimum among those feasible solutions. It is called as an optimal solution.

Depending on the number of objectives, optimization can be categorized into three types, and they are as follows:

P.S. De (✉) · B.S.P. Mishra
School of Computer Engineering, KIIT University, Bhubaneswar, Odisha, India
e-mail: praloy007@gmail.com

B.S.P. Mishra
e-mail: mishra.bsp@gmail.com

L.C. Jain et al. (eds.), *Intelligent Computing, Communication and Devices*,
Advances in Intelligent Systems and Computing 308,
DOI 10.1007/978-81-322-2012-1_12

Definition 1 When an optimization involves only one objective function which is meant to be optimized, then we will call it as a single-objective optimization problem [1].

Mathematically, it can be represented as follows:

$$\text{Maximize/Minimize} f(x), \quad x_i^{(L)} \le x_i \le x_i^{(U)}$$
$$\text{for}\, i = 1, 2, \ldots, n$$

Definition 2 When an optimization problem involves more than one objective functions which are meant to be optimized, then we will call it as a multi-objective optimization. Generally, a multi-objective optimization problem consists of two objectives and associated with a number of inequality and equality constraints.

Mathematically, it can be presented as follows:

$$\text{Minimize/Maximize} f_m(x), \quad m = 1, 2, 3;$$
$$\text{Subjected to}\, g_j(x) < 0, \quad j = 1, 2, \ldots, J;$$
$$h_k(x) = 0, \quad k = 1, 2, \ldots, K;$$
$$x_i^{(L)} \le x_i \le x_i^{(U)}, \quad i = 1, 2, \ldots, n.$$

where x is a n-dimensional vector having n decision variables. Mathematically, the solutions to any multi-objective optimization problem can be expressed in terms of nondominated points. In a minimization problem, a vector x_1 is partially less than another vector x_2, (x_1 x_2), when no value of x_2 is less than x_1 and at least one value of x_2 is strictly greater than x_1. If x_1 is partially less than x_2, we say that the solution x_1 dominates x_2 or the solution x_2 is inferior to x_1 [7]. Any member of such vectors is nondominated or superior by any other member. If the objective is to maximize a function, then we will define a dominated point if the corresponding component is smaller than that of a nondominated point. Every optimal solution to any multi-objective optimization problem is nondominated solution [2].

Definition 3 When an optimization problem involves four or more objectives, then it is called as a many-objective optimization problem.

Mathematically, it can be represented as follows:

$$\text{Minimize/Maximize} f_m(x), \quad \mathrm{m} = 4, 5, \ldots \mathrm{M};$$
$$\text{Subjected to}\, g_l(x) < 0, \quad l = 1, 2, \ldots, \mathrm{L};$$
$$h_m(x) = 0, \quad m = 1, 2, \ldots, M;$$
$$x_i^{(L)} \le x_i \le x_i^{(U)}, \quad i = 1, 2, \ldots, n.$$

2 A Brief Review on Existing Work

In earlier days, classical search methods are used for solving any optimization problem. These methods were point-by-point approach where a single solution in every iteration was modified to different and hopefully better solution. But from 1960, the field of search and optimization were changed by the introduction of evolutionary algorithms (EAs). EA uses nature's evolutionary principles for optimization. The most striking difference of EAs with classical search and other traditional optimization techniques is that EAs use a population of solution in every iteration instead of one solution. If any problem has a single optimal solution, all EA's population members are expected to converge to that solution. If any problem has multiple optimal solutions, EAs are used to capture a multiple optimal solution in their final population. It is more suitable to create multiple trade-off solutions for any multi-objective optimization problems [1].

There are so many evolutionary algorithms to solve multi-objective problem. Some of them are elitist. (Elite-preserving operator is used to save the elites for the next generation.) Elitist multi-objective EAs are better because the good solution found early on the run will never be lost until the better solution is discovered for number of generations.

In 1996, Rudolph proved that GAs converge to the global optimal solution of some function in the presence of elitism [1]. Deb et al. [3] proposed an elitist multi-objective EA named nondominated sorted genetic algorithm II in 2000. It uses an explicit diversity-preserving mechanism. They published a paper on this algorithm in IEEE Transactions On Evolutionary Computation, 2002, where they gave a comparison with other two well-known algorithms named SPEA [4] and PAES [5]. They showed that NSGA-II is better for obtaining good solution in some context. Some researcher also showed that SPEA and PAES are better for some cases [9].

Pareto front is obtained for multi-objective optimization problem by multi-objective evolutionary algorithms (MOEAs) that produce nondominated optimal solutions. It is a difficult task for the decision maker to select a single better solution. In 2008, Ishibuchi et al. [15] published a paper on Evolutionary Many Objective Optimization where they have shown their improvement on NSGA-II. They also discussed about the difficulties when we use many-objective optimization problem. Firstly, the deterioration of the search ability of Pareto dominance was based on evolutionary multi-objective optimization algorithms (EMOs). For many-objective optimization problem, almost all solutions in each population become nondominated. Convergence becomes weak for this reason. Another problem is handling a large number of nondominated solutions, and visualization also becomes more difficult for the decision maker for selecting the final solution from the set. Even for MOEAs, it becomes a difficult task for the decision maker for selecting a better or significant solution from the obtained Pareto front.

Jain and Deb [17] published a paper on improvement of NSGA-II for many-objective optimization problem. Their improved algorithm is known as NSGA-III.

Köppen and Yoshida [16] proposed an algorithm which is an improvement of NSGA-II for handling many-objective optimization problems. They said that NSGA-II is not efficient for many-objective optimization problem due to the decreasing probability of having Pareto-dominated solutions in the initial external population. To overcome this problem, they replaced the crowding distance assignment by substituting distance assignment where the distance is based on the measurement of the highest degree to which a solution is nearly Pareto-dominated by any other solution. Their result and performance showed that their improvement on NSGA-II is strong enough for solving many-objective optimization problems.

From the above study, we found that when the number of objectives increases, it means that for many-objective optimization problem, NSGA-II is not an efficient algorithm. Improvement is necessary for handling this task.

There always needs a balancing of the convergence and diversity to build up an efficient front. It is because converging to a set of solutions that are not close to the true optimal set is not suitable and the solutions must be sparsely spaced in the Pareto-optimal region [1]. Functions can be convex or nonconvex in nature. For nonconvex functions, we face a problem for obtaining a true Pareto front. We have also seen that if we decrease the population size, a true Pareto front can be obtained for any type of problems. Some researchers used their own thresholds for obtaining a true Pareto-optimal front [11]. The difficulty comes when a decision maker has to select a single solution from the obtained Pareto front because all the solutions are nondominated to each other and optimal.

We have worked on NSGA-II and analyzed the post-Pareto problem. Our proposed work can help the decision maker to select a better solution from the obtained Pareto front in a faster way. Different researchers gave their contribution to find the better solution after obtaining the Pareto front. Most of the researchers have used the clustering methods in different ways to the nondominated sets obtained by Pareto front. Most of the methods are used to find the similarity of elements in the nondominated set based on their objective function values, and after that, those similar elements or solutions are removed from the set.

Morse et al. [12] described an application of cluster analysis to the nondominated set. Based on the thresholds which were defined by the decision maker, a solution was removed if it was indistinguishable from other solutions in the nondominated set. Seven hierarchical clustering methods were used. There are also other methods such as Ward's method [14], the group average method, and the centroid method which performed well.

Rosenman et al. [13] applied complete linkage hierarchical clustering for reducing the size of the Pareto-optimal set. Their method allowed control of the diameter to the resulting clusters. They distinguished those solutions whose objective function values are similar. The objective functions were successively taken in order to avoid the implicit aggregation in applying proximity measures. Solutions of the nondominated set were clustered using a single criterion. If any solution within a cluster dominated by another solution in the cluster on all criteria except the clustering criterion, then the dominated solution was eliminated from consideration. The process was repeated for each criterion until the nondominated set became small in size.

Generally, the traditional clustering algorithms cannot produce alternative solutions. Most of the clustering procedures fail to create the optimal number of clusters in dataset on which they work on. Hierarchical clustering can make the heuristic overview of the whole dataset, but this process cannot relocate objects from incorrect group of the earlier stage. Hierarchical clustering cannot tell the optimal number of clusters for the nondominated set. We need the clustering procedure which can give a best overview of the whole dataset by creating the optimal number of quality clusters. *K*-means clustering can produce the optimal number of clusters in a single run as it needs the predefined number of clusters. This procedure may also give local optimal solutions due to its local search strategy [18].

In 2010, Chaudhuri et al. [11] published a paper on how to compute the most significant solution from Pareto front. They introduced an interesting cluster named "knee" cluster and used *k*-means clustering. They have also used silhouette plots for determining the optimal number of clusters, and then for each cluster, they selected a good representative solution by the solution which is closest to its respective cluster centroid. Sudeng et al. [8] proposed a pruning algorithm for multi-objective optimization by applying adaptive angle strategy (ADA) on both NSGA-II and SPEA2 [6] in 2013. They had shown that for any biobjective problem, the pruned Pareto-optimal solutions are located in the knee region of the obtained Pareto front.

We have tested the optimization problem consisting of three or two objective functions by NSGA-II to make the decision maker select the better solution quickly. We have introduced a novel ranking system after clustering the results obtained from the Pareto front. Decision maker can choose the solution from the preferred clusters according to their ranks. In the next section, we will describe our proposed work on which we contributed our focus on the improvement of NSGA-II for post-Pareto analysis.

3 Proposed Methodology

First, we use NSGA-II algorithm to solve any multi-objective optimization problem. NSGA-II can produce the Pareto front in which all the solutions are optimal and nondominated to each other. Now, the problem to us is how to extract a single better solution from this front. For this post-Pareto problem, we have clustered the solutions by *k*-means. Then, we have assigned ranks to the clusters. The decision maker can prefer the clusters according to their ranks. The ranks will be assigned to the clusters according to the number of their containing solution. Same rank will be assigned for those clusters which are similar in size. Decision maker can prefer any of them then.

The proposed methodology is described below:

3.1 Algorithm Steps

Begin {

Step 1: Use NSGA-II for solving any multi-objective optimization problem.

Step 2: Use any clustering method for making the cluster of nondominated solution from the Pareto front. We used classic *k*-means clustering. Cluster index = *k*-means (array of containing NSGA-II results) /cluster index created by *k*-means algorithm.

Steps of NSGA-II

NSGA-II ()

{

$R_t = P_t \cup Q_t$ /combining the parent and offspring population R_t of size 2N.

F = fast nondominated sort (R_t) /F = (F_1,F_2,…), all are the nondominated fronts of R_t , which are to be sorted according to nondomination. Elitism is ensured for all the population.

P_{t+1}=*NULL and* $i = 1$ / P_{t+1} is the new population

until $|P_{t+1}| + |F_i| \leq N$ /until the parent population is filled.

Crowding distance assignment (F_i) /calculate crowding distance in F_i that includes ith nondominated front in the parent pop

$P_{t+1} = P_{t+1} + F_i$

$i = i+1$ /check the next front for inclusion

Sort$(F_i , <_n)$ / Sort in descending order using $<_n$ (crowded-comparison operator)

$P_{t+1} = P_{t+1} + F_i\,[1{:}(N-|P_{t+1}|)]$ /Choose the first (N-| P_{t+1}|) elements of F_i

Q_{t+1}=*make-new-pop*(P_{t+1}) /use selection, crossover, and mutation to create a new population Q_{t+1}

$t = t+1$ / increment the generation counter

}

Steps of *k*-means

{

1. *Choose the total number of clusters for the final result and set the number as k. Select k patterns in the whole databases as the k centroids of k clusters randomly.*
   All instances are assigned to their closest cluster center according to the Euclidian distance metric.
2. *Classify every pattern to the closest cluster centroid. The closest represents the data value that is similar. Other features are also considered.*
3. *Recompute the cluster centroids, and then, there have k centroids of k clusters as we do after Step 1 of k-means.*
4. *Repeat the iteration of Step 2 and Step 3 of k-means until the convergence criterion fulfilled. The typical convergence criterion are as follows: no*

reassignment of any data from one cluster to another or the minimal decrease in squared error.

}

Step 3: Assign the size of each cluster according to the number of their contained solutions.
Size($C_1, C_2, \ldots\ldots\ldots\ldots, C_n$)

Step 4: Sort the clusters according to their size.
Sort($C_1, C_2, \ldots\ldots\ldots\ldots, C_n$)

Step 5: Rank the sorted cluster size-wise. The cluster containing the highest number of solution, i.e., highest sized, will be assigned as a rank 1, the next highest among the remaining will be rank 2, and so on.
Rank($C_1, C_2, \ldots\ldots\ldots\ldots, C_n$)

Step 6: Decision maker will check the most nondominated solution in the first-ranked cluster based on user criterion and choose that solution as a final. (In real life, if you will buy something from any shop or market, you will obviously prefer the shop which has more number of options to meet your objective.) If the better solution is not found, then repeat this step for remaining clusters according to their ranks.
end
}

4 Implementation and Result Analysis

We have implemented our improvement on MATLAB. There are various types of test problems. Our implementation was done in a machine build with Intel Core i5 processor, 8-GB RAM, and the platform used was Windows 7. We used some of standard test problems such as ZDT and DTLZ problems [9, 10].

4.1 Test Problems

The ZDT6 problem can be represented as follows:

$$\left.\begin{aligned} f_1(x) &= 1 - \exp(-4x_1)\sin^6(6\pi x_1) \\ f_2(x) &= g(x)\left[1 - \left(\frac{f_1(x)}{g(x)}\right)^2\right] \\ g(x) &= 1 + 9\left[\left(\sum_{i=2}^{n} x_i\right)\Big/(n-1)\right]^{0.25} \end{aligned}\right\} \quad (1)$$

This problem has also been tested before by Zitzler et al. This problem is six-variable problem ($n = 6$) having a nonconvex Pareto-optimal set. The density of the solutions across the Pareto-optimal region is non uniform and the density toward the Pareto-optimal front lies within the two objective functions. All variables lie in the region (0, 1), i.e., the Pareto-optimal region corresponds to $0 \le x_i^* \le 1$ and $x_i* = 0$ for $i = 2, 3, \ldots, 6$. It is nonconvex in nature, and this causes difficulties for getting true Pareto-optimal front.

For the nonconvex problems, we are not getting the true Pareto-optimal front. Thus, we have taken another test problem ZDT1 which is convex in nature. The problem can be represented as follows:

$$\left.\begin{array}{l} f_1(x) = x_1 \\ g(x) = (1 + \frac{9}{n-1} \sum_{i=2}^{n} x_i) \\ f_2(x) = 1 - \sqrt{\frac{f1}{g}} \end{array}\right\} \quad (2)$$

In this problem, all variables lie in [0, 1] and the Pareto-optimal region corresponds to $0 \le x_i^* \le 1$ and $x_i = 0$ for $i = 2, 3, \ldots, 6$. We can achieve the continuous Pareto-optimal front in objective space. Only difficulty of this problem is that the number of decision variables is large.

Researches have also been done on more than two objectives for solving multi-objective optimization. But most of researches were restricted to two-objective problems because

1. Balancing for convergence near the Pareto-optimal front and a good diversity can be tested adequately with two objectives.
2. Graphical representation of trade-off solutions in multi-objective problem with more than two objectives is difficult to achieve.

Therefore, we have also used a problem consisting of three-objective function named as DTLZ2.

$$\left.\begin{array}{l} f_1(x) = (1 + g(x)) \cos(0.5\pi x_1) \cos(0.5\pi x_2) \\ f_2(x) = (1 + g(x)) \cos(0.5\pi x_1) \sin(0.5\pi x_2) \\ f_3(x) = (1 + g(x)) \sin(0.5\pi x_1) \\ g(x) = \sum_{i=3}^{12} (x_i - 0.5)^2 \\ \text{subject to } 0 \le x_i \le 1, \quad i = 3, 4, \ldots, 12. \end{array}\right\}. \quad (3)$$

Our improvement was done on a real-coded NSGA-II which uses simulated binary crossover (SBX) operator [1] and polynomial mutation [1]. The crossover probability of $P_c = 0.9$ and mutation probability of $P_m = 1/n$ where n is the number of decision variables for this algorithm. We use distribution indexes for crossover and mutation operators as $\eta_c = 20$ and $\eta_m = 20$. K-means clustering used to the results obtained from the Pareto front. After cluster ranking, we obtain

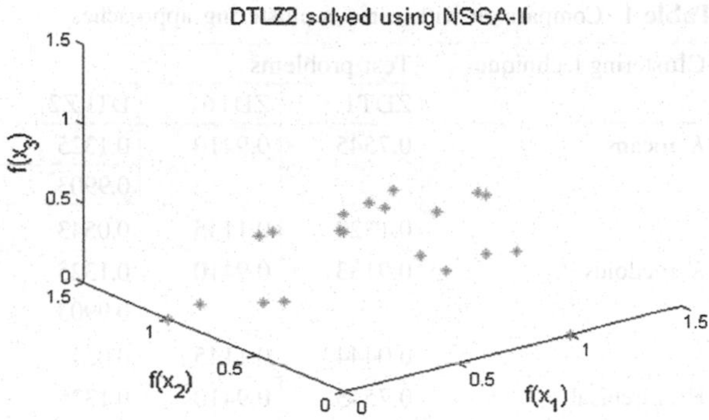

Fig. 1 NSGA-II result of DTLZ2

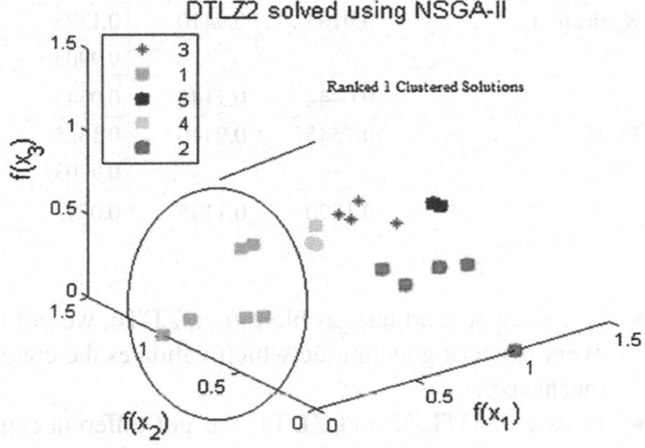

Fig. 2 Ranked clustered solutions obtained from the Pareto front

the following results for 2,000 generations. Some screenshots have been given for better understanding of our proposed work (Figs. 1 and 2).

From our resulting clusters in the "Result and Comparison with Other Clustering Methods" section, it is very clear that the good options can be found in the first-ranked cluster. We obtained some good results from the first-ranked cluster on the basis of our objectives or criterion.

4.2 Results and Comparison with Other Clustering Methods

We have used other clustering techniques such as *k*-medoids, *k*-means ++, hierarchical, and fuzzy *C*-means and compared the ranks with the used *k*-means results. We have taken a single solution obtained by *k*-means as a representative for other clustering techniques.

From Table 1, the following interesting observations are made:

Table 1 Comparison with various clustering approaches

Clustering technique	Test problems			Rank of clusters		
	ZDT1	ZDT6	DTLZ2	ZDT1	ZDT6	DTLZ2
K-means	0.7545	0.9410	0.1325	1	1	1
			0.9903			
	0.1320	0.1145	0.0543			
K-medoids	0.9133	0.9410	0.1325	1	1	1
			0.9903			
	0.0444	0.1145	0.0543			
Hierarchical	0.7545	0.9410	0.1325	1	1	2
			0.9903			
	0.1320	0.1145	0.0543			
K-means ++	0.9133	0.9410	0.1325	1	1	2
			0.9903			
	0.0444	0.1145	0.0543			
FCM	0.7545	0.9410	0.1325	2	1	1
			0.9903			
	0.1320	0.1145	0.0543			

- In case of second test problem, i.e., ZDT6, we got the same rank of clusters by every clustering technique which validates the concept of our proposed ranking mechanism.
- In case of DTLZ2 and ZDT1, we got different ranks in some clustering techniques. So, we believe this may happened due to the nature of clustering techniques.
- We found that for ZDT1, *k*-means ++ and *k*-medoids produced same result which is different from our representative solution. This is happened for the nature of both clustering techniques. In rank 1 cluster, we found that this is a better result from our representative solution which also exists there.

5 Conclusion and Future Work

We are facing difficulties for many-objective optimization problem. In future, we will try for NSGA-III and we will solve some real-life problems. As a researcher, we cannot say that our improvement is an efficient procedure, but our methodology can give a fair solution in a faster way. If we consider the worst time complexity of our whole process, it can be written as follows:

1. For NSGA-II, nondominated sorting complexity is $O(M(2N^2))$. Here, the domination checking is done for two times comparison. For this, we need $O(MN^2)$. Every solution will visit at most $N-1$ times before the domination count becomes zero. For $(N-1)$ solutions, we need $O(N^2)0.2N$ which is the total population size (R_t 's size). M is the total number of objectives. Crowding distance assignment is $O(M(2N)\log(2N))$, since M independent sorting of $2N$ solutions are required when all population members are in the front of non dominated solutions. Sorting on $<_n$ is $O(2N \log(2N))$. Overall complexity of NSGA-II is $O(MN^2)$.
2. For k-means clustering on the Pareto-optimal set $O(N)$, the computational complexity of k-means clustering is $O(N)$.
3. Size assigning complexity is $O(\mathrm{N})$, and sorting complexity is $O(N^2)$.
4. Ranking complexity is $O(N)$, and for domination check in each cluster, the complexity is $O(M(2N^2))$.
   So, we have found that the overall complexity $O(MN^2)$ is remained unaltered after the improvement on NSGA-II for post-Pareto analysis.

References

1. Deb, K.: Multi-objective optimization using evolutionary algorithm. Wiley, New York, ISBN 9814-12-685-3 (2005)
2. Srinivas, N., Deb, K.: Multi-objective optimization using non-dominated sorting in Genetic Algorithm. J. Evol. Comput. **2**(3), 221–248 (1994)
3. Deb, K., Pratap, A., Agarwal, S., Meyarivan, T.: A fast and elitist multi-objective genetic algorithm: NSGA-II. IEEE Trans. Evol. Comput. **6**(2), 182–197 (2002)
4. Zitzler, E., Thiele, L.: Multi-objective evolutionary algorithms: a case study and the strength pareto approach. IEEE Trans. Evol. Comput. **3**(4), 257–271 (1999)
5. Knowles, J., Corne, D.: The pareto archived evolution strategy: a new baseline algorithm for multi-objective optimization In: Proceedings of the 1999 Congress on evolutionary computation. NJ:IEEE Press, Piscataway, pp. 98–105 (1999)
6. Zitzler, E., Laumanns, M., Thiele, L.: SPEA2: Improving the strength pareto evolutionary algorithm, pp. 1–21. Department of Electrical Engineering, Swiss Federal Institute of Technology (ETH), Zurich, Switzerland (2001)
7. Tamura, K., Miura, S.: Necessary and sufficient conditions for local and global nondominated solutions in decision problems with multi-objectives. J. Optim. Theory Appl. **28**(24), August (1979)
8. Sudeng, S., Wattana Pongsakorn, N.: Pruning Algorithm for multi-objective optimization. In: 10th International joint conference on computer science and software engineering (JCSE), IEEE, pp. 70–75 (2013)
9. Huband, Simon, Hingston, Philip, Barone, Luigi, While, Lyndon: A review of multi-objective test problems and a scalable test problem toolkit. IEEE Trans. Evol. Comput. **10**(5), 477–506 (2005)
10. Deb, K., Mohan, M., Mishra, S.: A fast multi-objective evolutionary algorithm for finding well-spread pareto optimal solutions. KanGAL report number 20032002, pp. 1–18 (2003)
11. Chaudhari, P.M., Dharaskar, R.V., Thakare, V.M.: Computing the most significant solution from Pareto front obtained in multi-objective evolutionary. Int. J. Adv. Comp. Sci. Appl. (IJACSA) **1**(4), 63–68 (2010)

12. Morse, J.N.: Reducing the size of the nondominated set pruning by clustering. Comput. Oper. Res. **7**, 55–56 (1980)
13. Rosenman, M.A., Gero, J.S.: Reducing the Pareto optimal set in multicriteria optimization (with applications to Pareto optimal dynamic programming). Eng. Optim. **8**, 189–206 (1985)
14. Ward Jr., J.H.: Hierarchical grouping to optimize an objective function. J. Am. Stat. Assoc. **58**(301), 236–244 (1963)
15. Ishibuchi, H., Tsukamato, N., Nojima, Y.: Evolutionary many objective optimization: a short review. In: Proceedings of 2008 IEEE congress on evolutionary computation, pp. 2424–2431. Hong Kong, June 1–6, (2008)
16. Köppen, M., Yoshida, K.: Substitute Distance Assignment in NSGA-II for Handling Many Objective Optimization Problems, EMO 2007, LNCS 4403, pp. 727–741. Springer, Berlin (2007)
17. Jain, H., Deb, K.: An Improved Adaptive Approach for Elitist Nondominated Sorting Genetic Algorithms for Many-objective Optimization, EMO 2013, LNCS 7811, pp. 307–321. Springer, Berlin (2013)
18. Liu, L., Özyer, T., Alhajj, R., Barker, K.: Cluster validity analysis of alternative results from multi-objective optimization. In: Proceedings of the fifth SIAM international conference on data mining, pp. 496–500, Newport Beach, Canada (2005)

Test Case Creation from UML Sequence Diagram: A Soft Computing Approach

Ajay Kumar Jena, Santosh Kumar Swain and Durga Prasad Mohapatra

Abstract Unified modeling language (UML) is used to design the tests in various levels of testing. To create the test cases from the source code using traditional methods is becoming very difficult and cumbersome in cluster levels. UML artifacts provide a lot of facts which help the user to navigate through the flaws from the designed documents. In this work, we propose a method for creating the test cases using sequence diagram of UML models. As the testing can be started from the design process at the beginning phase, we preferred this approach. We proposed a model to generate a sequence flow chart (SFC) from sequence diagram and then tried to convert it to message control flow graph (MCFG). By using message sequence path coverage criterion, we traversed the MCFG and the test paths are generated. The test cases from these paths are created subsequently. Finally, genetic algorithm has been applied to generate test cases and also to optimize them. The model is implemented on a case study of ATM withdrawal system.

Keywords Sequence diagram · Sequence flow chart · Genetic algorithm

1 Introduction

Software testing is becoming more crucial and challenging due to increasing size and complexity of the present-day software. In complex software development process, more than 50 % of the cost and time are spent on testing, which is the

A.K. Jena (✉) · S.K. Swain
KIIT University, Bhubaneswar, Odisha, India
e-mail: ajay.bbs.in@gmail.com

S.K. Swain
e-mail: sswainfcs@kiit.ac.in

D.P. Mohapatra
National Institute of Technology, Rourkela, Odisha, India
e-mail: durga@nitrkl.ac.in

L.C. Jain et al. (eds.), *Intelligent Computing, Communication and Devices*,
Advances in Intelligent Systems and Computing 308,
DOI 10.1007/978-81-322-2012-1_13

most principal activity carried out in software development process [1]. Creation of test cases is the most difficult step in testing [2]. So, to design a large number of test cases and to test them is a very difficult and time-consuming task. Test cases created automatically can reduce the development cost by eliminating costly manual test case design. It also helps to achieve reliability through increased test coverage.

A test case is a triplet [S, I, O] where I is the inputted data to the system, S is the state of the system, and O is the expected output [1]. The output data produced by the execution of the software with a particular test case provide a specification of the actual program behavior. Normally, the test cases are designed based on the source code [2]. Though the code can be generated after the analysis and design, it is difficult to test at prior stages. In component-based software where the code is not available to the developer, it is difficult to generated test cases at integration level. To avoid wasting of time consumption and cost utilized in code-based system, it is desirable to generate the test cases at the design level, which increases the reliability of the software. Model-based testing is more efficient and effective than code-based approach as it is the mixed approach of source code and specification requirements [3]. The intermediate artifacts between source code and requirement specifications are models. Models preserve the essential information from requirement specification and are base for the final implementation. Unified modeling language (UML) is the modeling language, achieving a great attention as the industrial de facto standard for modeling object-oriented software. UML accomplishes the visualization of software at early stage, which helps in building the confidence of both software engineer and the end user on the system. It helps in making the proper documentation of the software and maintains the consistency in between the specification and design document. Out of 14 diagrams, UML sequence diagram is used for describing the behavior by modeling the sequence of messages in a system. Sequence diagram is a tuple (L, O, M, E) where L is the set of lifelines, O is the set of objects, M is the set of messages to be passed, and E is the set of execution specifications to be performed. Sequence diagrams describe how objects, or group of objects, interact within a system [4]. Most of the code-oriented structures are available in the sequence diagram, as a result it is very nearer to the code [4]. Test coverage criteria [5] are a set of rules that guide to decide the messages to be considered to adequate the test case design. Many testing approaches [6] have been proposed in literature to consider different test acceptable criteria for sequence diagrams. We use all messages coverage criterion in which signals for each message on the link connecting two objects should appear at least once in an acceptable test. This criterion confirms that all the messages between any two objects occur in the test. Generating test cases from the design documents has the advantage of permitting test cases to be available in the software development life cycle (SDLC), which makes the test plan more effective. Second advantage of design-based testing is to compliance the test of implementation with the documentation. Generating tests during design also allows testing activities to be shifted to an earlier part of the development process that allows for more planning of the test cases. Another advantage is that the test data are independent of any particular implementation. A tremendous improvement can be achieved by automating the process [3, 5]. We can get an optimal number of test cases of better quality by eliminating the redundant ones. Most of the researchers have not given their proper attention in using optimized set of test case generation using model-based technique which gives a motivation to us.

The rest of the paper is organized as follows. A review of related works is presented in Sect. 2. Section 3 presents our approach for the proposed work to generate test cases. In Sect. 4, the model is implemented in a case study of ATM withdrawal system. Section 5 presents the future work and conclusion.

2 Related Works

Three types of techniques have been illustrated in different works to generate test cases, i.e., code based, specification based, and model based. We presented a review of UML-based testing approaches reported in the literature.

Doungsa-ard et al. [7] proposed a framework for generating test data from software specifications. Test data are generated automatically using control flow graph with a deterministic algorithm. State machine diagram of UML was used in their work. A tool and an approach for generating test data from the software specification are used. To find the best test data, heuristic technique GA was applied. There is a transition problem and looping problem in the approach. In their approach, when the lengths of chromosomes vary, the result will vary. Prasanna et al. [8] have proposed a model-based approach for the generation of automated test cases. The test cases are derived by analyzing the dynamic behavior of the objects due to internal and external stimuli. A general tree is created from the object diagram. Genetic algorithm is applied on general tree to achieve tree structure. Tree crossover has been proposed to bring out all possible test cases. The scope of the paper has been limited to the object diagrams taken from the UML model of the system. No discussion has been mentioned for system-level faults and also behaviorial diagram. Shirole et al. [9] have presented an approach for automatic generation of feasible test paths using GA. The proposed architecture produces a set of test cases from UML state machine diagram. The authors transformed the state machine diagram to extended finite state machines and subsequently to extended control flow graph. They applied GA to generate test cases satisfying the coverage specified (all-definition cover, all-du path cover). It was observed by the authors that the feasibility of path is affected by the sequence of transitions.

3 Proposed Work

We propose our work to generate optimized test cases from specification document using sequence diagram of UML. UML is used as a de facto standard in both industry and academics. It is widely used as a modeling language, designed to

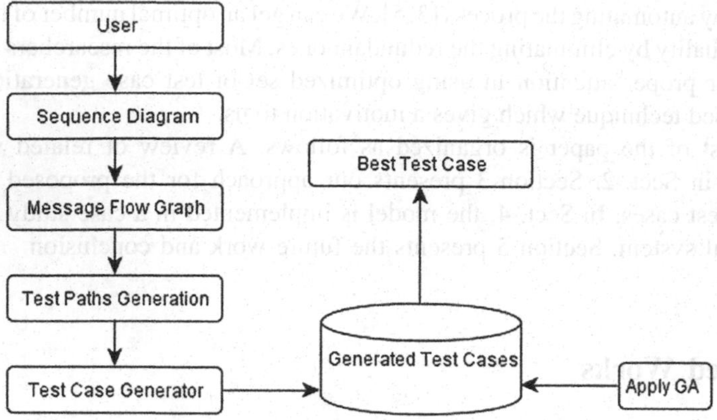

Fig. 1 Proposed framework

specify, construct, and document the artifacts which supports the dynamic and structural aspects of the software system.

3.1 Proposed Framework

The system that we are proposing will work as mentioned below:

1. Elaborate a problem statement for the system.
2. Generate the sequence diagram (SD) of the system.
3. Generate a graph MCFG, G = (V, E), where G represents the graph, V represents the nodes (messages), and E represents the set of connecting edges (flow).
4. Possible test paths are generated by using depth first search (DFS) method that guarantees visiting all the possible nodes, while removing the redundant ones.

3.2 UML Diagram for Design Document

In our approach, we considered the sequence diagram for ATM withdrawal system. The dynamic aspects of the object can be constructed, visualized, specified, and documented by the sequence diagram. A sequence diagram models behavior by specifying the sequence of actions and the conditions for coordinating actions. Figure 2 represents the sequence diagram for the ATM withdrawal system.

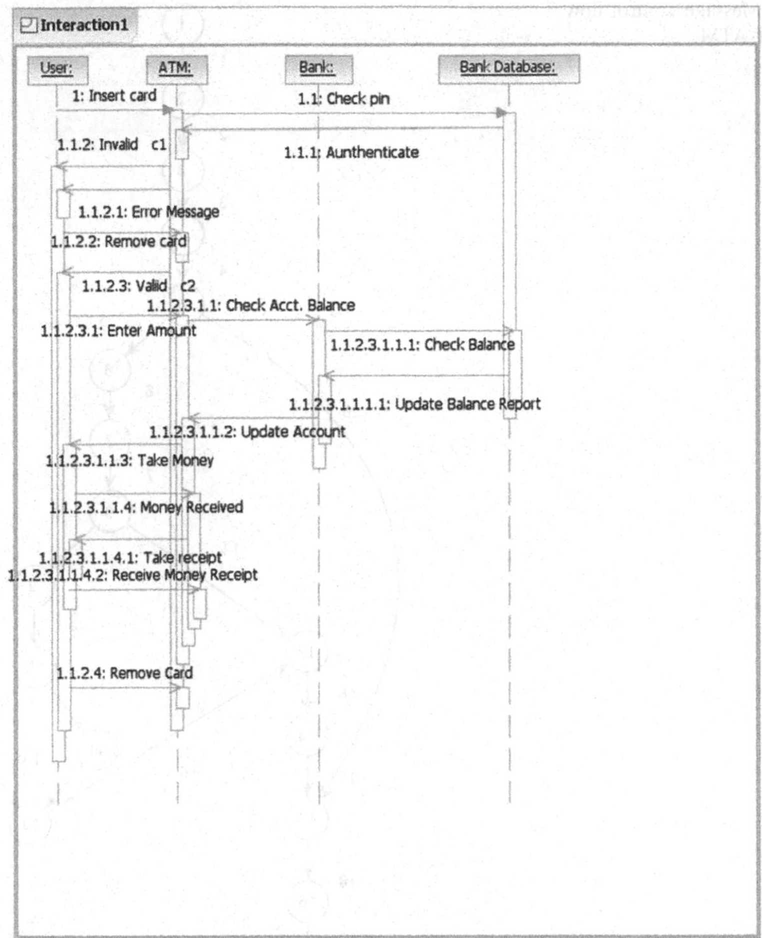

Fig. 2 Sequence diagram for ATM

3.3 Message Control Flow Graph

Control flow graph (CFG) is a static representation of the model that represents an alternative of the control flow. Cycle represented in a control flow graph implies the presence of a loop in the code. Possible paths from the start node to the end node of the CFG are given by control flow path (CFP). It shows different paths the program may flow during process. A MCFG is a directed graph, while its construct is represented by each node in the flow graph. Figure 3 represents the MCFG of ATM withdrawal system.

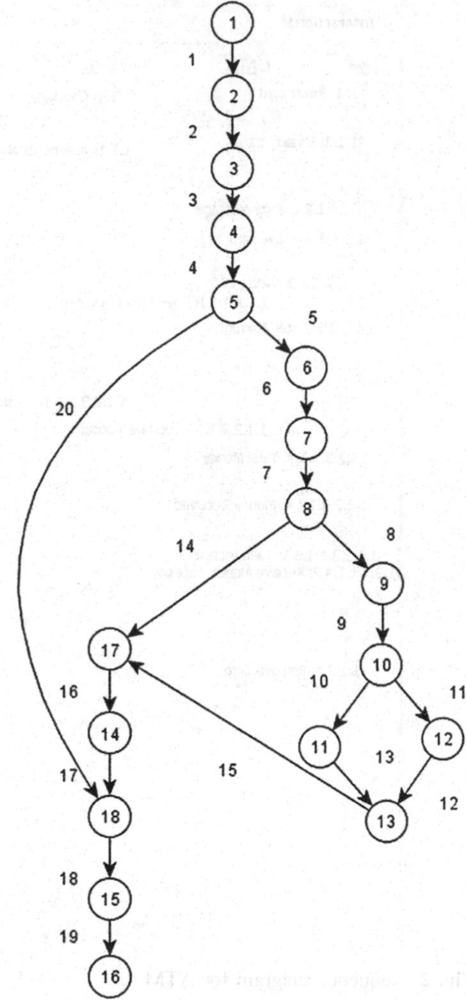

Fig. 3 Message control flow graph of ATM

3.4 Algorithm to Create Test Cases Using GA

1. *Draw the sequence diagram.*
2. *Convert the sequence diagram to message control flow graph.*
3. *Assign the weights to all the edges to the message flow graph.*
   - *Assign value one to the first edge*
   - *Increment the value of the next edge connecting to the next node and so on.*
   - *If the node is a decision node, then increment the value of the true path first and so on.*
4. *Generate the unique paths from the starting node to the end node.*
5. *Calculate the fitness value of the paths*

- *The weights of the paths are calculated as the sum of the weights from the starting node to the final node of the paths.*
- *The fitness function is defined as F(x) =* X^2

6. *The crossover can be done by choosing the first two strings as the first mate and second two strings as the second mate and so on. The offspring can be generated from the parents.*
7. *Mutate the genes in the population.*
8. *Again evaluate fitness after each generation.*
9. *Repeat the process until all the paths are covered or the maximum number of generations is reached.*
10. *Best test path generated*
11. *End*

4 Implementation and Results

In this section, we discuss the results obtained after implementation of our approach. We have taken the ATM withdrawal system as the case study. The sequence diagram is prepared by using the tool rational software architecture (RSA) and is shown in Fig. 2. After preparation of the diagram, it is converted to XMI (XML Meta Interface) code whose snapshot is given in Fig. 4. The message flow paths are generated by using Java under NetBeans IDE. Figure 3 shows the graph generated from the XMI code. The test cases for the ATM withdrawal system are shown in Table 1.

All possible test paths are generated before the generation of the test cases from the MFG. Our approach traverses the graph using DFS method which guarantees visiting all possible nodes of MFG. This helps us to achieve the message path coverage criterion. All possible test paths generated are as follows with weights.

Test Path 1 1 → 2 → 3 → 4 → 5 → 18 → 15 → 16 = 67
Test Path 2 1 → 2 → 3 → 4 → 5 → 6 → 7 → 8 → 9 → 10 → 11 → 12 → 13 → 17 → 14 → 18 → 15 → 16 = 207
Test Path 3 1 → 2 → 3 → 4 → 5 → 6 → 7 → 8 → 9 → 10 → 12 → 11 → 13 → 17 → 14 → 18 → 15 → 16 = 207
Test Path 4 1 → 2 → 3 → 4 → 5 → 6 → 7 → 8 → 17 → 14 → 18 → 15 → 16 = 112

4.1 Optimization of Test Paths Using GA

In our case study, we have population of four paths. So we select all the best paths for the first generation.

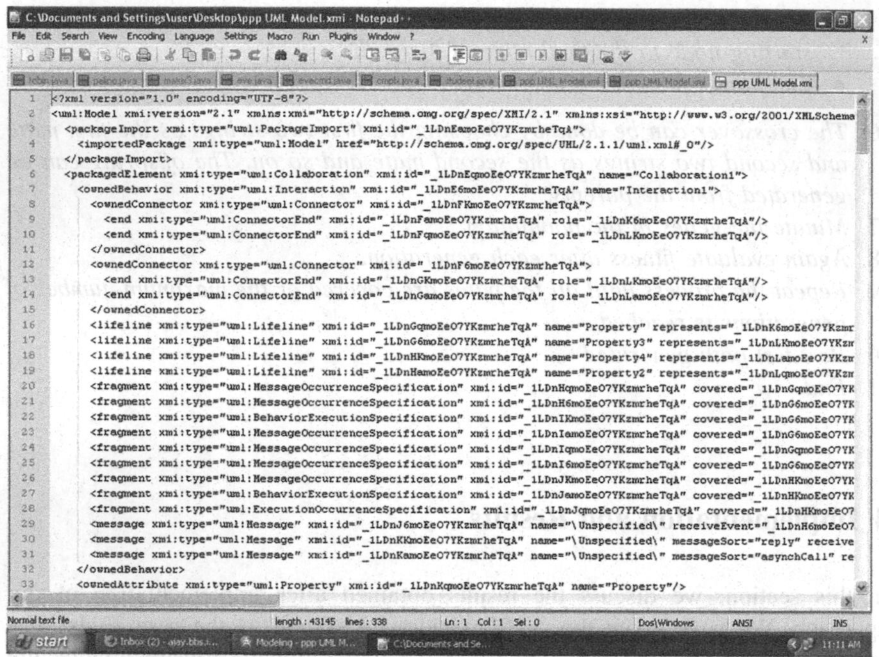

Fig. 4 Snapshot of the XML code

First Generation:

After calculating the weights of each test path, we apply the fitness function $f(k) = k^2$. The probability of each individual is calculated (Prob_ind) by $P(k) = f(k) / \sum f(k)$.

For the new generation, the selection of individuals is done by generating a random number between 0 and 255, as we consider an 8-bit number for our initial population size. We perform crossover by mating the chromosomes using multi-point crossover. The calculation of first generation is presented in Table 1.

Before crossover and after crossover

First pair:

Test Path 1 = 010|000|11 Test Path 5 = 01001111

Test Path 2 = 110|011|11 Test Path 6 = 11000011

Before crossover and after crossover

Second pair:

Test Path 3 = 110|011|11 Test Path 7 = 11010000

Test Path 4 = 011|100|00 Test Path 8 = 01001111

Second Generation:

In the second generation, the paths, Path 5, Path 6, Path 7, and Path 8, are considered.

We observed that the fitness value has been changed from 102,731 to 93,771 (second generation) in one generation. We noticed that the new populations Test

Table 1 First-generation data

String no.	Initial population	*k*-value	$f(k) = k^2$	Prob_ind	Expected count	Actual count
Test Path 1	01000011	67	4,489	0.04	0.17	0
Test Path 2	11001111	207	42,849	0.42	1.67	2
Test Path 3	11001111	207	42,849	0.42	1.67	2
Test Path 4	01110000	112	12,544	0.12	0.49	0
Sum			102,731.00	1.00	4.00	4.00
Avg			25,682.75	0.25	1.00	1.00
Max			42,849.00	0.42	1.67	2.00

Table 2 The observed test cases

Test case no.	PIN no.	Amount entered	Amount in account	Expected result	Actual result
1	5,107	N.A.	100,000	Invalid pin/ unsuccessful	Invalid pin/ unsuccessful
2	5,014	5,000	25,000	Success/Dispense cash Print receipt update account	Success/Print receipt Update account
3	5,014	3,000	20,000	Warning message show balance	Show balance
4	5,014	5,000	4,000	Insufficient funds show balance	Unsuccessful Show balance
5	5,014	1,050	6,000	Enter amount	Enter amount/ Multiple of 100

Path7 and Test Path8 (two new paths) are parts of Test Path2. By further calculations, we observed that Test Path2 is the optimized test path for the solution.

4.2 Test Case Creation

After the creation of all the test paths, we tried to generate the test cases from the paths. For the generated test paths, we can generate test cases on the basis of covering the messages. The test cases are shown in Table 2. Five different test cases are considered denoted in column 1.

The PIN number entered by the user is given in column 2, and column 3 represents the amount to be entered by the user. Balance in account and ATM is dealt by the bank side, which is described in column 4. Expected result and actual result are described in columns 5 and 6, respectively.

5 Conclusion

The test cases from the sequence diagram are generated by using our approach. Flaws can be detected from the model during the analysis level. So it reduces the cost and time elapsed. Message flow coverage criteria are used in our approach. The proposed model covers maximum messages in the graph. It is applied in many different sequence diagrams of different domains. Furthermore, the test cases are also reduced. The approach is not automated. An automated tool for the proposed approach can be developed. This approach may also be extended for other diagrams of UML for the purpose of test case generation. Finally, GA plays a significant role in optimizing the test cases.

References

1. Mall, R.: Fundamentals of software engineering. Prentice Hall of India, 3rd edn. (2009)
2. Abdurazik, A., Offutt, J.: Using UML collaboration diagrams for static checking and test generation. In: Proceedings of the 3rd International Conference on the UML, Lecture Notes in Computer Science, Springer GmbH, vol. 1939, pp. 383–395, New York, U.K. (2000)
3. Ali, S., Briand, L.C., Jaffar-ur-Rehman, M., Asghar, H., Zafar, Z., Nadeem, A.: A state based approach to integration testing based on UML models. J. Inf. Softw. Technol. **49**(11–12), 1087–1106 (2007)
4. Rumbaugh, J., Jacobson, I., Booch, G.: The Unified Modeling Language Reference Manual, Addison-Wesley (2001)
5. Binder, R.V.: Testing object-oriented systems: Models, patterns and tools (Series—Addison-Wesley Object Technology Series), (1999)
6. Jorgensen, P.C.: Software testing: a craftsman's approach, CRC Press, 2nd edn. (2002)
7. Doungsa-ard, C., Dahal, K., Hossain, A., Suwannasart, T.: Test data generation from UML state machine diagrams using gas. In: International Conference on Software Engineering Advances, IEEE Computer Society (ICSEA 2007)
8. Prasanna, M., Chandran, K.R.: Automatic test case generation for UML object diagrams using genetic algorithm. Int. J. Adv. Soft Comput. Appl., vol. 1 (2009)
9. Shirole, M., Suthar, A., Kumar, R.: Generation of improved test case from UML State diagram using genetic algorithm, ISEC '11, pp. 125–134, ACM (2011)

Solving Nonlinear Constrained Optimization Problems Using Invasive Weed Optimization

Y. Ramu Naidu and A.K. Ojha

Abstract Many real-world problems are constrained optimization problems. In solving nonlinear constrained optimization problems, penalty function method has been the popular approach. The performance of invasive weed optimization (IWO) with multistage penalty function is discussed in this paper. The proposed IWO is performed for six well-known problems, and results are reported. The obtained results demonstrate that IWO outperformed than other evolutionary algorithms.

Keywords Invasive weed optimization · Multistage penalty function · Nonlinear constrained optimization problems

1 Introduction

Constrained nonlinear optimization problems are frequently appeared in the fields of engineering and science [1–4]. Consider the following nonlinear constrained problem

$$\begin{aligned}&\min_x f(x) \quad x \in D \subset \mathbb{R}^n\\&\text{Subject to } g_i(x) \le 0 \quad i = 1, 2, \ldots m\\&h_j(x) = 0 \quad j - 1, 2, \ldots k\\&x_i^l \le x_i \le x_i^u\end{aligned} \tag{1}$$

where $g_i(x)$ are m inequality constraints, $h_j(x)$ are k equality constraints, x_i^l is a lower limit, and x_i^u is an upper limit.

Y. Ramu Naidu (✉) · A.K. Ojha
School of Basic Sciences, Indian Institute of Technology Bhubaneswar, Bhubaneswar, India
e-mail: y.ramunaidu@gmail.com

A.K. Ojha
e-mail: akojha57@iitbbs.ac.in

L.C. Jain et al. (eds.), *Intelligent Computing, Communication and Devices*,
Advances in Intelligent Systems and Computing 308,
DOI 10.1007/978-81-322-2012-1_14

Equality constraints are reformatted as

$$|h_j(x)| \leq \varepsilon$$

where ε is a positive tolerance value.

Solving constrained linear and nonlinear programming problems has been extensively studied for the past two decades. Several methods are proposed to solve constrained problems with the use of evolutionary algorithms. One such meta-heuristic method is called invasive weed optimization (IWO) proposed by Mehrabian and Lucas [5]. This algorithm is based on growing of weeds in crop fields. IWO is inspired by weed colonization in nature, and it is based on weed biology and ecology. Its important features include the allowing of all the weeds to participate in reproduction process and the reproduction of weeds that occurs without mating. IWO has been successfully applied in various engineering applications.

Generally, constrained optimization problems (CPP) are being solved with the use of penalty function methods [6–9]. In penalty function methods, the infeasible solutions are penalized so that the CPP is transformed into an unconstrained optimization problem. The main disadvantage of a penalty method is assigning degree of penalty to constraints. From the literature, a high degree of penalty puts more selective pressure on optimization algorithms for obtaining a feasible solution.

The organization of the paper is as follows: Following introduction, the basic ideas of IWO have been discussed in Sect. 2. Multistage penalty function has been presented in Sect. 3 and discussed the numerical illustrations in Sect. 4; finally, some conclusions have been included in Sect. 5.

2 Invasive Weed Optimization

The steps involved in the IWO are presented in the below:

- Initialization: A finite number of seeds are initialized randomly and distributed uniformly over the given search space.
- Reproduction of seeds: A seed after growing into a flowering plant produces new seeds. The number of seeds of a weed depends on its fitness value, minimum fitness value, and maximum fitness value. The number of seeds of all weeds linearly decreases from $S_{\max}$ to $S_{\min}$ as the following Eq. (2).

$$n(w_i) = \frac{S_{\max}(\max \text{fit} - \text{fit}(w_i)) + S_{\min}(\text{fit}(w_i) - \min \text{fit})}{\max \text{fit} - \min \text{fit}} \tag{2}$$

where n (w_i) = number of producing seeds of ith weed, and maximum number of seeds (S_{max}) and minimum number of seeds (S_{min}) are predefined parameters. fit(w_i) is the fitness value of ith weed, and max fit and min fit are maximum and minimum fitness values of the colony of weeds, respectively.

- Spatial dispersal: The produced seeds of the group are distributed normally with a mean of parent plant position and standard deviation (SD) is followed by the below Eq. (3).

$$\sigma_{\text{gen}} = \frac{(\text{gen}_{\max} - \text{gen})^{\text{mi}}}{\text{gen}_{\max}^{\text{mi}}}(\sigma_{\max} - \sigma_{\min}) + \sigma_{\min} \tag{3}$$

  where σ_{gen} is the SD at the present generation, and $\sigma_{\max}$ and $\sigma_{\min}$ are the maximum and minimum SD, predefined parameters. $gen_{\max}$ is the maximum generations. *mi* is the nonlinear modulation index, and necessity of *mi* is generated seeds could be close to the parent weed.
- Competitive exclusion: If the total number of weeds exceeds the maximum number of weeds (P_{max}) of the colony, the weeds which have worst fitness value are eliminated from the colony such that the constant number of weeds of the colony is maintained.
- Stopping condition: If the present generation number is equal to the maximum number of generations, the above process will be stopped and print the minimum fitness value of the colony of weeds.

3 Multistage Penalty Function

In constrained optimization problems, the constraints define a feasible region, i.e., if the candidate solution satisfies all constraints, we say that it belongs to the feasible region, and if not, then it is infeasible solution. In constrained optimization problems, a feasible solution is preferable to an infeasible solution. In the penalty function method, a constrained problem will be transformed into an unconstrained problem by incorporating a penalty function. Moreover, to obtain the optimal solution, a sequence of unconstrained optimization problems is solved and a highest value of a penalty function makes the minimization algorithm to converge to a local minimum. Instead a low value of the penalty function, the algorithm can hardly detect the feasible optimal solution.

The multistage penalty function is defined as [3, 7, 10]

$$F(\vec{x}) = f(\vec{x}) + h(ite)H(\vec{x}), \quad \vec{x} \in S \subset R^n \tag{4}$$

where $f(x)$ is the given objective function, $h(ite)$ is a penalty value changing dynamically, *ite* is the present generation number of the algorithm, and $H(x)$ is defined as [10].

$$H(\vec{x}) = \sum_{i=1}^{m} \theta(q_i(\vec{x}))q_i(\vec{x})^{\gamma(q_i(\vec{x}))} \tag{5}$$

$$\theta(q_i(\vec{x})) = \begin{cases} 10, & \text{if } q_i(\vec{x}) < 0.001 \\ 20, & \text{if } q_i(\vec{x}) < 0.1 \\ 100, & \text{if } q_i(\vec{x}) < 1 \\ 300, & \text{otherwise} \end{cases}$$

$$q_i(x) = \begin{cases} \max(0, g_i(\vec{x})) & 1 \le i \le m \\ \text{abs}(h_j(\vec{x})) & 1 \le j \le k \end{cases}$$

$$\gamma(q_i(\vec{x})) = \begin{cases} 1, & \text{if } q_i(\vec{x}) < 1 \\ 2, & \text{otherwise} \end{cases}$$

$$h(ite) = \sqrt{ite} \text{ for problem 1} \quad h(ite) = ite\sqrt{ite}$$

4 Results and Discussion

The proposed IWO is applied to six benchmark problems. All experiments are done in MATLAB (R2010a). All simulations are performed on a PC operating at 2 GB of RAM, Intel (R) Core (TM) i7-2600 processor at 3.00 GHz. The test problems are taken from [1, 2, 10–12]. For all experiments, the proposed IWO uses the following parameters: Nonlinear modulation index ($mi = 3$), initial number of seeds = 5, minimum generated seeds of the colony of weeds ($S_{\min}$) = 1), maximum generated seeds of the colony of weeds ($S_{\max}$) = 3, and the maximum number of weeds of the colony ($P_{\max}$) = 10. The maximum SD ($\sigma_{\max}$) depends on the dynamic range of design variables. The IWO works well if the maximum SD $\sigma_{\max}$ is set around a small percent (3 or 5 %) of the dynamic range of each design variable and the minimum SD $\sigma_{\min} = 10^{-12}$. The violation tolerance value of constraints is 10^{-5}. All experiments are performed for a maximum of 5,000 generations. The statistical results, i.e., mean, SD and the best solution, are reported for 10 runs in Table 1.

The obtained results are compared with those of the three variants of PSO [10], namely

i. PSO with inertia weight, PSO-In
ii. PSO with constriction factor, PSO-Co
iii. PSO with both factors, PSO-Bo.

Table 1 shows the comparison of the proposed IWO with the variants of PSO. For problem 1, all variants of PSO obtained infeasible solutions due to equality constraint. In comparison, the proposed IWO obtained feasible solution (E-10 considered as a zero). For problems 2, 4, and 5, the PSO variants obtained

Table 1 Comparison of the proposed IWO- and PSO-based results

Problem	Method	Best solution (Sum V.C)	Mean solution (Sum V.C)	Standard deviation
1	IWO	1.393461 (1.14E-13)	1.393461 (1.14E-13)	0.00
	PSO-In	1.393431 (0.000020)	1.394006 (0.000014)	0.0015
	PSO-Co	1.393431 (0.000020)	1.393431 (0.000020)	0.00
	PSO-BO	1.393431 (0.000020)	1.393431 (0.000020)	0.00
2	IWO	−6,961.813 (0.00)	−6,961.8103 (0.00)	0.0019
	PSO-In	−6,961.798 (0.000008)	−6,960.866 (0.000003)	0.608
	PSO-Co	−6,961.837 (0.000019)	−6,961.836 (0.000019)	0.0011
	PSO-BO	−6,961.837 (0.000019)	−6,961.774 (0.000013)	0.14
3	IWO	680.632 (0.00)	680.635 (0.00)	0.003
	PSO-In	680.639 (0.000019)	680.671 (0.000008)	0.034
	PSO-Co	680.635 (0.00130)	680.663 (0.00034)	0.050
	PSO-BO	680.636 (0.0000)	680.683 (0.000015)	0.014
4	IWO	−30,665.539 (0.00)	−30,665.539 (0.00)	0
	PSO-In	−31,543.484 (1.311)	−31,526.304 (1.297)	18.037
	PSO-Co	−31,542.578 (1.311)	−31,528.289 (1.326)	12.147
	PSO-BO	−31,544.459 (1.311)	−31,493.190 (1.331)	131.67
5	IWO	−31,025.559 (0.00)	−31,025.559 (0.00)	0
	PSO-In	−31,544.036 (0.997)	−31,523.859 (0.958)	17.531
	PSO-Co	−31,543.312 (0.996)	−31,526.308 (0.965)	19.153
	PSO-BO	−31,545.054 (0.999)	−31,525.492 (0.968)	23.392
6	IWO	−212.999 (0.00)	−212.999 (0.00)	0
	PSO-In	−213.0 (0.00)	−213.0 (0.00)	0
	PSO-Co	−213.0 (0.00)	−213.0 (0.00)	0
	PSO-BO	−213.0 (0.00)	−213.0 (0.00)	0

infeasible solutions. For problem 3, the variant of PSO-Bo was successful in obtaining a feasible solution. In problem 6, the PSO variants achieved feasible solution. Figure 1 represents the typical curves of IWO and PSO. In all problems, the proposed IWO successfully obtained feasible solutions. For all problems, IWO outperformed PSO variants as well as other evolutionary algorithms.

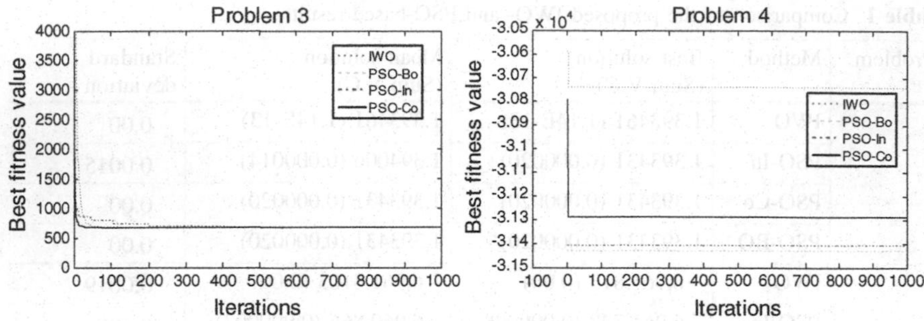

Fig. 1 Convergence curves of fitness value versus iterations for problem 3 and 4

5 Conclusions

In this paper, we discussed the performance of the IWO with multistage penalty function. The obtained results demonstrate that IWO with multistage penalty function is a good alternative algorithm to handle nonlinear constraints. In all cases, IWO achieved superior results than those obtained by other methods PSO and GA. Our future work is investing the best optimal solution with other IWO models.

References

1. Floudas, C.A., Pardalos, P.M.: A collection of test problems for constraints global optimization algorithms. In: Goos, G., Hartmanis, J. (eds.) LNCS, vol. 455. Springer, Heidelberg (1990)
2. Hock, W., Schittkowski, K.: Test examples for nonlinear programming codes. In: Beckmann, M., Kunzi, H. P. (eds.) LNEMS, vol. 187. Springer, Heidelberg (1981)
3. Yang, J.M., Chen, Y., Horng, J.T., Kao, C.Y.: Applying family competition to evolution strategies for constrained optimization. In: Peter, J.A., Robert, G.R., John, R.M., Russ, E. (eds.) USA 1997. LNCS, vol. 1231, pp. 201–211. Springer, Heidelberg (1997)
4. Pappula, L., Ghosh, D.: Large array synthesis using invasive weed optimization. In: IEEE conference on microwave and photonics, pp. 1–6. IEEE Press, India
5. Mehrabian, A.R., Lucas, C.: A novel numerical optimization algorithm inspired from weed colonization. Ecol. Inform. **1**, 355–366 (2006)
6. Homaifar, A., Lia, A.H., Qi, X.: Constrained optimization via genetic algorithms. Simulation **2**, 242–254 (1994)
7. Joines, J.A., Houck, C.R.: On the use of non-stationary function to solve nonlinear constrained optimization problems with GA's. In: IEEE Conference on evolutionary computation, pp. 579–584. IEEE Press, Orlando, Florida (1994)
8. Mezura, E.: Alternative to handle constraints in evolutionary optimization. Ph.D. thesis, CINVESTAV-IPN, Mexico (2004)
9. Rao, S.S.: Optimization: Theory and Applications. Wiley Eastern Limited, New York (1977)

10. Parsopoulos, K. E., Vrahatis, M.N.: Particle swarm optimization method for constrained optimization problems. In: 2nd euro-international symposium on computational intelligence, pp. 214–220. IOS Press, Kosice (2002)
11. Himmelblau, D.M.: Applied Nonlinear Programming. McGraw-Hill, New York (1972)
12. Michalewicz, Z.: Genetic Algorithms + Data Structures = Evolution Programs. Springer AI Series, New York (1992)

ESEC: An Ideal Secret Sharing Scheme

Greeshma Sarath, S. Deepu, Sudharsan Sundararajan and Krishnashree Achuthan

Abstract Secret sharing techniques have been used to ensure data confidentiality and enable secure reconstruction of data. In this paper, we propose a novel secret sharing scheme called ESEC based on Elliptic Curve Cryptography (ECC). The proposed scheme is to distribute a secret to a non-intersecting group of participants with different privilege levels. In our implementation, we have three phases involving share generation, distribution, and secret regeneration. The threshold of different privilege levels is different and for reconstruction of the secret all participants will have to be from the same privilege level. The proposed scheme is a perfect and ideal secret sharing scheme.

Keywords Secret sharing · Elliptical curve cryptography · Multilevel secret sharing · Cryptography

---

G. Sarath (✉)
Department of Computer Science and Engineering,
Amrita Vishwa Vidyapeetham, Kollam, India
e-mail: greeshmasarath@am.amrita.edu

S. Deepu · S. Sundararajan · K. Achuthan
Department of Cyber Security Systems and Networks,
Amrita Vishwa Vidyapeetham, Kollam, India
e-mail: deepusthatta@gmail.com

S. Sundararajan
e-mail: sudharsan77@gmail.com

K. Achuthan
e-mail: krishnashree.achuthan@gmail.com

L.C. Jain et al. (eds.), *Intelligent Computing, Communication and Devices*,
Advances in Intelligent Systems and Computing 308,
DOI 10.1007/978-81-322-2012-1_15

1 Introduction

Secret sharing scheme is a popular mechanism used to secure a common secret by distributing it among more than one participants in such a way that each participant gets at least one share but an individual share does not divulge any details about the original secret. The concept of threshold secret sharing was first discussed by Adi Shamir, where he demonstrated how a secret can be split into n shares out of which any "t" of them can be combined to reconstruct the secret. The key property of this scheme was that any $t - 1$ number of shares will not disclose any information about the secret. Shamir's scheme relies on a polynomial-based mechanism for secret sharing. Recently, Elliptic Curve Cryptography (ECC) which is a secure asymmetric encryption scheme is also used for secure secret sharing. ECC maps the shares onto elliptical curve points of a secure curve over a finite field Fp.

In our multilevel secret sharing scheme, each participant is assigned by a privilege level depending on their position in the hierarchy. The secret is divided among all the participants in the same privilege level, so that shareholders with same privilege level can unite together to generate the secret. Same secret can be regenerated by another group without interfering with each other. The threshold of shares needed depends on the privilege level of the shareholder, but threshold of a group is the number of shareholders in that group, i.e., all members with same privilege level has to participate in reconstruction phase. This scheme is a perfect secret sharing scheme, i.e., any subset of unqualified participants does not infer any information about the secret.

The whole secret sharing scheme is based on ECC [1]. Here the secret is mapped to points on elliptic curve. The length of all shares is same as that of the secret. Such secret sharing schemes are said to be ideal. The access structures of this scheme are multipartite or compared, means the participants are divided into several levels of all participants in the same level play same role in the whole process.

ECC is one of the most advanced and promising techniques in the field of public key cryptography. It offers many advantages over other cryptographic techniques which uses integer factorization or discrete logarithmic approach. The hardest problem in which ECC is built upon is elliptic curve discrete logarithmic problem. ECDLP is based on the infeasibility in computing discrete logarithms on elliptic curves over finite fields. It gives ECC, a greater strength-per-key-bit. It uses arithmetic with much shorter numbers 160,256 bits instead of 1,024, 2,048 bits and provides same level of security.

First we developed a non-threshold secret sharing scheme with all participants treated in equal manner. The secret is distributed equally as shares to all. It is an ideal, perfect secret sharing scheme using ECC. This scheme is then extended to incorporate independent users with different privilege levels. This scheme is referred as multilevel secret sharing. It is useful in situations where participants are arranged in a hierarchy. Each level in the hierarchy is independent from each other and the number of participants in each level is different. But the privilege of each member in a particular level is same. The share distribution relays on privileges. The extended version is also ideal and perfect secret sharing scheme.

2 Related Work

Secret sharing concept was introduced by Shamir [2] and similar concept was proposed by Blakley in [3] during the same period. Shamir [2] was based on polynomial interpolation while [3] was based on finite geometry. In [4], the authors discuss a variable threshold-based secret sharing scheme where the threshold value for regeneration of the secret can be changed. Another threshold-based secret sharing scheme based on RSA digital signature is described in [5]. Both the schemes can regenerate the secret using lesser number of shares than the total number of shares of the secret and they verify the integrity of the shares also. Sun et al. [6] discusses another secret sharing area named proactive secret sharing in which shares for the same secret are updated in accordance with a time interval. This scheme is based on bilinear pairing and ECC. Multilevel secret sharing based on ECC and bilinear pairing, where participants are assigned different privilege levels is discussed in [7]. Here the secret sharing scheme uses ECC based on some access structures. Setting access structures for different participants is a difficult problem. Scheme [8] is related to access structures of ECC-based secret sharing using algebraic-geometric does and their application in secure multiparty computation. Here the share size with respect to the secret is one of the issues. Just as in [9] where the authors propose an ideal secret sharing scheme where the size of the share is equal to the size of the secret our work too is an ideal secret sharing scheme. An ideal hierarchical secret sharing scheme suggesting a natural definition for the family of hierarchical access structures is discussed in [10]. The authors of [11] have presented a survey on different secret sharing mechanisms and discussed their properties. The scheme proposed in this paper is different from those mentioned above in that its simple and uses basic secret sharing and ECC to blend multilevel secret sharing and ideal secret sharing property to achieve perfect secrecy.

3 Non-threshold Secret Sharing Scheme Using ECC

We represent the secret as a point on an elliptical curve. We have to choose a secure curve in the set up phase to avoid curve specific attacks. If there are "n" participants, choose a secure elliptical curve $E\ (F_p)$ and its generator, g. Now, the secret can be represented as a sum of n points. Choose "n" random numbers r_1, $r_2, \ldots, r_n$, then the secret can be represented as $S = r_1{\cdot}g + r_2{\cdot}g +, \ldots, + r_n{\cdot}g$ that is, "S" itself is a point in the elliptic curve. These points $r_1{\cdot}g$, $r_2{\cdot}g$, …, $r_n{\cdot}g$ are the shares for each participants. These shares are distributed to all shareholders. To enhance the security, we encrypt these shares using ECElgamal encryption scheme, before the distribution. The shareholders will accept the shares, and stored in the encrypted form for future use.

When any participants request for reconstruction of secret, the regeneration unit broadcast a request for shares to all participants. After accepting the request from regeneration unit, each participant will decrypt shares using its own private key and encrypt it with public key of reconstruction unit. They use ECElgamal for encryption and decryption. The regenerator receiving all the shares and decrypt it and regenerating the secret, in such a way by simply Point-Addition of shares. The regenerated point is one of the points in the curve which represent secret. If any one of the shareholder sends invalid share then the secret cannot be regenerated. Which means the generated secret is entirely different one and the intension of malicious participant is not in use.

This scheme is not a threshold secret sharing because it needs coordination of all the participants for the reconstruction. But it is one of the efficient and fastest secret sharing schemes. The security aspect of this scheme is high, because it uses ECC for representing the shares and secret. It pivots on ECDLP which is hard to break. Also this scheme takes advantages of ECElgamal encryption while transmitting the shares. It can be used in different types of application.

Algorithms

CREATE-SECRET (n)

n = number of participants
Secret = α, point at infinity, share = 0

1. If(n < 0)
2. return 0
3. Else
4. While(share < n)
   Perform Point-Addition (Secret, Point-Multiplication(CREATE-SHARE(),g)
5. Share++
6. Return secret

CREATE-SHARE ()

1. Share = 0
2. share = select a random of fixed size
3. return share

RECONSTRUCTION-SECRET (n, shares)

1. secret = α, point at infinity
2. for i = 1 to n
3. perform point-addition (secret, Point-Multiplication(shares[i],g))
4. return secret

4 A Multilevel Secret Sharing Scheme

This is an extended version of our first method. Here each participant is assigned with a privilege level. The number of participants in each privilege level is assigned with an increase in the privilege. Each participant is assigned a privilege such that all participants at same level in the hierarchy enjoy same privilege. The privilege level is represented as $K = \{1, 2,\ldots, k\}$. Here also we have to choose a secure curve in the set up phase to avoid curve specific attacks. This scheme represents the secret and shares as elliptic curve points.

The participants in the lowest privilege level are first granted with shares. Let "*m*" is the number of participants in the lowest privilege level that is level with maximum number of participants. The generation unit select "*m*" random numbers $r_1, r_{2,} \ldots, r_m$ and assign $r_1 \cdot g, r_2 \cdot g, \ldots r_m \cdot g$ as shares to "*m*" participants in the lowest privilege level. The secret is generated as $S = r_1 . g + r_2 . g + , \ldots, + r_m \cdot g$. For all participants in higher privilege level assign m^i shares as $m_1 = r_1^1 \cdot g$, $m_2 = r_2^1 \cdot g$, … and $m_1^i = (S - (r_1^1 + r_2^1 + \ldots + r_1^{i-1}))g$, i.e., generate "*m*" shares if k_ith level contains "*m*" number of shares. After generating sufficient number of shares for each privilege level encrypt and distribute to corresponding participant.

At reconstruction phase, any of the participant send request for reconstruction to regeneration module. This module identifies the privilege level of particular participant and broadcast a request for shares to all the participants with same privilege. More precisely, the generator unit identifies the particular participant belongs to which level in the hierarchy and send request to all members in the group for their individual share. Now each member accepts the request, encrypt their share, and send back to regeneration unit. The generation unit acquires all the shares from the group and regenerate the secret without any interruption to other members in the rest group. Here also, if any one of them send invalid share then the secret cannot be regenerated. The valid share submitted by a participant with different privilege level is also worthless for reconstruction.

This multilevel secret sharing scheme takes the advantages of ECDLP and ECElgamal to get maximum security while creating and transmitting the shares. Even the groups having different privilege level, each member gets elliptical curve point of same size. So it is difficult to identify the privilege level from the share. Each group is capable of reconstructing secret which means the groups works independently without interrupt other group. It can be used in different types of application where low power conception, low computational power, etc., using the advantage of ECC.

Algorithms

CREATE-SHARE (n, l)
n = participant id
l = number of participants in a level

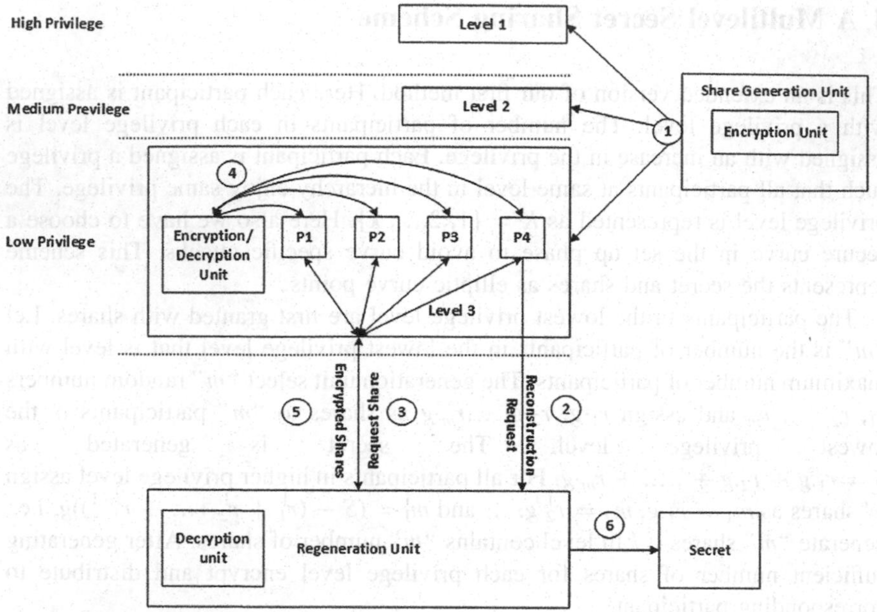

Fig. 1 Multilevel secret sharing scheme

1. sum = 0;
2. if (n < (l − 1))
3. select a random number share, $0 < \text{share} < 2^{192}$
4. sum = sum + share
5. return share
6. If (n = (l − 1))
7. secret-sum
8. return share

RECONSTRUCTION-SECRET (k, shares [])

1. ind = 0, resecret = α, a point at infinity.
2. While (ind < k)
3. perform Point-Addition(resecret, Point-Multiplication(shares[ind],g))
4. ind++
5. If (resecret == secret)
6. return "Secret can be regenerated"
7. Else
8. return "Secret cannot be regenerated"

Figure 1 shows our multilevel secret sharing scheme. The figure specifies the steps in order. In the 1st Step shares are created based on privilege level by share generation unit. The shares are encrypted and distributed in a hierarchy by the same unit. In 2nd Step, any participant can requested for regeneration to regeneration unit. Regeneration unit will identify the privilege level of requester at 3rd Step and also sends a request for shares to all members with same privilege.

In 4th Step, all members those who got request will decrypt their shares and encrypt it with public key of regeneration unit. Each member sends their encrypted share to regeneration unit at 5th Step. Regeneration unit accepts all the shares and decrypt it and regenerate the secret by elliptical curve point operations, i.e., at Step 6 we regenerate the secret.

5 Implementation and Analysis

5.1 Implementation

Our schemes are implemented in java 1.7.0 using the development environment Eclipse 2.0.1. The schemes are working according to our design. It contains process of generation, distribution, and reconstruction. The communication is through socket programming in java. We use elliptic curves recommended for Federal government use by NIST standard. We implemented with curve P-192 and it is applicable to other curves also. We did our implementation as separate modules in order to use it either at local host or at network. Each participant is identified by socket, so each participant will get an identity.

5.2 Analysis

The testing is performed on computer having Intel core i7 processor, 8 GB RAM, and 500 HDD. We run only our system on the computer and analyzed. The execution time with varying number of participants is analyzed. The scheme is analyzed with different curves also. The observations are tabulated and make graph in Fig. 2. It shows our system execution time is increasing gradually with slight variation with respect to increase in participants. Also when key bit size increased the curve shows same feature. From this, we can say that our system is directly proportional to key bit size and number of participants.

This multilevel secret sharing scheme distributes secret as shares to each participant at each level. The paper [12] discussed about access structures for secret sharing. In our scheme, any subgroup having same privilege could reconstruct the secret. An access structure from one privilege level combined with another privilege level never leads to secret. So the secret is purely safe from any unqualified subgroup access structure.

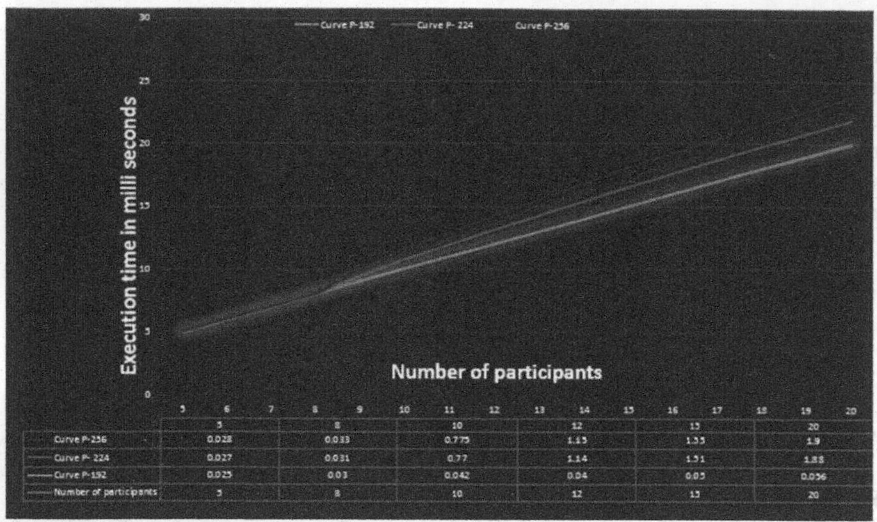

Fig. 2 Number of participants versus execution time for different key sizes in bits

6 Conclusion

The non-threshold-based scheme and multilevel secret sharing scheme is designed and developed. Both schemes are implemented in java with each participant as a separate module and communication is through socket programming.

Both Schemes are purely based on elliptical curve cryptography. They are ideal and perfect secret sharing schemes. Secret distributions made secure by using ECElgamal encryption scheme. Our multilevel encryption scheme can be used in applications where the participants belongs disjunctive groups. All the groups share a common secret and it can be reconstructed by a group without disturbing others.

References

1. Hankerson, D., Menezes, A., Vanstone, S.: Guide to elliptic curve cryptography, published by Springer, Berlin (2004)
2. Shamir, A.: How to share a secret. Commun. ACM **22**, 612–613 (1979)
3. Blakley, G.R.: Safeguarding cryptographic keys. In: Proceedings of the AFIPS National Computer Conference, pp. 313–317 (1979)
4. Yamamoto, H.: On secret sharing systems using (k, L, n) threshold scheme. IECE. Trans., vol. **J68**–A(9), pp. 945–952, 1985 (in Japanese). English translation: Electronics and Communications in Japan, Part I, vol. **69**(9), pp. 46–54, Scripta Technica, Inc., 1986
5. Xu, C., Xiao, G.: A threshold multiple secret sharing scheme. Acta Electronica Sin. **10**(32), 1387–1689 (2004)

6. Sun, H., Zheng, X., Yu, Y.: A proactive secret sharing scheme based on elliptic curve cryptography 1st International Workshop on Education Technology and Computer Science, vol. 2, pp. 666–669 (2009)
7. Hua, S., Aimin, W.: A multi-secret sharing scheme with general access structures based on elliptic curve (2010)
8. Chen, H., Ling, S., Xing, C.: Access structures of elliptic secret sharing schemes (2008)
9. Farras, O., Padŕo, C.: Ideal secret sharing schemes for useful multipartite access structures IEEE, Springer, Berlin (2007)
10. Farràs, O., Padró, C.: Ideal hierarchical secret sharing schemes. IEEE Trans. Inf. theor. **58**(5) (2012)
11. Beimel, A.: Secret-sharing schemes: A survey, Springer LNCS 6639, pp. 11–46 (2011)
12. Chen, H., Ling, S., Xing, C.: Access structures of elliptic secret sharing schemes. IEEE Trans. Inf. Theor. **54**(2) (2008)

Extended Service Registry to Support I/O Parameter-Based Service Search

H.N. Lakshmi and Hrushikesha Mohanty

Abstract The increasing availability of Web services within an organization and on the Web demands for an efficient search mechanism to find services satisfying user requirements. Universal Description, Discovery, and Integration (UDDI) is an industry standard for Service Registries, developed to solve the Web service search problem. However, UDDI offers limited search functionalities which may return a huge number of irrelevant services. Also, often consumers may be unaware of precise keywords to retrieve the required services satisfactorily and may be looking for services capable of providing certain outputs. In this paper, we propose a new system called ESR for extended and efficient service search using an Object Relational Database. The experimental results demonstrate the efficiency of service search in our extended service registry (ESR) and the variety of user queries supported.

Keywords Service Registries · Service search · UDDI · I/O parameters

1 Introduction

Web services are self-contained, self-describing, modular applications that can be published, located, and invoked across the Web. As growing number of services are being available, selecting the most relevant Web service fulfilling the requirements of a user query is indeed challenging. Various approaches can be used for service search, such as searching in Universal Description, Discovery, and

H.N. Lakshmi (✉)
University of Hyderabad, Hyderabad, India
e-mail: hnlakshmi@gmail.com

H.N. Lakshmi · H. Mohanty
CVR College of Engineering, Hyderabad, India
e-mail: hmcs_hcu@yahoo.com

L.C. Jain et al. (eds.), *Intelligent Computing, Communication and Devices*,
Advances in Intelligent Systems and Computing 308,
DOI 10.1007/978-81-322-2012-1_16

Integration (UDDI), Web and service portals. In this paper, we take up the issue of searching at Service Registries for its practicality in business world as providers would like to post their services centrally, as searching there is less time-consuming than searching on World Wide Web.

Basically, current technology supports service search by name, location, business, bindings, or TModels and binds two services based on composability of their protocols. The search API is limited by the kind of information that is available and searchable in UDDI entries and do not provide any support for complex searches like I/O parameter-based search and automatic composition of Web services. Also, our experiments show that average response time of current UDDI implementations increase with a substantial increase in number of services registered.

The above shortcomings of UDDI have motivated us to build an extended service registry (ESR) system capable of offering powerful and efficient search operations. We propose the use of Object Relational Database as repository of Web services. Information about the Web services, extracted from their WSDLs, is stored in tables and relational algebraic operators are used for service search. This work is an extension of our previous work [1], where we proposed a RDBMS approach for Service Registries.

Often consumers may be unaware of exact service names that are fixed by service providers. Rather consumers being well aware of their requirements would like to search a service based on their commitments (inputs) and expectations (outputs). Based on this concept, we have explored the feasibility of I/O-based Web service search in our proposed ESR system, to support varying requirements of the consumer. Utility of such an I/O-based Web service search for composition of Web services is shown in our previous work [1].

The rest of the paper is organized as follows. In Sect. 2, we describe the Object Relational Schema used for service registry. This is an extension of our previous work [1]. Section 3 discusses our experimental results. In Sect. 4, we essay the related work. We conclude our work in Sect. 5.

2 Object Relational Database for Service Registry

In this section, we first give an overview of how Web services are registered in UDDI. We then argue for the need of Object Relational Database for storing Web service information in a registry. We then propose an Object Relational Schema for our ESR.

2.1 WSDL and UDDI Relationship

The Web Services Description Language (WSDL) is a XML language for describing Web services. It contains service interface defined as a set of operations and messages, their protocol bindings, and the deployment details. UDDI (V3) offers the industry a specification that is used for building flexible, interoperable XML Web services registries useful in private as well as public deployments [2]. If offers clients and implementers a comprehensive and complete blueprint of description and discovery foundation for a diverse set of Web services architectures.

There are four primary data types in a UDDI registry—businessEntity: used to represent the business or service provider within UDDI, businessService: used to represent business descriptions for a Web service, bindingTemplate: contains the technical information associated to a particular service, and tModel : used to define the technical specification for a Web service. A businessEntity can contain one or more businessServices. A businessService can have several bindingTemplates and each bindingTemplate contains a reference to one or more tModels.

WSDL document contains six major elements that defines Web Services—types: provides data-type definitions, message: provides an abstract definition the data being and consists of logical parts, portType: defines a set of abstract operations each referring to an input message and output message, binding: specifies concrete protocol and data format specifications for the operations and messages defined by a particular portType, port: specifies an address for a binding and service: aggregates a set of related ports.

A summary of the mapping is as follows:

1. WSDL portType element is mapped to UDDI tModel.
2. WSDL binding element is mapped to UDDI tModel.
3. WSDL port element is mapped to UDDI bindingTemplate.
4. WSDL service element is mapped to UDDI businessService.

2.2 Object Relational Database as Service Registry

We observe that there are many sub-elements in WSDL that are not mapped to UDDI types such input message, output message, and message parts. Although these details can be added in tModel of the service, there are no supporting search functionalities provided in UDDI standards. This leads to the limited and poor search functionalities supported by UDDI. We, hence, propose to include these details in an ESR, enabling the registry to support additional search functionalities.

Each operation in a service has an input and output message, each of which in turn may have one or more message parts. For example, a HotelBooking service may take {Period, City} as input and provide {HotelName, HotelCost} as output. Thus, we require a strategy to store these multi-valued inputs and output messages in the registry. Such a structure clearly outs the principles of normalization.

Since current UDDI implementations store the UDDI data types in a Relational Database, a first thought would be to include the message details in a Relational database. A normalized Relational database solution to this requires that we store input and output message parts of each operation across multiple rows. With such a design, it takes multiple SQL Join operations to access and display the input or output message parts of a single operation. This implies reduced database performance. To avoid the need for multiple joins and to speed up the query, we propose to use multi-valued attributes for storing input and output message parts in our database design and, hence, use an Object Relational Database for our proposed ESR. Advantage of this approach is that it supports querying for services efficiently and also supports complex queries involving input and output parameters.

2.3 Extended Service Registry

In this section, we shall describe the Object Relational Schema for storing Web services in the registry. The relationship between the various tables in the proposed schema is depicted in the ER diagram in Fig. 1. This is an extension of our previous work [1].

1. Each Web service is given a unique ID (WSID) and stored with its name (WSName), port address (WSPortAdd), and operation name (OPName) in a Web service table (WSTable).

   WSTable: { WSID, WSName, WSPortAdd, OPName}

2. The parameters' table (ParTable) contains all the parameters, which take part as either input or output in any of the Web services in the registry, with their names in (PName). Each parameter is given a unique ID, (PID).

   PartTable : {PID, PName}

3. The Web service input/output table (WSInOutTable) lists all Web services in the registry with their respective input parameters (InPars) and output parameters (OutPars). InPars and OutPars are of collection type (ParList) and are stored as nested tables.

   WSInOutTable: {WSID, InPars, OutPars}

4. The schema also includes a query table for storing the user query (QueryT) and a table to store services matching the user query (MWSTable).

   - QueryT: {PNo, OutPars} where OutPars are the PIDs of output parameters specified in user query.
   - MWSTable: {WSIDs}

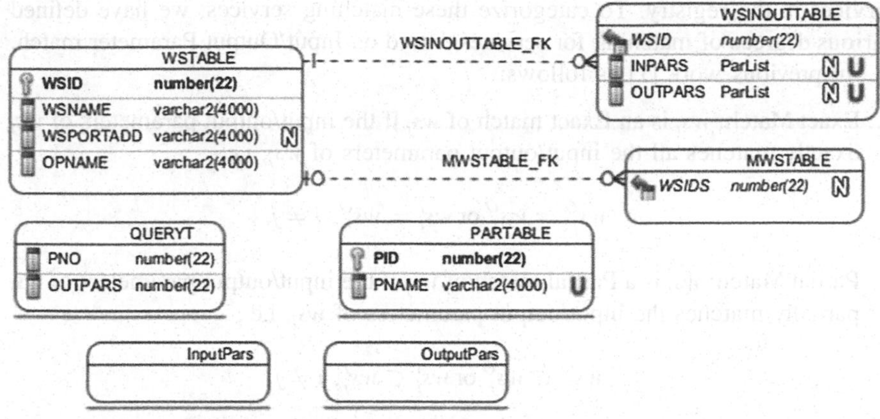

Fig. 1 ER diagram for service repository

Algorithm 1: Output Parameter-based Service Search

Input: Q^O, WSInOutTable: table
Output: MWSTable: table
foreach *ParName in* Q^O **do**

Select PID form ParTable where PName = *ParName*
INSERT PID into the QueryT table

foreach *ws in WSInOutTable* **do**
if $ws^O = Q^O$ *then*

INSERT *ws as* Exact Match in MWSTable

else if $ws^O \subset Q^O$ **then**

INSERT *ws as* Partial Match in MWSTable

else if $ws^O \supset Q^O$ **then**

INSERT *ws as* Super Match in MWSTable

else

continue

2.4 I/O Parameter-Based Service Search

A Web service, *ws*, has typically two sets of parameters—set of inputs ws^I and set of outputs ws^O. When a user searches for a service providing a requested set of outputs and/or accepting a requested set of inputs, there may be many matching services in the registry. To categorize these matching services, we have defined various degrees of matching for services, based on Input/Output Parameter match, in our previous work [1] as follows:

1. Exact Match: ws_i is an Exact match of ws_j if the input/output parameters of ws_i exactly matches all the input/output parameters of ws_j, i.e.,

$$ws_i^O = ws_j^O \text{ or } ws_i^I = ws_j^O,\ i \neq j.$$

2. Partial Match: ws_i is a Partial match of ws_j if the input/output parameters of ws_i partially matches the input/output parameters of ws_j, i.e.,

$$ws_i^O \subset ws_j^O \text{ or } ws_i^I \subset ws_j^I,\ i \neq j.$$

3. Super Match: ws_i is a Super match of ws_j if the input/output parameters of ws_i is a superset of the input/output parameters of ws_j, i.e.,

$$ws_i^O \supset ws_j^O \text{ or } ws_i^I \supset ws_j^I,\ i \neq j.$$

2.5 Process of Parameter-Based Service Search

To empower the Service Registries with additional search capabilities, we define algorithms for I/O parameter-based service search. Algorithm 1 presents pseudocode for output parameter-based service search. Q^O represents output parameters specified in user query. Similar procedure is used for input parameter-based service search.

3 Experimental Results

The effectiveness of the OR Databases approach is shown by conducting two sets of experiments:

1. Performance of basic keyword search in Service Registries—jUDDI and ESR approach.
2. Scalability of Service Registries.

3.1 Experimental Setup

To evaluate the response time of UDDI-based service directories, we query a service directory that is implemented in jUDDI. Apache jUDDI (pronounced "Judy") is an open source Java implementation of the Universal Description, Discovery, and Integration (UDDI v3) specification for Web services [3]. We conducted experiments on WSC Dataset [4]. We varied the number of Web services that were stored in Service Registries implemented in jUDDI and ESR, from 100 to 500 and then to 1,000. Five queries were submitted to the service directory in both the sets of experiments, and the response times of the service directory were computed. We ran our experiments on a 1.3-GHz Intel machine with 4-GB memory-running Microsoft Windows 7. Our algorithms were implemented using Oracle 10g and JDK 1.6. Each query was run 5 times and the results obtained were averaged, to make the experimental results more sound and reliable.

3.2 Comparison of Basic Keyword Search in Registries—JUDDI Versus ESR Approach

Web service search is performed in UDDI-based service directories using predefined APIs. find service() function in inquiry API, of jUDDI, is used to locate specific services in service registry and returns a serviceList structure that matches the conditions specified in the arguments. The various arguments supported are authInfo, businessKey, findQualifiers, and name. The default behavior of UDDI with respect to matching is exact match.

In the first set of experiments, we published 500 Web services in jUDDI and ESR. We then queried jUDDI with five different user requirements. Experiments were done for both approximateMatch and exactMatch Qualifiers. Same set of requirements were then fed to ESR and their response times were recorded. The experiment was repeated by publishing 1,000 Web services in both the registries. Figures 2 and 3 shows the performance results for Queries Q1 to Q5 on both the Service Registries for approximateMatch and exactMatch Qualifiers respectively.

In the second set of experiments, we varied the number of services published from 100 to 1,000 in jUDDI and ESR. We then queried jUDDI with four different user requirements. Experiments were done for both approximateMatch and exactMatch Qualifiers. The same sets of requirements were then fed to ESR, and their response times were recorded. Figure 4 depicts how the execution time for Query Q1 varies on both the Service Registries, with an increase in the number of services registered.

From the results obtained, we can infer that the time taken for search with approximate match is far lesser in our ESR approach when compared to the time taken in jUDDI. Though the time taken for service search with exact match appears to be close to the time taken in jUDDI, it is so only for the cases when

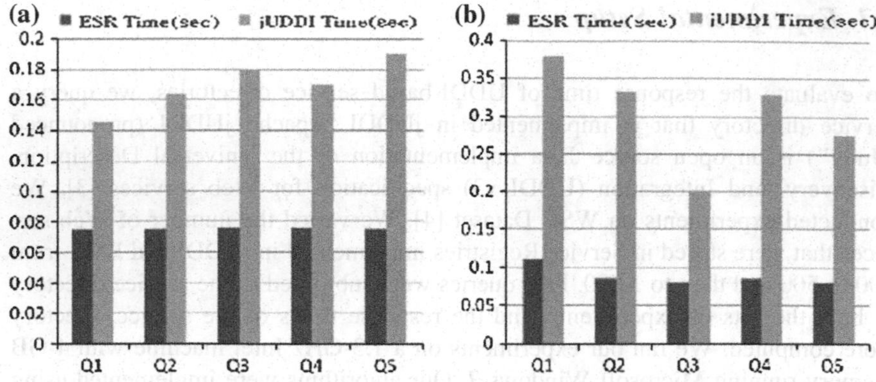

Fig. 2 jUDDI versus ESR for 500 services. **a** Exact match. **b** Approximate match

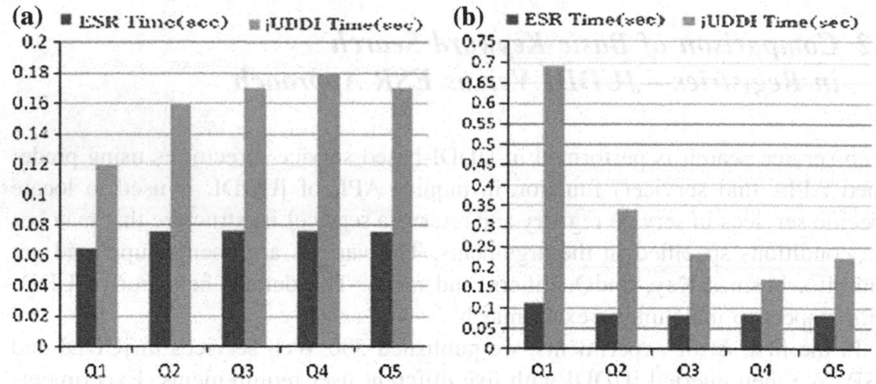

Fig. 3 jUDDI versus ESR for 1,000 services. **a** Exact match. **b** Approximate match

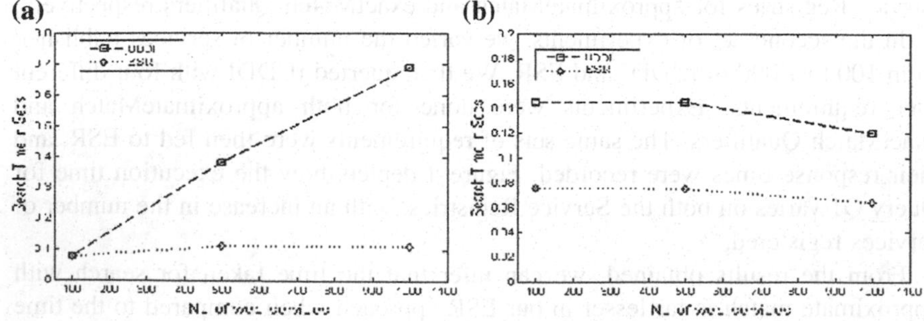

Fig. 4 Service search time in jUDDI versus ESR. **a** Exact match. **b** Approximate match

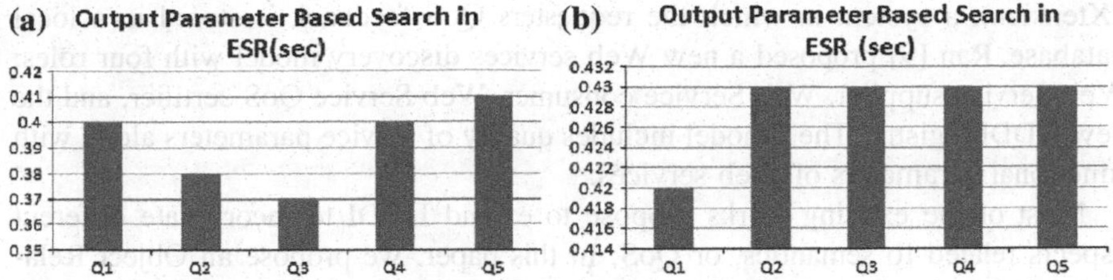

Fig. 5 Output parameter-based search in ESR. **a** Registry with 500 services. **b** Registry with 1,000 services

search results have very few matching services, meaning, for search results that have many matching services, our approach works faster than jUDDI, both in the case of approximate match and exact match. Also, the search time becomes more efficient in our approach with a substantial increase in number of services in the registry. Thus, we can finally conclude that service search in ESR is more scalable and efficient than searching in UDDI.

3.3 Performance of Output Parameter-Based Service Search

To empower the ESR with additional search capabilities, we defined algorithms for I/O parameter-based service search as discussed in Sect. 2.5. To include I/O parameter-based service search, we implemented Algorithm 1 and evaluated its performance using different output parameter patterns as user queries. In the first set of experiments, we published 500 Web services in ESR. The experiment was then repeated by publishing 1,000 Web services in ESR. Figure 5 shows the performance of output.

4 Related Work

In this section, we survey current e orts related to UDDI extensions and clustering Web services. Many efforts have been made to extend UDDI to improve service search, either by storing additional information about Web services typically by extending the WSDL format or by extending the API of UDDI registries to support additional functionalities.

Goodwin et al. [5] propose an architecture for semantic sensor matchmaker to make the service search and integration more efficient. To support addition of new query types, Mili et al. [6] proposed a generic extension framework to UDDI registries, by adding a middle tier acting as a broker between standard UDDI registries and clients. Juric et al. [7] incorporated version information into the businessService and tModel data structures. Zhou et al. [8] proposed UX (UDDI eXtension), a system in which the requesters QoS feedback is stored in a local database. Ran [9] proposed a new Web services discovery model with four roles: Web Service supplier, Web Service consumer, Web Service QoS certifier, and the new UDDI registry. Their model includes quality of service parameters along with functional parameters of Web services.

Most of the existing works propose to extend UDDI to incorporate different aspects related to semantics, or QoS. In this paper, we propose an Object Relational Schema for storing services in registry, to enable the registry to include multi-valued elements of WSDL and also to support I/O parameter-based service search.

5 Conclusion

Query-based Web service search is an important issue, especially for non-semantic Web services. Traditional UDDI-based service search lacks the ability to recognize all the features described in WSDLs. This leads to limited and poorly designed search operations that these registries offer. Experiments show that the performance of such service directories deteriorates with a substantial increase in number of services. Also, service search operations need to be extended to support varying requirements of the consumer. In order to improve the scalability and to empower Service Registries with additional search capabilities, we propose the use of Object Relational Databases as repository of Web services. We have simulated the algorithms on WSC Dataset [4]. The experimental results demonstrate the benefits of our ESR approach on varying user queries.

In the future work, we would like to work on a strategy for choosing the best matching service among the many candidate services. We further plan to integrate the current proposal with our previous work [1], to generate all possible compositions for a given user requirement, when expressed in terms of I/O parameters.

References

1. Lakshmi, H.N., Mohanty H.: RDBMS for service repository and composition. In: Fourth International Conference on Advanced Computing (ICoAC), 13–15 Dec 2012
2. UDDI Specifications: http://uddi.org/pubs/uddi_v3.html
3. Apache jUDDI:http://juddi.apache.org/index.html
4. The Web Service Challenge (WS-Challenge): http://www.ws-challenge.org/
5. Goodwin, J.C., Russomanno, D.J., Qualls, J.: Survey of semantic extensions to UDDI: implications for sensor services, SWWS, 16–22 (2007)
6. Mili, H., Tamrout, R.B., Obaid, A.: JRegistry: an extensible UDDI registry. Reports of NOTERE, pp. 115–128 (2005)
7. Juric, M.B., Sasa, A., Brumen, B., Rozman, I.: WSDL and UDDI extensions for version support in web services. J. Syst. Softw. **82**, 1326–1343 (2009)

8. Zhou, C., Chia, L., Lee, B.: QoS-aware and federated enhancement for UDDI. Int. J. Web Serv. Res. (IJWSR) **1**, 58–85 (2004)
9. Ran, S.: A model for web services discovery with QoS. ACM Sigecom Exchanges **4**(1), 1–10 (2003)

Realizing New Hybrid Rough Fuzzy Association Rule Mining Algorithm (RFA) Over Apriori Algorithm

Aritra Roy and Rajdeep Chatterjee

Abstract Association rules shows us interesting associations among data items. And the procedure by which these rules are extracted and managed is known as association rule mining. Classical association rule mining had many limitations. Fuzzy association rule mining (Fuzzy ARM) is a better alternative of classical association rule mining. But fuzzy ARM also has its limitations like redundant rule generation and inefficiency in large mining tasks. Rough association rule mining (Rough ARM) seemed to be a better approach than fuzzy ARM. Mining task is becoming huge now days. Performing mining task efficiently and accurately over a large dataset is still a big challenge to us. This paper presents the realization of new hybrid mining method which has incorporated the concepts of both rough set theory and fuzzy set theory for association rule generation and shows comparative analysis with Apriori algorithm based on test results of the algorithm over popular datasets.

Keywords Association rule mining · Fuzzy association rule mining · Fuzzy c-means clustering · Rough set theory · Rough association rule mining · Attribute reduction · Apriori algorithm

1 Introduction

Knowledge discovery in databases (KDD) is big procedure which consists of several subprocesses. Data mining is a subprocess of KDD process [1]. Data mining [2] discovers useful information and interesting patterns by the logical

A. Roy (✉) · R. Chatterjee
School of Computer Engineering, KIIT University, Bhubaneswar, India
e-mail: royaritra1990@gmail.com

R. Chatterjee
e-mail: cse.rajdeep@gmail.com

L.C. Jain et al. (eds.), *Intelligent Computing, Communication and Devices*,
Advances in Intelligent Systems and Computing 308,
DOI 10.1007/978-81-322-2012-1_17

analysis of a database. This derived information is very useful in intelligent systems such as Expert systems [3]. An Expert system uses heuristic mechanisms and knowledge to produce expert suggestion for decision making. Because of this need the idea of association rule mining has come. Association rule mining shows interesting association relationships among data items.

1.1 Data Mining and Association Rule Mining

We know that there is a large amount of data present in physical world. And useful knowledge is hidden within these data. Now extraction of knowledge from these dataset is a crucial thing because knowledge helps us in decision making. Data mining can be described as an important tool to discover the knowledge. Knowledge also can be termed as interesting patterns. So, the whole process of producing knowledge from raw data is called KDD. And data mining is just a part of KDD process. Association rules provide interesting correlations among data items. From these association rules, useful knowledge can be derived. Association rule mining is an aspect of data mining.

1.2 Fuzzy ARM Concepts and Rough ARM Concepts

Classical or crisp association rule mining uses the concept of sharp partitioning of dataset to convert numerical attributes into Boolean attributes. And it results in loss of information. Here, a user has to define the minimum support value which is again a problem because any wrong setting of minimum support value will cause errors in mining process. So, to eradicate these issues, the concept of fuzzy association rule mining (Fuzzy ARM) came. Fuzzy ARM process incorporates the concepts of fuzzy set theory [4] for the association rule generation. There is a huge number of fuzzy ARM algorithms is already present in research work. Some of them are interesting in terms of the mining strategy. In [5], we can see a variety of different fuzzy ARM techniques. In [6], we can see the use of fuzzy c-means (FCM) clustering [7, 8] technique for the preprocessing of the dataset. In [9], we found the automatic generation of minimum support value of each data item by the proposed algorithm. In [10], we can see a pruning mechanism to delete redundant rules by the concept of "certainty factor" [11, 12]. In [13], we can see again a pruning mechanism to eradicate redundant rules by the "equivalence concept". But the performance of these algorithms was not up to the mark as they were anticipated. Fuzzy ARM algorithms are not efficient in case of huge datasets. And also there is the chance of redundant rule generation. So, as an alternative of the fuzzy ARM, the concept of rough association rule mining (Rough ARM) came. Rough ARM incorporates the concepts of Rough set theory [14]. There is a significant number of research works already done in the area of Rough ARM. Some of them are interesting in terms of the mining strategy. In [15], we can find that rough set approach to association rule mining is a much easier technique than the maximal association method [16]. Here, rules generated in both methods are similar. In [17], we can see the use of rough set attribute reduction technique to reduce the size of the large dataset. In [18], we can find the use of the equivalence class concept for the mining task. Rough ARM seemed to be better than the fuzzy ARM. But the challenge of performing mining task efficiently on a large dataset, still remains. This paper represents a hybrid mining method which uses the concepts of both rough set theory and fuzzy set theory.

2 Theoretical Aspects

This section represents some theoretical aspects which are related to the proposed algorithm. These theoretical aspects are briefly discussed.

2.1 Rough Set Concepts and Reducts

Most of the time information which is available in the physical world is uncertain, imprecise, and incomplete. Performing mining task on such incomplete data can produce incomplete knowledge. So, it is very much needed to imperfect knowledge. And rough set [19] can remove this imperfectness. Let, U is finite set of objects and a binary relation $R \subset U \times U$ be given. The set U is called the *universe* and R is an *indiscernibility relation*. The discernibility relation describes our lack of knowledge about U. R is also can be represented as an equivalence relation. The pair (U, R) is termed as *approximation space*. Let X be a subset of U ($X \subseteq U$). Main objective is to represent set X with respect to R. $R(x)$ denotes the equivalence class of R determined by element x. Equivalence classes of the indiscernibility relation R is called *granules*. These granules are the fundamental parts of knowledge. And these granules of knowledge are understandable to us because of R. But by the indiscernibility relation, individual objects of U cannot be observed. From [20], we can get a simplified view of the rough set approximations. The set of all objects which can be with *certainty* classified as members of X with respect to R is called the *R-lower approximation* of a set X with respect to R, and denoted by $R_*(X)$. The set of all objects which can be only classified as *possible* members of X with respect to R is called the *R-upper approximation* of a set X with respect to R, and denoted by $R^*(X)$. The set of all objects which can be definitively classified neither as members of X nor as members of $-X$ with respect to R is called the *boundary region* of a set X with respect to R, and denoted by $RN_R(X)$.

So, from those relations, it can be easily anticipated that, set X is a *crisp set* with respect to R if and only if the boundary region of X is a null set. And set X is a *rough set* with respect to R if and only if the boundary region of X is not a null set.

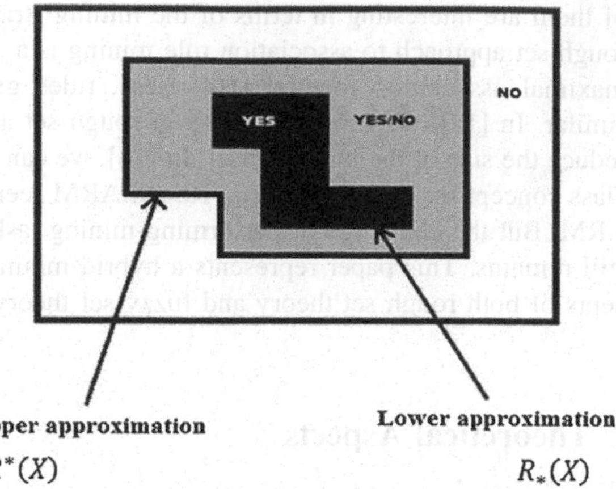

Fig. 1 Graphical illustration of the rough set approximations

The accuracy of approximation of a rough set X can be numerically described by the ratio of lower approximation and upper approximation of X. And this ratio is denoted by $\alpha_R(X)$. Here $|X|$ denotes the cardinality of $X \neq \emptyset$. The value of $\alpha_R(X)$ ranges between (0, 1) such that, $0 \leq \alpha_R(X) \leq 1$. If $\alpha_R(X) = 1$, then X is a *crisp set* with respect to R, means X is *definable* in U, and if $\alpha_R(X) < 1$, then X is a *rough set* with respect to R, means X is *indefinable* in U. Figure 1 clearly defines the rough set approximations.

Instead of set approximations, rough sets can be described also by using, a ***rough membership function*** proposed in [21]. Dimensionality reduction is a very important thing in case of large dataset. Here, dimensionality reduction means attribute reduction. Attribute reduction is done by the generation of reducts [20] which is a rough set concept. The key concept behind reduct is keeping only those attributes that preserves the indiscernibility relation as well as set approximations. So, reducts can be described as subsets of attributes which are minimal. The subset B' of B is a *reduct* of B if B' is independent and $IND_S(B') = IND_S(B)$.

2.2 Information System and Decision System

From [20], we can also get an idea about information systems. A dataset is represented as a table, of which each row is an object and each column is an attribute that can be measured for each object or simply provided by the user. Such table can be called as an information system. Let U is the universe and it is also a set of finite number of objects. And A is the finite set of the attributes which is nonempty. So, a dataset S will be represented as an information system $S = (U, A)$. A decision system is similar to an information system, with a little difference. In case of an information system, A is a nonempty finite set of attributes. But in case of decision system, we can see the presence of decision attributes. A decision attribute is a distinguished attribute by which knowledge can be expressed. And the values of decision attribute helps in evaluating an object. So, information systems having decision attributes present within them are called decision systems. In [17], a good example of a decision system is given. Which is given a decision system $S = \langle U, A, V, f \rangle$, here $U = \{x_1, x_2, \ldots x_n\}$ is the limited collection objects or samples. $A = C \cup D$ is set of finite number of attributes. $C = \{c_1, c_2, \ldots, c_m\}$ is the condition attribute set and $D = \{d_1, d_2, \ldots, d_j\}$ is the decision attribute set. $C \cap D = \varphi. f(x_i, c)$ is value of x_i in attribute c. And V is the value range of attribute set A.

2.3 Fuzzy c-Means Clustering

Clustering method is very useful in finding patterns in the dataset. Fuzzy c-means clustering method [22] is an extension of k-means algorithm where a data item can be a member of only one cluster. But in case of FCM clustering, a data item can be member of multiple clusters. Here, each data item has a degree of membership to be a member of each cluster. This clustering algorithm basically iteratively minimizes the following objective function,

$$J_m = \sum_{i=1}^{N} \sum_{j=1}^{C} u_{ij}^m \|x_i - c_j\|^2 \qquad (1)$$

where m is any real number such that $1 \leq \mathrm{m} < \infty$, u_{ij} is the degree of membership of x_i in the cluster of j, x_i is the ith d-dimensional measured data, c_j is the d-dimensional center of the cluster, and $\|*\|$ is any norm expressing the similarity between any measured data and the center. The fuzziness parameter m is an arbitrary real number ($m > 1$). We can see the relative overlapping of clusters in FCM clustering. Here, has to define the number of clusters and the minimum support value. And FCM clustering algorithm is not a deterministic algorithm.

3 Proposed Algorithm

This section represents the description of the proposed algorithm, pseudo code, and the analysis of the algorithm.

3.1 Algorithm Description

The proposed algorithm works this way. Firstly, algorithm will reduce the attributes by the using rough set. Here algorithm calculates the reducts [23]. The attributes, which does not belong to a reduct, are unnecessary attributes and therefore can be dropped. So, as the number of attributes is reduced, obviously the dataset will be also reduced. After that algorithm converts the crisp dataset into fuzzy dataset and applies FCM clustering technique on fuzzy dataset to create fuzzy clusters. Next, algorithm applies classical Apriori algorithm [24, 25] on each cluster to generate set of association rules from each cluster. And lastly, algorithm has a mechanism for the aggregation of the generated association rules. Algorithm works in four basic steps. Those are as follows,

Step 1:
Given a dataset $S = (U, A)$, a decision system where U is the finite set of objects and A is the attribute set. Calculate B indiscernible set, $[x]_B$. Here, $[x]_B$ calculate Blower approximation. Here, $X \subseteq U$ and $B \subseteq A$. Calculate positive region of the partition *U/D* with respect to *B*. *D* and *B* be the subset of *A*. Calculate γ for all possible subset *B* of *A*. *D* depends on *B* with degree of γ. Obtain *RED(B)*, it is the set of all reducts of *B* where, $RED(B) \subseteq A$ *and* $A' = RED(B) \bigcup D$. Here A' is the reduced attribute set. Dataset after attribute reduction is $S' = (U, A')$.

Step 2:
Convert crisp dataset S' into fuzzy dataset F using fuzzy MF (Membership function). Apply FCM clustering algorithm on F. Obtain set of clusters $C = \{C_1, C_2, \ldots C_j\}$. Here, j = number of clusters.

Step 3:
Apply classical Apriori algorithm on each cluster $C_i \in C$, where $i = 1, 2, 3, \ldots, j$. Obtain set of association rules R_i from C_i, where rule $r_i^l \in R_i$. Here $i = 1, 2, \ldots, j$ and $l = 1, 2, \ldots, m$. m = number of rules in each cluster and $0 < l \leq m$.

Step 4:
Initially $E = \emptyset$, Aggregated set of rules.
iff,

$$r_i^l \cap E = \emptyset$$

then,
$E = E \cup r_i^l$ where, $r_i^l \in R_i$
Finally obtain E, set of association rules from S'.

3.2 Pseudo Code

1. //Generate reducts from the given dataset
2. //Calculate γ for all possible subset B of A
3. $POS_B(D) = \bigcup_{X \subseteq U/D} \underline{B}(X)$
4. Obtain $A' = RED(B) \bigcup D$

5. Dataset after attribute reduction, $S' = (U, A')$
6. //Convert crisp dataset S' into fuzzy dataset F
7. //Apply FCM clustering algorithm on F
8. Obtain set of clusters C
9. //Apply classical Apriori algorithm on each cluster C_i
10. Obtain association rule set R_i from each cluster
11. //Rules aggregation
12. Initially aggregated set of rules, $E = \emptyset$
13. iff,
14. $r_i^l \cap E = \emptyset$
15. then,
16. $E = E \cup r_i^l$ where, $r_i^l \in R_i$
17. Finally obtain E

3.3 Analysis

The algorithm was applied to the iris dataset, wine dataset, and breast cancer Wisconsin (Diagnostic) dataset. These state-of-the-art datasets are available in the UCI machine learning repository [26]. Iris dataset is a dataset of iris plant which has 150 instances divided into 3 classes (3 classes of 50 instances each). It is multivariate, real valued dataset which does not have any missing value. And there are four attributes in this dataset, sepal length, sepal width, petal length, and petal width. The second dataset is wine dataset which is made from the chemical analysis of wines. And this dataset has 178 instances and 13 attributes. These attributes are alcohol, malic acid, ash, alkalinity of ash, magnesium, total phenols, flavanoids, nonflavanoid phenols, proanthocyanins, color intensity, hue, OD280/OD315 of diluted wines, and proline. Breast cancer Wisconsin (Diagnostic) dataset has 569 instances and 32 attributes. Two attributes are not necessary for mining purpose. One is ID number and the other one is class attribute. Rest 30 attributes are necessary for mining purpose. And the configuration of the system in which all the experiments are performed is given below,

- Intel Core i5-3360M CPU (2.80 GHz), 8.00 GB DDR3 RAM and SATA 7,200 rpm HDD

Initially, we have applied Apriori algorithm to the three datasets and also calculated the average confidence for the same. In our proposed algorithm, we need to reduce the attributes using rough set theory. In our experiment, we have used Rosetta [27] which is a rough set analysis tool. This is also the first step in our four-step algorithm. Before proceeding further, we have again applied the Apriori algorithm to the reduced datasets. Before applying FCM clustering technique in out reduced datasets, we are using Gaussian MF to transform it into fuzzy datasets. After that we have used Apriori algorithm to find out the rules from the clusters.

Table 1 Object-attribute information

Dataset name	Before reduction		After reduction		After applying FCM clustering			
					Cluster 1		Cluster 2	
	No. of objects	No. of attributes	No. of objects	No. of attributes	No. of objects	No. of attributes	No. of objects	No. of attributes
Iris	150	4	150	3	85	3	65	3
Wine	178	13	178	10	90	10	88	10
Breast cancer	569	32	569	13	332	13	237	13

Data transformation and association rule generation are done on MATLAB (R2011a) [28]. Table 1 gives us the complete information regarding attribute reduction as well as clusters.

We observe that the rough set technique performs well for large attributes dataset such as wine and breast cancer. And also we find the FCM clustering technique groups objects based on their overlapping nature and similarity.

It is obvious that applying Apriori algorithm to these clusters give us much lesser overhead in terms of computational complexity. Following Figs. 2, 3, and 4 represents the result of FCM clustering part of the proposed algorithm on these datasets.

Table 2 shows our extensive empirical results from which we can justify our proposed algorithm in terms of average confidence of the generated rules. In the comparative analysis, three types of experimental results are presented. In the first section, the average confidence of generated association rules directly from the main datasets is given. For all the experiments, we keep the minimum support value to 10 % and the minimum confidence threshold to 80 %.

We keep two issues in our mind while proposing our algorithm-first reduction of computational complexity and second generation of rules with high confidence. Except breast cancer data, the other two datasets give higher average confidence for reduced datasets over the original one. In breast cancer data in the process of reduction few attributes are discarded, which may involve in frequent item set calculation. It may be avoided by the use of other rough set techniques such as neighborhood rough set and variable precision rough set which allows certain degree of flexibility in the process of reduct computation. It is a scope for further research in the future. We are using two cluster centers and FCM further groups the objects based on their similarity (sometimes overlapping in nature).

The C_{A} reflects average confidence near to the average confidence of C_{O}, though C_2 gives best average confidence for all the data in all the cases. It is an interesting observation that C_{A} for breast cancer data is higher than its C_{R}. This particular observation justifies that our proposed algorithm (mainly the clustering step) improves the generation of high confidence rules. Also in wine data, we find C_{A} gives better confidence than its C_{O}. It validates our algorithm (RFA) which produces rules with high confidence values in much lesser computation.

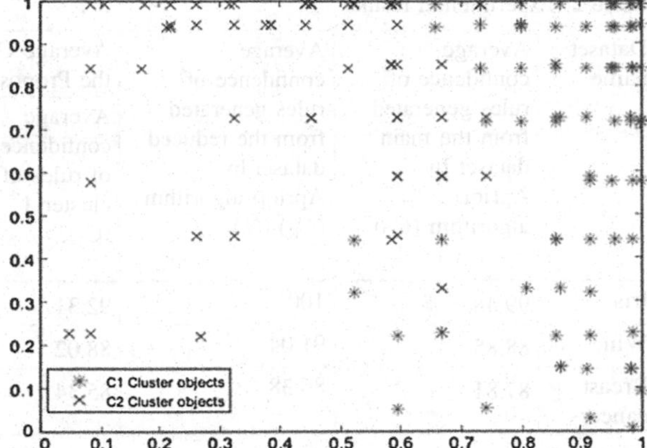

Fig. 2 Iris dataset clusters

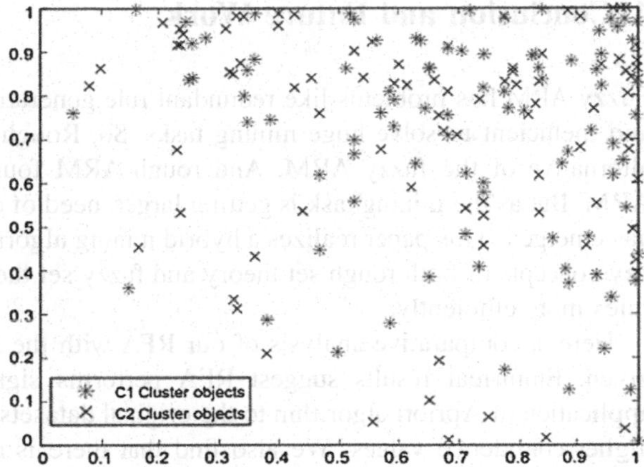

Fig. 3 Wine dataset clusters

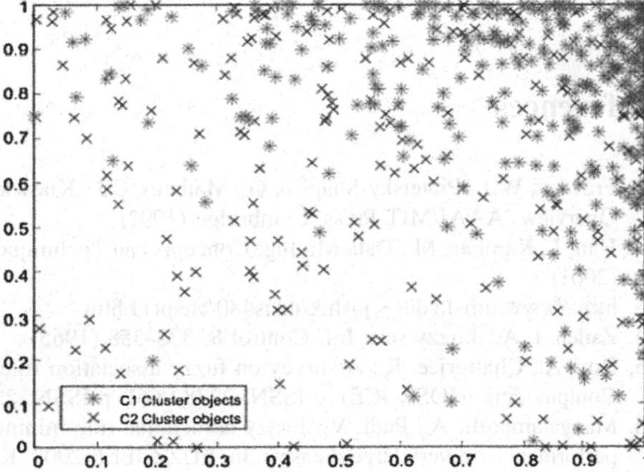

Fig. 4 Breast cancer Wisconsin (diagnostic) dataset clusters

Table 2 Experimental results

Dataset name	Average confidence of rules generated from the main dataset by Apriori algorithm (C_O)	Average confidence of rules generated from the reduced dataset by Apriori algorithm (C_R) (%)	Average Confidence of rules generated by the Proposed Algorithm (RFA)		
			Average confidence of rules of cluster 1 (C_1)	Average confidence of rules of cluster 2 (C_2)	Average of the average confidence of rules of cluster1 and 2 (C_A) (%)
Iris	99.48	100	92.31	100	96.15
Wine	88.85	91.04	88.02	91.39	89.7
Breast cancer	87.81	86.38	85.34	88.07	86.7

4 Conclusion and Future Work

Fuzzy ARM has problems like redundant rule generation, incomplete knowledge, and inefficient to solve huge mining tasks. So, Rough ARM concept came as an alternative of the fuzzy ARM. And rough ARM found to be better than fuzzy ARM. But as the mining task is getting larger, need of developing better algorithm has emerged. This paper realizes a hybrid mining algorithm which incorporates the key concepts of both rough set theory and fuzzy set theory to generate association rules more efficiently.

Here, a comparative analysis of our RFA with the Apriori algorithm has been given. Empirical results suggest RFA performs significantly well against the application of Apriori algorithm to the original datasets, i.e., all possible rules with higher confidence values. We also find that there is a need of further study for using other rough set techniques as well as other association rule mining algorithms. We will examine those prospects in our future research work.

References

1. Frawley, W.J., Piatetsky-Shapiro, G., Matheus, C.J.: Knowledge Discovery in Databases: An Overview. AAAI/MIT Press, Cambridge (1992)
2. Han, J., Kamber, M.: Data Mining: Concepts and Techniques. Morgan Kaufmann, Los Altos (2001)
3. http://www.umsl.edu/~joshik/msis480/chapt11.htm
4. Zadeh, L.A.: Fuzzy sets. Inf. Control **8**, 338–358 (1965)
5. Roy, A., Chatterjee, R.: A survey on fuzzy association rule mining methodologies. IOSR J. Comput. Eng. (IOSR-JCE), e-ISSN: 2278-0661, p-ISSN: 2278-8727, **15**(6), 1–8 (2013)
6. Mangalampalli, A., Pudi, V.: Fuzzy association rule mining algorithm for fast and efficient performance on very large datasets. In: FUZZ-IEEE 2009, Korea, ISSN: 1098-7584, E-ISBN: 978-1-4244-3597-5, pp. 1163–1168, 20–24 Aug 2009

7. Bezdek, J.C.: Pattern Recognition with Fuzzy Objective Function Algorithms. Kluwer Academic Publishers, Norwell (1981)
8. Hoppner, F., Klawonn, F., Kruse, R., Runkler, T.: Fuzzy Cluster Analysis, Methods for Classification, Data Analysis and Image Recognition. Wiley, New York (1999)
9. Mahmoudi, EV, Aghighi, V, Torshiz, MN, Jalali, M., Yaghoobi, M.: Mining generalized fuzzy association rules via determining minimum supports. In: IEEE Iranian Conference on Electrical Engineering (ICEE), E-ISBN: 978-964-463-428-4, Print ISBN: 978-1-4577-0730-8, pp. 1–6 (2011)
10. Watanabe, T.: Fuzzy association rules mining algorithm based on output specification and redundancy of rules. In: IEEE International Conference on Systems, Man, and Cybernetics (SMC), ISSN: 1062-922X, Print ISBN: 978-1-4577-0652-3, pp. 283–289 (2011)
11. Delgado, M., Marin, N., Martin-Bautista, M.J., Sanchez, D., Vila, M.-A.: Mining fuzzy association rules: An overview. In: Studies in Fuzziness and Soft Computing, vol. 164/2005, pp. 351–373. Springer, Berlin (2006)
12. Delgado, M., Marin, N., Sanchez, D., Vila, M.-A.: Fuzzy association rules: general model and applications. IEEE Trans. Fuzzy Syst. **11**(2), 214–225 (2003)
13. Watanabe, T., Fujioka, R.: Fuzzy association rules mining algorithm based on equivalence redundancy of items. IEEE Trans. Syst. Man Cybern. E-ISBN: 978-1-4673-1712-2, Print ISBN: 978-1-4673-1713-9, 1960–1965 (2012)
14. Pawlak, Z.: Rough sets. Int. J. Comput. Inf. Sci. **11**(5), 341–356 (1982)
15. Guan, J.W., Bell, D.A., Liu, D.Y.: The rough set approach to association rule mining. In: Proceedings of the Third IEEE International Conference on Data Mining (ICDM'03), Print ISBN: 0-7695-1978-4, pp. 529–532 (2003)
16. Feldman, R., Aumann, Y., Amir, A., Zilberstain, A., Kloesgen, W., Ben-Yehuda, Y.: Maximal association rules: a new tool for mining for keyword co-occurrences in document collection. In: Proceedings of the 3rd International Conference on Knowledge Discovery, pp. 167–170 (1997)
17. Chu-xiang, C., Jian-jing, S., Bing, C., Chang-xing, S., Yun-cheng, W.: An improvement apriori arithmetic based on rough set theory. In: Third Pacific-Asia Conference on Circuits, Communications and System (PACCS), Print ISBN: 978-1-4577-0855-8, pp. 1–3 (2011)
18. Jiao, X., Lian-cheng, X., Lin, Q.: Association rules mining algorithm based on rough set. In: International Symposium on Information Technology in Medicine and Education, Print ISBN: 978-1-4673-2109-9, Vol 1, pp. 361–364 (2012)
19. Pawlak, Z.: Rough Sets Theoretical Aspects of Reasoning about Data. Kluwer Academic Publishers, Dordrecht (1991)
20. Suraj, Z.: An introduction to rough set theory and its applications: a tutorial. ICENCO, (2004)
21. Brown, E.M.: Boolean Reasoning. Kluwer Academic Publishers, Dordrecht (1990)
22. Dunn, J.C.: A fuzzy relative of the ISODATA process and its use in detecting compact well separated clusters. J. Cybern. Syst. **3**, 32–57 (1974)
23. Komorowski, J., Pawlak, Z., Polkowski, L., Skowron, A.: Rough sets: a tutorial. In: Pal, S.K., Skowron, A. (eds.) Rough Fuzzy Hybridization. A New Trend in Decision-Making, pp. 3–98. Springer, Berlin (1999)
24. Agrawal, R., Imielinski, T., Swami, A.: Mining association rules between sets of items in large databases. In: Proceedings of the 1993 ACM SIGMOD International Conference on Management of Data (1993)
25. Agrawal, R., Srikant, R.: Fast algorithms for mining association rules in large databases. In: Proceedings of the 20th International Conference on Very Large Data Bases, VLDB, pp. 487–499 (1994)
26. UCI machine learning repository: https://archive.ics.uci.edu/ml/datasets.html
27. Rosetta: http://www.lcb.uu.se/tools/rosetta/
28. Matlab: http://www.mathworks.in/products/matlab/

Modal Analysis of Hand-Arm Vibration (Humerus Bone) for Biodynamic Response Using Varying Boundary Conditions Based on FEA

Ashwani Kumar, Deepak Prasad Mamgain, Himanshu Jaiswal and Pravin P. Patil

Abstract The main objective of hand-arm vibration (HAV) study is to identify the effect of vibration on human hand. Humerus bone is a long bone in the arm or forelimb that connects the shoulder to elbow. Free vibration analysis was performed to know the natural frequencies and natural vibration modes and identify the fracture location of the bone through the computer simulation based on FEA. Finite element analysis is an approximation technique used for the analysis of complex objects and geometries. The humerus bone analysis is subjected to free–free and fixed–fixed boundary conditions. For these two different boundary conditions, natural frequencies and natural vibration modes were identified. The mode shape shows that the natural frequency of free–free boundary condition varies from 0 to 1,185.3 Hz and for fixed–fixed boundary condition 943.36 to 7,703.9 Hz. On the basis of these two boundary conditions, mode shape is determined and fracture location can be easily notified. To prevent the fracture of humerus bone, external excitation frequency must be avoided to coincide with these natural frequencies. The results were compared with experimental results available in literature. For the design of humerus bone model, SOLID EDGE software is used and the model is imported in ANSYS R 14.5 (FEA based software) for the free vibration analysis.

Keywords Humerus bone · FEA · Vibration mode · Natural frequency · Fracture · Boundary conditions

A. Kumar (✉) · D.P. Mamgain · H. Jaiswal · P.P. Patil
Department of Mechanical Engineering, Graphic Era University, Dehradun 248002, India
e-mail: kumarashwani.geu@gmail.com

P.P. Patil
e-mail: pravinppatil2004@gmail.com

L.C. Jain et al. (eds.), *Intelligent Computing, Communication and Devices*,
Advances in Intelligent Systems and Computing 308,
DOI 10.1007/978-81-322-2012-1_18

1 Introduction

Human body subjected to any type of vibration is called human vibration. The main reason of human vibration study is to reduce the health risks and increase the level of comfort. The human vibration can be divided into two types known as hand-arm vibrations (HAV) and whole-body vibrations (WBV). HAV are induced via the hands, and this is the main cause of circulatory disorder, bone, joint, or muscle diseases. WBV are induced via the back and the feet of a person, and it may cause harm to the spinal column. The human body and each organ have its own natural frequencies that can resonate with vibration excitation received at their natural frequencies and this resonance may cause adverse health effects. The bone between the shoulder and the elbow is called the humerus bone. It is the longest bone of the human hand arm. The upper end of humerus bone fits into a socket in shoulder; it is like a cup structure. The bottom of the humerus bone is connected to the elbow; it is like a small cup. The radius of shoulder cup is bigger than that of elbow cup. These two cups are connected by a shaft cylindrical in shape. The joint of shaft to the cups is the structural weakness and fracture point. Our objective is to study humerus bone vibration with varying boundary conditions. Human body vibration has been studied for more than 50 years. Many researchers and authors have contributed a lot.

Tiemessen et al. [1] have performed a review study work for drivers to reduce the whole-body vibration exposure and try to find the solution to reduce the whole-body vibration. There are various factors responsible for WBV such as design consideration, posture, duration of vibration and amplitude of vibration. Musculoskeletal disorders (MSD) and low back pain is the resultant of whole-body vibration. A driver is subjected to whole-body vibration for long duration, so chances of MSD and low back pain are maximum. Ingólfsson et al. [2] have performed a literature review for pedestrian-induced lateral vibrations of footbridges and explore the various factors for footbridges design to reduce the retrofit cost because of vibration. Lings et al. [3] have performed the epidemiological literature review (1992–1999) for WBV and low back pain. They have concluded that there is a direct relationship between WBV and low back pain and suggested methods to reduce the WBV.

Huiskes et al. [4] have studied the importance of computer simulation for bio mechanics application. Researchers have been studying the vibration characteristic of femur bone from 1980. Khalil et al. [5] obtained natural frequencies and mode shapes of femur bone using experimental and analytical methods. The experimental measurements were based upon Fourier analysis of transfer function. For analytical solution, a mathematical model of 59 elements was analyzed using transfer matrix method. The first 20 experimental natural frequencies vary from 250 to 7,300 Hz. Khalil et al. [5] have considered the case of femur bone. It is the longest bone of human body. In continuation of this work, we have considered the case of humerus bone, in structure and length it is more similar to the femur bone. So in this work, we have considered two boundary conditions for checking the result. The result for natural frequency of the free–free boundary condition is very low starting from 0 Hz. Only in modes 9 and 10, natural frequency was in the range of experimental result. For fixed-fixed boundary condition, the natural frequency varies from 943.36 to 7,703.9 Hz, which is accurately in range with the experimental natural frequency of (250–7,300) Hz.

2 The CAD Model

The humerus bone model was constructed from CT scan data and reconstructed using software SOLID EDGE [6]. A three-dimensional model is required for a quantitative vibration analysis. To simplify the bone geometry, bone shaft is considered as a cylindrical shaft and the joints are made by hemi spherical cups. The radius of shoulder cup is bigger than that of elbow cup. The actual design of bone is a complex process so to simplify the geometry, the regular shapes available in modeling software are used, but properties are similar to actual bone. So analysis results predict the actual bone analysis. The dimension of Scale 1.0 humerus bone is as follows: the length (L) of Humerus bone shaft is 250 mm and cylindrical shaft radius (r) is 8.33 mm. The radius of shoulder cup (R_1) is 33.32 mm and elbow cup (R_2) is 24.99 mm, $L/r = 30$, $R_1/r = 4$, and $R_2/r = 3$ [7]. The CAD software SOLID EDGE is selected to prepare the solid model of the humerus bone. The CAD model is shown as Fig. 1. After the completion of the model, the *.IGES file is imported from the SOLID EDGE to ANSYS 14.5 [8] software for the analysis. The meshed model of humerus bone is shown in Fig. 2. The meshed model consists of 24,644 nodes and 8,647 elements. Linear tetrahedral elements were used for meshing.

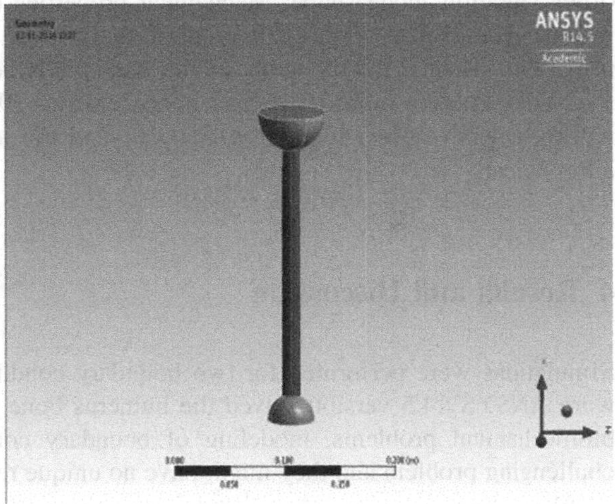

Fig. 1 Cad solid model of humerus bone

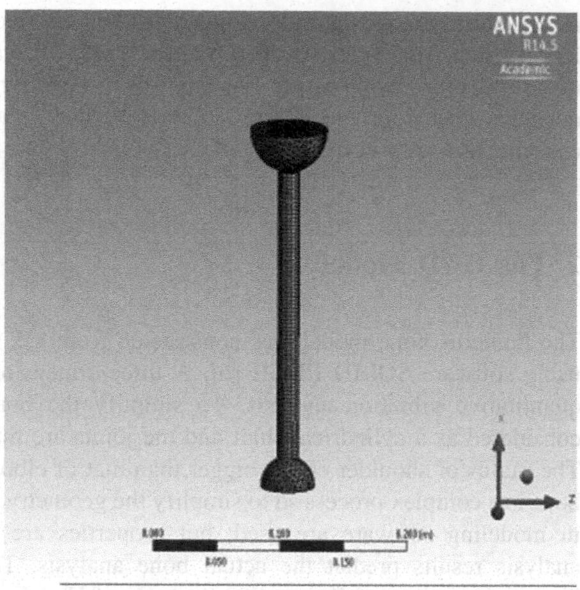

Fig. 2 Meshed model

3 Boundary Conditions and Material Properties

A humerus bone can be studied on two different boundary parameters. In free–free boundary conditions, all DOF of boundaries are subjected to variations. In fixed–fixed boundary conditions, all DOF are constrained in boundaries. For free–free condition, both the cups at joints are free. In fixed–fixed boundary condition, the cup is constraint to move. When the results have been compared with the literature result, fixed–fixed boundary condition is more appropriate and describes precisely human hand-arm biodynamic behavior. We have taken into account for first 10-order vibration mode shape. Mechanical properties (Young's modulus, Poisson ratio, and bone mass density) are required to study the humerus bone. The material properties selected for the study of the humerus bone were Young's modulus—17.2 GPa, Poisson ratio—0.30, and bone density—1900 kg/m^3 [7]. ANSYS 14.5 workbench is selected for modal analysis, and the load is selected by program automatically.

4 Results and Discussion

Simulations were performed for two boundary conditions. The FEM-based software ANSYS 14.5 version solved the humerus bone modal analysis problem. In biomechanical problems, modeling of boundary conditions and joints is very challenging problem and they might have no unique results. Through in our study,

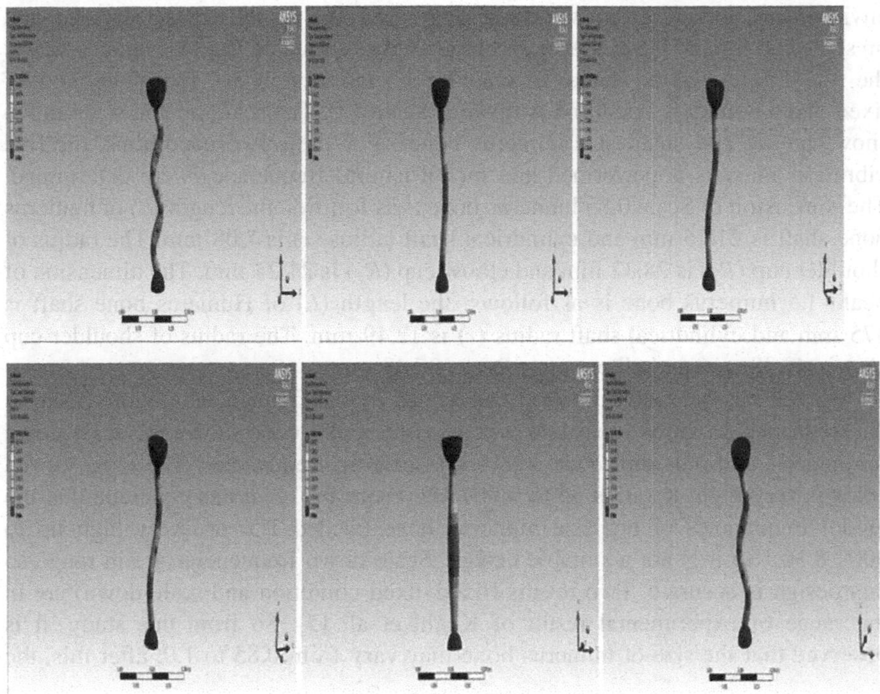

Fig. 3 Six different mode (1, 2, 3, 4, 5, and 6) shapes of the humerus bone model (fixed–fixed boundary condition)

we have found the natural frequencies and first ten modes' shape of humerus bone as shown in Fig. 3. Using ANSYS implicit code, the modal analysis has been performed. From analysis for free–free boundary condition, frequency range variation is 0–1185.3 Hz, and for fixed–fixed boundary condition, frequency range varies in between 943.36 and 7703.9 Hz. From the literature results, the fixed–fixed boundary condition resonant frequency lies in between the experimental result. So the fixed–fixed boundary condition is more precisely described humerus bone vibration condition. So we will discuss about this boundary condition only. The five sinusoidal lateral buckling modes (Mode 1, Mode 3, Mode 6, Mode 7, and Mode 10) and others localized expanded modes are shown in Fig. 3. External excitation or continuous exposure to hand vibration on humerus bone may cause fracture to the bone. The reason for the fracture is pairing of external excitation frequency to natural frequency or modal frequency of the humerus bone. The modal analysis results were satisfied with experimental work done by Khalil et al. [5].

After finding the accurate boundary condition, the humerus bone CAD model is scaled down by 0.85 and scaled up by 1.5. The scaling is used to check the CAD model suitability because the size of humerus bone may vary person to person. Size of humerus bone depends on various factors such as body growth, physical environment, and nutrition. While scaling by 0.85, the model size is decreased, and on scaled up by 1.5, the model size is increased as shown in Fig. 4. Figure 4 shows the five different mode shapes of scale-up 1.5 humerus bone. The model size of fixed–fixed boundary condition is rated as Scale 1.0. The scale-up and scale-down show the big and small size humerus bones. For these two conditions, the free vibration analysis is performed and modal natural frequencies were determined. The dimension of Scale 0.85 humerus bone is as follows: the length (L) of humerus bone shaft is 212.5 mm and cylindrical shaft radius (r) is 7.08 mm. The radius of shoulder cup (R_1) is 28.32 mm and elbow cup (R_2) is 21.24 mm. The dimension of Scale 1.5 humerus bone is as follows: the length (L) of Humerus bone shaft is 375 mm and cylindrical shaft radius (r) is 12.49 mm. The radius of shoulder cup (R_1) is 49.98 mm and elbow cup (R_2) is 37.48 mm.

A graph has been plot between two-scaled modal frequencies. Figure 5 shows the frequency variation in scaling process. Blue-color code shown the scale-down frequencies and red-color code shows the scale-up frequencies. The range of red color is very high; it varies up to 9,007 Hz. From Fig. 5, it can conclude that the modal frequencies of big size humerus bone (scaled 1.5) are very high up to 9007.8 Hz. So it is not a suitable design. Scale-down frequencies are in range so this design is accurate. Two results (fixed–fixed condition and scale-down) are in the range of experimental result of Khalil et al. [5]. So from this study, it is observed that the size of humerus bone may vary from 0.85 to 1.0; after this, the

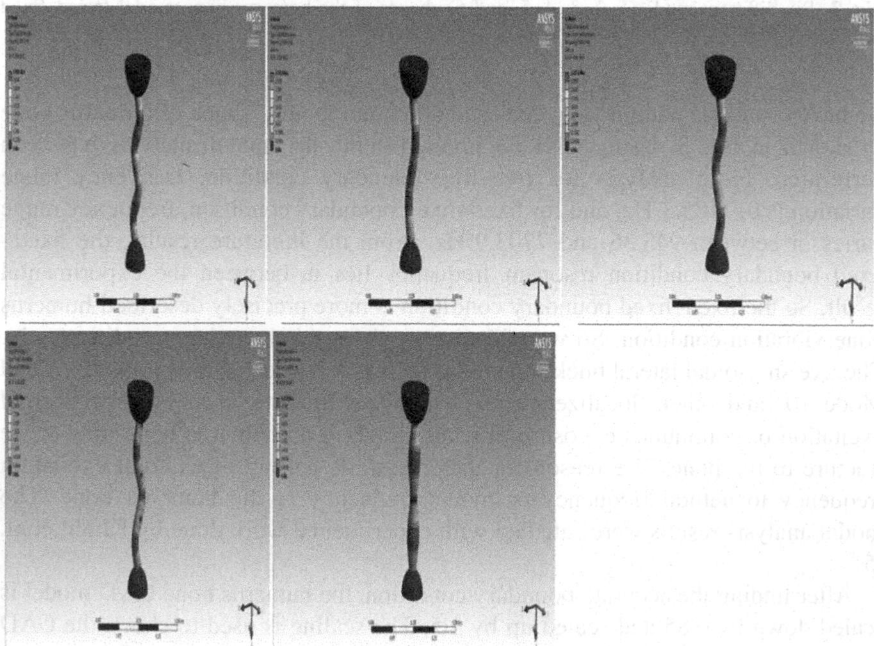

Fig. 4 Five different mode shapes (1, 3, 6, 7, and 9) of the humerus bone model scaled by 1.5

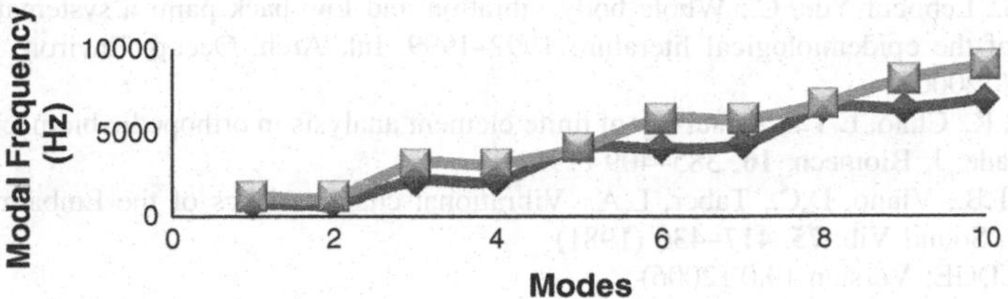

Fig. 5 Frequency variations in scaling

size will be large and the frequencies are very high. In order to prevent the fracture, external excitation frequency should not match with the natural frequency of humerus bone. During external excitation condition, the fracture location can be finding out by seeing the particular mode shape.

5 Conclusion

It is observed that road accident, sudden load application, and continuous HAV excitation are the main reasons for humerus bone failure. The results of this research work show that the chances of bone fracture are through bone shaft and joint (red-color coding). The natural frequency and first ten mode shapes of humerus bone were determined using fixed–fixed boundary condition. The scaling-up and scaling-down are used to determine the maximum size variation of humerus bone. According to the result, the size varies between 0.85 and 1.0 times of designed geometry dimension. The results of this study were verified by the experimental results available in literature. ANSYS 14.5 software has excellent analysis capabilities, and SOLIDEDGE software has a good function of solid modeling. They are suited for finite element analysis of complex shapes. The 3D solid model was prepared by using SOLIDEDGE software and is transferred to ANSYS 14.5. In this research work, we have considered the vibration problem of the humerus bone to explain the HAV using FEA method. Finite element analysis offers satisfactory results with additional ability to calculate regional mode and natural frequency with fracture locations during external excitation condition.

References

1. Tiemessen, I.J.H., Hulshof, C.T.J., Frings-Dresen, M.H.W.: An overview of strategies to reduce whole-body vibration exposure on drivers: a systematic review. Int. J. Ind. Ergon. **37**, 245–256 (2007)
2. Ingólfsson, E.T., Georgakis, C.T., Jönsson, J.: Pedestrian-induced lateral vibrations of footbridges: a literature review. J. Eng. Struct. **45**, 21–52 (2012)

3. Lings, S., Leboeuf-Yde, C.: Whole-body vibration and low back pain: a systematic, critical review of the epidemiological literature 1992–1999. Int. Arch. Occup. Environ. Health **73**, 290–297 (2000)
4. Huiskes, R., Chao, E.Y.S.: A survey of finite element analysis in orthopedic biomechanics: the first decade. J. Biomech. **16**, 385–409 (1983)
5. Khalil, T.B., Viano, D.C., Taber, L.A.: Vibrational characteristics of the Embalmed human femur. J. Sound Vib. **75**, 417–436 (1981)
6. SOLIDEDGE: Version 19.0 (2006)
7. Zadpoor, A.A.: Finite element method analysis of human hand arm vibration. Int. J. Sci. Res. **16**, 391–395 (2006)
8. ANSYS R 14.5: Academic, structural analysis guide (2013)

Preserving Privacy in Healthcare Web Services Paradigm Through Hippocratic Databases

Rekha Bhatia and Manpreet Singh

Abstract As is the case with every other area of digital life, privacy is a major concern in health care also, since online Web services in healthcare domain are increasingly becoming the need of society. The patients' sensitive personal information (PI) in such an environment is more at the risk of inadvertent disclosure. Safeguarding this PI from malicious users is critical to such systems. The existing standards of privacy policy enforcement like platform for privacy preferences (P3P) given by World Wide Web consortium (W3C) and enterprises privacy authorization language (EPAL) of IBM are not sufficient to protect sensitive PI of users shared online through Web services where multiple such unknown heterogeneous services collaborate to carry out the intended task. The user no longer will be interested in those services in which their privacy is at stake. This trend is hampering the online transactions-based business of many large corporate giants. The need of the hour is to integrate privacy policies along with traditional access control policies in order to address the sensitive information disclosure issue. In this paper, we have suggested how Hippocratic Databases can be efficiently used for dealing with privacy disclosure in healthcare scenarios.

Keywords Privacy · Access control · Web services · PI · Hippocratic Databases

1 Introduction

As organizations offering Web services that collect a huge amount of individuals PI, the users of these services are at greater risk of privacy breaches. Some of the known incidences of privacy breaches as reported by [1] include the retrieval of

R. Bhatia (✉)
Punjabi University Regional Centre, Mohali, India
e-mail: r.bhatia71@gmail.com

M. Singh
Punjabi University, Patiala, India
e-mail: msgujral@yahoo.com

L.C. Jain et al. (eds.), *Intelligent Computing, Communication and Devices*,
Advances in Intelligent Systems and Computing 308,
DOI 10.1007/978-81-322-2012-1_19

health records of two states of USA by a researcher at the Carnegie Mellon University from an anonymous database of state health insurance records and creation of a private company by Boston University to sell the personal data collected as part of a study. Similarly, there are several cases of selling personal health records of patients having specific allergies or records pertaining to substance abuse by private healthcare centres to pharmaceutical companies in order to benefit them.

Unless the organizations follow good privacy practices to save users' sensitive private information from inadvertent disclosure, the users will not be willing to transact with them through their online services. The organizations engaged in Web services-based business should be able to specify the extent of information disclosure a user has to forgo with in order to access the services, and the users of these services should be able to specify the type of access for their shared PI. The databases, which include responsibility for privacy preservation as a fundamental ideology, are known as Hippocratic Databases (HDBs). These databases are inspired from secured databases in which sensitive information is securely transmitted through secure channels and must be stored securely [1, 2]. HDBs get their name from the Hippocratic oath which doctors take to practice medicine in an ethical way. HDBs can be extended as privacy preserving access control systems for minimum disclosures of privacy-related information [3]. The efficiency considerations in these databases are secondary to privacy considerations. These HDBs are based on fair information practices inspired by the US Privacy Act (1974) [4, 5]. These practices include purpose specification, consent, limited collection, use, disclosure and retention of PI collected about individuals [6].

Recent interest of researchers in HDBs shows a promising direction for enforcing privacy policies in health care-related Web services. In [1], the authors proposed an architecture for HDBs in which organizations can specify the purpose of PI collection, to whom this PI can be disclosed and the retention period. In these HDBs, authorization tables with privacy information are also created which every user of the service can access. The drawback of these HDBs is that they do not distinguish between the suitability of various means available to carry out the requestor's intended task. As an example, billing information of our online purchases can be received through mobile phone, through email facility, through courier or through post. Each of these means of accessing service requires different pieces of PI disclosure.

In Web services paradigm, different heterogeneous Web services have to collaborate with each other in order to serve the requests of the users. The organizations offering these services may split the access request into sub-parts depending upon whether the organization is able to serve the request on its own or has to be reliant on other service providers to complete the intended request. In our view, the users of these services should be given a choice about which collaborating service partner they would prefer based on their trust value for each of them. This trust value can be generated through consolidation of their previous interactions with these service providers and through testimonies of the third parties. The detailed trust calculation process is shown in our previous work [7]. The paper is organized as follows. In Sect. 2, we briefly explain the related work in privacy preserving access control using HDBs. In Sect. 3, we present an example scenario to be used throughout our paper. In Sect. 4, we explain and apply a multi-criteria decision-making technique called analytical hierarchy processing (AHP) to calculate an optimal path of service delivery based on user's privacy preferences. Section 5 concludes the paper.

2 Related Work

With the emergence of open healthcare services, the issue of privacy preservation is becoming a top priority problem and is a very active research area. In paper [8], the authors specified a novel approach for policy creation, interpretation, and implementation in healthcare scenarios. They used the concept of model integrated computing and logic programming to successively reduce the gap between requirements and the privacy policies. The research community has developed formalisms covering different facets of the policy formalization problem [9–14]. While all these researchers made significant contributions, none is able to achieve the practical realization of those formalisms outside the research community. The authors in [15] presented an automated approach to derive the optimal path for reaching the required authorization level in order to fulfil the requests of the users. They organized the primary purpose of access into purpose directed graphs in order to collaborate with those partners who offer optimal services to complete the request of the users. Our work is inspired from this work and basically a reorientation of the thought process in a healthcare scenario.

3 Example Scenario

A primary healthcare centre which provides online healthcare services to its patients along with its collaborating secondary partners offering various specialized facilities such as specialized surgery and diagnostic facilities. In order to get treated in the primary healthcare centre, patients have to disclose their PI such as name, age, address and disease. The online payment to obtain nursing services at home requires credit information from the banker of the user. The diagnostic reports can be received through email, courier or speed post. Each of these different means to use the healthcare Web services requires different attributes of PI disclosure. Depending upon the choice of means to get the service, the service provider should ask the users only minimum required PI attributes to serve their purpose. For example, one user is ready to disclose his email address to receive the diagnostic reports, but another user having online facility only at his workplace may want to receive the report by a courier company. Still another user wants to get the service through postal department because in his views, this is the safest method of delivery. From this example, it is clear that privacy is a matter of personal choice.

The main parameters which should be considered while designing privacy aware healthcare services are privacy disclosure extent and trust value. The objectives of such privacy aware design are to minimize disclosure extent and maximize trust value. The solutions based on HDBs do not consider these parameters. Our work is oriented towards personalization of healthcare services based on the user's choice of collaborating service partner and chosen mean to receive that service. We specify a method to automatically get to an optimal path required to approach a service from the service provider, if multiple paths having different weights are available to receive the intended service. From weight, we mean a function depending upon two parameters:

i. Privacy disclosure extent
ii. Trust degree with collaborating service partner

Using HDBs, we allow each service user to specify his trust degree associated with collaborating partners of the primary service provider. Then, the weight combining privacy disclosure extent and trust degree for each collaborating partner is calculated for each alternate path. Consequently, an optimal path according to user's preferences is selected.

4 Enhancement of Hippocratic Databases Through Analytical Hierarchy Processing

The central concept of HDBs is the storage of a special attribute called purpose of data collection along with other attributes in every Table [1] as shown in Table 1.

Here, external recipients are the users/services to whom the specified PI attribute can be disclosed. Retention period is the valid duration for usage of PI. After the expiry of this period, PI attributes should not be retained. Authorized users are those users who are allowed to access the specified PI attribute. According to privacy metadata schema shown in Table 2, the purpose is stored in both the privacy policies table and the privacy authorizations table. For the services, where dynamic collaborations are not required, HDBs can be effective means of privacy control. But, for the emerging Web services scenarios in which dynamically the various services have to interact in order to fulfil the request of the users, the HDBs have to be enhanced in some way so as to include the cases like users do not have any trust in a collaborating partner of the service or users want to opt for a particular service delivery method when there is sufficient choice.

Table 1 Database schema

Table	PI attributes
Patient	Purpose, patient id, name, age, address, disease
Report	Purpose, patient id, report id, report data

Table 2 Privacy metadata schema

Table	PI attributes
Privacy policies	Purpose, table, PI attribute, {external recipients}, {retention period}
Privacy authorizations	Purpose, table, PI attribute, {authorized users}

If a purpose p can be decomposed into sub-purposes $p1, p2, \ldots, p_n$ and all of these should be satisfied in order to satisfy the primary purpose p, it is said to be AND decomposition of a purpose. Similarly, if only one among $p1, p2, \ldots, p_n$ must be satisfied in order to satisfy p, it is said to be OR decomposition of a purpose [1, 15–17]. This purpose hierarchy can be specified in the form of AND–OR graph as shown in Fig. 1.

In Fig. 1, diagnostic healthcare service provider A collaborates with credit card company B to check the balance in the requestor's card. This is shown by AND nodes (without arc) in the graph. The service provider A collaborates with three delivery partners for diagnostic report delivery, that is, speed post C, email D and courier E. Any of these can be selected by service user depending on the extent of his PI disclosure and trust in the delivery partners. This is shown by OR nodes (with arc) in the graph.

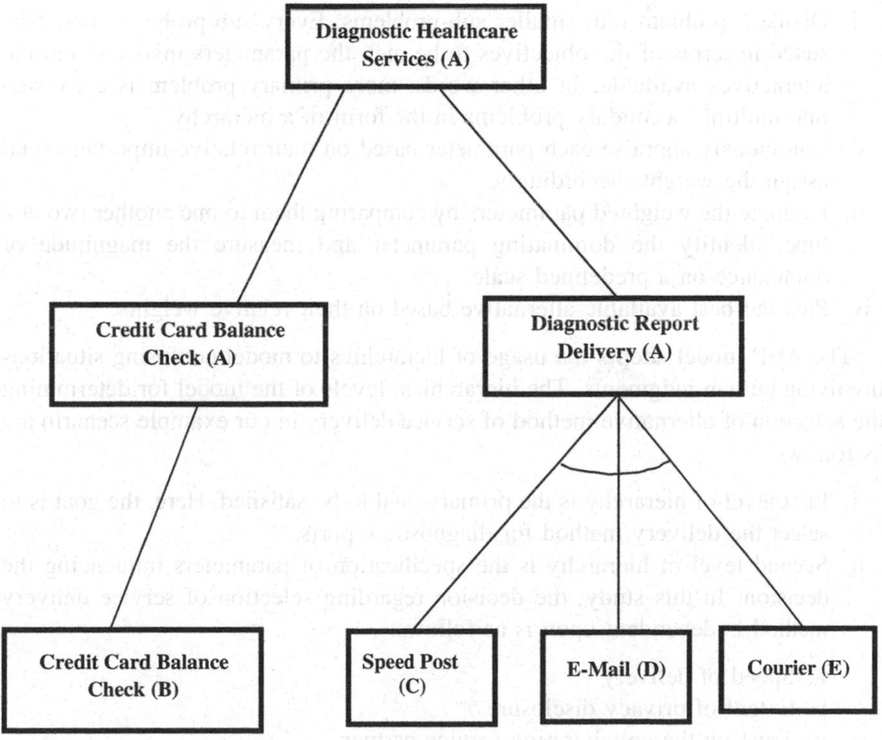

Fig. 1 AND–OR graph of purpose hierarchy for diagnostic healthcare services

To calculate the weights of the various paths as mentioned earlier, we use a well-known multi-criteria decision-making technique (MCDM), namely (AHP). By applying AHP, we can arrive at a much better quality decisions for scenarios requiring decisions based on multiple criteria and for confusing situations where everything is not measurable quantitatively [18]. AHP analyses the decision based on multiple criteria and structures those criteria depending upon their comparative significance. For measuring comparative significance, the weights are assigned to the multiple criteria with a great care.

The hierarchical levels in AHP are based on the various decision parameters which are of the same order of significance. Before making hierarchies, the decision problem should be clearly understood by the decision-maker and all the parameters on which the solution is dependent should be recognized. By successively dividing the decision-making problem in multiple sub-problems at different hierarchical levels based on decision parameters, the decision-maker can clearly conceptualize the issues involving comparison of homogeneous elements [19]. AHP has provision for accommodating small inconsistency in judgment as human beings are not always consistent. The consistency ratio scales are derived from the principal Eigenvectors and consistency index is derived from the principal Eigenvalue.

Applying AHP technique involves the following steps:

i. Divide a problem into smaller sub-problems. Every sub-problem should be stated in terms of the objectives to be met, the parameters involved and the alternatives available. In other words, more primary problem is converted into multiple secondary problems in the form of a hierarchy.
ii. Consciously appraise each parameter based on their relative importance and assign the weights accordingly.
iii. Evaluate the weighted parameters by comparing them to one another two at a time, identify the dominating parameter and measure the magnitude of dominance on a predefined scale.
iv. Pick the best available alternative based on their relative weights.

The AHP model adopts the usage of hierarchies to model confusing situations involving human judgments. The hierarchical levels of the model for determining the selection of alternative method of service delivery in our example scenario are as follows:

i. First level of hierarchy is the primary goal to be satisfied. Here, the goal is to select the delivery method for diagnostic reports.
ii. Second level of hierarchy is the specification of parameters influencing the decision. In this study, the decision regarding selection of service delivery method is dependent upon is as follows:

    a. Speed of delivery
    b. Extent of privacy disclosure
    c. Trust on the collaborating service partner

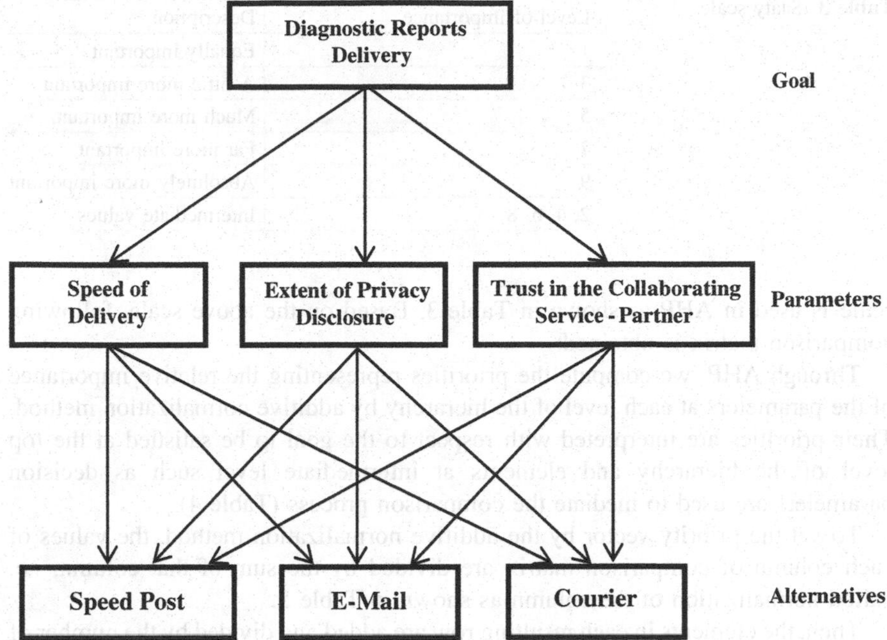

Fig. 2 AHP hierarchy for choice of collaborating service partner

iii. The last level of hierarchy describes the available alternatives. The alternatives available for service delivery in our case are as follows:

   a. Speed post
   b. Email
   c. Courier

Figure 2 describes the AHP hierarchy obtained by arranging the information about goals to be achieved, criteria influencing the decision and the available choices for service delivery.

In AHP, human judgements are used to calculate the relative rankings of the decision parameters involved, e.g. assume that for a service user of the diagnostic healthcare services:

a. Limiting extent of disclosure is five times as important as compared to speed of delivery.
b. The trust in the collaborating service partner is seven times as important as speed of delivery.
c. The trust in the collaborating service partner is three times as important as extent of disclosure.

Using pairwise comparisons, the relative importance of one parameter over another can be expressed. For assigning weights to the parameters, the Saaty rating

Table 3 Saaty scale

Level of importance	Description
1	Equally important
3	A little more important
5	Much more important
7	Far more important
9	Absolutely more important
2, 4, 6, 8	Intermediate values

scale is used in AHP as shown in Table 3. Based on the above scale, following comparison matrix is obtained:

Through AHP, we compute the priorities representing the relative importance of the parameters at each level of the hierarchy by additive normalization method. Their priorities are interpreted with respect to the goal to be satisfied at the top level of the hierarchy and elements at intermediate level such as decision parameters are used to mediate the comparison process (Table 4).

To get the priority vector by the additive normalization method, the values of each column of comparison matrix are divided by the sum of that column, i.e. called normalization of the column as shown in Table 5.

Then, the elements in each resulting row are added and divided by the number of values present in the corresponding row as illustrated in Table 6. Priority vector is actually the normalized Eigenvector of the matrix. For simplifying calculations, we are using an approximation method (additive normalization) to get approximately the same results. This method is valid for three or lesser number of decision parameters.

The next step is to find out a consistency ratio (CR) to check how consistent the judgments are, relative to the large samples of purely random judgments. AHP calculations are based on the assumption that the decision-maker is rational, i.e. if *A* is preferred to *B* and *B* is preferred to *C*, then *A* is preferred to *C*. This property is called transitive property. If the CR is greater than 10 % (0.01), then the judgments are interpreted as untrustworthy and the process must be repeated.

According to Saaty [18], for consistent reciprocal matrix, the largest Eigenvalue (λ_{max}) is equal to the number of comparisons (n). A measure of consistency, called consistency index (CI), as degree of consistency was also suggested by him, using the formula:

$$\text{CI} = (\lambda_{max} - n)/(n\text{-}1)$$

Table 4 Reciprocal matrix for the decision parameters

	Speed of delivery	Extent of privacy disclosure	Trust in the collaborating service partner
Speed of delivery	1/1	1/5	1/7
Extent of privacy disclosure	5/1	1/1	1/3
Trust in the collaborating service partner	7/1	3/1	1/1

Table 5 Pairwise comparison matrix for the parameters involved

	Speed of delivery	Extent of privacy disclosure	Trust in the collaborating service partner
Speed of delivery	1.0000	0.2000	0.1429
Extent of privacy disclosure	5.0000	1.0000	0.3333
Trust in the collaborating service partner	7.0000	3.0000	1.0000
Total	13.0000	4.2000	1.4762

Table 6 Calculation of priority vector from comparison matrix

	Speed of delivery	Extent of privacy disclosure	Trust in the collaborating service partner	Priority vector
Speed of delivery	0.0769	0.0476	0.0968	0.0738
Extent of privacy disclosure	0.3846	0.2381	0.2258	0.2828
Trust in the collaborating service partner	0.5385	0.7143	0.6774	0.6434
Total	1.0000	1.0000	1.0000	1.0000

The final step is to calculate the consistency ratio (CR) by using the Table 7 given by Saaty. The upper row in the table is the order of the random matrix, and the lower row is the respective consistency index for random judgments.

Each of the numbers in the RI table is the average of consistency indexes derived from a sample of randomly selected reciprocal matrices.

To verify the consistency of judgments in the diagnostic reports delivery example, first, we have to calculate the principal Eigenvalue (λ_{max}).

Consider $[AX = \lambda_{max}X]$ where A is the reciprocal matrix and X is the Eigenvector.

$$\begin{bmatrix} 1 & 0.2 & 0.1429 \\ 5 & 1 & 0.3333 \\ 7 & 3 & 1 \end{bmatrix} \begin{bmatrix} 0.0738 \\ 0.2828 \\ 0.6434 \end{bmatrix} = \lambda_{\mathbf{max}} \begin{bmatrix} 0.0738 \\ 0.2828 \\ 0.6434 \end{bmatrix}$$

From the above matrix equations, we calculate $\lambda_{max=}$ 3.06 and CI = 0.03. Using Table 7, we calculate CR = CI/0.58 which comes out as 0.05 which is <0.1. So the evaluations are consistent.

Table 7 Random consistency index (RI) [18]

n	1	2	3	4	5	6	7	8	9	10
RI	0	0	0.58	0.9	1.12	1.24	1.32	1.41	1.45	1.49

Table 8 Reciprocal matrix for the decision parameter—speed of delivery

	Speed post	Email	Courier
Speed post	1/1	1/3	7/1
Email	3/1	1/1	5/1
Courier	1/7	1/5	1/1

Table 9 Reciprocal matrix for the decision parameter—extent of privacy disclosure

	Speed post	Email	Courier
Speed post	1/1	3/1	7/1
Email	1/3	1/1	5/1
Courier	1/7	1/5	1/1

Table 10 Reciprocal matrix for the decision parameter—trust in the collaborating service partner

	Speed post	Email	Courier
Speed post	1/1	2/1	5/1
Email	1/2	1/1	3/1
Courier	1/5	1/3	1/1

The next step in AHP is to calculate the rankings of alternatives in terms of three decision parameters. We assume that the user under our study has the preferences shown in Tables 8, 9 and 10 for the available alternatives in terms of the decision parameters.

Again following the same method for the three alternative delivery methods, as we followed for decision parameters, we calculate the priority vectors for these alternatives as shown in Table 11

The matrix containing priority vectors of alternatives is multiplied by the priority vector of decision parameters to arrive at the final decision as shown below:

$$\begin{bmatrix} 0.3324 & 0.6434 & 0.5813 \\ 0.5870 & 0.2828 & 0.3091 \\ 0.0806 & 0.0738 & 0.1096 \end{bmatrix} * \begin{bmatrix} 0.0738 \\ 0.2828 \\ 0.6434 \end{bmatrix} = \begin{bmatrix} 0.5805 \\ 0.3222 \\ 0.0973 \end{bmatrix}$$

Table 11 Priority vectors for the three alternatives

	In terms of speed of delivery	In terms of extent of privacy disclosure	In terms of trust in the collaborating service partner
Speed post	0.3324	0.6434	0.5813
Email	0.587	0.2828	0.3091
Courier	0.0806	0.0738	0.1096

From the above equation, we can conclude that the most important alternative service delivery method in our example healthcare services scenario from the point of view of the selected user is speed post followed by email. The least important alternative service delivery method for the user is courier.

5 Conclusions

AHP technique can extend the HDBs to provide for alternative path selection based on user's trust in the collaborating partner as well as his preference for the extent of privacy disclosure. Using this technique, we can quantify the user's qualitative preferences to arrive at a decision more judiciously. Many legal hassles concerning privacy disclosures can be avoided by the service providers if they incorporate above technique in HDB before disclosing the sensitive PI of the customers to the third parties.

References

1. Agrawal, R., Bird P., Grandison, T., Kiernan, J., Logan S., Rjaibi, W.: Extending relational database systems to automatically enforce privacy policies. In: Proceedings of the 21st International Conference on Data Engineering, ICDE '05, pp. 1013–1022. Washington, DC, USA (2005)
2. Agrawal, R., Kiernan, J., Srikant, R., Xu, Y.: Hippocratic databases. In: 28th International Conference on Very Large Databases, Hong Kong (2002)
3. Bayardo, R., Grandison, T., Johnson, C., Agrawal, R., Asonov, D., Kiernan, J.: Managing disclosure of private health data with Hippocratic databases. IBM Research White Paper (2005)
4. Rotenberg, M.: The Privacy Law Sourcebook 2000, United States Law, International Law, and Recent Developments. Electronic Privacy Information Center, Washington, DC (2000)
5. Rotenberg, M.: Fair information practices and the architecture of privacy. Stanford Technology Law Review (2001)
6. U.S. Department of Health, Education, and Welfare: Records, computers and the Rights of Citizen, Report of the Secretary's Advisory Committee on Automated Personal Data Systems, xx–xxiii edn (1973)
7. Bhatia, R., Singh, M.: Trust based privacy preserving access control in web services paradigm. In: the Second IEEE International Conference on Advanced Computing, Networking and Security, ADCONS, pp. 243–246 (2013)
8. Nadas, A., Frisse, M.E., Sztipanovits, J.: Modeling privacy aware health information exchange systems. In: 1st International Workshop on Engineering EHR Solutions (IWEES), Amsterdam Privacy Conference (2012)
9. Barth, A., Mitchell, J., Datta, A., Sundaram, S.: Privacy and utility in business processes. In: Computer Security Foundations Symposium, CSF '07, 20th IEEE, pp. 279–294 (2007)
10. Datta, A., Franklin, J., Garg, D., Kaynar, D.: A logic of secure systems and its application to trusted computing. In: 30th IEEE Symposium on Security and Privacy, pp. 221–236 (2009)

11. Lam, P.E., Mitchell, J.C., Sundaram, S.: A formalization of HIPAA for a medical messaging system. In: Proceedings of the 6th International Conference on Trust, Privacy and Security in Digital Business, Berlin, pp. 73–85. Springer, Heidelberg (2009)
12. Becker, M.Y., Fournet, C., Gordon, A.D.: SecPAL: design and semantics of a decentralized authorization language. J. Comput. Secur. **18**(4), 619–665 (2010)
13. Craven, R., Lobo, J., Lupu, E., Ma, J., Russo, A., Sloman, M., Bandara, A.: A Formal Framework for Policy Analysis, Imperial College London, Technical Report (2008)
14. Simko, G., Sztipanovits, J.: Active monitoring using real-time metric linear temporal logic specifications. In: HEALTHINF, pp. 370–373 (2012)
15. Li, M., Sun, X., Wang, H., Zhang, Y.: Optimal privacy-aware path in hippocratic databases. In: 14th International Conference on Database Systems for Advanced Applications Brisbane, pp. 441–455, Australia (2009)
16. Nilsson, N. J.: Problem Solving Methods in AI. Mc Graw-Hill, New York (1971)
17. Rich, E., Knight, K., Nair, S.B.: Artificial Intelligence. Mc Graw-Hill, New York (2009)
18. Saaty, T.L.: The Analytic Hierarchy Process. McGraw-Hill, New York (1980)
19. Saaty, T.L.: Fundamentals of Decision Making and Priority Theory with the Analytic Hierarchy Process. RWS Publications, Pittsburg (2000)

Generation of Array Passwords Using Petri Net for Effective Network and Information Security

S. Vaithyasubramanian, A. Christy and D. Lalitha

Abstract Over the years, information and network security in the field of computer is an everlasting troublesome area under discussion. Information and network security instigates with the user. The important feature of security is 'Password Authentication.' A human-created password comes from a small domain. It is just a matter of time for hackers to break security measures with the available computer power and tools. Online services can be accessible by using distinct passwords with varying strengths. Cracking or decoding the password has created serious challenges or threats in providing security of the information. Highly secured password generation therefore has become a challenging task. In this work, Petri net-based array password generation has been done. The methodology adapted in this paper is novel and more immune.

Keywords Petri net · Authentication · Password · Alphanumeric · Graphical · Biometric · Limitations · Information security · Network security

1 Introduction

In this digital world, accesses to the available resources for a wide range of purpose people are using numerous computing devices. Computing devices have become omnipresent and to access people use networks. The foremost distress with the development of the latest standards and applications in the field of

S. Vaithyasubramanian (✉) · A. Christy · D. Lalitha
Sathyabama University, Chennai, India
e-mail: discretevs@gmail.com

A. Christy
e-mail: christy_a1@hotmail.com

D. Lalitha
e-mail: lalkrish_24@yahoo.co.in

L.C. Jain et al. (eds.), *Intelligent Computing, Communication and Devices*,
Advances in Intelligent Systems and Computing 308,
DOI 10.1007/978-81-322-2012-1_20

computer networks is their security. The leading uncertainties in this area are hacking the system and cracking the passwords. In general, passwords are the first and probably only protection against intrusion. A password affords the first line of security against unfair right of entry to computer [1]. Thought-provoking problem in this field is the security of the passwords [2]. People have done some work in this area to improve the security. Still, there exists a requirement of better methods to overcome password cracking. The password is an inert secret string composed of keyboard characters. User name gives the identity about the user while the password authenticators that he/she is the authorized user. The password is used as the authentication key. Authentication is only one feature of security. Information and network security starts with the user. Human-generated passwords come from a small domain. They are easy to guess, common passwords, short, and weak passwords. Password that is defiant to guessing attack, hybrid attack, brute force guessing, and dictionary attacks are called as strong password. Most strong passwords are computer-generated, hard to remember, and not user-friendly. Maintaining written catalogs of passwords on scraps of paper, or in a text document on the desktop or mail, is insecure and is effortlessly viewed by snooping eyes [1]. Using the unchanged password over and over again across an ample range of systems, Web sites form the nightmare scenario [3]. When a password cracker cracks out the password of a particular user, now have access to every part of that user's life such as system, e-mail, retail, financial, and work.

2 Types of Existing Authentication Methods

Passwords are fundamental in this computing world. A usual computer user has passwords for numerous functions: logging into e-mail accounts, accessing social networks, booking online tickets, net banking, accessing applications, and even to read the newspaper online. From using a password to sign into the operating system to passwords for various communications on the Internet, passwords can be classified as a 'necessary evil.' Right of entry to a computer system is based on alphanumeric password or graphical password or biometric authentication.

Alphanumeric password is a secret word, string, an expression, or a combination of various characters, and numerical that authenticates the identity of the user. For remembrance, user can create patterns of string passwords using CFG, CSG, and strong random password using Markov chain [4, 5]. Alternate to alphanumeric password came in the form of biometric authentication and graphical passwords during late 1970 and 1996s, respectively. Recognition of users finger print, face, voice, iris scan is used as users' unique biometric identification. In graphical password [6–9], the user can choose a pass point or image as their password. Picture-based and drawn-based graphical password classifications are used. More research is going on this area.

3 Limitations of Existing Methods

Human trend in creating password makes them vulnerable, and they are subject to various cyber attacks. The problem with an alphanumeric password arises largely from boundaries of humans 'long-term memory' [10], which forces them to choose weak password such as common password, dictionary word, obvious password, favorites, and easily guessable password [11]. By violating the policies of the service provider, user creates their password as short and weak passwords [3, 12]. Cracking those passwords is very easy. They are cracked by various attacks such as Guessing, Brute force attack, Dictionary Attack, and Hybrid attack [13]. The general issue in the graphical password input is to click outside the tolerance and the users need to understand the degree of precision needed. And they are subject to regular means of attacks such as Shoulder surfing, Intersection Analysis, Social Engineering, and Spyware attack [7, 14, 15]. Foremost problems in biometric authentications are direct hazards. The major issues concerning biometric are false rejection/acceptation rate [16, 17].

4 Need of a Good Password

Need of a good password is another vital anxiety. The password problem has led to innovations to improve passwords. As an alternate to the traditional alphanumeric passwords, the past decades have seen an emerging attention by means of graphical password and biometric authentication. These authentication techniques have their own strength and limitations. The problems with these techniques have led to the innovations of improving passwords. One innovation is array password, i.e., passwords that are based on the generation of array rather than alphanumeric strings, graphical password, and biometric authentication. The basic idea is that using arrays will lead to greater security and decrease the tendency to choose an insecure password. This, in turn, should increase overall security. Several array password systems described in the forthcoming section have been developed.

5 Petri Net

Petri net [18] is a graphical and mathematical modeling tool which can be applied to many systems. Tokens are used to simulate the dynamic and concurrent activities of systems. Petri net was introduced by Carl Adam Petri in 1962. The Petri net graph is a directed, weighted, bipartite graph consisting of two kinds of nodes, called places and transitions. Arcs are either from a place to a transition or from a transition to a place. In the graphical representation, places are represented by circles and transitions are represented by bars. Arcs are labeled with their

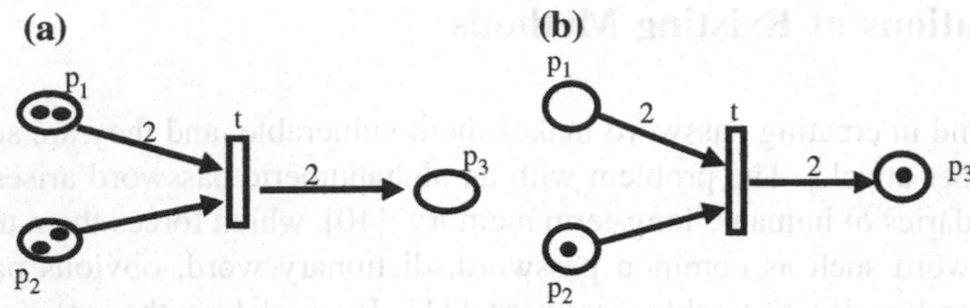

Fig. 1 **a** Position of tokens before the transition fires, **b** position of tokens after the transition fires

weights (positive integers). Labels for unit one is generally omitted. The weights can also be represented by drawing parallel arcs.

A marking assigns to each place a nonnegative integer. If the marking assigns to a place *p* a nonnegative integer *k*, then the place *p* is said to have *k* tokens. Pictorially, *k* black dots are placed in *p*. A marking is denoted by *M* a m-vector, where *m* is the number of places in the net. The *i*th component of *M* denotes the number of tokens in p_i.

In modeling conditions and events, places represent conditions and transitions represent events. A transition has a certain number of input places and output places representing the preconditions and post-conditions of the event, respectively. The presence of a token in a place is interpreted as the condition associated with the place. When a transition fires (an event takes place), marking of the net changes. Marking changes according to the firing rules which are given below. (i) A transition t is enabled if each input place is marked with at least $w(p, t)$ tokens, where $w(p, t)$ is the weight of the arc *p* to *t*. (ii) An enabled transition may or may not fire (depending on whether or not the event actually takes place). (iii) Firing an enabled transition *t* removes $w(p, t)$ tokens from each input place *p* of *t* and adds $w(t, p)$ tokens to each output place *p* of *t*, where $w(t, p)$ is the weight of the arc from *t* to *p*.

The graph given below illustrates the firing rules. Figure 1a shows the marking before firing transition *t* which is enabled, and Fig. 1b shows the marking after firing transition *t*, where *t* is disabled. Since the weight of the arc from p_1 is two, firing *t* removes two tokens from p_1. After firing *t*, two tokens are put in p_3 since the weight of the arc to p_3 is two.

The formal definition of a Petri net follows:

Definition A Petri Net structure is a four tuple $C = (P, T, I, O)$, where $P = \{p_1, p_2, \ldots, p_n\}$ is a finite set of places, $n \geq 0$, $T = \{t_1, t_2, \ldots, t_m\}$ is a finite set of transitions $m \geq 0$, $P \cap T = \emptyset$, $I: T \to P^{\infty}$ is the input function from transitions to bags of places, and $O: T \to P^{\infty}$ is the output function from transitions to bags of places.

Note: A bag, like a set, is a collection of elements over some domain. Unlike sets, bags allow multiple occurrences of elements. When the weight of an arc from a transition to a place is more than one, then the place is repeated in the bag as many times. The number of tokens put in the place will depend on the number of times it occurs in the bag. If the weight of an arc from a place to a transition is more than one, then again the place is repeated in the set as many times. The number of tokens required in the place, for the transition to be enabled, will depend on the number of times it occurs in the bag.

Definition A Petri Net marking is an assignment of tokens to the places of a Petri Net. The tokens are used to define the execution of a Petri Net. The number and position of tokens may change during the execution of a Petri Net.

After the basic definitions of Petri net, the Petri net structure that generates rectangular arrays is defined.

6 Array Token Petri Net Structure

This section defines a Petri Net structure [19, 20] to generate rectangular arrays. The basic notations used are first explained.

6.1 Basic Notations

Σ^{**} denotes the arrays made up of elements of Σ, and Σ^{++} denotes nonempty arrays made up of Σ. If A and B are two arrays having same number of rows, then $A \oplus B$ is the column-wise catenation of A and B. If two arrays have the same number of columns, then $A \ominus B$ is the row-wise catenation of A and B. $(x)^n$ denotes a horizontal sequence of n 'x', and $(x)_n$ denotes a vertical sequence of n 'x', where $x \varepsilon \Sigma^{**}$. $(x)^{n+1} = (x)^n \oplus x$, and $(x)_{n+1} = (x)_n \ominus x$. $\oplus$ denotes either $\oplus$ or $\ominus$.

The Array Token Petri Net Structure retains the four components of C as given in Definition 5.1. The tokens positioned in places are taken as rectangular arrays over a given alphabet. Some of the transitions are labeled in this net structure. Two types of labels have been used: (i) a designated input place and (ii) catenation rule. Column catenation and row catenation are the two types of catenations that are possible with rectangular arrays. These catenations take place provided that the condition for catenation is satisfied. When two arrays have the same number of rows, then the arrays can be catenated column-wise. When two arrays have the same number of columns, then the arrays can be catenated row-wise.

6.2 Catenation Rule

(i) Column catenation rule as a label for transition

Column catenation rule is in the form $A \oplus B$. Here, the array A denotes the m x n array in the input place of the transition. B is an array language whose number of rows will depend on 'm' the number of rows of A. The number of columns of B is fixed. For example, $A \; \Phi \; (x\,x)_m$ adds two columns of x after the last column of the array A which is in the input place and $(x\,x)_m \; \Phi \; A$ would add two columns of x before the first column of A. Here, 'm' would denote the number of rows of the input array A.

For example, if A is the array $\begin{matrix} a & a & a \\ a & a & a \\ a & a & a \end{matrix}$, then $A \; \Phi \; (x\,x)_m$ would be the array $\begin{matrix} a & a & a & x & x \\ a & a & a & x & x \\ a & a & a & x & x \end{matrix}$ and $(x\,x)_m \; \Phi \; A$ would be the array $\begin{matrix} x & x & a & a & a \\ x & x & a & a & a \\ x & x & a & a & a \end{matrix}$.

(ii) **Row catenation rule as a label for transition**

Row catenation rule is in the form $A \,\Theta\, B$. Here again, the array A denotes the $m \times n$ array in the input place of the transition. B is an array language whose number of columns will depend on 'n' the number of columns of A. The number of rows of B is always fixed. For example, $A\Theta \begin{pmatrix} x \\ x \end{pmatrix}^n$ adds two rows of x after the last row of the array A which is in the input place. But $\begin{pmatrix} x \\ x \end{pmatrix}^n \Theta\, A$ would add two rows of x before the first row of the array A. Here, 'n' would denote the number of columns of the input array A. For example, if A is the array $\begin{matrix} a & a & a \\ a & a & a \\ a & a & a \end{matrix}$, then $A\,\Theta \begin{pmatrix} x \\ x \end{pmatrix}^n$ would be the array $\begin{matrix} a & a & a \\ a & a & a \\ a & a & a \\ x & x & x \\ x & x & x \end{matrix}$ and $\begin{pmatrix} x \\ x \end{pmatrix}^n \Theta\, A$ would be the array $\begin{matrix} x & x & x \\ x & x & x \\ a & a & a \\ a & a & a \\ a & a & a \end{matrix}$.

Now, the firing rules are listed out in the array generating Petri net structure, where arrays are taken as tokens.

6.3 Firing Rules

In this Petri Net structure, three types of transitions are enabled and can be fired.

(i) When all the input places of a transition have the same array as token,

- Each input place should have at least the required number of arrays.
- Firing t removes arrays (according to the weight of the arc) from all the input places and moves the array to all its output places.

Figure 2a shows the position of the arrays before firing t, where it is enabled, and Fig. 2b shows the position of the arrays after firing t, where it is disabled.

(ii) When all the input places of a transition have different arrays as token,

- The transition has a designated input place as label.
- The designated place has the same array as tokens.
- Each input place should have at least the required number of arrays.
- Firing t removes arrays from all the input places and moves the array from the designated input place to all its output places.

Figure 3a shows the input place p_1 of the transition t which has two tokens of array A and another array A_1 in the input place p_2. The transition is enabled since the label designates the input place which has the required number copy of the same array. Figure 3b shows the position of the arrays after firing t, where it is disabled. The array from the designated place p_1 is moved to the output place p_3. The array A_1 is consumed in this process.

(iii) When all the input places of t (with catenation rule as label) have the same array as token,

- Each input place should have at least the required number of arrays.
- The condition for catenation should be satisfied.
- Firing t removes arrays from all the input places p and the catenation is carried out in all its output places.

An example to explain row catenation rule is given below. Figure 4a shows the position of the arrays before firing t where it is enabled, and Fig. 4b shows the position of the arrays after firing t where it is disabled. Since the transition is labeled with a catenation rule it takes place in p_3.

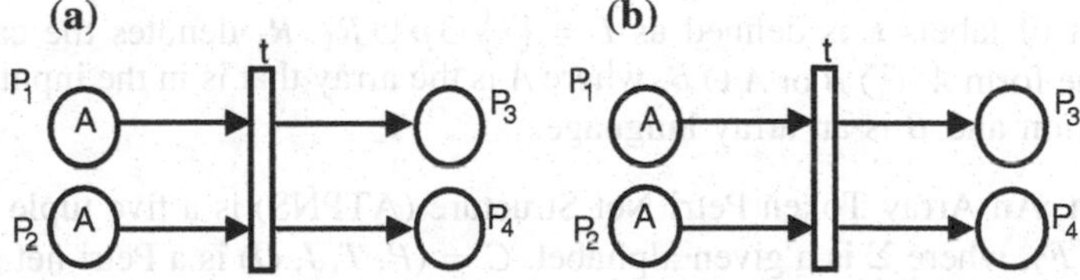

Fig. 2 **a** Transition without label before firing, **b** transition without label after firing

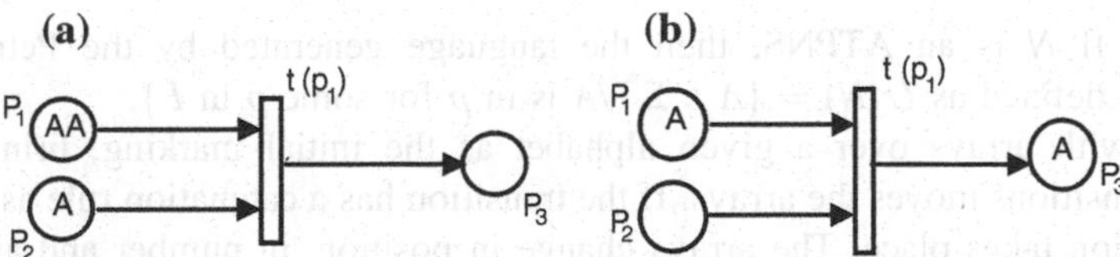

Fig. 3 **a** Transition t with a designated place as label before firing, **b** firing t puts the array in p_1, into the output place

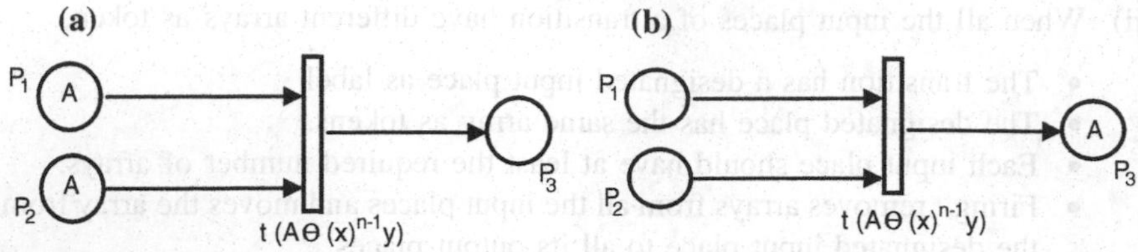

Fig. 4 **a** Transition t with row catenation rule as label before firing, **b** row catenation carried out in the output place p_3

If A is the array $\begin{matrix} a & a & a \\ a & a & a \\ a & a & a \end{matrix}$, the number of columns of A is 3, $n-1$ is 2. Firing t adds the row $x \quad x \quad y$ after the last row of A. Hence, A_1 is the array $\begin{matrix} a & a & a \\ a & a & a \\ a & a & a \\ x & x & y \end{matrix}$.

6.4 Set of Labels

Three types of labels are assigned to transitions in this model

(i) λ—when no label is attached to the transition
(ii) p_i—when one of the input places of the transition is designated as a label
(iii) R_1—when a catenation rule of the form $A \; ⦶ \; B$ or $A \ominus B$ is used as a label

The set of labels L is defined as $L = \{\lambda\} \cup p \cup R_1$. R_1 denotes the catenation rules of the form $A \; ⦶ \; B$ or $A \ominus B$, where A is the array that is in the input place of the transition and B is an array language.

Definition An Array Token Petri Net Structure (ATPNS) is a five tuple $N = (\Sigma, C, M_0, \sigma, F)$, where Σ is a given alphabet, $C = (P, T, I, O)$ is a Petri net structure with arrays of Σ^{**} in certain places of P as initial markings, M_0: $P \rightarrow \Sigma^{**}$, σ: $T \rightarrow L$ a mapping on the set of transitions to the set of labels and a finite set of final places $F \subset P$.

Definition If N is an ATPNS, then the language generated by the Petri net structure is defined as $L(N) = \{A \; \varepsilon \; \Sigma^{**} / A$ is in p for some p in $F\}$.

Starting with arrays over a given alphabet as the initial marking, firing the enabled transitions moves the arrays. If the transition has a catenation rule as label the catenation takes place. The arrays change in position, in number and also in size. All arrays reaching the final place or a set of final places is collected as the language generated by the Array Token Petri Net Structure.

7 Model Formulation and Generation of Array Passwords

In this section, we discuss the model formulation and firing sequence for the generation of array password. Our model generates up to 2×2 array passwords with two input symbols, and this can be extended. The number of possibilities and running time required to encrypt are listed in the following table. The table gives estimates of running time [2] required on a PDP-11/70 to test all possible characters of length n chosen from various sets of characters. The time estimation has been given only for the character but not for the order of the array. The major difficulty for the crackers will be (i) identification of the order the array and (ii) what type of character used.

7.1 Petri Net to Generate All Possible 2 × 2 Arrays Over the Alphabet {a, b}

ATPNS generating 2×2 array is as follows: Let the start arrays be $S_1 = B_1$ and $S_2 = B_2$. With $\sum = \{a, b\}$; $P = \{p, p', p_1, p_2, p_3, p_4\}$; $T = \{T, T', t_1, t_2, t_3, t_4, t_6, t_7\}$; $F = \{p_4\}$; $\sigma\,(T) = \lambda$; $\sigma\,(T) = \lambda$; $\sigma\,(T') = \lambda$; $\sigma\,(t_2) = A\ \theta\ B_1$; $\sigma\,(t_3) = A\ \theta\ B_2$; $\sigma\,(t_4) = A\ \Phi\ B_3$; $\sigma\,(t_5) = A\ \Phi\ B_4$; $\sigma\,(t_6) = A\ \Phi\ B_5$; $\sigma\,(t_7) = A\ \Phi\ B_6$. Let the arrays involved in the rules associated with the transition be given as follows.

$$B_1 = a,\ B_2 = b,\ B_3 = \begin{matrix} a \\ a \end{matrix},\ B_4 = \begin{matrix} a \\ b \end{matrix},\ B_5 = \begin{matrix} b \\ a \end{matrix},\ B_6 = \begin{matrix} b \\ b \end{matrix}$$

7.2 Firing Sequences

The arrays S_1 and S_2 are the start arrays of the net. In this example, it is the element in the a_{11} position of the array. Firing T will push the start array S_1 into the output place P_3. Firing $T'\ t_1$ will push the start array S_2 into the place P_3. A copy of the array also remains in P and P'. Firing t_2 or t_3 adds an element below, so that the array reaching P_3 will be of size 2×1. Hence, the firing sequence Tt_2, Tt_3 generates the arrays $\begin{matrix} a \\ a \end{matrix}$ and $\begin{matrix} a \\ b \end{matrix}$, respectively. The firing sequence $T'\ t_1\ t_2$' $T'\ t_1\ t_3$ generates the arrays $\begin{matrix} b \\ a \end{matrix}$ and $\begin{matrix} b \\ b \end{matrix}$, respectively. Firing t_4 adds the column $\begin{matrix} a \\ a \end{matrix}$. Firing t_5 adds the column $\begin{matrix} a \\ b \end{matrix}$. Firing t_6 adds the column $\begin{matrix} b \\ a \end{matrix}$, and firing t_7 adds the column $\begin{matrix} b \\ b \end{matrix}$ (Fig. 5).

Thus, there are 16 possible firing sequences $T\ t_2t_4$, $T\ t_2t_5$, $T\ t_2t_6$, $T\ t_2t_7$, $T\ t_3t_4$, $T\ t_3t_5$, $T\ t_3t_6$, $T\ t_3t_7$, $T'\ t_1t_2t_4$, $T'\ t_1t_2t_5$, $T't_1\ t_2t_6$, $T'\ t_1t_2t_7$, $T'\ t_1t_3t_4$, $T'\ t_1t_3t_5$, $T'\ t_1t_3t_6$,

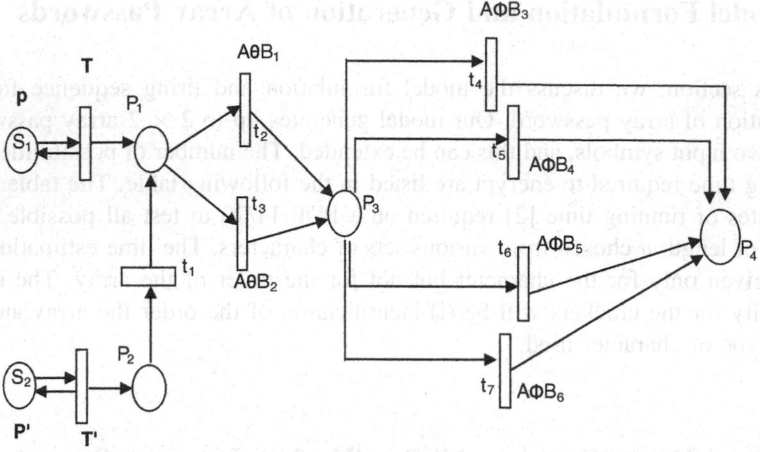

Fig. 5 Petri net Model generating 2 × 2 array passwords

Table 1 Number of possibilities and estimation time for 95 printable characters

S. no	Order/size	Number of possibilities for two input symbols and option of character		Running time required
1	1 × 1	2	95	120 ms
2	1 × 2	4	95^2	11 s
3	2 × 1	4	95^2	11 s
4	3 × 1	8	95^3	17 min
5	1 × 3	8	95^3	17 min
6	**2 × 2**	**16**	$\mathbf{95^4}$	**28 years**
7	**2 × 3**	**64**	$\mathbf{95^6}$	**29 years**
8	**3 × 2**	**64**	$\mathbf{95^6}$	**29 years**
9	**33**	**512**	$\mathbf{95^9}$	–

and T' $t_1t_3t_7$ generating the 16 possible combinations of 2 × 2 arrays over the alphabet $\{a, b\}$. Example of generating a 2 × 2 array is illustrated below.

$$\left[b \overset{t_1}{\Rightarrow} b \overset{t_2}{\Rightarrow} \begin{matrix} b \\ a \end{matrix} \overset{t_6}{\Rightarrow} \begin{matrix} b & b \\ a & a \end{matrix} \right]$$

Example1 Illustration generating a 2 × 2 Array.

The following table gives the clear picture about the order of the array, the number of possibilities for 2 input symbols, option of character, and running time required to estimate the characters (Table 1).

8 Conclusion

In this paper, we recommend a new way of creating array password using Petri nets. With only 2 input symbols and for 3×3 array, there are 512 possibilities and 95^9 character options. For 3×2 array, there are 64 possibilities and running time required is 29 years; the running time has been calculated for character selection not for the order of the array. This paves a new way of valuable and protected mechanism for Web logins. Our approach can be effectively and securely used as the authentication mechanism for the public and un-trusted terminals. To a great extent, further research and user studies are necessary for these password techniques to inclusive advanced levels of development and efficiency. True security, however, is an attribute of the entire human–computer environment, not just what is stored digitally. Upcoming effort in this password generation method should not leave the human out of the equation. With the immense growth in computer power, complexity increasing every day, today's secure applications will not be so safe tomorrow.

References

1. Gehringer, E.F.: Choosing passwords: security and Human factors. In: IEEE 2002 International Symposium on Technology and Society (ISTAS'02), pp. 369–373, ISBN: 0-7803-7284-0 (2002)
2. Morris, R., Thompson, K.: Bell Laboratories "Password Security: A case History" Communication of ACM, vol. 22, pp. 594–597, Nov 1979
3. Hong, J.: Passwords getting painful, computing still blissful. Commun. ACM **56**(3), March 2013
4. Vaithyasubramanian, S., Christy, A.: A Practice to Create user friendly secured password using CFG. In: Accepted for International Conference on Mathematical and Engineering Sciences—2014 (ICMES 2014)
5. Vaithyasubramanian, S., Christy, A.: A scheme to create secured random password using markov chain. In: Accepted for International Conference on Artificial intelligence and Evolutionary Algorithms in Engineering Systems—2014 (ICAEES 2014)
6. Sabzevar, A.P., Stavrou, A.: Universal multi-factor authentication using graphical passwords. In: Proceedings of the 2008 IEEE International Conference on Signal Image Technology and Internet Based Systems, pp. 625–632 (2008)
7. Gao, H., Jia, W., Ye, F., Ma, L.: A survey on the use of graphical passwords in security. J. Softw. **8**(7), July 2013
8. Adebola, O., Ithnin, N., Jali, M.Z., Akosu, N.: Graphical password Schemes design: enhancing memorability features using Autobiographical Memories. J. Theor. Appl. Inf. Technol. **53**(1), 10 July 2013
9. Sonkar, S.K., Paikrao, R.L., Kumar, A.: Graphical password authentication scheme based on color image gallery. Int. J. Eng. Innovative Technol. (IJEIT) **2**(4), October 2012
10. Yan, J.J., Blackwell, A.F., Anderson, R.J., Grant, A.: Password memorability and security: empirical results. IEEE Secur. Priv. **2**(5), 25–31 (2004)
11. Florencio, D., Herley, C.: A large-scale study of web password habits. In: Proceedings of the 16th International Conference on the World Wide Web, Acm Digital Library, pp. 657–666 (2007)

12. AlFayyadh, B., Thorsheim, P., Jøsang, A., Klevjer, H (2012) Improving usability of password management with standardized password policies. In: The Seventh Conference on Network and Information Systems Security—SAR-SSI 2012 Cabourg, ISBN: 978-2-9542630-0-7, May 2012
13. http://resources.infosecinstitute.com/dictionary-attack-using-burp-suite
14. Wiedenbeck, S., Waters, J., Birget, J.C., Brodskiy, A., Memon, N.: Authentication using graphical passwords: basic results. In: Human-Computer Interaction International (HCII) (2005)
15. Sarohi, H. K., Khan, F. U.:Graphical password authentication schemes: current status and key issues. Int. J. Eng. Innovative Technol. (IJEIT) **10**(2), No 1, March 2013
16. Ahmad, S.M.S., et al.: technical issues and challenges of biometric applications as access control tools of information security. Int. J. Innovative Comput. Inf. Control **8** (11), 7983–7999, ISSN 1349-4198, November 2012
17. Bhatnagar, M., Jain, R.K., Khairnar, N.S.,: A survey on behavioral biometric techniques: mouse vs. keyboard dynamics. In: IJCA Proceedings on International Conference on Recent Trends in Engineering and Technology, pp. 27–30 (2013)
18. Peterson, J.L.: Petri Net Theory and Modeling of systems. Prentice Hall Inc., Englewood Cliffs (1981)
19. Lalitha, D., Rangarajan, K.: Column and row catenation petri net systems. In: Proceeding of 5th IEEE International Conference on Bio-Inspired Computing: Theories and Applications, pp. 1382–1387 (2010)
20. Lalitha, D., Rangarajan, K., Thomas, D.G.: Rectangular Arrays and Petri Nets, Combinatorial Image Analysis, vol. 7655, pp. 166–180. LNCS, London (2012)

Application of Genetic Algorithm for Component Optimization to Deploy Geoprocessing Web

Sujit Kumar Behera, Lalit Kumar Behera, Payodhar Padhi and Maya Nayak

Abstract Spatial data infrastructures served through the Web combined with the ever increasing network and telecommunication capabilities, and made geospatial data largely available over the last few decades. In addition, providing semantic specifications to geospatial information, data sharing and interoperability have also been achieved. Consequently, effective and efficient implementation of the Web processing, data processing methods for geospatial information extraction, and knowledge discovery over the Web are a major challenge for various domains. This paper provides a basic framework to optimize the various components associated with the geoprocessing implementation using genetic algorithm

Keywords Geoprocessing · Component optimization · Genetic algorithm

1 Introduction

With the introduction of advanced technology of Earth observation and sensing, the volume of geoscientific data has increased tremendously in the past decade and is expected to keep growing continuously. Interoperability and accessibility of geoprocessing resources improve the application of geospatial data in various domains and help to increase the geospatial knowledge available to society. This interoperability is achieved by common standards, whereas accessibility to

---

S.K. Behera (✉)
Aricent, Bengaluru, India
e-mail: mesujit@gmail.com

L.K. Behera · P. Padhi
Konark Institute of Science and Technology, Jatni, Bhubaneswar 752050, Odisha, India

M. Nayak
Orissa Engineering College, Jatni, Bhubaneswar 752050, Odisha, India

L.C. Jain et al. (eds.), *Intelligent Computing, Communication and Devices*,
Advances in Intelligent Systems and Computing 308,
DOI 10.1007/978-81-322-2012-1_21

particular resources is enabled by the Web. Both aspects are supported by Web services technology.

Web-based distributed geospatial computing and large networks of collaborating applications are the next step in the evolution of geoprocessing [1]. To address the demands of geoprocessing in distributed environments like the Web, the combination of conventional analysis functions and advanced computing technologies requires new technical infrastructure, domain-specific models, and methodologies to support advanced data-mining tools and online community collaborations. Generally, the geo-processing Web consists of lightweight protocols, crowd-sourcing capability, and the capability to process real-time geospatial data sources provided by sensors. It enables distributed, interoperable, and collaborative processing of geospatial data for information and knowledge discovery. The geo-processing Web provides architecture, standards, and tools to meet these requirements. Some are service-oriented architecture (SOA), lightweight protocols, crowd-sourcing capability, and the capability to process and deliver real-time geospatial data provided by sensors.

2 Approaches for Geoprocessing

Geoprocessing is used to manipulate available spatial data. A common geoprocessing includes spatial processing (e.g., network analysis and coordinate transformation), thematic processing (e.g., classification and geocoding), temporal processing (e.g., change detection and temporal sub-setting), and metadata processing (e.g., geographic annotation and statistical calculation). The advances in information technology have immense contribution for the development of the geoprocessing paradigm. Existing geoprocessing paradigms can be classified into following types [2, 3]:

(1) Traditional geoprocessing applications are performed in a stand-alone environment like a desktop computer. Data storage, visualization, and processing are tightly coupled in a software repository. Thus, geospatial users often access, transform, and visualize data using one software package.
(2) Client/server data storage and processing functions are performed at the remote servers, while data presentation functions are performed by local clients such as Web browsers [4].
(3) Distributed object When the component-based software engineering principle is applied, geoprocessing functions can be provided by different software vendors by following the interoperable application programming interface

(API). These functional components can be assembled to accomplish a complex geoprocessing task, even though they may be provided by different software packages.

(4) Web services SOA provides an interoperable computing infrastructure for conducting advanced distributed geoprocessing tasks. These tasks use Web protocols and formats encoded in the extensible markup language. In the context of SOA, geoprocessing functions are provided as loosely coupled Web services and coordinated to execute complex geoprocessing tasks collaboratively as workflows. The research and development on new geoprocessing infrastructures in the past several years have been occupied predominantly with the employment of technology for improving discovery, performance, and workflows. It is, therefore, necessary to set up a research agenda clarifying the promise of geoprocessing over the Web. This is the aim of the geoprocessing Web [5, 6, 7].

3 The Geoprocessing Web and Component Optimization

As SOA is emerging as the basis for distributed computing and an interoperable framework of collaborating applications, an increasing amount of geospatial processing functions and applications are built upon it. Users can collect, analyze, and derive geospatial data, information, and knowledge over the Web. The geoprocessing Web covers the conceptual, methodological, technical, and managerial issues that facilitate distributed and collaborative geoprocessing over the Web.

Figure 1 illustrates a three-layer framework of the geo-processing Web. The first layer is the geoprocessing resource layer that provides sensors, geospatial data, and geoprocessing facilities. The second layer is the management platform aimed to provide basic utilities to manage geospatial data, services, and models. In particular, the basic utilities include some services for the retrieval, process, and visualization of geospatial data; the sensor services focuses on discovery of and accessing to sensor observations. In the third layer, a workbench or portal facilitates users to discover and understand geospatial resources, and develop geoprocessing models and applications. Geoprocessing models, after going through a collaborative peer review, can be registered in the model warehouse as a type of knowledge. The security and privacy across three layers ensures the resource sharing rules and conditions for different types of users. The framework adopts the service-oriented view in which each component is developed as services following standardized interfaces and protocols.

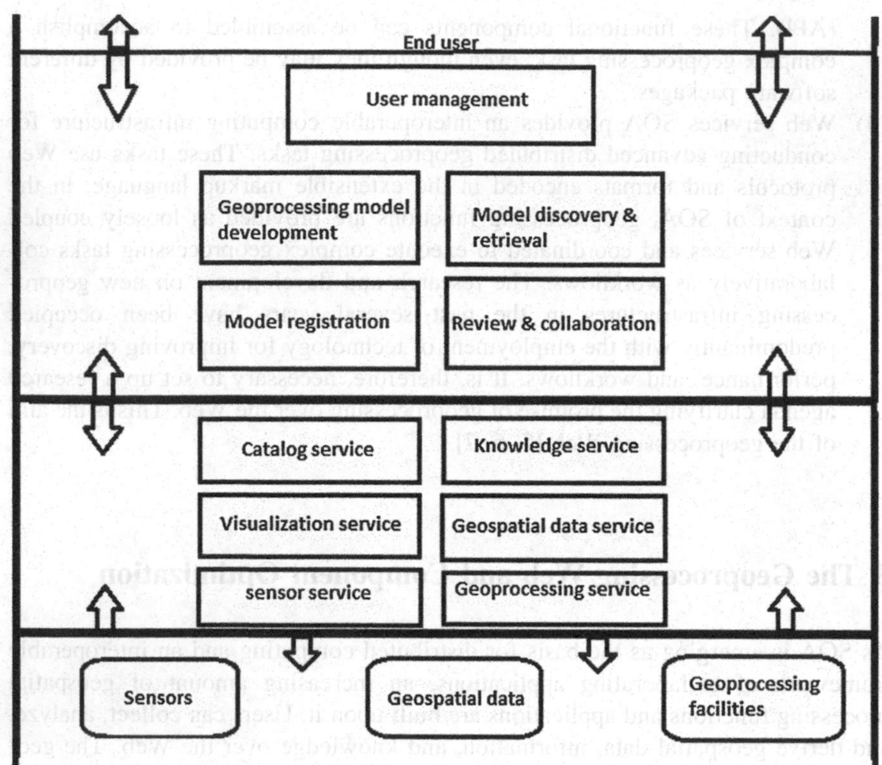

Fig. 1 Framework for geoprocessing Web

4 Proposed Work

Standards enable the interoperability of service components, support the easy integration of distributed components, and ensure the components being accessible by the large public. But the quantification of the various components at different layers is required for the cost optimization and implementation. The efficiency of a particular layer is the cumulative performance of all the components of that layer. By taking the field requirement values on a hundred-point scale rating of various components in different deployments for first layer, the objective function has been obtained with the help of regression analysis which is given in Eq. 1. In this equation, E represents the efficiency of the installation and c (i) various

Table 1 Fitness values in each generation and best and mean values

Sl.no.	f-count	Best f(x)	Mean f(x)	Stall generation
1	100	0.6952	0.6996	0
2	150	0.6949	0.6985	0
3	200	0.6947	0.6977	0
4	250	0.6938	0.6982	0
5	300	0.6934	0.6971	0
6	350	0.6930	0.6971	0
7	400	0.6929	0.6967	0
8	450	0.6927	0.6960	0
9	500	0.6927	0.6961	1
10	550	0.6923	0.6953	0
…………………………………………				
40	2050	0.6887	0.6887	0
41	2100	0.6887	0.6887	0
42	2150	0.6887	0.6888	0
43	2200	0.6887	0.6888	1
44	2250	0.6887	0.6888	0
45	2300	0.6887	0.6887	0
46	2350	0.6887	0.6888	1
47	2400	0.6887	0.6888	0
48	2450	0.6887	0.6888	1
49	2500	0.6887	0.6887	2
50	2550	0.6887	0.6887	0

X = 71.0000 75.0000 74.0000 56.8171 84.6711
Fval = 0.6887

components. When this objective function has been processed by the genetic algorithm [8, 9] the optimized values for different components have been obtained. A sample output of genetic algorithm is shown in Table 1, and the best and mean efficiency are shown in Fig. 2 which will help in cost-effective deployment.

$$E = ((0.0070 * c(1)) + (0.0030 * c(2)) + (0.0007 * c(4)) + (-0.0012 * c(6)) + (-0.0002 * c(7))) \quad (1)$$

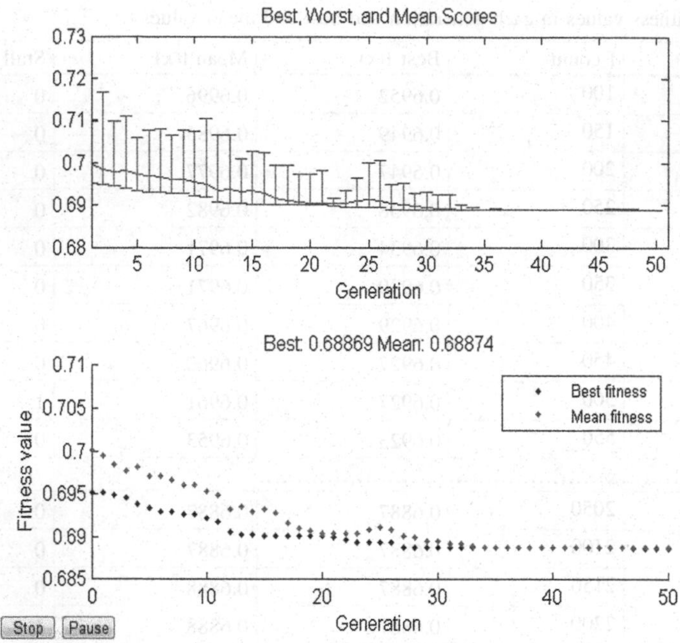

Fig. 2 Best and mean efficiency values and the fitness values in each generation

5 Conclusions

With the continuous increment of the available amount of spatial data sets, science, industry, and administration require Web-based geo-information concerning storage, availability, and processing. The development of spatial data infrastructures (SDIs) bring about the Web-based sharing of large volumes of distributed geospatial data and computational resources. A powerful, dependable, and flexible information infrastructure is required to process heterogeneous and distributed data into information and knowledge. This paper discusses the component optimization in the geo-processing Web. Our study reveals that the geo-processing Web implementation process can be improved by component optimization using genetic algorithm for enhancing the way in which geospatial applications and systems are designed, developed, and deployed.

References

1. Kiehle, C., Greve, K., Heier, C.: Requirements for next generation spatial data infrastructures-standardized web based geoprocessing and web service orchestration. Trans. GIS **11**(6), 819–834 (2007)

2. Tsou, M., Buttenfield, B.: A dynamic architecture for distributing geographic information services. Trans. GIS **6**(4), 355–381 (2002)
3. Zhao, P., Di, L., Yu, G.: Building asynchronous geospatial processing work-flows with web services (2011)
4. Abel, D., Taylor, K., Ackland, R., Hungerford, S.: An exploration of GIS architectures for internet environments. Comput. Environ. Urban. Syst. **22**(1), 7–23 (1998)
5. Berners-Lee, T., Hall, W., Hendler, J., Shadbolt, N., Wietzner, D.: Creating a science of the web. Science **311**(5788), 769–771 (2006)
6. Zhao, P., Di, L., Yu, G.: Building asynchronous geospatial processing work-flows with web services. Comput. Geosci. **39**(2), 34–41 (2012)
7. Zhao, P., Yu, G., Di, L.: Geospatial web services. In: Hilton, B. (Ed.), Emerging Spatial Information Systems and Applications, pp. 1–35. Idea Group Publishing, Hershey (2006)
8. Ji, Z., Li, Z., Ji, Z.: Research on genetic algorithm and data information based on combined framework for nonlinear functions optimization. Proc. Eng. **23**, 155–160 (2011)
9. Mezura-Montesa, E., Coello Coello, C.A.: Constraint-handling in nature-inspired numerical optimization: past, present and future. Evol. Comput. **1**, 173–194 (2011)

DNA-Based Cryptographic Approach Toward Information Security

Abhishek Majumdar, Tanusree Podder, Atanu Majumder, Nirmalya Kar and Meenakshi Sharma

Abstract Cryptography is a science of encoding a secret message in a manner so that any other person other than the receiver and the sender cannot be able to decode it, whereas the steganography is the science of concealing the information in any other cover media so that an attacker cannot recognize the presence of the secret message inside the cover media. This paper proposes a method using combined technologies of cryptography and steganography by following the concept of genetic engineering based on the DNA. This approach will enhance the message security, which is the main concern in today's world for message transmission. In this proposed algorithm, a long and strong 256-bit key is used for encryption with a strong new method of encryption, generation of DNA sequence, and a new method to conceal the encrypted DNA sequence to a cover image, which provides better security in the message against the intruders attack.

Keywords DNA sequence · Key · Nucleotide · Steganography · Cryptography

---

Please note that the LNCS Editorial assumes that all authors have used the western naming convention, with given names preceding surnames. This determines the structure of the names in the running heads and the author index.

A. Majumdar (✉) · M. Sharma
Computer Science and Engineering Department, SSCET, Badhani, Punjab, India
e-mail: abhishekmajumdar91@gmail.com

M. Sharma
e-mail: hod.csebadhani@srisaigroup.in

T. Podder · A. Majumder · N. Kar
National Institute of Technology Agartala, Agartala, Tripura, India
e-mail: tanusreepodder29@gmail.com

A. Majumder
e-mail: atanu.cse21@gmail.com

N. Kar
e-mail: nirmalya@nita.ac.in

L.C. Jain et al. (eds.), *Intelligent Computing, Communication and Devices*,
Advances in Intelligent Systems and Computing 308,
DOI 10.1007/978-81-322-2012-1_22

1 Introduction

Security of information during the transmission is the most required thing in this era. As a result, the concept of cryptography and steganography was introduced. The cryptography is the technology where a known text can be converted into certain coded format that will not be understandable to human whereas steganography came after cryptography to enhance its security, where the human cannot recognize the existence of the secret message. But nowadays, for better results in data transmission security, both of these technologies are used together that can guarantee a highly secured environment. Various encryption and steganographic tools have been used since a decade, but the human DNA-based encryption and steganographic approach is the most emerging and promising area among all due to the complex structures and several special features of the DNA. Practically, the cryptography and the biological genetic molecules do not have any direct connections to each other. But in the field of information security, the nature of the DNA can be adopted to enhance the security and reliability of the information. It is quite a new area and still it is not reached to a mature position. So many researchers, nowadays, take interest regarding this study.

In this paper, a modified DNA substitution is proposed using the various features of the DNA. This encryption algorithm used here is based on the concept of both the DNA and the conventional cryptography. In this, a large 256-bit key is used that works on the block cipher in a randomly generated ordered manner with several rounds. Moreover, the proposed steganographic approach is much more different and better as compared to the traditional ones. In this approach, we have used the concept of DNA primers that was taken from the large set of EBI database by maintaining its length depending upon the cipher DNA sequence, and moreover, primers are appended to the cipher DNA sequence with a special manner that enhance the security of the message.

2 Brief Idea of DNA

The DNA actually stands for deoxyribo nucleic acid. Biologically, it is the means of all kind of life that contains a molecule as the carrier of the genetic information. According to Watson and Crick, the DNA is a double helix made up of two strands of four basic nucleotide bases adenine (A), guanine (G), cytosine (C), and thymine (T). The construction of these two DNA strand is such that two bases can be paired together to join the strands to each other. But the key point is that adenine (A) will always paired with thymine (T) and the cytosine (C) will always paired with the guanine (G). The concepts about DNA that are generally followed in encryption are the codon, primer, and the large variation of the DNA sequence.

Codons are the sequence of three adjacent nucleotides and, for every codon, there exist a corresponding unique amino acid that is used for protein synthesis.

The primer is also a DNA sequence that contains a number of nucleotides. Primers are appended on the both sides of the DNA sequence and it is treated as keys. Through the use of primer the security of the DNA sequence can be enhanced.

Another strong feature of the DNA is its variation in sequence. A DNA sequence can also get from the large set of EBI or NCBI database where out of 163 million targets, any one can be chosen. The NCBI has the responsibility to provide the GenBank DNA sequence database since 1992. So, it is a great challenge or simply an impossible task to the attacker to guess the correct DNA sequence.

3 Literature Review

In this era, a large number of researchers are working on information security and they had implemented several algorithms based on DNA concept. Among them, some had used DNA computing in their algorithms, while in the other side, some have incorporated DNA characteristics into their algorithms. From the large set of research works on this field, some are as follows:

Majid Babaei proposed a Chaos theory-based text and image encryption method using DNA computing and also enhanced the performance of the one-time pad algorithm [1].

Mohammad Reza Abbasy, Azizah Abdul Manaf et al. described a method where a common DNA sequence is shared between the sender and the receiver. Then, the occurrences of the DNA representation of the plain text in the reference DNA sequence are listed by using an indexing method where every couple of nucleotides in DNA reference sequence is given an index number [2].

H.Z. Hsu and R.C.T. Lee et al. discussed three methods namely, the insertion method, the complementary pair method, and the substitution method. They secretly select a reference DNA sequence for each of these three methods and incorporate the message into the DNA sequence, using reverse binary coding method [3].

Bibhash Roy, Atanu Majumder et al. have given a new method of encryption, which has two levels of encryption that was concerned with how DNA sequencing can be used in cryptography [4]. Moreover, a new scheme of key generation and key sharing was also proposed by them.

The researchers are implementing this to enhance the security of the cipher text by appending extra coded information with it at different location of the cipher text [5]. Various kinds of extra codes that are padded can be authentication code, integrity code, starting primer, ending primer, etc.

Amal Khalifa and Ahmed Atito have derived a method where the plaintext is transformed into the collection of amino acids and encrypted using DNA-based playfair cipher [6]. After that, DNA complementary substitution is used to encrypt the message in greater extent.

Ashish Gehani, Thomas LaBean, and John Reif et al. have given the concept that how can the researchers nowadays use DNA as a medium for ultra-scale computation and for ultra-compact information storage [7].

On the other hand, researchers like Xing Wang, Qiang Zhang et al. proposed a new way that how cryptography can work with DNA computing, and how to transmit message effectively and in a secure manner. They have incorporated the asymmetric key cryptography RSA algorithm along with DNA computing theory [8].

Along with the cryptography, the steganography approach based on DNA is also taking as an area of interest by the researchers. Mohammad Reza, Najaf Torkaman, Pourya Nikfard et al. have given a concept regarding a DNA steganography method to hide the session key, and they used a DNA sequence from the wide range of publicly available DNA sequence databases [9].

On the otherhand, Suman Chakraborty, Sudipta Roy et al. have given a method where they have used the idea of soduko solution matrix to perform some computations on behalf of the message [10].

4 Proposed Method

The method proposed in this paper provides a reliable and secured data transmission. The overall method is carried out through two phases; these are DNA-based encryption of data and embedding of the resultant encrypted DNA sequence into an image.

4.1 Encryption

The encryption method used here is somehow different from the other encryption algorithms till now implemented. In the earlier encryption schemes, very basic concepts regarding the DNA were used such as simple EX-OR operation, primer addition, using complementary rules of DNA, etc. But here, the proposed DNA-based encryption algorithm is little different from others.

Here, the total encryption phase is based on the two sub-phases, one for the key selection and another is message encryption.

In the first sub-phase that is in key selection phase, a 256-bit key is chosen randomly to do the further encryption. This 256-bit key is then divided into four 64-bit blocks, which will be acted as sub-keys at the time of encryption. Each block of sub-key is labeled with the DNA base namely, A, T, C, G. Then, randomly select any combination of these four bases out of the possible 24 combinations without repetition, such as {A, C, T, G}; {G, A, C, T}, etc.

In the second sub-phase that is in the message encryption phase, the ASCII values of the plaintext are retrieved and after that, it will be transformed to its equivalent 8-bit binary form. Then, the plain text is divided into 256-bit blocks.

Table 1 Randomly generated ASCII table corresponding to each 8-bit binary

8-bit binary	ASCII	8-bit binary	ASCII
00000000	15	–	–
00000001	74	–	–
00000010	103	–	–
00000011	204	–	–
00000100	134	–	–
00000101	25	–	–
00000110	67	–	–
00000111	89	11111111	82

Each block of plain text will be the part of the encryption process. Now, each 256-bit blocks of plain text is divided into four 64-bit blocks. Now perform 64-bit EX-OR operation between the 256-bit plain text having four 64-bit blocks and the randomly chosen 256-bit key of four DNA bases each of having 64 bits. This key is acted as the round 1 key.

In round 2, the key order of DNA bases is right shifted to 1 block that is round 1 key is right shifted to 64 bits. This produced key will be EX-ORed with the result of the first round. Then, its result is go for the third round with the 1 block-shifted key of the second round, and this process will be continued for 4 rounds for the 4 keys generated by shifting of 1 block (64 bit) in each round. For example, if the randomly chosen key combination is ACGT where each base represents each block of 64 bit. So, in the second round, the key will be TACG since a block is shifted right; in third and fourth round, it will be GTAC and CGTA, respectively. As a result in each round of encryption, the plain text is encrypted in more and more depth by four different keys operating in each round, whereas a single was only chosen. The rest of the four keys were induced in each round from its previous one.

After that, generate a random table of ASCII values in the range of 0–255 corresponding to each 8-bit binary combination as per Table 1. So, in this approach, the main fact is that any 8-bit binary value not necessarily to be its actual decimal value, it may vary each time when a random table is generated. For example, the 8-bit binary 00000011 may correspond to the ASCII value 200 or 123 or anything else in each time the random table is generated. Then, the resultant ASCII values are transformed to their actual binary form and each two bit of them is converted to the DNA bases based on the randomly chosen manner to represent the bases from the Table 2.

After all of these the technologies of appending the primers are used in a separate way rather than the traditional ways to enhance the security of the DNA sequence. In this proposed method, a primer of length 2 × (1/100 × length of the cipher text(DNA sequence)) is taken from the EBI or NCBI databases from where among millions of targets a single DNA sequence of the desired length measured

Table 2 The possible DNA bases 2-bit binary value

Index	2 bit Binary values				Index	2 bit Binary values			
	00	01	10	11		00	01	10	11
1	A	T	C	G	13	C	T	A	G
2	A	T	G	C	14	C	T	G	A
3	A	C	T	G	15	C	A	T	G
4	A	C	G	T	16	C	A	G	T
5	A	G	C	T	17	C	G	A	T
6	A	G	T	C	18	C	G	T	A
7	T	A	C	G	19	G	T	C	A
8	T	A	G	C	20	G	T	A	C
9	T	C	A	G	21	G	C	T	A
10	T	C	G	A	22	G	C	A	T
11	T	G	C	A	23	G	A	C	T
12	T	G	A	C	24	G	A	T	C

by the above formula is chosen. Then, split it into two parts and append those as a front and end primer of the DNA sequence. As a result, the actual size of the primer may vary each time for different plain texts. Hence, the security of the message is very much hiked.

4.1.1 Algorithmic Steps

For the encryption phase, the steps that are followed are listed below:

Step 1: Read the byte values of the plain text and transform each byte value into 8-bit binary representation.
Step 2: Divide the plain text into a number of 256-bit blocks.
Step 3: Randomly choose a key of length 256-bit.
Step 4: Divide the 256-bit key into four 64-bit blocks namely, A, T, C, G.
Step 5: Randomly choose a combination of the 4 blocks of the key, e.g., TGCA.
Step 6: For each 256-bit of plain text, repeat the steps 7 to 9.
Step 7: Split the 256-bit block into four 64-bit blocks namely, PT1, PT2, PT3, PT4.
Step 8: Perform 64-bit EX-OR operation between the four blocks of plain text and the chosen key.
Step 9: Shift the key to right by 1 block, i.e., 64 bit.
Step 10: Randomly generate table of ASCII value ranging from 0–255 corresponding to each 8-bit binary.

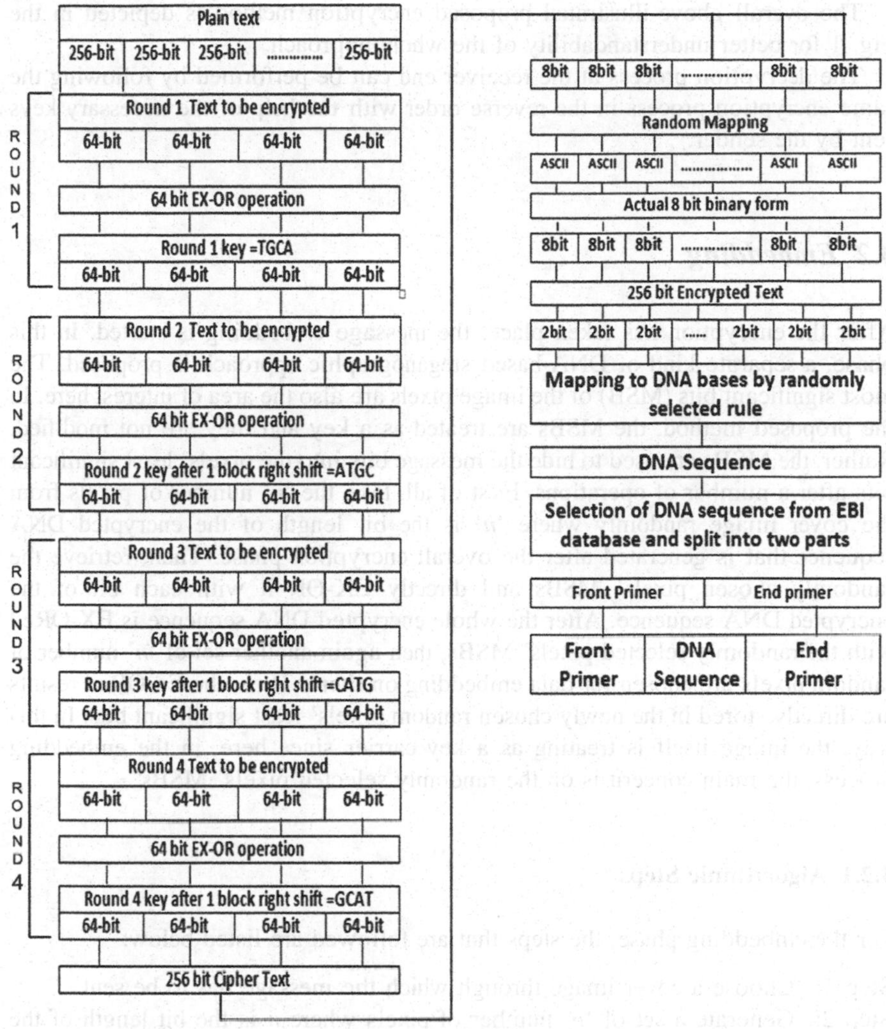

Fig. 1 Overall proposed encryption approach

Step 11: Transform the result of Step 10 to their actual 8-bit binary representation.

Step 12: Map each two bit of binary into a DNA bases by choosing any rule from the Table 2.

Step 13: Take a DNA sequence from the EBI database of length [$2 \times 1/100 \times$ length of the resultant DNA sequence generated after Step 12].

Step 14: Split the DNA sequence into two equal size sequences and append those as front and end primers, respectively.

The overall above-illustrated proposed encryption method is depicted in the Fig. 1 for better understandability of the whole approach.

The decryption process at the receiver end can be performed by following the same encryption process in the reverse order with the help of the necessary keys sent by the sender.

4.2 Embedding

After the encryption has taken place, the message embedding is started. In this phase, a separate kind of DNA-based steganographic approach is proposed. The most significant bits (MSB) of the image pixels are also the area of interest here. In the proposed method, the MSBs are treated as a key and they are not modified. Rather, the MSBs are used to hide the message bits into the pixels' least significant bits after a number of operations. First of all, take the '*n*' number of pixels from the cover image randomly where '*n*' is the bit length of the encrypted DNA sequence that is generated after the overall encryption phase. Then, retrieve the randomly chosen pixels' MSBs and directly EX-OR it with each bit of the encrypted DNA sequence. After the whole encrypted DNA sequence is EX-ORed with the randomly selected pixels' MSBs, then again another set of '*n*' number of random pixels are chosen for data embedding on them. Then, the Ex-ORed results are directly stored in the newly chosen random pixels' least significant bits. In this way, the image itself is treating as a key carrier since here, in the embedding process, the main concern is on the randomly selected pixels' MSBs.

4.2.1 Algorithmic Steps

For the embedding phase, the steps that are followed are listed below:

Step 1: Choose a cover image through which the message has to be sent.
Step 2: Generate a set of '*n*' number of pixels where n is the bit length of the encrypted DNA sequence to be hide.
Step 3: Retrieve MSBs of all the pixels in the set.
Step 4: Perform EX-OR operation between the MSBs and the encrypted DNA sequence bit representation.
Step 5: Again, generate another set of pixels and retrieve their LSBs.
Step 6: Replace each LSB with the each bit of the resulting bit sequence after Step 4.

5 Algorithmic Presentation

5.1 Key Selection

In this phase, a key of size 256 bit is selected randomly. Let, K is the key, K= '1000 1010 0011 0001 1100 0100 1011 0001 0111 0000 0010 0100 1111 1100 1111 0011 1110 1110 1011 1011 0100 1011 1111 0000 1011 1100 1101 1011 0011 1010 0001 1110 0001 1000 0011 0110 1010 1111 1000 1110 0001 0000 1001 1010 1001 1001 0010 0001 0000 1101 1010 1110 1000 1111 1000 1111 1101 1011 0011 1010 1111 0101 1111 1110'

Read the key values and generate four sub-keys of 64 bit each. Label the sub-keys with DNA bases (A, T, C, G) as follows:

A='1000101000110001110001001011000101110000001001001111110011110
011'

T='111011101011101101001011111100001011110011011011001110100001
1110'

C='000110000001101101010101111100011100001000010011010100110010010
0001'

G='0000110110101110100011111000111111011011001110101111010111111
110'

From the Table 1, let us randomly select DNA sequence with DNA bases be 'TGCA' then round 1 key, K1 = TGCA; round 2 key, K2 = ATGC; round 3 key, K3 = CATG; round 4 key, K4 = GCAT.

5.2 Message Encryption

As described in the above sections, the plain text is divided into a number of 256-bit blocks. Every 256-bit plain text block will go through the four round of encryption process. After each round, the resultant coded block will be operated with the each round key and get prepared for the next round. Split the 256-bit block into four 64-bit blocks namely, P11, P12, P13, P14, which is the input of the round 1. As taken, the sub-keys for round 1 are the K11 = T, K12 = G, K13 = C, K14 = A. The detailed flows of each round are as follows:

Round 1:
256-bit text PT1
64 bit blocks P11, P12, P13, P14
Round 1 sub-keys K11 = T, K12 = G, K13 = C, K14 = A
Compute cipher text CT1 = (P11 ⊕ K11) (P12 ⊕ K12) (P13 ⊕ K13) (P14 ⊕ K14)

Round 2:
256-bit text PT2 = CT1
64-bit blocks P21, P22, P23, P24
Round 2 sub-keys K11 = A, K12 = T, K13 = G, K14 = C
Compute cipher text CT2 = (P21 ⊕ K21) (P22 ⊕ K22) (P23 ⊕ K23) (P24 ⊕ K24)

Round 3:
256-bit text PT3 = CT2
64-bit blocks P31, P32, P33, P34
Round 3 sub-keys K31 = C, K32 = A, K33 = T, K34 = G
Compute cipher text CT3 = (P31 ⊕ K31) (P32 ⊕ K32) (P33 ⊕ K33) (P34 ⊕ K34)

Round 4:
256-bit text PT4 = CT3
64-bit blocks P41, P42, P43, P44
Round 2 sub-keys K41 = G, K42 = C, K43 = A, K44 = T
Compute cipher text CT4 = (P41 ⊕ K 41) (P42 ⊕ K42) (P43 ⊕ K43) (P44 ⊕ K44)

After the round 4, the generated cipher text CT4 is transformed into its binary form, from the randomly generated Table 1, each ASCII value corresponding to every 8-bit binary information is retrieved. After that, the actual 8-bit binary value of each ASCII value is found out. Then, having every two bit binary map into its corresponding DNA base as per the randomly selected rule from the Table 2. Hence, a DNA sequence M is produced. On the basis of the length of this DNA sequence from the EBI database, a DNA sequence of twice of the 1/100th of the length of the *M* is selected and treat it as a primer. Spit the primer into equal two parts and pad them on the both side of the *M*. Now the resultant DNA sequence becomes *M*′.

At last, an image is chosen and select a set of *n* pixels randomly, where '*n*' is the bit length of *M*. Retrieve the MSB of every selected pixel and perform EX-OR operation with each bit of the *M*′. After that another set of *n* pixels is selected randomly and simply replace their LSBs with the EX-ORed results of the previous.

6 Conclusion

In this paper, a method of message transmission is proposed using the combined technology of cryptography with the DNA-based cryptography and steganography. In this approach, the secret information has been hidden in more depth so that it will be almost impossible to an attacker to know about the message. Here, a long-size 256-bit key is used in different way each time in 4 rounds to encrypt the message. Moreover, the random allocation of the 256 ASCII values to each 8-bit binary and the use of randomly chosen binary values to each DNA bases from a 24 possible ways make our method more secure and reliable. The most strong point in this method is the way of the primer selection. Its size may vary each time for different plain texts. Moreover, the primer is selected out of 163 millions of targets, so it is clear that for the intruder, it is totally impossible to guess the correct primer from this sea of possible DNA sequences. Finally, the encrypted DNA sequence is hidden into the image in very secure manner rather than to hide the information directly into each consecutive pixel value. The proposed method will be a better and secure method since the image itself here acting as a key carrier through a set of pixels' MSBs. For the attacker, the understanding of the bit information after embedding the message into the image will be a tedious job.

References

1. Babaei, M.: A novel text and image encryption method based on chaos theory and DNA computing. Nat. Comput. **12,** 101107 (2013). doi: 10.1007/s11047-012-9334-9. Springer Science + Business Media, Berlin
2. Abbasy, M.R., Manaf, A.A., Shahidan M.A: Data hiding method based on DNA basic characteristics. In: Ariwa, E., El-Qawasmeh, E. (Eds.): DEIS 2011, CCIS 194, pp. 5362. Springer, Berlin, Heidelberg (2011)
3. Hsu, H.S., Lee, R.C.T.: DNA based encryption methods. In The 23rd Work—shop on Combinatorial Mathematics and Computation Theory, National Chi Nan University Puli, Nantou Hsies, Taiwan 545, April 2006
4. Roy, B., Majumder, A.: An improved concept of cryptography based on DNA sequencing. Int. J. Electron. Commun. Comput. Eng. **3**(6), 1264–1267 (2012)
5. Roy, B., Rakshit, G., Singha, P., Majumder, A., Datta. D.: An improved symmetric key cryptography with DNA based strong cipher. In: International Conference on Device and Communication, BIT Mesra, Ranchi, Jarkhand, India, Feb 2011
6. Khalifa, A., Atito, A.: High-capacity DNA-based steganography. In: The 8th International Conference and informatics and Systems (INFOS2012), IEEE, May 2012
7. Gehani, A., LaBean, T., Reif, J.: DNA-based cryptography. DI—MACS DNA based computers V, American Mathematical Society (2000)
8. Wang, X., Zhang, Q.: DNA computing-based cryptography. In: Proceedings of the IEEE International Conference, ISBN: 978-1-4244-3867-9/09 (2009)
9. Torkaman, M.R.N., Nikfard, P., Kazazi, N.S., Abbasy, M.R., Tabatabaiee, S.F.: Improving hybrid cryptosystems with DNA steganography. In: Ariwa, E., El–Qawasmeh, E. (eds.) DEIS 2011, CCIS 194, pp. 4252, 2011. Springer, Berlin, Heidelberg (2011)
10. Chakraborty, S., Roy, S., Bandyopadhyay, S.K.: Image steganography using DNA sequence and sudoku solution matrix. Int. J. Adv. Res. Comput. Sci. Softw. Eng. **2**(2), (2012)

Design a Fuzzy Logic Controller with a Non-fuzzy Tuning Scheme for Swing up and Stabilization of Inverted Pendulum

A.K. Pal and J. Chakrabarty

Abstract In this paper, a new non-fuzzy self-adaptive scheme is proposed for optimal swing up control of the inverted pendulum system. Further, the system stabilization is compared against an automatic fuzzy-based self-tuning technique. Our work proposes a twin fuzzy control scheme for effective control of cart position and inverted pendulum angle. A comparative analysis through simulation demonstrates the feasibility and reliability of the proposed approach.

Keywords Adaptive control · Fuzzy control · Tuning · Inverted pendulum

1 Introduction

The inverted pendulum is an interesting topic to researchers for its strong degrees of nonlinearity and inherent instability and yet simplicity of structure. So it can efficiently be used for determining the effectiveness of a control algorithm. Its basic structure consists of one/two flexible pendulum arms mounted to a cart on a moving rail. In the absence of a stabilizing controller, the pendulum arms, which have its center of mass above its pivot point, are unable to maintain their upright position. Thus, the control objective is to apply a force to move the cart so that the pendulum arms remain in the vertical unstable position while simultaneously being driven to the desired cart location. Furthermore, it should also include a swing up control aspect if initially the pendulum hangs freely in the vertical position. Control of an inverted pendulum is a very common control engineering problem based on flight simulation of rockets and missiles during the initial stages of flight,

A.K. Pal (✉) · J. Chakrabarty
Department of A.E.I.E, HIT, Kolkata, India
e-mail: arabindakumarpal@gmail.com

J. Chakrabarty
e-mail: jayshreechakraborty.1407@gmail.com

L.C. Jain et al. (eds.), *Intelligent Computing, Communication and Devices*,
Advances in Intelligent Systems and Computing 308,
DOI 10.1007/978-81-322-2012-1_23

wherein the aim was to stabilize the inverted pendulum such that the position of the carriage on the track was controlled quickly and accurately. The pendulum should remain erected in its inverted position during such movements.

The proportional integral derivative (PID) controller though known to operate on a majority of industrial process and control applications, they exhibit poor performance in case the system under study is nonlinear and ill-defined. Moreover, conventional controllers require system model, which may not always be available. Researchers have thus employed newer approaches of genetic algorithms, fuzzy logic [1, 2], neural networks [3], or a combination of any of the above to solve this nonlinear control problem. A few of the earlier methods involved the use of various linearization techniques to account for the nonlinearities of the system model [4]. However, application of fuzzy logic for inverted pendulum control still by far remains most popular, as they do not require precise knowledge of the system parameters. Attempts are persistent in the research world to obtain a superior optimal controller, as choosing the correct set of rules and scaling factor is not an easy task in order to fine tune the fuzzy logic controller (FLC) [5]. A number of approaches have been proposed to implement hybrid control structures that combine the robustness of conventional controllers with the intelligence of fuzzy logic techniques to control the nonlinear systems. Performance of PI-type FLC for higher order nonlinear systems is not satisfactory due to large overshoot, excessive oscillation, and slower response. PD-type FLCs are suitable for a certain class of nonlinear systems [6, 7]. PID-type FLCs are rarely used due to the difficulties in generation of a complicated three-dimensional rule base [8].

As already stated, the inverted pendulum model is nonlinear in nature and its parameters may vary with time. Different operator defined fuzzy tuning schemes have been proposed earlier to control such complex and nonlinear systems [9–12]. This type of fuzzy-based tuning scheme usually decreases the overshoot, but the system response may be delayed for its additional rule base. To overcome this shortcoming, in this paper, we propose a non-fuzzy self-adaptive scheme for tuning of fuzzy PD-type controllers.

2 Inverted Pendulum Description

Figure 1 shows the schematic of an inverted pendulum moving on a track of length 1 m [13]. The cart has a shaft to which two pendulums are attached and are able to rotate freely. The cart is driven by a DC motor attached at the end of the rail. The force with which the cart is pulled is controlled by applying a voltage to the motor. This voltage is our control signal u. The two variables that are read from the pendulum (using optical encoders) are the pendulum angle θ and the cart position on the rail x. The system is interfaced to a personal computer by means of a data acquisition card and is driven by MATLAB-/Simulink-based real-time software. The equations of motion for the inverted pendulum system are given by Eqs. 1 and 2.

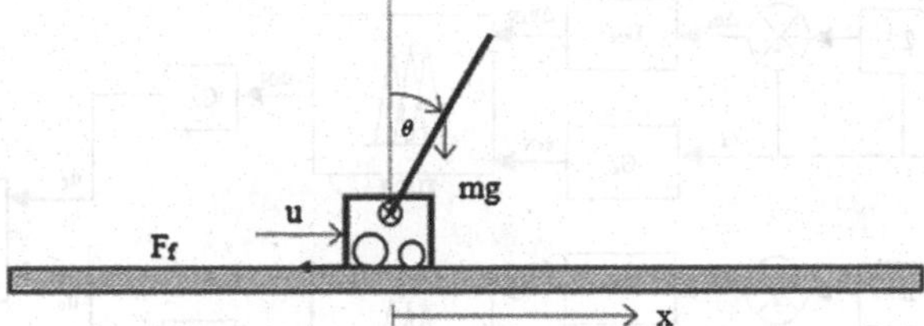

Fig. 1 Schematic of the inverted pendulum system

$$(M+m)x'' + kx' = u + (\text{mlsin}\theta)\theta'^2 - (\text{mlcos}\theta)\theta'' \quad (1)$$

$$(I+\text{ml}^2)\theta'' = \text{mglsin}\theta - (\text{mlcos}\theta)x'' - b\theta' \quad (2)$$

3 Controller Design

The proposed hybrid controller, for efficient control of cart position and inverted pendulum angle, is a PD-type fuzzy logic controller (FPDC). As already mentioned, the inverted pendulum control problem is divided into two separate control schemes. First is the swing up control, which allows the pendulum to reach the upright position with as minimal angular velocity as possible; and second is the inverted pendulum stabilization around the equilibrium point within a specified accuracy. Both these control objectives have to be attained simultaneously. A logical switch is used to change over between the swing up control mode and the inverted pendulum stabilization mode. The stabilization mode comes into effect once the pendulum reaches [−20°, +20°] of the final vertical upright position. The control action begins in the swing up control mode if initially the pendulum arms are hanging free. Here, zone detection signals in combination with a position controller act on the inverted pendulum system to achieve desired sway of the arms. Once in the stabilization mode, the system is acted upon by twin controllers, viz. position controller and angle controller (Fig. 2).

Each of the FLC used for performing the control action consists of five symmetrical triangular membership functions (MFs) of equal base and 50 % overlap, hence generating a set of $5^2 = 25$ fuzzy rules (given by Table 1). The inputs to the controller are the amount of deviation in the cart position or pendulum angle from the desired set point at a sampling instant k (denoted by e) and their corresponding change of error (denoted by Δe). Depending upon these two input parameters, each of the FLC is capable of generating a control voltage (denoted by u) that will effectively drive the system toward stability. The various input and output scaling factors play a role very similar to the gain coefficients in a conventional controller. The appropriate selection of these scaling factors is done based on the operator's

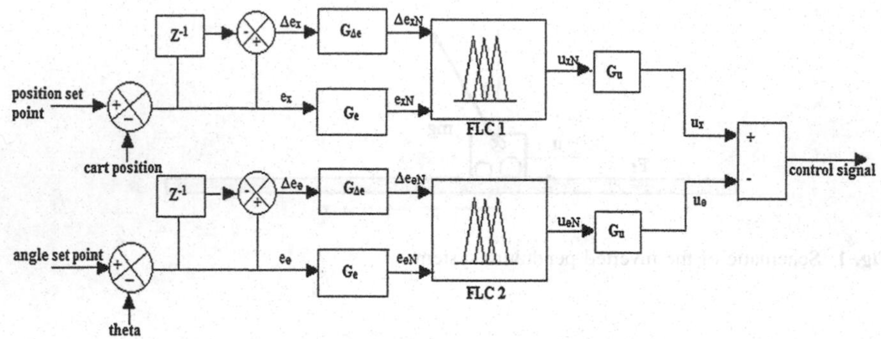

Fig. 2 Block diagram of FPDC twin controller for inverted pendulum stabilization

Table 1 Fuzzy rules for position and angle control

$\Delta e/e$	NB	NM	ZE	PM	PB
NB	NB	NB	NB	NM	ZE
NM	NB	NB	NM	ZE	PM
ZE	NB	NM	ZE	PM	PB
PM	NM	ZE	PM	PB	PB
PB	ZE	PM	PB	PB	PB

knowledge of the system under study to obtain the best performance of the control system. The relationships between scaling factors ($G_e, G_{\Delta e}, G_u$) and the input and output variables (e, Δe, ue, Δe, u) are given by Eqs. 4–6. The MFs of e_x, Δe_x, u_x for position controller are given by Fig. 3 and MFs of e_θ, Δe_θ, u_θ for angle controller are given by Fig. 4. Figure 5 shows the block diagram of FPDC for swing up control.

$$\Delta e(k) = e(k) - e(k-1) \tag{3}$$

$$e_N = G_e * e \tag{4}$$

$$\Delta e_N = G_{\Delta e} * \Delta e \tag{5}$$

$$u = G_u * u_N \tag{6}$$

The output scaling factor (G_u) should be determined very carefully for the proper implementation of control logic, since it is directly related to the stability of the system. Authors in [9–12] have proposed a self-tunable fuzzy-based inference system to adjust the gain G_u online according to the current states of controlled processes. A very similar scheme is applied here for tuning of the FPDC position controller which is expected to give a better swing up control by modifying the

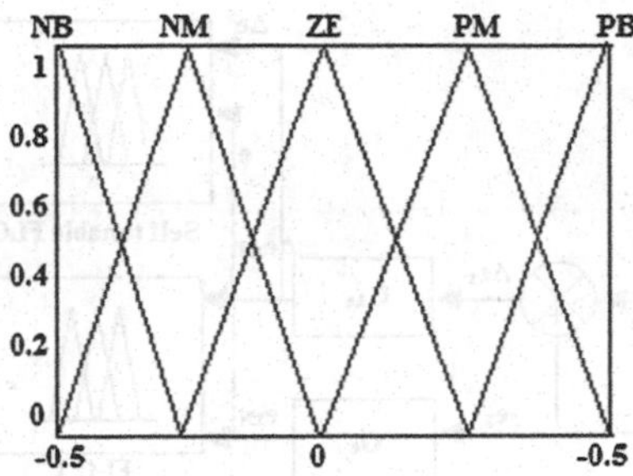

Fig. 3 MFs of $e_x, \Delta e_x, u_x$

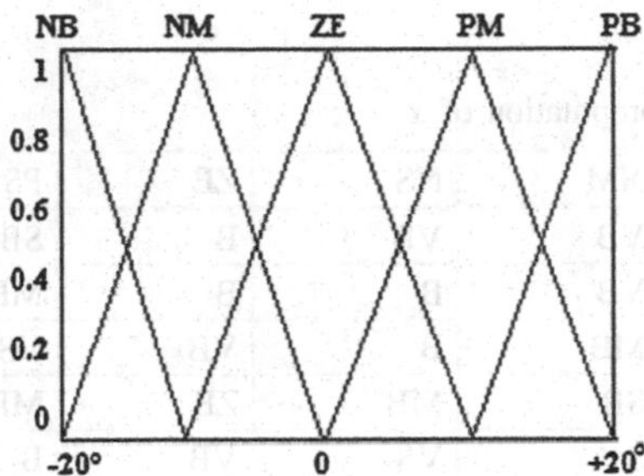

Fig. 4 MFs of $e_\theta, \Delta e_\theta, u_\theta$

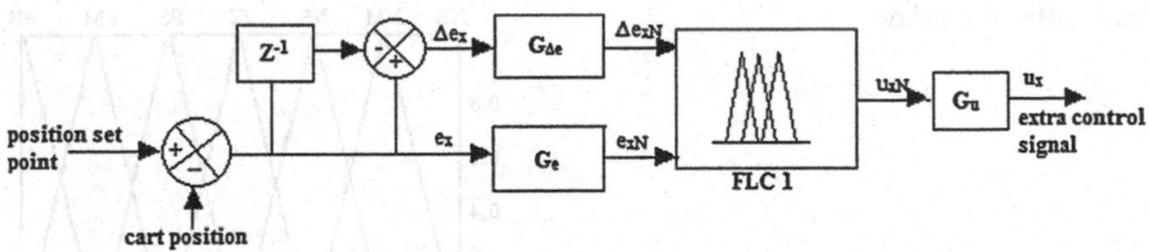

Fig. 5 Block diagram of FPDC for swing up control

controller gain by a factor α, which is now given by Eq. 7. This self-tuning mechanism for FPDC will be denoted as STFPDC in our discussion henceforth (see Fig. 6).

$$u = \alpha \cdot G_u * u_N \tag{7}$$

where α is a function of error(e) and change of error(Δe) of the system response. The fuzzy rule base for computation of the gain updating factor α consists of seven symmetrical triangular MFs of equal base and 50 % overlap, hence generating a set of $7^2 = 49$ fuzzy rules given by Table 2. The MFs of e, Δe, and α are given by Figs. 7 and 8, respectively.

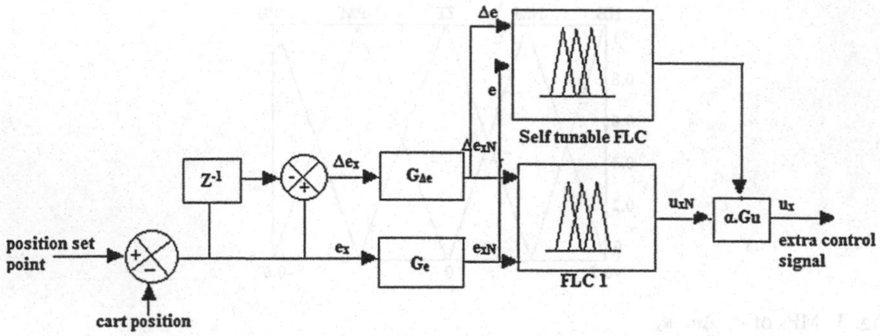

Fig. 6 Block diagram of STFPDC for swing up control

Table 2 Fuzzy rules for computation of α

$\Delta e/e$	NB	NM	NS	ZE	PS	PM	PB
NB	VB	VB	VB	B	SB	S	ZE
NM	VB	VB	B	B	MB	S	VS
NS	VB	MB	B	VB	VS	S	VS
ZE	S	SB	MB	ZE	MB	SB	S
PS	VS	S	VS	VB	B	MB	VB
PM	VS	S	MB	B	B	VB	VB
PB	ZE	S	SB	B	VB	VB	VB

Fig. 7 MFs of e and Δe

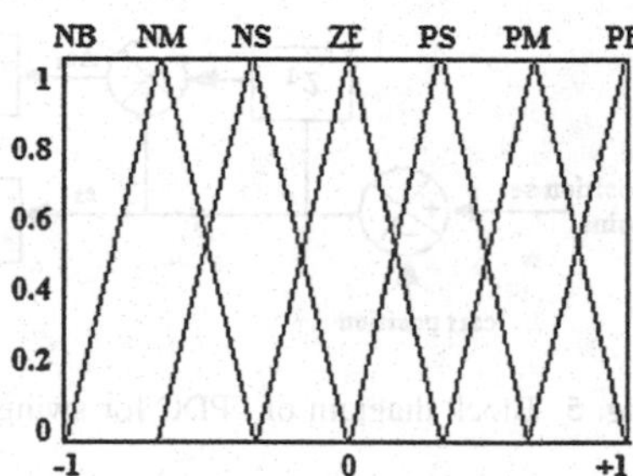

Fig. 8 MFs for computation of α

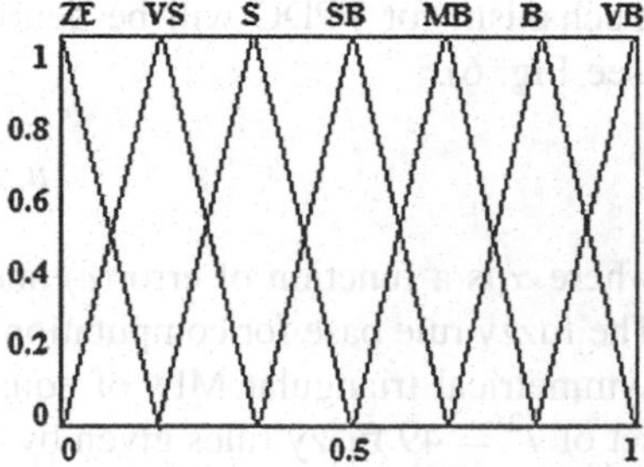

4 Proposed Tuning Scheme

Our idea here is to implement a certain non-fuzzy adaptive module to alter the output scaling factor online, to achieve a better performance. Such a scheme will reduce the design complexity by eliminating the need of multiple FLCs and the number of MFs while being able to exercise a similar control on the system [14]. The proposed scheme given by Fig. 9 is a self-adaptive PD-type fuzzy logic controller (SAFPDC) where a dynamic multiplying factor β is incorporated to adjust the controller gain (given by Eqs. 8 and 9).

$$u = \beta \cdot G_u * u_N \tag{8}$$

$$\beta = \delta \cdot e^{\mathrm{sqrt}(\Delta e)} \tag{9}$$

The value of β is computed at each sampling instant depending on the value of Δe which then automatically operates on the effective gain and thus on the system response. δ in the above expression is a factor that takes positive integer values to be decided by the operator, depending on the design specifications.

5 Results

In this section, we present the simulation results of each of the control scheme discussed above and perform a comparative study to illustrate the effectiveness of our proposed scheme over the others. Figure 10 shows the plot of pendulum cart position (in meter) against time (in second). The results show that the pendulum cart is stabilized within approximately 8 s in case of SAFPDC, while it takes much longer time in case of FPDC and STFPDC. The pendulum cart oscillation is also limited to 0.02 m above the set point during swing up. The various performance indices of the system response such as IAE, ISE, and ITAE have significantly reduced for SAFPDC when compared to STFPDC and FPDC. Figures 11, 12, and 13 show the plot of the inverted pendulum angle (in radian) against time

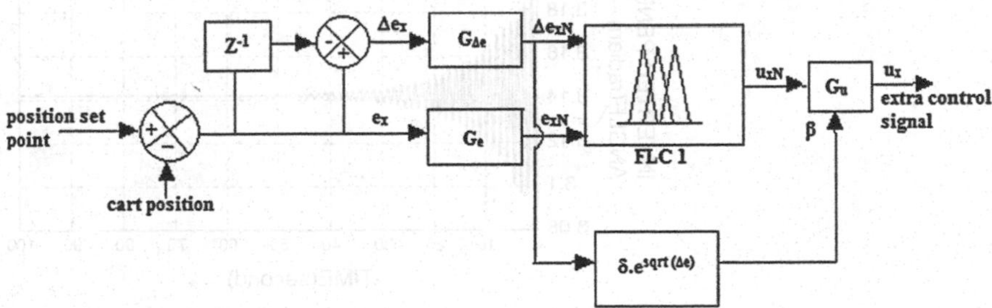

Fig. 9 Block diagram of SAFPDC for swing up control

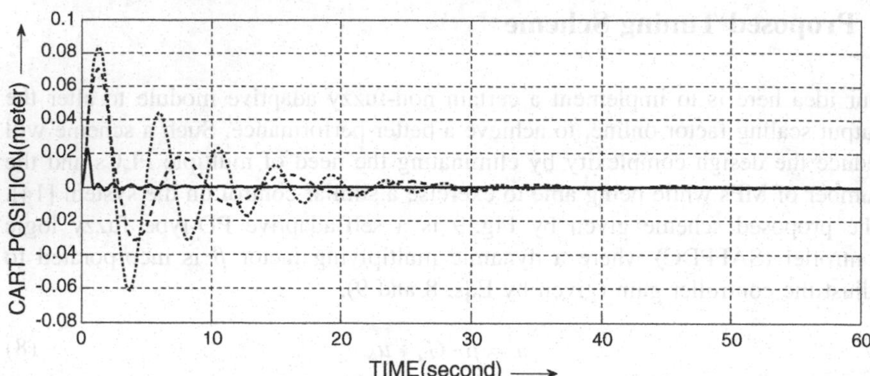

Fig. 10 Plot of pendulum cart position against time (*dotted* FPDC; *dashed* STFPDC; *solid* SAFPDC)

Fig. 11 Plot of inverted pendulum angle against time for FPDC

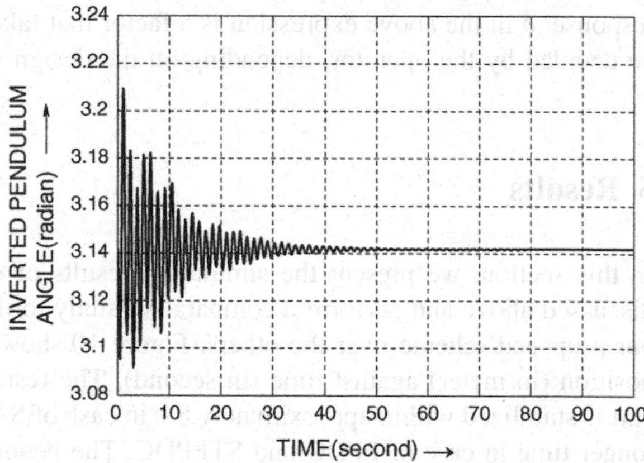

Fig. 12 Plot of inverted pendulum angle against time for STFPDC

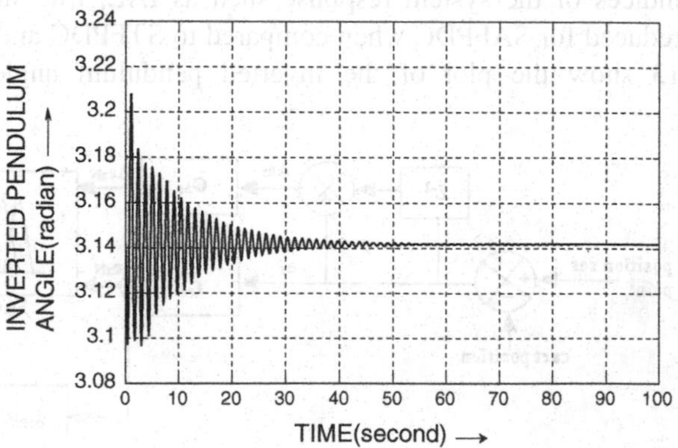

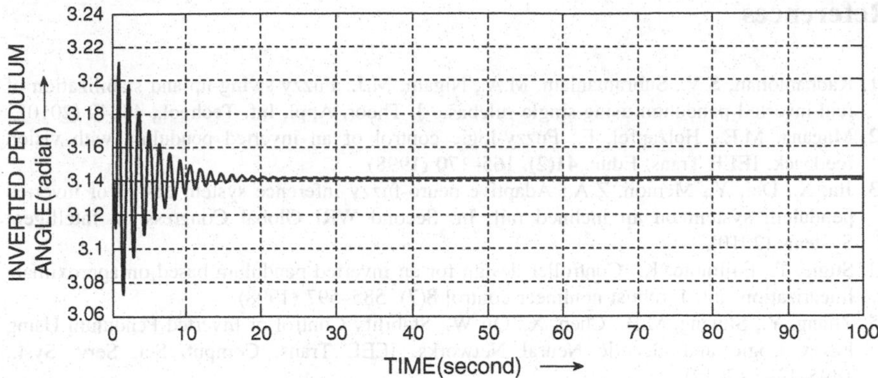

Fig. 13 Plot of inverted pendulum angle against time for SAFPDC

Table 3 Performance analysis of the proposed controllers for inverted pendulum control

Type of controller	t_p(s)	t_s(s)	IAE	ISE	ITAE
FPDC	1.3	28.33	4.4979	0.1721	33.3369
STFPDC	1.3	35.71	3.1523	0.0903	28.8326
SAFPDC	0.5	8.12	0.1751	0.0019	0.4249

(in second) for each of the control scheme separately for the sake of easy analysis. They very clearly reveal that in case of SAFPDC, the system takes the least time to reach and remain in its vertical upright position. The performance indices and the dynamic response characteristics have been tabulated in Table 3.

6 Conclusion

In this paper, we proposed a simple non-fuzzy self-adaptive scheme for PD-type FLCs. Here, the controller gain (output SF) has been updated online through a gain modifying parameter β defined on the change of error (Δe). Our proposed SAF-PDC scheme exhibited effective and improved performance compared to its conventional fuzzy counterpart and also outperforms controllers based on fuzzy self-tuning scheme. The proposed twin control scheme for inverted pendulum control reduces the computational complexity. By applying the proposed non-fuzzy tuning method and dual control scheme, the control objectives have been rightly achieved.

Acknowledgment The work was supported by the All India Council of Technical Education under Research Promotion Scheme (RPS File No. 8023/RID/RPS-24/2010-11).

References

1. Radhamohan, S.V., Subramaniam, M.A., Nigam, M.J.: Fuzzy swing-up and stabilization of real inverted pendulum using single rulebase. J. Theor. Appl. Inf. Technol., 43–49 (2010)
2. Magana, M.E., Holzapfel, F.: Fuzzy-logic control of an inverted pendulum with vision feedback. IEEE Trans. Educ. **41**(2), 165–170 (1998)
3. Jia, X., Dai, Y., Memon, Z.A.: Adaptive neuro-fuzzy inference system design of inverted pendulum system on an inclined rail. In: Second WRI Global Congress on Intelligent Systems (2010)
4. Sugie, T., Fujimoto, K.: Controller design for an inverted pendulum based on approximate linearization. Int. J. robust nonlinear control **8**(7), 585–597 (1998)
5. Zhang, Y., Shuang, M.B., Chen, X., Qi, W.: Stability Control of Inverted Pendulum Using Fuzzy Logic and Genetic Neural Networks. IEEE Trans. Comput. Sci. Serv. Syst., 1495–1498 (2012)
6. Huang, Y.W., Tung, P.C.: Fuzzy PD system in adaptive control systems having input saturation. J. Control Intell. Syst. **35**(3), 217–222 (2007)
7. Malki, H.A., Li, H., Chen, G.: New design and stability analysis of fuzzy proportional-derivative control systems. IEEE Trans. Fuzzy Syst. **2**, 245–254 (1994)
8. Pal, A.K., Mudi, R.K.: An adaptive fuzzy controller for overhead crane. In: IEEE Trans. Adv. Commun. Control Comput. Technol., 300–304 (2012)
9. Pal, A.K., Mudi, R.K.: Self-tuning fuzzy PI controller and its application to HVAC Systems. Int. J. Comput. Cogn. **6**(1), 25–30 (2008)
10. Mudi, R.K., Pal, N.R.: A robust self-tuning scheme for PI- and PD-type fuzzy controllers. IEEE Trans. Fuzzy Syst **7**(1), 2–16 (1999)
11. Pal, A.K., Mudi, R.K.: Speed control of DC motor using relay feedback tuned PI, fuzzy PI and self-tuned fuzzy PI controller. Control Theor. Inf. **2**(1), 24–32 (2012)
12. Pal, A.K., Mudi, R.K.: Development of a self-tuning fuzzy controller through relay feedback approach. In: Proceedings CCPE, Springer-Verlag (2011)
13. Feedback Instruments Ltd.: Manual on Digital Pendulum Control Experiments. In: Manual no. 33-936S Ed01 122006
14. Pal, A.K., Mudi, R.K.: An Adaptive PD-Type FLC and Its Real Time Implementation to Overhead Crane Control. Int. J. Emerg. Technol. Comput. Appl. Sci. **6**(2), 178–183 (2013)

Use of Machine Learning Algorithms with SIEM for Attack Prediction

E.T. Anumol

Abstract In the recent years, organizations face the ever growing challenge of providing security in the network infrastructure. An intrusion detection system is essentially a spruced up, intelligent variant of a firewall which does deep packet analysis which generate alerts but cannot predict multistep attacks. In this work, we propose an intrusion prediction system (IPS) with the extension of a commercial SIEM framework, namely open source security information management (OSSIM), to perform the event analysis and to predict future probable multistep attacks before they pose a serious security risk. Security information and event management (SIEM) framework affirms network protection by the correlation and management of network log files. Data mining techniques are used for processing of all normalized data from OSSIM and also for classification.

Keywords SIEM · OSSIM · Rapidminer · SVM · Network log files

1 Introduction

The amount of data in the industry is exploding. Similarly, the importance of these data is demanding higher security measures for them. Intrusion detection is the process of identifying that an intrusion has been attempted or has occurred. This necessity led to intrusion prediction system (IPS) which can alert even before the attack.

Attack prediction process is defined as the sequence of elementary actions that should be performed in order to recognize the attack strategy [1]. Recently in intrusion prediction community interest has been growing applying machine learning techniques to get high performance in execution time and classification

E.T. Anumol (✉)
TIFAC Core in Cyber Security, Amrita Vishwa Vidyapeetham, Coimbatore
Tamil Nadu, India
e-mail: anu.edayanikkattu@gmail.com

L.C. Jain et al. (eds.), *Intelligent Computing, Communication and Devices*,
Advances in Intelligent Systems and Computing 308,
DOI 10.1007/978-81-322-2012-1_24

accuracy. For collecting large amount of data for the training process, we are using the security information and event management (SIEM) system.

SIEM systems are collect logs from different sources such as Router/Switch, Firewalls, and IDS. SIEM system collects and combines network activity data, logs, security events, and external threat data into a powerful management dashboard that intelligently correlates, normalizes, and prioritizes them and this way greatly improving the remediation and response times and greatly enhancing the effectiveness of the system. The main objective of SIEM system was to reduce the high false positive rate and centralize the event analysis and produce an effective report by the correlation process.

2 Related Work

Different data mining algorithms were applied on the intrusion datasets to predict the attack risk of the network environment and compare their performances [2]; they were evaluated based on the predictive accuracy and ease of learning.

Intrusion detection system's accuracy was improved by correlating data among different logs [3, 4]. Different attacks were reflected in different logs and argued that some attacks were not evident when a single log was analyzed. This proposed system can identify attacks but cannot be predicted by examining log from the single source.

In [5] introduces a method for statistical approach to classify and identify distributed denial of service (DDoS) attacks using UCLA dataset. Then, it investigates the procedure of DDoS attacks; apply the K-Nearest Neighbor (K-NN) by selecting variables based on these features of the attack to classify the network status into each phase of DDoS attack.

3 Frame Work

In today's world, attacks are most commonly detected after they have occurred. So the idea of prediction is introduced where the training data will be generated based on the previously occurred events, and using real time data, we hope to predict attacks before they occur [6]. In the proposed system, SVM is used as the Machine Learning algorithm for classification.

3.1 OSSIM

Open source security information management (OSSIM) is an open source SIEM system which integrating a selection of tools such as Snort, OSSEC, and Ntop Ngios. OSSIM adopts several protocols: Rsyslog, FTP, SQL, SAMBA, OSSEC, and among others.

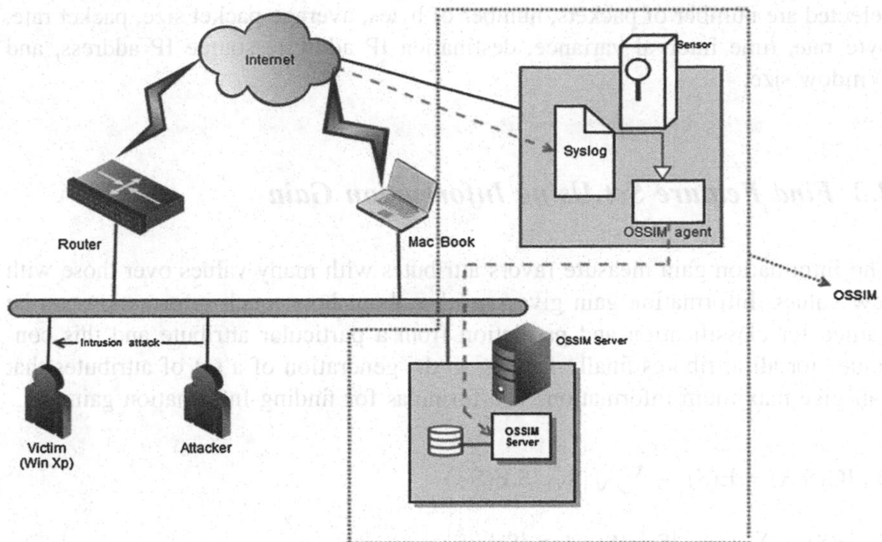

Fig. 1 How the OSSIM sensors collect data

OSSIM consists of five different key components: the server, agent, sensors, database, and framework.

Figure 1 shows how the OSSIM sensors collect data. The attack scenario is created to generate dataset for the attack. A number of network attacks can be simulated but the proposed system focuses on distributed DDoS, Smurf, and IP-Sweep attacks only. The attacker performs an intrusion attack on the victim and the corresponding log will be stored on the Syslog of the OSSIM agent. Our motive is to create features from these logs and developed some higher level features that distinguish normal from anomalous. The actions of OSSIM can be described in the following steps

1. Extract information from the events generated by the source devices deployed over the network infrastructure
2. Apply a policy and execute the correlation process to perform risk assessment
3. Raise an alert message to the administrator.

3.2 Dataset and Feature Selection

The training data is created by analyzing a network and capturing the network traffic. The attack scenario is created to generate datasets for attack. The features selected are number of packets, number of bytes, average packet size, packet rate, byte rate, time interval variance, destination IP address, source IP address, and Window size.

3.3 Find Feature Set Using Information Gain

The information gain measure favors attributes with many values over those with few values. Information gain gives an idea about how much information can be gained for classification and prediction from a particular attribute and this continues for all attributes finally leading to the generation of a set of attributes that can give maximum information. The formulas for finding Information gain are

4. $IG(S,A) = E(S) \; - \sum_{vV} |S_V|/S.E(S_V)$

5. $E(S) = \sum_{c \in C} -|S_c|/|S| * \log_2|S_c|/|S|$

3.4 Classification and Prediction

- Collects training data from different sources using SIEM and passed over to the SVM algorithm for classification
- The model for each attack based on the selected features are patterned
- Compares features of test data with the model of attack generated
- On comparison supporting to results attacks are predicted
- The flow chart for the propose system is (Fig. 2).

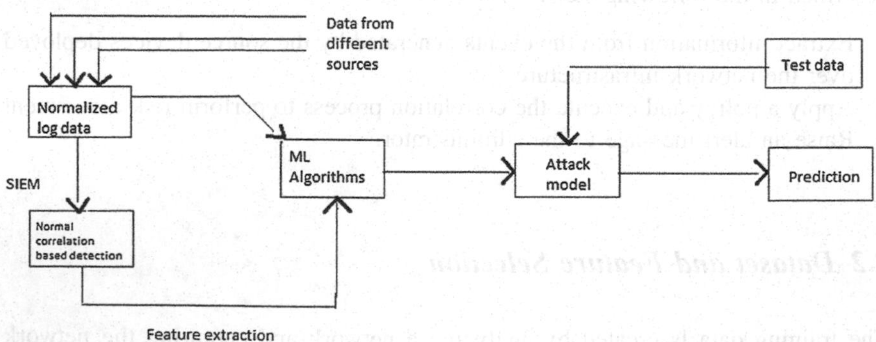

Fig. 2 The flow chart for the propose system

4 Conclusion

Network log files are essential for analysis in finding as any issues in complex computer systems. The OSSIM system was used for collecting log files from various kinds of sources. Various attributes are created out of which the best features with maximum gain ratio are selected for classification. The dataset is first passed through SVM. This training data serves as the model for prediction of test data. The system then classifies the test data based on what it has learnt from the training data.

References

1. Liu, S., Zhou, Z., Zhan, M.: Toward intelligent intrusion prediction for wireless sensor networks using three-layer brain-like learning. Int. J. Distrib. Sens. Netw. **2012**(243841), 14 (2012)
2. MeeraGandhi, G.: Machine learning approach for attack prediction and classification using supervised learning algorithms. Int. J. Comput. Sci. Commun. 1(2), July–December (2010)
3. Abadyz, C., Taylory, J., Senguly, C.: Log Correlation for Intrusion Detection: A Proof of Concept
4. Zopc, A.R., Vidhate, A.: Data minding approach in security information and event management. J. Future Comput. Commun. (2013)
5. Oo, T.T., Phyu, T.: A statistical approach to classify and identify DDoS attacks using UCLA dataset. Int. J. Adv. Res. Comput. Sci. Technol. (IJARCET) 2(5) (2013)
6. Hu, W., Liao, Y., Vemuri, V.R.: Robust Anomaly Detection Using Support Vector Machines

Performance Comparison of Single-Layer Perceptron and FLANN-Based Structure for Isolated Digit Recognition

Aryapriyanka Samal, Jagyanseni Panda and Niva Das

Abstract Isolated handwritten digit recognition is still a difficult task for a computer although a lot of research has been done on this topic for over two decades. The main difficulties arise due to a large number of variations and styles of digit patterns. Literature reveals that a great amount of research has been made to solve the problem. A variety of feature selection methods and classification methods have been proposed to improve the performance of the recognition system. In this paper, we have studied the functional link artificial neural networks (FLANN) and single-layer perceptron network for the task of classification, where the aim is to classify the input numeral as one of two classes. In contrast to multilayer perceptron network which increases the complexity of the network, we have compared the performance of two single-layer feedforward structures for the task of digit recognition. Using the functionality-expanded features, FLANN overcomes the nonlinearity nature of the problem which is commonly encountered in single-layer perceptron. Experimental results demonstrate that on a database of 30 digit patterns written by 30 people, FLANN-based classifier exhibits a recognition rate of 90 % when compared with perceptron-based classifier which exhibits only 30 % accuracy.

Keywords ANN · FLANN · Single-layer perceptron · Multilayer perceptron · Recognition

A. Samal (✉) · J. Panda · N. Das
Department of Electronics and Communication Engineering,
ITER, Sikha 'O' Anusandhan University, Odisha, India
e-mail: aryasamal@gmail.com

J. Panda
e-mail: jagyansenipanda@yahoo.co.in

N. Das
e-mail: nivadas@gmail.com

L.C. Jain et al. (eds.), *Intelligent Computing, Communication and Devices*,
Advances in Intelligent Systems and Computing 308,
DOI 10.1007/978-81-322-2012-1_25

1 Introduction

Over the years, computerization has taken over large number of numeral operations, and one such example is offline handwritten numeral recognition. Automatic handwritten English digit recognition still remains a challenging task for a computer although a lot of research has been done on this topic. The aim of handwritten digit recognition system is to classify the input digit as one of many classes. Automatic digit recognition system usually follows two steps: feature analysis and pattern classification.

Feature analysis involves extraction of relevant information for pattern classification from the input digit. The pattern classification task labels the digit as one of many classes using the class models. Considerable research work has been carried out in this area, and various methods have been proposed for the classification of handwritten digits. The methods include principal component analysis (PCA), support vector machine (SVM), nearest neighbor classification, neural computing, and fuzzy-based approaches [1, 2]. Literature reports some work related to recognition of handwritten digits of Indian scripts [3–6]. Technique for recognition of handwritten Hindi numerals based on modified exponential membership function fitted to fuzzy sets is presented in [7].

Literature reveals that still a lot of scope exists to design a robust system for recognition of handwritten digits. Among the various approaches, artificial neural network-based methods are very attractive since these need no precise mathematical model and have the learning ability. In this paper, we have proposed two single-layer ANN structures: functional link artificial neural networks (FLANN) and single-layer perceptron for the automatic classification of handwritten digit. In the first stage of recognition task, feature selection is done. Gradient features have been considered for this purpose. These features are then used to train the two different single-layer neural networks.

This chapter is organized into five sections. Sect. 2 presents a brief overview of data collection and feature extraction method. Section 3 deals with the classification task using two single-layer ANN structures. The simulation results are given in Sect. 4. Section 5 presents the conclusion.

2 Data Set and Feature Extraction

Data set of English handwritten numerals 0–9 is created by collecting the handwritten documents from writers. Data collection is done on a sheet specially designed for data collection. Writers from different professions were chosen including students, clerks, teachers, and vendors and were asked to write the numerals. No constraints were imposed on the use of ink or pen except that they have to write the numerals in the boxes of the sheets provided to them (Fig. 1).

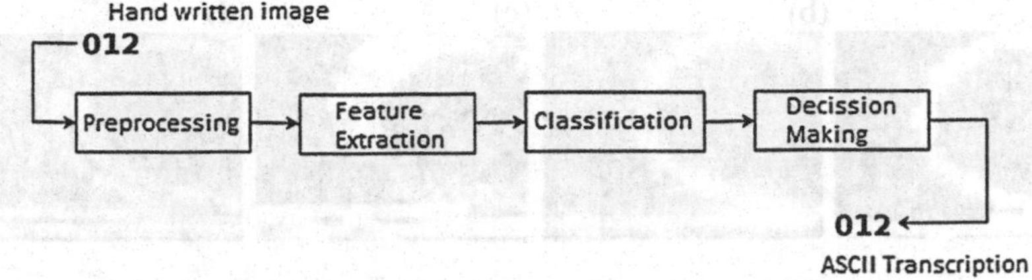

Fig. 1 General architecture of handwritten character recognition system

The feature extraction process consists of procedures for gradient calculation, feature vector generation, and dimension reduction of the feature vector. Each procedure is described in the succeeding subsections.

2.1 Calculation of Gradient [13]

A grayscale image is generated from an input binary image, and the gradient is calculated as described below.

1. Size normalization is applied to a binary character image so that the image has standard width and height (Fig. 2a).
2. Mean filter of size 2 × 2 is repeatedly applied r times to obtain a grayscale image (Fig. 2b).
3. The grayscale image is normalized so that the mean and the maximum of the gray scale are 0 and 1, respectively.
4. Roberts filter [8, 9] given by Eqs. (1) and (2) is applied to each pixel $g(i;j)$ of the normalized image to calculate the gradient (Fig. 2c and d).

$$\Delta u = g(i+1, j+1) - g(i, j)$$

$$\Delta v = g(i+1, j) - g(i, j+1)$$

$$\text{Direction: } \theta(i, j) = \tan^{-1}\left(\frac{\Delta v}{\Delta u}\right). \quad (1)$$

$$\text{Strength:} f(i, j) = \sqrt{\Delta u^2 + \Delta v^2} \quad (2)$$

2.2 Generation of Feature Vector [13]

A feature vector is composed of the strength of gradient accumulated separately in different directions as described below:

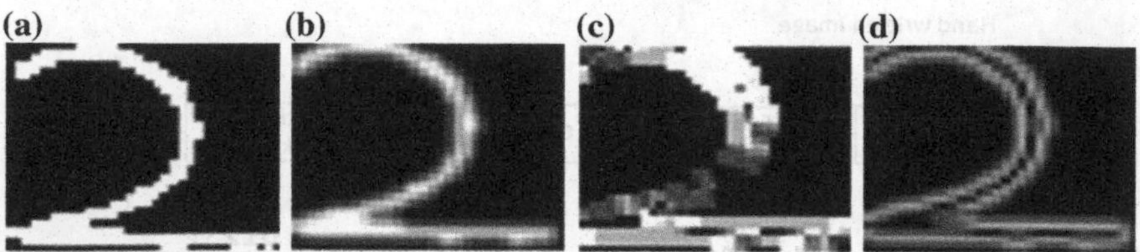

Fig. 2 Normalized binary image (**a**), grayscale image (**b**), direction of gradient (**c**), and strength of gradient (**d**)

1. The direction of gradient detected is quantized to 32 levels with $\frac{\pi}{16}$ interval.
2. The normalized character image is divided into 81 (9 horizontal $\times$ 9 vertical) blocks.
3. The strength of the gradient is accumulated separately in each of 32 directions, in each block, to produce 81 local spectra of direction.
4. The variable transformation ($y = x$^0.4) is applied to make the distribution of the features Gaussian-like [10, 11].

2.3 *Dimension Reduction*

The required processing time and storage can be reduced by the dimension reduction employing the PCA (KL transform). The PCA is a typical dimension reduction procedure based on the orthonormal transformation which maximizes the total variances and minimizes the mean square error due to the dimension reduction. It is shown that the dimensionality can be reduced to $\frac{1}{4}$ without sacrificing the recognition accuracy in handwritten numeral recognition employing the feature vector of size 400 detected from the gradient of the gray scale [12, 13].

3 Classification

ANN has been considered as one of the powerful classifiers for character and digit recognition. The most common architecture of ANN is the multilayer feedforward network. But due to its multilayered structure, the training speeds are typically much slower as compared to other single-layer feedforward networks [14]. Hence, in this paper, we have proposed a FLANN-based scheme and a single-layer-perceptron-based scheme for the recognition task.

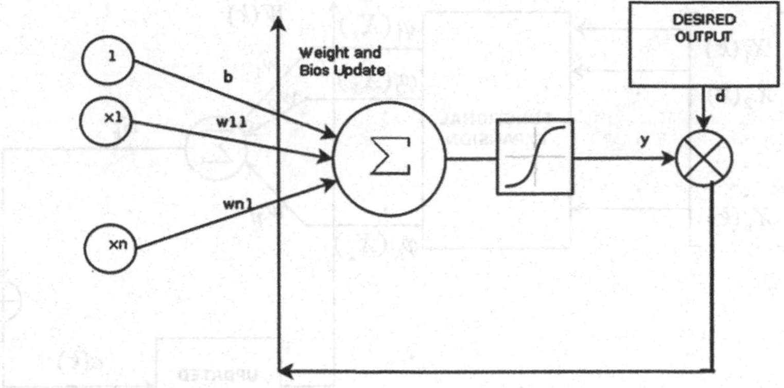

Fig. 3 Single-layer perceptron network

3.1 Single-Layer Perceptron Structure

$$x_j = \sum_{l=1}^{k} w_{ij} y_i \tag{3}$$

The basic structure of a single-layer perceptron is given in Fig. 3. It can be viewed as a single "neuron" with multiple inputs that generates an output signal. The value of this output depends on the relative strengths of weighted input signals. The perceptron output can be expressed as follows:

$$y(n) = f\left[w^T(n)x(n) + b\right] \tag{4}$$

where $w(n) = [w_1(n), \ldots w_N(n)]$ is the adaptive weight vector, $w(n) = [w_1(n), \ldots w_N(n)]^T$ is the input signal vector, and b is the bias term. The most commonly used activation functions are sigmoid and hard limiter. The perceptron weights are updated according to

$$w(n+1) = w(n) + \eta[d(n) - y(n)]x(n) \tag{5}$$

where η is the learning rate parameter less than 1. $d(n)$ is the desired output or target.

3.2 FLANN Structure

The FLANN approach removes the hidden layer from the ANN architecture to reduce the architectural complexity. It is a single-layer ANN capable of forming arbitrarily complex decision regions by generating nonlinear decision boundaries.

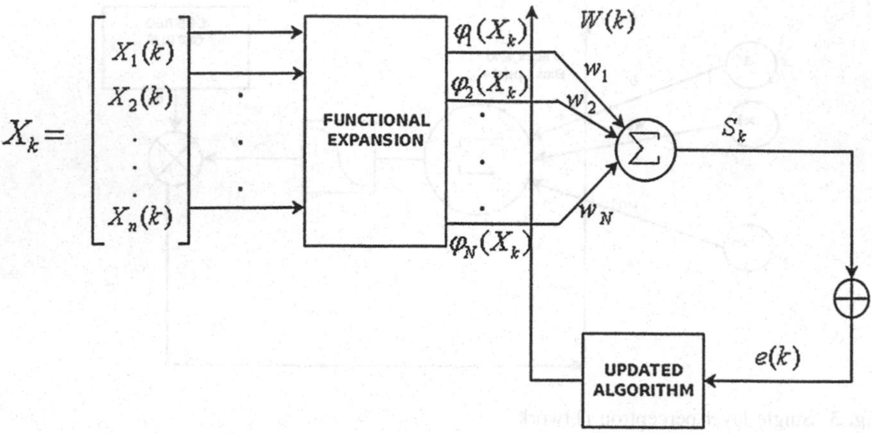

Fig. 4 FLANN structure

To bridge the gap between the linearity in the single-layer neural network and the highly complex multilayer neural network, the FLANN is used. The FLANN uses a single-layer feedforward neural network, and to overcome the linear mapping, it functionally expands the input vector. The basic structure of FLANN is given in Fig. 4. The functional expansion block makes use of a functional model comprising of a subset of orthogonal sine and cosine basis functions. The functional expansion effectively increases the dimensionality of the input vector, and hence, the hyperplanes generated by the FLANN provide greater discrimination capability in the input pattern space. These expanded input patterns are then fed to the single-layer neural network, and the network is trained to obtain the desired output (Figs. 5, 6, 7 and 8).

4 Simulation Result

To test the performance of the proposed FLANN-based scheme and single-layer-perceptron-based scheme, simulations were carried out on a MATLAB 2011b platform. Hundred isolated handwritten English digits from 0 to 9 were collected for the purpose of training and testing. First, the gradient feature was calculated using the procedure in Sect. 2. For each data set, a feature vector consisting of 2,592 features was generated. Then, using PCA, the features were reduced to 70 numbers. The feature vector comprising of 70 features was used to train both the ANN-based structures.

For the single-layer perceptron, each training set comprises of 70 features as input and appropriate target. The 70 number of weights are initialized with small random values between 0 and 1. Sigmoidal activation function is used as the nonlinearity. Training was carried out for 1,000 iterations, and learning curve for

Fig. 5 Data set of English numerals

Fig. 6 Data set after normalization

Fig. 7 Direction of gradient

Fig. 8 Strength of gradient

the same was plotted in Fig. 9. After training, the error reduces to marginal value and convergence is achieved. Thirty data sets were chosen for the purpose of testing. It was observed that out of 30 isolated digits, only 8 digits have been recognized correctly, showing an accuracy of 30 % only. The single-layer ANN being linear in nature exhibits poor recognition performance.

In the FLANN-based scheme, the 70 number of features act as input to the FLANN structure, which get expanded to 350 inputs. The 350 number of weights were initialized with small random values between 0 and 1. No activation function was used. The weight updation follows the same procedure as in case of single-layer perceptron. Training was carried out for 1,000 iterations, and learning curve was plotted accordingly in Fig. 10. The learning curve shows much quicker convergence than that in case of single-layer perceptron. Testing was carried out for 30 data sets. Out of 30 isolated digits, 27 digits were identified correctly, thereby demonstrating an accuracy of 90 %. Though the database was small, the FLANN-based scheme could produce 90 % correct result which clearly depicts the superior recognition performance of the scheme over the perceptron-based scheme.

The testing results for the perceptron-based scheme and the FLANN-based scheme are shown in Tables 1 and 2, respectively.

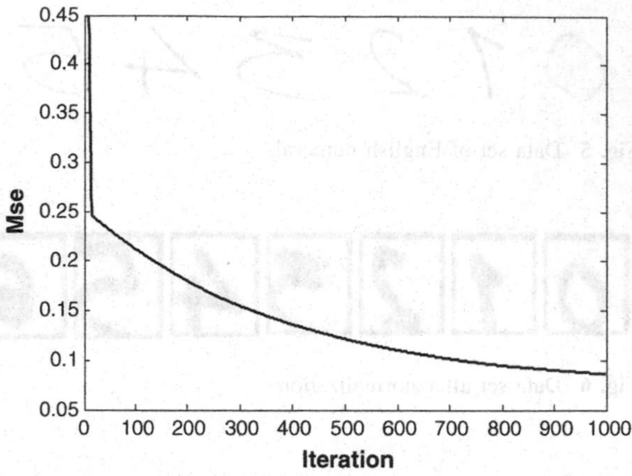

Fig. 9 Learning curve for single-layer perceptron network

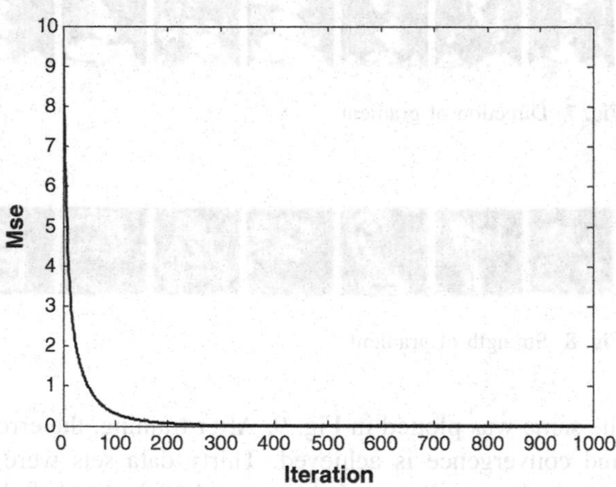

Fig. 10 Learning curve for FLANN

Table 1 Testing results of FLANN

Collected numerals	Pattern used in testing	Results	
		Classified	Not classified
0	3	2	1
1	3	3	
2	3	2	1
3	3	3	
4	3	3	
5	3	2	1
6	3	3	
7	3	3	
8	3	3	
9	3	3	

Table 2 Testing results of single-layer perceptron network

Collected numerals	Pattern used in testing	Results	
		Classified	Not classified
0	3	1	2
1	3		3
2	3		3
3	3	1	2
4	3	1	2
5	3	1	3
6	3	1	2
7	3	1	2
8	3		3
9	3	2	1

5 Conclusion

In this paper, we have considered the gradient features and used these features for the recognition of isolated handwritten digits. These features are then reduced using the PCA. The reduced features are then used for training the single-layer perceptron network and FLANN-based structure adopted for the classification task. The FLANN model functionally expands the given set of inputs which are then fed to the single-layer feedforward neural network. The network is trained as in the single-layer perceptron network, and the weights are adapted using simple LMS algorithm. Different sets of data were prepared for training and testing. Both single-layer networks are trained with 70 blocks of data. Learning curves have been drawn for both classifiers. The learning curves show that convergence is much faster and better in case of FLANN compared to single-layer perceptron. Both networks were tested for 30 sets of data. The FLANN-based classifier could identify 27 digits correctly, while the single-layer-perceptron-based network could identify only 8 digits. The experimental results confirm the superior recognition performance of FLANN-based classifier in comparison with the perceptron-based network. We also claim that the recognition accuracy will certainly improve if the database size is considerably increased. The improved performance of FLANN-based structure for the purpose of digit recognition also opens up scope for study of complex FLANN structures for achieving higher accuracy.

References

1. Lecun, Y., Bottou, L., Bengio, Y., Haffner, P.: Gradient based learning applied to document recognition. Proc. IEEE **86**(11), 2278–2324 (1998)
2. Vapnik, V.N.: The Nature of Statistical Learning Theory. Springer, Berlin (1995)
3. Pal, U., Chaudhuri, B.B.: Automatic recognition of unconstrained off-line Bangla Handwritten Numerals. In: Proceedings of Advances in Multimodal Interfaces. Lecture Notes on Computer Science (LNCS-1948), pp. 371–378. Springer, Berlin (2000)
4. Tripathy, N., Panda, M., Pal, U.: A system for Oriya handwritten numeral recognition. In: Smith, E.H.B., Hu, J., Allan, J. (eds.) Proceedings of SPIE, vol 5296, pp 174–181
5. Wen, Y., Lu, Y., Shi, P.: Handwritten Bangla numeral recognition system and its application to postal automation. Pattern Recogn. **40**(1), 99–107 (2007)
6. Rajput, G.G., Hangarge, M.: Recognition of isolated handwritten Kannada numerals based on image fusion method. In: Pattern Recognition and Machine Intelligence, LNCS 4815, pp. 153–160 (2007)
7. Bajaj, R., Dey, L., Choudhuri, S.: Marathi Numeral Recognition by Combining Decision of Multiple Connectionist Classifiers, vol. 27(1), pp. 59–72 (2002)
8. Roberts, L.G.: Machine perception of three-dimensional solids. In: Tippett, J.T. (ed.) Optical Electro-Optical Processing of Information, pp. 159–197. MIT Press, Cambridge (1965)
9. Russ, J.C.: The Image Processing Handbook, 2nd edn. CRC Press, Boca Raton (1995)
10. Wakabayashi, T., Tsuruoka, S., Kimura, F., Miyake, Y.: Increasing the feature size in handwritten numeral recognition to improve accuracy, systems and computers in Japan (English edition). Scripta. Technical. **26**(8), 35–44 (1995)
11. Fukunaga, K., Rheinboldt, W., Siewiorek, D. (eds.): Introduction to statistical pattern recognition, 2nd edn. Academic Press, New York (1990)
12. Kimura, F., Wakabayashi, T., Miyake, Y.: On feature extraction for limited class problem. In: Proceedings of the 13th ICPR, vol. II, pp. 191–194 (1996)
13. Shi, M., Fujisawa, Y., Wakabayashi, T., Kimura, F.: Handwritten numeral recognition using gradient and curvature of gray scale image. Pattern Recogn. **35**, 2051–2059 (2002)
14. Patra, J.C., Pal, R.N.: A functional link artificial neural network for adaptive channel equalization. Sig. Process. **43**, 181–195 (1995)

Computational Techniques for Predicting Cyber Threats

Ekta Gandotra, Divya Bansal and Sanjeev Sofat

Abstract With the increasing usage of Internet and computing devices with network competence, the Internet crimes and cyber attacks are increasing exponentially. Most of the existing detection and protection systems rely on signature based methods and are unable to detect sophisticated and targeted attacks like advanced persistent threats (APTs). In order to protect Internet users and cyber infrastructure from various threats, proactive defense systems are required, which have the capability to make intelligent decisions in real time. This paper reviews various computational techniques used in the literature for predicting cyber threats. It also highlights the challenges, which can be explored by researchers for future studies.

Keywords Cyber attacks · Cyber threats · Prediction · Intelligence

1 Introduction

The adoption of IT-based products/services in all the sectors demands for secure cyberspace. Existing detection and protection systems like firewalls, intrusion detection systems are unable to detect the new generation threats, which are targeted, persistent, stealthy, and unknown [1]. Recent cyber attacks like Aurora cyber-operation [2], Stuxnet against Iranian Nuclear Program [3], information-

---

E. Gandotra (✉) · D. Bansal · S. Sofat
Department of Computer Science and Engineering, PEC University of Technology, Chandigarh, India
e-mail: ekta.gandotra@gmail.com

D. Bansal
e-mail: divya@pec.ac.in

S. Sofat
e-mail: sanjeevsofat@pec.ac.in

L.C. Jain et al. (eds.), *Intelligent Computing, Communication and Devices*, Advances in Intelligent Systems and Computing 308,
DOI 10.1007/978-81-322-2012-1_26

stealing virus called "Dexter" [4], and many others are intended to emerge cyber warfare at various levels and lead to a large amount of losses in various countries. According to a report published by Internet Crime Complaint Centre (IC3) [5], in 2012, around 289,874 consumers complaint the IC3 with a loss of about 5,254 million dollar, which is 8.3 % up in reported losses since 2011.

In order to protect the Internet users and cyber infrastructure from various threats, adaptable and robust cyber defense systems are needed that have the ability to make intelligent decisions for detecting and forecasting various threats/attacks. The information so obtained can be shared with Computer Emergency Response Teams (CERTs) or other enterprises in real time so that they have better awareness of latest cyber threats and can take preventive measures to block them before they actually cause the damage. This paper aims to present the various computational techniques used in the literature for predicting cyber threats.

2 Prediction Techniques

A number of studies have been carried out for predicting cyber attacks/threats in last few years and can be categorized broadly into two types: statistical and algorithmic [6]. This section provides a description of each of the techniques along with examples from the literature.

2.1 Statistical Modeling

It is the most widely used method and employs variants of ordinary least square regression, logistic regression, time-series approaches, auto regression, etc. Classical time-series models use time domain method to predict future value from some combination of the past values, usually with a focus on reducing the systematic error in the model to white noise [7]. Some of the standard time-series forecasting methods are: moving average (MA), exponential smoothing (ES), exponential moving average (EMA), extrapolation, auto-regressive moving average (ARMA), etc. The details on these methods can be found in [7]. These are able to forecast intrusions after observing significant changes in values like traffic volume while attacks are launched. This approach is useful in doing short-term predictions and can be combined with probabilistic approaches or data mining techniques.

Park et al. [8] proposed a forecasting mechanism called FOrecasting using REgression analysis (FORE) through a real-time analysis of the randomness in the network traffic. They claim that their method can take action against unknown worms 1.8 times faster than the early detection mechanism.

Pontes et al. [9] presented a collaborative architecture of IDS with prediction approaches. Their forecasting system makes use of five techniques: simple MA

(SMA), exponential weighted MA (EWMA), combined SMA, combined EWMA, and Fibonacci sequence. In order to reduce a huge amount of alerts (false positives), they [10] proposed a two-stage system making use of multi-correlation for improving forecasting. In the first stage, an event analysis system (EAS) is employed for making multi-correlation between alerts from an IDPS with the logs of operating system. In the second stage, the forecasting techniques are applied on the data generated by EAS.

Fachkha et al. [11] proposed a distributed denial of service (DDoS) forecasting model for predicting (within minutes) the attacks' impact features like intensity rate and size. Ongoing DDoS attack is understood to recognize the similar situations in future for predicting short-term trends. A number of forecasting techniques namely MA, weighted MA, ES, and LR are used on three real DDoS case studies.

Watters et al. [12] suggested that in the field of cyber security, a predictive model should be used as a basis for policy making. They proposed a model named as Cyber Attacker Model Profile (CAMP) for analyzing the ethnographic properties of cyber crime, which in turn aims to explain a particular profile of attackers in qualitative terms. They used their approach to find the relationship between various social, economic, and demographic factors in the Baltic States and Eastern Europe using regression and correlation analysis.

2.2 Algorithmic Modeling

Algorithmic computational modeling includes probabilistic modeling, data mining, and machine learning approaches [6].

2.2.1 Probabilistic Modeling

This method provides an insight on the risk associated with a threat on probability scale [13]. Out of various technologies belonging to this modeling approach, the most widely used in network security includes Bayesian and Markov Chain method. The Bayesian method works by calculating posterior probability of specific events from prior probabilities obtained from historical data. This method is able to deal with complex distributions in network traffic and its results are easy to interpret but the problem with this method is that it is not easy to obtain prior distributions of a normal and an attack state. The computations involved in Markov chain model are very simple. However, there is apprehension in constructing the state profile in complex systems as all transition probabilities between possible states need to be calculated.

Wu et al. [14] suggested a cyber attacks prediction model based on Bayesian network. In addition to the vulnerabilities in the network that is captured by using attack graph, three environment factors (i.e., usage condition of the network, the value of assets in the network, and the attack history of the network) are considered. After incorporating these factors with attack graph, the attack probability of each node is computed using Bayesian network probability algorithm.

Kim et al. [15] pointed out that the threat trend is not directly revealed from the time-series data because it is normally implicit in its nature. They proposed a model for cyber threat trend analysis by making use of hidden Markov Model as well as incorporating the environmental information. This model is examined over a case study of Agobot and Mybot and found to be adaptive to dynamic environments.

Man et al. [16] proposed a method that makes use of ARMA and Markov Model for predicting network security situations. The prediction results of both the models are combined together with appropriate weight values to optimize the prediction.

Cheng [17] proposed an intrusion prediction technology based on Markov chain. A dynamic load-balancing algorithm is used in this prediction model, which can avoid packet loss and false negatives in high-performance network while handling heavy traffic loads in real-time.

Lim et al. [18] believed that the main pre-symptoms of cyber threats are activity and propagation of botnets. They proposed a prediction model capable of estimating the degree of botnet-based threats by monitoring their size, activity, and propagation.

Fava et al. [19] pointed out that attack projection usually rely on a priori network knowledge and system configurations and thus not suitable for diverged and changing nature of system settings. They proposed a variable-length Markov Model (VLMM), capable of capturing past and current attack sequences and can predict the likely future actions without having specific network information. They used the simulation results to demonstrate the performance of VLMM predictor.

2.2.2 Data Mining and Machine Learning

This method has been widely used in the areas like weather forecasting, earthquake prediction, stock market predictions, etc. It relies on extracting useful information and patterns from the large data sets [20]. The relationships so obtained can be exploited for proactive decision making. The computational complexity associated with data mining methods is usually high, and it is difficult to understand the current network security situation when an alarm is issued.

Thonnard and Dacier [21] proposed a data mining methodology to discover actionable knowledge related to Internet threats. The method consists of two parts: In the first part, an unsupervised clique-based clustering is applied to complex patterns derived from data for finding the groups of highly similar patterns with respect to single property each time. In the second step, the different sets of cliques are combined to form the groups called concepts (having common similarity pattern), which can best describe the real-world phenomenon to have better insight into the emerging Internet threats and attacks. They validated their work on a large data set collected through a global distributed honeynet.

Farhadi et al. [22] proposed a correlation intrusion attack prediction system consisting of two major components. In the first component, an attack scenario extraction algorithm (ASEA) is introduced for mining the alert stream for attack scenarios. The second component involves using hidden Markov Model to predict the next intrusion attack called plan recognition, which is then evaluated for both supervised and unsupervised learning algorithms using DARPA 2000 data set.

Tang et al. [23] proposed a network security situation prediction method, which is based on dynamic back propagation neural with covariance. They obtained the situation sequences using situation assessment model and used it as input training data and implemented the self-learning dynamic modification of selected parameters' values.

Kim et al. [24] proposed a framework for intrusion forecasting system that consists of three modules: data collection module, data analysis module, and a reporting module. Three forecasting techniques, i.e., time-series analysis, a probabilistic modeling, and data mining method are integrated in the data analysis module to predict DDoS attack in DARPA 2000-specific data set, and it is concluded that the false alarm rate can be reduced significantly if two or more methods are combined together for forecasting cyber attacks.

Algorithmic modeling has a lot of advantages over statistical modeling approaches. First, machine learning algorithms are best suited to handle big data sets as compare to the statistical approaches. Second, the algorithms have less dependency on rigid assumptions about the data generating process. Finally, unlike some statistical models, machine learning algorithms give better prediction results [6].

3 Challenges

The field of cyber security prediction needs a lot of attention in context to data sources and the computational techniques. Some of the challenges being faced in obtaining intelligence are as follows:

- Everyone looking for incident data realizes that it is incomplete and does not provide true predictions. Thus, there is a need of using several sources of data.
- Existing approaches for predicting cyber threats provide inadequate information, which is hardly scaled to the real-world scenario. An integrated technique capable of correlating information from different type of data sources is required.
- A huge amount of network data is streaming continuously in cyber infrastructure that requires new principles, methodologies, and algorithms for transforming raw data into the useful information.
- The evolving cyber threat landscape (full of volume, variety, and velocity of threats) needs the adoption of big data security analytics to meet better prediction and performance requirements.

The predictions or intelligence obtained need to be shared among geographically located analysts or forecasters so as to further improve and validate the results.

4 Conclusion

Violations caused by advanced cyber attacks require the continued insight and operational capabilities in the field. A future trend outlook regarding cyber attacks/threats may influence the decisions concerning the security provision before the incidents actually happen. This paper highlights the existing prediction techniques applied in cyber security field along with the major challenges. Though the prediction techniques being applied in the field of cyber security seem to be capable of enhancing cyber security measures, but still require a lot of research to deal with influx of new threats in a large computer networks.

References

1. Gandotra, E., Bansal, D., Sofat, S.: Malware analysis and classification: a survey. J. Inf. Secur. **5**, 56–64 (2014)
2. Thomas, T.: Google confronts China's "Three Warfares". In: Parameters: U.S. Army War College, vol. 40(2), p. 101 (2010)
3. Shakarian, P.: Stuxnet: Cyberwar revolution in military affairs. Small Wars J. (2011)
4. Sleuths detect virus at card swipe terminals. http://indianexpress.com/article/business/business-others/sleuths-detect-virus-at-card-swipe-terminals/
5. Internet Crime Report (2012) http://www.ic3.gov/media/annualreport/2012_IC3Report.pdf
6. Subrahmanian, V.: Handbook of Computational Approaches to Counterterrorism. Springer, Berlin (2013)
7. Brockwell, P., Davis, R.: Introduction to Time Series and Forecasting. Springer, Berlin (2010)
8. Park, H., Jung, O., Lee, H., In, H.: Cyber weather forecasting: forecasting unknown internet worms using randomness analysis. In: Gritzalis, D., Furnell, S., Theoharidou, M. (eds.) Information Security and Privacy Research, AICT, vol. 376, pp. 376–387. Springer, Heidelberg (2012)
9. Pontes, E., Guelfi, A.: IFS: intrusion forecasting system based on collaborative architecture. In: 4th IEEE International Conference on Digital Information Management, pp. 1–6. IEEE Press, Ann Arbor (2009)
10. Pontes, E., Guelfi, A., Silva, A., Kofuji, S.: Applying multi-correlation for improving forecasting in cyber security. In: 6th International Conference on Digital Information Management, pp 179–186. Melbourne (2011)
11. Fachkha, C., Harb, E., Debbabi, M.: Towards a forecasting model for distributed denial of service activities. In: 12th IEEE International Symposium on Network computing and Applications, pp. 110–117. Cambridge, MA (2013)
12. Watters, P., McCombie, S., Layton, R., Pieprzyk, J.: Characterising and predicting cyber attacks using the cyber attacker model profile (CAMP). J. Money Laundering Control **15**, 430–441 (2012)

13. Feller, W.: An introduction to probability theory and its applications. Willey, New York (1971)
14. Wu, J., Yin, L., Guo, Y.: Cyber attacks prediction model based on bayesian networks. In: 18th International Conference on Parallel and Distributed Systems, pp. 730–731. IEEE Press, Singapore (2012)
15. Kim, D., Lee, T., Jung, O., Peter, H.: Cyber threat trend analysis model using HMM. In: 3rd International Symposium on Information Assurance and Security, pp. 177–182. IEEE Press, Manchester (2007)
16. Man, D., Wang, Y., Wu, Y., Wang, W.: A combined prediction method for network security situation. In: International Conference on Computational Intelligence and Software Engineering, pp. 1–4. IEEE Press, Wuhan (2010)
17. Chenq, C.: A High-efficiency intrusion prediction technology based on Markov chain. In: Computational Intelligence and Security Workshop, pp. 518–521. IEEE Press, Harbin (2007)
18. Lim, S., Yun, S., Kim, J., Lee, B.: Prediction model for Botnet-based cyber threats. In: International conference on Convergence, pp. 340–341. IEEE Press, Jeju Island (2012)
19. Fava, D., Byers, S., Yang, S.: Projecting cyberattacks through variable-length Markov models. IEEE Trans. Inf. Forensics Secur. **3**, 359–369 (2008)
20. Maloof, M.: Machine learning and data mining for computer security: methods and applications. Springer, New York (2006)
21. Thonnard, O., Dacier, M.: Actionable knowledge discovery for threat intelligence support using a multi dimensional data mining methodology. In: IEEE International Conference on Data Mining Workshops, pp. 154–163, IEEE Press, Pisa (2008)
22. Farhadi, H., AmirHaeri, M., Khansari, M.: Alert correlation and prediction using data mining and HMM. ISC Int. J. Inf. Secur. **3**, 77–101 (2011)
23. Tang, C., Xie, Y., Quang, B., Wang, X., Zhang, R.: Security situation prediction based on dynamic BP neural with covariance. In: Advanced in Control Engineering and Information Science, pp. 3313–3317 (2011)
24. Kim, S., Shin, S., Kim, H., Kwon, K., Hen, Y.: Hybrid intrusion forecasting framework for early warning system. In: IEICE transaction on information and systems, ACM, E91-D, pp. 1234–1241 (2008)

Multi-objective Discrete Artificial Bee Colony Based Phasor Measurement Unit Placement for Complete and Incomplete Observability Analysis

K. Mahapatra, M.R. Nayak and P.K. Rout

Abstract The paper presents a multi-objective discrete artificial bee colony (DABC)-based optimization algorithm for deciding placement sites for phasor measurement units (PMU) in a power system for both complete and incomplete observability studies. The influence of depths of unobservabilities of some of the buses on the number of optimal PMU placement sites is investigated in this study. Limited availability of communication facilities around the power network is considered as constraints in the optimization problem. The simultaneous optimization of the two conflicting objectives such as minimization of the number of PMUs and minimization of the number of unobserved buses are performed, and the Pareto optimal solutions are obtained using the non-dominated sorting DABC with crowding distance mechanism. The formulation is extended to solve the pragmatic-phased installation of PMUs for complete observability studies. The performance of the suggested approach is tested on three IEEE test systems. The results obtained demonstrate the technique that provides utilities with systematic approaches for incrementally placing PMUs in the best possible way, thereby cushioning their cost impact.

Keywords Observability · Phasor measurement units · Discrete ABC

K. Mahapatra (✉) · M.R. Nayak · P.K. Rout
Siksha 'O' Anusandhan University, Bhubaneswar, India
e-mail: kaverimahapatra@gmail.com

M.R. Nayak
e-mail: manasnk72@gmail.com

P.K. Rout
e-mail: pkrout_india@gmail.com

L.C. Jain et al. (eds.), *Intelligent Computing, Communication and Devices*,
Advances in Intelligent Systems and Computing 308,
DOI 10.1007/978-81-322-2012-1_27

1 Introduction

A phasor measurement unit (PMU) is a device which measures the electrical waves to estimate the phasor of voltage and current at the bus on an electricity grid, using a common time source for synchronization. The multifold application of PMU for monitoring, protection, state estimation, and control considered the device as one of the most promising tool for the future smart grid revolution. However, the overall cost of the installation limits the number of PMUs and that poses the problem of finding optimal placement of PMUs considering various related operational issues.

This paper presents a new multi-objective PMU placement problem formulation for incomplete observability analysis of a power system network. Incomplete observability analysis is first introduced in [1] which presents the concept of depth of unobservability. Until now, incomplete observability analysis-based PMU placement was considered as a single objective optimization problem whose objective is to minimize the number of PMUs while attaining a desired depth of unobservability by minimizing the deviation from considered depth of unobservability. A weighted sum method was used in [1] for solving the problem. In [2], integer linear programming is used for solving the same by considering it as single objective type of problem where constraints represent the unobservability criteria. In this work, a multi-objective problem is formulated whose one objective is to minimize the number of PMUs and the other is to minimize the unobservable buses while allowing only a particular depth of unobservability or less depth of unobservability to coexist in the network. Instead of minimizing the deviation from the desired depth of unobservability, the main focus is kept on minimization of the number of unobservable buses by which number of states required to be estimated from the obtained measured data gets reduced which implies better estimation accuracy. It is because the presence of more estimated states increases the propagation of error.

In this work, non-dominated sorting discrete artificial bee colony optimization (NSDABC) is applied for the solution of the presented multi-objective optimization problem. The concept of 'depth of unobservability' is utilized, and its effect on the number of PMU placements is discussed. The resulting PMU placement scheme guarantees a near uniform distribution of PMUs around the network. It limits the distance between unobserved and observed buses.

The rest of the paper is organized as follows. Section 2 describes the incomplete observability of the system. Section 3 deals with the PMU placement problem formulation for incomplete observability analysis. The observability analysis procedure is presented in Sect. 4. Section 5 details the proposed multi-objective DisABC optimization algorithm. Section 6 discusses the results of the proposed algorithm when applied to PMU placement problem. Section 7 concludes the paper.

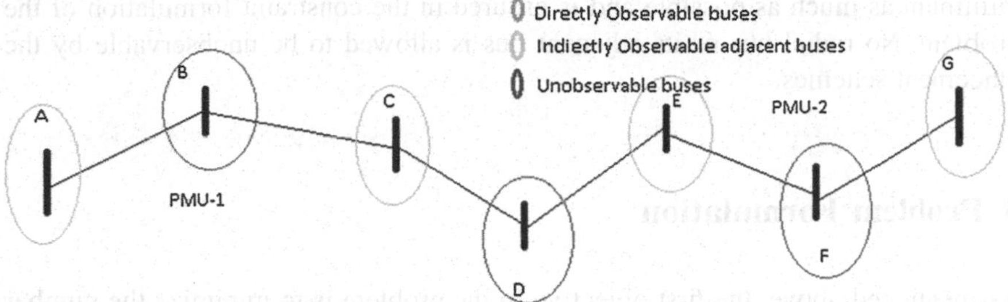

Fig. 1 Incomplete observability of a network

2 Incomplete Observability

Incomplete observability refers to the PMU placement scenario when the number and location of the PMUs are not sufficient to determine the complete set of bus voltages of a power system (the state of the power network) [1]. Figure 1 of [1] in this work represents an incompletely observed system. PMU-1 and PMU-2 measure the voltage phasors at buses *B* and *F*, respectively. *A*, *C*, *E*, and *G* bus voltages are calculated using the measured voltage and current values obtained from the PMU-placed buses. This placement scheme results in an unobservable bus *D*.

2.1 Depth-of-One Unobservability

A placement scheme is said to satisfy the depth-of-one unobservability criteria when any unobserved bus in a network is surrounded with observable neighboring buses whose phasors are estimated through the installed PMUs. The placement scheme must ensure not more than one unobservable buses to be connected to each other. Also, the unobservable group must not include any radial bus. Otherwise, it would become difficult to estimate its voltage and current phasors with better accuracy via only one neighboring bus.

2.2 Depth-of-Two Unobservability

A placement scheme is said to satisfy the depth-of-two unobservability criteria by allowing any two connected unobservable buses in the system. The placement scheme must ensure not more than two or unobservable buses to be connected to each other. The deviations from the desired depth of unobservability is kept minimum as much as possible and is ensured in the constraint formulation of the problem. No radial bus or its adjacent bus is allowed to be unobservable by the placement schemes.

3 Problem Formulation

As mentioned above, the first objective in the problem is to minimize the number of PMUs. If **X** is the binary solution vector representing a PMU placement scheme for an N bus system, the objective function value of the first objective for **X** is calculated by,

$$\min f_1 = \sum_{i=1}^{N} x_i \tag{1}$$

where x_i is a binary variable which shows the presence or absence of a PMU on bus i. The other objective of the problem is to minimize the number of unobservable buses or maximize the number of observable buses without getting deviated from a desired depth of unobservability. The number of unobservable buses in a system resulting from a PMU placement scheme can be determined from the observability analysis procedure [2]. If **Obs** is the vector which represents the observability information about for any solution vector X, then the total number of zero positions in the Obs vector indicates the number of unobservable buses.

$$\min f_2 = \sum_{i=1}^{N} \mathbf{not}(\mathbf{Obs_i}) \tag{2}$$

when the PMU placement set is determined for attaining a particular depth of unobservability say 'm', the problem formulation must allow less depth of unobservabilities such as $m-1, m-2, \ldots 2, 1$ to exist in the system but not higher such as $m+1$. In [1], the results show that authors have allowed even a higher depth of unobservability to exist in the system which is strictly restricted in this work. Each solution vector has to satisfy the feasibility constraint which examines any solution vector and considers it as valid only when the placement scheme does not violate any criteria defined for obtaining the desired depths of unobservability.

- It is valid, and the constraint satisfaction value (CV) for the vector is set to 1 if the solution vector leads to full observability.
- It is valid and the CV for the vector is set to 1 if there is not any deviation of more depth of unobservability from the desired depth of unobservability. Less depth of unobservabilities than the desired one is not a problem in this context. But the number of those groups must be kept as low as possible. That can be

ensured by adding a penalty term to the function value f_2. The value of penalty term is kept within 0–0.5. For example, when the desired depth of unobservability is 2, depth-of-one unobservability is allowed in the network. If any particular solution vector results in maintaining 'm' depth of unobservability while giving two deviations of '$m - 1$' depth of unobservability, the penalty function value is then $(2/N) * 0.5$.

- Any solution vector leading to more depth of unobservability than the desired is considered as an infeasible one, and a higher objective function value is assigned to it. The purpose of taking a higher function value is to discard it in the selection process of a minimization problem. CV for the vector is assigned a 0 value.

Two types of depth of unobservabilities are taken into consideration in this work.

1. Case 1: Depth-of-One unobservability
2. Case 2: Depth-of-Two unobservability.

4 Observability Analysis

The determination of observed and unobserved buses in a network resulting from a placement set is determined through the observability analysis procedure. There are two methods available for observability analysis. These are numerical and topological analysis. Topological observability analysis is used in this work in order to avoid the large number of calculations involved in the other one. The rules for topological observability analysis are as follows.

4.1 Direct Measurements

When a PMU is placed on a bus, it measures the voltage phasor of that bus and the current phasor of all its adjacent buses.

4.2 Pseudo-Measurements

When the voltage phasor of two terminal buses of a line is known, it is possible to calculate the current phasor of that line. Similarly, when the voltage phasor of one of the terminal buses of line and the current phasor of the line are known, the voltage phasor of the other terminal bus can be calculated.

4.3 Extension Measurements

The voltage phasor of zero injection (ZI) bus or one of its adjacent buses can be determined using the following rules:

- When the voltage phasor of all the adjacent buses to a ZI bus is known except the ZI bus itself, it can be readily determined using KCL equations.
- When in a group of a ZI bus containing the ZI bus and its adjacent buses, if all the bus voltage phasors except one are known, it can be determined making that bus observable using KCL equations.
- In a group of ZI buses connected to each other, if the number of unknown bus voltage phasors is equal to or less than the number of ZI buses, then those unobserved bus voltage phasors can be calculated.

5 Optimization Algorithm

Artificial bee colony optimization is a population-based stochastic algorithm which resembles the intelligent foraging behavior of the honey bees in order to solve complex, nonlinear, and non-convex mathematical optimization problems. The main difference between a continuous ABC and DABC [3] algorithm is in the type of mutation performed on any solution vector of population in the employed bee phase and in the onlooker phase. In this section, the complete algorithm steps are explained mathematically.

Step 1: Initialization of the optimization problem and the parameters

A binary vector $\boldsymbol{X}_i$ denotes a possible solution to the optimization problem. Initially, a set of *NP* uniformly distributed random binary vectors, each of size $N \times 1$ for an *N* bus system, is generated and stored in the population of bees $\boldsymbol{X}$. *NP*is same as the number of bees in the population.

$$\boldsymbol{X}_i = \{x_{i,1}, x_{i,2}, \ldots, x_{i,d}, \ldots x_{i,N}\} \tag{3}$$

$$\boldsymbol{X} = \{\boldsymbol{X}_1, \boldsymbol{X}_2, \ldots \boldsymbol{X}_i, \boldsymbol{X}_{i+1}, \ldots, \boldsymbol{X}_{NP}\} \tag{4}$$

The iteration number 't' is set to 1 and the maximum iteration number ($T_{\max}$) is specified along with the local search probability factor (L_P) and local search operation number (L_N). For each vector $\boldsymbol{X}_i^t$ of tth iteration, the objective function values are evaluated using Eqs. (1) and (2).

Step 2: Pareto Non-dominated Sorting

Each of the individuals is compared with every other individual $\boldsymbol{X}_i^t$ of the population in terms of their objective function values and is assigned to a front through the non-dominated sorting operation. A crowding distance-based ranking mechanism is used to add individuals from the population to any Pareto front. This is repeated until the population matrix $\boldsymbol{X}$ for the next iteration is filled with *NP* number of individuals from the Pareto set.

Step 3: Mutation Operation

A binary mutation operation is performed on every individual $\boldsymbol{X}_i^t$ in the population to generate child population. For every individual in the population, a corresponding random individual is chosen from the rest of the population for mutation. Both are considered as parent individuals. The NBSG algorithm is utilized based on the degree of dissimilarity which in turn depends upon the degree of similarity between two binary vectors and is proposed by Jaccard in [4]. The degree of dissimilarity between two vectors $\boldsymbol{X}_i$ and $\boldsymbol{X}_j$ is calculated as follows:

$$\text{Dissimilarity}\left(\boldsymbol{X}_i, \boldsymbol{X}_j\right) = 1 - \frac{M_{11}}{M_{01} + M_{10} + M_{11}} \tag{5}$$

where M_{01} is the total number of bits with $\boldsymbol{X}_i$ having a value of 0 and $\boldsymbol{X}_j$ having a value of 1. M_{10} is the total number of bits with $\boldsymbol{X}_i$ having a value of 1 and $\boldsymbol{X}_j$ having a value of 0. M_{11} is the total number of bits with both $\boldsymbol{X}_i$ and $\boldsymbol{X}_j$ having a value of 1. NBSG is used for generating the child population. In the first step after obtaining *dissimilarity* between two vectors, 'AD' is calculated using Eq. (6) with a random scaling factor '*r*' as a fixed decimal number.

$$\text{AD} = r \cdot \text{Dissimilarity}\left(\boldsymbol{X}_i, \boldsymbol{X}_j\right) \tag{6}$$

In the second step, a mathematical programming model is formulated in order to minimize the degree of dissimilarity of the old vector with the new solution vector to be generated. Minimization of this difference depends upon their M_{01}, M_{10}, M_{11} values which are calculated by solving the problem given below. Its mathematical model is expressed as

$$\min\left|1 - \frac{M_{11}}{M_{01} + M_{10} + M_{11}} - \text{AD}\right| \tag{7}$$

subject to the following constraints

$$array*20l M_{01} + M_{11} = n_1 M_{10} \leq n_0 M_{01}, M_{10}, M_{11} \geq 0\} \tag{8}$$

where n_1 and n_0 are the number of 1s and number of 0s, respectively, in vector $\boldsymbol{X}_i$. The total enumeration (TE) scheme employed here uses an exhaustive search technique to determine the optimum value of M_{01}, M_{10}, M_{11} for the new solution vector.

In the third step, the new solution V_i is generated for X_i using the following procedure. The V_i is first initialized to a vector of zeros. In the inheritance phase, any M_{11} number of bits are randomly picked from the positions of X_i which contains 1s and those in V_i are changed to 1s.

In the disinheritance phase, any M_{10} number of zero bits of V_i are picked randomly from the positions which contain 0s in X_i and are then changed to 1s. At the end of this step, all the newly generated solution vectors are stored in a matrix X^t_{new1} which is referred to as the first child population and their corresponding objective function values are evaluated.

Step 4: Local Search Step

Any L_N number of random vectors from the child population matrix X^t_{new1} is chosen with a probability L_P for swapping. Newly generated vectors are then stored in the sub-child population X^t_{new2}, and their corresponding objective function values are evaluated.

Local Search Algorithm

For $k = 1, k \leq L_N, k++$

Choose a vector $X^t_{p^k}$ of a random position p^k from the population X^t

If random number $< L_P$

choose any one bit position p^{lk} of vector $X^t_{p^k}$ for swapping

If $X^t_{p^k,p^{lk}}$ bit of the vector $X^t_{p^k}$ is equal to 1

$$X^t_{p^k,p^{lk}} \leftarrow 0$$

Else

$$X^t_{p^k,p^{lk}} \leftarrow 1$$

End if

End if

End For

Step 5: Mixture of Parent, Child Populations, and Non-dominated Sorting

Both the child populations X^t_{new1} and X^t_{new2} with the current population X^t (parent) of the current (tth) iteration are stored along with their corresponding objective values in a matrix P^t called as mixture population. Non-dominated sorting operation is applied on P^t by which all the solutions are sorted into different Pareto fronts which is then followed by a ranking based on crowding distance mechanism. Only the best NP number of vectors are stored in the updated population P^{t+1} for the next iteration $t+1$. At the end of this step, 't' is incremented by 1.

Step 6: Termination Criteria

Steps 3–5 are repeated until 't' reaches the maximum limit $T_{\max}$. The population obtained in the final iteration contains the Pareto optimal set.

Step 7: Best Compromised Solution

The best compromised solution is selected by a fuzzy rule which assigns a degree of satisfaction DS_i for each objective function to any solution vector presents in the Pareto optimal set according to the maximum and minimum objective function values obtained in that iteration.

$$\mathrm{DS}_i = \begin{cases} 1 & \text{if } f_i \le f_{i,\min} \\ \frac{f_{i,\max} - f_i}{f_{i,\max} - f_{i,\min}} & \text{if } f_{i,\min} < f_i < f_i \le f_{i,\max} \\ 0 & \text{if } f_i \ge f_{i,\max} \end{cases} \tag{9}$$

where $i = 1,\ 2$ since two objectives are considered in this work. The total degree of satisfaction of the solution vector for both the objectives is then calculated by

$$\mathrm{DS} = \frac{1}{\text{Total Number of Objectives}} \sum_{i=1}^{\substack{\text{TotalNumber}\\ \text{ofObjectives}}} \mathrm{DS}_i \tag{10}$$

After this, the best compromised solution is determined by selecting the solution vector of highest satisfaction degree among the set of Pareto optimal solutions.

6 Result Analysis

The proposed NSDABC algorithm is applied for solving the PMU placement problem in order to attain a desired depth of unobservability. Pareto fronts for three standard test systems, i.e., IEEE 14 bus, 30 bus, and 39 bus test systems, are obtained for depth-of-one and depth-of-two unobservability. Also, the placement set is obtained for complete system observability. All the results are obtained with consideration of the effect of ZI buses in the observability analysis procedure. The results can be shown in Figs. 2, 3 and 4 for depth-of-one unobservability and Figs. 5, 6 and 7 for depth-of-two unobservability for various test systems. These are compared with the results of [1] in Tables 1 and 2. It is also observed that the some of the results obtained through the proposed method can lead to less unobservable buses while ensuring a desired depth of unobservability. However, the placement schemes vary from each other for any change in depth of unobservability.

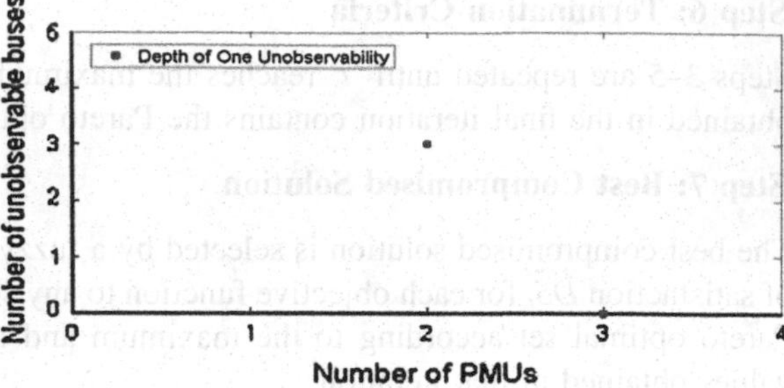

Fig. 2 Pareto front for IEEE 14 bus system for depth-of-one unobservability

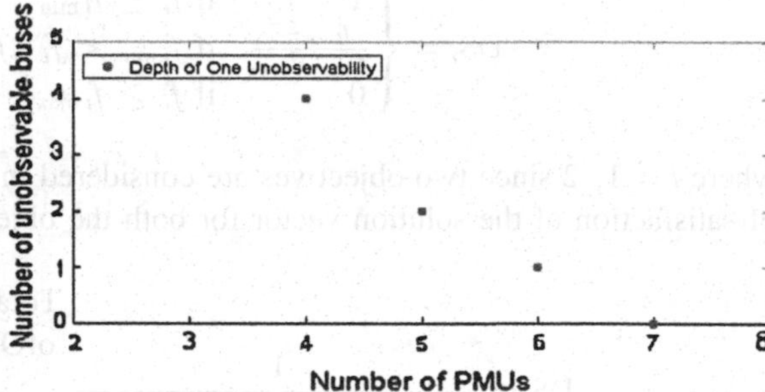

Fig. 3 Pareto front for IEEE 30 bus system for depth-of-one unobservability

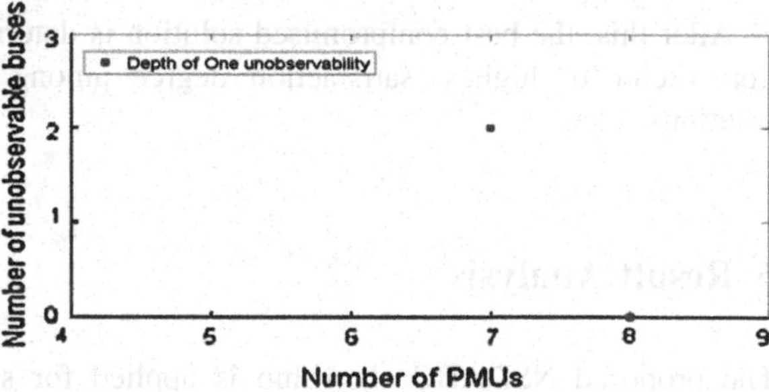

Fig. 4 Pareto front for IEEE 39 bus system for depth-of-one unobservability

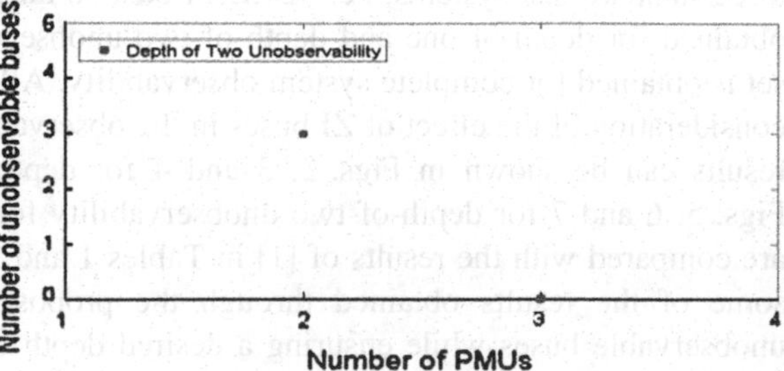

Fig. 5 Pareto front for IEEE 14 bus system for depth-of-two unobservability

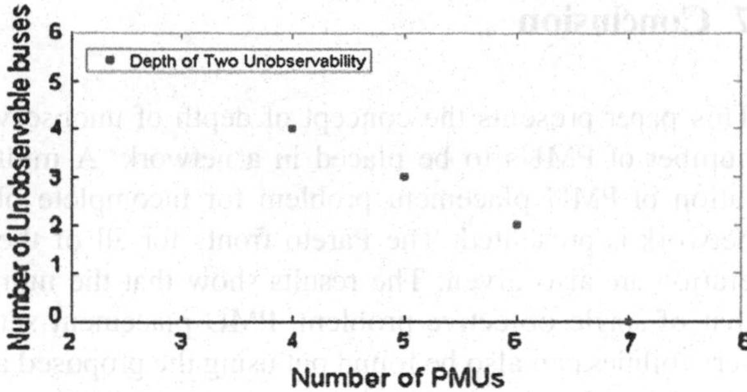

Fig. 6 Pareto front for IEEE 30 bus system for depth-of-two unobservability

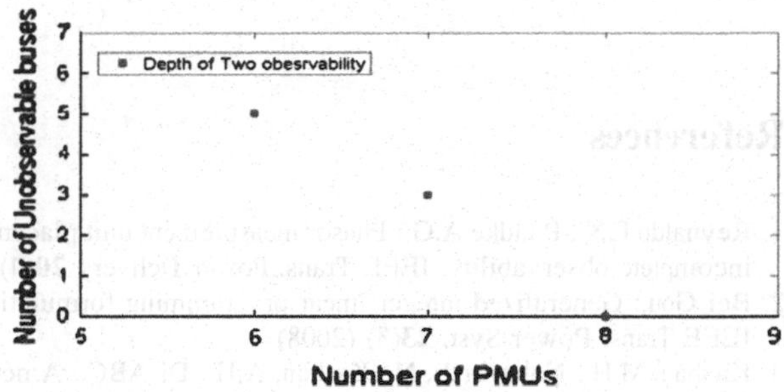

Fig. 7 Pareto front for IEEE 39 bus system for depth-of-two unobservability

Table 1 Number of PMUs for depth-of-one unobservability and complete observability

IEEE test systems	14 bus	30 bus	39 bus
Complete observability [1]	3	7	8
Depth-of-one unobservability [1]	2	7	–
Complete observability (proposed)	3	7	8
Depth-of-one unobservability (proposed)	2	7	7

Table 2 Number of PMUs for depth-of-two unobservability and complete observability

IEEE test systems	14 bus	30 bus	39 bus
Complete observability [1]	3	7	8
Depth-of-two unobservability [1]	2	3	–
Complete observability (proposed)	3	7	8
Depth-of-two unobservability (proposed)	2	3	6

7 Conclusion

This paper presents the concept of depth of unobservability and its effect on the number of PMUs to be placed in a network. A multi-objective problem formulation of PMU placement problem for incomplete observability analysis of any network is presented. The Pareto fronts for all of the test systems under consideration are also given. The results show that the number of PMUs matches with that of single objective problem. PMU placement sets for higher depth of unobservabilities can also be found out using the proposed approach. Also, evolutionary algorithms for this problem are easier and take less computational time to find a solution as compared to any classical techniques for large-scale networks.

References

1. Reynaldo F.N., Phadke A.G.: Phasor measurement unit placement techniques for complete and incomplete observability. IEEE Trans. Power Delivery **20**(4) (2005)
2. Bei Gou.: Generalized integer linear programming formulation for optimal PMU placement. IEEE Trans. Power Syst. **23**(3) (2008)
3. Kashan, M.H., Nahavandi, N., Kashan, A.H.: DisABC.: A new artificial bee colony algorithm for binary optimization. Appl. Soft Comput. **12**, 342–352 (2012)
4. Karaboga, D., Basturk, B.: A powerful and efficient algorithm for numerical function optimization: artificial bee colony (ABC) algorithm. J. Glob. Optim. **39**(3), 459–471 (2007)

Modified Ant Colony Optimization Algorithm (MAnt-Miner) for Classification Rule Mining

Sarbeswara Hota, Pranati Satapathy and Alok Kumar Jagadev

Abstract Classification rule mining is an important task of data mining. Ant colony optimization (ACO) algorithms are applied successfully to various optimization problems. Earlier Ant-Miner, an ACO algorithm was used to discover the classification rules and predictive accuracy was determined. In this paper, modified ant colony optimization (MAnt-Miner) is proposed to generate the classification rules and to enhance the predictive accuracy. This method is applied on breast cancer data set, and the experimental result showed that the predictive accuracy of MAnt-Miner is better than Ant-Miner.

Keywords Classification · Heuristic · Pheromone · Swarm intelligence

1 Introduction

Out of the several data mining tasks, classification is an important task. Classification problems can be viewed as optimization problems, where the goal is to find the best model that represents the predictive relationships in the data [1]. Actually, a classification problem consists of generating a small set of rules with predefined

S. Hota (✉) · A.K. Jagadev
Siksha O Anusandhan University, Bhubaneswar, India
e-mail: sarbeswarahota@soauniversity.ac.in

A.K. Jagadev
e-mail: alokjagadev@soauniversity.ac.in

P. Satapathy
Department of IMCA, Utkal University, Bhubaneswar, India
e-mail: satapathy.pranati@gmail.com

L.C. Jain et al. (eds.), *Intelligent Computing, Communication and Devices*,
Advances in Intelligent Systems and Computing 308,
DOI 10.1007/978-81-322-2012-1_28

class labels. The rules are generated from the training data set. The generated rules are then used to classify, i.e., to predict the class labels of unknown instances. The classification rules are of the form

$$\text{IF } \langle\text{condition}\rangle \text{ THEN } \langle\text{class}\rangle.$$

where condition is expressed as term1, term2, and so on, and each term is a triplet ⟨attribute, operator, value⟩.

The classification rule generation problem is NP-hard because for n number of attributes, the number of conditions is $O(2^n)$ that is exponential in nature. So many researchers have applied biologically inspired algorithms on these types of NP-hard problems to get the solutions.

Ant colony optimization (ACO) algorithm, one of the biologically inspired algorithms has been used extensively in classification task of data mining. Parpinelli et al. [2] are the first to propose ACO algorithm for discovering classification rules, known as Ant-Miner. Later on, this algorithm has been studied extensively by many researchers and modifies in various forms. The objective of our work is to propose a new heuristic value and pheromone updating strategy in the Ant-Miner.

The remainder of this paper is organized as follows. Section 2 introduces the ACO algorithms. Section 3 introduces the Ant-Miner. The proposed model of our work and the modifications in the Ant-miner are presented in Sect. 4. The experimental results obtained on one benchmark data set are presented in Sect. 5, and Sect. 6 concludes this paper.

2 Ant Colony Optimization Algorithm

ACO [3] is a branch of newly developed form of artificial intelligence called swarm intelligence. Swarm intelligence is a field that studies the collective intelligence of a swarm, i.e., a group of insects, which live in colonies such as ants, bees, etc. A single ant can do some simple tasks on its own, but the intelligent behavior shown by the colony lies in their cooperative work. Dorigo and Maniezzo presented the cooperative behavior of ant colonies. ACO algorithms [3] are based on the following ideas:

- The path traversed by an ant refers to a solution of a problem.
- The pheromone amount deposited on a path represents the strength of solution.
- The probability of choosing a path depends on the amount of pheromone laid by an ant.

In brief, the design of an ACO algorithm involves the following specification:

- Formulation of a problem based on a problem-dependent heuristic function and probability calculation of the edges between the vertices.
- A heuristic function (η) that represents the quality of the sub-solutions.
- The pheromone updating method, which increases or decreases pheromone trail (τ).
- A probability function that build the solution of the problem.

The basic pseudocode of ACO is as follows.

```
Procedure ACO
{        While (termination condition is not reached)
         {        Construct_Solution ()
                  Perform_Actions ()
                  Update_Pheromone ()
         }
}
```

3 Ant-Miner

The purpose of Ant-Miner is to discover classification rules from data set [2].

```
Algorithm Ant-Miner
{        Training_set= all training cases;
         DiscoveredRuleList=[ ];
         WHILE (No. of cases in the training set > max_uncovered_cases)
         {        i=0;
                  REPEAT
                  i=i+1;
                  Ant_i incrementally constructs a classification rule;
                  Prune the just constructed rule;
                  Update the pheromone of the trail followed by Ant_i ;
         UNTIL (i>= No_of_Ants) or (Ant_i constructed the same rule as the previous
         No_Rules_Converg-1 Ants)
         Select the best rule R_best among all constructed rules;
         Add rule R_best to DiscoveredRuleList;
         Remove the cases correctly covered by the selected rule from the training set;
         }
}
```

In Ant-Miner, initially the DiscoveredRuleList is empty and the training set is all the training cases. On each execution of WHILE loop that incorporates a number of executions of nested REPEAT-UNTIL loop, one classification rule is generated. The generated rule is appended to the DiscoveredRuleList, and the training cases that are correctly classified by this rule (i.e., instances satisfying the rule conditions and having the class predicted by the rule conclusion) are removed from the training set. This process is iterated while the number of uncovered training cases is greater than a user-specified threshold, called Max_uncovered_cases.

Each execution of the REPEAT-UNTIL loop of Ant-Miner algorithm comprises of these basic processes,

- Pheromone Initialization (τ) and pheromone updating strategy
- Problem-dependent heuristic value calculation (η)
- Rule construction using probability calculation of adding a term

4 Proposed Model and Modified Ant-Miner

The primary purpose of this paper is to discover the classification rules in a particular data set and classify with test data. Thus, it computes the classification accuracy. It performs the rule mining with the modified Ant-Miner. The general model of this work is described with the following steps as:

Step 1: Identification of categorical attributes and number of classes of a training data set
Step 2: Study of ACO algorithm for classification rule and its modifications
Step 3: Discover the classification rule list on the given data set
Step 4: Calculate the predictive accuracy of the rule list on the test data set and compare with Ant-Miner

In Ant-Miner, the ants select terms on the basis of pheromone amount and heuristic function, which measures the predictive power of a term. But in this method, the pheromone of each term is changed after an ant constructs a rule, while heuristic function is always the same, so that the next ant tends to choose terms used in the previous rule, whose pheromone is increased and is unlikely choose unused terms, whose pheromone is decreased. Consequently, the ants converge to a single constructed rule too quickly. This prevents to produce other potential rules.

Hence, in this paper, we use the existing pheromone initialization and probability calculation functions as that of Ant-Miner but a different heuristic function and pheromone updating method to find the most predictive terms in a rule.

4.1 Pheromone Initialization

Each term$_{ij}$ (term corresponding to attribute i and value j) corresponds to a segment in some path that can be followed by an ant. At each iteration of the WHILE loop of the Ant-Miner algorithm, all term$_{ij}$ are initialized with the same amount of pheromone, so that when the first ant starts its search, all paths have the same amount of pheromone. The initial amount of pheromone deposited at each path position is inversely proportional to the number of values of all attributes and is defined as

$$\tau[i,j](t=0) = \frac{1}{\sum_{i=0}^{a} b[i]} \tag{1}$$

where a *is* the total number of attributes, and $b[i]$ is the number of possible values that can be taken on by attribute A_i.

4.2 Heuristic Function

Heuristic function $H(W|A_i = V_{ij})$ of term$_{ij}$ computation of Ant-Miner [2] is based on information entropy and normalization, but this value is always the same irrespective of the contents of the rule in which the term occurs. Consequently, the ants likely converge to a single constructed rule too quickly. This avoids producing alternative potential rules. In order to overcome these shortcomings, we use a simple heuristic function [4], which is defined as follows:

$$\eta[i,j] = \frac{-\sum_{w=0}^{k} \text{freq } T_{ij}^{w} \log_2 \text{ freq } T_{ij}^{w}}{|T_{ij}|} \tag{2}$$

where k is the number of classes;
$|T_{ij}|$ is the total number of cases in partition T_{ij};
freq T_{ij}^{w} is the number of cases in partition T_{ij} with class w.

4.3 Probability Calculation

As specified in the Ant-Miner, the probability that term$_{ij}$ is chosen to be added to the current partial rule is given as

$$\text{Prob}[i,j] = \frac{\eta[i,j] \cdot \tau[i,j](t)}{\sum_{i=0}^{n} x[i] \cdot \sum_{j=1}^{bi} (\eta[i,j] \cdot \tau[i,j](t))} \quad (3)$$

where

- a is the total number of attributes,
- x_i is set to 1 if the attribute A_i was not yet used by the current ant, or to 0 otherwise.,
- b_i is the number of values in the domain of the ith attribute.

4.4 Pheromone Updating Strategy

After the rule construction by an ant, the amount of pheromone associated with each term that occurs in the constructed rule is updated [5] and the pheromone of unused terms is updated by normalization. For pheromone updating, we use the following method

$$\tau[i,j](t) = (1-\rho) + \left(1 - \frac{1}{1+Q}\right) \cdot \tau[i,j](t-1) \quad (4)$$

where ρ is the pheromone evaporation rate and Q is the quality of the constructed rule. In this paper, we take $\rho = 0.1$. The quality of a rule is computed by the formula: $Q = \mathit{sensitivity} \cdot \mathit{specificity}$, defined as:

$$Q = \frac{\text{TP}}{\text{TP + FN}} \cdot \frac{\text{TN}}{\text{FP + TN}} \quad (5)$$

where:

- TP (true positives) is the number of cases covered by the rule that has the class predicted by the rule.
- FP (false positives) is the number of cases covered by the rule that has a class different from the class predicted by the rule.
- FN (false negatives) is the number of cases that are not covered by the rule but that has the class predicted by the rule.
- TN (true negatives) is the number of cases that are not covered by the rule and that do not have the class predicted by the rule

The pheromone of unused terms is updated by normalization given as

$$\tau[i,j](t) = \frac{\tau[i,j](t-1)}{\sum \tau[i,j](t-1)} \quad (6)$$

5 Experimental Results

The performance of the modified Ant-Miner was evaluated using one public-domain data set from the UCI repository, i.e., the breast cancer data set having 659 cases, 9 categorical attributes, and 2 class labels. In our experiment, the parameters were set as follows:

- *No_of_ants* = 200.
- *Min_cases_per_rule* = 30 % of the total training data set.
- *Max_uncovered_cases* = 20 % of the total training data set.
- *No_rules_converg* = 10.

The comparison between Ant-Miner and modified Ant-Miner was carried out based on the predictive accuracy of the discovered rule lists. The number right after the "±" symbol is the standard deviation of the corresponding predictive accuracies rates (Tables 1 and 2).

From the above comparison, it is concluded that proposed Ant-Miner discovered rules with a better predictive accuracy than Ant-Miner in the data set. In addition, we can use the confusion matrix for analyzing the performance of both the algorithms. The confusion matrices for the two classes are shown in Tables 3 and 4.

Using confusion matrix as a tool for analyzing the performance, the accuracy rate can be calculated as

$$\text{Accuracy rate} = \frac{\text{TP} + \text{TN}}{\text{TP} + \text{FN} + \text{FP} + \text{TN}} \quad (7)$$

By using confusion matrix tool for analyzing the performance of the above two algorithms, it is concluded that the accuracy rate of the modified Ant-Miner is

Table 1 Test runs for predictive accuracy

Run number	Ant-Miner	Modified Ant-Miner
1	92.3695	97.1888
2	93.1727	98.7952
3	93.9759	98.7952
4	92.3695	97.1888
5	94.7791	96.7791
6	95.5823	96.3855
7	96.3855	97.1888
8	93.9759	97.9920
9	94.7791	97.1888
10	92.3695	95.5823

Table 2 Mean accuracy rate with standard deviations

Valuation item	Ant-Miner	Modified Ant-Miner
Accuracy rate (%)	94.2168 ± 1.6481	97.5100 ± 1.2672

Table 3 Confusion matrix for Ant-miner

	Class1	Class2	Total
Class1	194	5	199
Class2	9	42	51
Total	203	47	250

Table 4 Confusion matrix for modified Ant-Miner

	Class1	Class2	Total
Class1	195	4	199
Class2	2	49	51
Total	197	53	250

Table 5 Accuracy rate comparison

	Ant-Miner (%)	Modified Ant-Miner (%)
Accuracy Rate	94.4	97.6
Misclassification Rate	5.6	2.4

higher than that of the Ant-Miner and the misclassification rate of the modified Ant-Miner is lower than that of the Ant-Miner as shown in Table 5.

6 Conclusion

In this paper, we have modified the Ant-Miner with different heuristic function and pheromone updating method. The modified Ant-Miner was used to discover the classification rules from the breast cancer data set, and the predictive accuracy was compared with the predictive accuracy of Ant-Miner. It is inferred that for the given data set, the modified Ant-Miner works better that Ant-Miner. This classification rule mining problem can be designed as a multi-objective problem and multi-objective ACO algorithm can be used to get the set of Pareto solutions.

References

1. Fayyad, U.M., Piatetsky Shapiro, G., Smyth P.: From data mining to knowledge discovery: an overview. In: Advances in Knowledge Discovery and Data Mining, pp. 1–34. AAAI/MIT, Cambridge (1996)
2. Parpinelli, R.S., Lopes, H.S., Frietas, A.A.: Data mining with an Ant Colony Optimization Algorithm. IEEE Trans. Evol. Comput. **6**(4), 321–332 (2002)
3. Dorigo, M., Colorni, A., Maniezzo, V.: The Ant System: optimization by a colony of cooperating agents. IEEE Trans. Syst. Man Cybern. Part B **26**(1), 29–41 (1996)
4. Jiang, W.J., Xu, Y.H., Xu, Y.S.: A Novel Data Mining Algorithm based on Ant Colony System. In: Proceedings of the Fourth International Conference on Machine Learning and Cybernetics, pp. 18–21 (2005)
5. Liu, B., Abbass, H.A., Mckay, B.: Classification rule discovery with Ant Colony optimization. IEEE Comput. Intell. Bull. **3**(1) (2004)

An Intelligent Method to Test Feasibility Predicate for Robotic Assembly Sequence Generation

M.V.A. Raju Bahubalendruni and B.B. Biswal

Abstract Determination of a feasible assembly sequence with optimum assembly cost is significant for manufacturing industries to minimize the overall cost of manufacturing process. Since feasibility predicate is an essential qualifying criterion in the area of assembly sequence generation, an efficient method is developed to test the feasibility predicate of robotic assembly for a defined sequence. The correctness of the methodology is proven by integrating the method with the 3D solid CAD models. The automation of method avoids the human intervention and does not require any skill from the engineer.

Keywords Feasibility predicate · Robotic assembly · Assembly sequence

1 Introduction

Optimized assembly sequence of a product always plays key role in manufacturing industry in terms of cost effectiveness. It also supports in reducing the lead time of the product and improving the final product quality as well. Most of the researchers worked on finding out the at least one feasible solution to assemble the product and to find out the optimized assembly sequence from the feasible and stable assembly sequences with considerable approximations and assumptions. Feasibility predicate can be tested by precedence relations for an assembly. The feasibility test will be done after establishing the liaisons between the components; first of all, liaison matrix must be extracted automatically from the CAD

M.V.A. Raju Bahubalendruni (✉) · B.B. Biswal
Department of Industrial Design, National Institute of Technology Rourkela, Rourkela 769008, Odisha, India
e-mail: 512id1006@nitrkl.ac.in

B.B. Biswal
e-mail: bbbiswal@nitrkl.ac.in

L.C. Jain et al. (eds.), *Intelligent Computing, Communication and Devices*,
Advances in Intelligent Systems and Computing 308,
DOI 10.1007/978-81-322-2012-1_29

environment. There exist algorithms to obtain the liaison matrix from CAD data that is briefed by Bahubalendruni et al. [1]. Bourjault [2] depicted a method to obtain precedence relations; establishment conditions are used to answer the questions generated for each liaison. These questions can be generated through a computer program. However, the method of generation establishment conditions is dependent on the skill of the engineer and answering the questions based on the establishment conditions is also a skillful task. The correctness of methodology is dependent on the user provided answers, and the moreover, the method is manual process. De Fazio et al. [3] developed similar concept but modified the type of question with reduced count.

These methods ensure that whether a part can be disassembled/assembled for a given product and does not discuss about the direction along which the part can be disassembled/assembled. There exist methods to predict the collision between the parts. Cohen et al. [4], Ponamgi et al. [5], and Hubbard [6] described a collision detection method, Fu and Liu [7] proposed an algorithm for finding collision-free path among polyhedral obstacles and Gilbert and Foo [8]. And Zeghloul et al. [9] described a method to compute the distance between general convex objects in three-dimensional space. These methods are helpful in finding out the collision-free path for a part to disassemble.

In the current research, a method is developed to estimate the distance between the components and to check for a collision-free path to disassemble a component efficiently.

2 Overview on Feasibility

Feasibility predicate is true when a part can bring into contact to create an assembly/subassembly though collision-free path. Based on the assumptions that the assembly components are rigid and no destructive operation is done during the assembly, if a part can be disassembled from the product through collision-free path, hence the part also can be assembled to the product. There also exist many possible directions to disassemble the part from product; the optimal path must be found to ensure the optimal cost.

3 Simple Mechanism

In robotic assembly system, typically, parts are moved along six directions considering positive and negative sides (±X, ±Y, ±Z), assuming the components to be assembled on base part (disassembling in Z-direction is ignored). Each part is moved along a specific direction, if there exist interference with any component in the assembly, then the possibility of disassembling the part should be checked in other directions.

```
Method:
Get the assembly sequence and get the total number of parts in the assembly " n"
For i=n to 1
   For direction 1 to 5
      Move the part along the direction
      If i^th part interfere with any other part in the assembly then
         Change direction
      End if
      If the part can be remove along the direction without any collision then
         Capture the distance moved by the part along the direction
      End if
      Get the direction, in which the part moves short distance
   End for
End for
```

Though first direction may be the optimal and feasible, the disadvantage is the method that checks for other directions, also for the feasibility, and hence consumes lot of computational time. The distance to be moved by the part along a direction should be specified by the user that also raises the human effort.

4 Bounding Box Method

The bounding box method creates a three-dimensional cuboid for each component and considers the diagonal points to represent the part and to obtain the distances to be moved by any component in all the directions. Once the distances are obtained along all directions, these will be arranged in ascending order to test for the feasibility.

4.1 Distance Measurement Though Bounding Boxes Method

Consider gear assembly composed of seven parts as shown in Fig. 1, in three-dimensional environment, bounding box for each part and assembly can be created and represented as shown in Fig. 2.

The distance to be moved by a part in five directions can be obtained by using the diagonal coordinates of the bounding box of each part and assembly using Table 1.

The directions must be arranged in ascending order based on the distance to be moved, and checking for feasibility in the same order minimizes the time and efficient. The bounding box corner coordinates for the gear assembly shown in

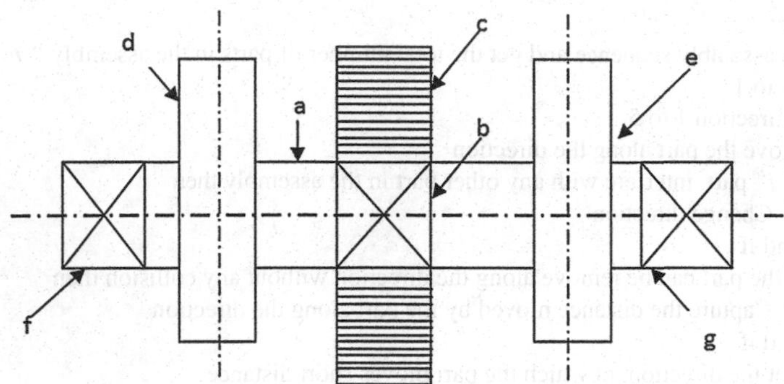

Fig. 1 Representation of seven-part gear assembly (*a* shaft; *b* bearing; *c* gear, *d* Arm, and *e* Arm, *f* nut, *g* nut)

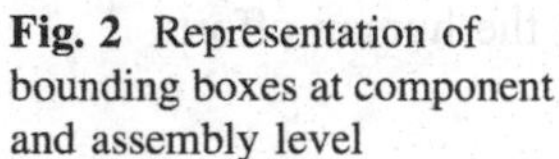

Fig. 2 Representation of bounding boxes at component and assembly level

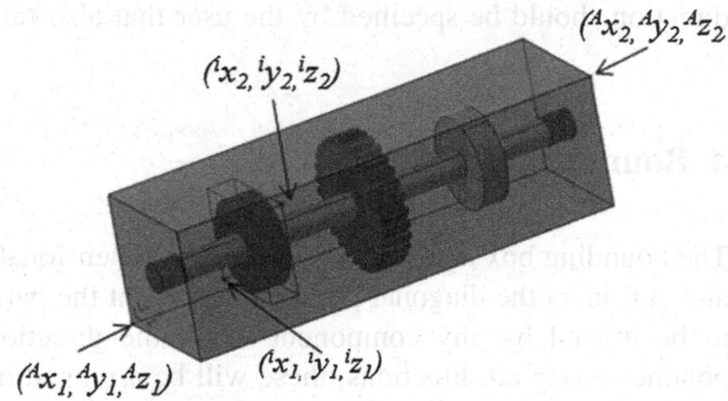

Table 1 Distance to be moved by the components to assemble/disassemble

	Part "*i*"				
Disassemble directions	X+	X−	Y+	Y−	Z+
Assemble directions	X−	X+	Y−	Y+	Z−
Distance to be moved	${}^Ax_2-{}^ix_1$	${}^ix_2-{}^Ax_1$	${}^Ay_2-{}^iy_1$	${}^iy_2-{}^Ay_1$	${}^Az_2-{}^iz_1$

Fig. 2 are listed in Table 2, and the distances to be moved by part D from the product is listed in Table 3.

From Table 3, it is efficient that to check the feasibility to disassemble the part "D" in the following directions "Z+, Y+, Y−, X−, X+". Though the distance to be moved is same in "Z−" direction and "Y±" directions, priority will be given to the "Z" direction as gravity force adds the significance.

Table 2 Component- and assembly-level bounding box corners

	x_1	y_1	z_1	x_2	y_2	z_2
Assembly	0	0	0	340	100	100
A	0	40	40	340	60	60
B	160	30	30	180	70	70
C	160	0	0	180	100	100
D	70	15	15	90	85	85
E	250	15	15	270	85	85
F	0	40	40	10	60	60
G	330	40	40	340	60	60

Table 3 Distance to be moved by the components to assemble/disassemble

	Part 4				
Disassemble directions	X+	X−	Y+	Y−	Z+
Assemble directions	X−	X+	Y−	Y+	Z−
Distance to be moved	${}^{A}x_2-{}^{4}x_1$	${}^{4}x_2-{}^{A}x_1$	${}^{A}y_2-{}^{4}y_1$	${}^{4}y_2-{}^{A}y_1$	${}^{A}z_2-{}^{4}z_1$
	270	**90**	**85**	**85**	**85**

Algorithm to extract bounding box and extracting distances

```
Open an assembly in 3D cad environment
Choose a plane (XY, YZ, or ZX) and repeat the steps for the two other planes
Create a plane1 offset to XY plane
    If there exist clash with the assembly then
        Move the plane1 till it results contact with the assembly
    End if
    Create a plane2 offset to the plane1 at unit distance
    If plane results clearance with the assembly then
        swap the direction
    End if
    Move the plane2 till it results contact.
Create a cuboid intersecting volume by six planes.
Schematic representation of bounding box is shown in Fig. 3.
```

4.2 Method to Test Feasibility

Once bounding boxes are created for each part and assembly, the distances are measured for each part based on Table 1.

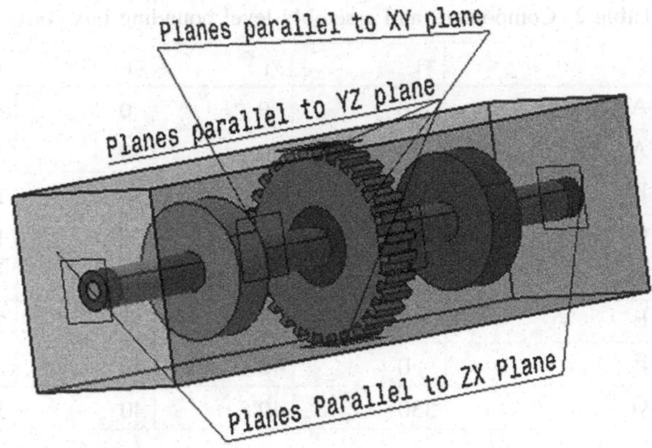

Fig. 3 Assembly bounding box generation representation

Methodology to test the feasibility predicate

```
for i= part n to 2.
  for j= 1 to 5 (directions arranged in ascending order for ith part)
    for k=0 to distance along j direction
       move the part to a distance "k" along jth direction & perform contact analysis
       if there exist interference then
         if j=5 then
           assembly sequence is not feasible
         end if
         change the direction (go to next j value)
       end if
       if k= distance along j direction then
         go to the next part
       end if
    end for
  end for
end for
```

There will not be any feasibility check for the last part, since it can be disassemble in all the possible directions, and the lowest distance direction will be given to it. The feasible assembly sequences will be transferred to the next phase for energy computation for the assembly process.

5 Conclusions

A new method called bounding box method is proposed to obtain the location of the part in the assembly to obtain the distance to move along the principle directions. An efficient method to test the feasibility predicate is described. Program for the integration of method with the CAD environment is described, which will be helpful in robotic assembly sequence generation.

References

1. Bahubalendruni, M.V.A.R., Biswal, B.B.: Computer aid for automatic liaisons extraction from cad based robotic assembly. In: Proceedings of the 8th International Conference on Intelligent Systems and Control (ISCO), pp. 42–45 (2014)
2. Bourjault, A.: Contribution a uneApprocheMethodologique de L'AssemblageAutomatise: Elaboration Automatique des Sequences Operatoires (Contribution to the methodology of automated assembly: automatic generation of operations sequences). Ph.D. Thesis, Université de Franche-Comté, Besançon, France (1984) (in French)
3. De Thomas, F., Whitney, D.E.: Simplified generation of all mechanical assembly sequences. IEEE J. Robot. Autom. **3**(6), 640–658 (1987)
4. Cohen, J.D., Lin, M.C., Manocha, D., Ponamgi, M.: I-COLLIDE: an interactive and exact collision detection system for large-scale environments. In: Proceedings of ACM International 3D Graphics Conference, pp. 189–196 (1995)
5. Ponamgi, M., Manocha, D., Lin, M.C.: Incremental algorithms for collision detection between solid models. In: Proceedings of the third ACM Symposium on Solid Modeling and Applications, pp. 293–304 (1995)
6. Hubbard, P.M.: Interactive collision detection. In: Proceedings of IEEE Symposium on Research Frontiers in Virtual Reality, pp. 24–31 (1993)
7. Fu, L.C., Liu, D.Y.: An efficient algorithm for finding a collision-free path among polyhedral obstacles. J. Robot. Syst. **7**(1), 129–137 (1990)
8. Gilbert, E.G., Foo, C.-P.: Computing the distance between general convex objects in three-dimensional space. IEEE Trans. Robot. Autom. **6**(1), 53–61 (1990)
9. Zeghloul, S., Rambeaud, P., Lallemand, J.P.: A fast distance calculation between convex objects by optimization approach. In: IEEE Proceedings on International Conference on Robotics and Automation, pp. 2520–2525 (1992)

A Probabilistic Method Toward SLAM for Mobile Robotic Systems

R.S. Anoop, T. Gireeshkumar and G. Saisuriyaa

Abstract Simultaneous localization and mapping (SLAM) problem helps a mobile robot in identifying its own position by providing an autonomously built map. This work proposes a software and hardware approach for online mobile robotic systems, which is capable of performing SLAM. The mapping of unknown environment with low-cost sensors, incorporating probabilistic method, is the highlight of this work. The hardware system comprises of a multisensor mobile robot developed on the ARM Cortex platform. The software part mainly incorporates pose graph data structure blended with mixture model, which is further optimized by stochastic gradient descent method

Keywords Simultaneous localization and mapping (SLAM) · Probabilistic robotics · Pose graph data structure · Gaussian mixture model

1 Introduction

Simultaneous localization and mapping (SLAM) in mobile robotic systems is a computational challenge, as the robot needs to continuously map its environment based on its current location, which can be resolved only when the information about the surroundings is known [1]. In planetary exploration, disaster environ-

R.S. Anoop (✉) · T. Gireeshkumar · G. Saisuriyaa
Amrita Vishwa Vidyapeetham University, Coimbatore 641112, India
e-mail: anooprajendran24@gmail.com

T. Gireeshkumar
e-mail: gireeshkumart@gmail.com

G. Saisuriyaa
e-mail: gsaisuriyaa@gmail.com

L.C. Jain et al. (eds.), *Intelligent Computing, Communication and Devices*,
Advances in Intelligent Systems and Computing 308,
DOI 10.1007/978-81-322-2012-1_30

ments, and military sectors, the use of mobile robots for mapping unknown environment is desirable. Here, huge amount of uncertainty occurs due to many factors such as unstructured environment, sensor inaccuracy, robot actuation errors, computational timing of processor, etc. [2]. To overcome all factors of uncertainty, a probabilistic approach for SLAM is needed. This approach represents ambiguity and degree of belief efficiently through mathematical models.

In this work, the sensor data obtained from different locations are used to form a spatial relationship between them. The sensor data are taken from an IR rangefinder, which is mounted on a servo motor fixed on the mobile robot. The sensor readings with respect to each servo angle are perceived to formulate a pose graph data structure [3, 4]. This pose graph is then used to generate probabilistic mixture model [5], and this model is further optimized to find the best optimal solution needed for SLAM. The algorithm for this purpose is proposed to be developed in the ARM Cortex platform, which is the core of multisensor mobile robot. The proposed algorithm, when implemented, does not obstruct the control action of the mobile robot.

The rest of the paper is organized as follows. Section 2 summarizes the literature survey. Section 3 shows the framework for the work. Finally, Sect. 4 encloses the conclusion and final remarks.

2 Literature Survey

Different methods have been developed in the past two decades in the field of probabilistic robotics for SLAM. Fox et al. [6] suggested a probabilistic approach for building large-scale indoor map using mobile robot. It puts forward an algorithm for maximum likelihood estimation that finds a solution for map generation problem. The work shows the experimental results in large cyclic environments and proves the approach to be appropriate and robust. The proposed work in this paper adopts certain features of the work of Pfingsthorn and Birk [3, 4], who extended the pose graph data structure with multimodal Gaussian distribution functions. Further, the optimization methods derived are compared with traditional state-of-the-art optimization methods such as stochastic gradient descent method, Levenberg–Marquardt method, etc. Finally, it was proved that prefilter stochastic gradient descent method and the prefilter Levenberg–Marquardt method to perform best.

3 Framework for the System

In this work, the mobile robot perceives environmental data from different points in its path. The data collection happens at periodic intervals by means of a servo actuated technique to construct the pose graph data structure. Using this pose graph, multivariate Gaussian distributions are generated that corresponds to each point at 300 ms time interval. After 1,500 ms, a set of distribution models are obtained. This collection of models is then summed up to get a mixture model. An optimization method has to be applied to make this data more accurate. Stochastic gradient descent method works well in this situation, since it optimizes the random variables obtained from a probability distribution function.

3.1 Software Design

Probabilistic approaches represent data as probabilistic distributions over a whole set of possible hypotheses. The idea of probabilistic approach toward SLAM, in this work, is formulated in three steps. Firstly, the real-time sensor data are used to develop a pose graph data structure. Secondly, a probabilistic mixture model is generated from this pose information. Further, this mixture model is filtered to obtain a single probability distribution and then optimized using stochastic gradient descent method. A simple schematic of the software model of the system is shown in Fig. 1.

Pose Graph Data Structure

Pose graph data structure can be denoted as a vector containing elements of vertices (*V*) and edges (*E*). The vertices are the positions where the sensor readings are obtained. Edges connect two consecutive vertices depending on constraint c_k [3, 4]. The spatial relationship between sensor data is obtained from the edges. The vertices v_i are denoted as:

$$v_i = (x_i, z_i) \tag{1}$$

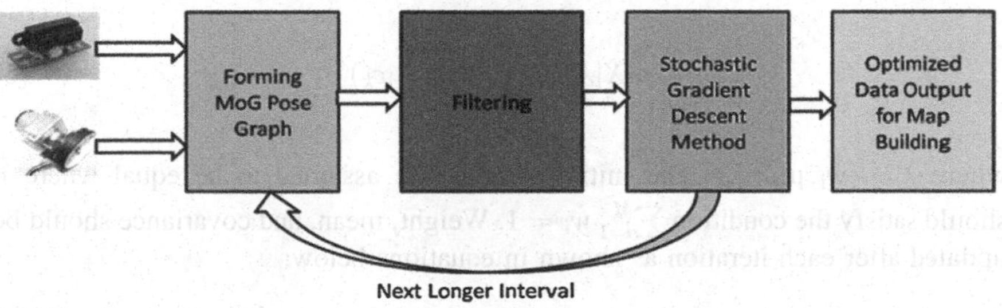

Fig. 1 Software model

where x_i is the rotation angle of servomotor and z_i is the sensor observation at x_i. The edges are denoted as:

$$e_k = \left(v_i, v_j, c_k\right) \tag{2}$$

where v_i is the current vertex, v_j is the next vertex, and c_k is the constraint, which is the distance between the vertex v_i and v_j. The vertices and edges can be expressed in such a way that v_i belongs to a whole set V and e_k belongs to a whole set E.

Generate Mixture Model

The mixture models can represent the ambiguities in sensor readings in an efficient way. For the sake of simplicity, the probability distribution function is modeled as Gaussian, which satisfies central limit theorem. The vector V is modeled as Gaussian; hence multivariate Gaussian distribution is needed. The equation for multivariate Gaussian distribution function [5] is as follows:

$$g(d_x|c_k) = \frac{1}{(2\pi)^{\frac{1}{2}} \cdot \sum_i \frac{1}{2}} \exp\left\{\frac{-1}{2}(d_x - \mu_k)^T \sum\nolimits_i^{-1} (d_x - \mu_k)\right\} \tag{3}$$

where d_x is the pose difference between the vertices v_i and v_j. The mean and covariance of the model are calculated initially by the following relation:

$$\sum\nolimits_{jk} = \frac{1}{N-1} \sum_{i=1}^{N} (d_x - \overline{x_j})(d_x - \overline{x_k}) \tag{4}$$

$$\bar{x} = \frac{1}{N-1} \sum_{i=1}^{N} d_x \tag{5}$$

where N is the number of observations. The mixture of Gaussian model is obtained by the weighted addition of the individual Gaussian models generated previously, as shown below:

$$p(X|\lambda) = \sum_{i=1}^{M} w_i g(d_x|c_k) \tag{6}$$

where $\lambda = \left(w_i, \mu_i, \sum_i\right)$. The initial weights are assumed to be equal where it should satisfy the condition $\sum_{i=1}^{M} w_i = 1$. Weight, mean, and covariance should be updated after each iteration as shown in equations below:

$$\overline{w_i} = \frac{1}{T}\sum_{t=1}^{T} P_r(i|x_t, \lambda) \tag{7}$$

$$\overline{\mu_i} = \frac{\sum_{t=1}^{T} P_r(i|x_t, \lambda) \cdot x_t}{\sum_{t=1}^{T} P_r(i|x_t, \lambda)} \tag{8}$$

$$\overline{\sigma_i^2} = \frac{\sum_{t=1}^{T} P_r(i|x_t, \lambda) \cdot x_t^2}{\sum_{t=1}^{T} P_r(i|x_t, \lambda)} - \overline{\mu_i}^2 \tag{9}$$

where

$$P_r(i|x_t, \lambda) = \frac{w_i g(d_x|c_k)}{\sum_{k=1}^{M} w_k g(d_x|c_k)} \tag{10}$$

Filtering and Optimizing

The generated mixture model has to be averaged to a single Gaussian component. For this purpose, expectation maximization method [7] is used. The component with highest weighted probability is chosen in this method. Further, the resulting Gaussian component is optimized using stochastic gradient descent (SGD) method.

3.2 Hardware Design

Since the sensor data have high importance in mapping, the hardware needs to be designed to satisfy the proper action and observation of the mobile robot. The core of the mobile robot is the MBED Board, which is developed with ARM Cortex M3 Microcontroller. The proposed hardware design of the system is divided into three perspectives: the sensor and actuation part, a communication module, and a remote station provided with graphical user interface. Figure 2 presents the schematic for the overall system architecture and the above-mentioned three modules. The sensor and actuation parts help in the robot motion and observation based on the microcontroller program. The communication module is selected on the basis of the required data rate and the distance between the mobile robot and the remote station. The remote station is a personal computer (PC). The installed control software in the PC allows the user to receive environmental information from the remotely located mobile robot, and this software is used to build the graphical user interface (GUI) showing the map of the robot environment.

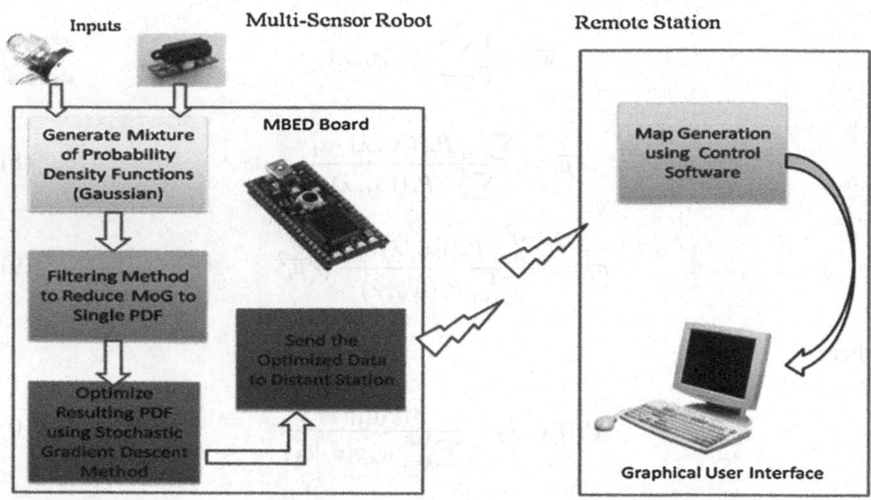

Fig. 2 Overall system architecture

4 Conclusion

An idea for SLAM using a probabilistic method is formulated in this paper. The advantages of various techniques adopted in the past were successfully included in the proposed work. The use of pose graph data structure and the mixture models helps to make a 2D map resembling the ground truth. The presented approach can be extended for a dynamic and unpredictable environment by imposing more probabilistic constraints from robot perception and action. Applications such as space exploration and military area exploration demand the best feasible path in the map from a starting point to destination. In these cases, the path planning algorithms can also be incorporated with the presented approach.

References

1. Frese, U.: A discussion of simultaneous localization and mapping. Auton. Robots **20**(1), 25–42 (2006)
2. Thrun, S., Burgard, W., Fox, D.: Probabilistic robotics. MIT press, Cambridge (2005)
3. Pfingsthorn, M., Birk, A.: Simultaneous localization and mapping with multimodal probability distributions. Int. J. Robot. Res. **32**(2), 143–171 (2013)
4. Pfingsthorn, M., Birk, A.: Handling local and global ambiguities via a generalized graph SLAM framework based on multimodal and hyperedge constraints. ICRA Workshop on robust and multimodal inference in Factor Graphs (2013)
5. Reynolds, D.: Gaussian mixture models. Encycl. Biometrics, 659–663 (2009)

6. Fox, D., Burgard, W., Thrun, S.: Probabilistic methods for mobile robot mapping. In: Proceedings of the IJCAI-99 Workshop on Adaptive Spatial Representations of Dynamic Environments (1999)
7. Thrun, S., Burgard, W., Fox, D.: A probabilistic approach to concurrent mapping and localization for mobile robots. Auton. Robots **5**(3–4), 253–271 (1998)

A Brief Survey on Concept Drift

V. Akila and G. Zayaraz

Abstract The digital universe is growing rapidly. The volume of data generated per annum is in the order of zeta bytes due to the proliferation of the Internet. Many real-world applications generate data that are continuous. This type of data is known as data streams. Examples of applications generating this kind of data are business transactions, Web logs, sensors networks, etc. The data stream is analyzed, and the underlying concepts are extracted to make predictions and decisions in real time. But as data streams evolve over time, they undergo concept drift. Concept drift means the statistical properties of the data stream change over time in unforeseen ways. This causes problems because the predictions based on the data streams become less accurate as time passes. To understand the behavior of data streams, it is important to investigate the changes of the data distributions and the causes of the changes. Therefore, periodic retraining, also known as refreshing, of any model is necessary. The survey covers the various techniques available in the literature to handle concept drift in data streams.

Keywords Data distributions · Data stream · Concept drift

1 Introduction

In today's information society, the emergent applications produce data streams at very high rates. These data streams can be explored to retrieve interesting patterns or to make predictions in real time. The data streams are dynamic, massive in size,

V. Akila (✉) · G. Zayaraz
Department of Computer Science and Engineering,
Pondicherry Engineering College, Pondicherry, India
e-mail: akila@pec.edu

G. Zayaraz
e-mail: gzayaraz@pec.edu

L.C. Jain et al. (eds.), *Intelligent Computing, Communication and Devices*,
Advances in Intelligent Systems and Computing 308,
DOI 10.1007/978-81-322-2012-1_31

and transient. Data streams are a sequence of time-stamped tuples that arrives in unbounded streams. It is not possible to store the data streams permanently due to their size. For the data streams to be useful, it has to be processed at the time it is generated and discarded later. Further, the underlying concept in the data streams also change. Concept drift refers to changes in the conditional distribution of the output (i.e., target variable) given the input (input features), while the distribution of the input may stay unchanged. Typical examples are customer preferences, spam mail, etc. Frequently, these changes make the model built over this data stream as obsolete. This problem known as concept drift is a complication in the field of incremental learning. The model can be relevant only if regular update is performed. The difficulty in the field of concept drift is distinguishing between concept drift and noise [1].

There are three types of concept drift that occur (i) sudden drift, (ii) gradual drift, and (iii) recurring drift. Sudden drift refers to change in class distribution. It occurs when a new class appears in the data stream and instances of the old classes disappear from the data stream. Gradual drifts are not as dramatic as sudden drifts. The class distribution is modified at a slower rate. Recurrent drifts are concept drifts that reappear with time.

In the given scenario, there is a need to stop, pause, and take stock of the concept drift handling techniques for data streams that are omnipresent in today's world. This paper presents an extensive survey of techniques that handle concept drift in data streams. The paper offers useful insights on the assumptions made by the concept handling techniques as well as the classification of techniques for concept drift handling. The assumptions are:

i. The underlying data stream is without impurities, which may not be the case in the real world.
ii. There is a single underlying concept that changes with time.

The classification of techniques is dealt in detail in the following heading.

2 Related Work

The classification of techniques that are used to handle concept drift derived from the survey is depicted in Fig. 1. The techniques can be broadly classified along these dimensions: (i) classifier type, (ii) drift type, (iii) data, and (iv) classifier update. The survey is presented on the perspective of drift type. Further, the techniques were compared along these directions:

i. Use of ensemble
ii. Learning type—online or batch
iii. Data—data distribution, class labels, or combination of the two
iv. Key feature employed in the technique.

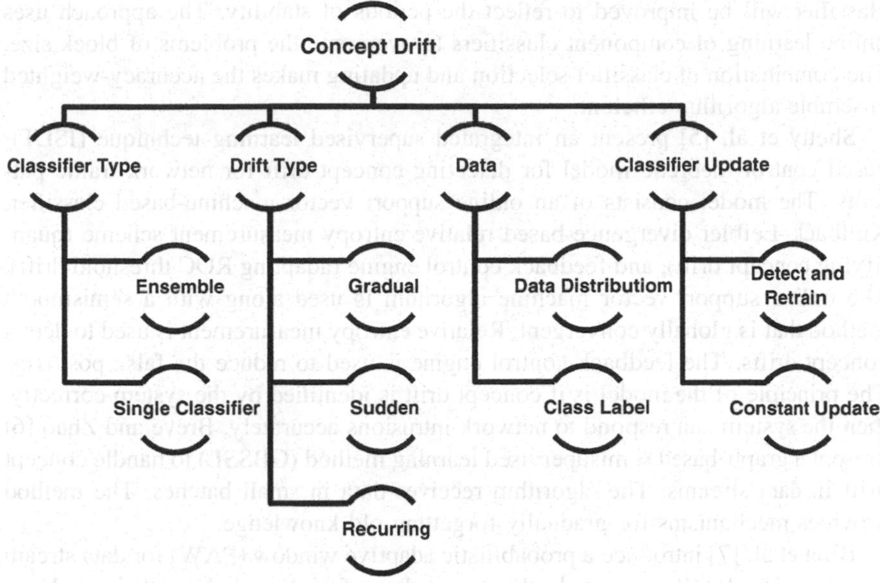

Fig. 1 Classification of concept drift handling techniques

2.1 Gradual Drift

The techniques handling gradual concept drift are analyzed below.

Masud et al. [2] make a threefold contribution: (i) an adaptive threshold for outlier detection, (ii) a probabilistic approach for class detection based on Gini Coefficient, and (iii) simultaneous multiple novel class detection. The system reduces the false alarm rate, increases the detection rate, and is applicable for multiclass problems. A new unified measure by name Neighborhood Silhouette Coefficient, which codifies the cohesion and separation of data, is presented. A flexible decision boundary for outlier detection is implemented by allowing slack space outside the decision boundary. This slack space is controlled by an adaptive threshold. The novel class instances are then detected by using a Gini Coefficient ψ. Alippi et al. [3] present a just-in-time adaptive classifier for gradual concept drifts. The method treats the drift as a sequence of stationary changes and uses a knowledge base management procedure to keep the classifier coherent with the current state of process. A change detection test is used to assess the variations in the trends. If there are no variations detected, then a polynomial regression is performed to modify the online knowledge base. If variations are detected, then the obsolete samples are deleted from the knowledge base.

Brzezinski and Stefanowski [4] recommend an accuracy-updated ensemble algorithm (AUE), which is an extension of accuracy-weighted ensemble algorithm. The proposed algorithm updates the classifier according to the data distribution. When no drift occurs between a sequence of blocks, then the component classifier will be improved to reflect the periods of stability. The approach uses online learning of component classifiers to overcome the problems of block size. The combination of classifier selection and updating makes the accuracy-weighted ensemble algorithm efficient.

Shetty et al. [5] present an integrated supervised learning technique (ISLT)-based control theoretic model for detecting concept drift for network traffic patterns. The model consists of an online support vector machine-based classifier, Kullback–Leibler divergence-based relative entropy measurement scheme (quantifying concept drift), and feedback control engine (adapting ROC threshold drift). The online support vector machine algorithm is used along with a semismooth method that is globally convergent. Relative entropy measurement is used to detect concept drifts. The feedback control engine is used to reduce the false positives. The principle of the model is if concept drift is identified by the system correctly, then the system can respond to network intrusions accurately. Breve and Zhao [6] present a graph-based semisupervised learning method (GBSSL) to handle concept drift in data streams. The algorithm receives data in small batches. The method provides mechanisms for gradually forgetting old knowledge.

Bifet et al. [7] introduce a probabilistic adaptive window (PAW) for data stream learning. The PAW stores only the recent data of high probability using a logarithmic number of instances. Turkov et al. [8] propose a Bayesian-based approach for pattern recognition problem. The time-varying parameters are treated as hidden random processes processing the Markov property. The approach uses an approximate dynamic programming-based algorithm to solve a two-class problem of spam emails. Minku and Yao [9] introduce an online ensemble learning approach called diversity for dealing with drifts (DDD). Concept drift is considered to be a change in distribution problem. The paper emphasizes on the diversity of ensemble to obtain generalization on a new concept. The paper presents an analysis of the effect of low- or high-density ensembles on the prequential accuracy depending upon the drift. It is shown that information from the old concept can help the learning of a new concept. This is done by training ensembles with high diversity on the old concept and using low diversity on the new concept. The DDD uses the diversity to handle the drift. The DDD system is robust against false alarms and quick in adapting to evolving concept.

A summary of the techniques analyzed for gradual drift is given in Table 1. From the table, it is evident that most of the techniques are based on batched learning techniques. Also the input stream is portioned into chunks before processing by the ensemble. The data analyzed are the data point distribution or the class label distribution or both.

2.2 Hybrid

The techniques handling gradual concept drift and sudden drifts together are analyzed below.

Table 1 Summary of techniques handling gradual drift

Technique	Ensemble	Learning		Data			Key feature
		Online	Batch	Data points	Class labels	Hybrid	
Adaptive threshold probabilistic approach for class detection based on Gini Coefficient and simultaneous multiple novel class detection	√		√		√		Adaptive threshold
Just-in-time adaptive classifier for gradual concept drifts			√	√			Knowledge management
AUE	√		√	√			Classifier update
ISLT		√		√			Adaptive threshold mechanism
GBSSL			√			√	Gradual forgetting
PAW	√				√		

The data stream is considered as a set of contexts by Gama et al. in [10]. The problem of drift detection is studied in the form of class probability detection. The instances of change in the context are identified as concept drift. The drift may be gradual or sudden. Three learning algorithms are used, perceptron, neural network, and decision tree. The system tries to control the online error rate of the algorithm. The key idea is, when the distribution of data is stationary, then error will decrease but when the distribution changes, then error will increase. Ross et al. [11] propose a method for detecting concept drift based on exponentially weighted moving average algorithm (EWMA) to calculate the error rate of the classifier. The method is used for two-class classification problem and makes only one pass over the data stream. The algorithm uses feedback called error stream. This error stream is viewed as a Bernoulli distribution with the Bernoulli parameter 'p' being the probability of misclassifying a point. The method is tested for sudden as well as gradual concept drift. A warning threshold is used to flag the beginning of the concept drift. After the warning flag is set the data stream is stored in the memory.

Brzezinski and Stefanowski [12] present the accuracy-updated ensemble (AUE2). This approach combines accuracy-based weighting mechanisms with the Hoeffding Trees. This hybrid approach allows AUE2 to perform well in sudden, gradual, recurring, short-term, and mixed drifts. Fixed block of input data stream with adaptive ensembles is used. For every new block, the classifiers are validated and the weakest of them is removed. The AUE2 system combines the best features of incremental learning with block-based ensembles. Susnjak et al. [13] advocate for a hybrid method that combines detect-and-retrain and constant-update of ensembles. A concept learning algorithm (CLA) is designed. The work is the extension of cascading training framework. The uniqueness of the approach is the development of layer confidence thresholds.

Deckert [14] presents a new framework, batch-weighted ensemble (BWE) for handling sudden and gradual concept drift. Equal-sized batches are used for the learning process. The ensemble does not build a new classifier on each batch. The classifiers' weight is computed by estimating the error rate of the recent batch. The proposed approach embeds in it a drift detection method by name batch drift detection method. The drift detection method builds a classification accuracy table, which is used to build a regression model. The regression model shows the direction of change as well as the level of change as warning and drift. If it is a warning level, then a new classifier is built on the batch and added to the ensemble. If the level is drift, then also a new classifier is built and added to the ensemble and the weights for the component classifier are established. Yeh and Wang [15] use a least square latent discriminant analysis (LSLDA) model. A novel rank one update method with a simplified class indicator matrix is proposed. The LSLDA model is useful in observing and modeling the data distribution changes. The proposed method is useful in recognizing newly added class labels. A forgetting factor is used to suppress the out of date data to enhance adaptively. The model is suitable for both sudden and gradual drifts.

Zhu [16] recommends a classification algorithm based on double window mechanism for handling concept drift streams (DWCDS). An N-basic classifiers is generated using N-cell sliding window. When the sliding window is full, the distribution changes of the data are checked to detect concept drifts. Voting mechanism is used with the ensemble classifier.

A summary of the techniques analyzed for gradual and sudden drift is given in Table 2. From the table, it is evident that most of the techniques are based on ensemble methods. The data streams are partitioned to chunks. To handle gradual drifts, the chunk size should be large and when handling sudden or abrupt drifts, the chunk size should small.

2.3 Recurrent Drift

The techniques analyzing the concept drifts that reappear in time are given below.

Table 2 Summary of techniques handling gradual and sudden drift

Technique	Context	Ensemble	Learning		Data			Key feature
			Online	Batch	Data points	Class labels	Hybrid	
Perceptron, neural network, and decision tree	√		√		√			Error rate
EWMA						√		Error stream
AUE2		√	√	√	√			Combine incremental with block-based ensemble
CLA		√						Layer confidence thresholds
BWE		√		√				Drift detector
LSLDA							√	
DWCDS		√			√			Double window

Alippi et al. [17] present a just-in-time classifier for recurrent concept drift. This system not only monitors the distribution of data points but also the class labels. As soon as a concept drift is detected, concept isolation is created and compared with the stored concepts. If a match occurs, then the just-in-time classifier uses the recurrent concept. The base classifiers used are *k*-NN classifier, NB classifier based on Gaussian distributions, and SVM classifier. Li et al. [18] present recurring concept drifts and limited labeled data (REDELLA), a decision tree approach for handling recurrent concept drift. The system works on unlabeled data. A *k*-means clustering algorithm is used to produce concept clusters, and unlabeled data are labeled using majority class at leaves method. A pruning mechanism is installed after a detection period to avoid over fitting. The prediction results are evaluated periodically to track the performance. A context-aware learning from data streams (CALDS) exploits the association between context information and learned concepts [19]. The method employs the error rate of the learning algorithm and relation between the context and the recurring concept. The paper presents a method for model similarity, representing context, model management, learn relation between context and concepts, and comparison of contexts and concepts.

Gomes et al. [20] present an ensemble-based approach for recurring drift. Stable concepts are identified using change detection methods based on error rate. Context information is given as a set of attribute value variables, which is integrated with the learned concepts to create a hierarchy. The context is represented

Table 3 Summary of techniques handling recurring drift

Technique	Context	Ensemble	Learning		Data			Key feature
			Online	Batch	Data points	Class labels	Hybrid	
Just-in-time (JIT) classifiers		√					√	
REDELLA			√			√		Unlabeled data
CALDS	√					√		Context
Context-aware learning system	√	√						

as an object in a multidimensional Euclidean space. The context similarity is derived using cosine similarity. The concept change detection is performed based on the error rate. To learn the relationship between the context attributes and underlying concepts, Naïve Bayes classifier is used. A summary of the techniques analyzed for recurrent drift is given in Table 3. From the table, it is evident that the techniques are based on ensemble methods. The data are analyzed based on class labels. The context in which the drift occurs is also taken to account.

3 Conclusion

Data streams are infinite in length and are continuous. The data streams can be used to infer knowledge to make real-time decisions. The underlying concept in the data stream is subject to change. The survey offers a macro-level view of the current scenario in concept drift handling methods.
The following observations are made from the survey:

- Many techniques are based on ensembles.
- Error rate is an important parameter used to detect the concept change.
- Determination of the appropriate data stream chunk size that is suitable for sudden as well as gradual drift is an open problem.
- The change in data point distribution as well as class label change contributes to concept drift. So, methods that consider both these parameters exhibit better performance.
- The number of passes made on the data stream contributes to the complexity of the drift handling techniques.
- Gradual drifts and a combination of gradual drift and sudden drift are explored extensively.
- Two-class classification problems are explored widely, whereas there is a dearth of techniques for multiclass problems.

The techniques that handle concept drift so far assume that there is a single underlying concept, and there are no uncertainties in the data stream. These assumptions do not hold true for real-world data streams. The survey brings attention to the fact that concept drift handling technique that can handle uncertain data streams with multiple change of concept is the need of the hour.

References

1. Tsymbal, A.: The problem of concept drift: definitions and related work. Technical report TCD-CS-2004-15, Trinity College Dublin, Ireland, pp. 1–7 (2004)
2. Masud, M., Chen, Q., Khan, L., Aggarwal, C., Gao, J., Han, J., Thuraisingham, B.: Addressing concept-evolution in concept-drifting data streams. IEEE International Conference on Data Mining, pp. 929–934 (2010)
3. Alippi, C., Boracchi, G.,Roveri, M.: An effective just-in-time adaptive classifier for gradual concept drifts. In: International Joint Conference on Neural Networks, pp. 1675–1682 (2011)
4. Brzezinski, D., Stefanowski, J.: Accuracy updated ensemble for data streams with concept drift. In: 6th International Conference on Hybrid Artificial Intelligent Systems. Lecture Notes in Computer Science, vol. 6679, pp. 155–162. Springer (2011)
5. Shetty, S., Mukkavilli, S.K., Keel, L.H.: An integrated machine learning and control theoretic model for mining concept-drifting data streams. In: IEEE International Conference on Technologies for Homeland Security (HST), pp. 75–80 (2011)
6. Breve, F., Zhao, L.: Particle competition and cooperation in networks for semi-supervised learning with concept drift. In: The 2012 International Joint Conference on Neural Networks (IJCNN), pp. 1–6 (2012)
7. Bifet, A., Read, J., Pfahringer, B., Holmes, G.: Efficient data stream classification via probabilistic adaptive windows. In: Proceedings of the 28th Annual ACM Symposium on Applied Computing, pp. 801–806 (2013)
8. Turkov, P., Krasotkina, O., Mottl, V.: The bayesian logistic regression in pattern recognition problems under concept drift. In: International Conference on Pattern Recognition, pp. 2976–2979 (2012)
9. Minku, L.L., Yao, X.: DDD: New ensemble approach for dealing with concept drift. IEEE Trans. Knowl. Data Eng. **24**(4), 619–633 (2012)
10. Gama, J., Medas, P., Castillo, G., Pedro Rodrigues, P.: Learning with drift detection. In: Advances in Artificial Intelligence, pp. 286–295. Springer, Berlin Heidelberg (2004)
11. Ross, G.J., Adams, N.M., Tasoulis, D.K., Hand, D.J.: Exponentially weighted moving average charts for detecting concept drift. Pattern Recognit. Lett. **33**(2), 191–198 (Elsiever) (2012)
12. Brzezinski, D., Stefanowski, J.: Reacting to different types of concept drift: the accuracy updated ensemble algorithm. IEEE Trans. Neural Netw. Learn. Syst. 25(1), 81–94 (IEEE) (2013)
13. Susnjak, T., Barczak, A.L.C., Hawick, K.A.: Adaptive ensemble based learning in non-stationary environments. In: International Conference on Neural Information Processing. LNCS, vol. 6443, pp. 438–445 (2010)
14. Deckert, M.: Batch weighted ensemble for mining data streams with concept drift. In: International Symposium on Methodologies for Intelligent Systems. LNAI, vol. 6804, pp. 290–299 (2011)
15. Yeh, Y., Wang, Y.F.: A rank-one update method for least squares linear discriminant analysis with concept drift. Pattern Recogn. **46**(5), 1267–1276 (2013)

16. Zhu, Q., Hu, X., Zhang, Y., Li, P., Wu, X.: A double-window-based classification algorithm for concept drifting data streams. In: IEEE International Conference on Granular Computing, pp. 639–644 (2010)
17. Alippi, C., Boracchi, G., Roveri, M.: Just-In-Time classifiers for recurrent concepts. IEEE Trans. Neural Netw. **24**(4), 620–634 (IEEE) (2013)
18. Li, P., Wu, X., Hu, X.: Mining recurring concept drifts with limited labeled streaming data. ACM Trans. Intell. Syst. Technol. **3**(2), 29:1–29:32 (ACM) (2012)
19. Gomes, J.B., Menasalvas, E., Sousa, P.A.: CALDS: Context-aware learning from data streams. In: Proceedings of the First International Workshop on Novel Data Stream Pattern Mining Techniques (StreamKDD'10), pp. 16–24 (2010)
20. Gomes, J.B., Menasalvas, E., Sousa, P.A.: Learning recurring concepts from data streams with a context-aware ensemble. In: Proceedings of the 2011 ACM Symposium on Applied Computing (SAC'11), pp. 994–999 (2011)

Live Virtual Machine Migration Techniques—A Technical Survey

T.Y.J. Naga Malleswari, G. Vadivu and D. Malathi

Abstract Cloud computing is achieved through virtualization. It means sharing of computing resources such as processors, memory, and I/O devices, thus making more utilization of computer systems. Virtual machines are simulated by virtual machine monitor (VMM) or hypervisor. Load balancing and power consumption are the two critical issues in cloud environment. This can be resolved by virtual machine migration. Virtual machine migration is the process of transferring a virtual machine from overloaded or under-loaded physical host to another physical host to balance the load or to reduce the consumption of resources or power. The important metrics to be focused in virtual machine migration are downtime and total migration time. So, any migration technique should transfer the virtual machine from one host to another host with minimum downtime and total migration time must be seamless. This paper describes types of virtualization, types of migration, several live migration techniques, their comparison, and the metrics that measure the performance of live migration.

Keywords Cloud computing · Virtualization · Live virtual machine migration · Gang migration · Memory compression · Deduplication · Shared storage

T.Y.J. Naga Malleswari (✉) · D. Malathi
Department of Computer Science and Engineering, SRM University, Chennai, India
e-mail: nagamalleswari.t@ktr.srmuniv.ac.in

D. Malathi
e-mail: malathi.d@ktr.srmuniv.ac.in

G. Vadivu
Department of Information and Technology, SRM University, Chennai, India
e-mail: vadivu.g@ktr.srmuniv.ac.in

L.C. Jain et al. (eds.), *Intelligent Computing, Communication and Devices*,
Advances in Intelligent Systems and Computing 308,
DOI 10.1007/978-81-322-2012-1_32

1 Introduction

Cloud Computing is a model having a pool of computing resources such as network servers, storage, applications, and services that can be shared by multiple users on demand by utilizing high speed network. The customer pays for the used resources and is called utility computing. Clouds allow the user to access virtualized resources such as hardware and development platforms by virtualization. Abstraction of computing resources such as storage, memory, processing power, and network or I/O is called virtualization. It is a system behaving more than one of the same systems [1].

Virtual machine migration service is the process of moving a virtual machine (VM) from one host server to another. There are various techniques of virtual machine migration such as live/hot migration or non-live/cold migration. All the system components such as CPU, networking, memory, and storage disks are virtualized; thus, the virtual machine state is captured as a set of easily moved data files [2].

In this paper, Sect. 1 gives the introduction of virtualization and VM migration. Section 2 tells virtualization types and Sects. 3–7 describes types of migration. Section 8 refers to basic techniques of live VM migration. In Sect. 9, other techniques of live VM migration are discussed. The metrics were described in Sect. 10. The techniques explained in Sect. 9 are compared in Sect. 12. The paper ends with findings and conclusion in Sects. 11 and 13 with the help of Sect. 12.

2 Virtualization

In Virtualization, a request for service is separated from the physical delivery of that service [3]. A hypervisor or virtual machine monitor (VMM) or virtualization layer is introduced in between operating system and hardware to provide virtualization.

2.1 Benefits

(i) On single computer several VMs can run with multiple OS instances concurrently.
(ii) It is less expensive and the demand of physical infrastructure is reduced by allowing several VMs on a high capacity server rather having on many small servers.
(iii) For testing applications on various platforms in software development process.
(iv) Dynamic partitioning and sharing of physical resources such as CPU, RAM, and I/O [4].
(v) Spending less time on provisioning, configuration, monitoring, and maintenance.

2.2 Types

(A) Hardware Virtualization

Figure 1 shows how the physical resource divided into multiple VMs with various workloads. Each VM assumes that the total underlying physical resources are owned by it. Hypervisor or virtualization layer [2] plays a vital role in it, and it manages scheduling and allocating physical resource. The OS on VM is called Guest OS.

(i) *Full Virtualization* emulates the bare machine hardware features such as I/O operations, memory access, instruction set, and interrupts into a VM. No modification required in Guest OS as it does not aware of being virtualized. For example VMWs are ESXi.
(ii) *Para Virtualization* [5] needs guest programs to be modified. It improves performance and efficiency by providing communication between the hypervisor and guest OS with the help of the software interface. It involves the OS kernel to be modified. Hyper call interfaces are provided by the virtualization layer to manage kernel critical operations such as memory management and interrupt handling. For example Xen.
(iii) *Hardware*-Assisted *Virtualization* [6] or accelerated virtualization. Xen refers it as hardware virtual machine (HVM). It introduces a new CPU execution mode in which hypervisor automatically trapped by the privileged and sensitive calls, thus removing the need of para virtualization. Hypervisor will run in new root mode.

(B) Memory Virtualization

CPUs have memory management unit (MMU) and translation look aside buffer (TLB) for virtual memory management. The guest OS maps the virtual addresses to the guest memory physical addresses. The hypervisor maps guest physical memory to the host actual memory using shadow page tables [3] and tables are updated dynamically.

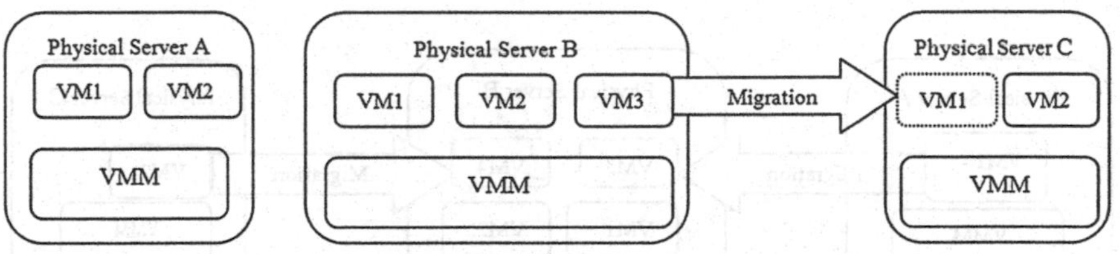

Fig. 1 Load balancing by migrating VM3 from physical server B to physical server C

(C) **Device and I/O Virtualization**

It manages and route the I/O requests between shared physical hardware and virtual devices using virtual NICs [3]. Physical hardware is virtualized and each VM is provided with a set of virtual devices by the hypervisor.

3 Virtual Machine Migration

VM migration is the process of transferring a VM from one physical host to another physical host to achieve load balancing, server consolidation to reduce the consumption of resources, power, fault tolerance, and online maintenance [7]. There are two methods of VM migration. They are cold migration and hot migration. Cold or non-live migration is migration of a powered off VM from one host to another [8]. The drawbacks of this method are VM status is lost and service interruption to the user [7]. Latter is the process of transferring running VM from one physical host to another without disconnecting the system. Storage, network connectivity, and memory of the virtual machine are transferred from the source machine to the destination machine. This is called live VM migration or hot migration. In this, the memory state and CPU registers are transferred to destination. As the VM is in running, data cannot be lost while migration. It does not require local disks to hold VM images rather it needs network attached storage (NAS) which acts as hard drive for the VMs and is accessed by physical machines [9]. Total migrating time is less in cold migration and down time is seamless.

3.1 Goals

(i) *Load Balancing*: Distribution of load across the physical servers improves the scalability. By migrating VMs from overloaded server to lightly loaded machines, the overall system load can be balanced [10].

(ii) *Server Consolidation/Energy Saving*: Maximization of resources and minimization of energy consumption is done by VM consolidation [11]. See in Fig. 2. Physical servers A and C have one VM each. These are consolidated

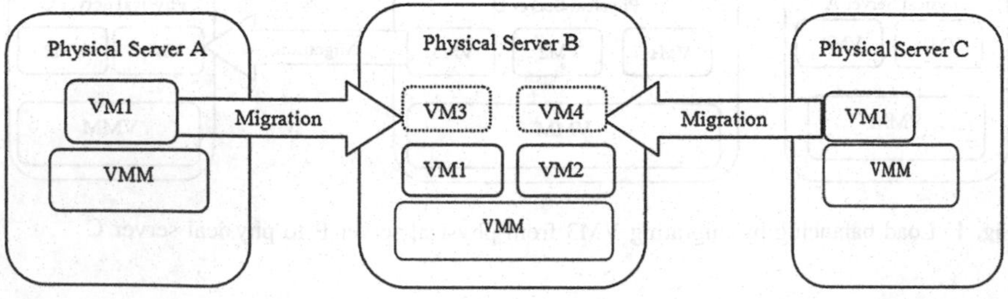

Fig. 2 Server consolidation by migrating VMs from physical host A and C to physical host B

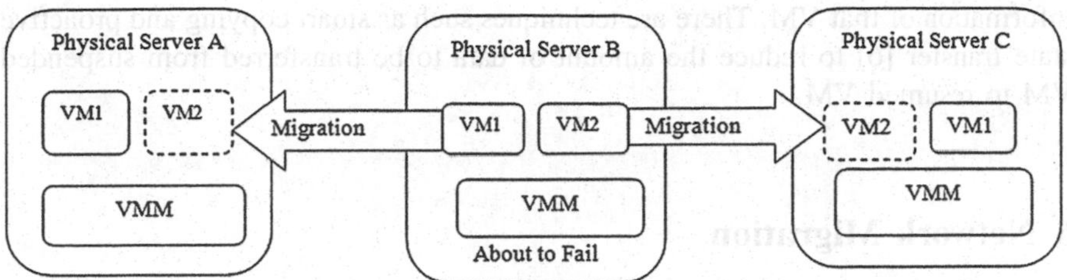

Fig. 3 Fault tolerance VMs migrated from host B–A and C as host B about to fail

and migrated to physical server B, and A and C can be switched off. Thus, reduces power consumption by servers.

(iii) *Fault Tolerance*: If any physical server predicted to be failed, VMs on this server migrated to other physical server, thus providing fault tolerance by means the availability of physical server is improved and avoids performance degradation [12] of applications. Figure 3 shows this.

(iv) *Online Maintenance*: VMs are migrated without disconnecting [7].

4 Memory Migration

It is process of migrating VM memory instance from source to destination. It has [13].

(i) *Push phase*: When VM on source still running the hypervisor transfers all memory pages to destination. The pages that are dirtied during transmission are resent again until the rate of recopied pages is more than dirtying rate to ensure consistency.

(ii) *Stop-and-copy phase*: The new VM is started after copying the memory pages of stopped source VM to destination.

(iii) *Pull phase*: When a required page is not found in VM after its execution starts on destination machine then page fault occurs, then the page is pulled from the source VM across the network.

5 File System Migration

A consistent and location-independent view of the file system is being available on all machines to support VM migration. Distributed file system is used to transfer the files of suspended VM state. The hypervisor stores the contents of each VM's virtual disks in local files and is transferred to destination along with other state information of that VM. There are techniques such as smart copying and proactive state transfer [6] to reduce the amount of data to be transferred from suspended VM to resumed VM.

6 Network Migration

Each VM having its own MAC address and virtual IP address to communicate with other remote systems. Mapping of virtual IP and MAC addresses to their corresponding VMs is done by the hypervisor. If the machines included in VM migration are connected with switched network, an unsolicited ARP reply is provided from the migrating host and that advertises the IP moved to a new location. Thus, future packets are sent to new location by reconfiguring all the peers. The migrating OS have actual MAC address to detect its move to a new port.

7 Device Migration

The physical hardware of each VM is virtualized and is presented in it with a set of standardized virtual devices. Device migration is possible through

(i) *A* device is emulated in software using *emulation*. (ii) Devices on the host system is virtualized using *virtualization*. (iii) All requests of a *nonmigratable device* driver are passed to the host machine, and as long as the device is in use migration is disallowed.

8 Basic Techniques

(A) Precopy Approach

It consists of push phase with stop-and-copy phase. The hypervisor sends a request to migrate OS from source host to destination host (*Premigration*). The destination host is checked for the availability of resources, if not found then VM on source host not affected. Otherwise the required resource is reserved for VM at destination host (*Reservation*). Then the source VM is stopped and all the pages are transferred to destination from source and are called as working set. During transferring, some pages may be dirtied. Repeat transferring dirty pages until the rate of recopied pages is not less than dirty page rate [14] (*Iterative Precopy*). VM on source host is suspended and further the network traffic is redirected to destination host (*Stop and Copy*). In case of any failure, copy of source host is used. An acknowledgment is sent to source host after receiving the OS image by destination host. Once the consistent copy of source VM is received, VM on source host is discarded (*Commitment*) and VM on destination host is activated or resumed (*Activation*) (Fig. 4).

(B) Post-Copy Approach

Push phase with stop and copy called as post-copy approach. Source VM is stopped and minimum processor state transferred to destination and VM is activated. Transferring memory pages to destination occurs on demand. The advantages over precopy approach are each memory page is moved to destination only once; total migration time and number of pages transferred are less [15] (Fig. 5).

Post-copy approach with improvements is listed as follows:

(i) *Post-copy* via *demand paging*: When the migrated VM on destination host started, and the pages it needs not in the memory then page fault occurs, it can be serviced by requesting the page over the network. The page is transferred and slows down the VM due to network traffic.

(ii) *Post-copy* via *active paging*: The pages are proactively pushed [16] to the VM while it is running. If any page faults, those can be serviced by demand

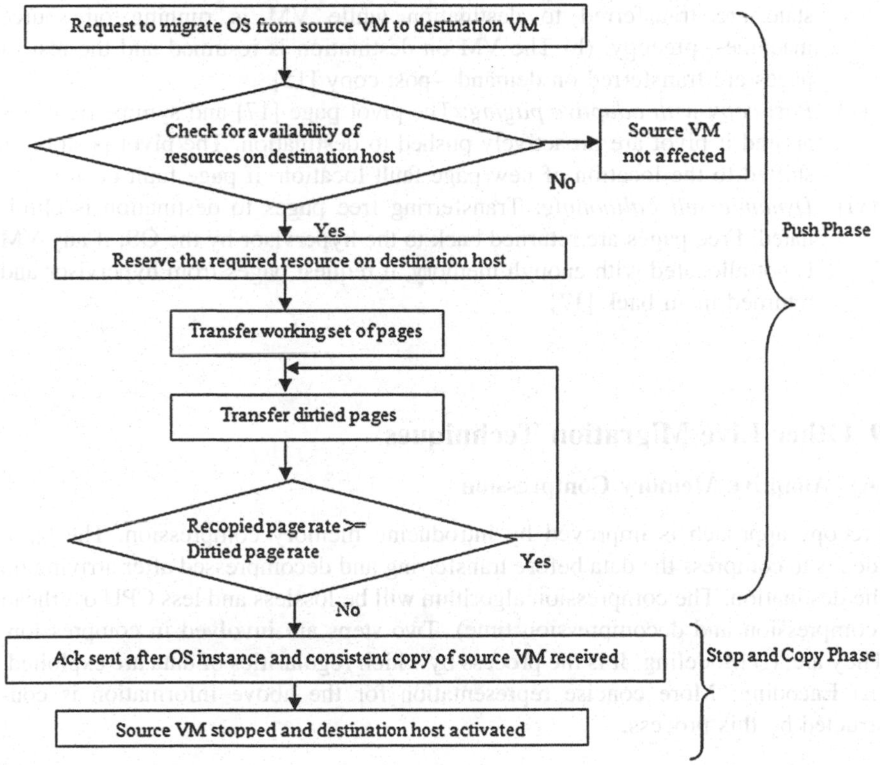

Fig. 4 Precopy approach

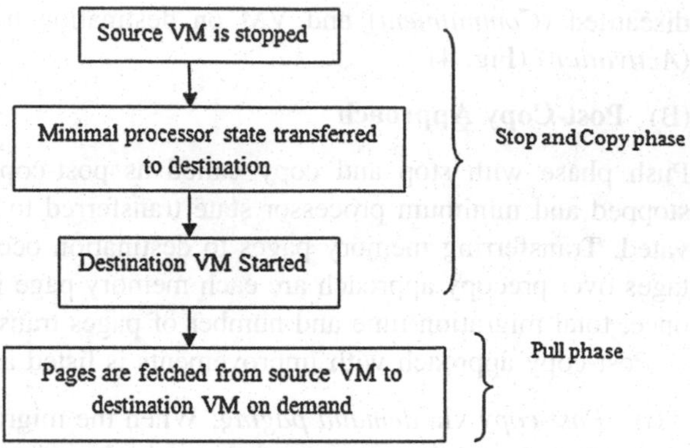

Fig. 5 Post-copy approach

paging. For the transferred pages, no page faults occur. Page faults given more priority than pushing pages.

(iii) *Post-copy* via *pre-paging*: The pages which are to be accessed in near future are transferred to the VM. This is like active paging technique but predicting the special locality of VM memory access pattern.

(iv) *Hybrid pre- and post-copy*: It includes: (a) The memory pages and processor state are transferred to destination while VM is running on source machine—precopy. (b) The VM on destination is resumed and the rest of pages are transferred on demand—post-copy [15].

(v) *Post-copy with adaptive paging*: The pivot page [17] and symmetric pages around it pivot are proactively pushed to destination. The pivot position is shifted to the location of new page fault location, if page fault occurs.

(vi) *Dynamic self ballooning*: Transferring free pages to destination is eliminated. Free pages are returned back to the hypervisor by the OS; if any VM is not allocated with enough memory, it request pages from hypervisor and returned them back [17].

9 Other Live Migration Techniques

(A) Adaptive Memory Compression

Precopy approach is improved by introducing memory compression. The basic idea is to compress the data before transferring and decompressed after arriving on the destination. The compression algorithm will be lossless and less CPU overhead (compression and decompression time). Two steps are involved in compression. They are: (i) Modeling: It is the process by which regularities of data are exploited. (ii) Encoding: More concise representation for the above information is constructed by this process.

Live VM migration shows better performance if the product of compression time (R_{cpr}) and compression ratio (ρ_{cpr}) is larger than available network bandwidth for migration (R_{tran}) [18].The formula written as

$$R_{cpr} \cdot \rho_{cpr} > R_{tran}$$

The compression ratio is defined mathematically as

$$\text{Compression Ratio, } \rho_{cpr} = \frac{\text{Uncompressed Size}}{\text{Compressed Size}}$$

Different compression algorithms are used based on the type of pages

(i) Pages with many zero bytes and nonzero bytes—maintaining the offset information of offset and values of nonzero bytes.
(ii) Pages with high similarity—using WKdm [19] compression algorithm which compresses memory data efficiently and quickly using dictionary and statistical techniques combined.
(iii) Pages with low similarity—any universal compression algorithm like LZO [20] with high compression ratio is used.

(B) Using Shared Storage

This uses precopy algorithm for live migration with an improvement. It sends the memory-to-disk mappings, i.e., a pair (*PFN, disk block*) where *PFN* is memory page in VM and *disk block* is the disk block index containing the data to the destination [21]. This technique works as

1. As it is using precopy of live VM migration, the memory pages are moved to destination in several iterations. In the meanwhile, memory pages are modified and dirtied. The changes are updated in page table entry (PTE) of MMU.
2. It brings changes in the iterative precopy phase of precopy algorithm instead of sending all dirty pages unique pages are transferred to the destination first.
3. A bit vector consists of dirty pages list and memory-to-disk mappings. If any block (duplicate block) is written to memory page *PFN* and become dirty in first iteration, then it is written back to the disk immediately before second iteration. There by maintains same (*PFN, disk block*) value for duplicate pages.
4. All the list of dirty pages and duplicate pages are sent to target. On receiving, the dirty pages are copied in destination VM.
5. NAS *fetch queue,* a back ground process fetches the contents of disk to target VM based on the duplicate pages list present in the queue.
6. Version number is assigned to each memory page. NAS fetch queue contains all outstanding requests (memory-to-disk mapping pairs). This request is deleted from the NAS fetch queue if any updated version of the same page is received.

7. The previous request for a memory page which is already in the queue is removed when it receives a new request for the same page.
8. During moving the data from NAS device to target, if the memory page is overwritten at the source host, then the data fetched from disk is discarded [22].
9. The target VM cannot start immediately after the last iteration. It restarts when all the NAS fetch queue requests are all processed.

(C) Exploiting Data De duplication

Migration with Data Deduplication (MDD) is used. With compression, lot of duplicated data compressed and transferred to the target; it takes more migration time even though the data is compressed. This is avoided by identifying similar memory pages by hash-based finger prints technology [23]. Data deduplicated using RLE encoding on the source host and on the target host. The steps are:

(i) *Identifying similar pages*:

It assumes 64-byte portions of a page as finger print. It finds hashes for blocks of those pages that are randomly selected locations on the page. The hashes are grouped as k groups with s hashes each. For each fingerprint c entries will be there in hash tables [22]. Each entry is called as reference page. Among this most suitable one is selected for Deduplication. SuperFastHash [24] is used as the hash function.

If the transferred reference pages are not there in the hash table then the reference page is cached. More memory is consumed. So merge the reference pages which have the same fingerprint (hash) but with different frame numbers. If any page having frame number *fn* modified in later rounds, reference page is found with modifications. Because of two cached reference pages with same frame number, inconsistency occurs. This can be eliminated using double hash FNHash and FPHash. Using FNHash, MDD finds the reference page in the hash table if found, MDD performs rehashing FPHash if needed, i.e., if any portion of this data changes and then do deduplication. If not found from FNHash using FPHash, it can be found then MDD do Deduplication. Thus, it ensures consistency.

(ii) *Dedeuplication of data*:

Let P_{trans} is the transferring page and P_{ref} is its reference page and data deduplication is found P_{parity} is find as $P_{parity} = P_{trans}\ XOR\ P_{ref}$

For P_{parity} duplicate portions, the above operation results in continuous zeros. Less information will be there in P_{parity}. Using RLE [22] encoding method, data transferred is further reduced. In destination, decoding is done to get P_{parity}. As P_{ref} is transferred to destination, P_{trans} is find reversely as $P_{trans} = P_{parity}$ XOR P_{ref}

MDD does not focused on the length of the encoded data because the probability of that data becoming longer is relatively small as P_{trans} *and* P_{ref} are highly similar. Much redundancy is eliminated by RLE encoding technique that accelerates migration very highly. To further accelerate the migration MDD uses multithreading to parallelize the Deduplication process.

(D) Energy aware virtual machine migration algorithm (EAM)

EAM [4] uses live migration. By calculating physical load and hit count of all the servers source server (victim) is selected. If any server has less load factor (PLi) than power off threshold (PoT) value is selected as victim server. For all servers, target threshold value (TST) is calculated. The target server is selected by using first fit algorithm. Then all the VMs running on the victim are to be migrated to the target server and source server is switched off to reduce power consumption. Underutilized servers are monitored by this algorithm and subsequently they will be switched off. If no target server is available in first round, it finds in next round using EAM. If overall OL [4] of any server is more than wake up threshold value (WoT) [4], then the server is switched on and some VMs from overloaded machine can be migrated to this server. It is assumed that 30 % of PoT, 85 %of WoT based on [25].

(E) Continual Migration

In continual migration, VM's states migrated to back up continually [26]. The internal states are migrated using live migration over the network. External states are shared with NAS. Migrated data is buffered on source host until current iteration of migration is over, and then it is moved to the backup host buffered block device with its length and check sum. These are verified before merging into migrated states (*Buffered Migration*) [26] ensures consistency. Memory pages are transferred uses traditional migration protocols in the first iteration. In later iterations, dirtied pages are monitored by continual migration and these dirty pages are moved using live migration (*Light Weight Migration*) [26]. By scheduling migrations using static intervals (*Scheduled Migration*) [26], the performance improved. Buffered block device has two modes, sync and async. *Sync mode*: Without translating or buffering any read or write operations are passed to native driver which is attached. *A sync mode*: Disk write operations are buffered in memory. Read operation finds and return the specific sector in the memory buffer. When switching from mode to mode, the entire buffered block device is flushed to the disk image file using cache write through policy. It introduces event channel to send the COMMIT messages from source to target. It contains 3 COMMITS.

(F) Asynchronous Replication and State Synchronization

Checkpointing/recovery and trace/replay technologies (CR/TR) [27] are used before stop-and-copy phase of precopy algorithm.

This technique contains the following phases [27]:

(i) *Initialization*: A VM is selected which is to be migrated.
(ii) *Reservation*: Resources on target host are reserved.
(iii) *Check Pointing*: The internal states of source VM are moved to target host. Device states and main memory are saved in checkpointing buffer [27] and moved to target using Copy-on-Write policy (COW).
(iv) *Iterative log transfer*: Nondeterministic events are recorded on source host in a log file. The check point data moved to target in first round. In further

rounds logged is moved. The target host replays the log files that are received from the checkpoint after VM resumes.

(v) *Waiting-and-Chasing*: After several rounds of log generates and transfer on source, it asks the target host whether to start the stop-and-copy phase. Target checks the cumulative unused file size in target and if it is less than the threshold value (Vthd) then it instructs source host to do stop-and-copy phase else to postpone the stop-and-copy phase.

(vi) *Stop-and-copy*: The Source VM is suspended and all the log files are moved to target. On target VM after last log file is received it is identical to source VM.

(vii) *Commitment*: Source VM is stopped after receiving the acknowledgment of commit message from target. From now, the network traffic is redirected to target. *Service taking over:* Target VM is activated and broadcasts it IP address.

(G) **Gang Migration**

A set of VMs from one set of physical machines are migrated to another set by gang migration using Deduplication, GMGD [28]. The cluster where a physical machine for VM resides is called as rack. In gang migration, similar pages are identified by a duplicate tracking mechanism [27], and they are suppressed by using of gang Deduplication among VMs existing in many clusters (*Duplicate Tracking Phase*). Similar memory pages among VMs within a single host are suppressed by using local Deduplication [27]. In VMM, *per-node* controller process by using content hashing similar memory pages are detected. Some of the pages become dirty while running. All shadow table entries of all the dirty pages are marked by the hypervisor. Rack wide content hashing is done for each rack and the hash values are exchange by the Deduplication servers of both the hosts. With the help of information present in Deduplication servers, the pages are sent or not are known. If the page status is *sent* retransmission of the same page is avoided as the Deduplication servers exchange the marked page information when all VMs are migrated to the other rack in parallel (*Migration Phase*). Sometimes the pages that are to be migrated are present in the VMs present in target rack. These cannot be transferred and care is taken by per-node controllers by synchronizing/forwarding this information to other hosts in the target rack (*Target-side VM Deduplication*) [28].

10 Metrics

Analysis of the performance of live VM migration is by the following parameters:

(i) *Application Degradation*: It is the degree of slowing down the performance of applications on virtual machines due to migration.

(ii) *Preparation Time*: The time required to finish transferring the VM's state from source to destination. In the meanwhile, memory pages are dirtied.

(iii) *Down Time*: During migration how much time the source VM is suspended.

(iv) *Resume Time*: The time between the end of VM migration on destination and restarting the VM. After VM restarts on destination, VM on source is destroyed [14].

(v) *Total Migration Time*: The time taken to complete the migration, i.e., preparation time, down time, and resume time.

(vi) *Network Traffic Overhead*: The usage of band width, additional traffic added [28].

(vii) *Pages Transferred*: The amount of pages (dirty, duplicates also) transferred.

11 Findings

Decision making to go with live migration techniques should be strategic. It is found

- Live migration via asynchronous replication and state synchronization check-pointing/recovery and trace/replay technologies (CR/TR) [28] is better in terms of downtime and network bandwidth.
- Gang migration migrates set of VMs to another rack of hosts with considerable amount of pages transferred. It reduces downtime and migration time more.
- Adaptive memory compression introduces the idea of memory compression.
- Migration with data Deduplication is best in terms of migration time.
- Live migration using shared storage reduces the migration time all most equal to other techniques but network bandwidth is a tradeoff.
- EAM reduces power consumption up to 28 % on static and 22 % on dynamic loads.
- Continual migration focused on fault tolerance (Fig. 6).

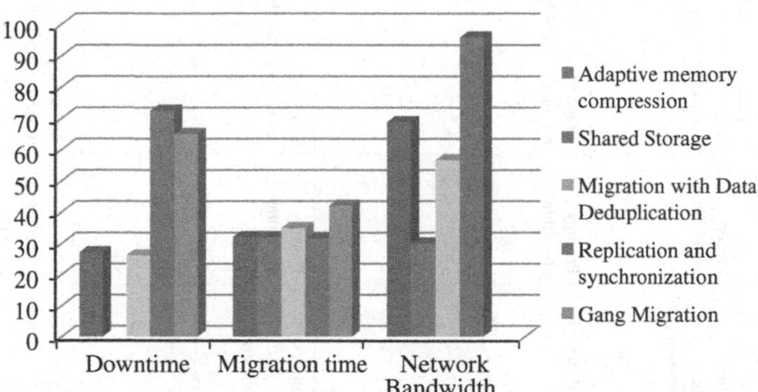

Fig. 6 Comparison of various live migration techniques

Table 1 Comparisons of different live migration technique

S. No.	Name of the technique	Basic migration technique	Event for migration	Selection of source physical server	Target physical server	Hypervisor used/ virtualization type/ implemented	Compared with	Metrics concentrated	Achievement/benefits	Drawbacks
1	Adaptive memory compression [18]	Precopy algorithm	Usage of resources > threshold	Based on SLA upper threshold	Based on SLA lower threshold (under-loaded)	Xen 3.1.0/ paravirtualisation	Unmodified Xen	Migration time	27.1 % downtime, 32 % total migration time, and 68.8 % number of pages transferred reduced	Compresses and transfers much duplicated data
								Downtime		
								Network traffic		
2	Shared storage [21]	Precopy algorithm	Usage of resources > threshold	Based on SLA upper threshold	Based on SLA lower threshold (under-loaded)	Xen 4.1/ paravirtualisation	Unmodified Xen	Migration time	Reduces	Increased downtime when more duplicated data on disk
								Downtime	32 % total migration time	
								Network traffic	Number of pages transferred is almost equal	Hosts involved in live migration should share storage
									37 % of downtime	
3	Migration with data deduplication [22]	Precopy algorithm	Usage of resources > threshold	Based on SLA upper threshold	Based on SLA lower threshold (under-loaded)	Xen 3.3.1 / paravirtualisation	Unmodified Xen	Migration time	Reduces	Not implemented in wide are networks
								Downtime	56.60 % of total data transferred during migration	
									34.93 % of total migration time	
									26.16 % of downtime on average	
4	Energy aware VM algorithm [25]	Live migration—precopy	Resource usage < threshold	PLi < PoT	Server with minimum TST	Simulator	Static and dynamic loads	Power consumption	Reduces	Not introduced in real data centers
									Power consumption	
									28 % on static load	
									22.0 % on dynamic load	

(continued)

Table 1 (continued)

S. No.	Name of the technique	Basic migration technique	Event for migration	Selection of source physical server	Target physical server	Hypervisor used/ virtualization type/ implemented	Compared with	Metrics concentrated	Achievement/benefits	Drawbacks
5	Continual migration [29]	Live migration—precopy	Detection failure of host	About to fail	Any VM	KVM/full virtualisation	Unmodified KVM	Fault tolerance	Hardware failure can be recovered less than one second	Not introduced in clusters
									Machine down time reduced	
									30 % overall performance	
									Ensures consistency	
6	Asynchronous replication and state synchronization	Live migration—precopy	Usage of resources > threshold	Based on SLA upper threshold	Based on SLA lower threshold (under-loaded)	UMLinux/Re-Virt-log and replay tool with modification in linux kernel	UMLinux/ Re-Virt-log and replay tool	Migration time	Reduces	Consistency not guaranteed
									72.4 % downtime	
								Downtime	31.5 % migration time	
								Network traffic fault tolerance	95.9 % network bandwidth	
									Applied to both LAN and WAN	
7	Gang migration	Live migration	Usage of resources > threshold	Based on SLA upper threshold	Based on SLA lower threshold (under-loaded)	QEMU/KVM	Unmodified	Migration time	Reduces	–
									42 % migration time	
							QEMU/ KVM	Network bandwidth	65 % network bandwidth	

12 Comparison

(Table 1)

13 Conclusion

Dynamic resource management is achieved through live virtual machine migration. A survey on virtualization, migration, and live migration techniques in which migration time, downtime, and total data transferred are focused parameters are presented in this paper. Some migration techniques provide fault tolerance and reliability. A comparative study of these techniques is also presented in this paper. Multiple VMs from one set of physical hosts can be migrated to another set using gang migration. It is found that gang migration reduces the network band width, down time, and migration time to a great extent by using Deduplication and migrating virtual machines in parallel.

References

1. Naga Malleswari, T.Y.J., Rajeswari, D.: A survey of cloud computing architecture and services provided by various cloud service providers. In: Proceedings of ICODC, 978-93-5087-502-5 201 (2012)
2. Buyya, R., Broberg, J., Goscinski, A.: Cloud computing: principles and paradigm
3. Understanding Full Virtualiztion, Paravirtualization and Hardware Assist, a white paper, VMWare
4. Al Shayeji, M.H., Samrajesh, M.D.: An energy-aware VM migration algorithm. In: ICACC (2012)
5. Li, Y., Li, W., Jiang, C.: A survey of virtual machine system: current technology and futrue trends. In: Third International symposium on Electronic Commerce and Security (2010)
6. Venkatesha, S., Sadhu, S., Kintali, S.: Survey of VM migration techniques (2009)
7. Kapil, D., Pilli, E.S., Joshi, R.C.: Live virtual machine migration techniques: survey and research challenges. 978-1-4673-4529-3/12- ©2012. IEEE
8. http://pubs.vmware.com/vsphere50/ index.jsp?topic=%2Fcom.vmware.vsphere.vcenterhost.doc_50 %2FGUID-326DEC3C-3EFC-4DA0-B1E9-0B2D4698CBCC.html
9. Strunk, A., Dargie, W.: Does live migration of VMs cost energy? 1550-445X/13 2013. IEEE
10. Mohan, A., Shine, S.: Survey on live VMM techniques. IJARCET **2**(1) (2013)
11. Graubner, P., Schmidt, M., Freisleden, B.: Energy-efficient virtual machine consolidation. IEEE, 1520-9202/13- ©2013
12. Leelipushpam, P.G.J., Sharmila, J.: Live VM migration techniques in cloud environment—a survey. In: Proceedings of 2013 IEEE Conference on ICT (2013)
13. Clark, C., Fraser, K., Hand, S., Hansen, J.G., Jul, E., Limpach, C., Pratt, I., Warfield, A.: Live migration of virtual machines. In: NSDI ’05: 2nd Symposium on Networked Systems Design and Implementation
14. Anala, M.R., Kashyap, M., Shobha, G.: Application performance analysis during live migration of virtual machines. IEEE, 978-1-4673-4529-3/12 /©2013

15. Sharma, S., Chawla, M.: A technical review for efficient virtual machine migration. In: International Conference on Cloud and Ubiquitous Computing and Emerging Technologies (2013)
16. Perez-Botero, D.: A brief tutorial on live VM migration from a security perspective
17. Hines, M.R., Gopalan, K.: Post-copy based live virtual machine migration using adaptive pre-paging and dynamic self-ballooning
18. Jin, H., Deng, L., Wu, S., Shi, X., Pan, X.: Live virtual machine migration with adaptive memory compression. IEEE, 978-1-4244-5012-1/09 ©2009
19. Wilson, P.R., Kaplan, S.F., Smaragdakis, Y.: The case for compressed caching in virtual memory systems. In: Proceedings of USENIX'99, pp. 101–116 (1999)
20. LZO Available: http://www.oberhumer.com/opensource/lzo/ (2009)
21. Jo, C., Gustafsson, E., Son, J., Egger, B.: Efficient live migration of virtual machines using shared storage. In: VEE'13, ACM 978-1-4503-1266-0/13/03 (2013)
22. Zhang, X., Huo, Z., Ma, J., Meng, D.: Exploiting data deduplication to accelerate live virtual machine migration. In: IEEE International Conference on Cluster Computing (2010)
23. Gupta, D., Lee, S., Vrable, M., Savage, S., Snoeren, A.C., Varghese, G., Voelker, G.M., Vahdat, A.: Difference engine: harnessing memory redundancy in VMs. In: Proceedings of the 8th USENIX Symposium on Operating Systems Design and Implementation (OSDI'08), pp. 309–322 (2008)
24. www.azillionmonkeys.com/qed/hash.html
25. Ferrari, D., Zhou, S.: An empirical investigation of load indices for load balancing applications. In: Courtois, P.-J., Latouche, G. (eds.) Proceedings of the 12th IFIP WG 7.3 International Symposium on Computer Performance Modelling, Measurement and Evaluation. North-Holland Publishing Co., Amsterdam, The Netherlands, pp. 515–528 (1987)
26. Cui, W., Ma, D., Wo, T., Li, Q.: Enhancing reliability for virtual machines via continual migration. In: 15th International Conference on Parallel and Distributed Systems (2009)
27. Liu, H., Jin, H., Liao, X., Yu, C., Xu, C.-Z.: Live virtual machine migration via asynchronous replication and state synchronization. IEEE Trans. **22**(12) (2011)
28. Deshpande, U., Schlinker, B., Adler, E., Gopalan, K.: Gang migration of VMs using cluster-wide deduplication. In: 13th IEEE/ACM International Symposium on Cluster, Cloud and Grid Computing 2013. IEEE

Author Biographies

T.Y.J. NagaMalleswari is a Research Scholar and assistant professor in Department of CSE, SRM University, Chennai. She did M. Tech in CSE from JNTU, Hyderabad in 2008. Her current research interests are cloud computing.

Dr. G. Vadivu is professor and HOD-IT, SRM University. She received doctorate in semantic Web SRM University, Chennai in 2013. She did masters in SRM University, in 2007. Her interests are semantic Web and data mining.

Dr. D. Malathi received A.M.I.E ECE from The Institution of Engineers, Calcutta in 1991, and M.E (CSE) from Madras University, Chennai, 1998, and PhD Information and Communication from Anna University, Chennai in 2010. She is a Professor, in Department of CSE, S.R.M University, Chennai. Her research interests include Artificial Neural Networks, Image Processing, Pattern Recognition, and Signal Processing.

A Design Phase Understandability Metric Based on Coupling and Cohesion for Object-Oriented Systems

Nikita Singh and Aprna Tripathi

Abstract As the software size grows, the maintenance become challenging. To make it easier, there is a need to measure some quality parameters in earlier phases of software development. Understandability has a major contribution to control the maintainability. Coupling and cohesion are two well-accepted parameters to measure the software quality parameters. In this paper, a model is proposed to measure the understandability that is based on coupling and cohesion.

Keywords Coupling · Cohesion · Software quality · Maintainability · Understandability

1 Introduction

When programmers try to reuse or maintain a software system developed by other programmers, the difficulty of understanding the system limits the reuses and makes maintenance more difficult. Counsell et al. [1] state that an important factor in the comprehension of a systems' architecture is an understanding of the systems' key classes and the coupling patterns. A study related to understandability of OO software was undertaken in [2]. Software quality model 9126 [3] defines understandability as one of the parameters to measure software usability. In an experiment of code inspection conducted by Porter et al. [4], 60 % of issues that professional reviewers reported were soft maintenance issues related to understandability.

N. Singh (✉)
Department of Computer Science and Engineering, SIET, Allahabad, India
e-mail: nikitasinghk@gmail.com

A. Tripathi
Department of Computer Science and Engineering, MNNIT Allahabad, Allahabad, India
e-mail: aprnatripathi@gmail.com

L.C. Jain et al. (eds.), *Intelligent Computing, Communication and Devices*,
Advances in Intelligent Systems and Computing 308,
DOI 10.1007/978-81-322-2012-1_33

2 State of the Art

Marcela Genero [4] determined the correlation between understandability and other parameter that are fetched from class diagram.

Nazir et al. [5] proposed a model to estimate the understandability of the class using design metric. Rajnish [6] used class complexity metric (CCM) to predict understandability. Shima et al. [7] proposes a software overhaul as a method for externalizing the process of understanding and presents a probability model to use process data of overhaul to estimate software understandability. Understanding attributes of cognitive processes can lead to new software metrics that allow the prediction of human performance in software development and for assessing and improving the understandability of text and code [8]. Cognitive information complexity measure (CICM) can be used to understand the cognitive information complexity and the information coding efficiency of the software [9, 10].

3 Proposed Approach

As defined by Nazir et al. [5], understandability is related with coupling, cohesion, and inheritance. In this paper, a new model of package-level understandability is established that depends on coupling, cohesion, and DIT. For coupling and cohesion, package-level coupling and cohesion metrics are used proposed in [11]. If there exist N number of methods in Class-B of package P_2 and Class-A of package P_2 calls M out of N number of methods, and then the weight of this sub-connection is defined as:

$$\text{Weight of sub-connection } W_{P_1C_A-P_2C_B} = \frac{\text{Number of methods called by class A}}{\text{Number of methods defined in class B}} = \frac{M}{N}$$

Here, Class-A and Class-B are represented by C_A and C_B

$$\text{Weight of connection } (P_{ij}) = \sum_{j=1}^{N} \text{weight of sub-connections from } P_j \text{ to } P_i \quad \text{for} \quad j \neq i$$

$$\text{Package-level coupling (PLC) for } P_i = \frac{\sum_{j=1}^{N} \text{weight of connections from } P_j \text{ to } P_i}{N-1}$$

where $P_j \in \{P - P_i\}$.

For any class C_i belonging to package P, let R_i be the number of relationships that are either directly or transitively related with this class C_i, N be the total number of classes in package P, and then, cohesion of class C_i is expressed as [11]:

$$\text{Cohesion of class } C_i(C_i\text{Coh}) = \frac{\text{Number of relationship (directly and/or transitively related with } C_i)}{(N-1)}$$

PLCoh of package P is normalized value of CiCoh for $i = 1$ to N. Mathematically, PLCoh is expressed as follows:

$$\text{Package-level cohesion (PLCoh)} = \frac{\sum_{i=1}^{N} C_i\text{Coh}}{N}$$

The DIT value for a class with in a generalization hierarchy is the longest path from the class to the root of the hierarchy. It is computed through an eclipse plug-in metrics [11]. To evaluate the proposed metric, six java-based large open source systems (OSS) are considered [11] that have multiple packages, since the proposed approach aims at package-level granularity to compute the PLCoh. Table 1 briefs about the OSS taken for experiment.

In order to establish a model for understandability, multiple linear regression technique has been used. Multivariate linear model is given as follows.

$$Y = a0 + a1x1 + a2x2 + a3x3\ldots \text{an } xn$$

where Y is dependent variable, $x1$, $x2$, $x3$,...xn are independent variables related to Y, and are expected to explain the variance in Y. $a1$, $a2$, $a3$,...an are the coefficients of the respective independent variables and a0 is the intercept.

$$\text{Understandability (proposed)} = 0.812 - 0.026 * \text{PLC} + 0.16 * \text{DIT} + 0.30 * \text{PLCoh}.$$

4 Result Analysis and Comparison

The results obtained by proposed approach are compared with the approach given by Nazir et al. To compare the results, the statistical correlation coefficient value is compared (Table 2).

Table 1 Source detail of project taken for empirical study

S. No.	Project name	URL
1.	Element construction set	http://jakarta.apache.org/ecs/
2.	Byte code engineering library	http://commons.apache.org/proper/commons-bcel/download_bcel.cgi
3.	JAKARTA-ORO	http://jakarta.apache.org/oro/
4.	XGEN	http://xgen.sourceforge.net/xgen/index.html
5.	Lamistra	http://sourceforge.net/projects/java-stratego/
6.	KEA	http://www.nzdl.org/Kea/download.html

Table 2 Correlation coefficient between different pairs

Parameters	Correlation coefficient	
	Proposed approach	Nazir et al. approach
PLC and understandability	−0.238	0.036
DIT and understandability	0.96	0.847
PLCoh and understandability	0.84	0.708

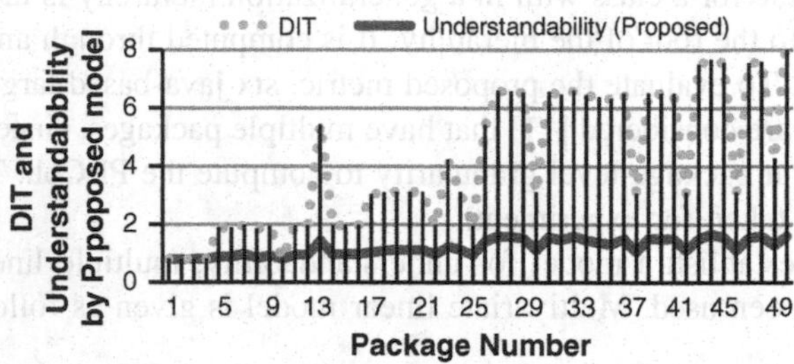

Fig. 1 Relationship between understandability (proposed) and DIT

5 Conclusion

Understandability of a software depends on the components relationship structure; paper proposes a model to measure the understandability that utilizes the coupling, cohesion, and depth of inheritance that are computed at package level. Result shows that the model proposed by Nazir et al. is strongly correlated with the proposed model. In the Nazir et al. approach, the correlation between understandability is 0.84, whereas in the proposed approach, the correlation value for the same parameters is 0.96.

Figure 1 represents the relationship between DIT and understandability. When DIT is lesser, lesser understandability is required.

References

1. Counsell, S., Newson, P., Mendes, E.: Architectural level hypothesis testing through reverse engineering of object-oriented software. In: 8th International Workshop on Program Comprehension (IWPC) (2000)
2. Harrison, R., Counsell, S.: Object oriented software understandability: An empirical investigation. In: Proceedings of 12 International Conference on Software and System Engineering and their Application (ICSSEA) (1999)
3. ISO/IEC 9126-1.: Software engineering—product quality—part 1: Quality model (2001)

4. Porter, A., Siy, H.P., Toman, C.A., Votta, L.G.: An experiment to assess the cost-benefits of code inspections in large scale software development. IEEE Trans. Softw. Eng. **23**(6), 329–346 (1997)
5. Nazir, M., Khan, R.A., Mustafa, K.: A metrics based model for understandability quantification. Int. J. Comput. **2**(4), 90–94 (2010)
6. Rajnish, K.: Class complexity metric to predict understandability. IJIEEB **6**(1), 69–76 (2014)
7. Shima, K., Takemura, Y., Matsumoto, K.: An approach to experimental evaluation of software understandability. In: Proceedings of the 2002 International Symposium on Empirical Software Engineering (ISESE'02). IEEE Computer Society, Washington, DC, USA, 2002
8. Mayrhauser, A., Vans, A.M.: Program understanding behavior during adaptation of large scale software. In: Proceedings of the 6th International Workshop on Program Comprehension (IWPC'98), Ischia, Italy, pp. 164–172, June 1998
9. Kushwaha, D.S., Misra, A.K.: Cognitive information complexity measure of object-oriented software: a practitioner's approach, SEPADS'06. In: Proceedings of the 5th WSEAS International Conference on Software Engineering, Parallel and Distributed Systems, pp. 174–179 (2006)
10. Kushwaha, D.S., Misra, A.K.: A modified cognitive information complexity measure of software. ACM SIGSOFT Software Engineering Notes, vol. 31(1), January 2006
11. Tripathi, A., Vardhan, M., Kushwaha, D.S.: Package level cohesion and its application. In: Fifth International Conference on Advances in Communication, Network, and Computing—CNC 2014, Elsevier, Chennai, 21–22 Feb 2014

Web Service Response Time Prediction Using HMM and Bayesian Network

Vani Vathsala Atluri and Hrushikesha Mohanty

Abstract Selection of suitable and efficient services by service consumers turns out to be a herculean task with the availability of abundant functionally similar services over the Web. In this scenario, quality of services being offered plays a crucial role in service selection by consumers. Response time has been a key factor influencing a consumer for selection of a Web service. Predicting it has been a major challenge for researchers. In this paper, we propose an approach for prediction of response time of Web services using hidden Markov model (HMM) and Bayesian networks.

Keywords HMM · Response time prediction · Web services · Bayesian networks

1 Introduction

A Web service is a piece of software that provides a service and is accessible over Internet. The role of Web services in today's world is continuously growing. This is not only due to the use of Web services by organizations having distributed operations but also for its operational ease. As the number of Web services that offer functionally similar services is growing enormously, service consumers have the luxury of selecting services that offer high quality of service within less time. In this context, predicting Web service response times would greatly help service users in selecting Web services of their interests.

V.V. Atluri (✉)
CVR College of Engineering , Hyderabad, India
e-mail: atlurivv@yahoo.com

V.V. Atluri · H. Mohanty
University of Hyderabad, Hyderabad, India
e-mail: mohanty.hcu@gmail.com

L.C. Jain et al. (eds.), *Intelligent Computing, Communication and Devices*,
Advances in Intelligent Systems and Computing 308,
DOI 10.1007/978-81-322-2012-1_34

For accurate response time prediction, various factors that add up to make response time should be predicted precisely. Hence, when a service consumer requests for response time prediction, prediction approach should consider predicting:

1. Instruction execution time: Time required for service execution. This time component is static in nature because time needed depends up on the structure of the service. The exact execution path taken is decided by the input parameters and their corresponding values.
2. Service waiting time: Time spent by the service request in queues waiting for Web server and other shared resources. This is a dynamic component of the response time. It depends on total number of requests currently pending at the Web server and other resources.
3. Network delay time: It is dynamic component of the response time. It is function of network traffic.

In this paper, we bring precision to our prediction by predicting all three components of response time stated above separately. Organisation of our paper is as follows: We compare various Web service response time prediction approaches in Sect. 2, and our prediction approach is elaborated in Sect. 3. Section 3.1 explains instruction execution time computation using input parameter values, and details on waiting time prediction using Bayesian networks. In Sect. 3.2 , we discuss network delay prediction using hidden Markov model. We present our conclusion and future work in Sect. 4.

2 Related Work

Prediction of response time and values for QOS attributes of Web services has been studied using different techniques [1–8]. Marzolla and Mirandola [2] propose an approach for performance assessment of Web service workflows based on BPEL constructs used. This requires knowledge of internal structure of the Web service and can be used by developers to predict performance of the service being developed.

Balogh et al. presented a knowledge-based approach in [6] for predicting execution time of stateful Web services. To predict the execution time of a Web service instance, it maintains a knowledge base of possible different past cases for different combinations of input parameters. Given a Web service instance, Euclidean distances are used to find out most similar past cases. The run-time for the given Web service instance is predicted to be the average output value of the most similar past cases. Cheung et al. [3] focuses on predicting the response time of a Web service from clients perspective. He proposes an approach that collects data from performance testing and applies interpolation at low workloads and extrapolation at high workloads, to predict the response time. Li et al. [1] and Chen [4] propose to employ user similarity metrics and Web services similarity statistics to predict Web service performance. However, these research works do not consider the dynamic nature of Web services for prediction, i.e. they do not predict time components 2 and 3 listed in the previous section.

Artificial neural networks have also proved themselves as reliable predictors. Gao and Wu [8] uses back propagation neural networks to predict the run-time of a given Web service. He uses availability, network bandwidth, response time, and reliability of a given Web service as inputs to the neural network that predicts execution duration as output. In all these cases, prediction is by mapping the current request to similar requests in the past and computing the expected response time. We present an approach which models Web service environment and predicts by considering dynamic nature of platform behaviour as it has considerable impact on execution time of Web services.

3 Web Service Response Time Prediction

Response time of a service may vary from one service request to another as processing at a given time depends on number of requests pending at the Web server, input parameter values, and sharing of common resources. The network through which a service request is routed to the service provider also contributes to delay in response time of the service. Delay introduced by the underlying network may be highly attributed towards the traffic flowing through the network. Hence, we propose the following approach:

1. Predict service execution time for a given service request by modelling Web server environment.
2. Predict delay introduced by underlying network by modelling the underlying network through which service request is routed.

The framework for prediction is depicted in Fig. 1.

A service user should send her prediction request to the prediction component which performs the prediction task using two sub modules: execution time prediction module and network delay prediction module. This prediction module can be hosted on the same Web server (which hosts the required service) in case of lightly loaded servers, or on another server which has access to the logs written by the primary Web server, in case of heavily loaded servers.

3.1 Execution Time Prediction

A service processing time or execution time can be subdivided into instruction execution time and waiting time. We propose to predict both the subcomponents individually.

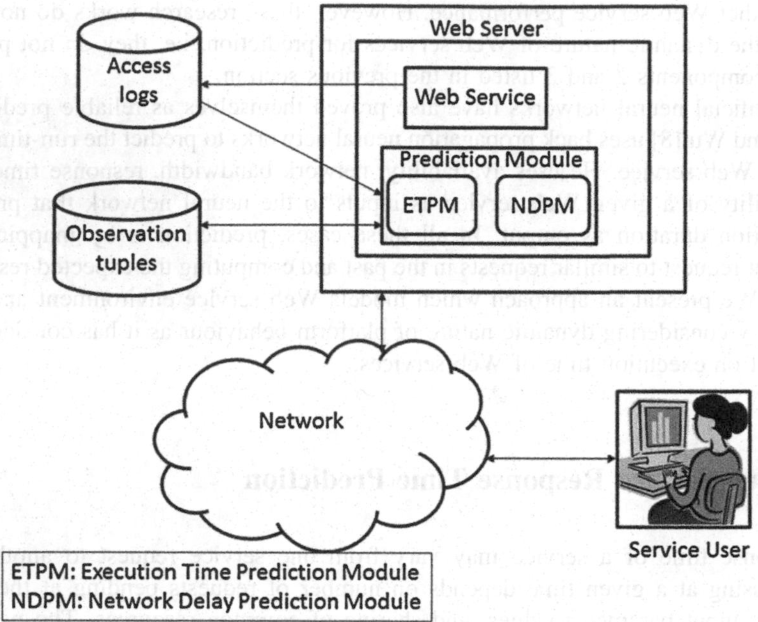

Fig. 1 Prediction framework

Predicting instruction execution time: Instruction execution time is the time taken to execute the code within a Web service. The most important factor that determines instruction execution time is the execution path taken by the current request. Distinct values for input parameters lead to different execution paths through a Web service which have totally different execution times. Hence, input parameter values determine the instruction execution time of a service request. Large input and output messages result in increased serialisation and deserialisation times which we include in instruction execution time.

Input vector *u* for a service request consists of two subcomponents *I* and sio. *I* is the set of input parameters, $I = \{i_1, i_2, \ldots, i_n\}$ where *n* indicates number of input parameters to the Web service call. sio stands for total size of input and output messages of the service request and reply. Thus, $u = \{\{i_1, i_2, \ldots, i_n\}$, sio. Prediction module has to be provided with all possible cases of input vector values that result in different instruction execution times. These execution times can be obtained by giving different input values to the Web service under no server load.

Predicting waiting time: A Web server may have several requests at a given time, made by different customers. Number of service requests of the same Web service at time *t* when a request is made affect the waiting time of the current Web service instance. Time spent waiting for other shared resources like database server also add on to the waiting time of a service request. In this paper, we consider only database server as a shared resource. The approach may be extended to other shared resources like file servers without loss of generality. Unlike instruction execution time, waiting time cannot be predetermined for a given set of input parameters as it depends on current server load and database server load. Prediction module has to estimate waiting time dynamically when it receives a prediction request. For a given input vector, server load and database server load, there might be several possible waiting times due to nonmeasurable factors like delay introduced by scheduling algorithm, etc. We propose to use Bayesian networks to predict waiting time which is conditionally dependent on the above-described factors. Maximum likelihood estimation is used for prediction. Predicting the waiting time involves finding out the expectation of waiting time values conditioned on measurable quantities such as server load, input vector values, etc. We propose to use Bayesian approach for estimating waiting time since conditional probabilities can be modelled naturally using Bayesian networks.

Prediction using Bayesian network: Let u_t represent a random variable that represents the input vector for service request submitted to the Web server in time interval t. y_t is a random variable that represents the server load in time interval t. n_t is another random variable that represents waiting time spent by the service request at the Web server. Random variable r_t represents database server load. The Bayesian network O computes waiting time probabilities conditioned on the input u_t, server load y_t, and database server load r_t. Let η be the expected value for probability distribution of waiting time η_t. Thus, η = expected value of probability distribution $Pr(n_t|u_t, r_t, y_t)$. Let $h(u_t, n_t, v, r_t) = Pr(n_t|u_t, r_t, y_t)$. Let D represent training data.

θ_j = represents probabilities of all k_j possible discrete values that n_t would take where $1 \leq j \leq m$ and m is number of different possible combinations of values of u_t, y_t, and r_t. These different possible discrete values for n_t are obtained using n observation tuples and IET. Thus, θ_j is a vector of values = $\{\theta_{j1}, \theta_{j2} \ldots\ldots \theta_{jkj}\}$.

N_{ji} is the number of cases in D that have jth possible discrete value represented by probability distribution θ_j. Thus, the expected value of the probability distributions η is computed as follows:

$$
\begin{aligned}
\eta_j &= \text{Expected value of } h(n_t, \, (u_t, y_t, r_t)^j) \\
&= \sum_i h(n_t^i, \, (u_t, y_t, r_t)^j).n_t^j
\end{aligned} \tag{1}
$$

where $(u_t, y_t, r_t)^j$ is jth possible combination of (u_t, y_t, r_t) in D where $(u_t = u, \; y_t = y, \; r_t = r)$.

$$
h(n_t^i, \, (u_t, y_t, r_t)^j) = \frac{N_{ji}}{N_j}.
$$

$$
N_j = \sum_{a=1}^{k_j} N_{ja}
$$

Table 1 Predictions for user id 1

WS Id	Matrix A	Matrix B	Predicted T (s)	Observed RT (s)
2	0.8667, 0.1333; 0.8600,0.1400	0.2500, 0.2813, 0.1563, 0.1250, 0.0625,0.1250; 0.2500, 0.2812 ,0.1429, 0.1384, 0.0625, 0.1250	0.250000	0.2580000
5	0.7333, 0.2667; 0.7214, 0.2786	0.9063, 0.0938; 0.8999, 0.1111	0.430000	0.438000
11	0.8000, 0.2000; 0.7999, 0.2111	0.8594, 0.1094, 0.0313; 0.8300, 0.1288, 0.0313	0.41000	0.695000

3.2 Network Delay Prediction

In this subsection, we present an HMM-based approach for inferring delay induced by networks based on observed metrics. A network can be in any one of the two possible states: **congested** and **noncongested**. Several factors have their role in determining current state of a network: Number of packets that are currently being routed through it, network bandwidth, routing and congestion control algorithms used etc. But prediction module has no direct access to all such details, and thus, network state remains hidden to prediction module. What could be measured is the delay that is introduced by the network in transmission of request and response messages between sender and receiver. When a new request (u, t) is submitted, network delay prediction module predicts the average network delay d_t in the time interval t. In order to predict network delay d_t in a time interval t, we make use of well-known HMM. In a regular Markov model, the state transitions are directly visible to an observer. In a HMM [9], the system being modelled is assumed to be a stochastic Markov process with unobservable or hidden states, but the output of the system in each of the time intervals, which is dependent on corresponding states of the system, is visible to the observer. We call this observable output of the system as emissions or observations O. State transition matrix A depicts the transition probabilities among states of the system. Each state has a probability distribution over the observation symbols. Emission probability matrix B governs probability distribution of the states over these emissions.

States of the underlying network are not visible to the prediction module and hence can be mapped to hidden states of a HMM. Since delay introduced by the network can be measured, and hence, are visible, they become emissions of the HMM. Network delay prediction module predicts average network delay in the time interval t by performing the following tasks:

1. Run the *EM Algorithm* on training data and train the HMM (i.e. determine the parameters of the HMM). This is a one-time task which is done initially.
2. Execution of the *Viterbi Algorithm* for network delay prediction upon receiving a request.

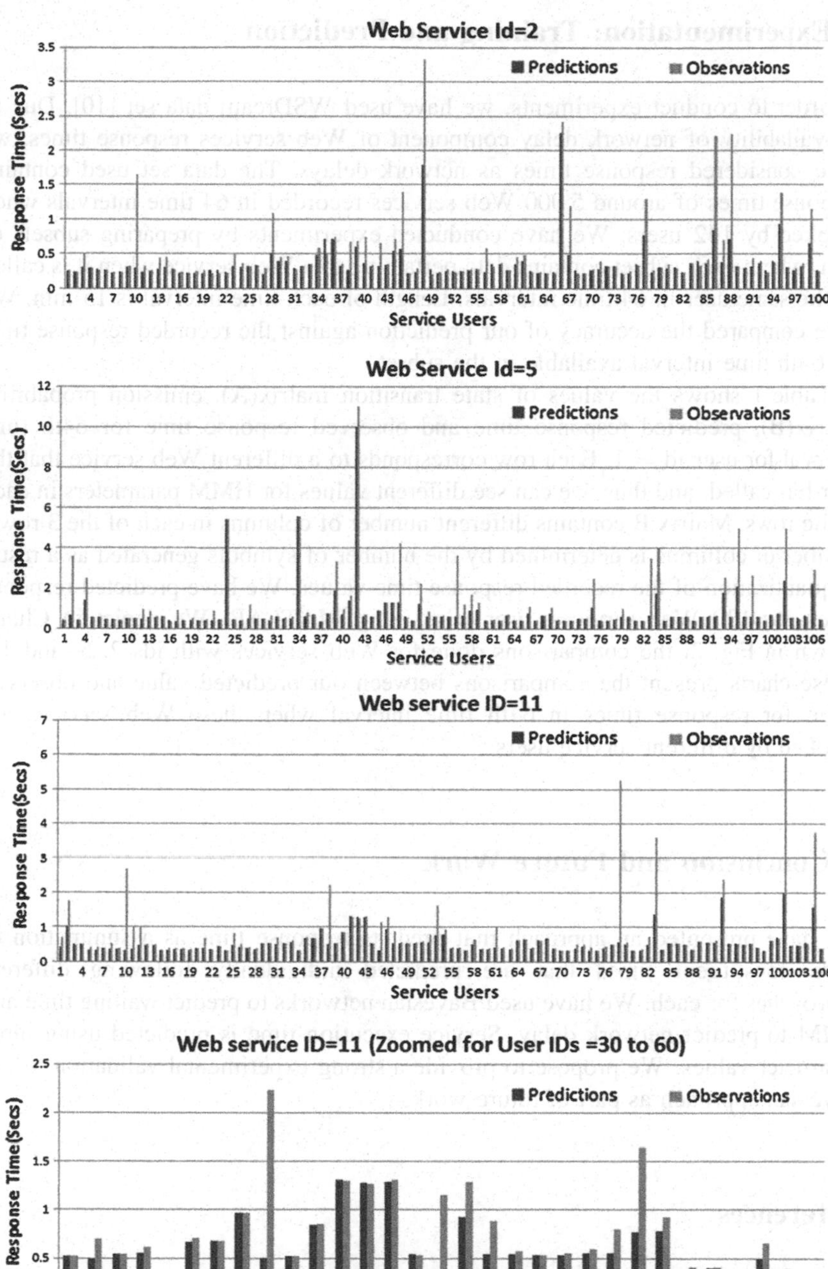

Fig. 2 Predictions versus observations

4 Experimentation: Training and Prediction

In order to conduct experiments, we have used WSDream data set [10]. Due to unavailability of network delay component of Web services response times, we have considered response times as network delays. The data set used contains response times of around 5,000 Web services recorded in 64 time intervals when invoked by 142 users. We have conducted experiments by preparing subsets of data, where each subset contains data pertaining to a Web service when it is called by the same user in 64 time intervals. Length of each time interval is 15 min. We have compared the accuracy of our prediction against the recorded response time for 64th time interval available in the subset.

Table 1 shows the values of state transition matrix(A), emission probability matrix(B), predicted response time and observed response time for 64th time interval for user id = 1. Each row corresponds to a different Web service that the user has called, and thus, we can see different values for HMM parameters in each of the rows. Matrix B contains different number of columns in each of the 3 rows; number of columns is determined by the number of symbols generated as a result of quantization of the recorded response time values. We have predicted response times for 100 Web services, user wise using MATLAB. We depict in Charts shown in Fig. 2, the comparisons done for Web services with ids 2, 5, and 11. These charts present the comparisons between our predicted value and observed value for response times in 64th time interval when these Web services are invoked by different service users.

5 Conclusion and Future Work

We have presented an approach that predicts response time as a summation of three constituent times that are predicted individually following different approaches for each. We have used Bayesian networks to predict waiting time and HMM to predict network delay. Service execution time is predicted using input parameter values. We propose to provide a strong experimental validation to our Bayesian approach as part of future work.

References

1. Li, Z., Bin, Z., Jun, N., Liping, H., Mingwei, Z.: An approach for web service QoS prediction based on service using information. International Conference on Service Sciences (ICSS), 2010 pp. 324–328
2. Marzolla, M., Mirandola, R.: Performance prediction of web service workflows. In: Proceedings of the quality of software architectures 3rd international conference on software architectures, components, and applications QoSA07, Springer Berlin, Heidelberg (2007)

3. Cheung, L., Golubchik, L., Sha, F.: A study of web services performance prediction: a client's perspective. Proceedings of the 2011 IEEE 19th Annual International Symposium on Modelling, Analysis, and Simulation of Computer and Telecommunication Systems. pp. 75–84 2011
4. Chen, L., Feng,Y.: An Enhanced QoS Prediction Approach for Service Selection. International Conference on Services Computing. pp. 727–728, 2011
5. Lelli, F., Maron, G., Orlando, S.: Client side estimation of a remote service execution. In: 15th International Symposium on Modeling, Analysis, and Simulation of Computer and Telecommunication Systems, Istanbul, Turkey, 2007
6. Balogh, Z., Gatial, E., Laclavik, M., Maliska, M., Hluchy, L.: Knowledge-based Runtime Prediction of Stateful Web Services for Optimal Workflow Construction. LNCS 3911, pp. 599–607, 2006. Springer Berlin Heidelberg 2006
7. Laranjeiro, N, Vieira, M, Madeira, H: Predicting Timing Failures in Web Services. Springer, Berlin (2009)
8. Gao, Z., Wu, G.: Combining QoS-based service selection with performance prediction. In: Proceedings of the 2005 IEEE International Conference on e-Business Engineering (ICEBE'05) 2005 IEEE
9. Rabiner, L.R.: A tutorial on hidden Markov models and selected applications in speech recognition. Proc. IEEE, **77**(2), 257–286 (1989)
10. Zhang, Y., Zheng, Z., Lyu, M.R.: WSPred. a time-aware personalized QoS prediction framework for web services. Proceedings of the 22th IEEE Symposium on Software Reliability Engineering (ISSRE 2011)

A Multi-agent-Based Framework for Cloud Service Description and Discovery Using Ontology

Manoranjan Parhi, Binod Kumar Pattanayak and Manas Ranjan Patra

Abstract The spectrum of cloud service providers has become large these days with increasing popularity of cloud computing as a service. These services appear to be similar in their approaches excepting that they vary in their key attributes such as price, computational power, and storage policy. As of now, there are no standardized specifications defined for a cloud service provider. Different cloud service providers use different vocabulary to specify similar operations. A number of quality of service (QoS) parameters may be associated with a service, and thus, identification of the appropriate parameter and its management is the vital issue that needs to be addressed by the service provider. In this context, the search engines such as Google, MSN, etc., do not represent the most effective and most efficient tools in order for determining the appropriate cloud service that could optimally meet the service requirements of a customer. It remains as a challenging issue for a customer requesting for a cloud service. Thus, there arises the necessity of a reasoning mechanism for service discovery that is capable of resolving the similarities among variety of services. In this paper, we address a semantic-based service description and discovery framework using multi-agent approach, where the cloud service descriptions that are automated based on shared ontology, contribute to optimal discovery of appropriate services as requested by consumers.

Keywords Cloud service description and discovery · Quality of service · Ontology · Multi-agent system

---

M. Parhi (✉) · B.K. Pattanayak
Department of Computer Science and Engineering, ITER, Siksha 'O' Anusandhan University, Bhubaneswar 751030, Odisha, India
e-mail: mrparhi@gmail.com

B.K. Pattanayak
e-mail: bkp_iter@yahoo.co.in

M.R. Patra
Department of Computer Science, Berhampur University, Berhampur 760007, Odisha, India
e-mail: mrpatra12@gmail.com

L.C. Jain et al. (eds.), *Intelligent Computing, Communication and Devices*, Advances in Intelligent Systems and Computing 308,
DOI 10.1007/978-81-322-2012-1_35

1 Introduction

Recently, popularity of cloud computing has facilitated the companies like Google, Amazon to provide reliable and cost-effective services [1]. Unlike semantic web, cloud computing does not possess a semantic base even if it is enriched with features like on-demand self-service, resource pooling, rapid elasticity, and so on. So, in order to provide an environment for automatic searching of services, across various cloud computing platforms, intelligent ontology-based cloud service registry is highly essential. As a whole, cloud services are published in the Internet by the service providers, which can be accessed by the consumers using Web portals. As of now, however, no efficient cloud discovery mechanisms are available to facilitate search across different cloud services that trigger the consumers of services to rely upon manual search while requesting for a cloud. Consumers may use the generic search engines that are available in the internet, but, however, these search engines may return the URLs in anticipation to a user's request those may not be possibly meeting the service requirements of the requester. Consequently, traversing all possible Web pages could be significantly time consuming. Even powerful search engines such as Amazon, Google, MSN, etc., do not possess the capability of reasoning about similar cloud services so as to determine the most appropriate service for a requester taking into account the specified service requirements.

An agent-based approach could be useful for dynamically searching for a cloud service. Agents can represent autonomous service consumers and providers as well as collaborate to dynamically configure and reconfigure service-based software applications.

In this paper, we present a multi-agent framework integrated with ontology for cloud service description and discovery. The rest of the paper is structured as follows: Sect. 2 describes the fundamentals of cloud computing. Cloud service life cycle management is explained in Sect. 3. Section 4 outlines literature survey on service discovery in cloud computing environment. Section 5 explains our proposed multi-agent framework based on ontology for cloud service description and discovery. Section 6 outlines the implementation details, and finally, Sect. 7 presents the conclusion and future work.

2 Cloud Computing

The idea of cloud computing as a service in the form of utility was foreseen by John McCarthy in 1960s [1–3]. Lack of standard definition of cloud computing may result in market hype that triggers the researchers to work on it. In this work, we use the definition of cloud computing as adopted by the National Institute of Standards and Technology (NIST) [4].

"Cloud computing is a model for enabling convenient, on-demand network access to a shared pool of configurable computing resources (e.g., networks, servers, storage, applications, and services) that can be rapidly provisioned and released with minimal management effort or service provider interaction."

Cloud computing incorporates a service-driven business model, wherein resources such as hardware, platform, and software as services are made available to requester on an on-demand basis. As practices reveal, the services offered by clouds can broadly fall into the categories: software as a service (SaaS), platform as a service (PaaS), and infrastructure as a service (IaaS).

- Software as a service: In this category, applications are made available to customers on demand. To the category of SaaS belong the applications like Salesforce.com, Rackspace, gmail, etc.
- Platform as a service: Platform-level resources such as operating system and varieties of software development frameworks can also be provided on demand. PaaS refers to providing platform layer resources, including operating system support and software development frameworks. Google App Engine, Microsoft Windows Azure, and Force.com, etc., are the examples of PaaS.
- Infrastructure as a service: IaaS refers to on-demand provisioning of infrastructural resources, usually in terms of VMs. The cloud owner who offers IaaS is called an IaaS provider. Examples of IaaS providers include Amazon EC2, GoGrid, and Flexiscale.

3 Cloud Service Life Cycle Management

The life cycle of a cloud service comprises of five separate phases like requirements, discovery, negotiation, composition, and consumption [5].

3.1 Service Requirements Phase

In this phase, the consumer needs to specify the technical, functional, and non-functional details of the requested cloud service. A request for service (RFS) can be issued only after the service requirements are identified and classified.

3.2 Service Discovery Phase

With reference to the specifications detailed in RFS along with the service descriptions, the provider can be discovered in the phase of service discovery.

3.3 Service Negotiation Phase

In service negotiation phase, the service provider and the consumer need to arrive to an agreement on the service delivered and the acceptance criteria of it.

3.4 Service Composition Phase

The service components from different service providers are integrated to a single service and delivered to the requester in the phase of service composition.

3.5 Service Consumption/Monitoring Phase

The service is delivered to the consumer based on the delivery mode agreed upon between the provider and the consumer that may be synchronous/asynchronous, real-time, batch mode, etc. Payment is made for the same after the service is delivered to the consumer, with respect to the pricing model specified in the SLA. The consumer then begins consuming the service.

4 Literature Survey

A large spectrum of literature related to service computing encompasses service discovery under a generic service-oriented framework [6]. Most of these can be applied in principle with suitable modifications for cloud service discovery since the technical foundations of cloud computing are the same as that of service-oriented architecture (SOA) [7]. In modern literature, variety of approaches and frameworks suitable for cloud service discovery using agents, QoS, and ontology have been proposed by researchers.

4.1 Agent

An agent represents a computational entity that acts on behalf of another entity (or entities) to execute a task or to achieve a specified goal. Although a single agent is capable of performing a given task, the agent paradigm was conceived as a distributed computing model where a set of agents (multi-agent) interact among each other by exchanging information and cooperating to perform complex tasks [8, 9]. *Agent-based cloud computing* [10, 11], a novel paradigm can provide agent-based solutions based on the design and development of software agents for improving cloud resources, service management, discovery, SLA negotiation, and service composition.

4.2 Ontology

Ontology, the knowledge representation technology for semantic web [12], facilitates the search for contents and information and improves crawling procedure too. A number of definitions for ontology can be found currently in the literature. As per Gruber [13], "An ontology is a formal and explicit specification of a shared conceptualization." In cloud ontology [14, 15], the hierarchical relations of cloud concepts are defined. For instance, the concept "Cloud system" has three different children nodes (IaaS, PaaS, SaaS). By consulting cloud ontology, similarity reasoning can be performed.

4.3 Quality of Service

Quality of service (QoS) includes the non-functional attributes (e.g., cost, response time, availability, security, etc.) that may influence the overall performance of any cloud service. QoS can be described in the user preferences to express their expectations. It can be included in a service advertisement when different service providers that present diverse versions of services to answer varying requirements of their customers. User preferences and service offerings have both functional and complex non-functional aspects that require to be matched against each other [16].

4.4 Related Work on Cloud Service Discovery

Sim [10], Kang and Sim [17] presented a cloud service discovery system (CSDS) that aims at enabling the cloud users in finding a cloud service over the Internet. The CSDS interacts with cloud ontology to identify the similarities among different services. It builds an agent-based discovery system that consults with ontology during information retrieval about cloud services. This work desires to launch a generic search engine to search required cloud services before reasoning suitable service for a client, which is time consuming. An ontology-based agent generation framework, proposed in [18], focuses at dynamically and fully automatically generating mobile agent for retrieving desired information on cloud platforms without user's intervention. However, this framework does not support the service discovery mechanism for cloud environment. A mobile agent-based service discovery framework integrated with ontology along with a prototype for service discovery for cloud is proposed by authors in [19]. It mainly assists users to discover suitable service on demand. Recall and precision are evaluated to test the accuracy of the system. However, this work lacks the implementation of service discovery to handle dynamic constraints and preferences as specified by the requesters. OWL-S-based semantic cloud service discovery and selection system is proposed by the authors in [20] that supports dynamic semantic matching of cloud services specified with complex constraints. This may pose challenges if part of the service descriptions is in free text, or if some cloud providers adopt custom ontologies. Therefore, ontology learning and alignment methods are required to address these issues. Ding et al. [21] proposed a hybrid technique based on syntactic and semantic of input and output of services for cloud service discovery. However, this approach is based on IO matching alone. Rehman et al. [22] and Wang et al. [23] introduced cloud service selection based on QoS parameters like cost, performance, response time, throughput, reputation, availability, reliability, etc. However, it is based on the typical QoS criteria adopted from Web service selection algorithms.

Based on the comparative study and analysis of different cloud service discovery approaches, we conclude that cloud service discovery technique should be automatic and dynamically providing ample scope to the customers to find the required services in ease by providing least intelligence and effort. Similarly, the service provider must get ample flexibility toward easy up gradation of their services from time to time. The discovery technique must ensure that neither any service provider nor any service consumer suffers due to the inefficient discovery technique used in the service discovery framework.

5 Proposed Framework

The overall architecture of the proposed system is presented in Fig. 1. The various agents used in the discussed framework are as follows:

5.1 Provider Agent

This agent helps the service provider to register a new cloud service or to upgrade an existing service. Provider agent has a direct communication with the cloud service provider and keeps track of cloud service popularity from user feedback. It can update the cloud service functionality from time to time dynamically.

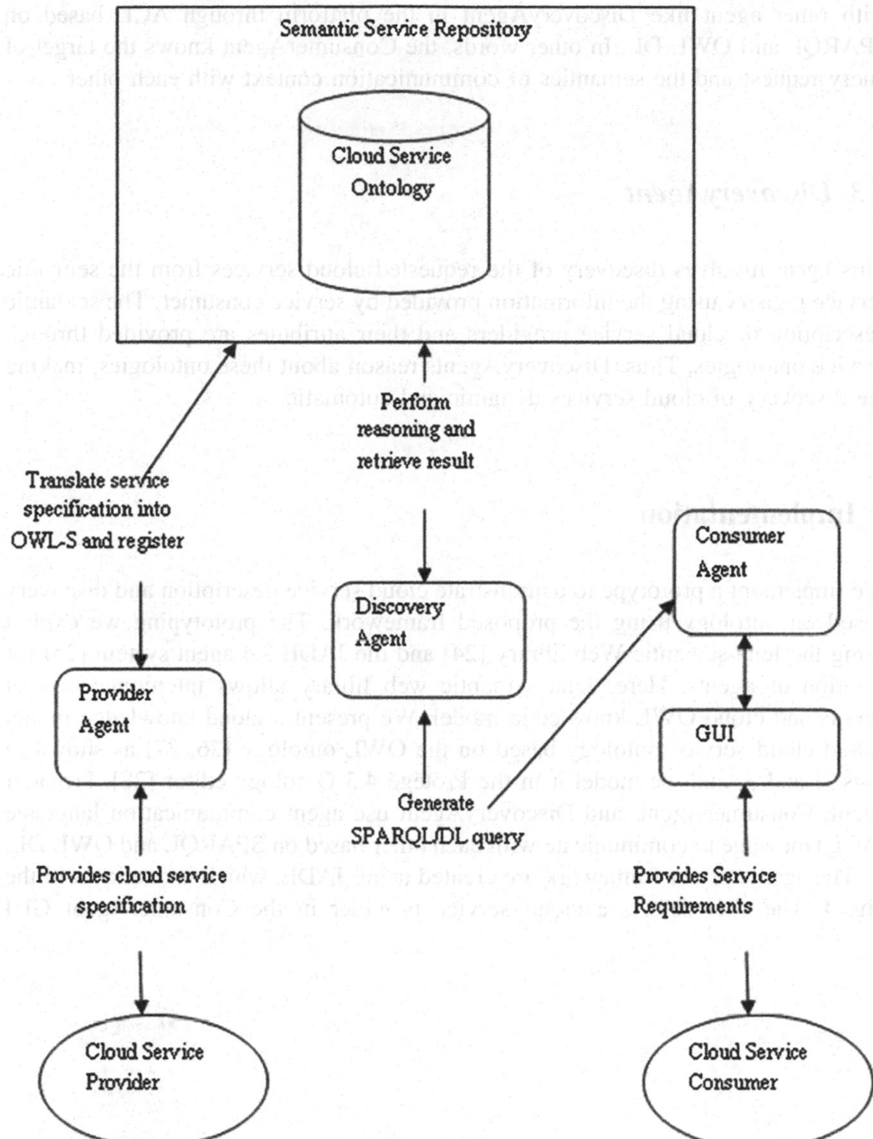

Fig. 1 Proposed multi-agent-based framework

5.2 *ConsumerAgent*

ConsumerAgent provides a user-friendly graphical user interface that can help user to select a query for cloud service. The ConsumerAgent knows how to interact with other agent like DiscoveryAgent in the platform through ACL based on SPARQL and OWL DL. In other words, the ConsumerAgent knows the target of query request and the semantics of communication context with each other.

5.3 DiscoveryAgent

This agent involves discovery of the requested cloud services from the semantic service registry using the information provided by service consumer. The semantic description of cloud service providers and their attributes are provided through service ontologies. Thus, DiscoveryAgents reason about these ontologies, making the discovery of cloud services dynamic and automatic.

6 Implementation

We implement a prototype to demonstrate cloud service description and discovery based on ontology using the proposed framework. The prototyping we exploit using the Jena semantic Web library [24] and the JADE 3.4 agent system [25] for creation of agents. Here, Jena semantic web library allows interconnection of agents and cloud OWL knowledge model. We present a cloud knowledge model called cloud service ontology based on the OWL ontology [26, 27] as shown in Figs. 2 and 3, and we model it in the Protégé 4.3 Ontology editor [28]. Provider agent, ConsumerAgent, and DiscoveryAgent use agent communication language (ACL) message to communicate with each other based on SPARQL and OWL DL.

The agents of our framework are created using JADE, which is as shown in the Fig. 4. The user selects a cloud service provider in the ConsumerAgent GUI

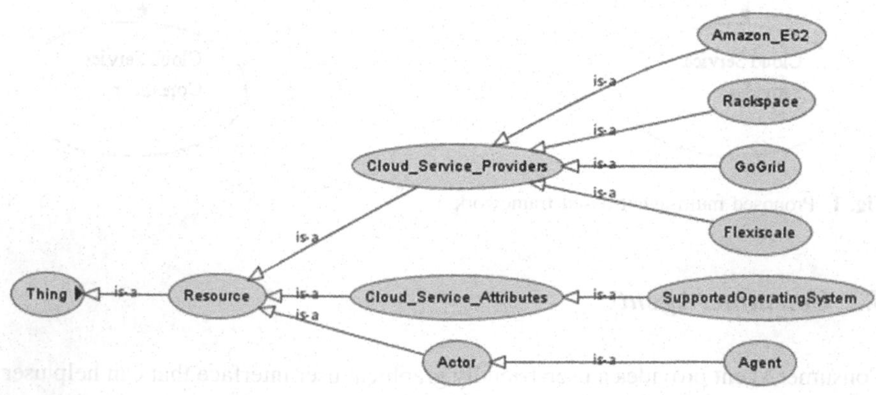

Fig. 2 Cloud service ontology

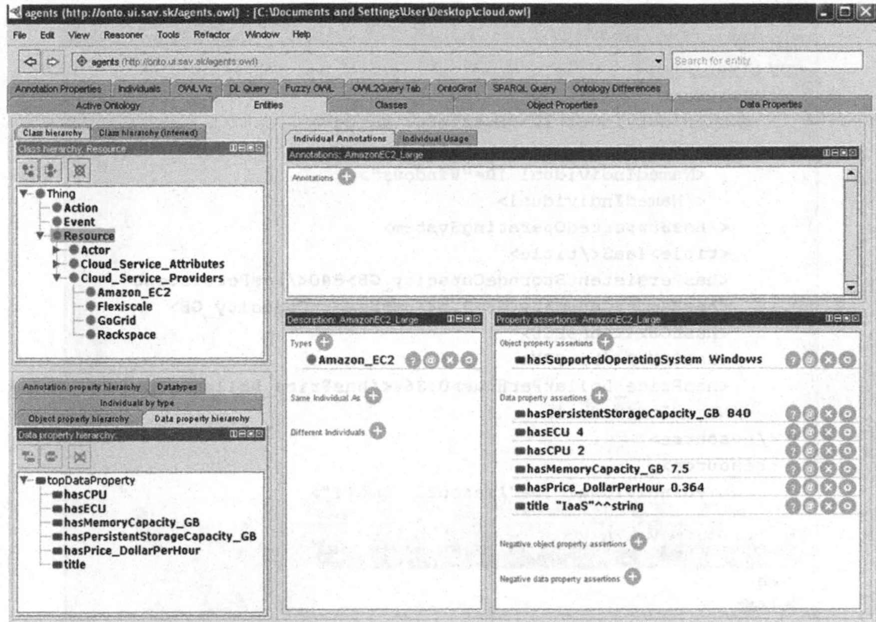

Fig. 3 Setting up the object properties and data properties of Amazon-EC2 cloud service provider

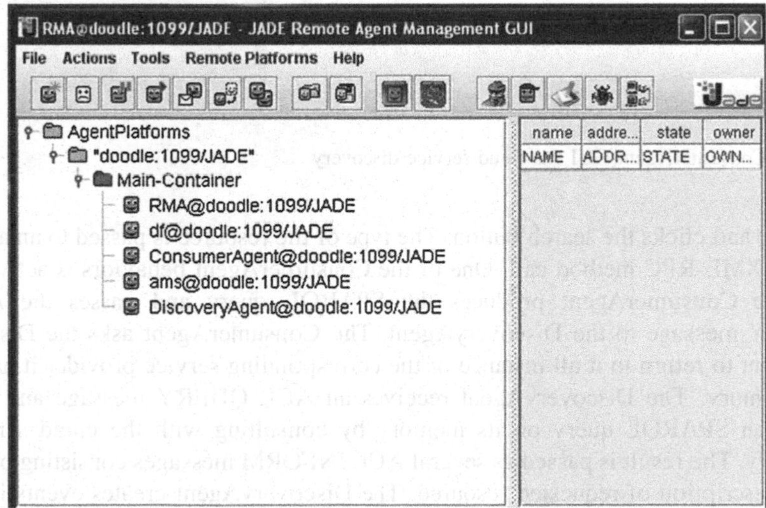

Fig. 4 Multi-agent-based platform of our framework

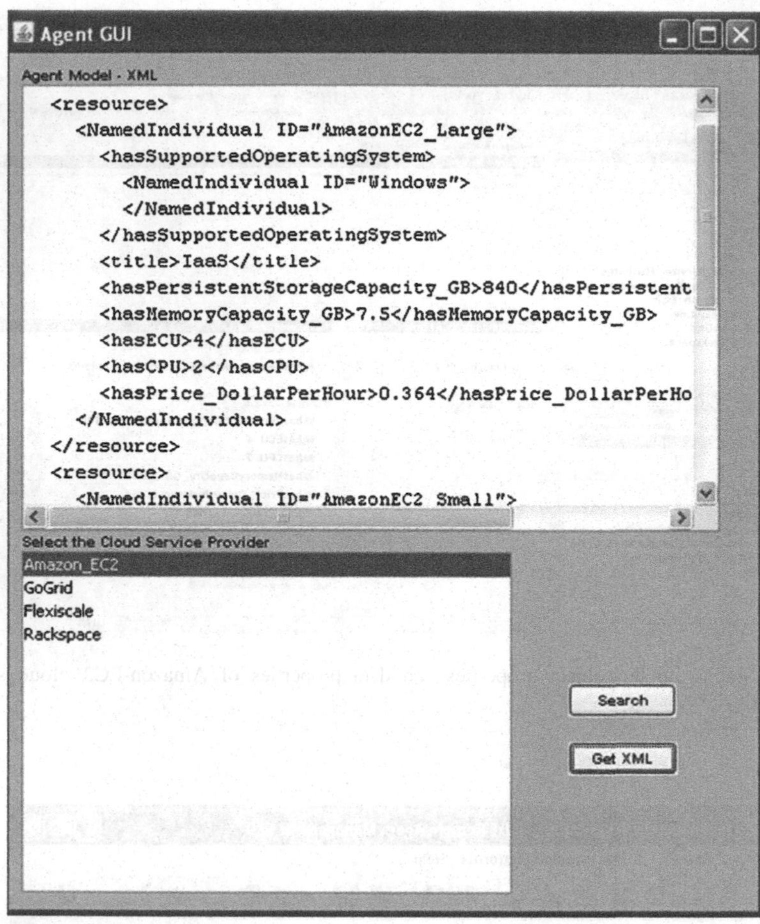

Fig. 5 ConsumerAgent GUI for cloud service discovery

(Fig. 5) and clicks the search button. The type of the resource is passed to an agent by the XML-RPC method call. One of the ConsumerAgent behaviors is activated and the ConsumerAgent produces the SPARQL query and passes the ACL QUERY message to the DiscoveryAgent. The ConsumerAgent asks the DiscoveryAgent to return to it all instance of the corresponding service provider it has in the memory. The DiscoveryAgent receives an ACL QUERY message and performs an SPARQL query on its memory by consulting with the cloud service ontology. The result is passed as several ACL INFORM messages consisting of the RDF description of requested resource. The DiscoveryAgent creates events in its memory that resources were sent to ConsumerAgent. This way the DiscoveryAgent keeps information about the environment. When the ConsumerAgent receives an ACL INFORM message, it stores its context into its memory model. Events about receiving resources are created in the ConsumerAgent memory. In addition, it adds references to returned resources to the DiscoveryAgent individual resource property. When a user clicks on the "getXML" button (Fig. 5) in the ConsumerAgent GUI, the DiscoveryAgent individual from the ConsumerAgent memory is returned in the XML format. In this XML, we can see cloud service provider instances along with their object and data properties, which are now in the ConsumerAgent memory.

7 Conclusion and Future Work

In the paper, we have proposed a framework that integrates multiple agents and ontology and developed a prototype for service description and discovery in cloud environment. The framework mainly assists in describing the cloud service providers and their attributes in a standardized way by using ontology and helps the users in discovering suitable service according to their requirements. User can select his request for discovering required service. The request will be automatically handled by the ConsumerAgent and then in turn the DiscoveryAgent perform the necessary reasoning based on predefined cloud service ontology and reasoning rules. Finally, the result is obtained in terms of XML in an understandable way. In the future, we will aim at implementing the entire system including cloud service composition and negotiation. In addition, we have to do more evaluations and comparison.

References

1. Zhang, Q., Cheng, L.: Cloud computing: State-of-the-art and research challenges. Journal of Internet Services and Applications -Springer **1**(1), 7–18 (2010)
2. Dillon, T., Chen, Wu., Chang, E.: Cloud computing: issues and challenges. In: Proceedings of 24th IEEE International Conference on Advanced Information Networking and Applications (AINA) pp. 27–33, April 2010
3. Buyya, R., et al.: Cloud computing and emerging it platforms: vision, hype, and reality for delivering computing as the 5th utility. J. Future Gener. Comput. Syst. Elsevier **25**(6), 599–616 (2009)
4. The NIST Definition of Cloud Computing available at: http://www.nist.gov/itl/cloud/upload/cloud-def-v15.pdf
5. Joshi, K.P., Yesha, Y., Finin, T.: Automating cloud services lifecycle through semantic technologies. IEEE Trans. Serv. Comput, pp. 1–14, 2012
6. Papazoglou, M.P.: Service-oriented computing: concepts, characteristics and directions. In: Proceedings of Fourth International Conference on Web Information Systems Engineering (WISE), pp. 3–12, Dec 2003
7. Wei, Y., Brian Blake, M.: Service-oriented computing and cloud computing challenges and opportunities. J. Int. Comput. IEEE **14**(6), 72–75 (2010)
8. Wooldridge, M.: An introduction to multiagent systems, 2nd edn. Wiley, New Jersey (2009)
9. Sycara, K.P.: Multiagent systems. AI Mag. **19**(2), 79–92 (1998)
10. Sim, K.M.: Agent-based cloud computing. IEEE Trans. Serv. Comput. **5**(4), 564–577 (2012)

11. Talia, D.: clouds meet agents: toward intelligent cloud services. J. Int. Comput. IEEE **16**(2), 1089–7801, 78–81 (2012)
12. Semantic Web Available at: http://www.w3.org/standards/semanticweb/
13. Gruber, T.R.: Towards principles for the design of ontologies used for knowledge sharing. Int. J. Hum. Comput. Stud. **43**(5–6), 907–928 (1995)
14. Youseff, L., Butrico, M., Silva, D.D.: Toward a unified ontology of cloud computing. Proceedings of IEEE Grid Computing Environments Workshop, Austin, TX, pp. 1–10, Nov 2008
15. Hoefer, C.N., Karagiannis, G.: Taxonomy of cloud computing services. In: Proceedings of IEEE GLOBECOM Workshop on Enabling the Future Service Oriented Internet, Miami, FL, pp. 1345–1350, Dec 2010
16. Yu. H, Reiff-Marganiec, S.: Non-functional property based service selection: a survey and classification of approaches. 2008
17. Kang, J., Sim, K.M.: Cloudle: a multi-criteria cloud service search engine. In: Proceedings IEEE Asia Pacific Services Computing Conference, Hangzhou, China. pp. 339–346, 2010
18. Chang, Y.S., Yang, C.T., Luo, Y.C.: An ontology based agent generation for information retrieval on cloud environment. J. Univers. Comput. Sci. **17**(8), 1135–1160 (2011)
19. Chang, Y-S., Juang, T-Y., Chang, C-H.: Integrating intelligent agent and ontology for services discovery on cloud environment. In: Proceedings of IEEE International Conference on Systems, Man, and Cybernetics, Seoul, Korea, pp. 3215–3220, 2012
20. Ngan, L.D., Kanagasabai, R.: OWL-S based semantic cloud service broker. In: Proceedings of IEEE 19th International Conference on Web Services (ICWS), pp. 560–567, June 2012
21. Ding, D., Liu, L., Schmeck, H.: Service discovery in self-organizing service-oriented environment. In: Proceedings. IEEE Asia-Pacific Services Computing Conference. Hangzhou, China. pp. 717–724, Dec 2010
22. Rehman, Z., Hussain, F.K., Hussain, O.K.: Towards multicriteria cloud service selection. In: Proceedings of Fifth International Conference on Innovative Mobile and Internet Services in Ubiquitous Computing (IMIS), Seoul, Korea. pp. 44–48, 2011
23. Wang, S., et al.: Cloud model for service selection, In: Proceedings IEEE Conference on Computer Communications Workshops (INFOCOM WKSHPS), Shanghai, China pp. 666–671, April 2011
24. HP Labs and Open Source Community: Jena Semantic Web Library. 2006, http://www.sf.net/
25. Telecom Italia: JADE (Java Agent Development Framework) Website. http://jade.cselt.it/, 2004
26. W3C: Web Ontology Language (OWL). http://www.w3.org/TR/owl-features/. 2006
27. Höfer, C.N., Karagiannis, G.: Cloud computing services: taxonomy and comparison. J. Internet Serv. Appl. **2**, 81–94 (2011)
28. Stanford University: Protégé Ontology Editor. http://protege.stanford.edu/. 2006

Enhancing Security in Cloud Computing Using Bi-directional DNA Encryption Algorithm

Ashish Prajapati and Amit Rathod

Abstract Cloud computing is the latest technology in the field of distributed computing. It provides various online and on-demand services for data storage, network services, platform services, etc. Many organizations are unenthusiastic to use cloud services due to data security issues as the data resides on the cloud services providers' servers. To address this issue, there have been several approaches applied by various researchers worldwide to strengthen security of the stored data on cloud computing. The Bi-directional DNA Encryption Algorithm (BDEA) is one such data security techniques. However, the existing technique focuses only on the ASCII character set, ignoring the non-English user of the cloud computing. Thus, this proposed work focuses on enhancing the BDEA to use with the Unicode characters.

Keywords Cloud computing · Data security issues · Bi-directional DNA encryption algorithm · DNA digital code

1 Introduction

Cloud computing has recently reached popularity and developed into a major trend in IT. We perform such a systematic review of cloud computing and explain the technical challenges facing in this paper. In Public cloud the "Pay per use" model is used. In private cloud, the computing service is distributed for a single society. In hybrid cloud, the computing services is consumed both the private cloud service and public cloud service. Cloud computing has three types of services. Software as

A. Prajapati (✉)
Computer Engineering, Parul Institute of Engineering and Technology, Vadodara, India
e-mail: er.ashishprajapati@gmail.com

A. Rathod
IT Department, Parul Institute of Engineering and Technology, Vadodara, India
e-mail: rathod.amit.h@gmail.com

L.C. Jain et al. (eds.), *Intelligent Computing, Communication and Devices*,
Advances in Intelligent Systems and Computing 308,
DOI 10.1007/978-81-322-2012-1_36

a Service (SaaS), in which customer prepared one service and run on a single cloud, then multiple consumer can access this service as per on demand. Platform as a Service (PaaS), in which, it provides the platform to create application and maintains the application. Infrastructure as a Service (IaaS), as per term suggest to provides the data storage, network capacity, rent storage, data centers, etc. It is also known as Hardware as a Service (HaaS).

2 Literature Surveys

In cloud computing, the major issue is to provide the security of data and this data security is prepared by the authentication, encryption and decryption, message authentication code, hash function, and digital signature and so on. So, here we discuss about some security problems and their solutions.

2.1 Use of Digital Signature with Diffie-Hellman Key Exchange and AES Encryption Algorithm to Enhance Data Security in Cloud Computing [1]

Rewagad and Pawar [1]. In this paper, the researcher using three way architecture protection schemes. Firstly, Diffie-Hellman algorithm is used to generate keys for key exchange step. Then digital signature is used for authentication, thereafter AES encryption algorithm is used to encrypt or decrypt user's data file. Diffie-Hellman key exchange algorithm is vulnerable to main in the middle attack. The most serious limitation is the lack of the authentication.

2.2 Union of RSA Algorithm, Digital Signature, and Kerberos in Cloud Security [2]

Hojabri and Heidari [2]. In this paper, the researcher first performs the concept of Kerberos authentication services. At the next step the authenticate server (AS) of Kerberos do verifies users and created the ticket granting ticket and session key and it sent to the users. The next step users send the ticket granting ticket and session key to ticket granting server (TGS) for getting the service. Then TGS sends ticket and session key for user. In final step, the users send the request service to cloud service provider for using the cloud service and also cloud service provides service to users. After doing this step, user can use the cloud service provider. But for more security they performed RSA algorithm for encryption and decryption and then they use digital signature for authentication.

2.3 Implementation Digital Signature with RSA Encryption Algorithm to Enhance the Data Security of Cloud in Cloud Computing [3]

Somani et al. [3]. In this paper, there are two enterprises A and B. An enterprise A has some data that are public data and enterprise has public cloud. Now B wants some secure data from A's cloud. So RSA algorithm and digital signature are used for secure communication. In this method, enterprise A takes data from cloud, which B wants. Now the data or document is crushed into little line using Hash code function that is called message digest. Then A encrypts the message digest within private key the result is in the digital signature form. Using RSA algorithm, A will encrypt the digital-signed signature with B's public key and B will decrypt the cipher text to plain text with his private key and A's public key for verification of signature.

3 Proposed Work

Previous section describes the study about the cloud computing, basics of cloud computing, and security problems occur in cloud. Then some study papers to solve these security problems. Here in this paper, the bi-serial DNA encryption algorithm is performing that providing the two level of security. The whole flow is described in next section.

3.1 DNA Digital Coding

In information science, the binary digital coding encoded by two state 0 or 1 and a combination of 0 and 1. But DNA digital coding can be encoded by four kinds of bases as shown in Table 1. That is ADENINE (A) and THYMINE (T) or CYTOSINE (C) and GUANINE (G). There are possibly 4! = 24 pattern by encoding format like (0123/ATGC) [4].

Table 1 DNA digital coding

Binary value	DNA digital coding
00	A
01	T
10	G
11	C

3.2 Encryption Process

In this part, user enters its message into Unicode plaintext. A Unicode plaintext is converted into its ASCII form then it converts into hexadecimal code and then converts into binary code using decimal convertor. Using binary code, find DNA digital coding from Table 1. And at last using primer pair, they find PCR amplification. And last amplified message is ready to send. The whole structure is shown in Fig. 1. Diffi-Hellman key exchange algorithm is used for key sharing.

Example:

Unicode:

આશિષ

ASCII:

\u0A86\u0AB6\u0ABF\u0AB7

Hexadecimal value:

5c 75 30 41 38 36 5c 75 30 41 42 36 5c 75 30 41 42 46 5c 75 30 41 42 37

Binary value:

0101110001110101001100000100000100111000001101100101110001110101
0011000001000001010000100011011001011100011101010011000001000001
0100001001000110010111000111010100110000010000010100001000110111

DNA Digital coding:

TTCATCTTACAATAATACGAACTGTTCATCTTACAATAATTAAGACTG
TTCATCTTACAATAATTAAGTATGTCTTACAATAATTAAGACTC

Primer 1: C, Primer 2: G, Generated Primer pair: CGC.

Amplified Message:

Using two prime pair, generate PCR amplification

TCGCTCGCCCGCACGCTCGCCCGCTCGCTCGCACGCCCGCACGCACGC
TCGCACGCACGCTCGCACGCCCGCGCGCACGCACGCCCGCTCGCGCGC
TCGCTCGCCCGCACGCTCGCCCGCTCGCTCGCACGCCCGCACGCACGC
TCGCACGCACGCTCGCTCGCACGCACGCGCGCACGCCCGCTCGCGCGC
TpTCGCTCGCCCGCACGCTCGCCCGCTCGCTCGCACGCCCGCACGCACGC
TCGCACGCACGCTCGCACGCCCGCGCGCACGCACGCCCGCTCGCG TCGC
TCGCCCGCACGCTCGCCCGCTCGCTCGCACGCCCGCACGCACGCTCGCAC
GCACGCTCGCACGCCCGCGCGCAC

Now message is ready to send.

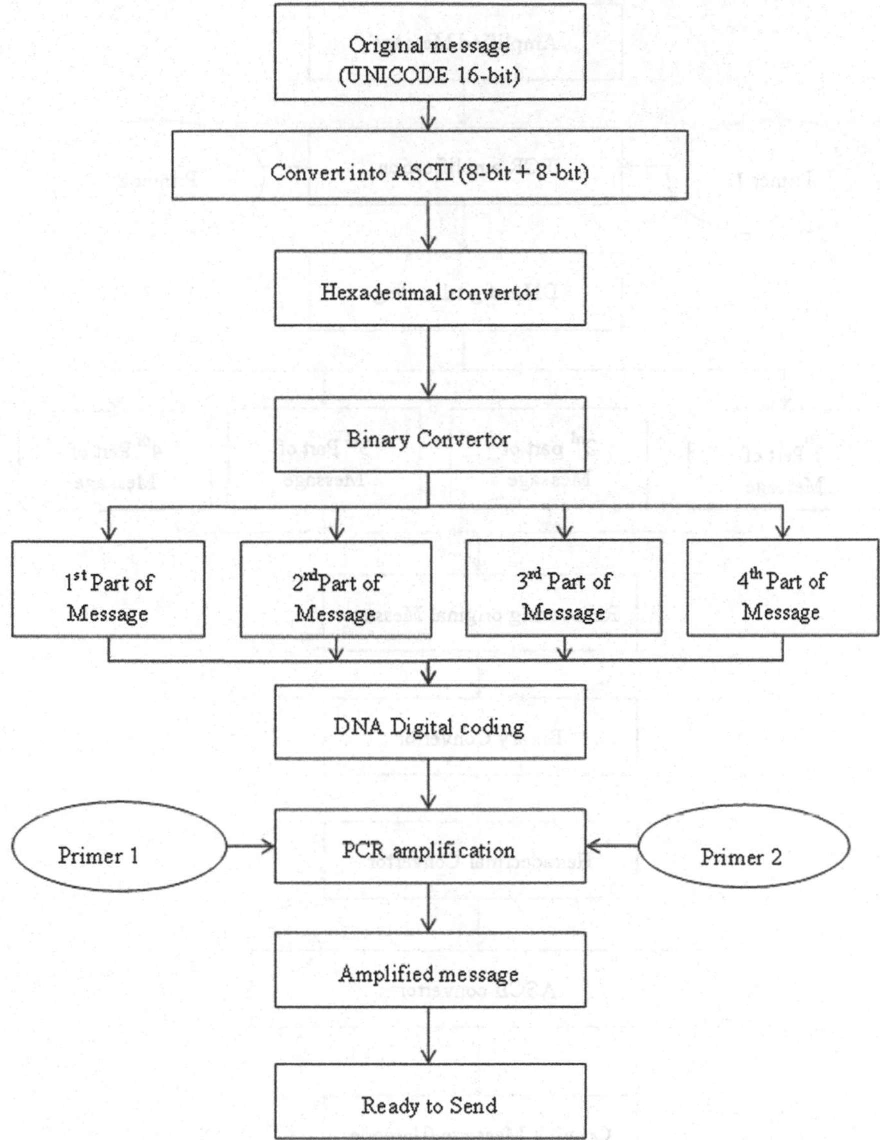

Fig. 1 Encryption of Bi-serial DNA algorithm

3.3 Decryption Process

From the receiver side, get the encrypted data using high compressed algorithm to recover compress data. Then using two correct primer pair, sender gets the DNA digital code. Then after DNA digital coding is converted into binary code then

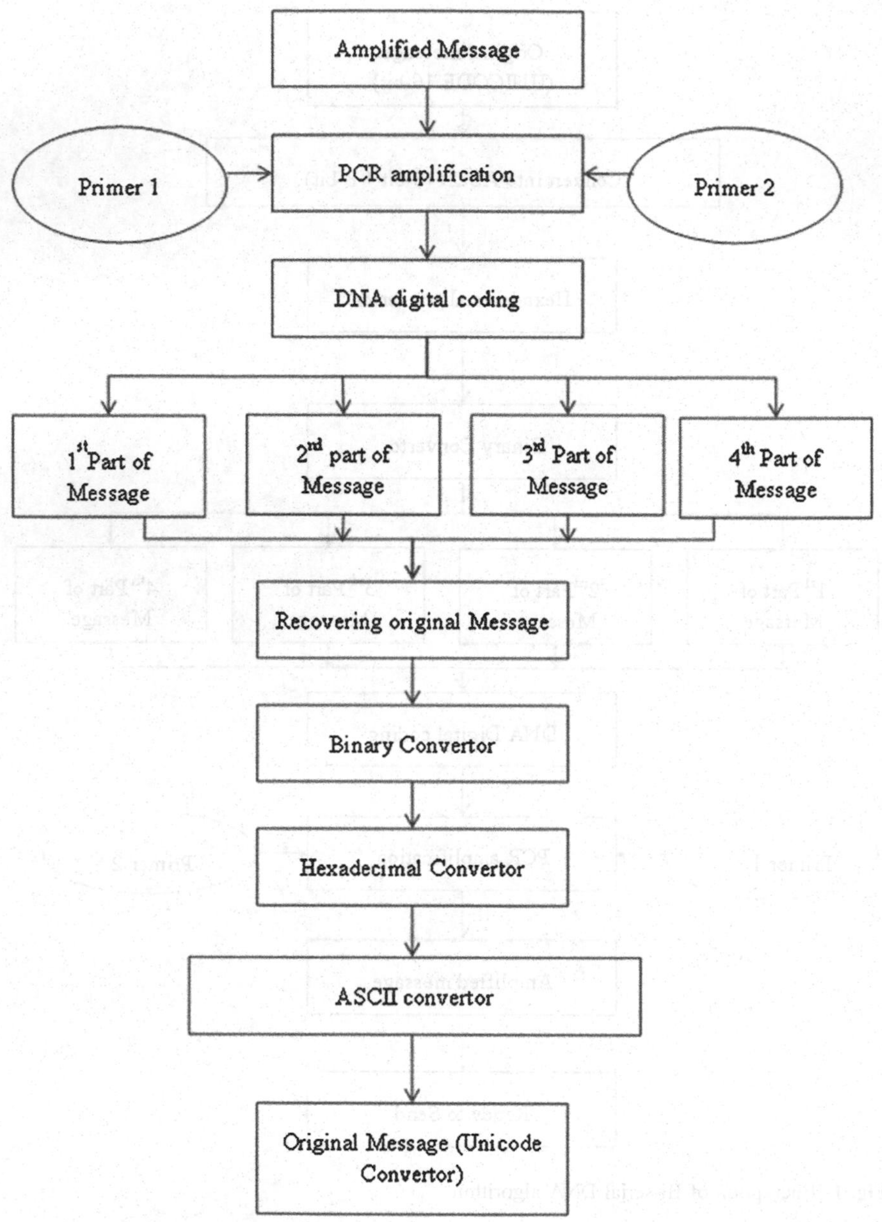

Fig. 2 Decryption of Bi-serial DNA algorithm

converted into hexadecimal code. Hexadecimal code is converted into normal plaintext by using decimal convertor. The whole flow is explained by Fig. 2 that is given below.

Example:
Primer 1: C, Primer 2: G, Received Primer pair: CGC.

Amplified Message:
TCGCTCGCCCGCACGCTCGCCCGCTCGCTCGCACGCCCGCACGCACGC
TCGCACGCACGCTCGCACGCCCGCGCGCACGCACGCCCGCTCGCGCGC
TCGCTCGCCCGCACGCTCGCCCGCTCGCTCGCACGCCCGCACGCACGC
TCGCACGCACGCTCGCTCGCACGCACGCGCGCACGCCCGCTCGCGCGC
TpTCGCTCGCCCGCACGCTCGCCCGCTCGCTCGCACGCCCGCACGCACGC
TCGCACGCACGCTCGCACGCCCGCGCGCACGCACGCCCGCTCGCG TCGC
TCGCCCGCACGCTCGCCCGCTCGCTCGCACGCCCGCACGCACGCTCGCA
CGCACGCTCGCACGCCCGCGCGCAC

DNA Digital coding:
TTCATCTTACAATAATACGAACTGTTCATCTTACAATAATTAAGACTG
TTCATCTTACAATAATTAAGTATGTCTTACAATAATTAAGACTC

Binary value:
0101110001110101001100000100000100111000001101100101110001110101
0011000001000001010000100011011001011100011101010011000001000001
0100001001000110010111000111010100110000010000010100001000110111

Hexadecimal value:
5c 75 30 41 38 36 5c 75 30 41 42 36 5c 75 30 41 42 46 5c 75 30 41 42 37

ASCII:
\u0A86\u0AB6\u0ABF\u0AB7

Unicode:
આશિષ

4 Conclusion and Future Scope

Data security is the main challenge for cloud usability. Various algorithms such as RSA, Diffie-Hellman, and DNA encryption are available to provide data security for the data stored on cloud. Digital signatures and Extensible Authentication Protocols are used for authentications. Using Bi-directional DNA Encryption Algorithm (BDEA), we achieve 2-layer security for ASCII character sets. The proposed system focuses on extending the BDEA algorithm to be used with Unicode character set. This can help reach to the wider community of the cloud users. The future work will focus on the possible attacks and cryptanalysis of the cipher text and measure its strength.

References

1. Rewagad, P., Pawar, Y.: Use of digital signature with Diffie-Hellman key exchange and AES encryption algorithm to enhance data security in cloud computing. In: 2013 International Conference on Communication System and Network Technologies (IEEE Computer Society)
2. Somani, U., Lakhani, K., Mundra, M.: Implementing digital signature with RSA encryption algorithm to enhance the data security of cloud in cloud computing. In: 2010 IEEE 1st International Conference on Parallel, Distributed and Grid Computing (PDGC-2010)
3. Hojabri, M., Heidari, M.: Union of RSA algorithm, digital signature and KERBEROS in cloud computing. International Conference on Software Technology and Computer Engineering (STACE-2012)
4. http://arxiv.org/ftp/arxiv/papers/1101/1101.2577.pdf

Service Composition Using Efficient Multi-agents in Cloud Computing Environment

Abhijit Bastia, Manoranjan Parhi, B.K. Pattanayak and M.R. Patra

Abstract In today's Internet world, end users need everything as a service (EaaS) being available to them throughout the world. Keeping users in the centre, developers are emplaning cloud computing environment which can satisfy needs of users virtually. Cloud computing is a collection of Web-accessible resources (i.e. Web services) that should be dynamically composed between service providers and brokers and virtualized based on consumer's needs on an on-demand basis. Mapping of the users' requirements should be done in an automated manner. But, distributed and constantly changing cloud computing environments pose new challenges to automated service composition such as: (i) dealing with incomplete information regarding cloud resources (e.g. location and providers), and (ii) dynamics of service providers, which set service fees on a supply-and-demand basis. To address these issues, we have proposed a multi-agent-based approach to compose services in multi-cloud environments for different types of cloud services: *one-time virtualized services, persistent virtualized services, vertical services*, and *horizontal services*. Cloud participants and resources are implemented and instantiated by agents. Previously the researchers have proposed self-organizing agents those make use of *Service Capability Table* and the *semi-recursive contract net protocol* (SR-CNP) to evolve and adapt cloud service compositions. To the existing work, we have planned to modify some of agents' behaviours, as a result we can reduce number of message passing by half in order to increase

---

A. Bastia (✉) · M. Parhi · B.K. Pattanayak
Department of Computer Science and Engineering, ITER, Siksha 'O' Anusandhan University, Bhubaneswar 751030, Odisha, India
e-mail: abhijit.bastia@gmail.com

M. Parhi
e-mail: mrparhi@gmail.com

B.K. Pattanayak
e-mail: bkp_iter@yahoo.co.in

M.R. Patra
Department of Computer Science, Berhampur University, Berhampur 760007, Odisha, India
e-mail: mrpatra12@gmail.com

L.C. Jain et al. (eds.), *Intelligent Computing, Communication and Devices*,
Advances in Intelligent Systems and Computing 308,
DOI 10.1007/978-81-322-2012-1_37

overall performance. Also we are planning a 2-layered (3 levels of multi-agents) self-organizing MAS that will establish a cloud service composition.

Keywords One-time virtualized services · Persistent virtualized services · Vertical services · Horizontal services · Agent behaviour · SR-CNP · SCT

1 Introduction

World Wide Web has become the most popular ammo in today's era of innovations and technologies. Cloud computing is an Internet-based computing, which typically involves the provisioning of dynamically scalable and often virtualized resources as services over the Internet. Some of the distinguishing characteristics of cloud computing are elasticity, scalability, hardware virtualization, fast service configuration, etc. [9, 10]. In cloud computing environments, variety of services can be provided which are broadly classified as infrastructure as a service (IaaS), platform as a service (PaaS), and software as a service (SaaS) as shown in Fig. 1. These cloud services can be composed into a value-added service termed as everything as a service (EaaS) to satisfy the dynamic needs of users. Various uncertainties can affect the correctness, availability, and reliability of service composition in cloud computing environments due to the heterogeneous, autonomous, and dynamic characteristics of cloud services. Therefore, how to ensure service compositions with multi-agents in the unpredictable cloud environments needs to be urgently addressed.

Now-a-days, the number of cloud providers is increasing, and the services offered by cloud providers have also increased. There are also increasing demands for cloud services from consumers. There are two major challenges to address. First, anticipating all of the possible required services is extremely difficult, particularly for software services. The second challenge is in selecting the optimum required single composite services, which are provided by different service providers with different quality of service (QoS) attributes; an optimal combination for forming a complicated service must be composed. Thus, there is a need for

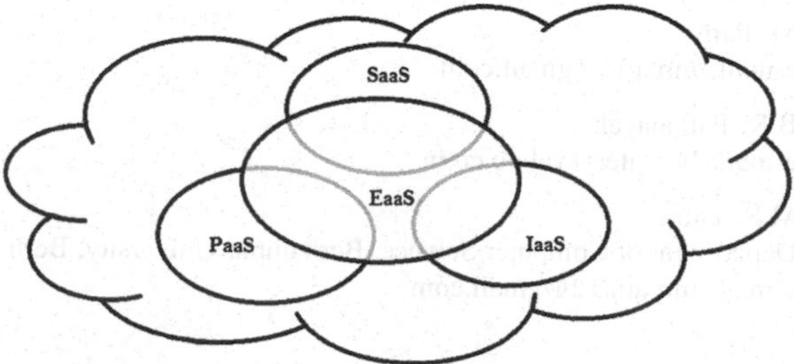

Fig. 1 Basic cloud service composition

dynamic and automated cloud service composition that can support everything as a service model capable of satisfying complex consumer requirements as they emerge and nullify human intervention by the use of agents. Literally, agent is an entity which acts on behalf of another entity. An agent is essentially a special software component that has autonomy that provides an interoperable interface to an arbitrary system and/or behaves like a human agent, working for some clients in pursuit of its own agenda. Agents are independent of problem solvers (e.g. cloud participants) that may collaborate to achieve a global objectives (e.g. service composition) while simultaneously considering both individual goals and constraints. Therefore, a multi-agent system is a collection of autonomous, interacting, and cooperative agents that react to events and may self-organize by means of interaction, negotiation, coordination, and collaboration.

The significance of this paper is that it is an effort in providing an efficient multi-agent-based approach for dealing with one-time, persistent, vertical, and horizontal cloud service compositions. This paper modernized some aspects of [1]. The rest part of this paper is distributed as follows. In Sect. 2, a literature study on cloud service composition model is discussed along with their merits and limitations. In Sect. 3, an efficient multi-agent-based cloud service composition model is proposed, along with the description of agents' job (Sect. 3.1), algorithms for proposed agents' behaviour (Sect. 3.2). In Sect. 3.3, the service capability tables (SCTs) for proposed model is discussed and semi-recursive contract net protocol (SR-CNP) is discussed in Sect. 3.4. In Sect. 4, implementation of the proposed model is discussed along with evaluation of both the models from speculative point of view is described. In Sect. 5, the future direction of this model is discussed along with the concluding remarks.

2 Literature Study

Several researches have been done on cloud service composition by using various approaches. In our review, we have considered the study of those papers which are related to cloud service composition.

A multi-agents-based cloud service composition model was proposed in [1, 3, 8] which is a 3-layered architecture. The positive points in [1] are as follows: (i) it is a multi-agent system that nullified the human intervention in cloud service composition, (ii) the implementation of service ontology which can be semantic or syntactic (which is another research area), and (iii) the distributive architecture of the model. Along with the above advantages, the paper lacks in some aspects like (a) service provider agents are cluster of homogeneous type of resource agents, i.e. an SPA only holds RAs of one-type services from SaaS, PaaS, and IaaS, (b) unnecessary message passing to rejected agents.

A QoS-based service composition using state transition machine is proposed in [2]. Positive point in this paper is that it composes service considering all optimal service resources, but the drawback is that it incurs more communication overhead to reach at a composite service. A QoS-based service composition considering network traffic is proposed in [4]. Though this paper considers optimal services available locally to incur less network propagation but it may lack in case of a heavy traffic arising in a particular locality resulting a bottleneck condition in a geographical area.

A QoS-based tree-pruning service composition is proposed in [5]. Though it is a QoS-based service composition technique but it is centralized approach. A QoS-based service composition based on service-level agreement is proposed in [7]. Here, agreement is done before delivering the services and the composite services are well tested. But the drawback is that a bottleneck situation may rise at cloud directory and also the testing result may vary for different trusted composition verifiers.

The basic aim of considering [1] is all its positivity as described earlier. Along with the above-discussed advantages, the paper lacks in some aspects like (a) service provider agents are cluster of homogeneous type of resource agents, i.e. an SPA only holds RAs of one-type services from SaaS, PaaS, and IaaS, (b) unnecessary message passing to rejected and/or to useless agents, and (c) inefficient implementation of consumer agent and broker agent levels which may give rise to delay. In our proposed model, we have tried to address the above issues.

3 Proposed Cloud Service Composition Model

In our proposed composition model, we have designed a 2-layered multi-agent-based cloud service composition model instead of a 3-layered model with an aim to reduce the processing time. The proposed model is given in Fig. 2. The service ontology considered here is a syntactic though semantic ontology can also be implementable as described in [6] which we will try to address in our future work. We have taken 3 agents namely (i) consumer agent (CA), (ii) service broker agent (SBA), and (iii) resource provider agent (RPA). Also we have taken an SBA of heterogeneous type, i.e. the SBA is a cluster of different types of services (i.e. RPAs). SBA will perform two tasks. Firstly, it will act as a service provider. Secondly, it will act as a broker agent if it fails to do the first task. The agents' behaviours are summarized in Table 1.

3.1 Agents' Job Description

The agents' job is defined from their functionality point of view, and the tasks that can be performed by agents are defined as their behaviour. Agents (CAs, SBAs, and RPAs) interact among each other to compose and manage cloud services by adopting diverse agent behaviours.

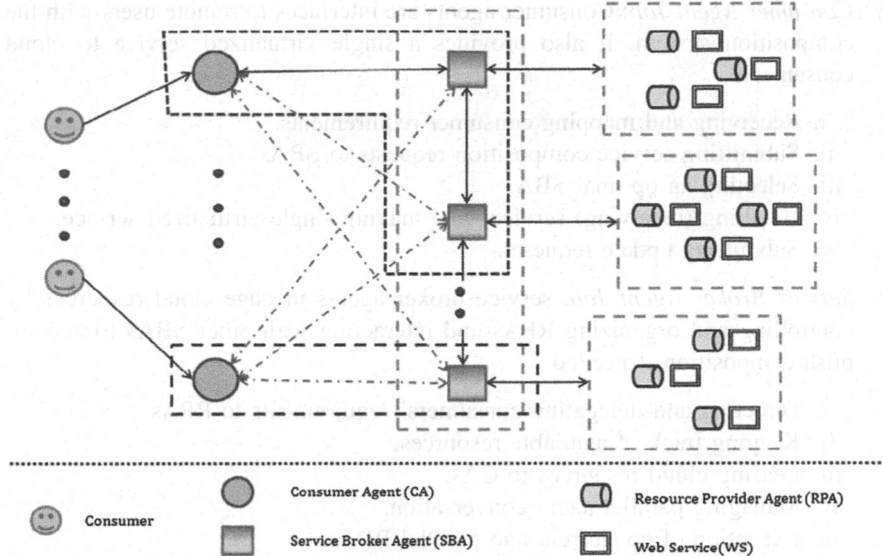

Fig. 2 Proposed cloud service composition model

Table 1 Agents' behaviour

Agent	Behaviour identifier	Main function
CA	*SR-CNPinitiatorCA*	To submit service composition call-for-proposals to SBAs
	ServiceAugmenterCA	To submit requests for incremental updates
	ServiceRevokerCA	To submit requests for subtractive updates
	ResultHandlerCA	To composite virtualized service
	ContractChangeMonitor	To receive expired contracts' notifications
SBA	*SR-CNPparticipantSBA*	To handle consumers' service composition requests from CAs or other SBAs
	SR-CNPinitiatorSBA	To submit call-for-proposals to resolve requirements to RPAs and service composition requests to other SBAs
	ReqAssignerSBA	To assign requirements to RPAs
	ResultHandlerSBA	To receive results from both RPAs and contracted SBAs
	IntermediarySBA	To handle requests to resolve requirements from RPAs
	ServiceRevokerSBA	To submit requests for subtractive updates
	ContractChangeMonitor	To receive expired contracts' notifications
RPA	*MainStructureRPA*	To resolve requirements by orchestrating a Web service
	ResourceHandlerRPA	To receive new web/cloud services and/or terminate web/ cloud services
	RequesterRPA	To request external requirements to SBAs
	ReleaserRPA	To release cloud resources

a. *Consumer Agent Job*: Consumer agents are interfaces to remote users with the composition system. It also provides a single virtualized service to cloud consumers.

   i. Receiving and mapping consumer requirements.
   ii. Submitting service composition requests to SBAs.
   iii. Selecting an optimal SBA.
   iv. Handling (receiving) services and making single virtualized service.
   v. Submitting update requests.

b. *Service Broker Agent Job*: service broker agents manage cloud resources by controlling and organizing RPAs and interacting with other SBAs to accomplish composition if needed.

   i. Directing and delegating consumers' requirements to RPAs.
   ii. Keeping track of available resources.
   iii. Leasing cloud resources to CAs.
   iv. Managing parallel agent conversation.
   v. Execution of concurrent and parallel RPAs.

c. *Resource Provider Agent Job*: resource provider agents arrange Web services and control the access to them. RPAs receive requests to resolve requirements from service brokers. Then, RPAs handle the requests via their associated Web service, returning the output to the SBAs.

   i. Arrange Web services and control access to them.
   ii. Process received requests from SBAs.

3.2 Algorithm for Agent Behaviour

We have defined about sixteen algorithms for these agent behaviours. Two of these algorithms are given in Tables 2 and 3 for readers' interest and rest are not given for the implementation point of view.

3.3 Service Capability Tables

SCTs are used by agents to register and consult information about other cloud agents. The records of SCTs are composed of (i) agent address, (ii) requirements fulfilled by agent, and (iii) last known status of the agent/service. The SCTs are (a) *dynamic*: because agents can be removed or added to the table, (b) *exact*: because agents' address and functionalities are well known, and (c) *incomplete*: because agent may unaware of the full list of existing agents. The SCT parameters for different agents are tabled in Tables 4 and 5.

Table 2 Algorithm for SR-CNPinitiatorCA

SR-CNPinitiatorCA Behaviour:
Input: (i) Consumer Requirements, (ii) SBAs' address
Output: Single Virtualized service
1. creating atomic requirements
2. CA sends *call-for-proposals* to resolve requirements to m
feasible SBAs' from SCT
3. if(ProposalReceived(Proposals,timeout1)) then
4. evaluate proposals 5. CA sends accept-proposal to *1 Best* SBA
6. create contract between CA and SBA
7. if(Receive(virtualized service,timeout2)) then
8. consume virtualized service
9. else
10. remove SBA and update status of SBA to *failed* in SCT
11. throw exception
12. else
13. update status of SBAs to unreachable in SCT
14. throw exception
Exception: Repeat steps again

3.4 Semi-recursive Contract Net Protocol

A problem whose solution can be obtained from the solution to smaller instances of the same problem is a recursive problem. A semi-recursive (i.e. to some extent recursive) agent interaction protocol is a protocol that attains its design objective (e.g. composing a set of cloud consumer requirements) by re-instantiating its communication pattern within itself to solve smaller instances of its design objective (e.g. composing a subset of the cloud consumer requirements). The SR-CNP follows a divide-and-conquer strategy. Agents in the CNP have two roles, i.e. initiator and participant. Here, CAs and SBAs can adopt the SR-CNP initiator role, but only SBAs and RAs can take the role of participant. Figure 3 shows the interaction of these agent behaviours among each other.

4 Implementation and Speculative Evaluation

To implement our framework, we have used Java Agent DEvelopment (JADE) platform. JADE is a very powerful middleware framework built with Java to design a multi-agent systems-based architecture. Consumer agents, service broker agents, and resource provider agents are created with JADE as shown in Figs. 4, 5, and 6. Figure 4 shows a JADE environment which consist of 3 proposed agents

Table 3 Algorithm for SR-CNP participant SBA

```
SR-CNPparticipantSBA Behaviour:
Input: (i) call-for-proposals from CAs to resolve Req
Output: (i) instantiation of SR-CNPinitiatorSBA Behaviour or (ii) refuse message
if(ProposalReceive(call-for-proposals(Req.))) then
reqRPA ← get requirements fulfilled by RPAs from Req
if(reqRPA == NULL) then
reqSBA ← get requirements fulfilled by other SBAs
end if
if(reqRPA ∨ reqSBA == Req) then
SBA sends Proposal
if(ProposalReceive(reply, timeout)) then
if(reply == accepted) then
if(reqRPA != NULL) then
contract RPA by SR-CNPinitiatorSBA
else if(reqSBA ! = NULL) then
contract SBA by SR-CNPinitiatorSBA
end if
end if
end if
else
SBA sends refuse message
```

Table 4 SCT for CA

About CA	About SBA
i. CAs' address	i. SBAs' address
ii. CAs' status (available, unreachable, and failed)	ii. SBAs' status (available, unreachable, and failed)
	iii. Requirements fulfilled by SBAs

Table 5 SCT of SBA

About SBA	About RPA
i. SBAs' address	i. RPAs' address
ii. SBAs' status (available, unreachable, and failed)	ii. RPAs' status (available, unreachable, and failed)
iii. Requirements fulfilled by SBAs	iii. Requirements fulfilled by RPAs

namely consumer agent, service broker agent, and resource provider agent with their agent IDs. Figure 5 shows a consumer interface which will give a composite service, and Fig. 6 shows a resource interface to register a resource.

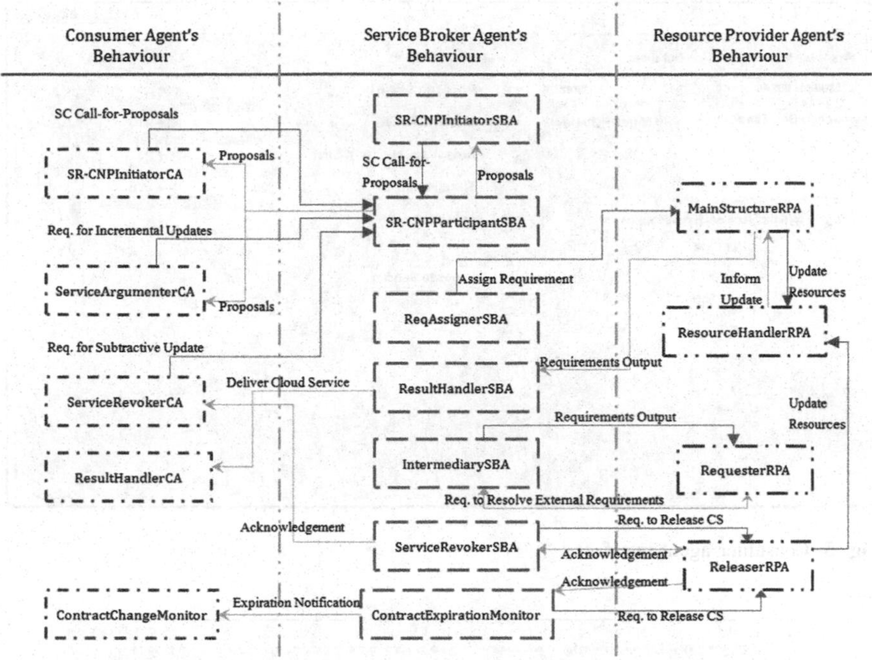

Fig. 3 Interaction of agent behaviours

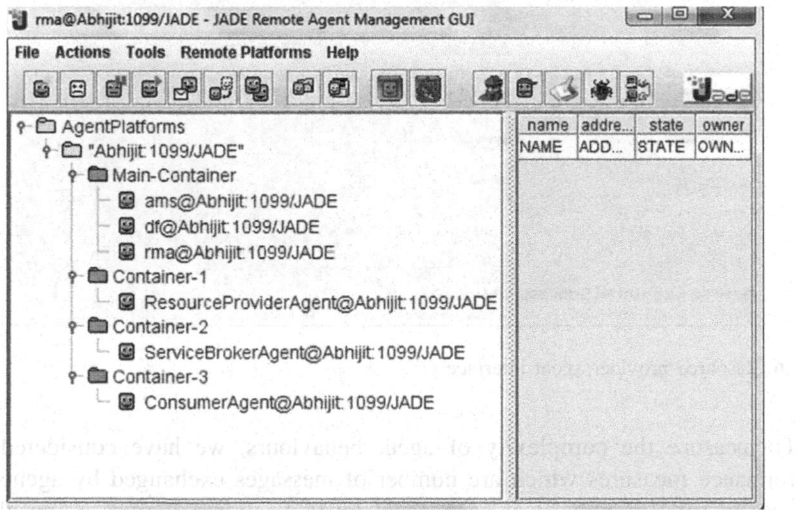

Fig. 4 JADE framework for proposed cloud service composition

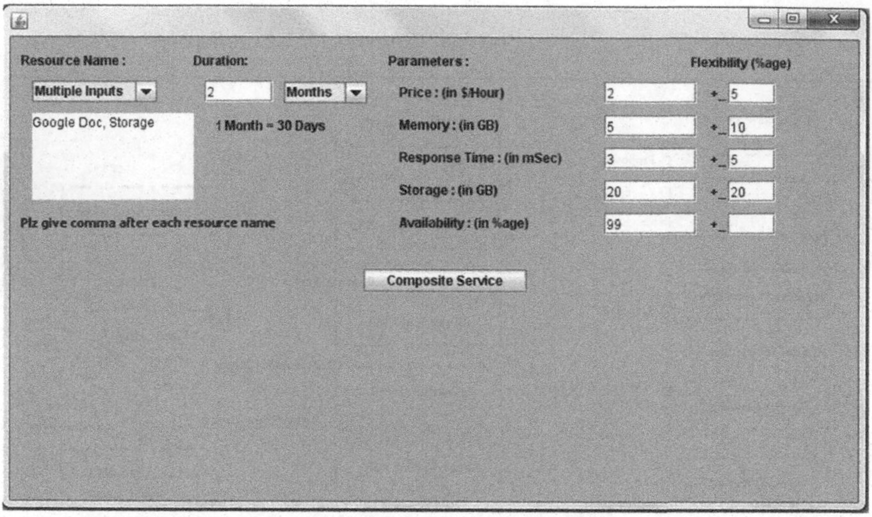

Fig. 5 Consumer agent interface

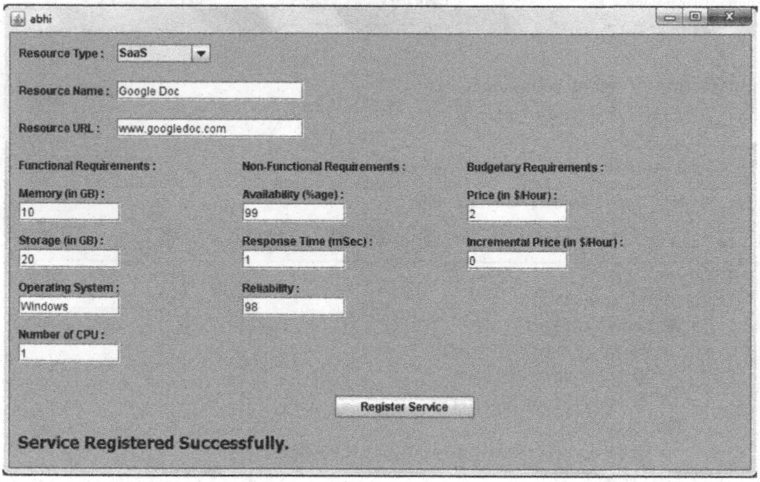

Fig. 6 Resource provider agent interface

To measure the complexity of agent behaviours, we have considered two performance measures which are number of messages exchanged by agents and processing time of request at each agent level. Section 4.1 gives a speculative evaluation on number of message exchanged, and in Sect. 4.2, the speculative evaluation on the processing of request is discussed.

Table 6 Comparative Speculative Evaluation

After getting the inputs, the CA sends q responses (1 accept-proposal message and $(q-1)$ reject-proposal messages). If the contracted SPAs fail, the failure is propagated to the BA. Then, the CA sends $(q-1)$ call-for-proposals messages to the remaining feasible BAs, and $(q-1)$ responses, and so on. Thus, in the worst case scenario the CA sends: $p(2q+2(q-1)+2(q-2)+\cdots+2(2)+2(1))$ $=2p(q+(q-1)+(q-2)+\cdots+(2)+(1))$ $=2p(q(q+1)/2)$ $=p*q(q+1)$ messages	After getting the inputs, the CA sends 1 response (1 accept-proposal message). (In this model, a timeout is set so that all other agents will consider themselves rejected if they do not get any accept-proposal within that timeout period.) If the contracted RPAs fail, the failure is propagated to the SBA. Then, the CA sends $(q-1)$ call-for-proposals messages to the remaining feasible SBAs, and so on. Thus, in the worst case scenario, the CA sends: $p((q+1)+((q-1)+1)+((q-2)+1)+\cdots)$ $=p((q+1)+q+(q-1)+(q-2)+\cdots)$ $=p((q+2)(q+1)/2)$ $=p*(q+2)(q+1)/2$ messages

4.1 Evaluation of Number of Message Exchanged

The number of message sent by an agent is dependent upon its connectivity. For both the models, we have considered similar test bed. The agents involved in evaluation were as follows: (i) for previous model, 25 CAs, 25 BAs, 30 SPAs, and 4,500 RAs (ii) for our proposed model, 25 CAs, 30 SBAs, and 4,500 RPAs. Both RAs and RPAs are randomly distributed to SPAs and SBAs, respectively. Weak connection has up to 33 % connectivity to next layer. Similarly, moderate has up to 66 % connectivity and strong has up to 100 % connectivity to next layer. The comparison shows a reduction of message passing by half through our proposed model which holds for each layer. A description on the speculative evaluation between the two approaches is discussed in Table 6. Figure 7 shows the evaluated number of message exchanged by previous model, and Fig. 8 shows the same for our proposed model.

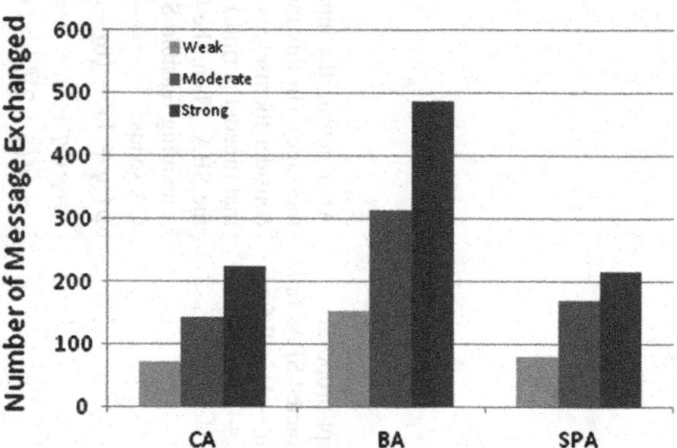

Fig. 7 Number of message exchanged by previous model

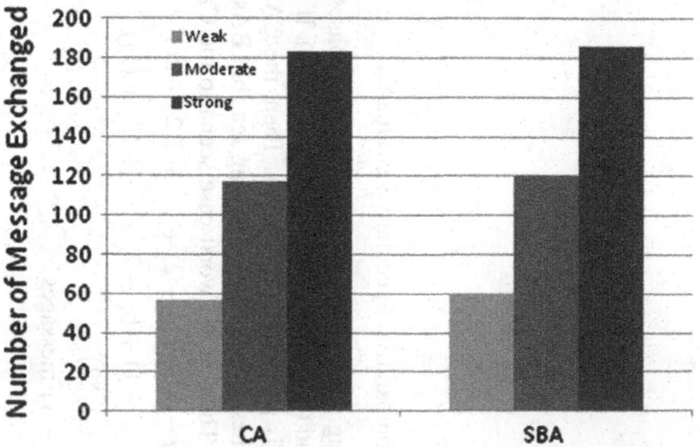

Fig. 8 Number of message exchanged by proposed model

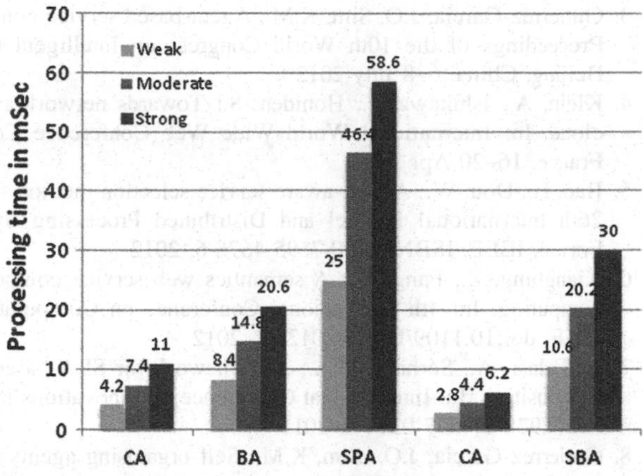

Fig. 9 Overall processing time of agents by both models

4.2 Evaluation of Processing Time of Request

To evaluate the processing time for each request, we have considered similar conditions, i.e. for sending a message 0.2 ms time is considered, and 1 ms time is considered for processing a request for both the models. Figure 9 shows the overall processing time needed to process single instance of a request by both the models, where CA, BA, and SPA belong to the previous model, and CA and SBA belong to our proposed model.

5 Conclusion and Future Work

In this paper, we have focused on demonstrating the effectiveness of adopting agent-based techniques for cloud service composition by implementing the desirable property that our agents can autonomously and successfully deal with changing service requirements through self-organization and collaboration. As an extension to the current work, we intend to incorporate the semantic ontology to our proposed model. We hope that our multi-agent-based framework can become more practical in real-world applications with full-phase implementation.

References

1. Gutierrez-Garcia, J.O., Sim, K.M.: Agent-based cloud service composition. Appl. Intell. doi:10.1007/s10489-012-0380-x, 5 Sept 2012
2. Liu, S., Xiong, G., Zhao, H., Dong, X., Yao, J.: Service composition execution optimization based on state transition matrix for cloud computing. In: Proceedings of the 10th World Congress on Intelligent Control and Automation, IEEE, Beijing, China, 6–8 July 2012

3. Gutierrez-Garcia, J.O, Sim, K.M.: Agent-based service composition in cloud computing. In: Proceedings of the 10th World Congress on Intelligent Control and Automation, IEEE, Beijing, China, 6–8 July 2012
4. Klein, A., Ishikawa, F., Honiden, S.: Towards network-aware service composition in the cloud. In: International World Wide Web Conference Committee, IW3C2, ACM, Lyon, France, 16–20 Apr 2012
5. Bao, H., Dou, W.: A QoS-aware service selection method for cloud service composition. In: 26th International Parallel and Distributed Processing Symposium Workshops and PhD Forum, IEEE, ISBN:978-0-7695-4676-6, 2012
6. Xiangbing, Z., Fang, M.: A semantics web service composition approach based on cloud computing. In: 4th International Conference on Computational and Information Sciences, IEEE; doi:10.1109/ICCIS.2012.43, 2012
7. Al Falasi, A., Serhani, M.A.: A framework for SLA-based cloud services verification and composition. In: International Conference on Innovations in Information Technology, IEEE, ISBN:978-1-4577-0314-0, 2011
8. Gutierrez-Garcia, J.O., Sim, K.M.: Self-organising agents for service composition in cloud computing. In: 2nd IEEE International Conference on Cloud Computing Technology and Science, doi:10.1109/CloudCom.2010.10, 2010
9. Zhang, Q., Cheng, L.: Cloud computing: state-of-the-art and research challenges. J. Internet Serv. Appl. **1**(1), 7–18 (2010)
10. Dillon, T., Chen, Wu., Chang E.: Cloud computing: issues and challenges. In: Proceedings of 24th IEEE International Conference on Advanced Information Networking and Application (AINA) pp. 27–33, Apr 2010

Comparative Study of DE and PSO over Document Summarization

Rasmita Rautray and Rakesh Chandra Balabantaray

Abstract With the exponential growth in the quantity and complexity of information sources, a number of computational intelligent-based techniques have developed in literature for document summarization. In this paper, a comparative study of two population-based stochastic optimization techniques has been proposed for document summarization. It specifies the relationship among sentences based on similarity and minimizes the weight of each sentence to extract summary sentences at different compression level. Comparison of both the optimization techniques based on fallout value of extracted sentences shows the good performance of PSO compared to DE on five different English corpus data.

Keywords Differential evolution · Particle swarm optimization · Sentence similarity · Summarization

1 Introduction

Automatic text summarization is the process to create compressed form of original document without losing the main content. It is a three-step process of analyzing features related to a document as input representation, converting input representation into a summary representation, and finally producing appropriate summary from summary representation. The important factor in good summary generation is the length of the summary, which is the ratio of summary length and

---

R. Rautray (✉)
Department of Computer Science and Engineering, SOA University, Bhubaneswar 751030, Odisha, India
e-mail: rashmitaroutray@soauniversity.ac.in

R.C. Balabantaray
CLIA Lab, Department of Computer Science, IIIT, Bhubaneswar, Odisha, India
e-mail: rakeshbray@gmail.com

L.C. Jain et al. (eds.), *Intelligent Computing, Communication and Devices*, Advances in Intelligent Systems and Computing 308,
DOI 10.1007/978-81-322-2012-1_38

original document length. The quality of summary is acceptable, when the compressed length is 10–30 %. Summarization can be carried out by two methods: extractive and abstractive. Selection of sentences with highest score from original document and putting it all together to form a summary is called extractive summary. Alternatively, summary generated using linguistic methods is called abstractive summary. The detailed theory of extractive and abstractive summary generation is available in [1, 2]. Summarization approach comprising in three steps: clustering the sentences based on semantic distances of sentences, calculating the accumulative sentence similarity and applying extraction rules to choose desired sentences, that has been proposed in [3–5]. To deal with the problem of improvising the performance of text selection in document summarization using statistical tools, a number of global optimization techniques have been proposed in literature. In [6, 7] a genetic algorithm-based document, summarization has been proposed to generate optimal summary by combining article sentences and query sentence. But solution based on GA needs a large numbers of parameters to be tuned. Most of the researchers claim that solution selection based on DE does not depend on their fitness value and fine-tuning of parameters [8]. Again in [9], PSO has shown it is superiority in terms of computational efficiency, small set of parameter tuning, and less number of function evaluations compared to GA.

In this paper, a comparative study of DE and PSO for single document summarization has been proposed based on inter-sentence relationship considering an English corpus data. Single document summarization can be viewed as an optimization problem that needs to minimize the redundancy among sentences, simultaneously keeping the sentences with high relativity with important sentences. Initially, according to the user specification, a compression level (i.e., the percentage of summary length needs to be retained in document summarization) is to be specified. Then, at different compression level, sentences are selected as summary sentences based on a threshold value. To extract the optimal sentences, a new fitness function taking the weighted average of important sentence feature (ISF) and minimal similarity (MF) has been taken in the optimization procedure. A fallout value (*F* value) obtained from the recall, and precision of extracted summary sentences has taken for comparing the performance of DE and PSO for document summarization. The rest of the paper is organized as follows: Sect. 2 details system overview. Section 3 describes the application of the proposed algorithm on a sample test collection and evaluation result. Finally, we end this paper with conclusion.

2 System Overview

Figure 1 illustrates an overview of sentence extraction-based summarization of single document as an optimization problem. The input to the system is a single document. The output is a concise summary providing the condensed information in the input document.

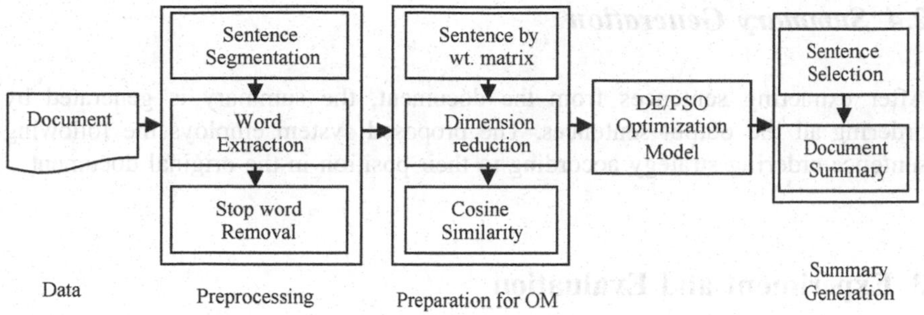

Fig. 1 Overall process of optimization model for text summarization

The proposed system can be decomposed into the following sub-processes: (1) preprocessing, (2) preparation for optimization model, (3) optimization algorithm, and (4) summary generation.

2.1 *Preprocessing*

Initially, document is segmented into sentences and words for each sentences is extracted. Then, the functional words or stopwords such as "a", "the", and "of" (frequently occurring insignificant words) are removed from the word list.

2.2 *Preparation for Optimization Model*

Feature is an important aspect of any text mining. So the different features such as term weight, sentence weight, and inter-sentence similarity for each sentence and formulation of objective function need to be prepared for input to the optimization model.

2.3 *Optimization Algorithm*

Objective function in text mining can be optimized by using different evolutionary algorithms. In this paper, such function is optimized using a differential evolution [10, 11] and particle swarm optimization algorithm [12, 13].

2.4 Summary Generation

After extracting sentences from the document, the summary is generated by ordering all the output sentences. The proposed system employs the following sentence ordering strategy according to their position in the original document.

3 Experiment and Evaluation

In this section, we conduct experiments on different data corpus of English articles and computed extracts at 10, 15, and 20 % levels. We compare the results with word summarizer. Below, we show a sample test and its summaries at 10 % levels generated by our scheme. We also show the summaries generated by word at different levels (Table 1).

Table 1 Generated summary at 10 % level

Word summarizer
Expert systems are software programs that store knowledge extracted from human experts
Expert systems thus appear to mimic human experts in a particular field or domain such as tax or auditing
Early expert systems focused on expert emulation, attempting to replicate the behavior and decisions of human
Expert systems are increasingly used in accounting
DE summarizer
Expert systems are software programs that store knowledge extracted from human experts
In contrast to artificial intelligence, ESs do not try to develop basic postulates and evolve these into intelligent behavior, but accept human knowledge/experience as its basics and attempt to formulate form of aggregate behavior
Expert systems are not only effective in responding to questions from a wide domain of knowledge and that have more than one answer but also effective in handling repetitive tasks in fuzzy domain of knowledge
Expert systems incorporate the knowledge of single or multiple human experts and are able to help accountants improve the quality of their service in the areas of audit planning, internal control evaluation, and identification of audit risk
PSO summarizer
Expert systems are software programs that store knowledge extracted from human experts
Expert systems thus appear to mimic human experts in a particular field or domain such as tax or auditing
Expert systems are increasingly used in accounting
Expert systems are not only effective in responding to questions from a wide domain of knowledge and that have more than one answer but also effective in handling repetitive tasks in fuzzy domain of knowledge

Table 2 Result of DE summarizer and PSO summarizer

Doc Id	Compression level (%)	DE summarizer			PSO summarizer		
		P	R	F	P	R	F
D1	10	0.25	0.25	0.25	0.5	0.5	**0.5**
	15	0.28	0.28	0.28	0.42	0.42	**0.42**
	20	0.5	0.4	0.44	0.62	0.5	**0.55**
D2	10	0.42	0.5	0.45	0.5	0.5	**0.5**
	15	0.5	0.5	0.5	0.75	0.75	**0.75**
	20	0.5	0.54	0.51	0.66	0.72	**0.68**
D3	10	0.14	0.14	0.14	0.28	0.28	**0.28**
	15	0.09	0.09	0.09	0.27	0.27	**0.27**
	20	0.14	0.14	0.14	0.5	0.5	**0.5**
D4	10	0.25	0.25	0.25	0.2	0.2	0.2
	15	0.5	0.42	0.45	0.33	0.28	0.302
	20	0.37	0.33	0.34	0.5	0.44	**0.46**
D5	10	0.28	0.33	0.302	0.28	0.33	0.302
	15	0.5	0.55	0.52	0.5	0.55	0.52
	20	0.33	0.36	0.34	0.5	0.54	**0.51**

The observations clearly indicate that the summaries generated by the PSO summarizer are better than DE summarizer at 10, 15, and 20 % in almost all test cases

The main approach for summary quality is the intrinsic content evaluation which is often done by comparison with an ideal summary. For sentence extraction, it is often measured by co-selection. It finds out how many ideal sentences the automatic summary contains. The main evaluation metrics of co-selection are precision, recall, and F-score [5].

For each document, a summary generated by Word summarizer has been considered as reference summary (denoted by S_{ref}). We then compare the candidate summary (denoted by S_{cand}) with the reference summary and compute the precision (P), recall (R), and F values as follows (Table 2):

$$P = \frac{|S_{\text{ref}} \cap S_{\text{cand}}|}{S_{\text{cand}}} \quad R = \frac{|S_{\text{ref}} \cap S_{\text{cand}}|}{S_{\text{ref}}} \quad F = \frac{2PR}{P+R}$$

4 Conclusion

Summary maintains information richness and diversity by preserving the topic-based information contained in the original documents. The requirement raises a fundamental problem: How important will a selected summary be to represent the whole documents? This paper presents a single document summarization model which extracts key sentences from given documents while reducing redundant information in the summaries. The model is represented as an optimization problem. Our approach uses the sentence-to-sentence relation to select salient sentences from given documents and reduce redundancy in the summary. The experimental results provide strong evidence that PSO rather than DE is a viable method for document summarization.

References

1. Alguliev, R.M., Aliguliyev, R.M.: Evolutionary algorithm for extractive text summarization. J. Intell. Inf. Manag. **1**, 128–138 (2009)
2. Alguliev, R.M., Aliguliyev, R.M., Mehdiyev, C.A.: Sentence selection for generic document summarization using an adaptive differential evolution algorithm. J. Swarm Evol. Comput. **1**(4), 213–222 (2012)
3. Alguliev, R.M., Aliguliyev, R.M., Isazade, N.R.: DESAMC+DocSum: differential evolution with self-adaptive mutation and crossover parameters for multi-document summarization. J. Knowl. Based Syst. **3**(5), 21–28 (2012)
4. Aliguliyev, R.M.: A new sentence similarity measure and sentence based extractive technique for automatic text summarization. J. Expert Syst. Appl. **36**, 7764–7772 (2009)
5. Chatterjee, N., Mohan, S.: Extraction—based single-document summarization using random indexing. In: IEEE International Conference on Tools with Artificial Intelligence, pp. 448–455 (2007)
6. Dixit, R.S., Apte, S.S.: Apte: improvement of text summarization using fuzzy logic based method. IOSR J. Comput. Eng. **5**(6), 5–10 (2012)
7. Hassan, R., Cohanim, B., Weck, O.D.: A comparison of particle swarm optimization and the genetic algorithm. In: American Institute of Aeronautics and Astronautics, pp. 1–13 (2004)
8. He, L., Shi, D., Wang, L.: An adaptive discrete particle swarm optimization for TSP problem. In: Second Asia- Pacific Conference on Computational Intelligence and Industrial Applications, pp. 393–396 (2009)
9. He, Y.X., Liu, D.X., Ji, D.H., Yang, H., Teng, C.: MSBGA: a multi document summarization system based on genetic algorithm. In: Proceedings of the Fifth International Conference on Machine Learning and Cybernetics, pp. 2659–2664 (2006)
10. Jones, K.S.: Automatic summarizing: the state of the art. Inf. Process. Manag. **43**(6), 1449–1481 (2007)
11. Karaboga, D., Akay, B.: A comparative study of artificial bee colony algorithm. J. Appl. Math. Comput. **214**, 108–132 (2009)

12. Kogilavani, A., Balasubramani, P.: Clustering based optimal summary generation using genetic algorithm. In: Proceedings of International Conference on Communication and Computational Intelligence, pp. 324–329 (2010)
13. Kowsalya, R., Priya, R., Nithiya, P.: Multi document extractive summarization based on word sequences. Int. J. Comput. Sci. Issues (IJCSI) **8**(2), 510–517 (2011)

Medical Image Thresholding Using Particle Swarm Optimization

Debashis Mishra, Isita Bose, Utpal Chandra De and Madhabananda Das

Abstract Image processing has been serving as one major part of medical science since 1980s as automation of image analysis offers better results in efficient time period to help specialists in diagnosis and eradication of diseases. Most frequently, medical fields face different cases of detecting tumors, kidney stones, fractures in bones, etc. through various images such as ultrasound images and X-ray images. But it is very difficult for identification of some particular structure in some medical images. Hence, such images need more improvement in terms of noise reduction and segmentation. Image thresholding is a kind of segmentation process which partitions the image into different objects. Particle swarm optimization (PSO) is one bio-inspired optimization technique which gets one optimized threshold value for image thresholding in this paper using proper fitness function.

Keywords Swarm intelligence · PSO · Image processing · Image segmentation · Thresholds · Image histogram · Medical images

1 Introduction

Image processing is a major section of research in all of the domains of science. It is one process of manipulating raw image for different purpose such as image enhancement, analysis. Segmentation is a vital image processing technique which

D. Mishra (✉) · I. Bose · M. Das
School of Computer Engineering, KIIT University, Bhubaneswar, Odisha, India
e-mail: debashis.engg@gmail.com

I. Bose
e-mail: isitabose89@gmail.com

M. Das
e-mail: mndas_prof@kiit.ac.in

U.C. De
School of Computer Application, KIIT University, Bhubaneswar, Odisha, India
e-mail: utpal@kiit.ac.in

L.C. Jain et al. (eds.), *Intelligent Computing, Communication and Devices*,
Advances in Intelligent Systems and Computing 308,
DOI 10.1007/978-81-322-2012-1_39

is used to partition the image into different regions to detect different objects. This can be applied in medical fields such as to locate stones or tumor in ultrasound images, radiography images, and study of anatomical structure. Nondestructive testing (NDT) is very common in medical science which consists of radiography technology, sonography technology, etc [10]. Ultrasonic imaging or sonography [8] is widely used in medical imaging. Generally, these images are observed by specialists and doctors to get important information, which may be affected due to noise in image or having different objects and elements in one image. Again manual analysis needs a lot of human efforts and may be biased for any cause. Noise in images can be eradicated by different de-noising technique [5]. But for better result, the image should be segmented into different sections to have more analyzed information.

Image segmentation can be achieved by many methods, explained in next section, among which image thresholding is the simplest one. In this method, the image is segmented by a threshold value. So finding an optimized threshold value is one strenuous job, due to which an advanced optimization technique, particle swarm optimization (PSO), explained later, has been implemented in this paper.

2 Background

2.1 Threshold-based Image Segmentation

In this type of segmentation, image is segmented into different regions based upon one or more threshold value. As we are talking about medical images such as radiography and ultrasonography images, the concentration is paid upon grayscale image only. It is required to find any abnormal dark or light spot in medical images to detect tumors or fractures in different kind of medical images. By the numbers and natures of threshold values, image thresholding can be of different types as given below.

- **Global Thresholding**: In this type of thresholding, only one gray-level value is chosen as global threshold [7, 9] and a binary image is produced on the basis of the chosen threshold value.
- **Semi-thresholding**: This can be told as modified global thresholding where one region takes the value of 0 and other one retains the original gray level [6].
- **Variable Thresholding**: Here, image is not segmented on the basis of a constant threshold value rather using a variable one which changes according to the surrounding pixels (i.e., regional) or to the position of pixel (i.e., adaptive) [7, 9].
- **Multiple Thresholding**: In multiple thresholding, the image is partitioned into number of segments using more than one threshold values [6, 7, 9].

2.2 Particle Swarm Optimization

PSO, [3] a swarm intelligence method, mimics the bird flying and searching for foods. It has initial population searching for optimized value, but it does not contain any kind of mutation or crossover like genetic algorithm. In PSO, each particle follows its own best value obtained yet denoted as personal best or 'pbest' and also follows the global best value among all the particles in a particular duration denoted as global best or 'gbest.' These best values are chosen by each particle's fitness value obtained by a function named 'fitness function.' After finding pbest and gbest values, the velocities and positions of each particle is updated using Eqs. 1 and 2 [3, 4].

$$V_i^{t+1} = V_i^t + K_1 \times \text{rand}() \times (P_i - X_i^t) + K_2 \times \text{rand}() \times (G^t - X_i^t) \tag{1}$$

$$\chi_i^{t+1} = \chi_i^t + V_i^{t+1} \tag{2}$$

Here, V_i^t is the velocity of tth particle at iteration 't', position of tth particle at iteration 't' is denoted as χ_i^t. P_i is the personal best (pbest) and G^t is global best (gbest). K_1 and K_2 are learning factors or speed factors (here, $K_1 = K_2 = 2$), and rand() is a random function [1, 3, 4]. After certain termination criteria or a number of iterations, its stops, and the gbest value found is considered as the optimized one.

3 Proposed Thresholding Method

Here, only grayscale medical images have been considered for thresholding purpose having gray level from 0 to L (say). Here, PSO technology is implemented to optimize the threshold value to partition the image into two regions, i.e., dark and bright regions. PSO can only be implemented by knowing the particle and the fitness function. It is obvious that the 'threshold' value is the particle.

Fitness function is a function of particle used to find appropriate value of the particle so that the fitness function will be optimized. Here, Kapur's entropy criterion method [2] is taken as the fitness function which is defined below. For the better segmentation result, a threshold value is chosen which will maximize the entropy criterion method. Let one 2D grayscale image contains N pixels with gray levels ranging from 0 to $L - 1$. Let the number of pixels for a particular gray level 'i' is denoted as $\eta(i)$, then the probabilistic chance of occurrence of gray level 'i' i.e., $\rho(i)$ in the image can be found by dividing $\eta(i)$ with total number of pixels N [2].

Kapur's entropy criterion method is illustrated in Eq. 3 which is a function of threshold; hence, single objective PSO system can be implemented [2].

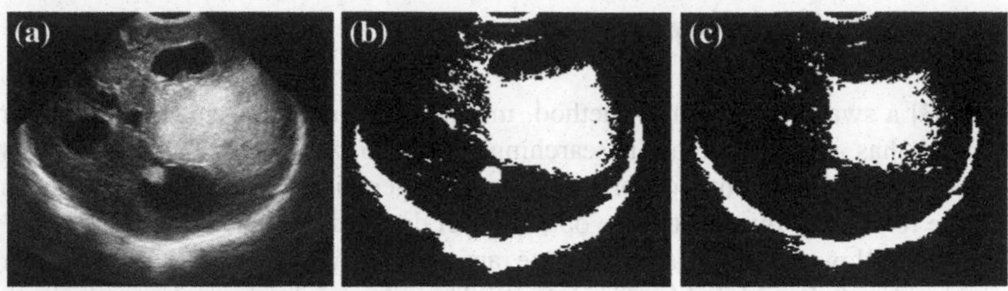

Fig. 1 a Ultrasound image. **b** Thresholding using PSO. **c** Classical thresholding

$$f(t) = F_0 + F_1 \tag{3}$$

$$F_0 = -\sum_{i=0}^{t-1} \frac{\rho_i}{w_0} \log_e \frac{\rho_i}{w_0}, \quad \text{where} \quad w_0 = \sum_{i=0}^{t-1} \rho_i \tag{4}$$

$$F_1 = -\sum_{i=t}^{L-1} \frac{\rho_i}{w_1} \log_e \frac{\rho_i}{w_1}, \quad \text{where} \quad w_1 = \sum_{i=t}^{L-1} \rho_i \tag{5}$$

3.1 Experiments and Results

A sample of 50 various ultrasound images are being tested with the proposed method with different swarm size and termination criteria. The system is applied to an ultrasound image (Fig. 1a) to detect brighter region with a swarm size 10 and for 30 iterations, and the result is shown in Fig. 1b. It is clear for the given figures that the system is capable of segmenting the brighter regions in the ultrasound image. Also one comparative approach with the traditional global thresholding and the proposed method is shown in Fig. 1c. It can be clearly visible that the proposed method is capable of detecting some fine brighter regions from the ultrasound image which will be more helpful for the doctors in the diagnosis process.

4 Conclusion

PSO is able to give optimized threshold value in effective time period in medical images. The proposed method helps the specialists or doctors in medical science for the detection of various objects such as tumor and calculi in ultrasound images or to detect fractures or ligament tears in any X-ray report. The future works focus on other advanced bio-inspired computation method or swarm intelligence technique such as ACO, CSO, ABC to solve the problem in cost- and time-effective manner.

References

1. Duraisamy, S.P., Kayalvizhi, R.: A new multilevel thresholding method using swarm intelligence algorithm for image segmentation. J. Intell. Learn. Syst. Appl. **2**, 126–138 (2010)
2. Kapur, J.N., Sahoo, P.K., Wong, A.K.C.: A new method for gray-level picture thresholding using the entropy of the histogram. Comput. Vis. Graph. Image Process. **29**, 273–285 (1985)
3. Kennedy, J., Eberhart, R.: Particle Swarm Optimization, pp. 1942–1948. IEEE (1995)
4. Kennedy, J., Eberhart, R.: Particle Swarm Optimization, Developments, Applications and Resources. IEEE (2001)
5. Mishra, D., Bose, I., Das, M.N., Mishra, B.S.P.: Detection and reduction of impulse noise in RGB color image using fuzzy technique. Distributed Computing and Internet Technology, LNCS, vol. 8337, pp. 299–310. ICDCIT, Springer, February 2014
6. Ritter, G.X., Wilson, J.N.: Handbook of Computer Vision Algorithms in Image Algebra, 2nd edn. CRC Press, Boca Raton (2001)
7. Sahoo, P.K., Soltani, S., Wong, A.K.C.: A survey of thresholding techniques. Comput. Vis. Graph. Image Process. **41**(2), 233–260 (1998)
8. Saini, K., Dewal, M.L., Rohit, M.K.: Ultrasound imaging and image segmentation in the area of ultrasound: a review. Int. J. Adv. Sci. Technol. **24**, 41–59 (2010)
9. Sezgin, M., Sankar, B.: Survey over image thresholding techniques and quantitative performance evaluation. J. Electron. Imaging **13**(1), 146–165 (2004)
10. Wang, X., Wong, B.S.: X-ray image segmentation based on genetic algorithm and maximum fuzzy entropy. Conference on Robotics, Automation and Mechatronics. IEEE, pp. 991–995 (2004)

Criterions for Databases in Cloud Computing Environment

Sarada Prasanna Pati and Prasant Kumar Pattnaik

Abstract Cloud computing is popularly used environment where many hardware and software resources are delivered as a service especially over the Internet. Since a majority of cloud applications require voluminous data processes ability at petabyte scale, a database management systems (DBMSs) catering to these applications forms a critical component in the cloud software stack. Cloud database has emerged as the amalgamation of distributed storage and virtualization technology. With growing market of distributed Web applications and rising trend of cloud computing, a need for a database suitable for cloud computing environment has drawn focus. This paper is intended to frame a set of criterions for database to be suitable for cloud computing environment which could provide us with much needed features such as high availability, flexible schema, lower cost of investment, scalability, elasticity, etc. We have also tried to evaluate the compliance of our proposed set of criterions with respect to some of the commercially used cloud databases. This proposed set of criterions would help to figure out the extent to which a database is suitable for cloud computing environment.

Keywords Cloud computing · Cloud databases · NoSQL databases · CDMS · MongoDB · Cassandra · NuoDB

S.P. Pati (✉)
Department of Computer Science and Engineering, ITER, Siksha 'O' Anusandhan University, Bhubaneswar, India
e-mail: saradapati@soauniversity.ac.in

P.K. Pattnaik
School of Computer Engineering, KIIT University, Bhubaneswar, India
e-mail: patnaikprasantfcs@kiit.ac.in

L.C. Jain et al. (eds.), *Intelligent Computing, Communication and Devices*, Advances in Intelligent Systems and Computing 308,
DOI 10.1007/978-81-322-2012-1_40

1 Introduction

A database is a collection of information that is organized so that it can easily be accessed, managed, and updated. It can be looked at as being a collection of *records*, each of which contains one or more *fields* (i.e., chunks of data) about some *entity* (i.e., object), such as a person, group, country, chemical, etc. Databases can be classified into various types depending on parameters such as data model, functionality, storage structure, modes of operation, etc.

The commercial database industry has seen many changes since its inception. Although many different types of database systems exist for decades, the *relational database* has been the pioneer. The relational database offers concrete solution to most of the data storage requirements in different areas ranging from real-world automation to Web applications. With the growth in distributed computing, also evolved the concept of distributed data storage and *distributed database*. It is a type of database configuration that consists of loosely coupled data repositories upon which distributed transactions run. *Mobile databases* are yet another category of commercially successful databases that are drawing the attention of most of us. A mobile database is a portable database that is physically separate from the central server and can be connected to by a mobile computing device over a wireless mobile network.

However, extensive growth in digital data on the Internet, requirements in novel data storage and access strategies, better broadband facilities and emergence of cloud computing have led to a new database paradigm, called *cloud database* [1]. Distributed storage and virtualization technology are considered as the back bone of cloud database. Cloud storage providers deliver economies of scale by using the same storage capacity to meet the needs of many organizations [2]. The popularity of the cloud database is increasing, and it is said that cloud database and services are going to rule the future database industry. There are many commercial cloud databases available in the market which share common features such as schema free, highly scalable, eventually consistency, user-friendly API, etc. On the other hand, these databases differ in many aspects [3] such as data model, database functionality, query interface, customized API, etc. For a better and constructive growth of these breed of databases, they need to be standardized like other conventional and traditional databases. This paper is an attempt toward this direction where we analyze the characteristics as well as requirements of commercial cloud databases and frame a set of twelve criterions to which these databases should comply to.

2 Related Work

Edgar F. Codd the famous mathematician proposed the concept of relational database management system (RDBMS) that is based on the relational model. Most of the popular commercial and open-source databases currently in use are based on the relational model. Codd introduced a set of 13 rules (numbered zero to twelve) to determine whether a database management systems (DBMSs) can be considered *relational* [4], i.e., RDBMS. The rules and their explanations can be found in [4].

To make the comparison of distributed databases easier, C.J. Date formulated 12 'commandments' or basic principles of distributed databases [5]. Although no current distributed database management system (DDBMS) complies with all of them, they constitute a useful objective. The commandments and their explanations can be found in [5].

The rapid growth in usage of mobile database is been attributed to the increased availability of powerful lightweight hand-held computing devices with more storage capacity and more powerful CPU, and low-cost mobile connectivity. The essential requirements and characteristics that these databases should satisfy for their better suitability in mobile computing environment can be found in [6, 7].

3 Cloud Databases

A cloud database [8] is a distributed database that has been conceived for a virtualized computing environment. A cloud database offers comprehensive database functionality by allowing its users to store and access data on a remote disk/datacenter at anytime from anywhere through Internet [3]. Cloud storage providers use the same storage capacity to meet the needs of many clients [2]. Putting the database in the cloud can be an effective way to support cloud-enable business applications as part of a wider software-as-a-service (SaaS) deployment by simplifying the processes required to make information available through the Internet. Cloud databases offer tremendous advantages over their traditional counterparts in terms of supporting unstructured and semi-structured data, increased performance, better accessibility, on the fly elastic scalability, multi-tenancy, fastest recovery, and failover with minimal downtime, support to change in storage requirements, and low-cost investment and maintenance of in-house hardware. Different researchers have classified the cloud databases based on different characteristics such as structured–unstructured, SQL based–NoSQL based, and ACID based–NoACID based [9].

3.1 NoSQL Databases

NoSQL is an emerging category of DBMS, which actually means '*Not Only* SQL,' is getting a wide spread acceptance in cloud database domain. Need of these databases arose when conventional relational databases were finding it difficult to scale and manage large data. Its main characteristic is its non-adherence to relational database concepts. These databases are usually distributed, do not follow a fixed schema, do not offer a SQL interface, usually avoids join operations, scales horizontally, and maybe available open source [10]. Moreover, NoSQL does not guarantee ACID properties always in order to achieve scalability and elasticity along with easy management of large data. These databases are designed to excel in speed and volume. Some of the popular examples of NoSQL databases are MongoDB, Cassandra, CouchDB, Redis, etc. Selection of a particular database purely depends on the client requirement.

3.2 Commercially Available Cloud Databases

Some of the prominent commercially available NoSQL cloud databases are MongoDB, Cassandra, CouchDB, Redis, etc. It is not that only NoSQL databases are suitable for the cloud computing environment, NuoDB, an SQL database, is also gaining popularity due to its potential of fitting into cloud computing environment effectively.

Cassandra: Cassandra [2] is an open-source distributed database management system. It is a top-level project by Apache Software Foundation designed to handle very large amounts of data that are spread across many commodity servers ensuring highly available service with no single point of failure. The basic fundamental of Cassandra is that it is a columnar database or rather a column-oriented distributed database. The data are stored in the form of columns, and it is uniquely marked using 'keyspace.'

MongoDB: MongoDB [11] is a cross-platform, schema free, document-oriented database system developed and supported by 10gen that provides high performance, easy scalability, and high availability. It is part of the NoSQL family of database systems. Unlike that of relational database that stores data in table structures, MongoDB stores structured data as JSON-like documents with dynamic schemas (MongoDB calls the format BSON) that makes the integration of data in certain types of applications easier and faster. Its features such as high availability, faster updates, auto-sharding, easy scalability, and rich query language make it one of the richest cloud databases.

NuoDB: NuoDB [12] is the world's first and only patented, elastically scalable, and SQL/ACID database. It is a tailor-made relational database with novelty to support the cloud's dynamic, asynchronous nature. It is a cloud data management system (CDMS) that if fully SQL supports ACID properties strictly and at the same time offers elastic scalability with least administrative hassles. NuoDB delivers high performance at Web scale with highly efficient and flexible resource utilization. NuoDB provides the ability to replicate data globally in real time. NuoDB is purpose built for the cloud from the ground up on an emergent architecture—a shared nothing, asynchronous, peer-to-peer design that is ideal for modern data centers yet delivers the power, reliability, and functionality of a traditional SQL database.

4 Proposed Set of Criterions for Cloud Database Management System

Here, we frame a list of criterions in the form a set of 13 rules (numbered zero to twelve) that any database must comply with in order to be suitable for use in cloud.

Rule 0: *Superset of Traditional Database*: The cloud database should be open to new standards which may contradict or differ as per the current SQL standards and should be open to non-SQL principles. CDMS should be at least able to do what traditional databases can do. A cloud database should not only provide basic relational database management functionalities such as ACID transactions, efficient query language but should also be capable of managing heavy load demands and handling structured as well as unstructured data blocks.

Rule 1: *Comprehend the Cloud*: A cloud database should be specialized for handling virtualization and abstraction. It should integrate itself in the cloud with minimum fuss and should have compatibility with cloud-oriented tools and software's. Its integration with cloud should bring a positive change in the performance and the performance scale should not be ordinary like existing conventional databases.

Rule 2: *Elastic Scalability*: A CDMS should be flexible and fast enough to add or delete resources (storage or computational) as per the need on a large scale. It should have the tendency to carry out millions of transactions per second and handle petabytes of data by adding real or virtual machines, networks and storage devices to a live database. Moreover, a CDMS must discard the resources when they are no longer needed.

Rule 3: *Efficient Geographical Distribution*: A cloud database must run concurrently on multiple datacenters in order to support faster disaster recovery, cent percent availability of applications, and to support geographically distributed workload. A CDMS must be able to deliver active operations with consistent semantics, work across and between wide area networks and understand how to localize activity or caches.

Rule 4: *Continuous Availability*: A CDMS must be robust, highly available and should not have a single point of failure. Moreover, it should handle system changes and diverse operations like network partitions with ease. It should remain available, or in case of failure should exhibit graceful degradation. A CDMS must support live up gradation of underlying software and hardware infrastructure and CDMS versions, and must support dynamic changes to schemas and other database administration tasks without compromising with CDMS availability.

Rule 5: *Run and Scale Anywhere*: A CDMS should be free from infrastructural constraints, that is, it should be able to run on any infrastructure like from single machines to private clouds, public clouds or even hybrid clouds. It must be able to run in a heterogeneous environment integrating different machines, virtual machines, operating systems, or network infrastructures. A CDMS should excel on enterprise and commodity hardware equally.

Rule 6: *Exclusive Logical Database*: Irrespective of the complexity of the application, a CDMS should provide its user a single, logical, consistent, and 100 % available database. A CDMS must guard users from using explicit partitioning, shading or caching techniques to achieve massive database scalability. The CDMS must encapsulate or prevent the occurrence of these complexities, so that a developer or administrator can focus on using the database irrespective of scale and complexity.

Rule 7: *Dynamic Multi-tenancy*: A CDMS must be dynamically multi-tenant. It must be able to manage large numbers of databases on a finite set of resources and be able to reassign resources to databases as needed. A CDMS must be able to hibernate inactive databases and wake them up on demand.

Rule 8: *Distributed Security*: A CDMS should ensure proper authentication and access control of database processes. There should be enterprise class security at system level and database level including access control and authentication of machine before being accepted by a body. Database-level security for users of the database is a must, and there should be encryption of all communications between machines.

Rule 9: *Ability to Store Data Anywhere*: A CDMS should store the data in storage system which is best suited for the given scenario and should be deployed locally, remotely in a datacenter or on a public or private cloud. It should be able to store all data redundantly in multiple locations, simultaneously and with transactional consistency, using a heterogeneous mix of storage locations and storage technologies.

Rule 10: *Metric and Billing Services*: Like any other cloud service, a cloud database must offer pay per use service, that is, it should charge as per the service used. Based on the varying user need, the plan should be custom-made in order to be an efficient and business-friendly cloud database.

Rule 11: *Performance Measuring and Self-Service Provision*: A user or client should be able to measure the performance of the service provided at the current instance. This might include inbuilt performance testing tools—such as load testing, node testing, etc. Cloud database clients should be able to have on-demand provisioning of IT resources for them like requesting additional amount of computing, storage, software, process, or more from the service provider without any hassles. After use of these resources is over, they should be automatically available for service.

Rule 12: *Should posses APIs (for Distinguishing Ability)*: Cloud database services consist of a database manager component that controls the underlying database instances using a service API. The service API is exposed to the end user and permits users to perform maintenance and scaling operations on their database instances. Cloud services should have standardized APIs, which provide instructions on how two application or data sources can communicate with each other.

Table 1 Compliance of SQL and cloud databases to the proposed set of rules

Proposed cloud database rules		SQL	NoSQL (Cassandra and MongoDB)	NUODB
0.	Superset of traditional database	Yes	No	Yes
1.	Comprehend the cloud	No	Yes	Yes
2.	Elastic scalability	No	Yes	Yes
3.	Efficient geographical distribution	No	Yes	Yes
4.	Continuous availability	No	Yes	Yes
5.	Run and scale anywhere	No	No	Yes
6.	Exclusive logical database	No	Yes	Yes
7.	Dynamic multi-tenancy	No	Yes	Yes
8.	Distributed security	No	Yes	Yes
9.	A CDMS must be able to store the data anywhere	No	Yes	Yes
10.	Metric and billing services	No	No	No
11.	Performance measuring and self-service provision	Partially	Partially	Partially
12.	Should posses APIs (for distinguish ability)	Yes	No	No

5 Analyzing Cloud Databases as Per the Proposed Criterions

Both Cassandra and MongoDB are popular cloud databases. NuoDB is considerably very young though, brings live a new category of cloud databases called NewSQL [12]. However, behavior of all these databases is different as per the load and deployment scenario. This is probably because of their structural difference and their modus operandi of attaining scalability and elasticity.

In Table 1, we are analyzing the compliance of the afore said databases with respect to our proposed set of rules.

6 Conclusion

All cloud DBMSs show varied properties set because of different scaling and elasticity techniques involved. Cloud databases such as MongoDB and Cassandra, the NoACID databases [9], attain their elasticity and scalability goal, thereby shifting from traditional database nature, whereas databases like NuoDB stick to their traditional form. NuoDB follows about 10 proposed criterions among the 13 proposed ones showing about 80.76 % compatibility with proposed criterions. NoSQL category having databases such as MongoDB and Cassandra show round about 65 % compatibility for the proposed criterions. Keeping acceptance pattern of proposed criterions in consideration, it is suggested that at least cloud database follows 8 criterions among the proposed 13 criterions to be effective as a cloud database management system.

References

1. Buyya, R., et al.: Cloud computing and emerging IT platforms: vision, hype, and reality for delivering computing as the 5th utility. Future Generation Comput. Syst. **25**(6), 599–616 (2009)
2. Wu, J., et al.: Recent advances in cloud storage. In: 3rd International Symposium on Computer Science and Computational Technology (ISCSCT'10), Jiaozuo, P. R. China, 14–15 Aug 2010, pp. 151–154 (2010)
3. Arora, I., Gupta, A.: Cloud databases: a paradigm shift in database. IJCSI **9**(4), 3 (2012). ISSN (Online) 1694-0814
4. Codd, E.F.: A relational model of data for large shared data banks. Commun. ACM **13**(6), 377–387 (1970)
5. Date, C.J.: Twelve rules for distributed database. Comput. World **21**(23), 75–81 (1987)
6. Sen, R.: DBMS techniques for lightweight computing devices. In: Proceedings of MobiDE, ACM Press 2011, pp. 1–8 (2011)
7. Panda, P.K., Swain, S., Pattnaik, P.K.: Review of some transaction models used in mobile databases. Int J Instrum. Control Autom. (IJICA) **1**(1), 99–104 (2011)
8. Bloor, R.: What is a Cloud Database. Technical report (2011)
9. Pasayat, S.K., Pati, S.P., Pattnaik, P.K.: Classification and live migration of data—intensive cloud computing environment. In: Intelligent Interactive Technologies and Multimedia Communications in Computer and Information Science, vol. 276, pp. 316–324. Springer, Berlin, Heidelberg (2013)
10. Agarwal, R., et al.: The Claremont report on database research. SIGMOID record (ACM) **37**(3), 9–19 (2008). ISSN 0163-5808
11. MongoDB: www.mongodb.org/
12. http:// www.nuodb.com/

Measuring Web Site Usability Quality Complexity Metrics for Navigability

Sandeep Kumar Panda, Santosh Kumar Swain and Rajib Mall

Abstract Now days, Web site design depends on a key feature such as navigability. This paper aims to find the usability, quality of web structure, based on the construction of hierarchical structure of the Web site, digit of strikeouts, and cyclomatic complexity of a Web site roadmap. The PowerMapper tool generates the roadmap for the league Web site. Route matrix is used to check the maximum digit of strikeouts to find the web page and Web site complexity is found by Web site structural cyclomatic complexity. The ease of use of the Web site, such as usability quality is calculated in 10-point scale and using some mathematical formula, the output suggests the improvement of Web site structure.

Keywords Cyclomatic complexity · Navigability · Web site roadmap

1 Introduction

Of late, there has been a proliferation of commercial Web sites due to the increased use of the Internet. There is a phenomenal increase in many organizations those are using the web for trading, marketing, promoting, and transacting products and services to consumers. Apart from firms and organizations, there seems to be a very large growth of the Internet by consumers for various purposes,

---

S.K. Panda (✉) · S.K. Swain
School of Computer Engineering, KIIT University, Bhubaneswar, Odisha, India
e-mail: skpanda00007@gmail.com

S.K. Swain
e-mail: sswainfcs@kiit.ac.in

R. Mall
Department of Computer Science and Engineering, IIT Kharagpur, Kharagpur, West Bengal, India
e-mail: rajib@cse.iitkgp.ernet.in

L.C. Jain et al. (eds.), *Intelligent Computing, Communication and Devices*,
Advances in Intelligent Systems and Computing 308,
DOI 10.1007/978-81-322-2012-1_41

including online shopping and information search. The consumer interest rise in online shopping is affecting the traditional retail sales. The rise in business to consumer electronic commerce has made many organization looks for new ways to understand online shopping behavior to attract and keep the consumers. Till today, the center of focus for web users is usability engineering.

To be successful, Web sites need to have good usability. Usability is an overall measure of how easy the user interface is to use [1–3]. Nielsen [3] stated that if users were unable to find a product, they would not buy it. Measuring the usability, navigation is one of the key attributes. We define how easily the real users find the desired information by linking through the Web site via navigability.

In the construction of Web site roadmap, navigation places an important role because it finds the route to be traversed to get a desired web page. The roadmap of the Web site looks like a tree starting from the home page as origin node. The Web site origin node such as the home page is constructed in a certain way that it should not consist of lots of routes. According to Benjamin Yen, in 2007 maximum digits of links in a route are 20 in a web page and there are only four strikeouts needed to reach a desired page [4]. At the time of designing a roadmap of a Web site, developer must consider these facts.

Section 2 represents a survey of related works. Section 3 describes the three metrics that are a road map of Web site structure, maximum digit of strikeout routes and web structure cyclomatic complexity. Section 4 presents the evaluation output. Section 5 concludes the paper with a critical analysis and interpretation of our work. Finally, we discuss the possible future extensions to our work.

2 Literature Survey

Web usability is the ease of use of a Web site [3]. Navigation is one of the important components of web pages that support the user in finding information and in browsing through the site's content. There is a convenient and obvious way to move between related pages and sections and also easy to return to the home page. The Web site structure relies on the efficiency of usability. The roadmap of the Web site should be in such a way that the user can easily interact Web site without any formal training. A Web site interface is a complex mix of text, links, formatting, graphic elements, and other aspects that affect the site's overall quality [5]. An effective web design is one that makes it easier for users to navigate through the different pages on the site [6]. The roadmap of a Web site appears as a directed graph where every single node serves as a web page and a route serves as a path to that page [7].

3 Procedure

To assess the usability quality of the Web site, the roadmap consists of three phases: construction of hierarchical structure of the web pages of the Web sites, calculation of route dimension metrics and finding the roadmap cyclomatic complexity of the Web site.

3.1 Construction of Hierarchical Structure of Web Site

A simple web software tool that is PowerMapper generates an HTML roadmap from a given uniform resource locator (URL). The search algorithm, breadth first search (BFS) is used to travel the Web site starting from the origin node such as a home page. All the pages are fetched recursively from the Web site. The tool generates Web site roadmap as a hierarchical tree with the origin node as a home page. In filter process, it removes all the multimedia and graphic files, because in Web site roadmap there is no significance of these files. The hierarchical tree structure of the Snapdeal.com commercial Web site is given in Fig. 1.

3.2 Calculation of Route Dimension Metric

A route dimension is needed to score the maximum digit of strikeouts per page. The route dimension of the tree is determined by the summation of all the level

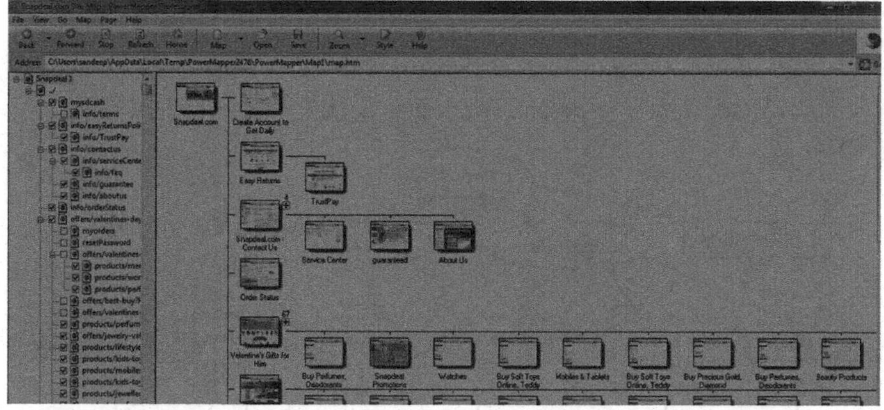

Fig. 1 Snapdeal.com Web site roadmap

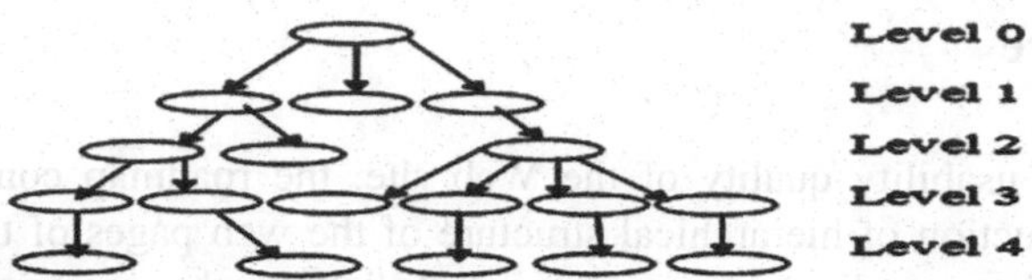

Fig. 2 A tree with four levels

nodes. In formulae 1, the route dimension is calculated as the summation of all level weights with the digit of nodes in each level. The maximum digit of strikeouts is calculated applying formulae 2.

$$\text{Route Dimension} = \sum Lj \cdot Nj \tag{1}$$

where Lj is level digit j, Nj is the digit of nodes at level j.

$$\text{Maximum digit of strikeouts} = \text{route dimension}/m \tag{2}$$

where m is the digit of nodes in the tree. An example tree is shown in Fig. 2.

Route dimension $= 0 \times 1 + 1 \times 3 + 2 \times 3 + 3 \times 6 + 4 \times 5 = 47$

Average digit of strikeouts = 47/18 = 2.61

3.3 Roadmap Complexity

In graph theory, the Web site structural cyclomatic complexity is described as follows. The origin node such as the home page is structured as a tree. In constructing a roadmap of a Web site such as a tree, one has to know the height and level of the tree. The origin node contains lots of sub nodes and dead nodes. An origin node structure with the sub nodes of a Web site is shown in Fig. 3.

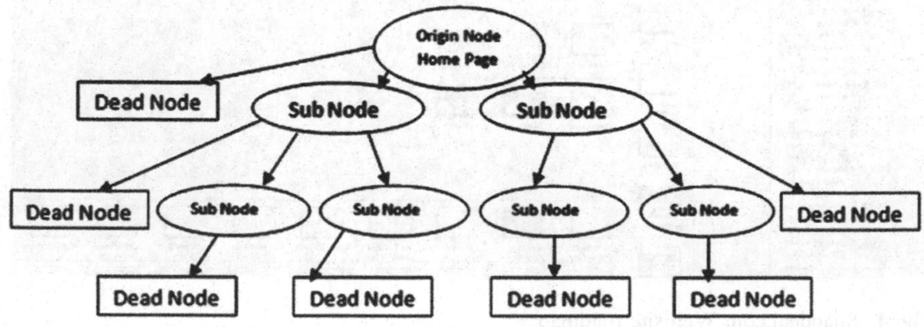

Fig. 3 A tree structure of a Web site

There are too many intermediate links are present in an origin, such as the home page of a Web site. In a roadmap, every single node appears as a web page, again the web pages are routed differently and at the least there is a dead node. In an origin node of a roadmap, at every single level, all sub nodes such as a web page, those do not have further routes ended with a dead node and a sub node those have further routes directs to the node up to the end level. We have evaluated the Web site structural cyclomatic complexity in formulae 3. According to McCab [8] in 1976, the cyclomatic complexity digit does not exceed 10.

$$\text{Web site Structural Cyclomatic Complexity} = (R - N + D + 1)/N \qquad (3)$$

R number of node routes
N number of nodes in the roadmap
D number of dead ends in the roadmap

4 Evaluation

The Web sites of more than 10 Indian electronics commerce online shopping Web sites are taken under assessment mode. The PowerMapper web tool fetches URL address of each e-commerce Web site and creates roadmap. The Web site roadmap combines all the web pages and looks like a tree structure at different levels. The maximum digit of strikeouts desired to retrieve a web page is calculated with route dimension metric applying formulae 1 and 2. The Web site structural cyclomatic complexity number is calculated from the Web site roadmap applying formulae 3. The roadmap of every single e-commerce Web site is computed in 10-point scale. Every single e-commerce Web sites are focused on organizational web pages in the roadmap, Web site structural cyclomatic complexity of the Web site and maximum digit of strikeouts from the 10-point scale value. The calculation process for 10-point scale value is given in Tables 1 and 2. The roadmap of the Snapdeal.com commercial Web site is given in the Fig. 3. The navigability value of the Snapdeal.com commercial Web site up to two levels is given in Table 3. The usability quality of 10 e-commerce Web site roadmap assessments are given in Table 4. Table 5 shows the 10 point scale assessment description.

Table 1 The roadmap and cyclomatic complexity calculated from 10-point scale

Sl. no.	Quality parameter	10-point scale value evaluation
1.	Number of links on web page of roadmap tree	If (number of links in the home page = total number of web pages) and (number of links in home page ≤20) then $k1$ = 10 else
		If (number of links in a page between 10 and 29) then $k1$ = 10 else
		If (number of links in a page = 9 or 21) then $k1$ = 9 else
		If (number of links in a page = 8 or 22) then $k1$ = 8 else
		If (number of links in a page = 7 or 23) then $k1$ = 7 else
		If (number of links in a page = 6 or 24) then $k1$ = 6 else
		If (number of links in a page = 5 or 25) then $k1$ = 5 else
		If (number of links in a page = 4 or 26) then $k1$ = 4 else
		If (number of links in a page = 3 or 27) then $k1$ = 3 else
		If (number of links in a page = 2 or 28) then $k1$ = 2 else
		If (number of links in a page = 1 or 29) then $k1$ = 1 else $k1$ = 0
2.	Cyclomatic complexity	If (cyclomatic complexity ≤1) then $k2$ = 10 else
		If (cyclomatic complexity ≤2) then $k2$ = 9 else
		If (cyclomatic complexity ≤3) then $k2$ = 8 else
		If (cyclomatic complexity ≤4) then $k2$ = 7 else
		If (cyclomatic complexity ≤5) then $k2$ = 6 else
		If (cyclomatic complexity ≤6) then $k2$ = 5 else
		If (cyclomatic complexity ≤7) then $k2$ = 4 else
		If (cyclomatic complexity ≤8) then $k2$ = 3 else
		If (cyclomatic complexity ≤9) then $k2$ = 2 else
		If (cyclomatic complexity ≤10) then $k2$ = 1 else $k2$ = 0
Value = average ($k1$, $k2$)		

Table 2 Maximum digit of strikeouts index calculated from 10-point scale

Sl. no	Quality parameter index	10-point scale value
1.	Average digit of strikeouts per web page	If average digit of strikeouts ≤2.5 then
		Value = value + 0.75 else
		If average digit of strikeouts ≤4 then
		Value = value + 0.5 else
		If average digit of strikeouts ≤5 then
		Value = value + 0.25 else

Table 3 The Snapdeal.com Web site structure quality evaluation

University name: Snapdeal.com			
Route dimension = 1,575			
Average digit of strikeouts = 2.90590			
Level number	Subtree in Web site structure	Number. of web pages in subtree	10-point scale value
1.	1	26	1
2.	1	1	1
	2	4	3
	3	17	4
	4	12	5
	5	6	7
	6	9	8
	7	9	8
	8	3	3
	9	1	1
	10	11	10
	11	1	1
	12	3	3
	13	22	1
	14	20	1
	15	04	4
	16	02	2
	17	08	5
	18	01	1
	19	02	2
3.112903			
Cyclomatic complexity = (R − N + D + 1)/N = (2,064 − 452 + 90 + 1)/452 = 3.365044		10-point scale value = 7	
Route dimension = 1,575		Average digit of strikeouts = 2.90590	
10 point scale value of snapdeal.com structure = 3.112903 + 7 + 0.50 = 10			

Table 4 Usability quality of various e-commerce Web site structure

Sl. no.	University name	*P*1	*P*2	Avg (*P*1, *P*2)	*R*	10-point scale value	Remarks
1.	Snapdeal.com	3.112903	7	5.056451	0.50	10	Very good
2.	Flipkart.com	4.21875	8	6.109375	0.50	12	Very good
3.	Ebay.com	5.666667	3	4.333333	0.25	8.9	Needs minor changes
4.	Homeshop18.com	2.55	6	4.275	0.25	8.8	Needs minor changes
5.	Quikr.com	4.380952	8	6.190476	0.5	12	Very good
6.	Jabong.com	2.833333	8	5.516667	0.75	11	Very good
7.	Myntra.com	3.058824	9	6.029412	0.75	12	Very good
8.	Futurebazaar.com	2.365854	6	4.182927	0.75	9.1	Good
9.	Naaptol.com	4.346154	5	4.673077	0.75	10	Very good
10.	Yepme.com	3.352941	6	4.676471	0.75	10	Very good

*P*1 = Road map calculation value, *P*2 = cyclomatic complexity value, *R* = route dimension metrics

Table 5 10-point scale value description

0–4	5–6	7	8	9	10
Very poor	Poor	Needs improvement	Needs minor changes	Good	Very good

5 Conclusion

We investigated three metrics of e-commerce Web site roadmap. Our study shows that a digit of strikeouts, web structure cyclomatic complexity, and navigability are the key dimensions toward Web site measure. The web developer concentrates on these key dimensions to measure the usability quality of the Web site. There are still limitations to the roadmap complexity, such as the structure of each page route, that will also an important issue of navigability. In the future, we have extended our work to find out the major and minor problems such as broken links of navigability on a Web site.

References

1. Bachiochi, D., Bernstein, M., Chouinard, E., Conlan, N., Danchak, M., Furey, T., Neligon, C., Way, D.: Usability studies and designing navigational aids for the World Wide Web. J. Comput. Netw. ISDN Syst. **29**, 1489–1496 (1997)
2. Najjar, L.: Designing e-commerce user interfaces. In: Proctor, R.W., Vu, K.-P.L. (eds.) Handbook of Human Factors in Web Design, pp. 514–527. Lawrence Erlbaum, Mahwah (2005)

3. Nielsen, J.: Usability 101: Introduction to usability. Alertbox: Current Issues in Web Usability (2003)
4. Yen, B., Hu, P.J.H., Wang, M.: Toward and analytical approach for effective website design: a framework for modeling, evaluation and enhancement. Electron. Commer. Res. Appl. **6**, 159–170 (2007)
5. Newman, M.W., Landry, J.A.: Sitemap, storyboards, and specifications: A sketch of web site design practice. In: Proceedings of Designing Interactive Systems: DIS 2000, Automatic Support in Design and Use, pp. 263–274. ACM Press, New York (2000)
6. Mendes, E., Mosley, N., Counsel, S.: Web metrics estimating, design and authoring effort. IEEE Multimedia **8**(1), 50–57 (2001)
7. Offutt, J.: Web Software Application Quality Attributes, pp. 187–198. Quality Engineering in Software Technology, Nuremberg (2002)
8. McCabe, T.J.: A complexity measure. IEEE Trans. Softw. Eng. **2**, 308–320 (1976)

Personalized Web Page Recommendation Using a Graph-Based Approach to Implicitly Find Influential Users

Ashish Nanda, Rohit Omanwar and Bharat Deshpande

Abstract In this paper, we propose a novel graph-based approach for modeling the browsing data of Web users in order to understand their interests and their relationship with other users in the network. The aim was to identify users who are more influential while recommending pages to a network of users with similar interests. We call these users influential users and assign them an influence score that indicates the extent to which similar users follow their recommendations. By monitoring the browsing activity of influential users, we can identify their interest profiles as well as relevant pages quickly, and recommend these pages to users with similar interests. We call our proposed graph-based model a recommendation network. In this graph, nodes represent users and an edge between users *u* and *v* expresses the fact that *u* and *v* have similar interests, in particular the weight of the edge is the degree to which the user interest profiles match. Based on the graph, we build a recommendation system for Web pages, taking into account the influence of users in a network. Experimental results that measure the precision, with which recommended Web pages are visited by users, indicate that our system performs significantly better than traditional collaborative filtering-based recommender systems.

Keywords User profile · Influential nodes · Web page recommendation · Web usage mining

A. Nanda (✉) · R. Omanwar · B. Deshpande
Birla Institute of Technology and Science, Goa Campus, Zuarinagar, Goa, India
e-mail: f2010175@goa.bits-pilani.ac.in; ashish.nanda.5591@gmail.com

R. Omanwar
e-mail: h2012060@goa.bits-pilani.ac.in

B. Deshpande
e-mail: bmd@goa.bits-pilani.ac.in

L.C. Jain et al. (eds.), *Intelligent Computing, Communication and Devices*,
Advances in Intelligent Systems and Computing 308,
DOI 10.1007/978-81-322-2012-1_42

1 Introduction

The Internet is a great source of information today, with millions of Web pages spanning across almost all topics at various levels of detail. The World Wide Web, thus serves as a great resource to millions of users looking to find specific information on various topics as well as keeping themselves informed and up to date by reading news articles, blogs, etc. With the exponential increase of information on the Web over the last decade, it has become even more difficult for users with varying amount of interest across different topics to find those Web pages that are most relevant to them. Sophisticated search engines and personalized Web page recommenders are of even more importance today, in order to filter through information and provide the user with Web pages that are most relevant to his interests. While search engines help users to find information related to a specific query or need, there are times when the user is not in need of any specific information, but is still looking to find content that is interesting to him, in the form of news, entertaining articles, pages on topics of his interest, or posts from friends on social media Web sites. Recommender systems aim at solving this problem of information filtering, by providing relevant suggestions to help users find content related to their interests. Traditional recommender systems such as those for movies, book, or online retail are based on user rating or purchase behavior, where a purchase or explicit rating is a clear indication of interest. However, designing a Web page recommender is much more challenging, since as compared to the traditional recommender systems; visiting a Web page is not as clear an indication of interest in that page or topic as say, purchasing an item online or explicitly rating a movie. Another challenge is that Web content is subject to high dynamicity, with old pages becoming unimportant very fast, and new pages appearing constantly. However, if the system were to successfully deal with such high levels of noise, personalized Web page recommenders would be an essential solution to the problem of information filtering, by helping users find relevant and interesting content among a plethora of information on the Web.

We introduce a novel and intuitive approach to make personalized recommendation of Web pages: given a users browsing history, we build an interest profile for the user across several categories, and in turn find users with highly similar interests whose visited pages can be recommended. However, unlike previous approaches, we do not just identify clusters of similar users, but user the browsing logs and similarity score to construct a weighted graph between users with matching interests. Further, we study the interaction between users in the network, particularly the precision with which Web pages recommended by a user are visited by similar users in the network. Users who have a high precision are called the influential nodes in the network, since their recommended Web pages are often found interesting to other users, and hence a recommendation by an influential user in the network could be assumed to be more appealing than that of a user who has been found to be less influential. Each user thus gets an influence score, depending on how successfully his recommended Web pages are visited, in order to predict the interestingness of a future recommendation. We take inspiration for our model from a similar phenomenon in social networks such as twitter and Facebook, in which users follow each other and are influenced by each other as information propagates through the network by posts and reposts of content. In information networks, influential users are better and faster at finding interesting content, and hence other users "follow" them. In a similar way, we can say influential nodes in our network are those users, whose recommendations are often found interesting and appealing by other similar users and can be said to "follow" their recommendations. However, unlike social networks, the influential nodes are implicit, and users have no knowledge of the behavior of others, but still similar interests between the users and common browsing behavior successfully explain the influence pattern in the network.

2 Related Work

Web usage mining and personalization: A user's Web history is a valuable source of information in order to understand a user's browsing behavior and interests. Bilenko and White [1] and White et al. [2] analyze users Web history to authoritatively recommend web pages for queries based on the Web sites, a user frequently visits after a query. Liu et al. [3] and Vivismo [4] tried mapping user queries to sets of categories so as to disambiguate words in a users query and assign a context for the same. In previous work, researchers have tried to personalize the search engine results using a user's interest in different categories [5–7]. For that the open directory project (ODP, dmoz.org) has been used as ontology. We too have used the same resource to create a vocabulary of keywords for different topics in the Web. Recommender systems: A recommender system works in a way that it analyzes a users past behavior and suggests products and content (e.g., books, movies, gadgets, news, etc.) that are relevant to a users interest. Das et al. [8] proposed a collaborative filtering-based approach to recommending news articles, while GroupLens research system [9] uses the *k*-nearest neighbor algorithm to recommend Usenet news articles. While GroupLens uses explicit ratings, Morita and Shinoda [10] exploit "time spent" as an implicit rating. Another example is Basu et al. [11] who use social- and content-based information in recommendations. Influence in social and information networks: Our method is also inspired from the importance of influential nodes in social and information networks. Some of the main research lies in the areas of measuring strength of social influence over the links in a network [12–14] and also in discovering influential users, as explained by Agarwal et al. [15] for discovering influential bloggers in a community, and Kempe et al. [16] on maximizing the spread of influence in a social network.

3 Approach to the Problem

In this section, we give an outline of our approach to personalized Web page recommendation. The overview of the system flow can be seen in Fig. 1. We first take the input through an application we developed, which records the browsing history of a set of users. This gives us the dataset D of records in the form (u, p, t) where u denotes the user, who visited the page p, and spent a duration of time t on the page. We propose a graph-based model to understand the interactions between users and find influential nodes in clusters of similar users. We use a weighted and attributed graph $G(U, E, \Omega, \beta, W)$, which is built from the records in dataset D and call it the recommendation network. The attributes of G are described as follows:

- Each node in the recommendation network corresponds to a user $u \in U$. Each user has an interest profile that is created based on his browsing history, with different amounts of interest across a set of topics $T = [1, N]$. The amount of interest of user in a topic is stored in the matrix Ω, where the amount of interest in a topic c for a user u is given by $\Omega(u, c)$.
- An edge in the graph G connects each user to each of his k-nearest neighbors found by calculating similarity in areas of interest. Thus, an edge $(u, v) \in E$ means that u and v have similar browsing patterns and interests.
- The weight of an edge (u, v) between two users is determined by the similarity between the two users' interest profile across the set of topics $T[1, N]$. The weight of the edge $W(u, v)$ is the similarity score calculated by the cosine similarity formula for the interest profiles of users u and v.

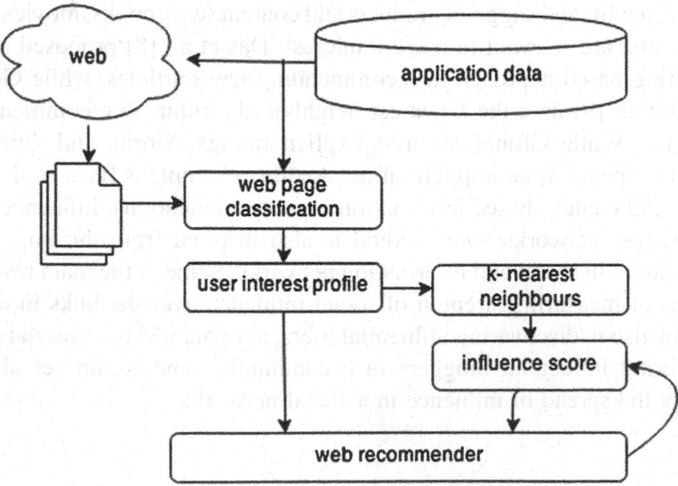

Fig. 1 System flow graph

- The value $\beta(u)$ is the influence score of a user $u \in U$ in his cluster of similar users. The influence score of a user implies the precision with which his network of k most similar users "follow" his recommendations.

The basic idea of creating the recommendation network is that: whenever a user finds a page p interesting, that page can be recommended to users with similar interest by passing information along the edges of the graph. The reverse process also takes place, wherein, information about a recommendation that is visited by others in a network, is propagated back to the user as implicit feedback and is kept track of through the users influence score. Although initially we have a cold start, over a period of time, not only do we establish the users interests across several topics, but also the influential recommenders in a network. We calculate a recommendation score $S(u, p)$, where a page P is recommended to a user u by the other users in the network. It takes into account the topic of the page P, and the amount of interest of the user in that topic, i.e., $\Omega(u, c)$. It also takes into account the weight of similarity $W(u, v)$ and the influence score $\beta(v)$ for the user v that recommended the page to u. The pages are then ranked for the user u by the score $S(u, p)$, and the top few pages are recommended. Although our approach is initially a cold start, and a visit to a page is much more noisy indication of interest than an explicit rating, we find that our method performs significantly better than traditional collaborative filtering approaches.

3.1 Creating User Profiles from the Dataset

Recommendation systems are designed to learn user interests from their past behavior in order to suggest them content that are related to their preferences. Thus, as mentioned above, we first use the Web logs in the dataset that are in the form (u, p, t), which indicates that a user u visited the page p for a time duration t. As shown in Fig. 2, we then use these records to understand users interests. We consider a set of 26 topics from the ODP directory, such as entertainment, sports, politics, computers, and arts and assume that a Web page can belong to one and only one topic. Thus given a topic c, we construct the vocabulary $V(c)$, which is a set of terms such as keywords and phrases that describe the topic and are typically associated with it. In order to construct the vocabulary for a topic c, we used the resources of ODP categories, Microsoft Research, as well as submitting keyword-based queries to popular search engines such as Google and examining the returned results. We then extract the most discriminative keywords from all documents that belong to a topic in order to construct its vocabulary, $V(c)$. All documents that belong to their respective topics were then parsed, and a hash table was created for each topic. The words in $V(c)$ served as the keys of the hash table for that topic, and the word occurrence frequency (N_k) across all documents in the topic c, for each word w_k, was stored as the values in the hash table. The total word count for each topic (N) was also calculated. Once a vocabulary was built for each topic, we then move

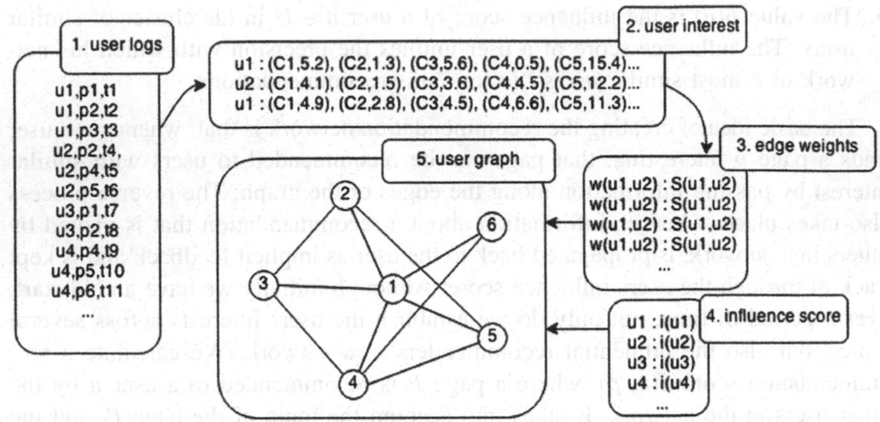

Fig. 2 Stepwise creation of influential user graph from dataset

on to the task of classifying the Web pages browsed by the user. For this purpose, given a page *p*, we create its bag-of-words representation *B*(*p*) which is made of terms from the content of the page. We normalize these terms by removing stop words, special characters, and by carrying out stemming. After creating *V*(*c*) for all topics *c* and *B*(*p*) for all pages *p*, in order to classify a page into a topic, we then use the Naïve Bayes classifier, which has been used successfully for such purposes in the past [17, 18].

3.2 Bayes Theorem

Consider $D = d1, d2, d3, \ldots dp$ to be a set of documents and $C = c1, c2, c3, \ldots cq$ be set of categories. Each of the *p* number of documents in *D* are classified into one of the *q* number of categories from set *T*. The probability of a document *d* to be in category *c* using Bayes theorem is given by:

$$P(c|d) = \frac{P(c)P(d|c)}{P(d)} \quad (1)$$

As *P*(*d*) is independent of the class, it can be ignored. Hence, the category to which *d* belong is given by

$$c_{\text{map}} = P(c)P(d|c) \quad (2)$$

Each document *d* has a bag-of-words representation, i.e., *B*(*d*). Assuming that the attributes (terms, i.e., $t_1, t_2 \ldots t_n$) are independent of each other,

$$P(d|c) = P(t_1|c)P(t_2|c)P(t_3|c)\ldots P(t_n|c) \tag{3}$$

$$c_{\text{map}} = P(c)\prod_{k=1}^{n} P(t_k|c) \tag{4}$$

Thus for our case of classification of Web pages p into the topic c:

$P(c)$ known as the prior probability is 1/26 since there are 26 categories and each of them is equally likely.

$P(t_k|c) = P(w_k|c)$ is the posterior probability, and is equal to the probability of occurrence of the word w_k in the category c. Thus

$$P(w_k|c) = N_k/N \tag{5}$$

where N_k and N for each w_k are stored in the hash table of the corresponding category. A word-topic combination that does not occur in the training data makes the entire result zero. In order to solve this problem, we use Laplace smoothing. The equation after adding Laplace correction becomes:

$$P(w_k|c) = (N_k + 1)/(N + |\text{Vocabulary}|) \tag{6}$$

Since the prior probability, $P(c) = 1/26$ for all categories, c_{map} only depends on the posterior probability. In this way, we classify the document d into that category c_i for which c_{map} has the maximum value:

$$c_{\text{map}} = P(c_i) = \prod_{k=1}^{n} P(w_k|c_i) \tag{7}$$

Once we classify all the pages into their respective categories, we can understand the users interests across different categories by keeping a count of the Web pages classified in different topics. For each page in the user history that is classified in a topic c, we increase the count of documents in that topic by one. Thus, the relative frequency of occurrence of Web pages in a topic determines the weight of that category in terms of the user's interest.

Thus, the amount of interest of a user u in a topic c is given by:

$$\Omega(u, c) = n_c/n \tag{8}$$

where:

n_c number of Web pages classified in topic c

n total number of Web pages across all topics

In this way, we calculate the interest profile of each user across all categories, as shown in step 2 of Fig. 2.

3.3 Edge Strength Between Users

An edge in the graph G connects each user to each of his k-nearest neighbors found by calculating similarity in areas of interest. Thus, an edge $(u, v) \in E$ means that u and v have similar browsing patterns and interests. We use the cosine similarity formula to find the similarity between the interest profiles of two users:

$$\text{similarity} = \cos(\Theta) = \frac{A \cdot B}{|A| \cdot |B|} \tag{9}$$

Here A and B are two vectors that represent the weights of interest in the 26 topics we consider. Using the cosine similarity formula, we weight all users with respect to similarity to the active user. We then select k users that have the highest similarity to the active user. These users form the neighborhood of similar users and an edge (u, v) is made between the active user u and each of these k users. The weight of the edge $W(u, v)$ is the similarity score calculated by the cosine similarity formula for the interest profiles of users u and v. Thus, the edges imply that a user interacts with his k most similar users, and also that the weight of that association depends on the amount of similarity between the interests of the users across the different topics.

3.4 Influence Score

The value $\beta(u)$ is the influence score of a user $u \in U$, in his cluster of similar users. The influence score of a user implies the precision with which the network of his k most similar users "follow" his recommendations. Thus to start with, all users have the same influence score, however, after a period of time, when the user has made several recommendations to similar users, we can estimate his influence of recommendation in the network by seeing the precision with which the pages he recommended were visited by others in his network of similar users. Thus by keeping track of the visited pages among those recommended by a user, we are able to establish to what degree similar users "follow" his recommendations, and hence can accordingly weight the recommendations from each user depending on their influence score. The influence score for a user u is given by:

$$\beta(u) = \frac{\text{number of recommended pages visited by users}}{\text{number of pages recommended to users}} \tag{10}$$

Thus, the higher the influence score of a user in the network, the more important his recommendations should be to others in the network, as compared to users with lower influence. This is because, a high influence score implies, that the user tends to discover more interesting pages and that to faster than others in his network, and hence users usually "follow" his recommendations. Thus, a newly recommended page by an influential user is also likely to be visited more by the network for the same reasons, than a page recommended by a user with a lower influence score.

4 Web Page Recommendation

Our Web page recommendation technique takes advantage of the information in the graph of the recommendation network, which allows for better recommendations than traditional collaborative filtering approaches. Since we recommend a user v pages from his neighbors itself, consider a user u who belongs to the neighborhood. We consider suggesting to user v pages that have been visited by user u and been found interesting. Based on our preliminary experiments, a user must view a page for at least 12 s for it to be considered interesting. To improve the relevance of our recommendations, we rank recommendations by considering the influence score $\beta(u)$ of the user u from whom the recommendation originates, as well as the edge weight $W(u, v)$ that reflects the strength of the connection and similarity in interests between u and v. Additionally, we use page topics to boost scores of pages whose topics match the interests of the user v, i.e., $\Omega(v, c)$. Additionally, we also take into account the unique visits to a page as a measure of popularity of the page, where V_p is the number of visits to page p, and $V_{\max}$ and $V_{\min}$ are the maximum and minimum visits to any page in the dataset. Overall, the recommendation score $S(v, p|u)$ of a page p recommended to v, given that the recommendation has originated by the user u is:

$$S(v, p|u) = \log\left(V_{\max} - V_{\min}/V_p - V_{\min}\right) \times \beta(u) \times W(u, v) \times \Omega(v, c) \quad (11)$$

We calculate this score for all pages p for all neighbors u of a user v. The pages are then ranked based on $S(v, p|u)$ and recommended to the user v.

5 Experimental Results

For evaluating our method, we conducted experiments on a group of 100 college students, by asking them to browse all their Web pages through our application for a period of 2 months. We used their browsing logs to construct their user profiles, a few of which can be seen in Fig. 3, which demonstrates how our Naïve Bayes classification technique generates unique and accurate interest profiles across the 26 categories. During this time, nearest neighbors and influence scores were also calculated for a user and his recommendation network. After this period, we began evaluating our results. The recommended pages are ranked by the score $S(v, p|u)$. The recommendation algorithm is evaluated by using precision-at-k ($p@k$) for $k = 1, 5, 10, 15$, which gives an indication of the fraction of recommended pages

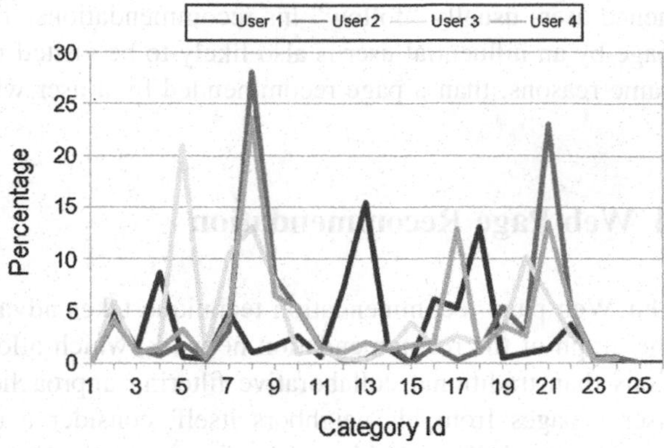

Fig. 3 Mapping user interest profiles across categories

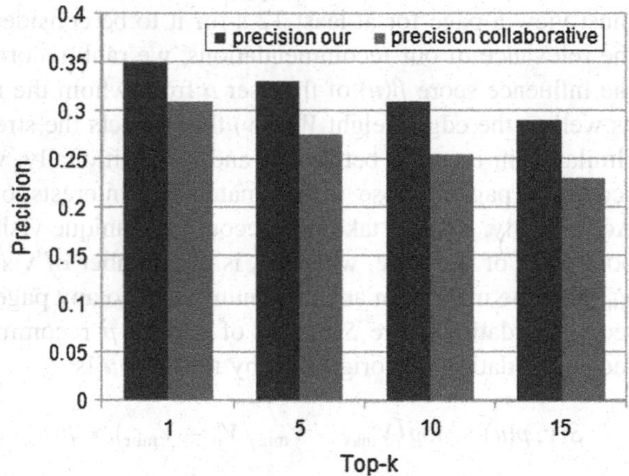

Fig. 4 Comparison of precision of our method *v/s* traditional collaborative filtering

that are actually visited by v. We compare our recommendation algorithm against traditional collaborative filtering approaches. We assume that a click on a page corresponds to a rating equal to 1 while a non-click corresponds to a rating equal to 0, and we compute user similarity with Pearson's correlation coefficients. We present the results we achieved in Fig. 4 and as we can see, our approach significantly outperforms algorithms based on traditional collaborative filtering. In particular, the improvement is more than 10 % for $p@1$ and 20 % for $p@15$.

6 Conclusion

In this paper, we propose a novel graph-based approach to Web page recommendation, that makes use of user-browsing behavior data to construct a user profile, and a network of k-nearest neighbors with edges weighted by similarity.

We also discover implicitly the influence of each user as a recommender in the network and identify to which extent a network "follows" users recommendations. The general idea of our technique is to monitor the activity of influential users, and recommend pages discovered by them to users who follow their suggested pages. Experiments show, our graph-based model that takes into account user interests, as well as their influence as recommenders in a network, results in significant improvement over traditional recommender systems based on collaborative filtering. In the future, we also plan to investigate other possible measures of edge weights by considering different influence models, and also plan to apply our method to a much larger dataset.

References

1. Bilenko, M., White, R.W.: Mining the search trails of surfing crowds: Identifying relevant websites from user activity. In: Proceedings of the 17th International Conference on World Wide Web, WWW'08, New York, NY, USA, pp. 51–60. ACM (2008)
2. White, R.W., Bilenko, M., Cucerzan, S.: Studying the use of popular destinations to enhance web search interaction. In: Proceedings of the 30th Annual International ACM SIGIR Conference on Research and Development in Information Retrieval, pp. 159–166. ACM (2007)
3. Liu, F., Yu, C., Meng, W.: Personalized web search by mapping user queries to categories. In: Proceedings of the 11th International Conference on Information and Knowledge Management, pp. 558–565. ACM (2002)
4. Vivisimo: http://www.vivisimo.com
5. Bennett, P.N., Svore, K., Dumais, S.T.: Classification-enhanced ranking. In: Proceedings of the 19th International Conference on World Wide Web, WWW'10, New York, NY, USA, pp. 111–120. ACM (2010)
6. Collins-Thompson, K., Bennett, P.N., White, R.W., de la Chica, S., Sontag, D.: Personalizing web search results by reading level. In: Proceedings of the 20th ACM International Conference on Information and Knowledge Management, pp. 403–412. ACM (2011)
7. Dou, Z., Song, R., Wen, J.-R.: A large-scale evaluation and analysis of personalized search strategies. In: Proceedings of the 16th International Conference on World Wide Web, pp. 581–590. ACM (2007)
8. Das, A., Datar, M., Garg, A., Rajaram, S.: Google news personalization: Scalable online collaborative filtering. In: Proceedings of the 16th International Conference on World Wide Web, WWW'07, New York, NY, USA, pp. 271–280. ACM (2007)
9. Konstan, J.A., Miller, B.N., Maltz, D., Herlocker, J.L., Gordon, L.R., Riedl, J.: Grouplens: applying collaborative filtering to usenet news. Commun. ACM **40**(3), 7787 (1997)
10. Morita, M., Shinoda, Y.: Information filtering based on user behavior analysis and best match text retrieval. In: Proceedings of the 17th Annual International ACM SIGIR Conference on Research and Development in Information Retrieval, pp. 272–281. Springer, New York, Inc. (1994)
11. Basu, C., Hirsh, H., Cohen, W., et al.: Recommendation as classification: Using social and content-based information in recommendation. In: AAAI/IAAI, pp. 714–720 (1998)
12. Tang, J., Sun, J., Wang, C., Yang, Z.: Social influence analysis in large-scale networks. In: Proceedings of the 15th ACM SIGKDD International Conference on Knowledge Discovery and Data Mining, pp. 807–816. ACM (2009)

13. Goyal, A., Bonchi, F., Lakshmanan, L.V.S.: Learning influence probabilities in social networks. In: Proceedings of the 3rd ACM International Conference on Web Search and Data Mining, pp. 241–250. ACM (2010)
14. Saito, K., Nakano, R., Kimura, M.: Prediction of information diffusion probabilities for independent cascade model. In: 12th International Conference KES 2008, pp. 67–75. Springer, Berlin, Heidelberg (2008)
15. Agarwal, N., Liu, H., Tang, L., Yu, P.S.: Identifying the influential bloggers in a community. In: Proceedings of the 2008 International Conference on Web Search and Data Mining, pp. 207–218. ACM (2008)
16. Kempe, D., Kleinberg, J.M., Tardos, ÂE.: Maximizing the spread of influence through a social network. In: Proceedings of the 9th ACM SIGKDD International Conference on Knowledge Discovery and Data Mining, pp. 137–146. ACM (2003)
17. McCallum, A., Nigam, K.: A comparison of event models for Naïve Bayestext classification. In: AAAI/ICML-98 Workshop on Learning for Text Categorization, pp. 41–48 (1998)
18. Wang, Y. Hodges, J. Tang, B.: Classification of web documents using a Naïve Bayes method. In: Proceedings of the 15th IEEE International Conference on Tools with Artificial Intelligence, pp. 560–564. IEEE Computer Society, Washington, DC, USA (2003)

Ensuring Data Security and Performance Evaluation in Cloud Computing

Pourya Shamsolmoali and M. Afshar Alam

Abstract Cloud computing is threatened by unanswered security issues that are risky for both the cloud providers and users. Cloud is a computing design that manages large sets of distributed resources, of which scientists benefit from their convergence. The aim of this paper is divided into two parts: firstly a brief review on cloud computing mainly focusing on security and secondly offer a solution that eradicates possible threats. In particular, we proposed a new data security model that can efficiently protect the data whether in the cloud database or in transition. We start with an established authentication server and data server providing user authentication, user verification, and data support. The system follows SSL protocol for data encryption and protection, and secure deliver report (SDR) is used for data reliability and integrity of communication.

Keywords Cloud computing · Data protection · Security · Authentication · Encryption · Secure deliver report

1 Introduction

Cloud computing brought innovative feature to Internet and data storage. The cloud offers massive benefits to businesses by significantly reducing the purchase cost of hardware and software. A definition by Casola et al. [1] and Foster et al. [2] declares that cloud is a large-scale distributed computing that is taken by economics scale, in which a group of virtualized, managed computing power, platforms, storage, and services are delivered. NIST provided a definition which

P. Shamsolmoali (✉) · M. Afshar Alam
Department of Computer Science, Jamia Hamdard University, New Delhi, India
e-mail: pshams@jamiahamdard.ac.in

M. Afshar Alam
e-mail: aalam@jamiahamdard.ac.in

L.C. Jain et al. (eds.), *Intelligent Computing, Communication and Devices*,
Advances in Intelligent Systems and Computing 308,
DOI 10.1007/978-81-322-2012-1_43

depicts cloud computing as a model permitting convenient network access to a shared lake of computing resources (e.g., storage, servers, applications, and services) [3]. In comparing with traditional networking method, cloud computing can be quickly provisioned and released with minimal management endeavor or interaction of service provider. Service providers must ensure that all the security facets are functional. In case of failure, they are to be held responsible. Lots of advantages like lower costs, pay for use, fast deployment, scalability, ubiquitous network access, low-cost disaster, data recovery, data storage solution, and greater resiliency are offered by cloud.

As cloud computing is getting an increased reputation, concerns are being expressed about the security issues introduced through the approval of this new model [4]. In this paper, we concentrated on cloud computing with the main focus on security. The data security model is considered, and unique security requirements are documented. The remainder of this paper is structured as follows: Sect. 2 describes the cloud computing security. Section 3 describes data security in cloud. Section 4 covers the proposed model. Section 5 describes the security proof of proposed model. Section 6 covers the performance evaluation. As a final point, Sect. 7 documents some conclusions.

2 Cloud Computing Security

Security in cloud computing is controlling the access of unauthorized users to the system state. The main aspects of security are confidentiality, availability, and integrity [5]. Confidentiality means assuring the users that their information will not be disclosed without their authority. Availability is a process of ensuring the information is available to the end users whenever and wherever they need. Integrity is avoidance of the unauthorized modification or information deletion. Rick Blasidell has done a research on top threats to cloud computing; he recognized four threats for cloud "Threat no 1: Security threats, Threat no 2: Outages, Threat no 3: Malicious insiders, and Threat no 4: Lack of information" [6]. CSA also published another research work on the top threats to cloud computing in March 2010 [7]. The point of the research was to support cloud providers and consumers in identifying the main vulnerable points and major risks of cloud and also how cloud provider infrastructure has security from these risks. It is our contention that CSA on cloud computing security could be considered as pioneering work to direct aspiring future researchers to guide them in this area. We have found CSA's research on top threats [8] is latest among the distinguished research works in cloud computing security [9].

3 Data Security in Cloud

The main weakness that traditionally correlated with cloud computing is the lack of arrangement for data security and the perception of moving to the cloud. This makes critical data to be uncovered at the time of attack [10]. It depends on the form of cloud computing, policies, and management. Mackay et al. noted three specific data storage security issues in the cloud environment for service provider; suppose to attend; the first mechanism is encryption, whereby all the data in cloud is encrypted. The second issue is shared resources. The last issue to consider is the integrity of the data that migrate to cloud storage [11]. Rong et al. [12] pointed that the owner of data should have full control over authorization of data sharing. With authorization given by the data owner, selected users have access to the data stored in the cloud [13]. This action should not privilege the service provider any right to access the user's data. Khorshed et al. [9] noted there are some benefits in monitoring API in cloud-based centralized system, but Web application based on API generally shares more vulnerabilities.

4 Proposed Model

In this section, we present a framework that has been structured to offer absolute solutions to preserve the integrity, confidentiality, and authenticity of data. We applied multiple techniques such as verification of the digital signature and double authentication to protect the critical data. The system consists of three main parts:

Cloud Provider who manages, provides cloud storage services, and has high computation power.

Data Owner organizations or an individual customer who has vast data files to be stored in the cloud storage.

User who will register with data owner and uses or shares a data stored on cloud storage. The user has limitation on right to use data files.

In proposed model as shown in Fig. 1, when a user wants to access the data stored in the cloud, first of all the user needs to register with the data owner by getting a valid username and password through the application interface. Next, the data owner sends authentication message to the user. At the end, the owner forwards the registered ID to cloud to store it within the user directory of the authentication server.

The abstract flow in Fig. 2 illustrates the interaction between the four parties, and it consists of the following steps:

1. The user sends the ID and password to the data owner, first authentication.
2. The user replies the security question provided by data owner, second level of authentication.
3. The data owner redirects the user ID and digital signature to authentication server; therefore, cloud will be sure that the owner let user access of data.

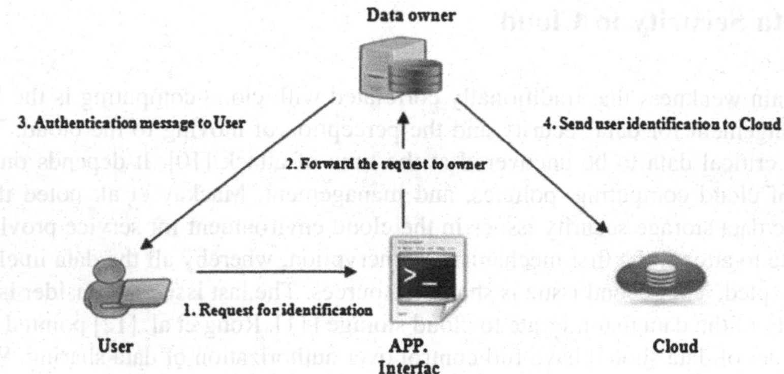

Fig. 1 User registration process

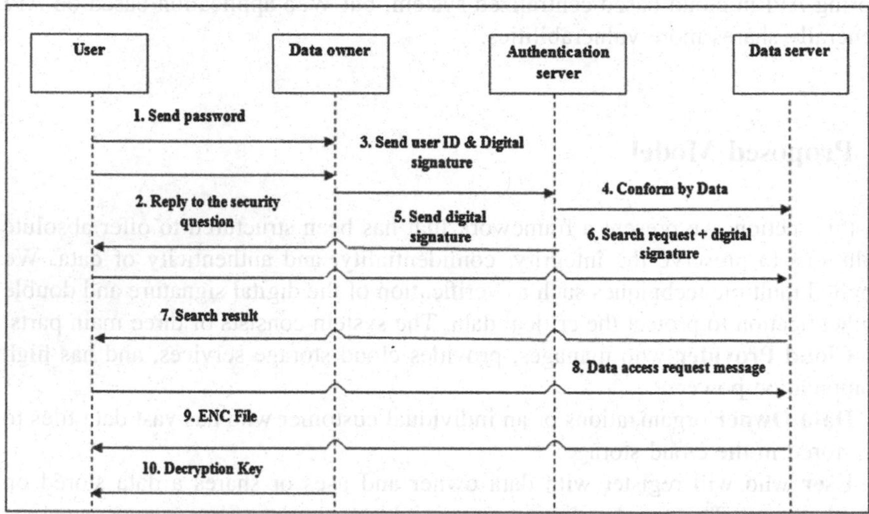

Fig. 2 Overall view of proposed model

4. The authentication server validates the user authorization grant. It also validates that the user is a trusted entity by data server and issues access permission.
5. The authentication server forwards the digital signature to the user. Then, the user can use it as an authentication token.
6. The user afterward sends the search request and asks for protected resources by presenting digital signature to data server.
7. The data server responds to search request and delivers the search result to user.
8. The user generates a request to the data server for retrieving encrypted data.

9. Afterward, the data server sends back the requested data in encrypted format.
10. Then, the data owner dispatches the decryption key to the user.

5 Security Proof

5.1 *Confidentiality of Data*

Secure traveling of data on network is a tough and highly complex issue, while the data threat is continuously rising. The cloud environment does not only require traffic protection. In addition, secure mechanism of communication is also essential [14]. To prevent the loss of data in transition, SSL protocol in our model is used. SSL generates end-to-end encryption by interacting between applications and the TCP/IP protocols to present authentication and an encrypted communication between data owner, server, and user. SSL protocol is embedded into every Web browser; this helps user from additional software installation on their system. To create secure communication between data server and user, first the data server sends the identification information to the user just after the connection is created and then sends the user a copy of its SSL certificate. The user verifies the certificate and replies to the data server. The data sever sends back a token to build SSL session.

5.2 *Limitation of Service Provider in Data Access*

When the data reside in a cloud database, the management and responsibility are handled by the service provider. Assume that the data in a cloud database are secure from any external party as the service provider uses strict security roles to protect his environment. The service provider can oppose the data owner. As the data in the cloud is not in direct control of the owner, any harm can be possible by cloud provider. For this reason, the best solution is applied in the proposed model. Privacy of data can be protected through encryption [14]. SSL protocol as we explained in previous section encrypts the data and builds secure communication over the public Internet.

5.3 *Secure Deliver Report (SDR)*

The data in the cloud is always vulnerable and under the threat of being interfered by any attack [15]. As all the precautionary methods such as double authentication, data encryption, and SSL protocol are used in the proposed model during transition

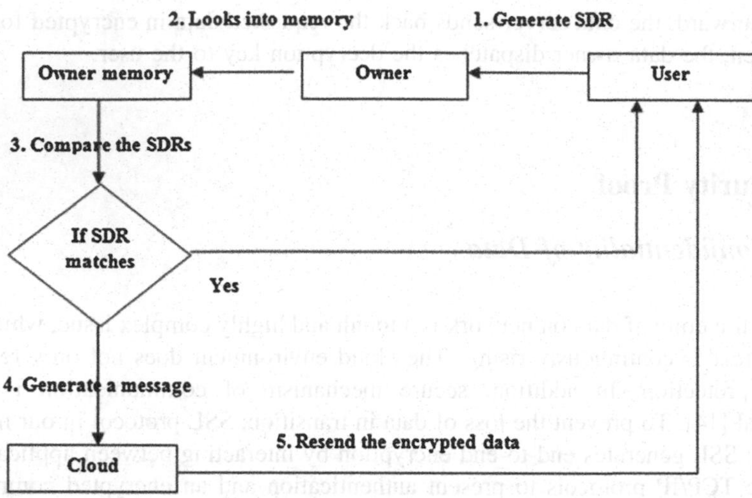

Fig. 3 Secure deliver report (SDR)

time, this does not guarantee the total security of the proposed model in respect to threats. The model has one more parameter called SDR. Firstly, SDR is generated by the data owner before sending the data to cloud; the owner keeps the SDR in his memory. On another side, the user can generate the SDR of received data and send it to the data owner, and the data owner compares the new SDR with the original one that he has. If both SDRs match, the user is assured that the data have not been interfered. In case the owner gets a mismatch SDR, he generates a message to cloud requesting to resend the data file to the same user as shown in Fig. 3.

6 Performance Evaluation

An effective data security model should be able to overcome all the existing issues of cloud computing to prevent the owner's data from all the risks associated. In Table 1, we have done a comparison between other data security models and the proposed model. The proposed model is evaluated with respect to implementation. This model is tested on CloudSim Simulator. Figure 4 represents the status of security after implementation of security parameters. Encryption provides additional security than identification and authentication, and it provides more security than SDR. Figure 4 shows the security evaluation of our proposed model. However, by taking the combination of all three security parameters, the security of data has the best efficiency.

Table 1 Comparison of data security

	Wang et al. [16]	Prasad et al. [17]	S.K. Sood [13]	Proposed model
Authorization	Yes	Yes	Yes	Yes
Confidentiality	Yes	Yes	Yes	Yes
Integrity	Yes	Yes	Yes	Yes
Encryption	Yes	Yes	Yes	Yes
Identification and authentication	Yes	Yes	Yes	Yes
Data security even after loss of user ID	No	No	Yes	Yes
User verification	No	No	No	Yes
Secure deliver checking	No	No	No	Yes

Fig. 4 Security evaluation

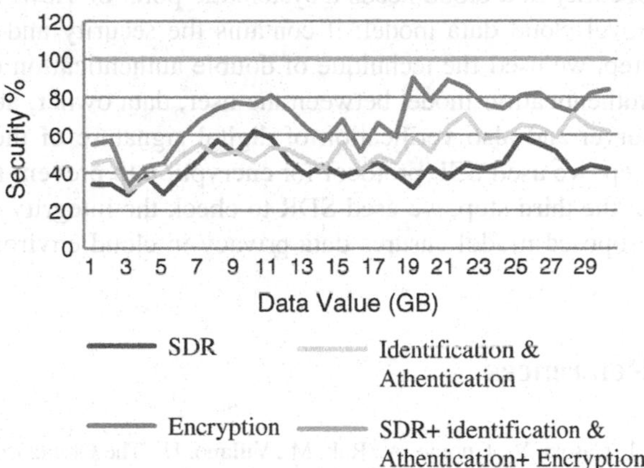

To create a virtual cloud environment, we have used a HP Proliant DL 580 G7 Server, with following features: dual 1.864 GHZ processors, 16 GB RAM, 5 × 300 GB SCSI Hard Drives. We also selected VMWare ESXi 5.0.0 Hypervisor as virtual machine manager (VMM) and Windows 7 as guest operating system. For monitoring load and performance testing, we used CloudTestLite by SOASTA. Previously, we showed the security performance of our proposed model. After implementation of all security parameters, performance of database connection pool usage in five minutes is illustrated in Fig. 5. The figures shows that the proposed model has a very good performance as compared to different models.

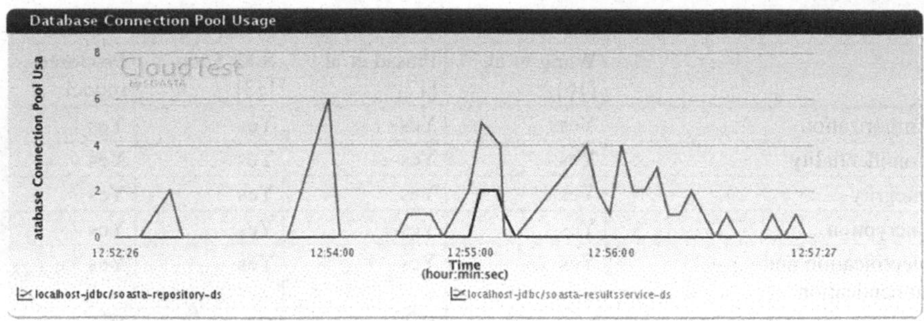

Fig. 5 Database connection pool

7 Conclusion

Security in a cloud needs a systematic point of view. In this paper, we proposed a novel cloud data model; it contains the security and performance issues. In first step, we used the technique of double authentication of user, we synchronized an authentication model between the user, data owner, authorization server, and data server and also verification of digital signature of the data owner. In the second step, we used SSL protocol for encryption to prevent the loss of data in transition. In the third step, we used SDR to check the integrity of data. We believe that our proposed model ensures data privacy in cloud environment.

References

1. Casola, V., Cuomo, A., Rak, M., Villano, U.: The CloudGrid approach: security analysis and performance evaluation. Future Gener. Comput. Syst. (2011). doi:10.1016/j.future.2011.08.008
2. Foster, I., Zhao, Y., Raicu, I., Lu, S.: Cloud computing and grid computing 360-degree compared. In: 2008 Grid Computing Environments Workshop, 2008, pp. 1–10. doi:10.1109/GCE.2008.4738445
3. NIST.: NIST cloud computing program. http://www.nist.gov/itl/cloud/. Retrieved 21 May 2011
4. Zissis, D., Lekkas, D.: Addressing cloud computing security issues. Future Gener. Comput. Syst. **3**(28), 583–592 (2012)
5. Teneyuca, D.: Internet cloud security: the illusion of inclusion. Inf. Secur. Tech. Rep. **3–4**(16), 102–107 (2011)
6. www.rickscloud.com/top-threats-for-cloud-computing/
7. Archer, J., Boehm, A.: Security guidance V 3.0, Cloud Security Alliance (2011)
8. Archer, J., Boehme, A., Cullinane, D., Kurtz, P., Puhlmann, N., Reavis, J.: Top threats to cloud computing, version 1.0 (2010). Cloud security alliance retrieved 7 May 2011
9. Khorshed, M.T., Ali, A.B.M., Wasimi, S.A.: A survey on gaps, threat remediation challenges and some thoughts for proactive attack detection in cloud computing. Future Gener. Comput. Syst. **6**(28), 833–851 (2012)

10. Zhou, M., Mu, Y., Susilo, W., Yan, J., Dong, L.: Privacy enhanced data outsourcing in the cloud. J. Netw. Comput. Appl. **4**(35), 1367–1373 (2012)
11. Mackay, M., Baker, T., Al-Yasiri, A.: Security-oriented cloud computing platform for critical infrastructures. Comput. Law Secur. Rev. **6**(28), 679–686 (2012)
12. Rong, C., Nguyen, S.T., Jaatun, M.G.: Beyond lightning: a survey on security challenges in cloud computing. Comput. Electr. Eng. **39**(1), 47–54 (2013)
13. Sood, S.K.: A combined approach to ensure data security in cloud computing. J. Netw. Comput. Appl. **6**(35), 1831–1838 (2012)
14. Gao, J., Xiao, Y., Liu, J., Liang, W., Chen, C.L.P.: A survey of communication/networking in smart grids. Future Gener. Comput. Syst. **2**(28), 391–404 (2012)
15. Noureddine, M., Bashroush, R.: An authentication model towards cloud federation in the enterprise. J. Syst. Softw. (2010). doi:10.1016/j.jss.2012.12.031
16. Wang, C., Wang, Q., Ren, K.: Ensuring data storage security in cloud computing. In: 17th International Workshop on Quality of Service (IWQoS). IEEE, pp. 1–9 (2009)
17. Prasad, P., Ojha, B., Shahi, R.R., Lal, R.: 3-dimensional security in cloud computing. Comput. Res. Dev. **3**, 198–208 (2011)

Context-Aware Computing Using Rich Contexts for Pervasive Environment

P. Rajasekaran, N. Prabakaran and V. Magudeeswaran

Abstract Applications in pervasive environment are, detect the environmental changes, further processing bountiful supply of information, and respond to the processed information by altering contexts overtly. It aims the interlinking of independent devices, whose purpose is to reinforce the contact between technology and intended human users. Many of the existing applications are not proactive and not in rich context too. The usual reminders will remind, even when the intended user is unavailable, become unknown, and finally expires. We propose rich-context conminder which is rich in context and supports events to be updated and expired automatically after completion of the task. Allure of our proposal is, to grant the reminder in the suitable real-time situation with respect to user, physical, and environmental behaviors. It is a deliberate attempt to provide a soft real-time system which proactively determines suitable timing to deliver them in the appropriate situation.

Keywords Pervasive computing · Context · Sensing · Soft real-time system · Rich context

1 Introduction

Pervasive computing environment has been greatly implemented in several fields, and it has been growing extremely with various areas such as context-aware computing, embedded systems, wireless sensor networks, and real-time system

P. Rajasekaran (✉) · V. Magudeeswaran
PSNACET, Dindigul, India
e-mail: p_rajasekar88@yahoo.co.in

V. Magudeeswaran
e-mail: magudeeswaran@gmail.com

N. Prabakaran
SCSE, VIT University, Vellore, India
e-mail: dhoni.praba@gmail.com

L.C. Jain et al. (eds.), *Intelligent Computing, Communication and Devices*,
Advances in Intelligent Systems and Computing 308,
DOI 10.1007/978-81-322-2012-1_44

designing [1, 2]. Generally, the pervasive refers, able to access the information constantly anywhere and anytime. Several groups of device are endeavoring to communicate among them in order to provide pervasiveness [3]. Existing reminders will present their information without knowing appropriate situation like unavailability of intended user and sometimes visible to everyone [4]. Reminders might use two kinds of styles such as descriptions and visual. If it is description, the intended user has to look upon for what purpose it reminds, but on the other hand, if it is visual, viewing and removing them manually after completion. Only a personal human assistant can remind the description with signal in the appropriate situation. It is not possible for everyone to have an assistant. Even this system has a flaw that if the assistant is not with us, it is hard to explain. We can include the audio/voice signal with the reminders in case of need and make them to present on the suitable situation as well as remove them automatically once completed [5]. Devices themselves should be capable of accessing the most effective form of connectivity with the surroundings, since commonly it is an invisible transaction and thus eliminates the overt in nature [6]. The remaining part of this paper is continued with related works, models and problems, proposed scheme, and evaluation. Finally, we conclude this paper.

2 Related Works

In the literature survey, survey works have been conducted on low-level contexts, rich-level contexts, and active–passive contexts used in pervasive computing environment. Pervasive computing involves dynamic tasks which are not provided by traditional computers [7, 8]. It introduces list of design challenges which are related to user interfaces and highly demanding context-based applications [9]. Context-aware computing meets the dynamic requirements of users, which add the contexts such as time, user location, and nearby user. We separate our work by four parts such as (i) context resolving, (ii) context possession, (iii) context progression, and (iv) context control. However, we discussed basic models and problems associated in designing the context modeling systems.

3 Models in Context Modeling System

3.1 Context Identification

Contexts are often extracted from sensors that generate huge volume of electrical signals which are converted as a computational data after processing [10]. This raw data are known as low-level context and transformed as a rich-level context after post-processing process [9]. For instance, Image pictured at home with my parents. Now the image includes what time the picture taken, when, where, and with whom.

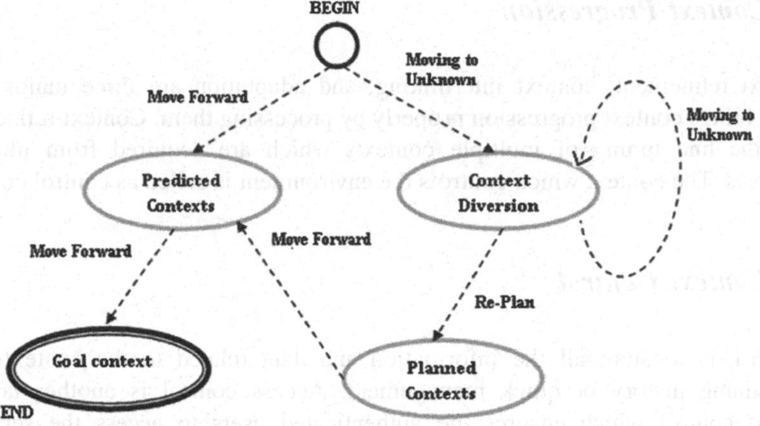

Fig. 1 Prediction of context deviation

3.2 Model and Sources of Contexts

Some rich-level contexts are originated from user, physical, and environment contexts. After sensing, these contexts are combined and further subjected to fine-refinement. As a result of this composite context, the more accurate context is generated [3]. This context is enhanced more than individual contexts and named as context refinement. To achieve the context more accurate, the search begins with moving. Move may fall either in planned path or in deviated path (Fig. 1).

If it falls in planned path, then the predicted current context is achieved via context refinement and further it accomplishes the goal context. It leads to context diversion if it is in deviated or unknown path. From the diversion, again it may go to deviation or re-plan. Re-planning make a path to goal context, but infinite deviation effects the context diversion unfortunately.

4 Rich Contexts for Context-Aware Computing

4.1 Context Resolving

Context resolving includes mainly three steps, those are acquiring user context, post-processing and filtering them finally. The context acquisition from user is directly or indirectly done by interaction mode. Raw contexts are now ready for processing, so this low level is processed to rich level by post-processing. Harmonize these contexts into more accurate context, so that common depiction can be accomplished. Here, the environment may be office, house, etc. so the range is fixed within the place.

4.2 Context Progression

Context refinement, context interlinking, and adaptation are three major parts which brings context progression properly by processing them. Context refinement does the fine tuning of multiple contexts which are acquired from multiple resources. The context which controls the environment is called as control context.

4.3 Context Control

Our task is to store all the information and data related to the contexts and maintaining history or quick management. Access control is another task of context control which ensures the authenticated users to access the services. Context management is accomplished through context processing and refinement.

5 Evaluation on Decision Making

The rich-context-aware alarm influences pervasive environment by using various sensors. Initially, the alarm is given with time, user status, and expiring time with irrespective to the environment. In our scenario, 'user contexts' specify the presence of the user, which is physically presence status. 'Environment contexts' refer the situation that exists with the intended user, where as environment situation may be meeting hall, home, outdoor, etc. 'Physical contexts' refer the time, date, etc. The decision tree shown below represents how the decision is made based on the requirements met.

Basically, 3 contexts are chosen as input and 23 combinations are possible as output. Let A, B, and C are the notations used for representing the user contexts,

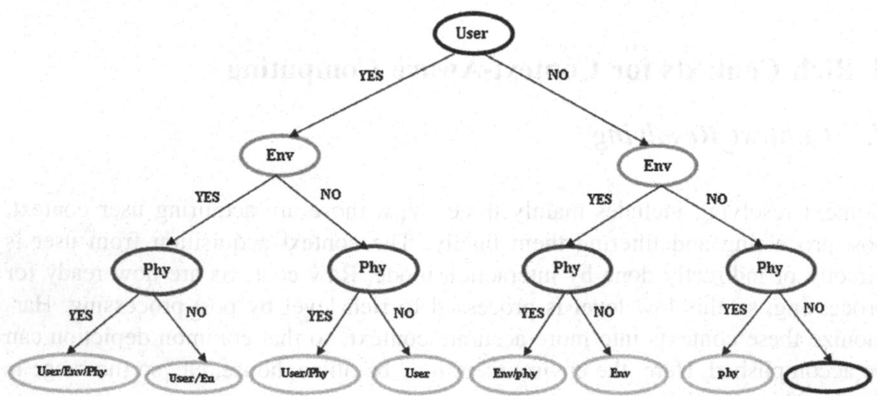

Fig. 2 Decision making tree

environmental contexts, and physical contexts, respectively. The decision tree shows the possible outcome from the sensor's decision tree (Fig. 2).

A-> Intended user is physically present. B-> Time and date are matched. C-> Existing environmental situation is observed. Now, the case becomes true for left extreme and false for right extreme. Similarly for all the respective situations, the context results are determined by using programmable logic with gates.

Once the conditions are satisfied, the task is activated based on the environment type. There is also an option for expiring them automatically once the task is completed. Figure 3 illustrates the cases and the corresponding conditions.

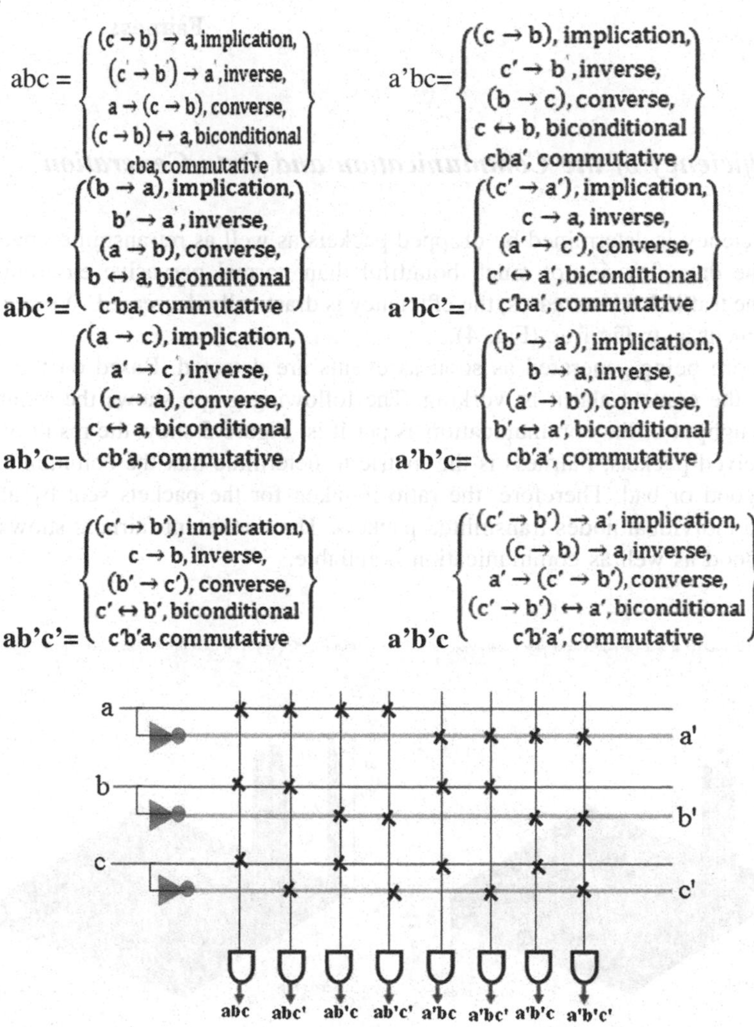

Fig. 3 User, physical, and environment contexts

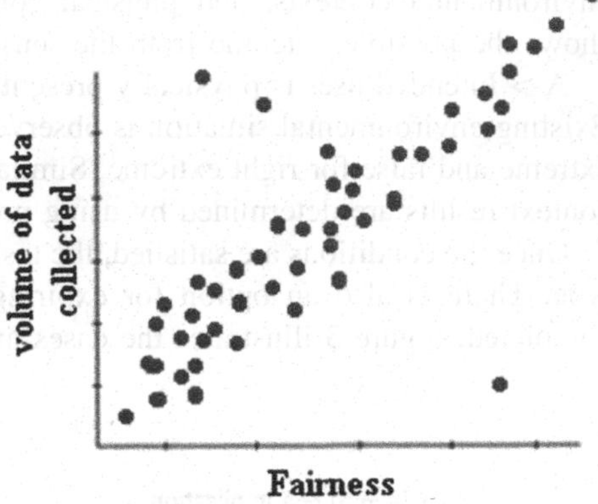

Fig. 4 Fairness of the communication

5.1 Efficiency of the Communication and Data Generation

The efficiency is determined by dropped packets as well as retransmitted packets. Here, the data flow is not much bountiful than normal pervasive environment. When the traffic load increases, the efficiency is drastically decreased. We keep the environment as traffic free (Fig. 4).

Data are being generated as soon as events are detected. Based on the comparison, the context alarm is working. The following graph shows the reliability and throughput of the communication as per it is. Figure 5 show the result of sent and received packets. Fairness is the metric to determine that the communication link is good or bad. Therefore, the ratio is taken for the packets sent by all the nodes to individual nodes transmitted packets. The consistent fairness shows that link is good as well as communication is reliable.

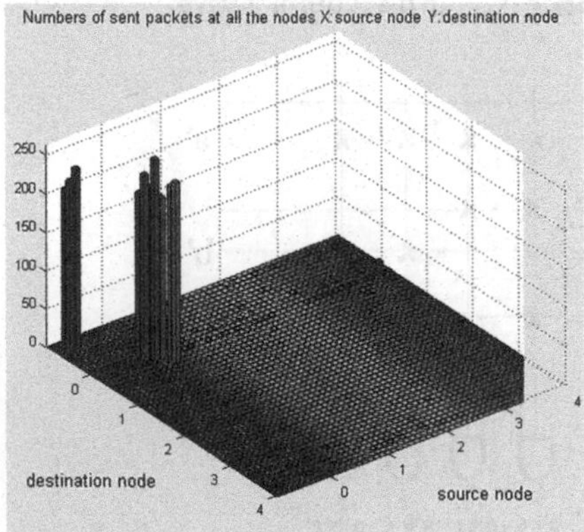

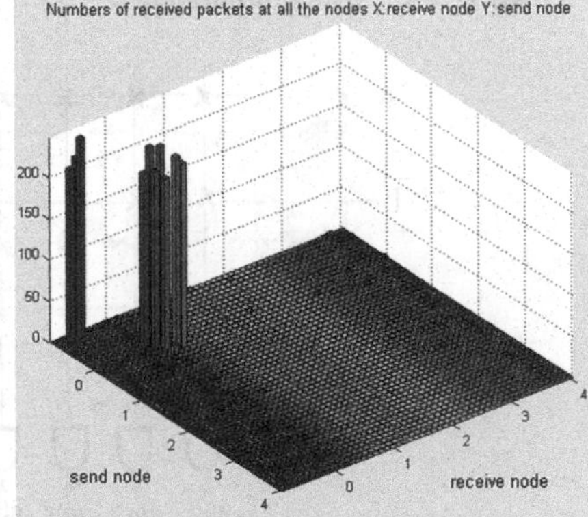

Fig. 5 Number of packets sent and received in the communication

6 Conclusion

We have proposed this scheme to conclude that the results of evaluations are general representations of context aware with the mixed level of low-level and rich-level contexts. We maintain context management, querying, and combining them in order to support flexible appropriate actions. The further results can be justified by implementing with the infrastructure in the form of high-level contexts in addition to context acquisition, querying, and management. This scheme reaches reasonable communication in the pervasive environment which shows better in terms of context resolving, progression, context possession, and control. Specially for improving rich-level context, it will assist to combine various basic low-level contexts. This scheme becomes reliable for making context alarm in the pervasive environment using sensor network communication.

References

1. Bergqvist, J., Ljungberg, F.: ComCenter: a person oriented approach to mobile communication. Extended abstract. In: Proceedings of CHI 2000, pp. 123–124 (2000)
2. Prabakaran, N., Geetha, K., Janani, K.: Open stream scheme for node level congestion control in wireless sensor network by International Conference and published in IEEE explorer (2011)
3. Dey, A.K., Abowd, G.D.: Towards a better understanding of context and context-awareness. In: CHI 2000 Workshop on the What, Who, Where, When, and How of Context-Awareness (2000)
4. Miyata, Y., Norman, D.A.: Psychological issues in support of multiple activities. In: Norman, D.A., Draper, S.W. (eds.) User Centered Design. pp. 265–284 (1986) (Chapter 13)
5. Byun, H.E., Cheverst, K.: Harnessing context to support proactive behaviours. In: ECAI2002 Workshop on AI in Mobile Systems, Lyon, (2002)
6. Prabakaran, N., Geetha, K., Janani, K.: Data propelling scheme for node level congestion control in wireless sensor network published in IEEE explorer (2012)
7. McFadden, T., Henricksen, K., Indulska, J.: Automating context-aware application development. In: UbiComp 1st International Workshop on Advanced Context Modelling, Reasoning and Management, September 2004, pp. 90–95
8. Nilsson, M., Hjelm, J., Ohto, H.: Composite capabilities/preference profiles: requirements and architecture. W3C Working Draft **21**, 2–28 (2000)
9. Kulik, J., Rabiner, W., Balakrishnan, H.: Adaptive protocols for information dissemination in wireless sensor networks. In: Proceedings of ACM MobiCom, Seattle, WA, Aug. 1999, pp. 174–185
10. Sankarsubramaniam, Y., Akan, O.B., Akyildiz, I.F.: ESRT: event to sink reliable transport in wireless sensor network. In: Proceedings of ACM MobiHoc'03

Financial Trading System Using Combination of Textual and Numeric Data

Jibendu Kumar Mantri and Braja B. Nayak

Abstract Forecasting stock return is a challenging concept that has attracted researchers' attention for many years. It involves an assumption that fundamental information publicly available in the past has some predictive relationships to the future stock returns. This study helps the investors to decide the better timing for buying or selling stocks based on the knowledge extracted from the historical prices of stock market using different data mining techniques. But this paper tries to provide a conclusive analysis based on the accuracies for stock market forecasting using the methods MLP and decision tree for buying and selling stocks.

Keywords MLP · Decision tree · BSE Sensex

1 Introduction

The financial markets around the global play an important role in the process of economic growth and developments by facilitating savings and channeling funds from savers to investors. In stock exchange, stock price mechanism is fully determined by market condition, caused mainly by changes in trading volume, modification in macroeconomic policies, shifts in investor tolerance of risk, and increased uncertainty. Over the past decades, many attempts have been made at understanding and predicting the future using different data mining methods. Among them, forecasting the price movements in stock market and making the

J.K. Mantri (✉)
Department of Computer Science and Engineering, North Orissa University, Baripada, Orissa, India
e-mail: jkmantri@gmail.com

B.B. Nayak
ECIL, Hyderabad, India
e-mail: brajabnayak@rediffmail.com

L.C. Jain et al. (eds.), *Intelligent Computing, Communication and Devices*, Advances in Intelligent Systems and Computing 308,
DOI 10.1007/978-81-322-2012-1_45

decision to buy and sell are considered as a major challenge as it is high-risk investment [1]. Investors have been trying to find a way to predict stock prices and to find the right stocks and right timing to buy or sell. To achieve those objectives, some researchers used the techniques of fundamental analysis, where trading rules are developed based on the information associated with macroeconomics, and industrial units. Recently, different new techniques such as decision trees, rough set approach, and artificial neural networks have been applied to this area [2–4]. Specially, decision trees have been found very effective for the classification of huge and frequently modifiable databases, e.g., stock market and shopping mall [5]. Also, decision tree concepts are used to discover rules and relationships by systematically breaking down and subdividing the information contained in data set [6]. For this context, this paper tries to represent the analysis of Indian stock market data (BSE) using two prominent techniques, i.e., multilayer perceptron and decision tree for buying and selling stocks.

2 Design

The applications of neural networks to financial forecasting have become very popular over the last few years due to nonlinearity existence in financial data [7]. Whether randomly or fully predictable, a correctly designed neural network will theoretically converge to an optimal result. Most of the other models do not have this property!

- The linearity assumption and normal distribution may not hold in mostly financial time series. Neural networks can model nonlinear systems and do not have any assumption about input probability distribution.
- ANNs are universal function approximation. It has been shown that a neural network can approximate any continuous function to any desire accuracy.
- ANNs can be generalized. After learning the data presented to them, ANNs can often correctly infer the unseen part of a population even if the sample data contain noisy information.
- Compared with GARCH model, neural networks are significantly more accurate.
- Heteroscedasticity phenomena can be captured by ANNs.

But the successful application of neural networks to data analysis in stock market is the multilayer perceptron (MLP) model. Multilayer perceptron models are nonlinear neural network models that can be used to approximate almost any function with a high degree of accuracy [8].

But decision trees embody also a supervised classification approach. The idea came from the ordinary tree structure which is made up of a root, nodes (the positions where branches divides), branches, and leaves.Decision tree is constructed from nodes that represent circles, and the branches are represented by the segments that connect the nodes. It starts from the root, moves downward, and generally is drawn from left to right. The node from where the tree starts is called a root node. The node where the chain ends is known as the “leaf” node. Two or more branches can be extended from each internal node, i.e., a node that is not leaf node. A node represents a certain characteristic, while the branches represent a range of values. These ranges of values act as a partition points for the set of values of the given characteristic. Also, the grouping of data in the decision tree is based on the values of attributes of the given data. A decision tree is made from the preclassified data. The division into classes is decided upon the features that best divide the data. The data items are split according to the values of these features. This process is applied to each split subset of the data items recursively. The process terminates as far as all the data items in current subset belong to the same class.

3 Data Preparation

These research data are collected from BSE Sensex from January 2005 to December 2013. At the beginning, the data collected contained nine attributes; this number was reduced manually to five attributes as the other attributes were found comparatively less important and having any direct effect on the study. Table 1 shows the five attributes selected with their descriptions and their possible values. The attribute “Action” is the investors’ overall response reflected in a concerned month (as the data are taken in month-wise interval) through the movement of the Sensex.

At the beginning, when the data were collected, all the values of the attributes selected were continuous numeric values. Data transformation was applied by generalizing data to a higher-level concept so as all the values became discrete. The criterion that was made to transform the numeric values of each attribute to discrete values depended on the previous month’s closing price of the stock. If the values of the attributes Open, High, Low, and Close were greater than the value of the previous attribute for the same trading month, the numeric values of the attributes were replaced by the value Positive as shown in Table 2. If the values of the attributes mentioned above were less than the value of the previous attribute, the numeric values of the attributes were replaced by Negative. The Action attribute is the investors’ action whether to buy or sell.

Table 1 Attribute description

Attribute	Description
Open	Current month open price of the stock
High	Current month maximum price of the stock
Low	Current month minimum price of the stock
Close	Current month close price of the stock
Action	The action taken by the investor on this stock

Table 2 Converted data for the implementation of decision tree

Month and year	Open	High	Low	Close	Action
Jan 2006	Positive	Positive	Positive	Positive	Buy
Feb 2006	Positive	Positive	Positive	Positive	Sell
Mar 2006	Positive	Positive	Positive	Positive	Sell
Apr 2006	Positive	Positive	Positive	Positive	Sell
May 2006	Positive	Positive	Negative	Negative	Buy
Jun 2006	Negative	Negative	Negative	Negative	Buy
Jul 2006	Positive	Positive	Positive	Positive	Buy
Aug 2006	Positive	Positive	Positive	Positive	Sell
Sep 2006	Positive	Positive	Positive	Positive	Sell
Oct 2006	Positive	Positive	Positive	Positive	Sell
Nov 2006	Positive	Positive	Positive	Positive	Sell
Dec 2006	Positive	Positive	Negative	Positive	Sell

Table 2 is a converted form of the attributes. The numeric values are replaced by indicators, e.g., positive or negative, and the actions are represented by buy or sale, depending upon the swing of the stock market as per its indices. Though the actual analysis comprises nine years of data extending from the year 2005 to year 2013, only the converted form of data for the year 2006 has been presented to maintain simplicity.

The decision tree of Fig. 1 is constructed by taking attribute "High" as the root node. The tree subsequently expanded to attribute "Low" via attribute "Close." The paths of the tree lead to the action of the investors at the decision node of the tree shown as a square.

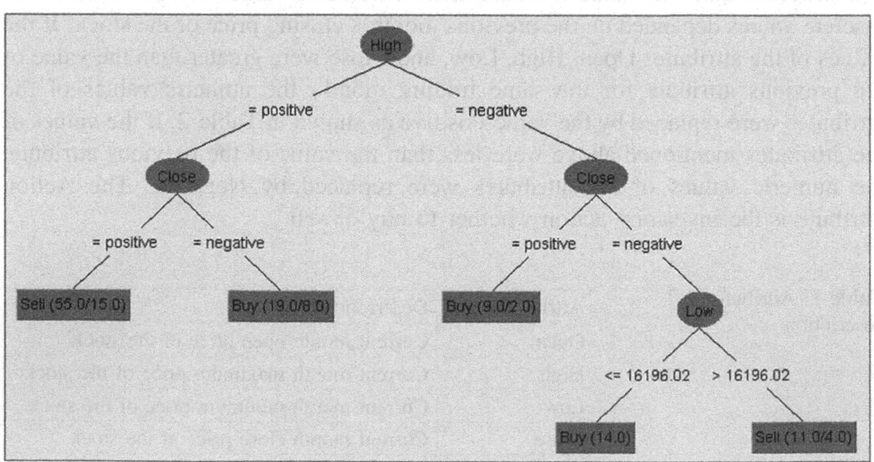

Fig. 1 Decision tree

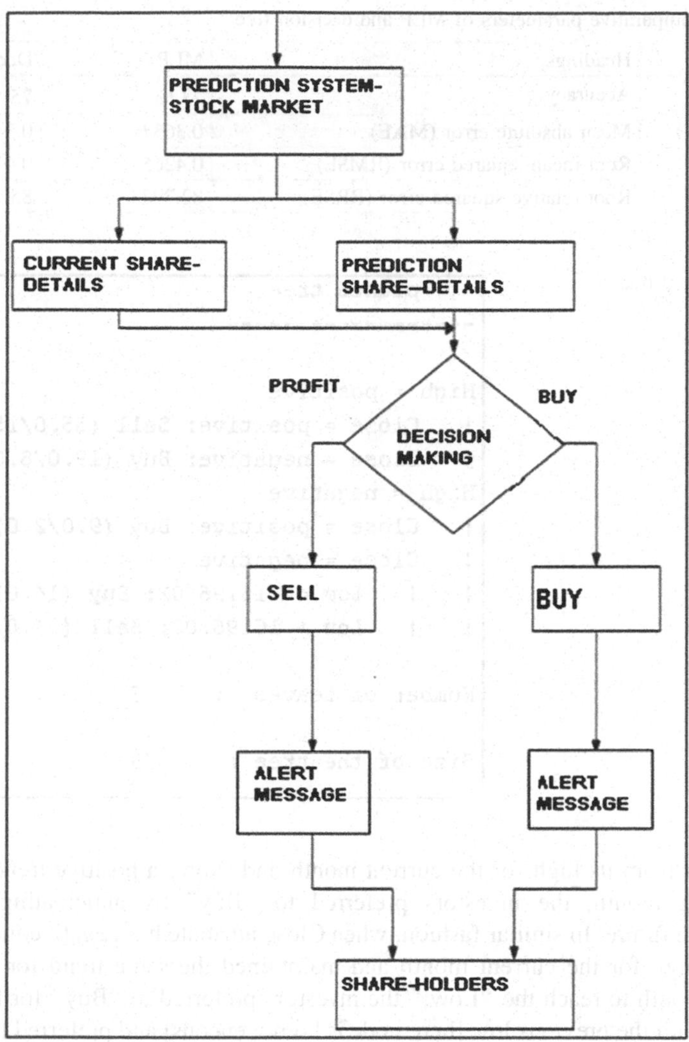

Fig. 2 Proposed model

A close observation of Table 2 and Fig. 1 will made the interrelationship between market movement and the investors' response (which is termed as "Action") understandable. It can be noticed that in most of the cases, when Close attribute has positive movement from its high for the current month and maintained the same trend for the succeeding month, the investors preferred to "Sell" for maximum gain. But when Close attribute has positive movement from its high for the current month and failed to maintain the same trend for the succeeding month, the investors preferred to "Buy" rather than "Sell" in the anticipation of raise in share prices in future. On the other hand, when Close attribute has negative

Table 3 Comparative parameters of MLP and decision tree

Sl. No	Headings	MLP	Decision tree
1	Accuracy	73 %	75 %
2	Mean absolute error (MAE)	0.3637	0.3472
3	Root mean-squared error (RMSE)	0.4265	0.4133
4	Root relative-squared error (RRSE)	82.7974	85.4237

Fig. 3 Pruned tree

```
pruned tree
------------------

High = positive
|   Close = positive: Sell (55.0/15.0)
|   Close = negative: Buy (19.0/8.0)
High = negative
|   Close = positive: Buy (9.0/2.0)
|   Close = negative
|   |   Low <= 16196.02: Buy (14.0)
|   |   Low > 16196.02: Sell (11.0/4.0)

Number of Leaves  :     5

Size of the tree :      9
```

movement from its high for the current month and shows a positive trend for the succeeding month, the investors preferred to "Buy" by anticipating market recovery in future. In similar fashion, when Close attribute has negative movement from its high for the current month and maintained the same trend for the succeeding month to reach the "Low," the investors preferred to "Buy" for less than or equal with the previous low [here node is homogeneous] and preferred to "Sell" otherwise. Considering the above facts, a model has been proposed as in Fig. 2 for a stock market prediction system.

4 Performance Comparison

A decision tree can be used as a model for a sequential decision problem under uncertainty. A decision tree describes graphically the decisions to be made, the events that may occur, and the outcomes associated with combinations of decisions and events. It is one of the methods which use the classification approach. Here, the C 4.5 algorithm developed by Quinlan is used to generate associations between different features and different trends. Table 3 shows the values of different parameters using MLP and decision tree.

The outputs given in Table 3 depict a comparative study of the three efficiency measures, e.g., MAE, RMSE, and RRSE of MLP and decision tree clearly indicate the relative efficiency of the models for the analysis of the data in the present study. The pruned tree generated by decision tree is shown in Fig. 3.

5 Conclusion

This study presents a proposal to use the decision tree classifier on the historical prices of the stocks to create decision rules that give buy or sell recommendations in the stock market than MLP. Such proposed model can be a helpful tool for the investors to take the right decision regarding their stocks based on the analysis of the historical prices of stocks in order to extract any predictive information from that historical data. The evaluation of a larger collection of learning techniques such as neural networks and genetic algorithms can represent a rich area for future investigation. Also, reconsidering the factors affecting the behavior of the stock markets such as trading volume, news, and financial reports which might impact stock price can be another rich field for future research.

Acknowledgment I am very much thankful to my esteemed teacher, Honorable Vice Chancellor of North Orissa University, Prof. Sanghamitra Mohanty for her help and advice for this new application to stock market.

References

1. Al-Debie, M., Walker, M.: Fundamental information analysis: an extension and UK evidence. J. Account. Res. **31**(3), 261–280 (1999)
2. Enke, D., Thawornwong, S.: The use of data mining and neural networks for forecasting stock market returns. Expert Syst. Appl. **29**, 927–940 (2005)
3. Quinlan, J.R.: Induction of decision tree. Mach. Lean. **1**, 81–106 (1986)
4. Wang, Y.F.: Mining stock price using fuzzy rough set system. Expert Syst. Appl. **24**, 13–23 (2003)
5. Agrawal, R., Lin, K.-I., Sawhney, H.S., Shim, K.: Fast similarity search in the presence of noise, scaling, and translation in time-series databases. In: Proceedings of 21st International Conference Very Large Data Bases (VLDB '95), Sept 1995, pp. 490–501 (1995)
6. Wu, M.C., Lin, S.Y., Lin, C.H.: An effective application of decision tree to stock trading. Expert Syst. Appl. **31**, 270–274 (2006)
7. Mantri, J.K., Gahan, P., Nayak, B.B.: Artificial neural networks—an application to stock market volatility. Int. J. Sci. Technol. **2**(5), 1451–1460 (2010)
8. Mantri, J.K., Mohanty, D., Nayak, B.B.: Design neural network for stock market volatility: accuracy measurement. Int. J. Comput. Technol. Appl. **3**(1), 242–250 (2012)

A Novel Trust-Based Privacy Preserving Access Control Framework in Web Services Paradigm

Rekha Bhatia and Manpreet Singh

Abstract Web users are increasingly becoming conscious about the personally identifiable information (PII) being collected and used by Web service providers. Users of these services are usually asked by the service providers to reveal their PII in order to access the services provided by them. While collecting this PII, the service providers must ensure their customers that the PII provided by them must be handled according to the privacy policies and laws. Currently, the enforcement of privacy policies and laws is done manually. This process is error prone and can leak information to the third parties which the information provider has never imagined. The automation of privacy policy enforcement is a must for Web service providers to deal with the privacy handling issue. This paper is an effort towards how to automate the privacy policy enforcement along with traditional authorization policies followed in legacy access control systems. As trust plays an important role in human life and we constantly update and upgrade our trust relationships with other people based on our outlooks in response to the changing situations, the dynamic nature of heterogeneous Web services collaboration is handled through a trust-based access control mechanism.

Keywords Privacy · Access control · Trust · Web services · PII

1 Introduction

In Web services paradigm, the legacy system of access control, and the privacy and trust-based access control mechanism should be integrated in a seamless manner. In traditional systems, the identities of the personally identifiable

R. Bhatia (✉)
Punjabi University Regional Centre, Mohali, India
e-mail: r.bhatia71@gmail.com

M. Singh
Punjabi University, Patiala, India
e-mail: msgujral@yahoo.com

L.C. Jain et al. (eds.), *Intelligent Computing, Communication and Devices*,
Advances in Intelligent Systems and Computing 308,
DOI 10.1007/978-81-322-2012-1_46

information (PII) requestors are used as enforcement mechanism. Such systems do not take into consideration, the privacy aspects like whether the PII they are requesting, are to be used for the specific purpose for which it was disclosed by the PII owner, whether the environmental conditions in which they are seeking this information are according to the satisfaction of the user and whether they are accessing this information during the valid retention period of PII mentioned by the PII owner. In order to extend traditional system of access control with privacy parameters as mentioned in Barker's taxonomy [1], the following requirements should be satisfied:

i. Integration of privacy authorization system along with conventional access control system.
ii. Since unknown Web services collaborate to carry out the requests of the information seekers, the trust factor should be introduced in the access control process.
iii. Preprocessing of PII submitted by information owners before storing in the repository is must.
iv. Integrated access control system should be simple to use (Fig. 1).

Our privacy aware access control model (PA^2CM) addresses all the above-stated issues. The model includes privacy parameters namely retention period, environmental conditions prevailing at the time of access, and granular levels for restricted access along with the traditional access control parameters. Besides, PA^2CM also includes the access control check module during the time access is occurring in order to capture the dynamically varying conditions during the course of access (Fig. 2).

The paper is organized as follows: in the next section, related work in privacy preserving access control is discussed. In Sect. 3, a description of privacy aware access control model is provided. In Sect. 4, the example of health care scenario used throughout this paper has been explained. In Sect. 5, we have introduced trust for controlling access to Web services while preserving privacy of the PII owners. In Sect. 6, we have concluded the paper stressing upon the need to integrate privacy policies into access control process of Web services.

2 Related Work

Previous research related to access control technologies has presented various access control models since the 1960s. These models include access matrix [2], discretionary access control (DAC) [3], mandatory access control (MAC) [3], and RBAC [4]. Most of these conventional access control models are specified in terms of subjects, objects, and the actions that can be performed on the objects. A subject is an entity that can initiate requests for a data resource to perform an action or a series of actions on objects. It can be users, automated agents, or processes. An object is an entity on which an action can be performed. An action is a privilege

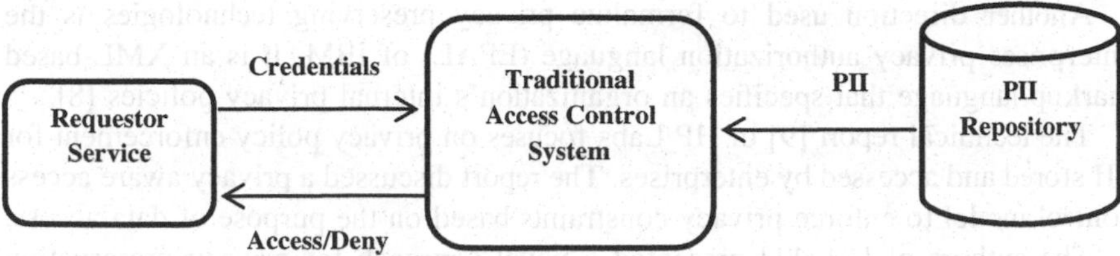

Fig. 1 Traditional access control system

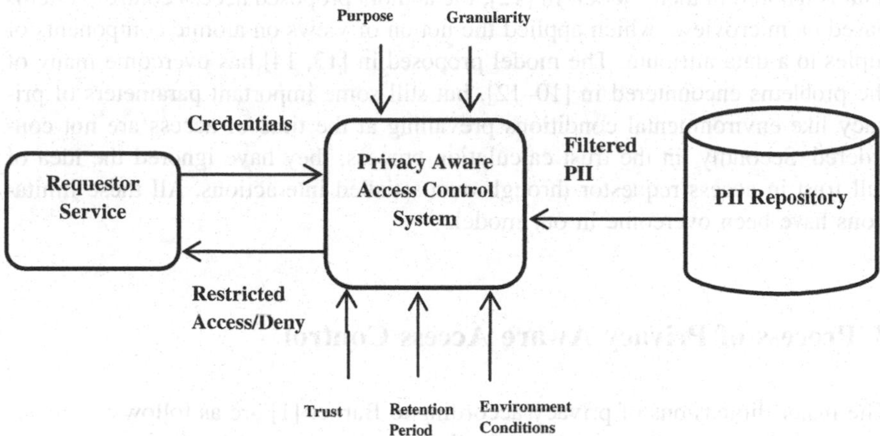

Fig. 2 Privacy aware access control system

invoked by a subject. Permissions are authorizations to perform some actions on the specified resources. Access rights to a resource are mainly regulated by a set of access control rules [5]. A general access control rule is as follows:

ALLOW [Subject] to perform [Action] on [Object].

All these conventional models have not mentioned of privacy policies and practices. The W3C's platform for privacy preferences project (P3P) is related to the development of P3P specifications for Websites to encode their privacy practices [6]. The P3P user agents enable users to automatically get information about Website practices and to automate actions in response to the Websites' privacy practices. P3P uses a language called 'A P3P preference exchange language (APPEL)' [7]. This language is used for expressing user's preferences for automated response generation regarding the acceptability of machine-readable privacy policies from P3P-enabled Websites. Although this framework is not primarily meant for supporting Web services privacy policies, we can see an enhanced version of P3P to incorporate the Web services privacy policy framework.

Another direction used to formalize privacy preserving technologies is the enterprises privacy authorization language (EPAL) of IBM. It is an XML-based markup language that specifies an organization's internal privacy policies [8].

The technical report [9] of HP Labs focuses on privacy policy enforcement for PII stored and accessed by enterprises. The report discussed a privacy aware access control model to enforce privacy constraints based on the purpose of data access.

The authors of [10, 11] presented a novel approach for privacy preservation access control based on the idea of purpose. In their model, the purpose information associated with a given data resource specifies the intended use of the data resource. Although their model protects the data privacy but the usability factor of data is ignored in their model. In [12], the authors proposed access control systems based on microviews which applied the notion of views on atomic components of tuples to a data attribute. The model proposed in [13, 14] has overcome many of the problems encountered in [10–12], but still some important parameters of privacy like environmental conditions prevailing at the time of access are not considered. Secondly, in the trust calculation process, they have ignored the idea of full trust in access requestor through own satisfied interactions. All these limitations have been overcome in our model.

3 Process of Privacy Aware Access Control

The major dimensions of privacy according to Barker [1] are as follows: purpose, granularity, and retention as shown in Fig. 3.

Purpose is the reason behind providing personal information to the data collector, for example, a person may provide detailed information about the problem he is suffering from, to a doctor in order to ensure receive necessary medical cure but not to any third party like his insurance company. Disclosing such sensitive personal information to insurance company can cause user to suffer heavy financial losses. This aspect of privacy is of huge importance in Web services paradigm as personal information can potentially be made available to such persons which the information provider has never imagined. The granularity defines the magnitude or degree of user's PII to be made available in response to a query, for example, information that a person has HIV AIDS can be made available to his close relatives but the exact name of the disease should not be revealed to the public at large. Retention period defines the amount of time period during which the collected PII for a specific purpose should be retained in the repositories of the service provider, for example, if user has provided 3 months (the period of his treatment in a medical care centre) as the retention period of his personal information, then his PII should not be retained by the service provider in his repositories beyond that time.

The conventional access control systems such as MAC [3], DAC [3], and RBAC [4] are based on which user is performing which action on which data object, whereas our privacy aware access control model (PA^{2}CM) is based on the information about which PII is used for which purpose, in which time period and

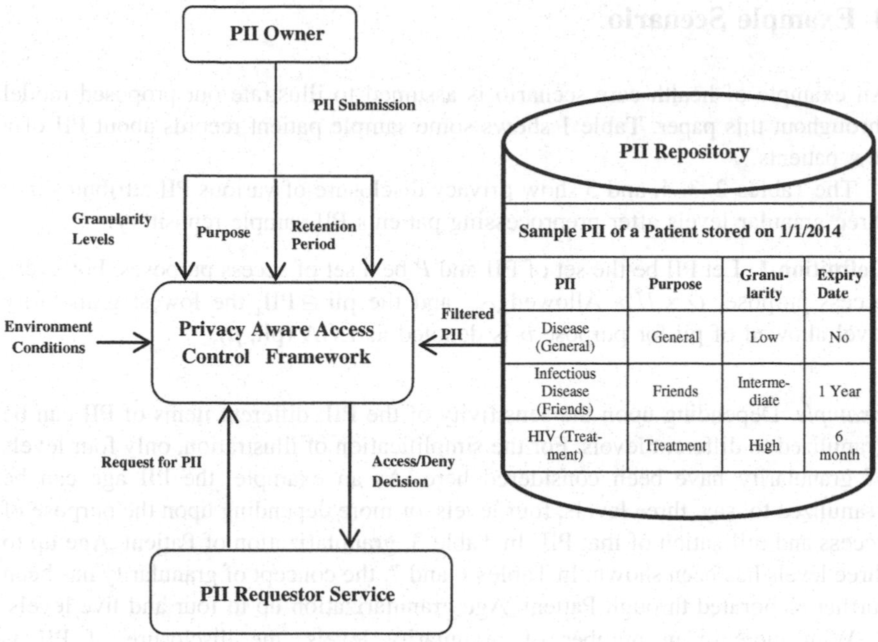

Fig. 3 Process of privacy aware access control

up to which degree. Our model is based on the magnitude of information revelation allowed for a particular user rather than only the permission for information revelation. Our model stresses upon the need to preprocess submitted PII before being stored in the repositories. Each data item is granulized to different levels, where each granularity level corresponds to a different privacy level. The main purpose of our model is to extend conventional access control models to include privacy parameters in order to minimize privacy disclosure and to maximize data usability as suggested in [14, 15].

To enforce privacy-based restrictions, a conventional access control model should be extended to include the privacy requirements along with PII in an integrated manner. The extended privacy aware access control model in Fig. 3 addresses the issue of enforcing privacy policies along with traditional access control policies in an integrated manner so as to prevent dubious users to violate the privacy rights of the PII owners. The model is based on the granularity of access so that data owners can control the degree of data disclosure based on the purpose of access. For example, a person having a sexually transmitted disease is generally unwilling to provide minute details about his disease to his family and friends, but to receive proper treatment, the exact minute disclosure of his disease is very much required by a doctor. Similarly, expiry date for retaining PII about exact name of the disease is limited to, say, 1 month according to the owner of PII but the information that the patient was having some disease in the past can be retained for some more time according to the privacy preferences of the owner of the PII.

4 Example Scenario

An example of health care scenario is assumed to illustrate our proposed model throughout this paper. Table 1 shows some sample patient records about PII of a few patients.

The Tables 2, 3, 4, and 5 show privacy disclosure of various PII attributes into three granular levels after preprocessing patients PII sample repository.

Definition 1 Let PII be the set of PII and *P* be a set of access purposes. For every access purpose, $O \times U \times \text{Allowed}_{\text{AC}_{\text{PII}}}$ and the pii $\in$ PII, the lowest granularity level allowed of pii for purpose *p* is denoted as LGA (pii, *p*).

Example Depending upon the sensitivity of the PII, different items of PII can be granulized to different levels. For the simplification of illustration, only four levels of granularity have been considered here. As an example, the PII age can be granulized to, say, three levels, four levels, or more depending upon the purpose of access and utilization of that PII. In Table 3, granularization of Patient_Age up to three levels has been shown. In Tables 6 and 7, the concept of granularity has been further elaborated through Patient_Age granularization up to four and five levels.

With increase in number of granularity levels, the disclosure of PII is decreasing (for the last level, PII disclosure is minimum). The maximum number of granularity levels of a particular PII should be decided by the owner of the PII depending upon the different purposes for which he wants to provide that PII. While deciding about the number of granularity levels, the most important thing to be considered is that the utilization of that information should not be lost in the process. As an example, the Patient_Age value when granularized to levels such as 'between 20 and 30', 'below 40', 'above 60' carries some meaning but when granularized to such level as 'below 100' carries no meaning for most of the purposes as the probability of going it above 100 is the least.

According to definition 1, if

PII = {Patient_Name, Patient_Address, Patient_Age, Patient_Disease} and P = {Treatment, Billing, Insurance} then, the lowest granularity level allowed of PII Patient_Age for billing purpose is: LGA (Patient_Age, Billing) = NR. Here, NR means, for billing purpose, the revelation of Patient_Age is not required. Similarly, the lowest granularity level allowed of PII Patient_Disease for insurance purpose is LGA (Patient_Disease, Insurance) = Low.

Tables 8, 9, and 10 show privacy revelations for maximizing utility while minimizing sensitive PII disclosure for treatment purpose, billing purpose, and insurance purpose, respectively.

Definition 2 Let PII be the set of PII and *P* be a set of access purposes. For every access purpose $p \in P$, the set REQD_PH $\subseteq$ PII denotes the required PII in order to satisfy access purpose *p*. The granularity level to be disclosed for *p* is: (i) For all pii $\in$ REQD_PH, the pii is permitted to be disclosed up to LGA (pii, *p*), where

Table 1 Patients PII repository (sample)

PII → values↓	Patient_Name	Patient_Age	Patient_Address	Patient_Disease
1	Revati Sharma	69	30, Miller Ganj, Ludhiana	HIV AIDS
2	Mala Kalra	21	564, Model Town, Delhi	Cancer
3	Venu Gopal	87	64D, Jawahar Nagar, Jaipur	Fever
4	Ritvik Mehra	35	12, Civil Lines, Ganga Nagar	Cancer
5	Meenu Singh	5	345, Sector 44B, Chandigarh	Fever

Table 2 Preprocessing of Patient_NamePII

Patient_Name	Granularity_Level
Revati Sharma	High
R. Sharma	Intermediate
R. S	Low

Table 3 Preprocessing of Patient_Age PII

Patient_Age	Granularity_Level
69	High
50–75	Intermediate
50–100	Low

Table 4 Preprocessing of Patient_Address

Patient_Address	Granularity_Level
30, Miller Ganj, Ludhiana	High
Miller Ganj, Ludhiana	Intermediate
Ludhiana	Low

Table 5 Preprocessing of Patient_Disease

Patient_Disease	Granularity_Level
HIV AIDS	High
Infectious disease	Intermediate
Some disease	Low

Table 6 Patient_Age PII granularity up to 4 levels

Patient_Age	Granularity_Level
69	1
60–70	2
60–75	3
60–80	4

Table 7 Patient_Age PII granularity up to 5 levels

Patient_Age	Granularity_Level
69	1
60–70	2
60–75	3
60–80	4
60–85	5

Table 8 Privacy disclosure for treatment purpose

PII	Granularity_Level
Patient_Name	High
Patient_Address	Low
Patient_Age	High
Patient_Disease	High

Table 9 Privacy disclosure for billing purpose

PII	Granularity_Level
Patient_Name	High
Patient_Address	Low
Patient_Age	NR
Patient_Disease	NR

Table 10 Privacy disclosure for insurance purpose to a third party

PII	Granularity_Level
Patient_Name	High
Patient_Address	High
Patient_Age	High
Patient_Disease	Low

LGA is the lowest granularity level allowed of pii for access purpose p. (ii) For pii $\notin$ REQD_PH, the pii is permitted to be disclosed up to granularity level NR.

Example According to definition 2, if PII = {Patient_Name, Patient_Address, Patient_Age, Patient_Disease} and P = {Treatment, Billing, Insurance} then, required PII (REQD_PII) for access purpose billing is:

REQD_PII = {Patient_Name, Patient_Address}

LGA (Patient_Name, Billing) = High
LGA (Patient_Address, Billing) = Low

Since (Patient_Age, Patient_Disease) $\notin$ REQD_PII for billing purpose, the granularity level specified for both these PII are NR as mentioned in Table 7. This example highlights the fact that introduction of granularity levels, in the PII submitted by a user, can maximize the utilization of that PII for the access purpose in hand, while at the same time can minimize the disclosure of sensitive PII of a user.

Definition 3 Let PII be the set of PII resources, U be the set of information users, AC be the set of actions allowed on resources, P be the set of purposes, GL be the set of granularity levels, RP be the set of retention period for various PII resources, and EC be the set of environmental conditions to be taken care of in a particular scenario. The new privacy aware authorization is a 7-tuple (pii, u, ac, *p*, gl, rp, ec) pii $\in$ PII, $u \in U$, ac $\in$ AC, $p \in P$, gl $\in$ GL, rp $\in$ RP, ec $\in$ EC.

The above 7-tuple states that the user *u* has been authorized to access pii resource for performing *ac* action under granularity level *gl* and purpose *p* if retention period *rp* and environmental conditions *ec* are valid. Besides, purpose of access and granularity levels, the other two important parameters of privacy in our model are retention period and environmental conditions prevailing at the time of access.

Example As an example, if retention period specified for Patient_Disease PII is 6 months from today, the expiry date for the information related to this PII will be (today + 6 × 30)th day. For simplification of discussion, the 30 days in a month is mentioned, but for real-time calculations, the exact number of days will be calculated. To illustrate the role of environmental conditions prevailing at the time of access, consider the emergency situation where delay in getting access rights to Patient_Disease PII for the doctor can prove fatal for his life.

5 Trust in Privacy Aware Access Control

Trust is the basis of human society in order to overcome fears and doubts while interacting within our social set-ups. We select a particular school for our kids because that school has the strong reputation of being trustworthy in the past. Similarly, we go for a particular brand of clothing, toys, and utility items because of their well built up reputation either through our own previous experience or through a word of mouth from others. In the online transactions, all heterogeneous Web services cannot be trusted to the same degree because most of the times, service provider and service requestor are unknown to each other. In order to support the process of collaboration among unknown Web services and to establish trust for privacy policy enforcement, there should be no human intervention. The process of collaboration should be fully automated so as to reduce the possibility of errors. The trust of the PII requestor service can be established in 2 ways [13–16]:

i. By feedback through service provider's own previous experience of transactions with the requestor service.
ii. By feedback, testimonies and recommendations of third party sources.

Figure 4 describes the integration of trust in our privacy aware access control model. Let *T* (Req) be the total trust of the requestor service with the provider service, *T*1 (Req) is the value of direct trust through previous experience with the requestor service, and *T*2 (Req) is the value of indirect trust obtained through recommendations and testimonies of other service providers. This trust calculation process has been mentioned in our previously published work [15].

Trust calculation by feedback through service provider's own previous interactions with requestor:

$$T1(\text{Req}) = \sum_{(i=1)}^{n} \text{DSF}(\text{Req}, i) * \text{WT}(\text{Req}, i) / \sum_{(i=1)}^{n} \text{WT}(\text{Req}, i),$$

where *i* stands for the number of previous interactions, DSF (Req, *i*) means the direct satisfaction value of the requesting service in *i*th interaction with the data provider service, and WT (Req, *i*) stands for the weight assigned to that particular interaction by the service provider depending upon the various privacy parameters specified earlier.

Trust calculation by testimonies and feedback through peer agents:

$$T2(\text{Req}) = \sum_{(j=1)}^{m} \text{DSF}(\text{Req}, j) * \text{WT}(\text{Req}, j) / \sum_{(j=1)}^{m} \text{WT}(\text{Req}, j),$$

where *j* stands for the number of testimonies of other data providing services, RSF (Req, *j*) means the satisfaction value of the *j*th service provider with the requestor service, and WT (Req, *j*) stands for the weight assigned to that particular interaction by the service provider depending upon the various privacy parameters specified earlier.

Total trust calculation:

If((<∀i : DSF(Req, i)==1) AND(T1(Req) > = Theta))
T(Req) = *T*1 (Req),
Else
T(Req) = fn (*T*1 (Req), *T*2 (Req))
EndIf

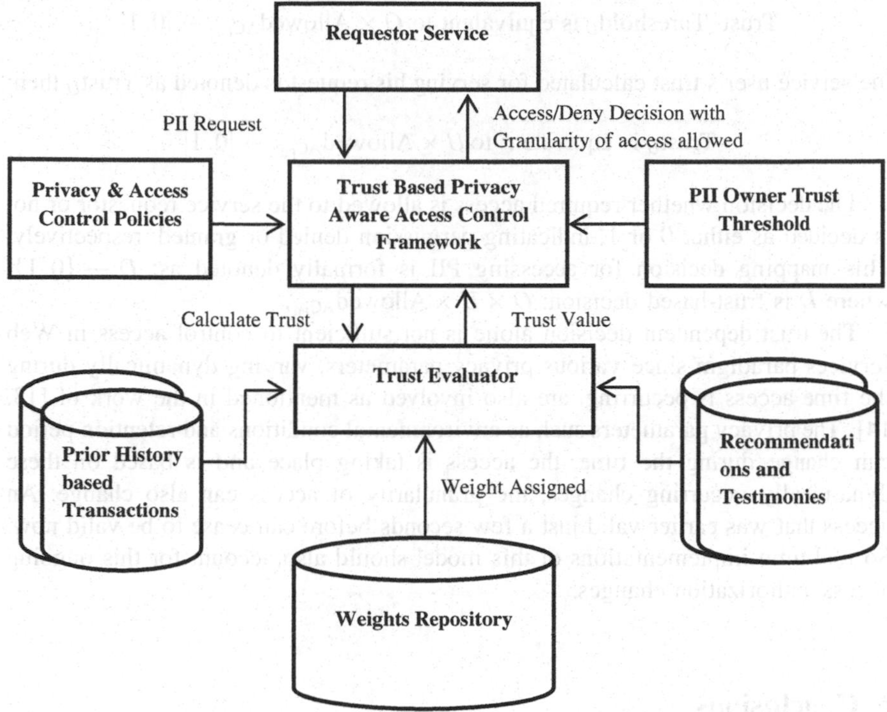

Fig. 4 Trust-based privacy aware access control framework

Here, theta stands for trust threshold value provided by the PII owner in order to allow access to his PII.

While calculating total trust value, some computational overheads are avoided in our method as there is no need to check for third party testimonies and recommendations if the service provider is fully satisfied with the requestor service in all its previous direct transactions and if those transactions have collective weighted value greater than or equal to the trust threshold provided by the PII owner. Otherwise, the total trust will be a function of both the direct and the indirect trust. Another important thing is that the requestor's trust is not always the same during the time the access is taking place, as he does not always behave in the same manner. The trust value of the requestor should be updated at regular intervals in order to reflect his current status.

The formal specification for incorporation of trust in the access control decision-making process is as follows:

Let O, U, PII, and AC are the set of service providers, service requestors, PII, and actions. Then, actions allowed on PII are designated by the subset: $\text{Allowed}_{\text{AC}_{\text{PII}}} \subseteq \text{AC} \times \text{PII}$. If service provider's trust threshold for performing an action on PII is denoted as Trust_Threshold_O, then:

$$\text{Trust\_Threshold}_O \text{ is equivalent to } O \times \text{Allowed}_{\text{AC}_{\text{PII}}} \rightarrow [0, 1]$$

the service user's trust calculated for serving his request is denoted as Trust_U then:

$$\text{Trust}_U \text{ is equivalent to } U \times \text{Allowed}_{\text{AC}_{\text{PII}}} \rightarrow [0, 1]$$

The decision whether required access is allowed to the service requestor or not is decided as either 0 or 1, indicating permission denied or granted, respectively. This mapping decision for accessing PII is formally denoted as: $D \rightarrow \{0, 1\}$, where D is trust-based decision: $O \times U \times \text{Allowed}_{\text{AC}_{\text{PII}}}$.

The trust-dependent decision alone is not sufficient to control access in Web services paradigm since various privacy parameters, varying dynamically during the time access is occurring, are also involved as mentioned in the work of [13, 14]. The privacy parameters such as environmental conditions and retention period can change during the time, the access is taking place and is based on these dynamically occurring changes, the granularity of access can also change. An access that was earlier valid just a few seconds before can cease to be valid now. So real-time implementations of this model should also account for this ongoing access authorization changes.

6 Conclusions

Challenging access control issues arise in heterogeneous services collaboration. Conventional technical solutions focus mainly on access control in a single organization or a single system, which are not suitable for dealing with the access control issues in complex unknown Web services collaboration scenarios. The PA^{2}CM model specifically addresses the need to integrate privacy policy enforcement in the legacy access control process of Web services, resulting into a single-integrated access control model. This novel model modifies the way access control technology is currently deployed in Web services paradigm so that information owners retain control of their sensitive PII and privacy policies even after sharing it online.

References

1. Barker, K., Askari, M., Banerjee, M., Ghazinour, K., Mackas, B., Majedi, M., Pun, S., Williams, A.: BNCOD, pp. 42–54 (2009)
2. Lampson, B.W.: Dynamic protection structures. In: Proceedings of American Federation of Information Processing Societies conference, Las Vegas, pp. 27–38. Nevada, USA (1969)
3. D.T.C.S.E.C. (TCSEC), DoD 5200.28-STD Foundations, MITRE Technical Report 2547 (1973)

4. Bell, D.E., LaPadula, L.J.: Secure computer systems: mathematical foundations. The MITRE Corp, vol. 1–111, Bedford, Mass (1973)
5. Louwerse, K.: The electronic patient record; the management of access—case study: Leiden University Hospital. Int. J. Med. Inf. **49**(1), 39–44 (1998)
6. World Wide Web consortium (W3C), platform for privacy preferences (P3P). Available at: www.w3.org/P3P
7. APPEL, A P3P preference exchange language 1.0 (APPEL1.0) (Working Draft), World Wide Web consortium (W3C), April 2002. Available at: http://www.w3.org/TR/P3P-preferences/
8. IBM, the enterprise privacy authorization language (EPAL),EPAL 1.1 specification, 2004. Available at: http://www.zurich.ibm.com/security/enterprise-privacy/epal/
9. Casassa Mont, M., Thyne, R., Chan, K., Bramhall, P.: Available at: http://www.hpl.hp.com/techreports/2005/HPL-005-110.pdf (2005)
10. Byun, J. W., Bertino, E., Li, N.: Purpose based access control of complex data forprivacy In: Proceedings of SACMAT'05, pp. 102–110. ACM Press, New York (2005)
11. Byun, J.W., Bertino, E., Li, N.: Purpose based access control for privacy protection in relational database systems. Technical Report 2004-52, Purdue University (2004)
12. Byun, J.W., Bertino, E.: Micro-views, or on how to protect privacy while enhancing data usability: concepts and challenges. SIGMOD Rec. **35**(1), 9–13 (2006)
13. Li, M., Wang, H., Ross, D.: Trust-based access control for privacy protection in collaborative environment. In: The 2009 IEEE International Conference on e-Business Engineering, pp. 425—430. Macau, China (2009)
14. Li, M., Wang, H.: Protecting information sharing in distributed collaborative environment. In: 10th Asia-Pacific Web Conference Workshop, pp. 192–200. Shenyang, China (2008)
15. Bhatia, R., Singh, M.: Trust based privacy preserving access control in web services paradigm. In: the Second IEEE International Conference on Advanced Computing, Networking and Security, ADCONS, pp. 243—246 (2013)
16. Wang, Y., Vassileva, J.: Trust and reputation model in collaborative networks. In: Proceedings of 3rd IEEE International Conference Collaborative Computing, pp. 150–157 (2003)

A Novel Low-Power Domino Logic Technique Providing Static Output in Evaluation Phase for High Frequency Changing Inputs

Sumit Sharma and Kamal Kant Kashyap

Abstract Domino logic is widely used for high switching speed and high-performance circuits. Domino logic consists of an inverter used between two stages of dynamic logic. Robustness of domino logic degrades with scaling down as leakage power increases. This paper presents a new proposed domino logic circuit with improved speed. Present work proposed domino logic scheme which gives static output also in evaluation phase with high frequency inputs. Conventional domino styles do not provide static output with changing inputs during evaluation phase. This proposed circuit is designed by making use of current mirror circuit and modified keeper circuitry. The proposed circuit has low power-delay product as compared to conventional domino logic styles. All the circuits have been simulated in cadence virtuoso 180 nm technology. According to simulation results obtained, the circuit shows more than 50 % better performance as compared to conventional domino styles.

Keywords Domino logic · Dynamic gates · Evaluation phase · Pre-charge phase · Static gates · Robustness

1 Introduction

CMOS gates are basically designed using static logic and dynamic logic. Static logic gates comprises of both pull-up network and pull-down network. This leads to number of transistors required for N-input gate to be equal to 2N. Also, short circuit

---

S. Sharma (✉)
School of VLSI Design and Embedded Systems, NIT Kurukshetra,
Kurukshetra, Haryana 136119, India
e-mail: sumit1207sharma@gmail.com

K.K. Kashyap
ECE Department, NIT Kurukshetra, Kurukshetra, Haryana 136119, India
e-mail: kamalkant7185@gmail.com

L.C. Jain et al. (eds.), *Intelligent Computing, Communication and Devices*,
Advances in Intelligent Systems and Computing 308,
DOI 10.1007/978-81-322-2012-1_47

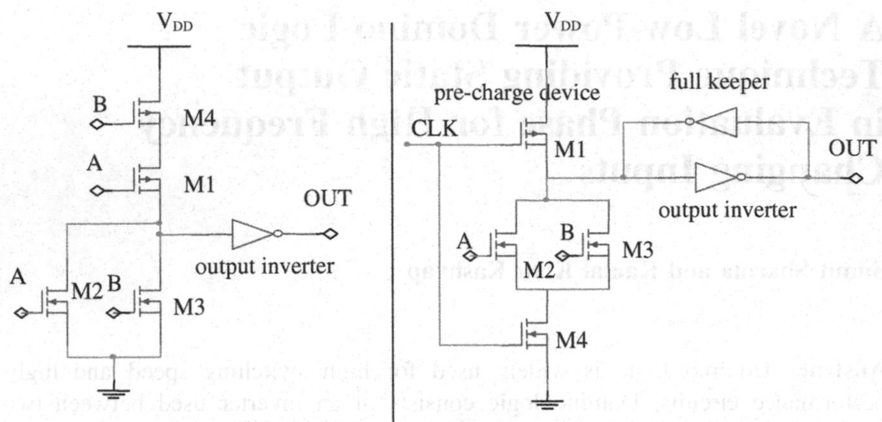

Fig. 1 Implementation of OR gate using (i) Static CMOS logic (ii) Domino logic

power consumption is increased especially for large fan-in gates. Pseudo-NMOS logic overcomes drawback of more area requirement of static CMOS as it comprises of a grounded PMOS transistor in PUN and PDN performs the evaluation function. The numbers of transistors required for N-input gate reduces to $N + 1$. But this leads to increase in static power consumption. By considering the advantages of low-power consumption of static CMOS and less area requirement for Pseudo-NMOS logic, dynamic logic circuits designed by introducing clock in the circuit.

Dynamic logic needs $N + 2$ numbers of transistors for a gate of fan-in N and operates in two phases: pre-charge phase and evaluation phase [1]. Dynamic logic basically suffers from the problems of charge sharing, charge leakage, and capacitive coupling, etc. [2]. Also problem arises when cascading two dynamic stages. To overcome this problem in cascading of dynamic gates, Domino logic is introduced.

Domino logic is basically an inverter placed between the two cascading stages. Domino logic is used in digital signal processing and high-speed applications such as data-path in microprocessor [3].

Upsizing the keeper circuit improves robustness [4]. Among conventional domino techniques, high-speed domino logic and conditional keeper domino logic [5] provide the most robust output. They use of delayed version of clock to increase robustness.

All the conventional domino logic techniques do not allow the charging of dynamic node again in evaluation phase if it is discharged once until next pre-charge phase comes. Thus, circuit does not give correct output if input changes during evaluation phase. In this paper, we propose a domino technique which gives static output in evaluation phase for high frequency changing inputs during clock period. This is done by modifying the keeper circuit, using footer transistor for preventing charge leakage during pre-charge phase and by using current mirror circuit. The proposed circuit offers less power dissipation and better power-delay product.

2 Previous Work

The dynamic node is in floating state in evaluation phase, which may lead to distorted output because of noise. The typical domino logic style includes footless domino logic [6], footer domino logic [7], high-speed domino logic, conditional keeper logic, etc. Figure 1 shows the circuit for OR gate in static CMOS logic as well as in domino logic. Figure 2 shows the (a) footless domino logic style (b) footer domino logic style for high fan-in OR gate. The footer domino dissipates less power as compared to footless domino style because of stacking effect.

In case of footer domino logic style, a footer transistor is used as shown in Fig. 2b and clock is applied to its gate terminal. When clock goes high in evaluation phase then footer transistor turns ON and provides a discharge path according to logic function performed by the evaluation network, otherwise turns OFF in pre-charge phase so reduces the leakage [8] hence the power dissipation. In case of footless domino style because of no footer transistor, there is more power dissipation. If high fan-in gate are used, leakage is more because of increase in number of transistors.

So, for high-speed applications, weak keeper circuit is required and for highly robust circuits, strong keeper circuit is used.

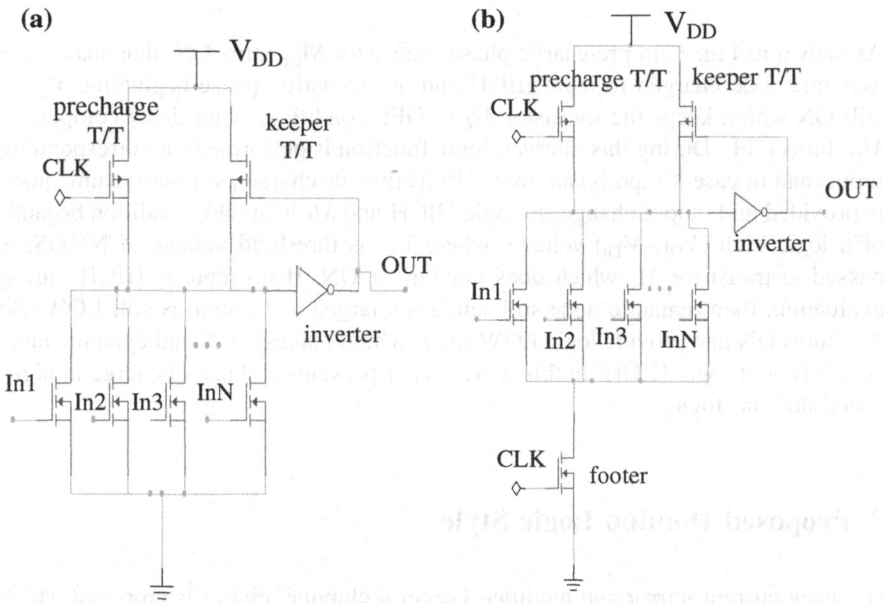

Fig. 2 High fan-in OR gate implementation using **a** footless domino logic **b** footer domino logic

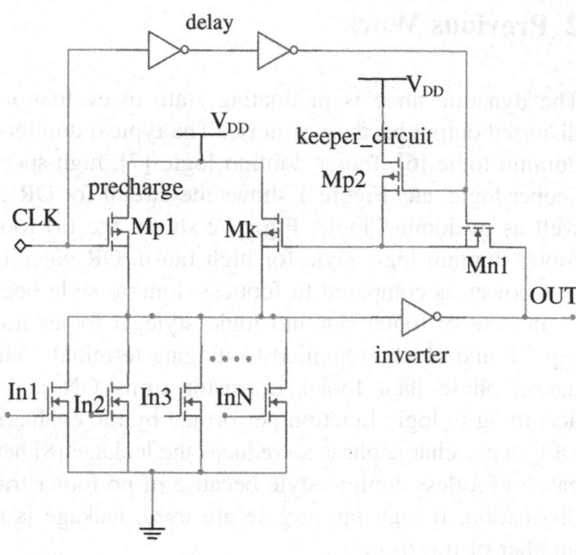

Fig. 3 High-speed domino logic [5]

2.1 High-Speed Domino Logic

As shown in Fig. 3, in pre-charge phase, transistor M_{p1} turns ON, thus makes the dynamic node charged to logic HIGH and in evaluation phase beginning, M_{p2} is still ON which keeps the transistor M_k to OFF condition. After delay completion M_{p2} turns OFF. During this interval, logic function is performed for corresponding inputs and in case if input(s) are logic HIGH then discharge path for dynamic node is provided and output changes to logic HIGH and M_k is in OFF condition because of a logic high (V_{DD}–V_{Tn}) voltage, where V_{Tn} is threshold voltage of NMOS, is passed to transistor M_k which does not turn it ON. If no input is HIGH during evaluation, then dynamic node still remains charged and output is still LOW. So M_{n1} turns ON and provide logic LOW to M_k which makes it ON and dynamic node is still kept at logic HIGH. In this way, keeper prevents leakage of charge in high-speed domino logic.

3 Proposed Domino Logic Style

By using current mirror and modified keeper technique, circuit is proposed which gives static output in evaluation phase for high frequency changing inputs as shown in Fig. 4. Current mirror [9] works on the principle that if the same gate-source voltage applied to two similar transistors then the current flowing through the drains of those two transistors will be equal. The use of current mirror circuit

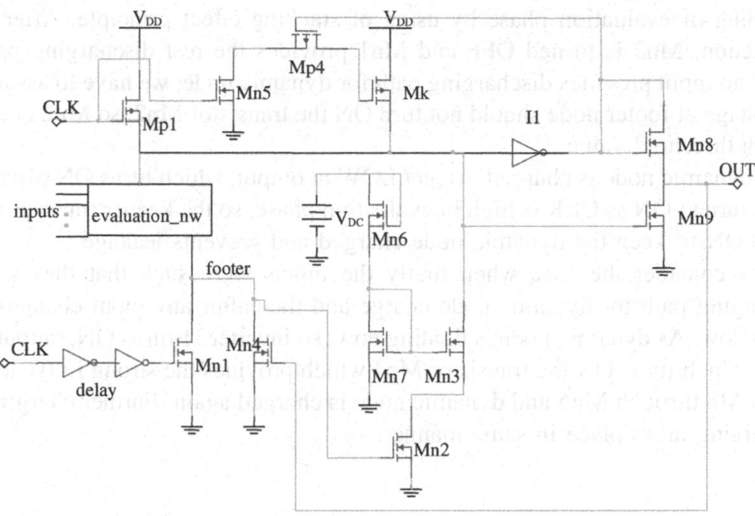

Fig. 4 Proposed domino logic style

improves performance by making the circuit less sensitive to the variations in power supply.

The keeper circuit is modified by using transistors Mn8 and Mn9. The inverter I1 used at dynamic node is of low threshold NMOS such that it turns ON even with a small voltage applied to it. Two inverters introduce delay in CLK to make the footer Mn1 remains OFF in beginning of evaluation phase. During this period, evaluation network performs its operation. Transistor Mn4 and Mn2 are of high threshold voltage. Mn4 turns ON when output is strong enough. It prevents further discharge through Mn2. The leakage reduction is done by making use of transistors Mn6, Mn7, and Mn3 in form of current mirror. Transistors Mn8 and Mn9 are also used for improving the circuit by leakage reduction [10] also lead to reduction of delay and power dissipation. In pre-charge phase, when clock is low, pre-charge transistor Mp1 become ON, dynamic node is charged. Transistor Mn5 is in OFF state. When the pre-charge phase begins, footer transistor Mn1 is ON because of delay applied to clock and it may lead to discharge dynamic node but as transistor Mp1 is ON, dynamic node still remains HIGH.

When evaluation phase begins, CLK goes HIGH. In the beginning of evaluation phase, the footer transistor Mn1 is OFF because of delay introduced in clock. In this time slot, evaluation network performs its logic function.

If any input combination tries to pull the dynamic node down, the high threshold transistor Mn2 turns ON and the dynamic node discharges through the transistors Mn3 and Mn2. This whole circuitry used prevents leakage in the beginning of evaluation phase by using of stacking effect principle. After delay completion, Mn2 is turned OFF and Mn1 provides the rest discharging path. In case if no input provides discharging path for dynamic node, we have to assure that the voltage at footer node should not turn ON the transistor Mn2, so Mn2 is chosen of high threshold value.

As dynamic node is charged, we get LOW at output, which turns ON Mp4. Mn5 is also turned ON as CLK is high in evaluation phase, so the keeper transistor M_k is turned ON to keep the dynamic node charged and prevents leakage.

Now consider the case when firstly the inputs were such that they provide discharging path for dynamic node charge and thereafter any input changes from high to low. As dynamic node is floating now, so inverter I1 turns ON, output turns to low which turns ON the transistor Mp4 which provides the strong LOW logic to keeper Mk through Mn5 and dynamic node is charged again. Further charging and discharging takes place in same manner.

4 Simulation Results

We simulated different topologies using cadence virtuoso 180-nm technology model. And 32 fan-in OR gates are taken as evaluation network for all topologies. The simulation results are tabulated here.

To compare different topologies, we use the parameters power consumption, delay, and power-delay product (PDP). It can be seen that proposed design shows significant improvement in performance. Table 1 shows the different values of power, delay, and PDP at $V_{DD} = 1.2$ V and $V_{DC} = 0.8$ V.

The outputs of footer domino style, footless domino style, high-speed domino style, and proposed domino circuit are shown further (Figs. 5, 6, 7, and 8).

Table 1 Comparative performance for different standard and proposed topologies

Parameters	Topologies				
	Footless	Footer	HSDL	CKL	PROPOSED
Power ($\times 10^{-5}$ W) dissipation	6.2	4.5	4.2	4.6	2.3
Delay (ns)	1.02	2.06	1.01	1.02	.51
PDP ($\times 10^{-14}$ J)	6.32	9.27	4.242	4.692	1.173

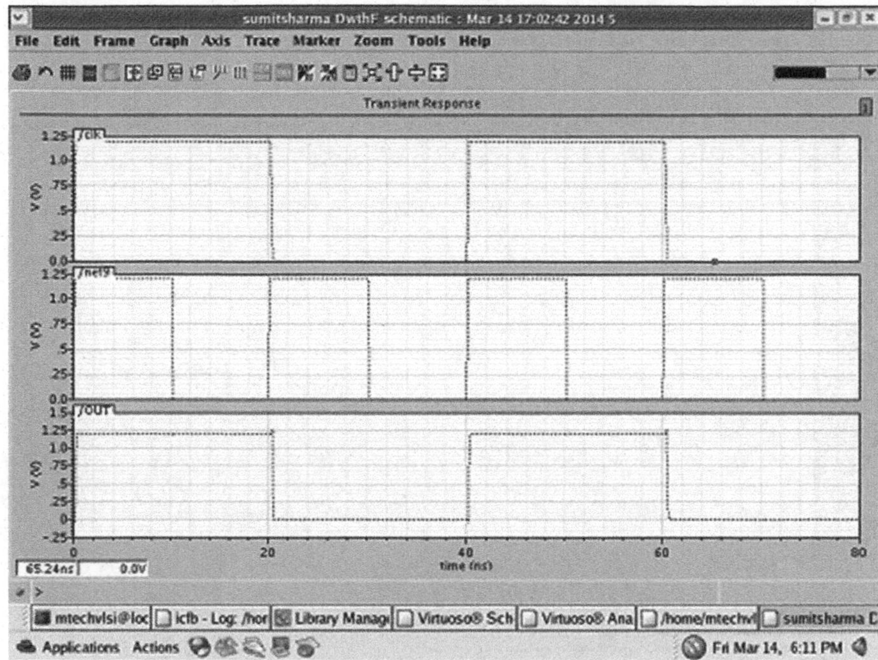

Fig. 5 Output wave form of footer domino style

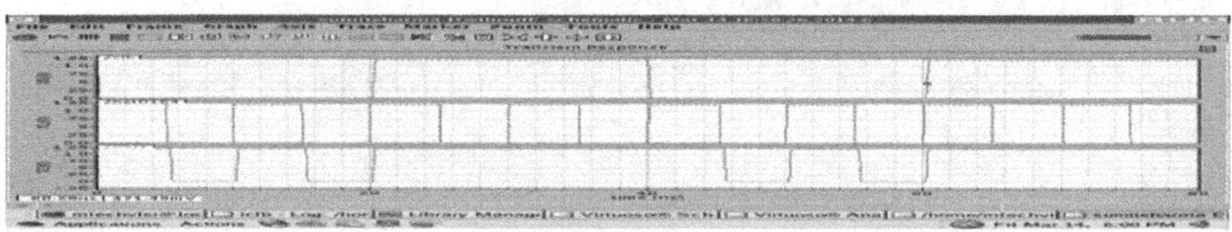

Fig. 6 Output wave form of footless domino style

These outputs are taken by using 32 fan-in OR gate as evaluation network. The proposed circuit has more area as number of transistors is increased. Also we applied high frequency input to one transistor and applied LOW to remaining inputs.

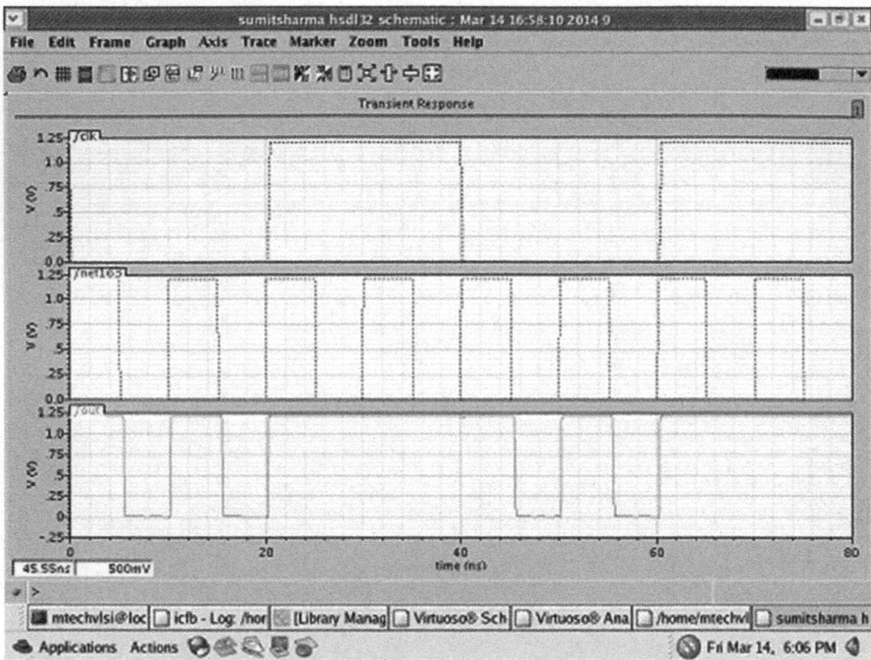

Fig. 7 Output wave form of high-speed domino logic

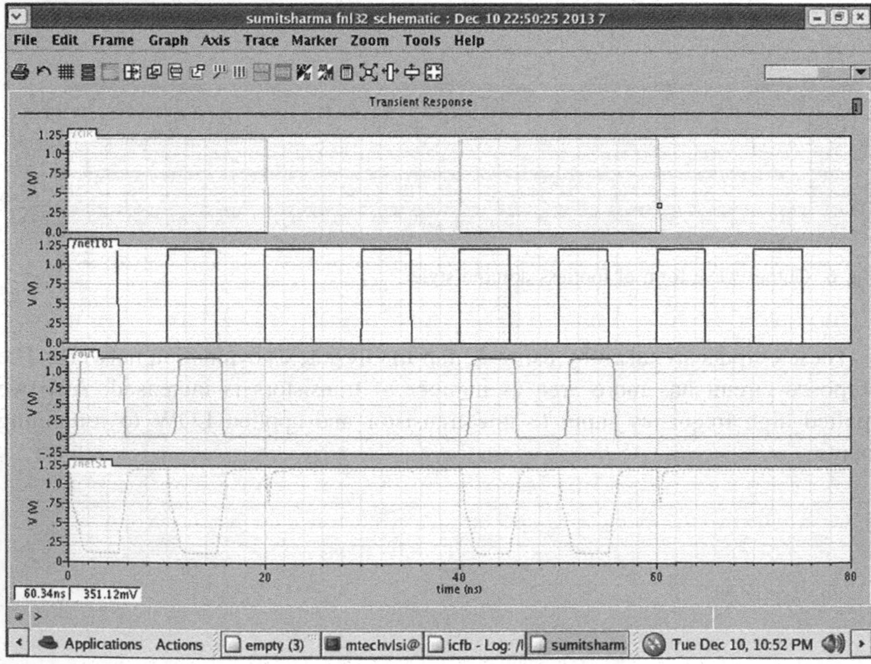

Fig. 8 Output wave form of proposed domino logic style

5 Conclusions

In this paper, we proposed new domino logic style that gives static output in evaluation phase for high frequency inputs. Proposed domino style also shows better performance compared to other conventional domino logic styles. Also, power dissipation, delay, and Power-delay product (PDP) is improved.

References

1. Rabaey, M., Chandrakasan, A., Nikolic, B.: Digital Integrated Circuits, 2nd edn, pp. 269–274. Prentice Hall, Englewood Cliffs (2002)
2. Neil, H.E., David, H.: Principle of CMOS VLSI design: a system perpective, 3rd edn. Addison-Wesley, Reading (2004)
3. Sung, R.J.-H., Elliott, D.G.: Clock-logic domino circuits for high-speed and energy-efficient microprocessor pipelines. IEEE Trans. Circuits Syst. II Express Briefs **54**(5), 460 (2007)
4. Das, K.K. et al.: Low-leakage integrated circuits and dynamic logic circuits. U.S. Patent 6933744 (2005)
5. Moradi, F., Vu Cao, T., Vatajelu, E., Peiravi, A., Mahmoodi, H., Wisland, D.: Domino logic designs for high-performance and leakage-tolerant applications. Int. VLSI J. **46**, 247–254 (2013)
6. Anis, M.H., Allam, M.W., Elmasry, M.I.: Energy-efficient noise-tolerant dynamic styles for scaled-down CMOS and MTCMOS technologies. IEEE Trans. Very Large Scale Integr. (VLSI) Syst. **10**, 71–78 (2002)
7. Oklobdzija, V.G., Montoye, R.K.: Design-performance tradeoffs in CMOS-domino logic. IEEE J. Solid State Circuits **21**(2), 304–306 (1986)
8. Ding, L., Mazumder, P.: On circuit techniques to improve noise immunity of CMOS dynamic logic. IEEE Trans. Circ. Syst. **12**, 910–925 (2004)
9. Gray, P.R., Hurst, P.J., Lewis, S.H., Meyer, R.G.: Analysis and Design of Analog Integrated Circuits. Wiley, New York (1984)
10. Kursun V, Friedman EG (2002) Low swing dual threshold voltage domino logic. In: Proceedings ACM/SIGDA Lakes Symposium on VLSI, pp. 47–52

The Impact of Gate Underlap on Analog and RF Performance of Hetero-Junction FET

Rohit Jana and Angsuman Sarkar

Abstract Due to enhanced carrier mobility, InP/InGaAs heterostructure double gate MOSFET evinced himself as an attractive candidate for applications in high performance digital logic circuits. In this paper, our aim was to analyze the impact of gate underlap on analog and RF performance of InP/InGaAs hetero-junction FET using TCAD device simulator. The analog and RF parameters of HFET such as drain resistance (R_o), transconductance (g_m), and unity-gain cutoff frequency (f_T) are studied for varying underlap length raging from 2 to 9 nm. It is shown that the analog and RF performance of hetero-junction FET is severely affected by amount of underlap and this effect can be moderated by an optimal underlap, which yields a trade-off between the analog and RF performance.

Keywords Hetero-junction FET · Gate underlap · Analog/RF performance · Unity-gain cutoff frequency · Transconductance · Drain resistance

1 Introduction

Gradually, as MOSFETs are scaled into sub 20-nm regime, the device performance starts to degrade due to the significantly increased short channel effects (SCEs) [1]. The new III–V heterostructure underlap DGMOS device provides higher ON current, lesser delay, lower energy-delay products, and lower DIBL than the silicon-based devices [2]. Now, the carrier injection velocity in turn depends on the effective mass (m^*) and low-field carrier mobility [3]. However, due to low density of states (DOS) in Γ-valley, they have small inversion charge (Q_{inv}) and consequent

---

R. Jana (✉) · A. Sarkar
ECE Department, Kalyani Government Engineering College, Kalyani, India
e-mail: rohitece08@gmail.com

A. Sarkar
e-mail: angsumansarkar@kgec.ac.in

L.C. Jain et al. (eds.), *Intelligent Computing, Communication and Devices*,
Advances in Intelligent Systems and Computing 308,
DOI 10.1007/978-81-322-2012-1_48

reduction in drain current (I_d) [4]. Due to higher value of mobility and carrier injection velocities III–V materials provide faster switching than silicon [2]. Also, materials having high permittivity and high mobility are generally responsible for higher leakage current and SCEs [5]. The quasi-planar structure of underlap devices has become more feasible [6] and this gate underlap in nanoscale MOSFETs can substantially reduce SCEs by reducing off state current and minimizing the gate delay [7]. The underlap devices also decreases the parasitic capacitances considerably, which, in turn, possess higher speed and lower power dissipation. Coupling between source and drain is reduced significantly, especially for a shorter channel device, eventually causes off state leakage to decrease. On the other side, channel resistance increases with underlap, thus degrading the on performance [8].

2 Device Description

The structure of III–V heterostructure (InGaAs/InP) underlap DGMOSFET device is shown in Fig. 1 having gate length (L_g) 18 nm with an lightly doped (10^{15} cm^{-3}) ultra-thin body (t_{bd}) of 6 nm. The device body consists of narrow-band $In_{0.53}Ga_{0.47}As$ layer(t_{ch}) of 4 nm and two wide-band InP barrier layers (t_b) of 1 nm each, i.e., $t_{bd} = t_{ch} + 2t_b$. The source and drain lengths are kept fixed at 5 nm, the top and bottom gate oxide thickness (t_{ox}) is 1.2 nm. The simulated device structure has source/drain regions with high doping concentration of 10^{20} cm^{-3}. In this new device, structure traditional dielectric silicon dioxide (SiO2) is replaced by high-K hafnium dioxide (HfO_2) to minimize leakages. The underlap length is altered from 1 to 9 nm for better device performance investigation. Physical properties of both III–V semiconductors (narrow-band $In_{0.53}Ga_{0.47}As$ and wide-band InP) are listed in Table 1.

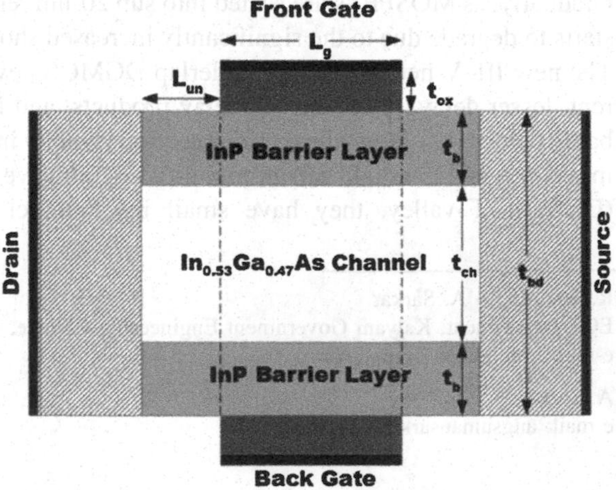

Fig. 1 Schematic cross-sectional diagram of III–V heterostructure underlap DG MOSFET

Table 1 Physical properties of $In_{0.53}Ga_{0.47}As$ and InP

Semi-conductor	E_g(eV)	CBO (eV)	VBO (eV)	μ_e (cm^2/V s)	μ_h (cm^2/V s)	ε_0	Lattice constant (Å)
$In_{0.53}Ga_{0.47}As$	0.74	0.22	0.38	12,000	300	13.9	5.868
InP	1.344	–	–	12.5	5.867	5,400	200

3 Device Simulation Result

A Two-dimensional numerical device simulator SILVACO ATLAS is used for the heterostructure MOSFET 2-D device simulations. Shockley-Read-Hall (SRH), Auger method along with Newton numerical technique has been used to obtain solutions of a set of coupled equations—such as current equations for electrons and holes, and continuity equations for electrons and holes which is employed in ATLAS. Figure 2 shows that transfer characteristics ($I_d - V_{GS}$) of III–V heterostructure (InGaAs/InP) underlap DGMOSFET device with constant $L_{un} = 5$ nm and $L_g = 18$ nm, for two different values of the drain bias. The simulated $I_d - V_{GS}$ device characteristics are compared with simulated results as reported in [2], for the purpose of simulator calibration in order to verify the correctness of the simulation models chosen.

4 Analog Performance

The variation of analog performance parameters such as Transconductance (g_m) and the output resistance (R_o) for varying amount of underlap are analyzed in this section.

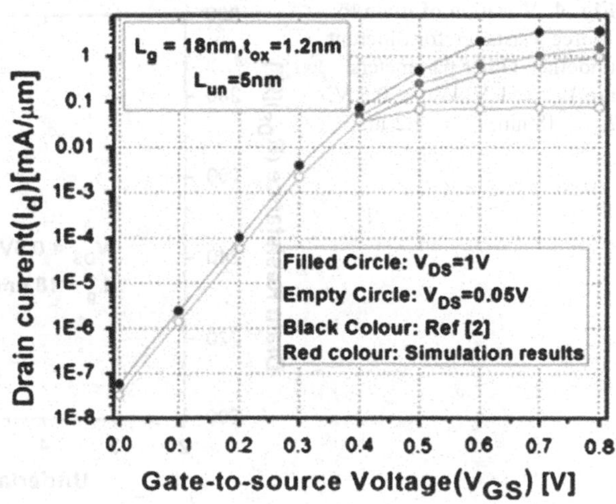

Fig. 2 Variation of drain current with gate-to-source voltage for two different drain-to-source voltages. Other parameters are $L_{un} = 5$ nm, $L_g = 18$ nm, $t_{ox} = 1.2$ nm

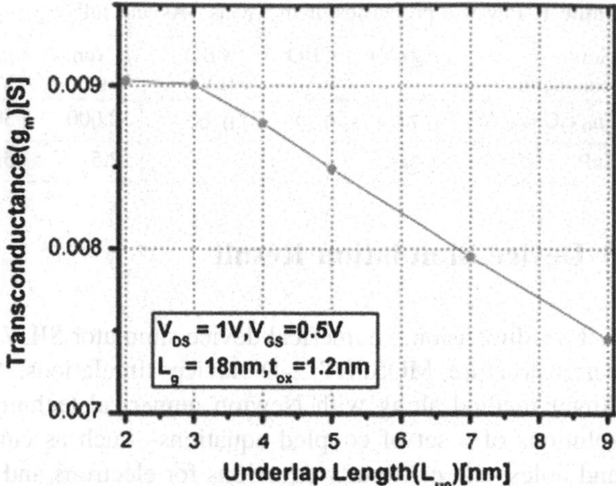

Fig. 3 Variation of Transconductance (g_m) with underlap, Other parameters are $V_{DS} = 1$ V, $V_{GS} = 0.5$ V, $L_g = 18$ nm, $t_{ox} = 1.2$ nm

Figure 3 shows the variation of the g_m with respect to amount of underlap for an applied drain voltage equals to 1 V. It is evident from Fig. 3 that as underlap increases; the transconductance reduces, leading to a decrease in the ON current I_{ON} as well as the OFF current I_{OFF}. This reduction in the drain current especially at the higher values of underlap can be attributed to the reduction of the effective channel length. Thus, beyond a certain amount of underlap (>3 nm), the transconductance is inversely proportional to the amount of underlap and decreases linearly with an increase in the amount of underlap. Figure 4 reveals that the output drain resistance (R_o) become saturated and attains a maximum value of 295 Ω above 7 nm underlap. Therefore, it is obvious that the intrinsic gain defined by the product of transconductance g_m and drain resistance R_o achieves its pick values for higher underlap length as it is dominates over the change in transconductance with respect to amount of underlap.

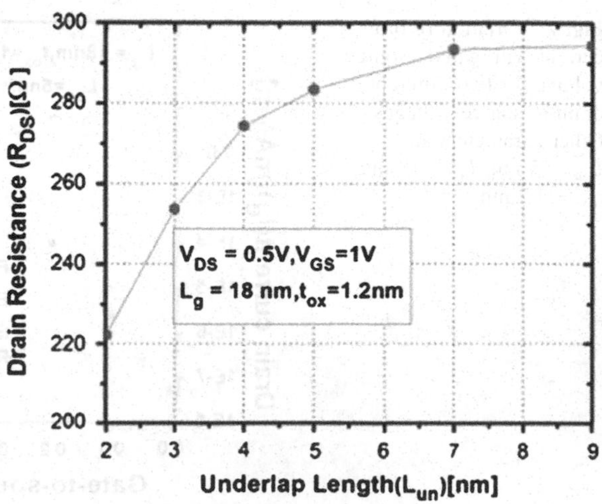

Fig. 4 Variation of drain-to-source resistance for different underlap. Other parameters are $V_{GS} = 1$ V, $V_{DS} = 0.5$ V, $L_g = 18$ nm, $t_{ox} = 1.2$ nm

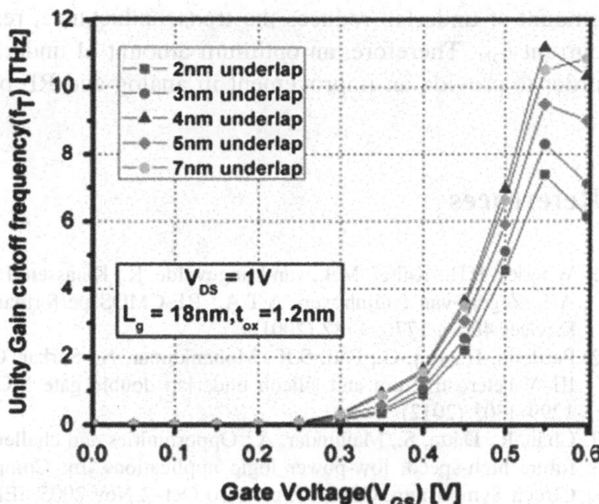

Fig. 5 Variation of unity-gain cutoff frequency for different gate voltage. Other parameters are $V_{DS} = 1$ V, $L_g = 18$ nm, $t_{ox} = 1.2$ nm

5 RF Analysis

In this section, we study on the RF performances of III–V heterostructure underlap DGMOSFET device. In high-speed digital applications (speed and high swing) f_T is an interesting parameter and is given by

$$f_T = \frac{g_m}{2\pi(c_{gs} + c_{gd})} \approx \frac{g_m}{c_{gg}} \tag{1}$$

where C_{gd}, C_{gs} and C_{gg} are the gate-to-drain, gate-to-source and total gate capacitances, respectively, and g_m is the transconductance.

Figure 5 plots the variation of unity-gain cutoff frequency (f_T) as a function of gate-to-source voltage (V_{GS}) for different amount of underlap. It is evident from the figure that the f_T will reach its peak value for an optimal amount of underlap equals to 4 nm. The III–V heterostructure underlap DGMOSFET device exhibits much higher values of f_T (in THz range) as compare to traditional MOSFET (in GHz), due to the enhanced value of g_m attributed by it higher mobility in such devices.

6 Conclusions

In this paper, the impact of gate underlap on the analog and RF performance of hetero-junction FET is investigated. We have shown that increasing the amount of underlap results in an increase in drain resistance (R_O). However, increasing the amount of underlap reduces the transconductance, resulting in a decrease in ON current I_{ON}. Therefore, an optimum amount of underlap can be quite effective in order to provide an improvement in analog and RF performance.

References

1. Woerlee, P.H., Knitel, M.J., van Langevelde, R., Klaassen, D.B.M., Tiemeijer, L.F., Scholten, A.J., Zegers-van Duijnhoven, A.T.A.: RF-CMOS performance trends. IEEE Trans Electron Devices **48**(8), 1776, 1782 (2001)
2. Pardeshi, H., Raj, G., Pati, S.K., Mohankumar, N., Sarkar, C.K.: Comparative assessment of III–V heterostructure and silicon underlap double gate MOSFETs. Semiconductors **46**(10), 1299–1303 (2012)
3. Chau, R., Datta, S., Majumdar, A.: Opportunities and challenges of III–V nanoelectronics for future high-speed, low-power logic applications. In: Compound Semiconductor Integrated Circuit Symposium, CSIC'05, p. 4, 30 Oct–2 Nov 2005. IEEE
4. Oktyabrsky, S., Ye, P.: Fundamentals of III–V Semiconductor MOSFETs. Springer, Berlin (2010)
5. Nainani, A., Yuan, Z., Krishnamohan, T., Saraswat, K.: Optimal design of III–V heterostructure MOSFETs. In: 2010 International Conference on Simulation of Semiconductor Processes and Devices (SISPAD), pp. 103, 106, 6–8 Sept 2010
6. Bansal, A., Paul, B.C., Roy, K.: Modeling and optimization of fringe capacitance of nanoscale DGMOS devices. IEEE Trans Electron Devices **52**(2), 256, 262 (2005)
7. Bansal, A., Roy, K.: Analytical subthreshold potential distribution model for gate underlap double-gate MOS transistors. IEEE Trans. Electron Devices **54**(7), 1793, 1798 (2007)
8. Trivedi, V., Fossum, J.G., Chowdhury, M.M.: NanoscaleFinFETs with gate-source/drain underlap. IEEE Trans Electron Devices **52**(1), 56, 62 (2005)

Low-Power, High-Speed, Indirect Frequency-Compensated OPAMP with Class AB Output Stage in 180-nm CMOS Process Technology

Subhrajyoti Das, Sushanta K. Mandal, Adyasha Rath and Sweta Padma Dash

Abstract In this paper, the design of low-power, high-speed, two-stage, indirect frequency-compensated operational amplifier is presented. The OPAMP employs split-length devices and class AB output stage. Split-length technique is employed both in load device as well as in differential-pair device. The split-length device enhances the phase margin (PM) and unity gain bandwidth (UGB) while maintaining lower supply voltage. The class AB output stage provides a faster settling time and reduced power dissipation. Simulations of the proposed circuits were carried out in cadence specter on 180-nm process technology at a supply voltage of 1.6 V. The proposed split-length current mirror load OPAMP exhibits power dissipation of 82 μW, UGB of 43.26 MHz, and PM of 79.25°. Similarly, the proposed split-length differential-pair OPAMP exhibits a power dissipation of 78 μW, UGB of 49.71 MHz, and PM of 85.34°.

Keywords OPAMP · Indirect frequency compensation · Split-length transistor · Class AB stage

1 Introduction

The OPAMP is generally considered as the basic building block of a myriad of analog circuits. The stability criterion and frequency compensation methods enhance the versatility of the OPAMP. The increasing demand of low power, low

S. Das (✉) · S.K. Mandal · A. Rath · S.P. Dash
School of Electronics Engineering, KIIT University, Bhubaneswar 751024, India
e-mail: subhrajyoti@live.in

S.K. Mandal
e-mail: sushantakumar@yahoo.com

A. Rath
e-mail: adyasharath.1690@gmail.com

S.P. Dash
e-mail: swetapadmadsh@gmail.com

L.C. Jain et al. (eds.), *Intelligent Computing, Communication and Devices*,
Advances in Intelligent Systems and Computing 308,
DOI 10.1007/978-81-322-2012-1_49

area, and the need to support a wide range of load capacitances motivates to design OPAMP using the frequency compensation technique.

In order to achieve stability in two-stage OPAMP, Miller compensation technique is used [1–3]. The Miller compensation technique is called as direct compensation technique in which a miller capacitor is used. This technique employs capacitance multiplication method that causes pole splitting, and dominant pole is compensated. But, there is a right-half plane (RHP) zero introduced in s-plane due to the compensating capacitor (C_c). The phase margin (PM) of OPAMP is hampered due to this RHP zero, which is at $z_1 = g_{m2}/C_c$ in s-plane. The PM can be improved by increasing the value of C_c but at the cost of reduced unity gain bandwidth (UGB), which is given by, $f_{un} = g_{m1}/2\pi C_c$ [2]. RHP zero can be removed by blocking the feed forward current. To block the feed forward current, either a zero null resistor is added in series with the C_c in feedback path or a voltage buffer is inserted in between the first and second stages of OPAMP [2–5]. Another method employs indirect frequency compensation in which compensating current feedback is indirectly given to the output of differential amplifier through a common gate stage [4, 6] to remove RHP zero. In this work, indirect frequency compensation is achieved with split-length devices while operating at low supply voltage [2, 7]. To improve the transient response and slew rate of OPAMP, a class AB output stage OPAMP is designed [8].

2 Indirect Compensation and Split-Length Composite Transistor

In indirect frequency compensation, the compensation capacitor (C_c) is connected in between the output node and an internal low-impedance node of the first stage. Through this C_c, the compensation current is allowed to feedback indirectly from output node to internal high impedance node of first stage [2].

Cascaded OPAMPs are normally used to achieve high open-loop gain with two gain stages. When first stage of a differential amplifier is cascaded, then a low-impedance node is easily realized. This low-impedance node is used for indirect compensation of OPAMP where C_c is connected to that node [1, 2]. Due to shrinking CMOS technology, the urge to scale down supply voltage arises. Thus, for nano-CMOS, process cascading is not suitable. In order to get rid of this problem, split-length composite transistor is used in the OPAMP. To create low-impedance node, splitting can be done on an NMOS or a PMOS transistor. For NMOS, lower transistor operates in triode region, but not in saturation region. Similarly, for PMOS, upper transistor operates in triode region, but not in saturation region as shown in Fig. 1 [2, 7].

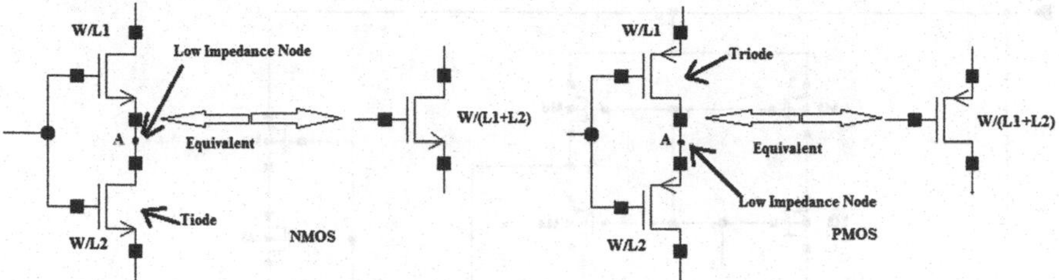

Fig. 1 Split-length NMOS and PMOS transistors [2]

3 Class AB Output Stage

The class AB output stage is used to obtain better transient response and higher slew rate as well as also helps in maintaining low power consumption. Class AB stage has the capability to generate output currents I_{out}, which is same as the biasing current [9–11]. Generally, class AB output stage is implemented by using a floating battery with value V_{battery} connected between the gate terminals of output transistor and operating as a common source amplifier [9] (Fig. 2).

4 Proposed Circuit

In this section, the two types of OPAMP are designed separately:

1. Spilt-length current mirror load (SLCM) OPAMP with class AB output stage
2. Spilt-length differential-pair (SLDP) OPAMP with class AB output stage.

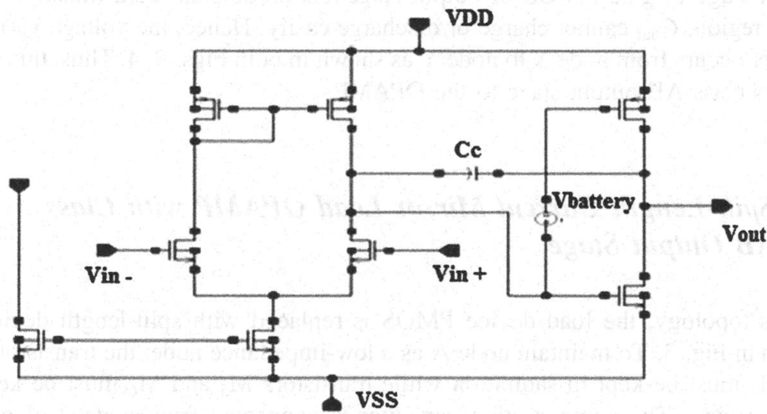

Fig. 2 Class AB output stage with floating battery

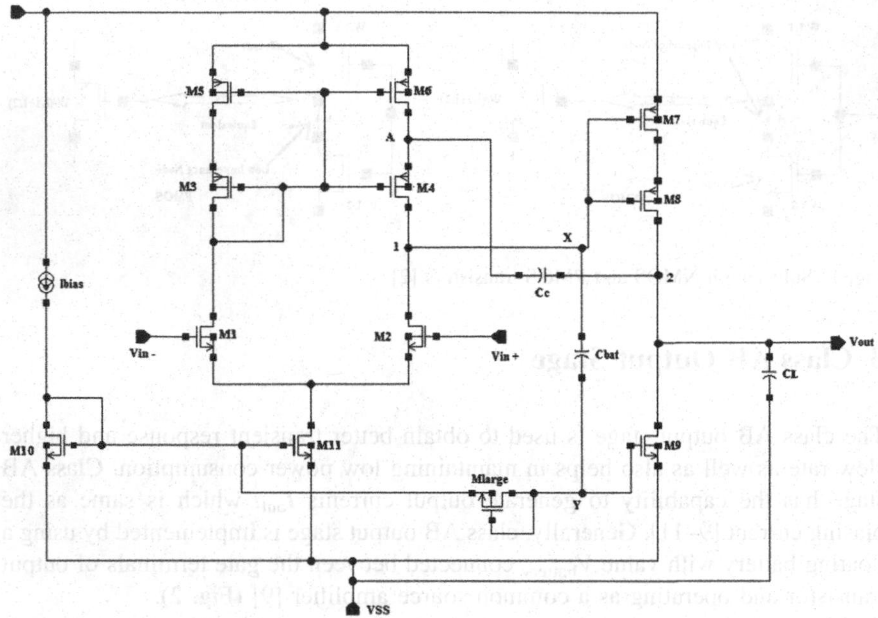

Fig. 3 Split-length load devices two-stage OPAMP with class AB stage

To create low-impedance node, either the length of differential-pair transistor (i.e., NMOS) or load transistor (i.e., PMOS) is splitted. The compensation capacitor is connected to this low-impedance node for indirect frequency compensation. For class AB output stage, a diode-connected NMOS and a capacitor is used instead of floating battery. A diode-connected NMOS transistor M_{large} is used as a resistive element (Figs. 3, 4), which is always in cutoff region. For floating battery, a capacitor (C_{bat}) with small capacitance value is connected to the output of first stage to gate NMOS of output stage. As diode-connected transistor is in cutoff region, C_{bat} cannot charge or discharge easily. Hence, the voltage variation transfer occurs from node X to node Y as shown in both Figs. 3, 4. Thus, this setup renders class AB output stage to the OPAMP.

4.1 Spilt-Length Current Mirror Load OPAMP with Class AB Output Stage

In this topology, the load device PMOS is replaced with split-length device as shown in Fig. 3. To maintain node A as a low-impedance node, the transistors M_3 and M_4 must be kept in saturation while transistors M_5 and M_6 must be kept in triode region. The compensation capacitor is connected from node 2 of output stage to node A for indirect frequency compensation. Here, we get UGB

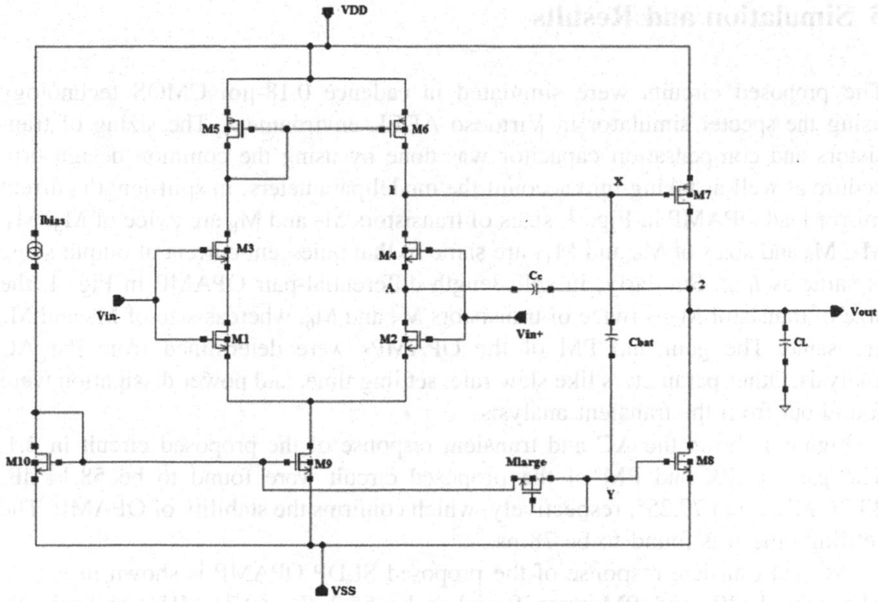

Fig. 4 Split-length differential-pair (*SLDP*) two-stage OPAMP with class AB stage

$f_{un} = g_{m1}/(2\pi(2C_c))$. The quiescent current in M_9 is equal to the current in M_{11} because both the transistors have same gate-to-source voltage and both are equally sized. The voltage variation between node X and node Y is very large at the slewing operation of output stage, while the OPAMP is in dynamic mode. This is because of slower charging and discharging of C_{bat}. By this arrangement, the output stage exhibits class AB or push–pull operation.

4.2 Spilt-Length Differential-Pair OPAMP with Class AB Output Stage

In this topology, the differential-pair device NMOS is replaced with split-length device as shown in Fig. 4. To make node A a low-impedance node, the transistors M_3 and M_4 must be kept in saturation region and transistors M_1 and M_2 must be kept in triode region. The compensation capacitor C_c is connected from node 2 of output stage to node A for indirect frequency compensation. The UGB is given by $f_{un} = 2g_{m1}/(2\pi C_c)$. The class AB stage operation is same as in SLCM OPAMP.

5 Simulation and Results

The proposed circuits were simulated in cadence 0.18-μm CMOS technology using the specter simulator in Virtuoso ADEL environment. The sizing of transistors and compensation capacitor was done by using the common design procedure as well as taking into account the model parameters. In split-length current mirror load OPAMP in Fig. 3, sizes of transistors M_7 and M_8 are twice of M_3, M_4, M_5, M_6 and sizes of M_9 and M_{11} are same so that quiescent current at output stage is same as I_{bias}. Similarly, in split-length differential-pair OPAMP in Fig. 4, the size of transistor M_7 is twice of transistors M_5 and M_6, whereas size of M_9 and M_8 are same. The gain and PM of the OPAMPs were determined from the AC analysis. Other parameters like slew rate, settling time, and power dissipation were found out from the transient analysis.

Figure 5 shows the AC and transient response of the proposed circuit in 4.1. The gain, UGB, and PM of the proposed circuit were found to be 58.14 dB, 43.26 MHz, and 79.25°, respectively, which confirms the stability of OPAMP. The settling time was found to be 78 ns.

AC and transient response of the proposed SLDP OPAMP is shown in Fig. 6. The gain, UGB, and PM were found to be 55.2 dB, 49.71 MHz, and 85.34°, respectively. The settling time of the SLDP OPAMP is calculated to be 50 ns from its transient response as shown in Fig. 6.

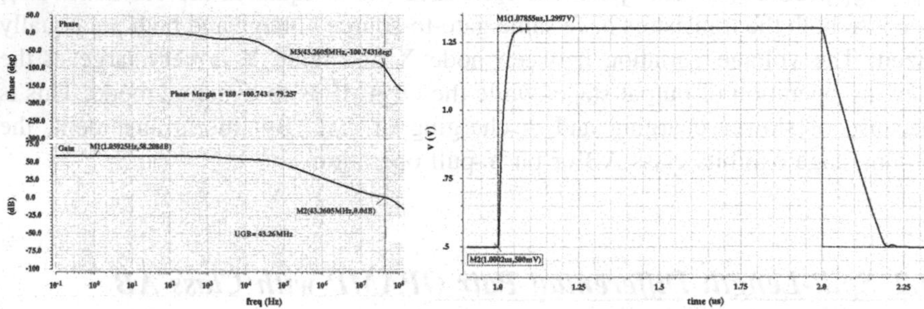

Fig. 5 AC response and transient response of SLCM load OPAMP

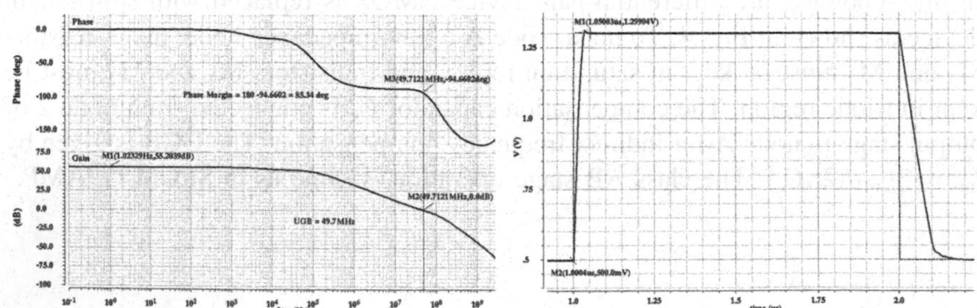

Fig. 6 AC response and transient response of SLDP OPAMP

Table 1 Performance comparison of the OPAMPs

OPAMP performance parameter	SLCM OPAMP in [7]	SLDP OPAMP in [7]	SLCM OPAMP with class AB stage	SLCM OPAMP with class AB stage
Technology (nm)	500	500	180	180
Supply voltage (V)	2.5	2.5	1.6	1.6
DC gain(dB)	66	60	58	55
UGB (MHz)	20	35	43	49
C_c (pF)	2	2	1.5	1
Phase margin	80	60	79	85
Settling time (ns)	60	75	78	50
+Slew rate (V/μs)			22	16
−Slew rate (V/μs)			15	10
Power (mW)	0.7	0.7	0.082	0.078

In Table 1, a comparative analysis of the performance of the proposed work is done with the existing OPAMP designs. From the table, we clearly see that the proposed OPAMPs gives better results in terms of unity gain bandwidth, PM, and power than the existing OPAMP designs as in [7].

6 Conclusion

Frequency compensation is necessary for multistage OPAMPs. In the proposed OPAMPs, indirect frequency compensation method is employed with split-length device, which gives faster and low-power OPAMP. Simulation result shows that the employment of class AB output stage improves the transient response of OPAMPs with low static power dissipation with less circuit complexity. For OPAMP designing in nano-CMOS process, these indirect compensated topologies with class AB stage are suitable.

References

1. Allen, P.E., Holberg, D.R.: CMOS Analog Circuit Design, 2nd edn. Oxford University Press, Oxford (2002)
2. Baker, R.J.: CMOS: Circuit Design, Layout, and Simulation, 2nd edn. Wiley Interscience (2005)
3. Tsividis, Y., Gray, P.: An integrated NMOS operational amplifier with internal compensation. IEEE J. Solid-State Circuits **11**(6), 748–754 (1976)
4. Ahuja, B.K.: An improved frequency compensation technique for CMOS operational amplifiers. IEEE J. Solid-State Circuits **18**, 629–633 (1983)

5. Hurst, P.J., Lewis, S.H., Keane, J.P.: Miller compensation using current buffers in fully differential CMOS two-stage operational amplifiers. In: IEEE Transactions on Circuits Systems I—Regular Papers, vol. 51, no. 2, Feb 2004
6. Palmisano, G., Paumbo, G.: A compensation strategy for two-stage CMOS opamps based on current buffer. IEEE Trans. Circuits Syst. I Fundam. Theory Appl. **44**(3), 257–262 (1997)
7. Saxena, V., Baker, R.J.: Compensation of CMOS Op-amps using split-length transistors. Submitted to IEEE International MWSCAS, Aug 2009
8. Yan, Z., Shen, L., Zhao, Y.: A low-voltage CMOS low-dropout regulator with novel capacitor-multiplier frequency compensation. In: Proceedings of IEEE ISCAS'08, May 2008, pp. 2685–2688
9. You, F., Embabi, S.H.K., Sanchez-Sinencio, E.: Low-voltage Class-AB buffers with quiescent current control. IEEE J. Solid-State Circuits **33**(6), 915–920 (1998)
10. Torralba, A., Carvajal, R.G., Martinez-Heredia, J., Ramírez Angulo, J.: Class-AB output stage for low voltage CMOS op-amps with accurate quiescent current control. Electron. Lett. **36**(21), 1753–1754 (2000)
11. Aloisi, W., Palumbo, G., Pennisi, S.: Design methodology of Miller frequency compensation with current buffer/amplifiers. IET Proc. Circuits Devices Syst. **2**(2), 227–233 (2008)

An Analytical Surface Potential Model of Surrounding Gate Tunnel FET

Soumen Paul and Angsuman Sarkar

Abstract In this paper, for the first time a new propitious surface potential model of tunnel field effect transistor (TFET) is presented. TFET uses band-to-band tunneling (BTBT) process and has a wide perspective in the field of nano-scale device suspending MOSFET as a switching device. The sub-threshold swing limitation of conventional MOSFET is minified by using TFET, thus establishing its own pathway and representing itself as an unprecedented device for low power application. In order to incorporate the advantages of both surrounding gate structure combined with TFET, an analytical model of surrounding gate TFET is presented. The surface potential model is developed using Gauss's law of Electrostatics and implemented on a cylindrical structure. Performance of the device is tested for variation using high gate dielectrics and also by altering gate oxide thickness of surrounding gate TFET channel and silicon body thickness.

Keywords BTBT · Gated *p-i-n* diode · Pseudo-2D · Surrounding gate · TFET · Surface potential

1 Introduction

Researchers are always in search of new and advanced technology. The evolution of tunnel field effect transistors (TEFTs) which is commonly known as TFET or surface tunnel transistors (STTs) is the consequences of such innovative research. MOSFET as a switching device was being employed extensively until and unless TFET came into existence. TFET works on band-to-band tunneling mechanism [1]

S. Paul (✉) · A. Sarkar
ECE Department, Kalyani Government Engineering College, Kalyani, India
e-mail: soumenpaul50@gmail.com

A. Sarkar
e-mail: angsumansarkar@kgec.ac.in

L.C. Jain et al. (eds.), *Intelligent Computing, Communication and Devices*,
Advances in Intelligent Systems and Computing 308,
DOI 10.1007/978-81-322-2012-1_50

and not only has it provided sub-threshold slope less than 60 mV/decade [2], but also shows promising result for low power application. Previous work on analytical modeling of TFET using double gate TFET in [3] indicated or evinced improved result for ON current while OFF current remains in fempto or pico ampere range. Double gate TFET is dependent on drain potential and these new findings is investigated in paper [4]. High gate dielectric and thin film structure boost ON current and the significance of using both in DG TFET is shown in [5, 6]. An Ultra-thin silicon body over insulator tunneling field effect transistor structure [7]. Now here in order to boost this ON current to a greater extent surrounding gate TFET using cylindrical structure is prepared and assumed ON current to be improvable.

2 Model Derivation

2.1 Device Structure

A schematic cross section of the surrounding gate n-channel TFET using cylindrical structure is shown Fig. 1. In this study, we consider $N_1 = 10^{20}$ cm^{-3}, $N_2 = 10^{17}$ cm^{-3}, and $N_3 = 5 \times 10^{18}$ cm^{-3} are doping concentration in p^+(p-type)source, intrinsic (i) region, and n^+(n-type) drain, respectively. Thickness of silicon body is taken as $t_{Si} = 10$ nm, whereas thickness of gate dielectric is considered as $t_{Ox} = 3$ nm. Dielectric constant of gate dielectric is $C_{Ox} = 3.9$ for SiO_2 (Silicon dioxide), $C_{Ox} = 25$ for HFO_2 (Hafnium dioxide), and $C_{Ox} = 29$ for

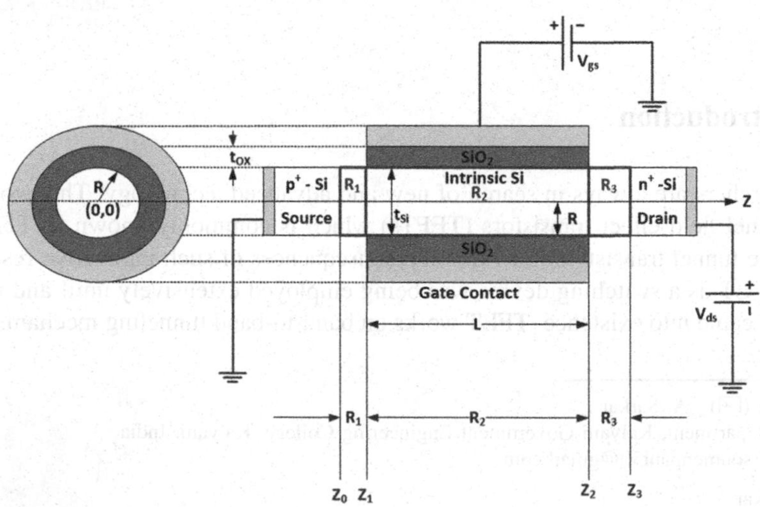

Fig. 1 Schematic cross section of the surrounding gate n-channel tunnel FET showing related variables

Zirconium dioxide (ZrO_2) is considered. Here polysilicon gate material is used having work function $W_M = 4.5$ eV, whereas work function for silicon channel is $W_{Si} = 4.68$ eV.

2.2 Model for Surface Potential and Electric Field

In our previous work, an analytical model for a cylindrical surrounding gate MOSFET is reported [8], which showed the use of Gauss's law on an elementary cylindrical volume and a model for surface potential was developed by equating the incoming and outgoing flux in different directions. The differential equation as previously obtained in [8] for cylindrical surrounding gate MOSFET is

$$\frac{\partial^2 \psi_s}{\partial z^2} - \lambda^2 \psi_S = \beta \tag{1}$$

In this paper, the same differential equation is utilized to model surface potential and electric field analytically for a SRGTFET.
where

$$\lambda = \sqrt{2C_{Ox}/ \in_{Si} R} \quad \text{and} \quad \beta = \frac{qN_S}{\in_{Si}} - \lambda^2 V'_{gs}$$

To model SRGTFET, let us consider N_1 is the doping concentration in p-type source region, $-N_3$ is the doping concentration in n-type drain, and $\pm N_2$ is the doping concentration in lightly doped region.

The general solution of (1) is given by

$$\Psi_{Si} = A_i e^{\lambda z} + B_i e^{-\lambda z} - \frac{\beta_i}{\lambda^2} \tag{2}$$

where A_i and B_i are constant coefficients that need to be derived. The general solution for lateral electric field is obtained by finding the derivative of surface potential with respect to z-direction (along the channel) and given as

$$E_{zi} = \lambda \left[A_i e^{\lambda z} - B_i e^{-\lambda z} \right] \tag{3}$$

Similar to the method employed by Bardon et al. [3], to model surface potential, the channel depletion region inside the source and the drain is also considered in our study. Taking this into consideration, the total channel length is divided into three regions namely R_1 which is depletion region inside source, R_2 is the intrinsic region, and R_3 which is depletion region inside drain with their boundaries on z-direction marked by z_0, z_1, z_2, and z_3. Thus,

$$z_0 \leq R_1 \geq z_1 \quad \text{for region } R_1$$
$$z_1 \leq R_2 \geq z_2 \quad \text{for region } R_2$$
$$z_2 \leq R_3 \geq z_3 \quad \text{for region } R_3$$

Ψ_0 and Ψ_3 is defined as surface potential at point z_0 and z_3 are the boundary values of potential at source side and drain side and are given by

$$\Psi_0 = -kT/q\ln(N_1/n_i) \tag{4}$$

$$\Psi_3 = kT/q\ln(N_3/n_i) + V_d \tag{5}$$

As surface potential is same (i) at the end of region R_1 and at the beginning of region R_2 and (ii) at the end of region R_2 and at the beginning of region R_3, A_1, B_1 and A_3, B_3 are obtained and the generalized expression is given as

$$A_i = \frac{\psi_i e^{-\lambda z_{i-1}} - \psi_{i-1} e^{-\lambda z_i} + \frac{\beta_i}{\lambda^2}\left[e^{-\lambda z_{i-1}} - e^{-\lambda z_i}\right]}{2\sinh(L_i)} \tag{6}$$

$$B_i = \frac{\psi_{i-1} e^{\lambda z_i} - \psi_i e^{\lambda z_{i-1}} - \frac{\beta_i}{\lambda^2}\left[e^{\lambda z_{i-1}} - e^{\lambda z_i}\right]}{2\sinh(L_i)} \tag{7}$$

where $L_i = \lambda(z_i - z_{i-1})$

$$\Psi_1 a_{11} + \Psi_2 a_{12} = E \tag{8}$$

$$\Psi_1 a_{21} + \Psi_2 a_{22} = F \tag{9}$$

Now applying continuity of electric field (i) at the end of region R_1 and at the beginning of region R_2 and (ii) at the end of region R_2 and at the beginning of region R_3 as a result of which two equations are formed and finally A_2, B_2 is obtained. To find the unknown surface potentials Ψ_1 and Ψ_2, let us substitute the values of the coefficient A_1, B_1, A_2, B_2 and A_3, B_3 in that two equations and two simultaneous equations are obtained as given above in the form of Eqs. (8), (9). The unknown surface potentials Ψ_1 and Ψ_2 are obtained by solving (8) and (9) using crammer's rule and is given by

$$\Psi_1 = \frac{\begin{vmatrix} E & a_{12} \\ F & a_{22} \end{vmatrix}}{D}, \quad \Psi_2 = \frac{\begin{vmatrix} a_{11} & E \\ a_{21} & F \end{vmatrix}}{D}, \quad D = \begin{vmatrix} a_{11} & a_{12} \\ a_{21} & a_{22} \end{vmatrix}$$

3 Results and Discussion

3.1 Plausibility of Surface Potential and Lateral Electric Field

Figure 2 shows the variation of surface potential against position along the channel for a SRGTFET with gate length $L_g = 50$ nm and length of source/drain region is equal to 30 nm for different gate bias voltages with a constant $V_{DS} = 1$ V. Surface potential increases in steps as gate voltage increases in intrinsic region from $V_G = 0$ V to $V_G = 1.4$ V.

Potential variation near and across the junctions causes changes in surface potential in source and drain region where it is constant in lightly doped region when fixed gate voltage is applied as shown in Figs. 2 and 3 plots the variation of lateral electric field as a function of position along the channel length for different gate bias voltages. Due to this varying potential across tunnel junction, a peak value of electric field is observed near tunneling junction, which needs to be maximized in order to obtain a large drain current.

3.2 High Gate Dielectric

In order to achieve higher gate controllability, the impact of the use of high-k gate dielectric material is studied. Figure 4 shows the variation of surface potential along length of the channel where gate material SiO_2 is replaced by a gate material having higher dielectric value equals to 21. The impact of the introduction of the

Fig. 2 Surface potential variation for different gate voltage increasing from 0 to 1.4 V (gate length = 50 nm, gate dielectric = 3.9, $t_{OX} = 3$ nm and $V_{DS} = 1$ V). Total length of the device is 110 nm

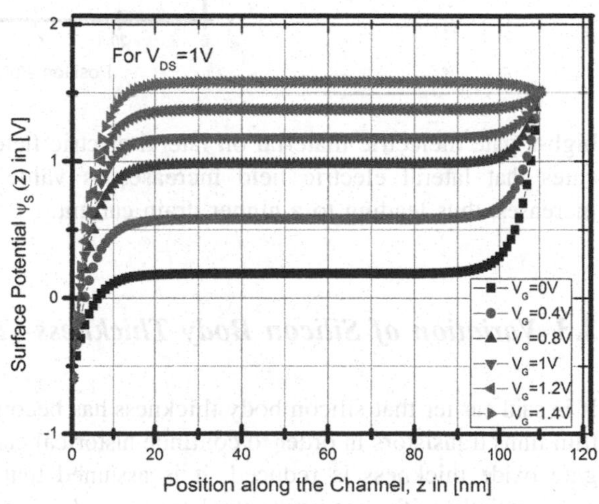

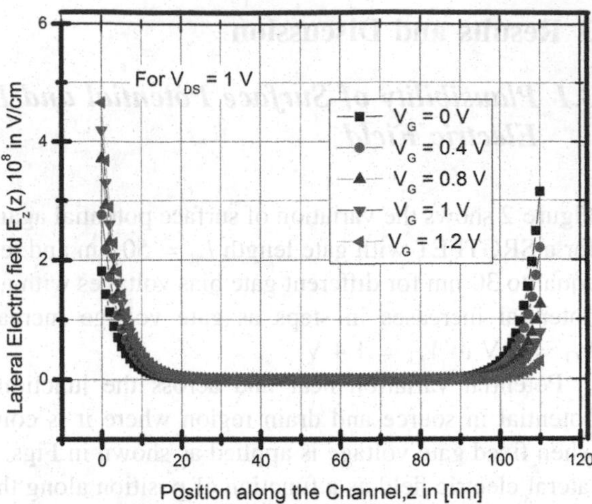

Fig. 3 Lateral electric field variation for different gate voltage increasing from 0 to 1.4 V (gate length = 50 nm, gate dielectric = 3.9, t_{OX} = 3 nm and V_{DS} = 1 V). Total length of the device is 110 nm

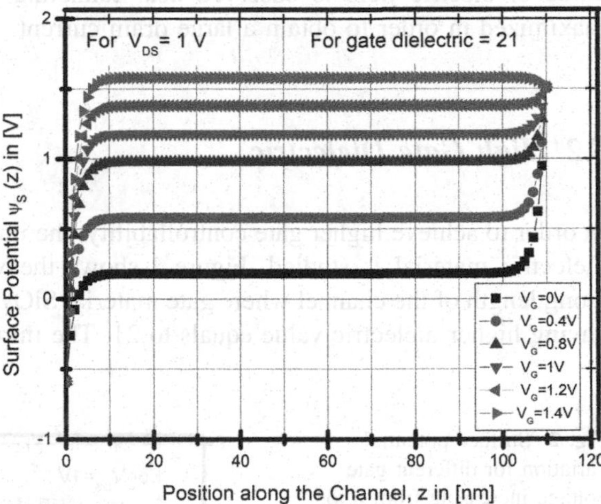

Fig. 4 Surface potential variation for different gate voltage increasing from 0 to 1.4 V (gate length = 50 nm, gate dielectric = 21, t_{OX} = 3 nm and V_{DS} = 1 V). Total length of the device is 110 nm

high-k gate dielectric material on lateral electric field is shown in Fig. 5. It indicates that lateral electric field increases as value of gate dielectric material increases, thus leading to a higher drain current.

3.3 Variation of Silicon Body Thickness (T_{Si})

It is said earlier that silicon body thickness has become a key parameter for ultra-thin films transistors in order to continue historical cadence of scaling. Now, when gate oxide thickness is reduced, it is assumed that tunneling current increases exponentially with gate oxide thicknesses as $I_{BTBT} \approx \exp(-\sqrt{t_{Si}})$. Figure 6 plots

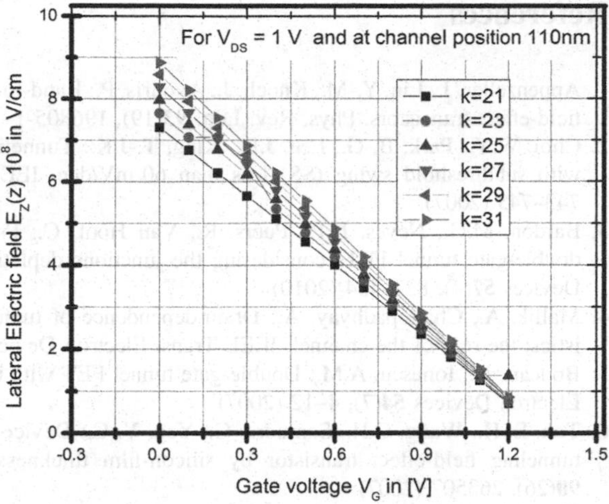

Fig. 5 Lateral electric field variation for different gate voltage increasing from 0 to 1.4 V (gate length = 50 nm, t_{OX} = 3 nm and V_{DS} = 1 V). Total length of the device is 110 nm

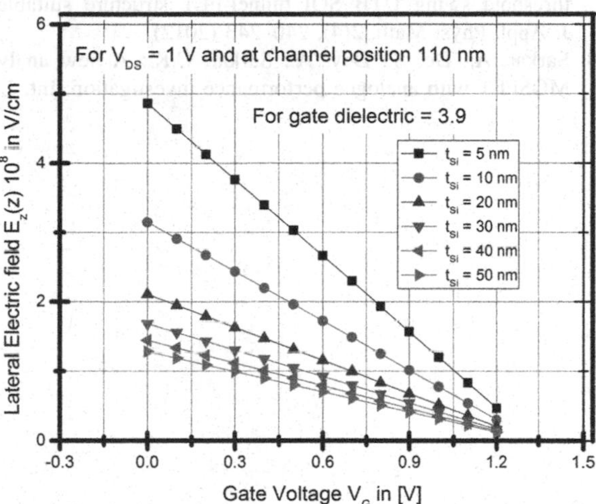

Fig. 6 Lateral electric field variation for different gate voltage increasing from 0 V to 1.4 V. (gate length = 50 nm, gate dielectric = 3.9, t_{OX} = 3 nm and V_{DS} = 1 V). Total length of the device is 110 nm

the lateral electric field distribution as a function of gate bias for different silicon body thickness. Large increase in the lateral electric field is observed when silicon body is reduced to 5 nm.

4 Conclusion

In this paper, the surface potential and lateral electric field of surrounding gate tunnel FET is analytically modeled using pseudo 2-D analysis and derived applying Gauss law. The impact of variation of silicon body thickness and high-k dielectric gate materials has also been scrutinized.

References

1. Appenzeller, J., Lin, Y.-M., Knoch, J., Avouris, P.: Band-to-band tunneling in carbon nanotube field-effect transistors. Phys. Rev. Lett. **93**(19), 196805-1–196805-4 (2004)
2. Choi, W.Y., Park, B.-G., Lee, J.D., King, T.-J.K.: Tunneling field-effect transistors (TFETs) with subthreshold swing (SS) less than 60 mV/dec. IEEE Trans. Electron Devices **28**(8), 743–745 (2007)
3. Bardon, M.G., Neves, H.P., Puers, R., Van Hoof, C.: Pseudo-Two-dimensional model for double-gate tunnel FETs considering the junctions depletion regions. IEEE Trans. Electron Devices **57**(4), 827–434 (2010)
4. Mallik, A., Chattopadhyay, A.: Drain-dependence of tunnel field-effect transistor characteristics: the role of the channel. IEEE Trans. Electron Devices **58**(12), 4250–4257 (2011)
5. Boucart, K., Ionescu, A.M.: Double-gate tunnel FET with high-k gate dielectric. IEEE Trans. Electron Devices **54**(7), 4–12 (2007)
6. Toh, E.-H., Wang, G.H., Samudra, G., Yeo, Y.-C.: Device physics and design of double-gate tunneling field-effect transistor by silicon-film thickness optimization. Appl. Phys. Lett. **90**(26), 263507 (2007)
7. Gupta, P.S., Kanungo, S., Rahaman, H., Sinha, K., Dasgupta, P.S.: An extremely low subthreshold swing UTB SOI tunnel-FET structure suitable for low-power applications. Int. J. Appl. Phys. Math. **2**(4), 240–243 (2012)
8. Sarkar, A., De, S., Dey, A., Sarkar, C.K.: A new analytical subthreshold model of SRG MOSFET with analogue performance investigation. Int. J. Electron. **99**(2), 267–283 (2012)

Design and Characterization of Ka-Band Reflection-Type IMPATT Amplifier

L.P. Mishra and M. Mitra

Abstract Till date, IMPact Avalanche Transit Time (IMPATT) has emerged as a most powerful semiconductor source in the range of microwave and millimetre wave for the application in high-range communication and RADAR. In this paper, IMPATT device has been designed for Ka-band reflection-type amplifier. The characterization of the amplifier has shown its efficiency as an initial high gain of about 17 dB for an input power of 10 μW. It is also being found that the gain decreases with the increase of input power and the gain becomes nearly 3 dB for an input power of 17 mW. It is being observed that the amplifier is very much stable over an input power range of 2–17 mW and can operate in CW mode.

Keywords IMPATT amplifier · CW amplifier · Ka-band amplifier · Reflection-type amplifier

1 Introduction

The first SDR (IMPATT) diode was developed by Jhonston [1], and the microwave oscillation was found by Lee [2] using the diode. Though lot of work has been done over the years regarding the development of IMPATT oscillator for higher-power output, higher efficiency, and higher frequency range of operation. The effect of negative resistance of IMPATT diode in the amplification of microwave signals was first observed by Hines [3]. It was found that nonlinear effects are dominant considerations in power amplifier design because of efficiency and economy considerations of the device. Later on, the development of IMPATT amplifier was

---

L.P. Mishra (✉)
Department of ECE, ITER, Siksha 'O' Anusandhan University, Bhubaneswar, Odisha, India
e-mail: laxmimishra@soauniversity.ac.in

M. Mitra
Department of E&TC, IIEST, Shibpur, West Bengal, India
e-mail: monojit_m1@yahoo.co.in

L.C. Jain et al. (eds.), *Intelligent Computing, Communication and Devices*,
Advances in Intelligent Systems and Computing 308,
DOI 10.1007/978-81-322-2012-1_51

made by Snider [4] in 1970 was used in 'X' band range as CW amplifier. Still a lot of development has been made in the field of IMPATT amplifier at lower frequency band. Hence, a Ka-band reflection-type IMPATT amplifier has been developed indigenously using a Ka-band IMPATT diode oscillator as an input signal source. These amplifiers become very much attractive due to its small size, simple arrangement, and sufficient power addition for various applications in the field of high-frequency communication and RADAR.

2 Experimental Set-up and Analysis of Amplifier

Figure 1 shows a suitable equivalent circuit of an IMPATT amplifier. It includes a circulator to separate input and output waves. A resonant circuit including a capacitor which is the junction capacitance of the diode as well as the circuit capacitance, a tuning inductor *L*, and a transformer is used to establish the desired loading conductance G_L. The schematic experimental set-up is shown in Fig. 2 where port 1 is connected with a Ka-band signal source. The Ka-band IMPATT amplifier is connected to the port 2 of the circulator, and the amplifier signal is reflected back to port 3 from which we will get the output power.

3 Related Theory

The amplifier has independence impedance control. The real part of the load impedance is determined by the physical dimensions of a two-section quarter-wave transmission-line impedance of a short-circuited coaxial stub placed in series with the transformer $X_L = Z_0 \tan \beta_l$.

The most common types of active diodes are the IMPATT diode and the transferred electron device (generally called a Gunn diode). These devices have high-frequency capability since the saturated velocity of an electron in a semiconductor is high (generally on the order of ~ 107 cm/sec), and the transit time is

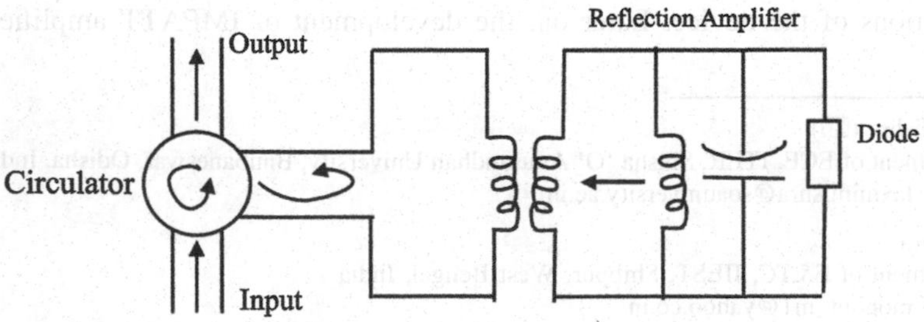

Fig. 1 Equivalent circuit of IMPATT amplifier

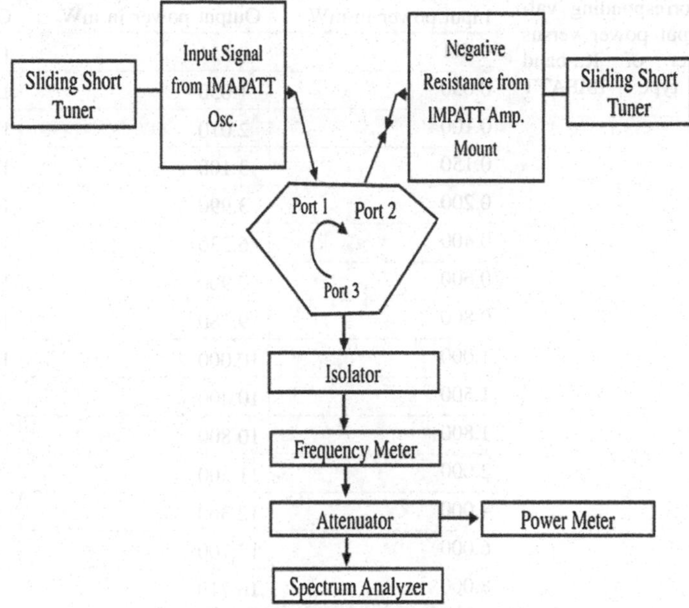

Fig. 2 Experimental set-up for CW-type silicon Ka-band reflection-type IMPATT amplifier

short since the length of the region over which the electron transits can be made on the order of a micron (i.e. 10–4 cm) or less. Oscillation frequency on the order of 400 GHz has been achieved with IMPATT diodes [5].

Military and commercial interest in Ka-band frequencies is motivated by low-atmospheric propagation loss in the frequency range surrounding 30 GHz. The lower loss is beneficial for communication and radar applications that require free-space transmission. Moreover, antennas designed at Ka-band are of reduced size and mass relative to antennas designed at lower frequencies with comparable gain. Several researchers have demonstrated Ka-band quasi-optical amplifier arrays [6–10].

If the resistance of the amplifying IMPATT diode is r_d, and the load resistance present in the circuit is R_L, and the IMPATT diode is kept just below the threshold of oscillation, the amplifier gain would be given by

$$A = \left| \frac{r_d - R_L}{r_d + R_L} \right|$$

when r_d is negative, the magnitude of 'A' becomes greater than unity, which indicates the IMPATT diode is behaving as an amplifier.

Table 1 Corresponding values for output power versus input power of Ka-band reflection type IMPATT amplifier

Input power in mW	Output power in mW	Gain in dB
0.011	0.551	17.000
0.050	1.050	13.220
0.100	2.010	13.030
0.150	3.100	13.150
0.200	3.990	13.000
0.400	6.330	12.000
0.600	7.900	11.200
0.800	9.180	10.600
1.000	10.000	10.000
1.500	10.400	8.400
1.800	10.800	7.780
2.000	11.500	7.590
4.000	12.360	4.900
6.000	17.700	4.700
8.000	16.710	3.200
10.000	20.410	3.100

4 Experimental Results and Discussion

The variation of input power and the corresponding amplifier output power for CW mode of operation was recorded in the form of a Table 1, and for each value of input power, the gain is calculated in dB. A plot of output power versus input power is shown in Fig. 3, and a plot of gain versus input power is shown in Fig. 4.

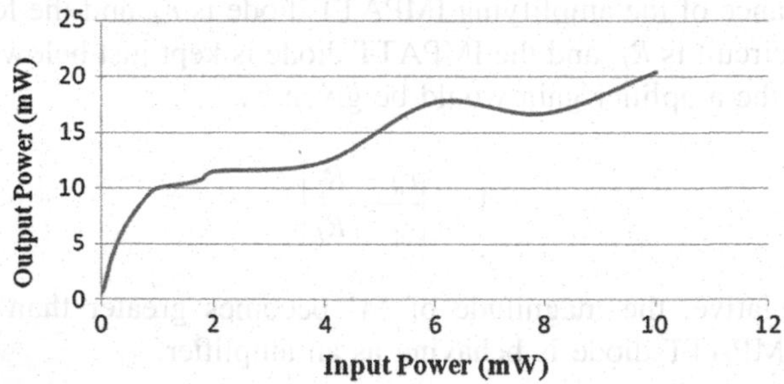

Fig. 3 Plot of output power versus input power of Ka-band reflection-type IMPATT amplifier operating at 36 GHz

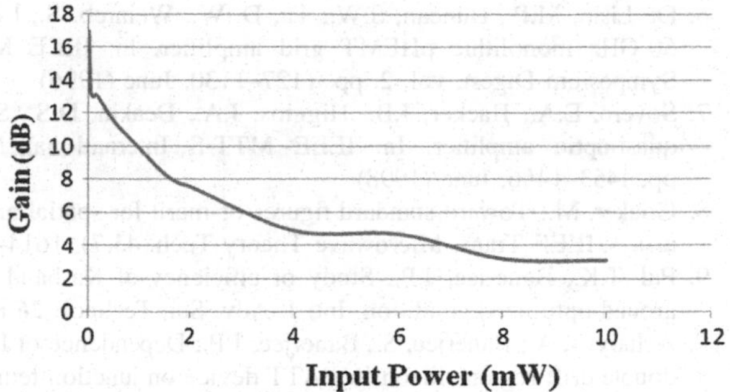

Fig. 4 Plot of gain versus input power of Ka-band reflection-type IMPATT amplifier operating at 36 GHz

From Fig. 3, it is being observed that the output power varies linearly with input power. Over a range of 10 μW to 10 mW of input power, output power can vary from 1 to 20 mW. The frequency of the signal is found 36 GHz. The plot of gain versus input power under Fig. 4 shows that the gain decreases with the increase of input power. The gain is about 17 dB at a power of 11 μW and a minimum of 3 dB at an input power of 10 mW. It is also being found that the gain becomes almost constant when input power level is 8 to 10 mW and these acts as a stable amplifier over the range.

5 Conclusion

Thus, the reflection-type IMPATT amplifier can be used in CW mode for a low-power input system. It can fulfil the requirement of Ka-band amplification. Modified versions also can be used to power combine MMICs, making cavity combiners.

References

1. Johnston, R.L., Loach, B.C., Cohen, B.G.: A silicon diode microwave oscillator. Bell Syst. Tech. J. **44**(2), 369–372 (1965)
2. Lee, C.A., Batdorf, R.L., Weignmann, W., Kaminsky, G.: Technological developments evolving from research on Read diodes. IEEE Trans. Electron Devices **13**(1), 175–180 (1966)
3. Hines, M.E.: Negative resistance diode power amplification. IEEE Trans. Electron Devices **17**(1), 1–8 (1970)
4. Snider, D.M.: A one watt CW high—efficiency X-Band Avalanche diode amplifier. IEEE Trans. Microwave Theory Tech. **18**(11), 963–967 (1970)
5. Liu, C.M., Sovero, E.A., Ho, W.J., Higgins, J.A., De Lisio, M.P., Rutledge, D.B.: Monolithic 40-GHz 670-mW HBT grid amplifier. In: IEEE MTT-S International Microwave Symposium Digest, pp. 1123–1126, June (1996)

6. De Lisio, M.P., Duncan, S.W., Tu, D.-W., Weinreb, S., Liu, C.-M., Rutledge, D.B.: A 44/60 GHz monolithic pHEMT grid amplifier. In: IEEE MTT-S International Microwave Symposium Digest, vol. 2, pp. 1127–1130, June (1996)
7. Sovero, E.A., Hacker, J.B., Higgins, J.A., Deakin, D.S., Sailer, A.L.: Ka-band monolithic quasi-optic amplifier. In: IEEE MTT-S International Microwave Symposium Digest, pp. 1453–1456, June (1998)
8. Gouker, M.: Toward standard figures-of-merit for spatial and quasi-optical power-combined arrays. IEEE Trans. Microwave Theory Tech. **43**(7), 1614–1617 (1995)
9. Pal, T.K., Banerjee, J.P.: Study of efficiency of Ka-band IMPATT diodes and oscillators around optimized condition. Int. J. Adv. Sci. Technol. **26** (2011)
10. Acharyya, A., Banerjee, S., Banerjee, J.P.: Dependence of DC and small-signal properties of double drift region silicon IMPATT device on junction temperature. J. Electron Devices **12**, 725–729 (2012)

Energy Transfer to Piezoelectric Component Through Magnetic Resonant Coupling

P.P. Nayak, D.P. Kar, S.N. Das and S. Bhuyan

Abstract In this paper, a non-contact energy transfer method has been experimentally investigated to drive a piezoelectric component operating in the thickness vibration mode. The energy transfer system uses a receiving coil connected to the piezoelectric component placed away from an energy-transmitting coil along its central axis. The impedance characteristics and frequency characteristics of the voltage developed across the piezoelectric component operating in the thickness vibration mode are experimentally investigated. It is observed that a mechanical resonance vibration is excited in the driven piezoelectric component through strongly coupled magnetic resonance between the coils as well as due to the mechanical resonance of the piezoelectric component. It has been found that the voltage developed across the piezoelectric component is maximum at its resonance frequency. It is also found that the energy received by the piezoelectric component connected to the receiving coil depends on the operating frequency, coil design, and distance between the coils.

Keywords Magnetic coupling · Piezoelectric component · Resonance · Energy transfer

1 Introduction

There has been rising interest in the application of piezoelectric devices such as actuation mechanism [1], sensing [2], precision positioning, micro-manipulations [3], miniature ultrasonic motors [4], and implantable biomedical electronics [5]. Piezoelectric devices can vibrate mechanically at their resonant frequency when excited by an applied voltage of the same frequency. While the idea of non-contact

P.P. Nayak (✉) · D.P. Kar · S.N. Das · S. Bhuyan
ITER, Siksha 'O' Anusandhan University, Bhubaneswar, India
e-mail: praveennayak@soauniversity.ac.in

L.C. Jain et al. (eds.), *Intelligent Computing, Communication and Devices*,
Advances in Intelligent Systems and Computing 308,
DOI 10.1007/978-81-322-2012-1_52

energy transfer is not entirely new, its use for operating piezoelectric devices has only been brought to the fore recently [6, 7]. While significant headway has been made in the field of non-contact energy transfer to piezoelectric devices, the amount of energy transferred in most cases is still not sufficient to operate the piezoelectric devices. In order to solve this issue, a non-contact energy transfer method has been investigated in this work. With the recent surge of miniature piezoelectric devices, non-contact energy transfer can provide a convenient alternative to traditional power sources used to operate piezoelectric actuators, sensors, and micro-manipulators.

2 Experimental Setup, Operating Conditions, and Mechanism

Figure 1 shows the schematic diagram of non-contact energy transfer to the piezoelectric component. An input power source is connected to a transmitting coil. The piezoelectric component is connected to the receiving coil. Both the transmitting and receiving coils are made of copper windings. A matching circuit is used to make the energy transfer system resonance. The transmitting coil is in resonance with a capacitor, while the receiving coil is in resonance with the clamped capacitance of the piezoelectric component.

The experiment is carried out under the following conditions. The piezoelectric component used in this experiment is made of lead zirconate titanate (PZT) ceramic material (C203-PZT, supplied by Fuji Ceramic Corporation) with a size of $30 \times 8 \times 2\ \text{mm}^3$ and is shown in Fig. 2. The piezoelectric component is poled along its thickness direction. The top and bottom surfaces of the PZT are covered with silver metal electrodes.

The medium is air for the non-contact energy transfer to the piezoelectric component. The copper wires are wound to form spiral transmitting and receiving

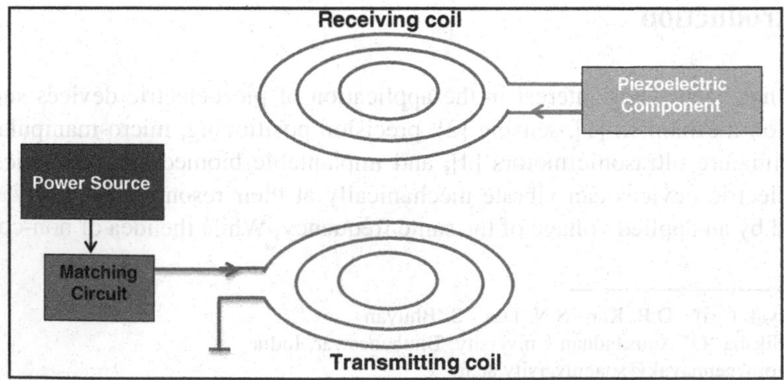

Fig. 1 Non-contact energy transfer system for driving piezoelectric device

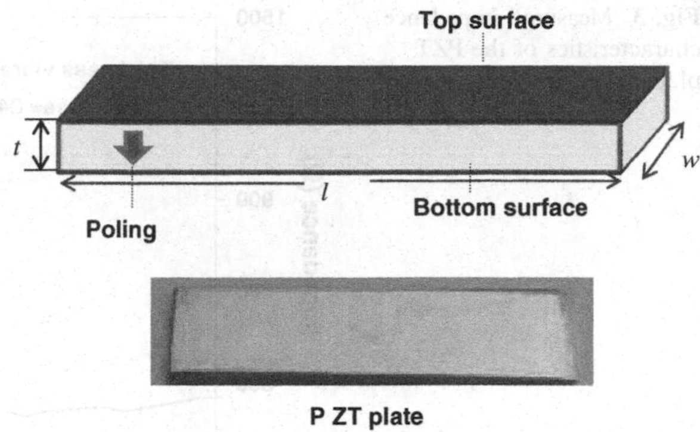

Fig. 2 Schematic diagram and photograph of the piezoelectric component

Table 1 Relevant property constants of the PZT material

Properties		Value
Electromechanical coupling factor	k_t	0.47
Frequency constants (mm KHz)	N_t	1,470
Relative dielectric constants	$\varepsilon_{33}^T/\varepsilon_0$	1,450
Piezoelectric coefficients (10^{-12} C/N)	d_{33}	325
Mechanical quality factor	Q_m	2,000
Dissipation factor (%)	tan δ	0.3

coils of radius of 6 cm. The separation distance between the coils is 3 cm. The input source voltage is 7 V_{rms}. The resonant capacitor connected to the transmitting coil is of the same value as the clamped capacitance ($C = 920$ pF) of the piezoelectric component operating in the thickness mode. The relevant material properties of the piezoelectric component are shown in Table 1.

3 Results and Discussion

The impedance characteristics of the piezoelectric component operating in the thickness vibration mode are depicted in Fig. 3. The impedance characteristic of the driven PZT component is measured by an impedance analyzer (HP4194A). From the frequency constant of the piezoelectric material and the impedance characteristics, it is found that the resonance frequency of the used piezoelectric component in thickness vibration mode is 772 KHz. The measured equivalent circuit parameters of the driven piezoelectric plate operating in the thickness vibration mode are shown in Table 2.

Figure 4 shows the frequency characteristics of the voltage developed across the piezoelectric plate driven by non-contact energy transfer. It is seen that the voltage across the piezoelectric plate attains a maximum at the resonance frequency of the piezoelectric component. The piezoelectric voltage drops suddenly

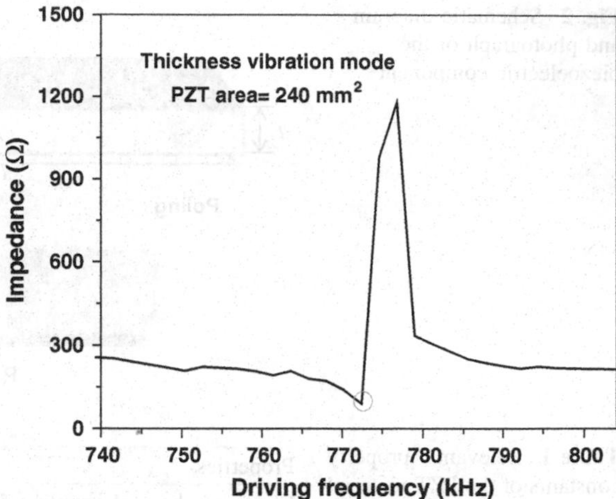

Fig. 3 Measured impedance characteristics of the PZT plate in the thickness mode

Table 2 Measured equivalent circuit parameters of the driven piezoelectric component

Piezoelectric component	Resonance frequency f_r(kHz)	Equivalent resistance R_m(Ω)	Equivalent inductance L_m(mH)	Equivalent capacitance C_m(pF)	Clamped capacitance C_d(pF)
PZT plate	772	42	16.3	2.52	712.2

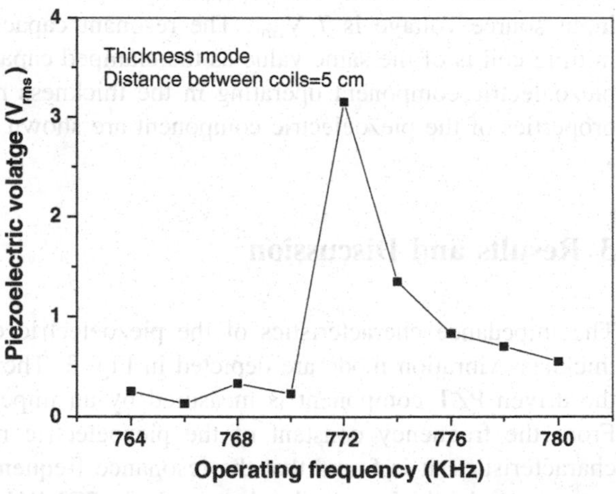

Fig. 4 Frequency characteristics of the voltage developed across the piezoelectric component operating in the thickness mode

if the operating frequency is detuned from the resonant frequency. When the operating frequency of the energy-transmitting source (both the transmitting and receiving coil's resonant frequency) is close to the mechanical resonance frequency of the piezoelectric plate, a mechanical resonance vibration is stimulated in the piezoelectric plate by the converse piezoelectric effect. This mechanical

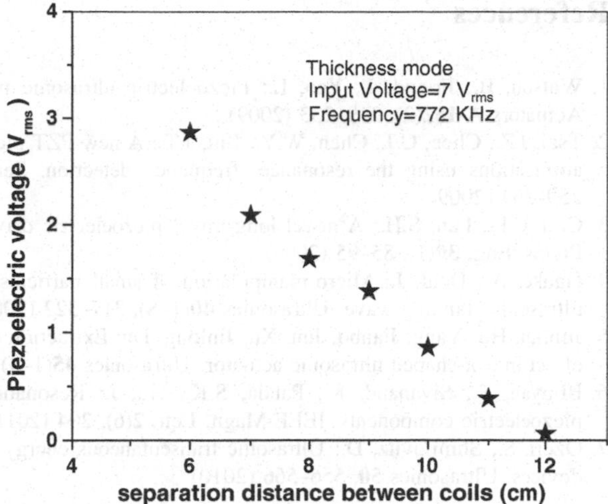

Fig. 5 Dependence of the separation distance between the coils on the voltage developed across the piezoelectric component operating in the thickness mode

resonance vibration generates a voltage across the piezoelectric plate due to the direct piezoelectric effect. Hence, the peak of the voltage across the piezoelectric plate is only due to the piezoelectric resonance.

The dependence of the non-contact energy transfer to the piezoelectric component on the vertical distance between the transmitting and receiving coils is shown in Fig. 5. The transmitting and receiving coils are coaxially aligned, and the vertical distance between them is varied. It is seen that the voltage developed across the piezoelectric plate decreases with the increase in the distance between the transmitting and receiving coils. This is because the magnetic field coupling between the transmitting and receiving coils decreases with the increase in the distance between the coils.

4 Conclusion

In this work, a non-contact energy transfer system has been experimentally investigated to drive the piezoelectric component. The effects of operating frequency and distance between the transmitting and receiving coils of the energy transfer system are analyzed. Experimentally, it is observed that the energy transfer system has its own strongly magnetic-coupled region that is an optimum frequency for which maximum energy transfer occurs. The energy transfer further reduces with the increase in the distance between the transmitting coil and receiving coils connected to the piezoelectric component of the non-contact energy transfer system. The analyses will give future guidelines for designing a compact non-contact energy transfer system to drive the piezoelectric devices used in sensing, precision positioning, micro-actuation mechanism, and chemical and biomedical engineering.

References

1. Watson, B., Friend, J., Yeo, L.: Piezoelectric ultrasonic micro/milli-scale actuators. Sens. Actuators A **152**(2), 219–233 (2009)
2. Tsai, J.Z., Chen, C.J., Chen, W.Y., Liu, J.T.: A new PZT piezoelectric sensor for gravimetric applications using the resonance- frequency detection. Sens. Actuators B: Chem. **139**(2), 259–264 (2009)
3. Chu, C.L., Fan, S.H.: A novel long-travel piezoelectric driven linear nanopositioning stage. Precis. Eng. **30**(1), 85–95 (2006)
4. Haake, A., Dual, J.: Micro-manipulation of small particles by node position control of an ultrasonic standing wave. Ultrasonics **40**(1–8), 317–322 (2002)
5. Junhui, Hu, Yang, Jianbo, Jun, Xu, Jinlong, Du: Extraction of biologic particles by pumping effect in a π-shaped ultrasonic actuator. Ultrasonics **45**(1–4), 15–21 (2006)
6. Bhuyan, S., Sivanand, K., Panda, S.K., Hu, J.: Resonance-based wireless energizing of piezoelectric components. IEEE Magn. Lett. **2**(6), 204 (2011)
7. Ozeri, S., Shmilovitz, D.: Ultrasonic transcutaneous energy transfer for powering implanted devices. Ultrasonics **50**, 556–566 (2010)

Interface Study of Individual and Stacked High-k/P-Si MOSCAPs by CV Technique

Milan Maitri Mishra, Gayatri Pradhan, Farida Ashraf Ali and Gouranga Bose

Abstract Interface trap charge density, threshold voltage, and flat band voltage of aluminum oxide (Al_2O_3), hafnium oxide (HfO_2), titanium oxide (TiO_2), and yttrium oxide (Y_2O_3) of different oxide thickness have been calculated after high-frequency CV simulation of MOSCAPs at room temperature, where P-type silicon is taken as substrate. The calculated D_{it} value is less in TiO_2 ($1.37 \times 10^{10}\ eV^{-1} cm^{-2}$) than Al_2O_3 ($4.81 \times 10^{10}\ eV^{-1}\ cm^{-2}$), HfO_2 ($1.58 \times 10^{10}\ eV^{-1}\ cm^{-2}$), and Y_2O_3 ($2.34 \times 10^{10}\ eV^{-1}\ cm^{-2}$) for 5 nm oxide thickness. The threshold and the flat band voltage of all oxide layers are found to be around 0.35 and −0.3 V, respectively, which match well with the experimentally reported values. Furthermore, the CV simulation, threshold, and flat band voltage calculations were done for 10-nm and 15-nm-thick individual oxides (Al_2O_3, HfO_2, TiO_2, and Y_2O_3) and compared. In addition, the interface trap charge densities, threshold voltages, and flat band voltages of stacked HfO_2/TiO_2 and Al_2O_3/TiO_2, Y_2O_3/TiO_2 are calculated for 2 nm oxide thickness by the same CV technique. It is found that the interface states of stacked HfO_2/TiO_2 ($1.18 \times 10^{10}\ eV^{-1}\ cm^{-2}$) are marginally less than stacked Al_2O_3/TiO_2 ($1.34 \times 10^{10}\ eV^{-1} cm^{-2}$) and Y_2O_3/TiO_2 ($1.29 \times 10^{10}\ eV^{-1}\ cm^{-2}$).

Keywords High-k · Interface traps · MOSCAPs · Threshold voltage · Flat band voltage

1 Introduction

Microelectronics industries used to use silicon dioxide as gate dielectric for more than three decades because of its excellent material properties which include chemical and thermal stability at high temperature, good interface quality with low

M.M. Mishra · G. Pradhan · F.A. Ali · G. Bose (✉)
Institute of Technical Education and Research (ITER), Siksha 'O' Anusandhan University, Bhubaneswar 751030, Odisha, India
e-mail: gourangabose@soauniversity.ac.in

L.C. Jain et al. (eds.), *Intelligent Computing, Communication and Devices*,
Advances in Intelligent Systems and Computing 308,
DOI 10.1007/978-81-322-2012-1_53

interface states, high melting point, resistivity, larger band gap, and many more [1]. The downscaling of device dimension associated with Moore's law enables to integrate more number of transistors in a chip at low cost and thereby improving the device performance [2]. This moves the thickness of oxide layer to nanometer range, which limits the functionality of silicon dioxide by increasing direct tunneling and leakage current density [3]. Also the ultrathin SiO_2 faces many problems like MOSFET capacitance density reduction, high defect density, and boron diffusion in case of poly silicon gate. Thus, an oxide layer with high dielectric constant than the conventional silicon dioxide was required to overcome the scaling-related difficulties [4]. The high-k materials to be used should have large band gap (>5 eV), high permittivity (>12), and low defect density to replace silicon dioxide [5, 6].

Aluminum oxide (Al_2O_3) is a promising candidate to use instead of SiO_2 because of its low cost and high band gap (~8.8 eV) although its permittivity is not so high (~9). Also Al_2O_3 provides low leakage current density due to high band gap, has good breakdown voltage, and is considered as a good reaction barrier between the high-k layer and the silicon substrate [1, 6]. From capacitance point of view, oxide with k value ranging from 25 to 30 is preferred. Thus, hafnium-based oxides (HfO_2) are considered as their dielectric constant is 25 and band gap is 6 eV and they have good thermal stability with silicon [1, 7]. Very high dielectric constant (~80) and higher breakdown strength make titanium oxide (TiO_2) as a better alternative for deep submicron technologies. Device with TiO_2 gate dielectric shows nearly ideal behavior although its band gap is low (~3.5 eV) [8]. Another oxide yttrium oxide (Y_2O_3) drags attention because of its high band gap of 6 eV, good dielectric properties (k–13 eV), and good thermal stability up to 2,000 °C [9]. Besides all these, hafnium and titanium oxide provide better results regarding low-power applications and interface trap densities. These two potential candidates, HfO_2 and TiO_2, can be used as stack to provide better results regarding D_{it} and threshold voltages [10]. Furthermore, the interface properties of stacked Al_2O_3/TiO_2 and Y_2O_3/TiO_2 were also analyzed [11]. The interface traps at the oxide/semiconductor interface is a prime factor behind the shifts the flat band voltage, threshold voltage, and decrease carrier mobility leading to circuit [12]. In this study, the high-frequency CV characteristic of the MOSCAP structure with individual high-k dielectric layer and stacked high-k layer has been analyzed and the interface trap charge density has also been calculated by Termann method [13].

2 Simulated Structure

MOSCAPs having P-type silicon substrate with non-uniform doping concentration, i.e., 5×10^{15} cm^{-3} near the interface and 5×10^{17} cm^{-3} at the bulk, a 5-nm-thick high-k dielectric layer of Al_2O_3, HfO_2, TiO_2, Y_2O_3, and Al gate have been simulated using high-frequency (1 MHz) CV technique at room temperature by the device

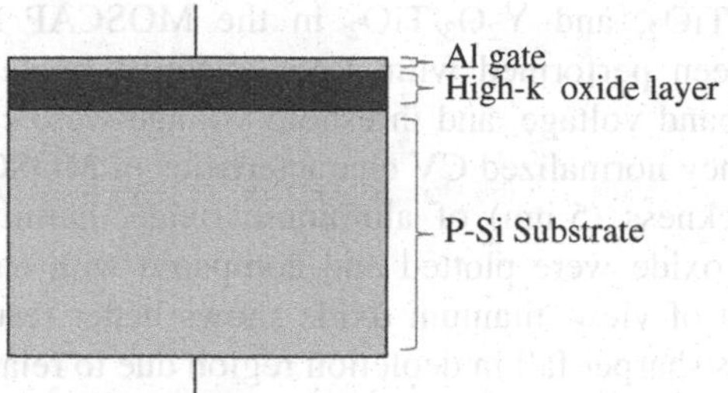

Fig. 1 Modeled MOSCAP structure with individual high-k layers

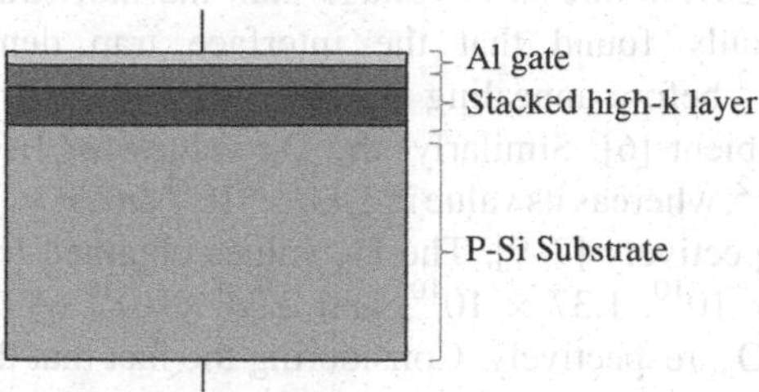

Fig. 2 Modeled MOSCAP structure with stacked high-k layers

simulator ATLAS of the TCAD simulation tool SILVACO. Thereafter, device parameters like capacitance and threshold voltages have been extracted. Shockley–Read–Hall recombination and CVT model along with numerical methods like Newton–Gummel-trap methods were also taken into account by the simulator ATLAS during CV simulation to extract the device properties [14] (Fig. 1).

Furthermore, stacked oxide layers provide better interface quality than individual oxide layer; thus, two high-k oxide layers with different oxide thickness (t_1 and t_2) and different permittivity (K_1 and K_2) have been taken into consideration to model the MOSCAP structures. The stacked combinations taken are HfO_2/TiO_2, Al_2O_3/TiO_2, and Y_2O_3/TiO_2 for 2 nm oxide thickness. The two high-k layers in stack act as two capacitors in series, and the equivalent oxide thickness can be obtained by Eq. (1) as in [15] (Fig. 2).

$$EOT = t_1 + \left(\frac{K_1}{K_2}\right) t_2 \tag{1}$$

3 Results and Discussion

By simulation along with numerical calculation, normalized capacitances, D_{it}, threshold voltages, and flat band voltages were calculated and compared. Taking different high-k oxide ($Al_2O_3/HfO_2/TiO_2/Y_2O_3$) with a thickness of 5 nm, simulations at room temperature have been done and compared. Again using stacked

HfO_2/TiO_2, Al_2O_3/TiO_2, and Y_2O_3/TiO_2 in the MOSCAP structures, the CV simulations have been performed with 2 nm effective oxide thickness and the respective D_{it}, flat band voltage, and threshold voltage were calculated.

The high-frequency normalized CV characteristics of MOSCAPs with constant dielectric layer thickness (5 nm) of aluminum oxide, hafnium oxide, titanium oxide, and yttrium oxide were plotted and compared with each other in Fig. 3. From material point of view, titanium oxide shows better result than other three oxides because of its sharper fall in depletion region due to relatively less interface trap charge density. Furthermore, the normalized capacitance with stacked HfO_2/TiO_2, Al_2O_3/TiO_2, and Y_2O_3/TiO_2 was compared in Fig. 4. From capacitance and threshold voltage point of view, the stack oxide layer yields better performance as it provides reduced device threshold voltage than the individual oxides.

It is experimentally found that the interface trap density for Al_2O_3 is 1×10^{12} eV^{-1} cm^{-2} before annealing and reduced to 4×10^{11} eV^{-1} cm^{-2} after annealing in H_2 ambient [6]. Similarly, the D_{it} values for HfO_2 are also around 6×10^{12} eV^{-1} cm^{-2}, whereas its value is 3.11×10^{11} and 9×10^{11} eV^{-1} cm^{-2} for TiO_2 and Y_2O_3, respectively [7–9]. The D_{it} values obtained from simulations are 4.81×10^{10}, 1.58×10^{10}, 1.37×10^{10}, and 2.34×10^{10} eV^{-1} cm^{-2} for Al_2O_3, HfO_2, TiO_2, and Y_2O_3, respectively. Considering the fact that the simulated results well match with the experimentally reported values, comparison of D_{it} profile with 5-nm-thick Al_2O_3, HfO_2, TiO_2, and Y_2O_3 high-k layer is shown in Fig. 5. In all the three cases, the D_{it} value increases with increases in gate voltage and its value is less in case of TiO_2 providing better interface quality than other three oxides.

The D_{it} profiles with stacked HfO_2/TiO_2, Al_2O_3/TiO_2, and Y_2O_3/TiO_2 have been compared in Fig. 6, which indicate the insignificant difference in interface trap charge density of stacked Al_2O_3/TiO_2 (1.34×10^{10} eV^{-1} cm^{-2}) Y_2O_3/TiO_2

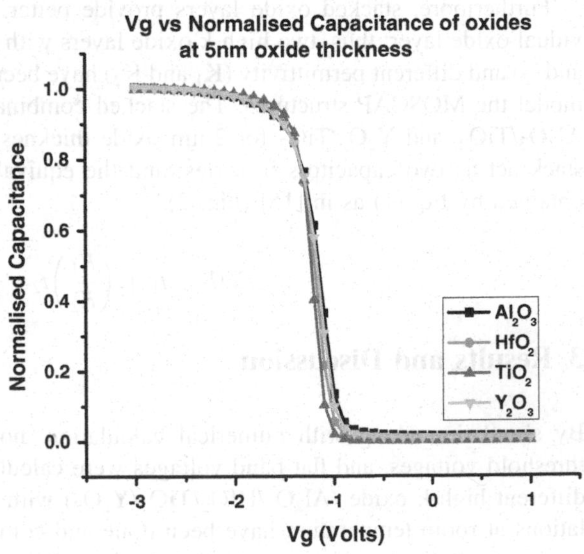

Fig. 3 CV characteristics with 5-nm-thick $Al_2O_3/HfO_2/TiO_2/Y_2O_3$ layers

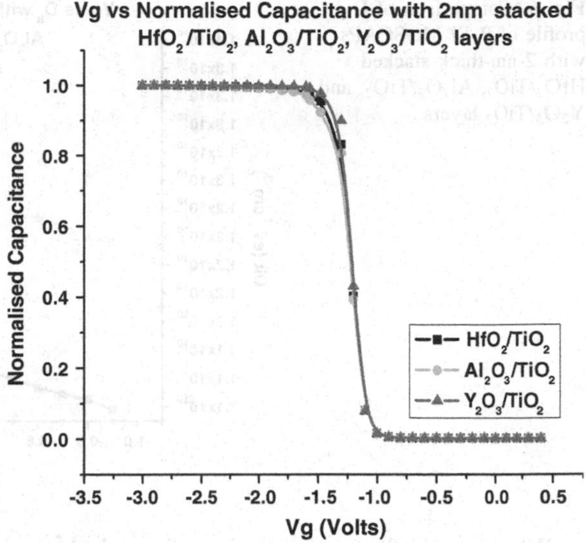

Fig. 4 CV characteristics with 2-nm-thick stacked HfO_2/TiO_2, Al_2O_3/TiO_2, and Y_2O_3/TiO_2 layers

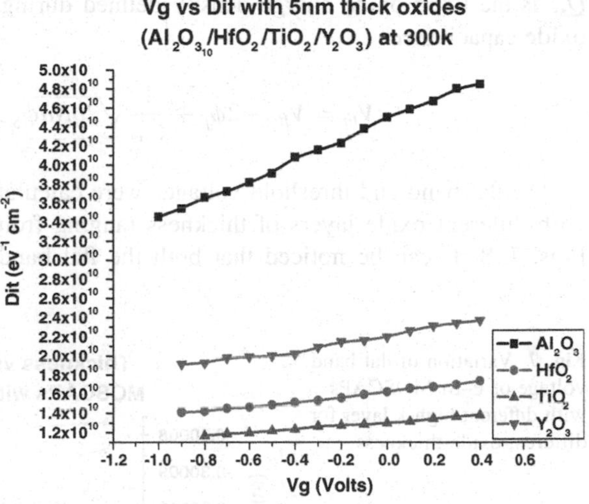

Fig. 5 Comparison of D_{it} profile of P–Si MOSCAPs with different high-k layers and 5 nm oxide thickness

(1.29×10^{10} eV^{-1} cm^{-2}) and HfO_2/TiO_2 (1.18×10^{10} eV^{-1} cm^{-2}). Among these, the hafnium oxide and titanium oxide stack can be considered to provide better interface quality than the other two oxides stacks.

The flat band and threshold voltages were also calculated by the given Eqs. (2) and (3) as in [16, 17]

$$V_{fb} = \left[\phi_m - \left(\mathcal{N} + \frac{E_g}{2e} + \phi_f\right)\right] - \frac{Q_{ss}}{C_{ox}} \tag{2}$$

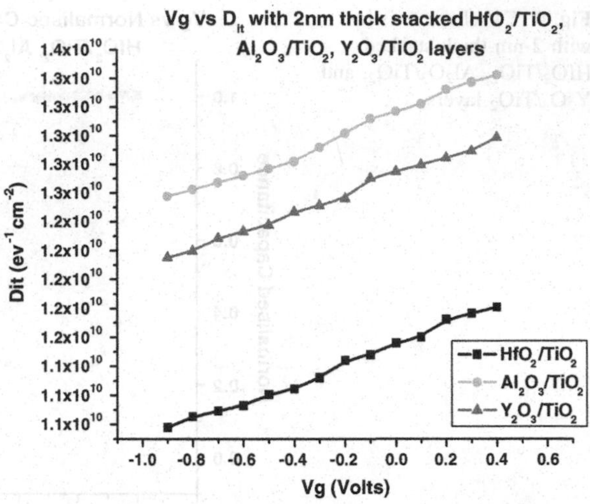

Fig. 6 Comparison of D_{it} profile of P–Si MOSCAPs with 2-nm-thick stacked HfO_2/TiO_2, Al_2O_3/TiO_2, and Y_2O_3/TiO_2 layers

Where ϕ_m is the work function of metal, $\mathcal{N}$ is the electron affinity of semiconductor, E_g is the band gap energy, ϕ_f is the Fermi potential of P-type substrate, Q_{ss} is the fixed oxide charge density defined during simulations, and C_{ox} is the oxide capacitance.

$$V_{th} = V_{fb} - 2\phi_f + \frac{1}{C_{ox}}\sqrt{2qNa\phi_{Si}}\sqrt{|-2\phi_f|} \tag{3}$$

The flat band and threshold voltages were calculated for MOSCAPs structures with different oxide layers of thickness ranging from 15 to 5 nm and plotted in Figs. 7, 8. It can be noticed that both the flat band and threshold voltages are

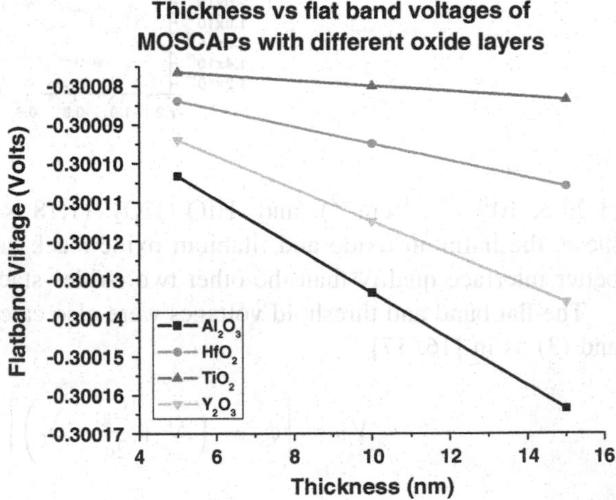

Fig. 7 Variation of flat band voltage of P–Si MOSCAPs with different high-k layer for different oxide thickness

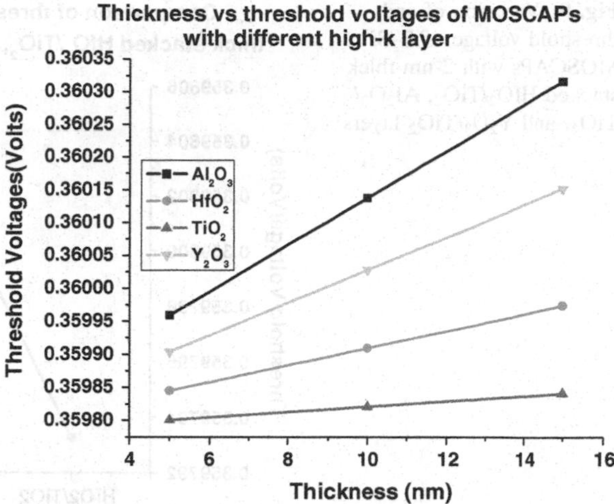

Fig. 8 Variation of threshold voltage of P–Si MOSCAPs with different high-k layer for different oxide thickness

linearly dependent on the oxide thickness. Threshold voltage is decreasing with downscaling and found to be least in case of 5 nm oxide thickness. Furthermore, threshold voltage is less in case of titanium oxide and gradually increasing toward hafnium oxide and yttrium oxide, respectively, and finally maximum in case of aluminum oxide. The experimentally reported flat band voltage for aluminum oxide is 0.54 V, varies from −1.0 to −0.5 V in case of hafnium oxide, 0.98 V for titanium oxide, and −0.5 V for yttrium oxide. Similarly, threshold ranges from 0 to 1 V in all the cases [6–9]. The obtained threshold voltage and flat band voltage are 0.359959, 0.35984, 0.359801, 0.359904 and −0.3001, −0.30008, −0.30007, −0.30009 V for Al_2O_3, HfO_2, TiO_2, Y_2O_3, respectively, which comes in the range of experimentally reported results. Differences between the experimental and simulated results may arise due to the factors related to fabrication.

From Fig. 9, it is realized that device threshold voltages are less in the case of HfO_2/TiO_2 stack than Al_2O_3/TiO_2 and Y_2O_3/TiO_2 layers of 2 nm oxide thickness. Therefore, while considering threshold voltage minimization HfO_2/TiO_2 stack combination can be preferred over others. Since the stack of two high-k oxide layers behaves like two capacitors in series, and their equivalent permittivity differs from the individual oxide. The permittivity of HfO_2/TiO_2 stack is more than that of Al_2O_3/TiO_2 and Y_2O_3/TiO_2. This is one of the most important requirements of gate oxide in MOS transistors, as threshold voltage has an indirect relationship with permittivity, for least threshold voltage.

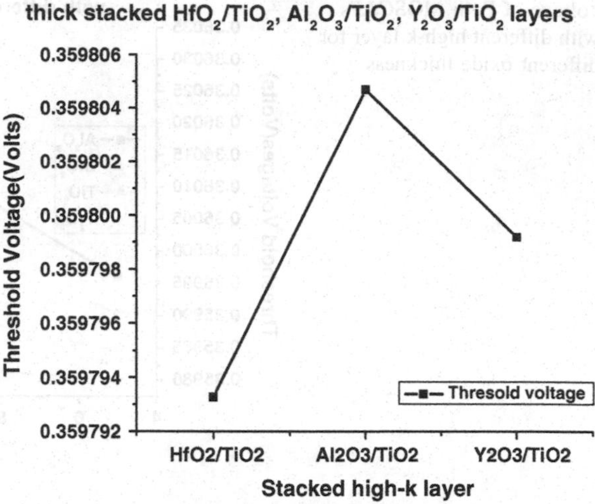

Fig. 9 Variation of and threshold voltage of P–Si MOSCAPs with 2-nm-thick stacked HfO_2/TiO_2, Al_2O_3/TiO_2, and Y_2O_3/TiO_2 layers

4 Conclusion

The high-frequency CV analysis of Al/high-k/Si MOSCAP structure was performed with Al_2O_3, HfO_2, TiO_2, and Y_2O_3 layers, and the interface trap charge densities, threshold voltages, and normalized capacitance were calculated by Termann method for 5 nm oxide thickness. Further analysis of the CV was also carried out by taking 2-nm-thick stacked HfO_2/TiO_2, Al_2O_3/TiO_2, and Y_2O_3/TiO_2 layers. From the results, it has been found that individually TiO_2 and among the stacks HfO_2/TiO_2 give better electrical properties. Additionally, D_{it} was calculated for both types of structures and compared. It was found that the interface trap charge density is about 3.5 times more in Al_2O_3 than TiO_2. The Y_2O_3 and HfO_2 have two and three times less D_{it} values than Al_2O_3. Furthermore, the stacked oxide provides less interface states than that of individual oxides and the HfO_2/TiO_2 combination has minimum interface states than that of other two combinations. Also better results were found for the stacked oxides in terms of threshold voltages and flat band voltages. Due to these reasons, we can propose the stacked high-ks for future nanoelectronic applications.

References

1. Chowdhury, M.H., Mannan, M.A., Mahmood, S.A.: Int. J. Emerg. Technol. Sci. Eng. **2**(2), (2010)
2. Okamotoa, K., Adachia, M., Kakushimab, K., Ahmeta, P., Sugiib, N., Tsutsuib, K., Hattoria, T., Iwaia, H.: IEEE Solid State Device Research Conference, 1-4244-1124-6 (2007)
3. Huang, A.P., Yang, Z.C., Chu, P.K.: Advances in Solid State Circuits Technologies, pp. 446 (2010). ISBN 978-953-307-086-5

4. Park, I.S., Lee, T., Ko, H., Ahn, J.: J. Korean Phys. Soc. **49**, S760–S763 (2006)
5. Lee, B.H., Oh, J., Tseng, H.H., Jammy, R., Huff, H.: Mat. Today, Elsevier Ltd. **9**(6), pp. 32–40 (2006)
6. Jeon, I.S., Park, J., Eom, D., Hwang, C.S., Kim, H.J., Park, C.J., Cho, H.Y., Lee, J.H., Lee, N.I., Kang, H.K.: Appl. Phys. Lett. **82**(7), (2003)
7. Chen, W.B., Xu, J.P., Lai, P.T., Li, Y.P., Xu, S.G.: Sci. Dir. **47**, 937–943 (2007)
8. Subham, K., Khan, R.U.: J. Nano- Electr. Phys. **5**(1) 01021(5 pp) (2013)
9. Roh, K., Yang, S., Hong, B., Roh, Y.: J. Korean Phys. Soc. **40**(1), 103–106 (2002)
10. Dutta, S., Sivaramakrishnan, R., Gopalan, S., Shankar, B.: World Acad. Sci. Eng. Technol. **34**, (2009)
11. Wu, Y.H., Lin, C.C., Hu, Y.C., Wu, M.L., Wu, J.R., Chen, L.L.: IEEE Electron. Dev. Lett. **32**(8), (2011)
12. Golam Sarwar, A.T.M., Siddiqui, M.R., Siddique, R.H., Khosru, Q.D.M.: TENCON 2009–2009 IEEE Region 10 Conference, 978–1–4244–4547–9/09 (2009)
13. Ong, D.W.G.: Modern MOS Technology, Processes, Devices and Design, Chap. 4, pp. 53–55, McGraw-Hill publishing Company Limited, New Delhi (1986)
14. Atlas User Manual Version: atlas 5.16.3.R, SILVACO International
15. Schroder, D.K.: Defects in Microelectronic Materials and Devices, Chap. 5. CRC Press, Florida (2008). ISBN: 978-1-4200-4376-1
16. Sze, S.M.: Physics of Semiconductor Devices, 2nd edn. Wiley, New York (1981)
17. Neemen, D.A.: Semiconductor Physics and Devices, 3rd edn., Chap. 10, pp. 438 Tata McGraw-Hill publishing Company Limited, New Delhi (2007)

Mesh-Type Split-Ring Resonator as Parasitic Radiator for SAR Reduction in Mobile Handset

Sarada Prasan Rout and Amlan Datta

Abstract The cellular personal communications system has marked a phenomenal and intensive development and its cons have been observed lately. A lot of work has not been dispensed to minimise the hazardous effects. Thus, the protection toward a person's head and safeguarding from the radiations of non-ionizing electromagnetic signals produced by cellular phones while in use is fetching one of the most concerned tasks to neutralize to an infinitesimally small amount. Since planar inverted-F antenna (PIFA) is a very favoured design in mobile applications, it is of prime importance to analyze the matching properties of such an antenna in real application scenarios (i.e., in close proximity to the head, grasped by the hand). Any matching deviation leads to the reduction of radiated power which should be accounted for in the system design. Different specifications of the proposed antenna are measured through simulations in free space as well as in the presence of the human head model. Specific absorption rate (SAR) is also computed for this antenna, and the interaction of the antenna with the human body is also investigated. The results of elaborated antenna design that have minimized radiation toward the user's head are presented in this paper.

Keywords Cellular personal communication · Non-ionizing · Electromagnetic signals · PIFA · Radiated power · SAR

S.P. Rout (✉) · A. Datta
School of Electronics Engineering, KIIT University, Bhubaneswar, India
e-mail: saradarout196@gmail.com

A. Datta
e-mail: amlandatta01@gmail.com

L.C. Jain et al. (eds.), *Intelligent Computing, Communication and Devices*,
Advances in Intelligent Systems and Computing 308,
DOI 10.1007/978-81-322-2012-1_54

1 Introduction

Nowadays, the public concerns regarding the negative health effect owing to constant exposure to electromagnetic waves (EM) have increased by an observable amount. When a container is filled with water, it holds up to a certain point, but once it reaches the rim of the container, it starts spilling. In the same manner, human body will also absorb radiation up to a certain limit. Hence, due to the concerned hazard, there is a need to take precautionary steps in this regard. Law making and many public corporations around the world monitor the safety guidelines/criteria and enforce them for protection against EM waves [1, 2]. Voltage-standing wave ratio (VSWR), radiation pattern, and antenna gain of a handset are influenced by the user. The separating distance from an antenna in the handset and the head has a proportional relationship with the SAR and antenna efficiency. The separating space between the mobile antenna and the user's head is said to be one of the important factors for deviation in antenna matching properties whether it is an external or built-in antenna setup [3, 4].

Organization of paper: Sect. 2 illustrates methodology used for the simulations, antenna design, and parameters. We present simulation results with relevant discussion in Sect. 3. Conclusion and Acknowledgment sections are at the end of this paper.

2 Methodology

The planar inverted-F antenna (PIFA) applications for mobile phones are increasing in a number of ways. The PIFA has the advantages of having a small and simple design, multiband resonant properties, lightweight, low cost, attractive radiation pattern, reliable performance, and typical low SAR properties [5]. SAR is the measure of electromagnetic energy being absorbed by the biological tissue mass when exposed to radiating devices. This power absorbed by the user is measured by a parameter called SAR.

$$\mathrm{SAR} = \frac{\sigma[E_t]^2}{\rho} \tag{1}$$

2.1 Antenna Design Parameters

The setup of the proposed antenna is composed of a main patch radiator at the top with a height of 8 mm from the ground plane of dimensions 45 × 30 mm. The proposed antenna is mounted on a metallic chassis. The main plate at the top layer, a ground plane at the bottom layer, and a folded stub plate are perpendicular to the two plates. The main plate and the folded stub share the common feeding strip

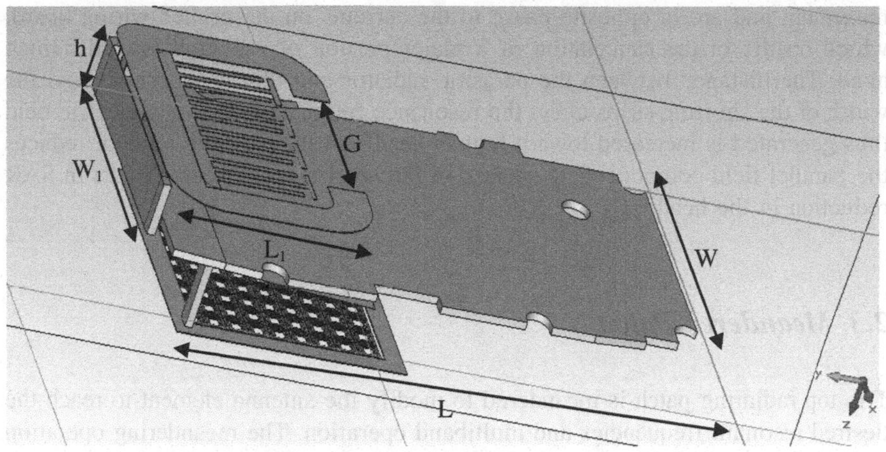

Fig. 1 Perspective view of PIFA model

connected to the transmission line etched on the back of the ground plane. The two patches share a common shorting strip, while the folded stub is not grounded. The stub is used to bring about a perfect matching at particular resonance frequency. The metal chassis supports both the Printed Wiring Board (PWB) and the ground plane. The ground is being connected by a shorting plate and at one end of the top resonating patch. A feed source is positioned on the ground plane and linked to the patch by a 2-mm large conductor. A shorting plate of 3 mm width is placed 6 mm away from the feed. Figure 1 shows the geometry of the PIFA element optimized for operating at two different frequencies. In this structure, the top radiating patch is of length 40 mm and the width is 20 mm. The height of the patch is 0.035 mm. Both the ground plane and the substrate plate are 1 mm thick.

2.2 Mesh-Type Parasitic Radiator with Matching Capacitive Stub

Parasitic radiator is a parallel-plate-like structure which is shorted at the upper corners of the PWB of handset. Parasitic radiator significantly reduces the coupling between the PWB in the handset and the head phantom. The basic principle regarding the operations of parasitic radiator is said to be canceling a part of the field near the head. An array of split-ring resonator is of size 1.9 × 1.9 mm and 2.9 × 2.9 mm for inner and outer cells, respectively. Ferrite is used as major building material for the mesh-type parasitic radiator. There will be a significant reduction in SAR values for a resonating length of the parasitic radiator with the array of split-ring resonators. The currents generated on the parasitic radiator are maximum and are in opposite phase to the currents on the printed wiring board, which results in the cancelation of a major portion of radiation toward human head. The distance between the parasitic radiator and the metal chassis and the width of the shorting strips affect the resonance frequency. The total electric field thus generated is increased toward human head, but the parasitic radiator reduces the parallel field components to surface of the head, which in turn results in SAR reduction in the head.

2.3 Meandered Patch

The top radiating patch is meandered to modify the antenna element to reach the desired resonant frequencies and multiband operation. The meandering operation makes effective uniform flow of current throughout the length and radiation as well (Figs. 2, 3, 4).

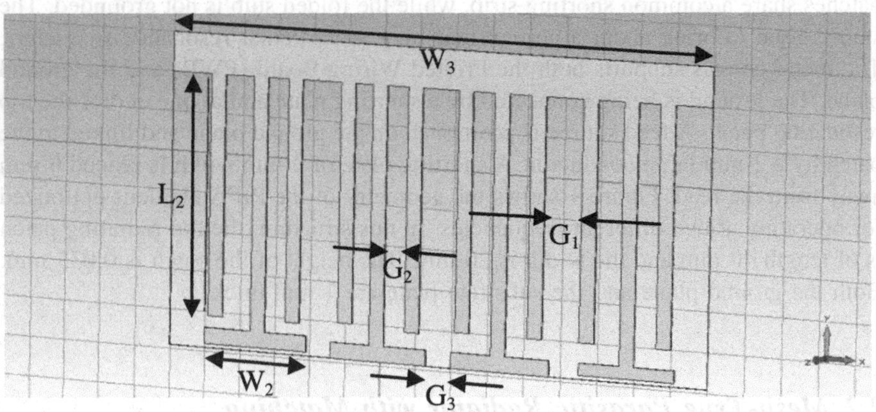

Fig. 2 Front view of PIFA meander element model

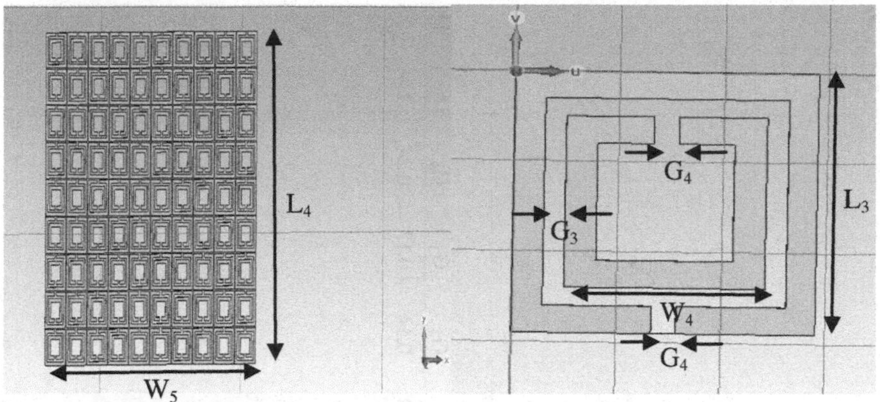

Fig. 3 Perspective view of an array of split-ring resonators

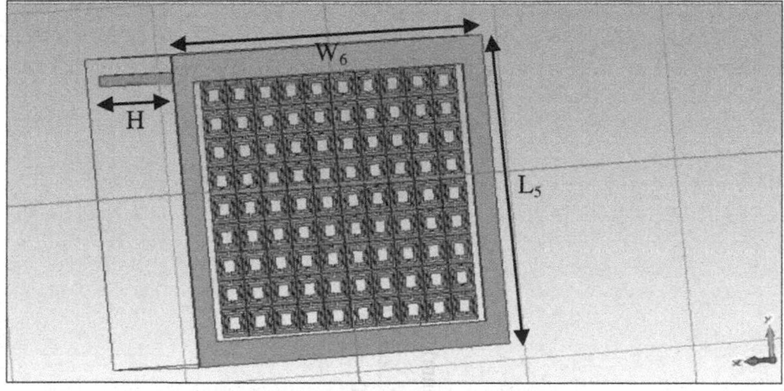

Fig. 4 Perspective view of mesh-type parasitic radiator

Mesh-type parasitic radiator is created and aligned on the upper edge of the PWB. The parasitic radiator is connected to the ground plane in its upper corner with 1.3-mm-wide shorting strips. The length of the parasitic radiator is varied between 3 and 53 mm, and the height from the PWB h is settled between 2.4 and 3.1 mm (Table 1).

Table 1 Dimensions of the PIFA model

Parameters with numeric values (all dimensions in mm)																		
L	L_1	L_2	L_3	L_4	L_5	W	W_1	W_2	W_3	W_4	W_5	G	G_1	G_2	G_3	G_4	h	H
100	20	12.5	2.9	26.1	32.1	40	30	0.88	22	1.9	29	12	0.28	0.14	0.12	0.24	2.4	4.8

3 Antenna Simulation Results

3.1 Return Loss and Radiation Pattern

The return loss has been evaluated for the design through CST Microwave Studio solver [6]. The simulation results for the handheld device is shown in Fig. 5 focus a reasonable value of return loss. The minimum return loss is at 1 GHz and 1.9 GHz, and the value is about −13 and −14 dB, respectively, indicating amount of power that is transferred to the load or amount of power reflected back.

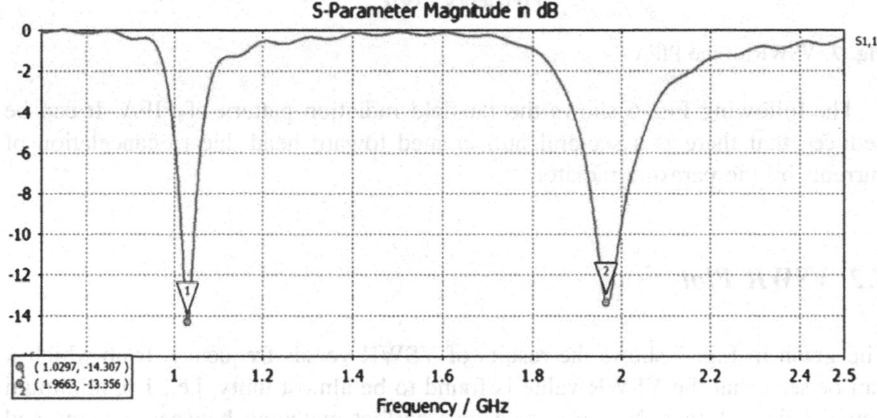

Fig. 5 Return loss (in dB)

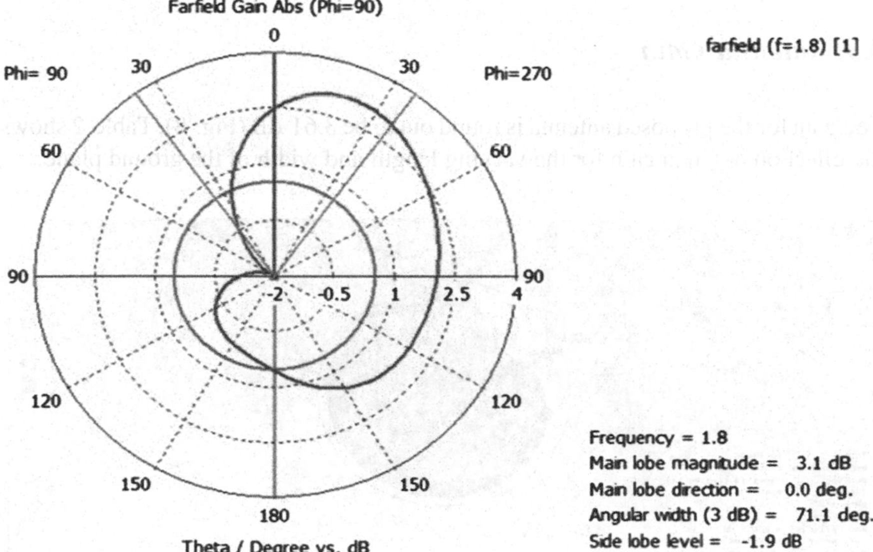

Fig. 6 Radiation pattern

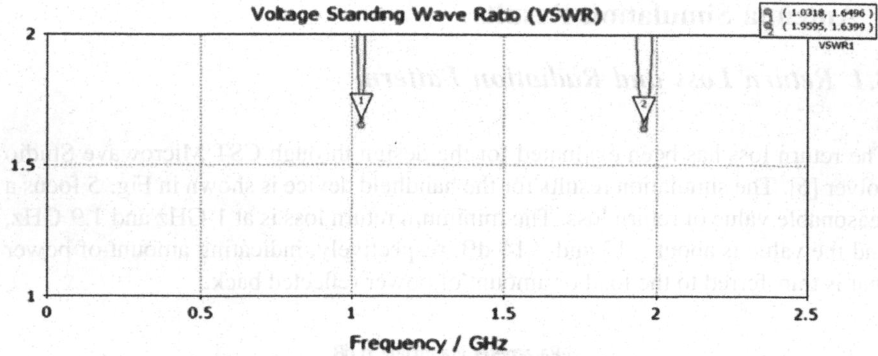

Fig. 7 VSWR of the PIFA

The following Fig. 6 shows the far-field radiation pattern of PIFA. It can be deduced that there is a sectoral null created toward head due to cancelation of currents by the parasitic radiator.

3.2 VSWR Plot

The graph in Fig. 7 shows the results of VSWR versus frequency, from which it can be seen that the VSWR value is found to be almost unity, i.e., 1.6, minimum power reflected back to source, and thus perfect matching between antenna and feeding system.

3.3 Antenna Gain

The gain for the proposed antenna is found out to be 3.61 dB (Fig. 8). Table 2 shows the effect on antenna gain for the varying length and width of the ground plane.

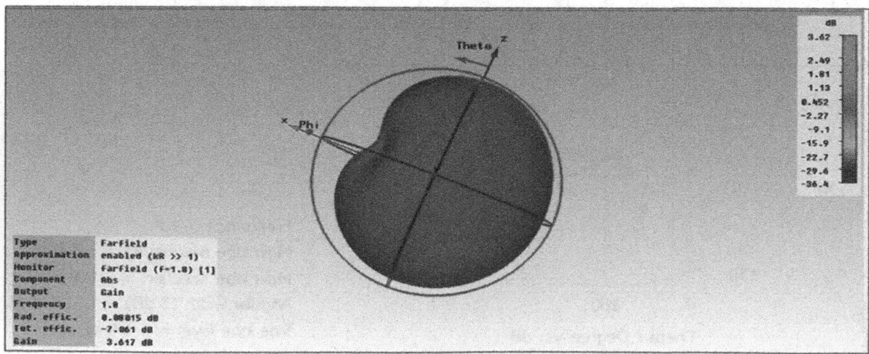

Fig. 8 Gain of the PIFA

Table 2 Effect of varying length and width of ground plane on antenna gain

S. No	Parameters	Dimensions in mm	Gain in dB	Material assigned
1	Varying ground thickness (length constant—40 × 20 mm)	0.8	2.09	1. Radiating patch: copper 2. PCB board: Rogers ultralam 3. Ground: copper 4. Parasitic radiator: ferrite sheets
		0.9	3.24	
		1	3.12	
2	Varying ground length (thickness constant—0.9 mm)	40 × 22.5	3.30	
		40 × 25	3.37	
		45 × 25	3.51	
		45 × 30	3.62	
		45 × 35	3.57	

3.4 SAR Calculations

The PIFA is embedded into the handset. The configuration is for the measurement of SAR as shown in Fig. 9.

The simulated results for the antenna designed can be used for cellular applications at 1 and 1.9 GHz band. The interaction between PIFAs with parasitic radiator and human head model has a significant effect on antenna gain, VSWR, and the radiation pattern. However, the simulation result of the proposed PIFA gives a low value for peak SAR almost equal to 0.46 W/kg averaged over 10 g of tissue and almost 0.65 W/kg averaged over 1 g of tissue where the FCC guidelines to sell a phone in the USA are 1.6 W/kg averaged over 10 g of tissue.

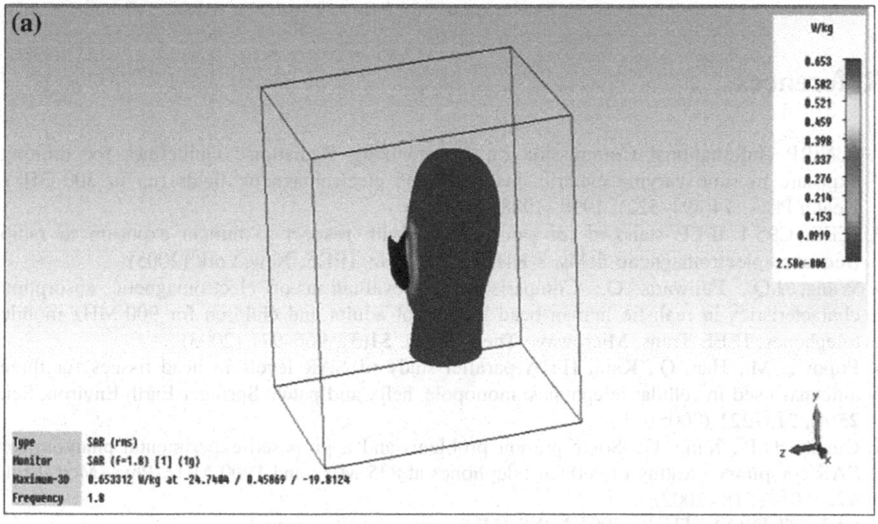

Fig. 9 SAR value. **a** Averaged over 1 g of tissue. **b** Averaged over 10 g of tissue

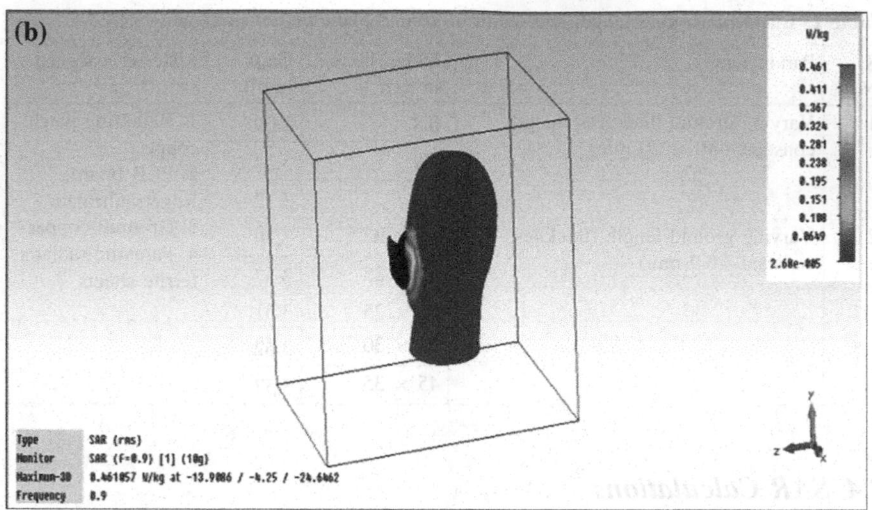

Fig. 9 (continued)

Electromagnetic radiation toward the user's head is dropped in the simulated antenna installed on the handset. It is also observed that the frequently used antennas can be replaced with the newly devised one, without any transformation in fundamental construction of cellular phone.

Acknowledgment I would like to thank Mrs. Lizina Khatua, Assistant Professor, KIIT University, and Mr. Asit Panda, Assistant Professor, NIST, Berhampur, for his support during all antenna simulations in CST Microwave Studio.

References

1. ICNIRP (International Commission on Non-Ionizing Radiation): Guidelines for limiting exposure to time varying electric, magnetic and electromagnetic fields (up to 300 GHz). Health Phys. **74**(494–522), 1998 (1988)
2. IEEE, C95.1 IEEE standard for safety levels with respect to human exposure to radio frequency electromagnetic fields, 3 KHz to 300 GHz, IEEE, New York (2005)
3. Wang, J.Q., Fujiwara, O.: Comparison and evaluation of electromagnetic absorption characteristics in realistic human head models of adults and children for 900 MHz mobile telephones. IEEE Trans. Microwave Theor. Tech. **51**(3), 966–971 (2003)
4. Popović, M., Han, Q., Kanj, H.: A parallel study of SAR levels in head tissues for three antennas used in cellular telephones: monopole, helix and patch. Springer Earth Environ. Sci. **25**(4), 215–221 (2005)
5. Gandhi, O.P., Kang, G.: Some present problems and a proposed experimental phantom for SAR compliance testing of cellular telephones at 835 MHz and 1900 MHz. Phys. Med. Biol. **47**, 1501–1518 (2002)
6. CST STUDIO SUITE^{TM} 2006 MWS (Microwave Studio) manual

On-Chip Improved Single Clock Swing Enhanced Charge Pump Circuit

Nikhil Deo, Rahul Roushan and Rohit Kumar Sah

Abstract In this paper, an improved single-clock swing-enhanced charge pump circuit is presented. It is an improved version of single-clock swing-enhanced charge pump (SCSECP). We have made certain modifications in the previous SCSECP circuit while keeping the general topology same, and as a result, higher output voltage is achieved. In general, charge pump circuits use the available low-voltage supply to generate a higher voltage, but in our case, the proposed six-stage charge pump gives an output voltage of 51.70 V at no-load for an input voltage of 1.2 V.

Keywords Charge pump · Clock-level shifter · Mixed clock · High voltage

1 Introduction

With rapid advancement in VLSI design, a large number of different digital as well as analog modules can be integrated on a single chip. These modules may require different supply voltages for their operation. In such cases, power converter can be used to generate multiple voltage levels from a single low-voltage supply. Either inductive boost converter or charge pump circuit can be used to design power converters. But for on-chip power converters, charge pump is preferred over inductive boost converters as charge pump only requires capacitors and MOS transistors, whereas boost converter requires an inductor which makes the design bulky and costly.

N. Deo (✉) · R. Roushan · R.K. Sah
North Eastern Regional Institute of Science & Technology, Itanagar,
Arunachal Pradesh, India
e-mail: nikh.deo@gmail.com

R. Roushan
e-mail: rhl.roushan@gmail.com

L.C. Jain et al. (eds.), *Intelligent Computing, Communication and Devices*,
Advances in Intelligent Systems and Computing 308,
DOI 10.1007/978-81-322-2012-1_55

The fundamental idea of the charge pump circuit was first proposed by Cockcroft and Walton [1]. But his proposed model was not feasible for on-chip implementation. Later, Dickson [2] proposed a model for charge pump similar to that of Walton Cockcroft's model that uses MOS transistor instead of diode as charge transfer device. A new topology of the charge pump circuit was proposed by Ding et al. [3] in which the intermediate voltage generated by the first-grade charge pump is used by a clock-level shifter to upgrade the clock voltage swing, which is then fed to second-grade charge pump to achieve high output voltage. Ansari et al. [4] proposed a single-clock charge pump circuit suitable for low-voltage application. Roushan et al. proposed a single-clock swing-enhanced charge pump (SCSECP) [5], which incorporates a single-clock charge pump and clock-level shifter circuit to achieve higher output voltage. Recently, Mondal and Paliy proposed a new enhanced single-clock charge pump [6] suitable for micro-scale energy-harvesting systems.

In this paper, an improved version of SCSECP circuit is presented and we have shown that our improved single-clock swing-enhanced charge pump (ISSCP) can achieve higher output voltage than the SCSECP circuit.

2 Charge Pump Circuit

This section gives a brief introduction to some of the charge pump that forms the basic building block for our charge pump.

2.1 Dickson Charge Pump

The Dickson charge pump [2] lays down the basic idea for most of the present day charge pumps. Figure 1 shows a twelve-stage Dickson charge pump; this circuit is fed with two anti-phase pumping clocks and a low DC voltage (V_{IN}). The circuit has only diode-connected PMOS and capacitor. The diode-connected PMOS allows the current to flow in only one direction and capacitors are charged and discharged successively during each clock cycle. The PMOS (P_F) and capacitor C_F at the output form a capacitive load [6].

2.2 Single-clock Charge Pump

Single-clock charge pump proposed by Ansari et al. [4] has proved to be more efficient than the Dickson charge pump. In this charge pump, the need for two anti-phase clocks is eliminated. A single stage of their charge pump consists of two NMOS, two PMOS, and two charging capacitors. We can cascade these stages in

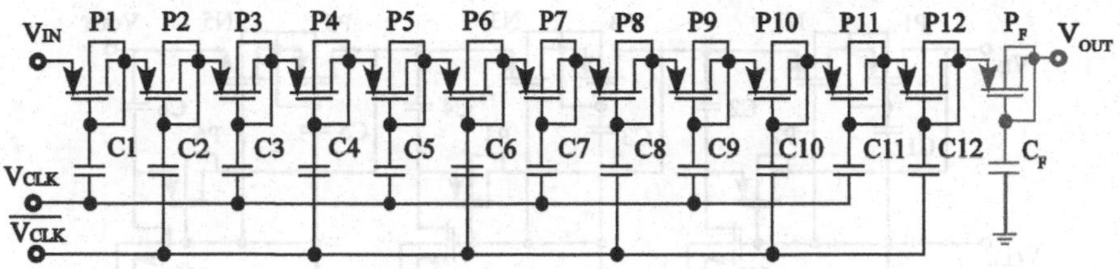

Fig. 1 Twelve-stage Dickson charge pump with capacitive load

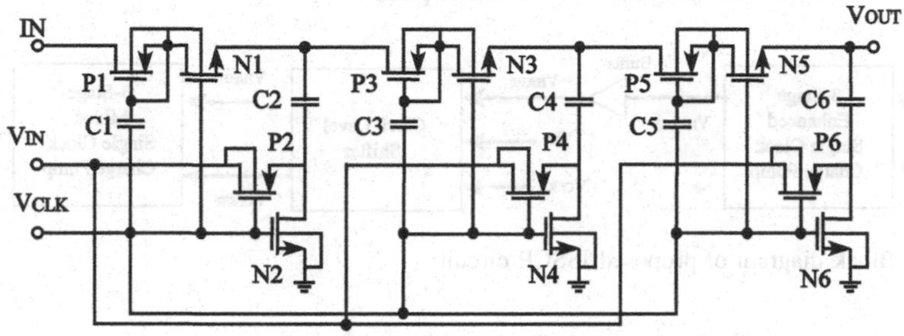

Fig. 2 Three-stage single-clock charge pump

series to attain a desirable voltage level. Figure 2 shows a three-stage single-clock charge pump.

2.3 Enhanced Single-clock Charge Pump

A three-stage enhanced single-clock charge pump with appropriate changes in single-clock charge pump circuit topology proposed by Mondal and Paliy [6] is shown in Fig. 3. In this charge pump, the internal voltage of each stage is used by the succeeding similar stages to supply voltage to both the PMOS transistors of that stage. This charge pump can achieve higher output voltage amplitude than the single-clock charge pump circuit of Fig. 2.

3 Proposed Improved Single-clock Swing-enhanced Charge Pump Circuit

We have proposed an ISSCP which can achieve higher output voltage than the SCSECP circuit. ISSCP is based on the same topology that was first used in [3] and later in [5]. ISSCP consists of three different blocks as shown in Fig. 4; the first block consists of three-stage enhanced single-clock charge pump. This block

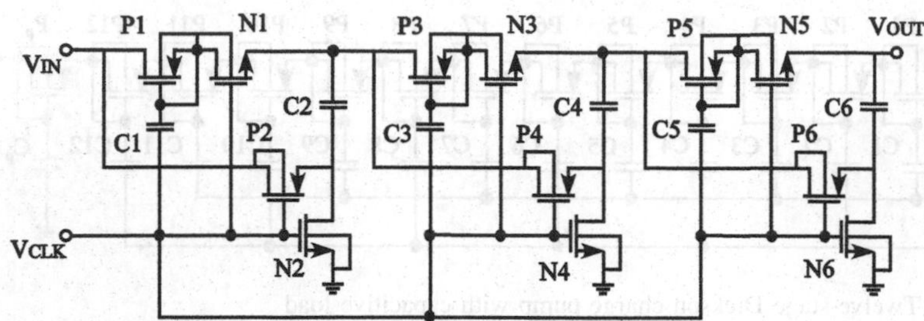

Fig. 3 Three-stage enhanced single-clock charge pump

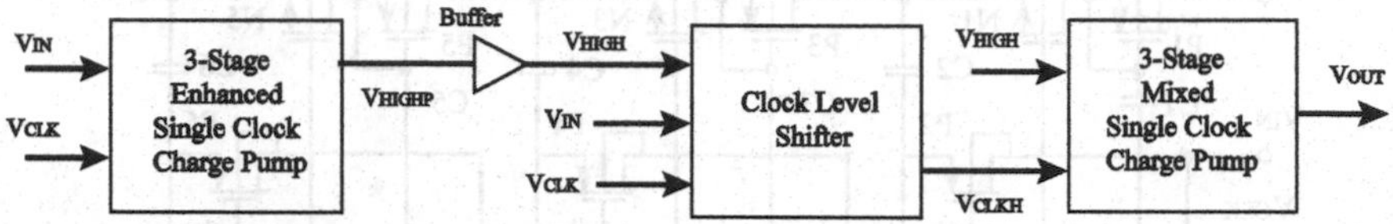

Fig. 4 Block diagram of proposed ISSCP circuit

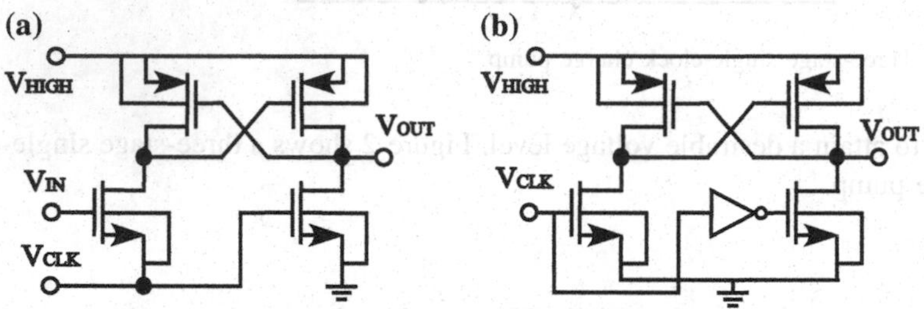

Fig. 5 **a** Clock-level shifter for ISSCP. **b** Clock-level shifter for SCSECP

produces a certain pulsating square voltage (V_{HIGHP}) having an amplitude higher than the input voltage (V_{IN}).

The second block consists of the clock-level shifter which is the heart of this charge pump. Figure 5a shows the clock-level shifter that is used in ISSCP, and it is the pass-transistor half-latch-level converter [7, 8]. Similarly, Fig. 5b shows the clock-level shifter used in the SCSECP circuit. As shown in Figs. 4 and 5, the clock-level shifter circuit has three inputs; the first input (V_{HIGH}) acts as the supply voltage for this clock-level shifter. The voltage V_{HIGH} is the output voltage of the buffer (Fig. 4). The second input to the clock-level shifter is V_{IN} which is a DC voltage, and the third input is the original clock signal V_{CLK}.

The clock-level shifter upgrades the voltage swing of the original clock from 0–V_{IN} to 0–V_{HIGH}, and hence, we get a new upgraded clock signal (V_{CLKH}) with a swing of 0–V_{HIGH}.

The third block is a three-stage mixed single-clock charge pump (Fig. 6). This circuit is a mixed-clock version of the circuit in Fig. 3. This idea is based on the

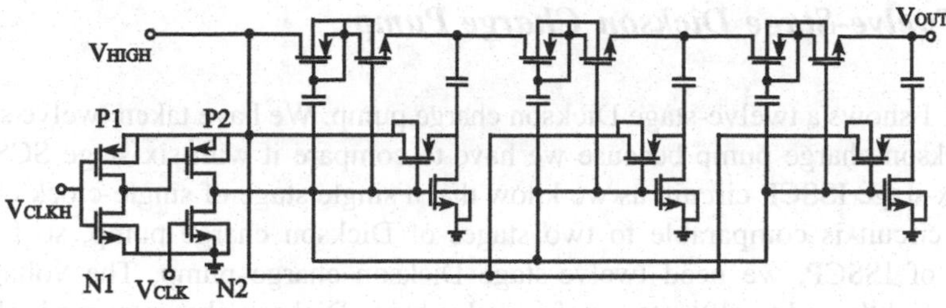

Fig. 6 Mixed single-clock charge pump

mixed-clock second-grade charge pump circuit used in [3], in which both the clock V_{CLK} and the new V_{CLKH} are used. It uses the voltage V_{HIGH} as its input voltage and V_{CLKH} as the clock signal for its operation and boosts the V_{HIGH} to a much higher level; hence, as a result, a higher output voltage (V_{OUT}) is generated.

For successful operation of this circuit, there are some issues that need to be resolved. These issues are also there in SCSECP circuit. Charge pump circuits are known to have poor output regulation, so to sustain a steady voltage at the output of the charge pump. One of the possible solutions as mentioned in [9, 10] is to use an 'energy efficient feedback control scheme,' and also as the output V_{HIGH} of the first block of Fig. 4 is pulsating in nature, a good regulator [11] is required to obtain a stable DC voltage V_{HIGH}.

To solve these issues of our charge pump, further research is required. In our design, we have placed a buffer block which will incorporate these control circuitries. As the buffer is not designed, so for our experimental setup, we assumed that the buffer output voltage is a steady DC voltage V_{HIGH} equals to the stable positive amplitude of the pulsating voltage V_{HIGHP} as obtained from simulation results, this is based on the same logic that was used for SCSECP in [5].

Hence, for simulation purpose, the clock-level shifter is connected to (a) DC voltage supply of 7.4 V for V_{HIGH}, (b) voltage supply of 1.2 V for V_{IN}, and (c) a clock supply voltage (V_{CLK}) of voltage swing (0–1.2 V).

4 Simulation Results

The charge pump circuits are implemented in 180-nm standard CMOS technology, and simulations were carried out using LT spice simulator. The input voltage (V_{IN}) is set to 1.2 V and the charge storage capacitors of 1 pF are used in each stage. The size of NMOS is 0.36 μm/0.18 μm and for PMOS is 0.72 μm/0.18 μm in each stage. The clock voltage amplitude (V_{CLK}) is 1.2 V, and clock frequency chosen is 1 MHz with 50 % duty cycle. To show a comparison, we have simulated a twelve-stage Dickson charge pump, a three-stage SCSECP circuit, and our three-stage ISSCP circuit. All of them were designed with 180-nm technology under no-load and capacitive-load conditions.

4.1 Twelve-Stage Dickson Charge Pump

Figure 1 shows a twelve-stage Dickson charge pump. We have taken twelve stages of Dickson charge pump because we have to compare it with six-stage SCSECP and six-stage ISSCP circuit, as we know that a single stage of single-clock charge pump circuit is comparable to two stages of Dickson charge pump, so for six stages of ISSCP, we need twelve-stage Dickson charge pump. The voltage at fourth, eighth, and twelfth stages of a twelve-stage Dickson charge pump is shown in Fig. 7. We get an output voltage of 13.12 V at the twelfth stage, i.e., at the drain of the P12 (PMOS) as shown in Fig. 1.

4.2 Six-stage Single-clock Swing-enhanced Charge Pump

Figure 8 shows the output waveform of last three stages, i.e., fourth, fifth, and sixth stages of a six-stage SCSECP circuit under no-load condition. A maximum output voltage of 47.10 V is achieved at the sixth stage, 33.67 V at the fifth stage, and 20.30 V at the fourth stage of the SCSECP circuit.

4.3 Proposed Improved Six-stage Single-clock Swing-enhanced Charge Pump

The simulation results for the proposed charge pump circuit under no-load condition are shown in Fig. 9. It shows the output voltages of fourth, fifth, and sixth stages. The maximum output voltage obtained at the output (i.e., sixth stage) is 51.70 V. This is higher than the above-mentioned charge pump circuits.

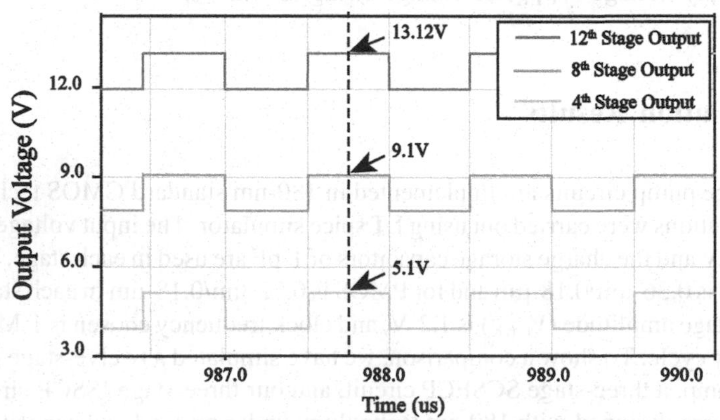

Fig. 7 Voltages at fourth, eighth, and twelfth stages of a twelve-stage Dickson charge pump

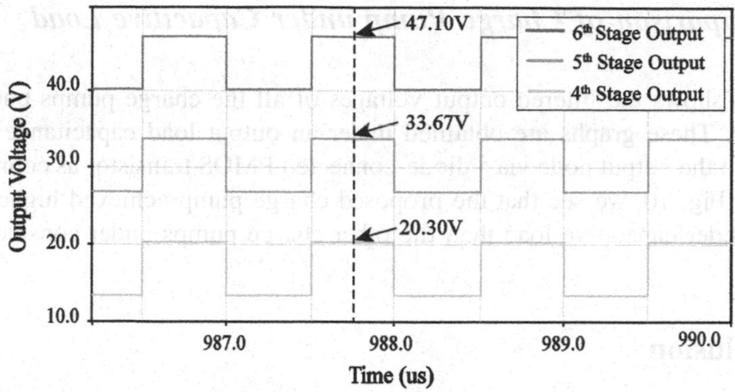

Fig. 8 Pulsating voltages at fourth, fifth, and sixth stages of a six-stage SCSECP circuit

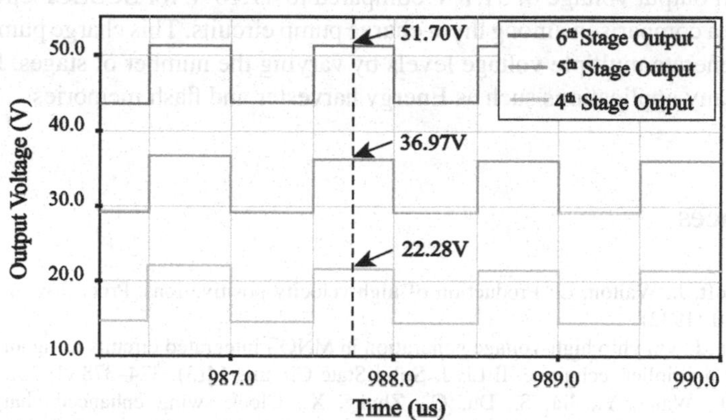

Fig. 9 Pulsating voltages at fourth, fifth, and sixth stages of the proposed ISSCP

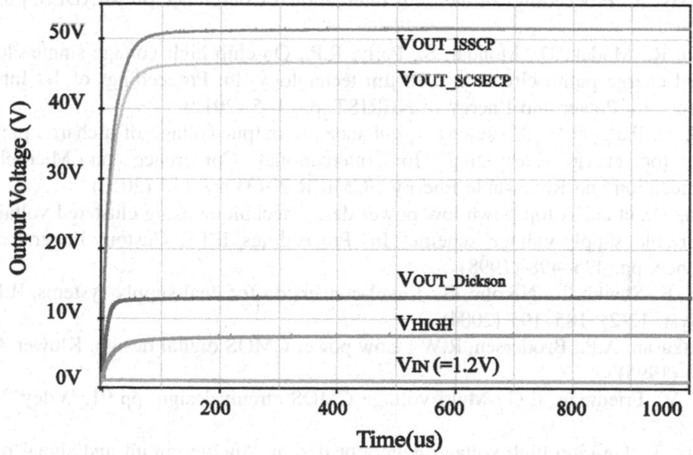

Fig. 10 Output voltages of 12-stage Dickson, 6-stage SCSECP and 6-stage ISSCP under load

4.4 Comparison of Charge Pump under Capacitive Load

Figure 10 shows the filtered output voltages of all the charge pumps under consideration. These graphs are obtained under an output load capacitance of 1 pF attached to the output node via a diode-connected PMOS transistor as connected in Fig. 1. In Fig. 10, we see that the proposed charge pump achieved higher output voltage under capacitive load than the other charge pumps under consideration.

5 Conclusion

Our circuit obtained a higher output voltage than the SCSECP circuit. We managed to get high output voltage of 51.7 V compared to 47.10 V for SCSECP circuit. We also made a comparison among these charge pump circuits. This charge pump can be used to generate multiple voltage levels by varying the number of stages. It can be used in many applications such as Energy harvester and flash memories.

References

1. Cockcroft, J., Walton, E.: Production of high velocity positive ions. Proc. Roy. Soc. A **136**, 619–630 (1932)
2. Dickson, J.: On-chip high-voltage generation in MNOS integrated circuits using an improved voltage multiplier technique. IEEE J. Solid-State Circuits **11**(3), 374–378 (1976)
3. Ding, J., Wang, Y., Jia, S., Du, G., Zhang, X.: Clock swing enhanced charge pump. International Conference of Electron Devices and Solid-State Circuits (EDSSC), pp. 1–2 (2011)
4. Ansari, M., Ahmad, W., Signell, S.: Single clock charge pump designed in 0.35 µm technology. In: Proceedings of the 18th International Conference on MIXDES, pp. 552–556 (2011)
5. Roushan, R., Modak, D., Mondal, S., Paily, R.P.: On chip high voltage single clock swing enhanced charge pump circuit in 0.18 µm technology. In: Proceedings of 1st International Conference on Power and Energy in NERIST, pp. 1–5 (2012)
6. Mondal, S., Paily, R.P.: A strategy to enhance the output voltage of a charge pump circuit suitable for energy harvesting. In: International Conference on Microelectronics, Communication and Renewable Energy (ICMiCR-2013) pp. 1–5 (2013)
7. Hamada, M., et al.: A top-down low power design technique using clustered voltage scaling with variable supply-voltage scheme. In: Proceedings IEEE Custom Integrated Circuits Conference, pp. 495–498 (1998)
8. Ishihara, F., Sheikh, F., Nikolić, B.: Level-conversion for dual-supply systems. IEEE Trans. VLSI Syst. **12**(2), 185–195 (2004)
9. Chandrakasan, A.P., Brodersen, R.W.: Low power CMOS digital design. Kluwer Academic, Norwell (1995)
10. Kursun, V., Friedman, E.G.: Multi-voltage CMOS circuit design, pp 91, Wiley, New York (2006)
11. Tanzawa, T.: On-chip high-voltage generator design. Analog circuit and signal processing, Springer, Berlin, p. 116 (2013)

Material-Based Vibration Characteristic Analysis of Heavy Vehicle Transmission Gearbox Casing Using Finite Element Analysis

Ashwani Kumar, Arpit Dwivedi, Himanshu Jaiswal and Pravin P. Patil

Abstract Transmission system is the most important part of the vehicle assembly. Truck transmission casing was subjected to vibration induced by the vehicle running, varying speed, and torque conditions. It is required to find the natural frequency and mode shape for accurate prediction of transmission casing life and prevent it from damage. The main objective of this research work is to find the influence of transmission casing material on natural frequency and mode shape of free vibration. Casing is made of material like cast iron, Al alloys, Mg alloys, structural steel, and composites. In this paper, the free vibration analysis of transmission casing has been performed by finite element simulation using ANSYS 14.5 software. The vibration patterns for first twenty modes were studied for all types of materials. The mode shapes show that the natural frequency of all materials varies in the range of 1,000–3,800 Hz. The analysis results were compared with experimental result available in the literature.

Keywords Transmission casing · Cast iron · FEA · Free vibration · Material influence · Modal frequency

1 Introduction

In automobile truck, the main function of engine is to generate power and transmission is used to transmit powers from engine to automobile wheels. This power transmission process is not as simple as it feels. The transmission components or

A. Kumar (✉) · A. Dwivedi · H. Jaiswal · P.P. Patil
Department of Mechanical Engineering, Graphic Era University, Dehradun 248002, India
e-mail: kumarashwani.geu@gmail.com

P.P. Patil
e-mail: pravinppatil2004@gmail.com

L.C. Jain et al. (eds.), *Intelligent Computing, Communication and Devices*, Advances in Intelligent Systems and Computing 308,
DOI 10.1007/978-81-322-2012-1_56

parts are subjected to varying loading conditions. These varying loading and boundary conditions produce noise and vibration. Various types of damping material like cast iron are used to eliminate and absorb the vibration waves [1, 2]. Noise and vibration are the two technical indexes for the transmission failure, so vibration is selected as a study parameter. Researchers have done various studies on dynamic response of transmission system since past two decades, but it is a very complex procedure in terms of design, measurement, or mathematical modeling.

Automobile transmission system is a combination of gears to meet the torque variation for the varying speed conditions. Transmission system can be classified into three types—automatic, manual, and continuously variable transmission. The simplest type of transmission is manual. Manual transmission is of two types: sliding mesh and constant mesh. The main reason of noise and vibration is wrong shifting of gears, uneven road surfaces, loose fixturing of transmission gears, components, and housing. Clashing is the general phenomena that occur during shifting of gears. Clashing is a loud noise produced during collision of gear tooth, and this collision leads the transmission failure. In other two types of transmission, automatic and continuously variable transmissions, there is less driver interaction [3, 4]. Slack in drive train mechanism produces high vibration known as transmission shock. Transmission shock is the high grade of vibration that may cause the failure of housing [5, 6].

Tuma [7] has studied the noise and vibration of transmission system. Author has solved the gear noise problem by introducing an enclosure to reduce radiated noise. TARA trucks have been selected as a research object. The Fourier transform is used for the analytical analysis. Analytical result is verified using experimental investigation. The extensive noise is produced during the tooth meshing or at structural resonance frequency. The natural frequency of vibration is varying in between 500 and 3,500 Hz at varying rpm. The severe vibration occurs at the frequency range of 500–2,500 Hz. Our results show that the natural frequency of vibration varies from 1,002.5 to 2,954.8 Hz, which is verified by Jiri Tuma results. Åkerblom [8] has performed a literature review and concluded that transmission error is an important excitation mechanism for gear noise and vibration. In addition to transmission error, friction and bending moment are another reason responsible for failure.

Nacib et al. [9] have studied the heavy gearbox of helicopters. To prevent breakdown and accident in helicopters, gear fault detection is important. Spectrum analysis and cepstrum analysis method is used to identify damage gear. Fourier analysis is used for analytical results. Gordon et al. [10] have studied the source of vibration. A sports-utility vehicle with sensor and data acquisition system is used to find the vibration source. This study was focused on vehicle vibration response from road surface features. Kar and Mohanty [11] have used motor current signature analysis (MCSA) and discrete wavelet transform (DWT) for studying the gear vibration. Load fluctuations on the gearbox and gear defects are two major sources of vibration.

2 Three-Dimensional Model of Gearbox Casing

Solid Edge [12] software has good geometric modeling capability. It has various features suited for complex geometry like transmission casing modeling. The geometrical dimensions were obtained from drawing in Dehradun with the permission of manufacturer of truck. We have constructed the geometry by drawing and taking measurements in workshop. Few parts have not been considered for designing to simplify the geometry, and it has no impact on vibration frequency. The CAD model is shown in Fig. 1. For free vibration analysis, finite element analysis (FEA) software ANSYS 14.5 [13] has been used as an analysis tool. Figure 2 shows the meshed model of gray cast iron transmission housing. ANSYS 14.5 has high-quality meshing facility. The meshed model consists of 377,697 nodes and 228,341 elements; linear tetrahedral elements were used for meshing (Fig. 3).

Fig. 1 CAD model

Fig. 2 Mesh model

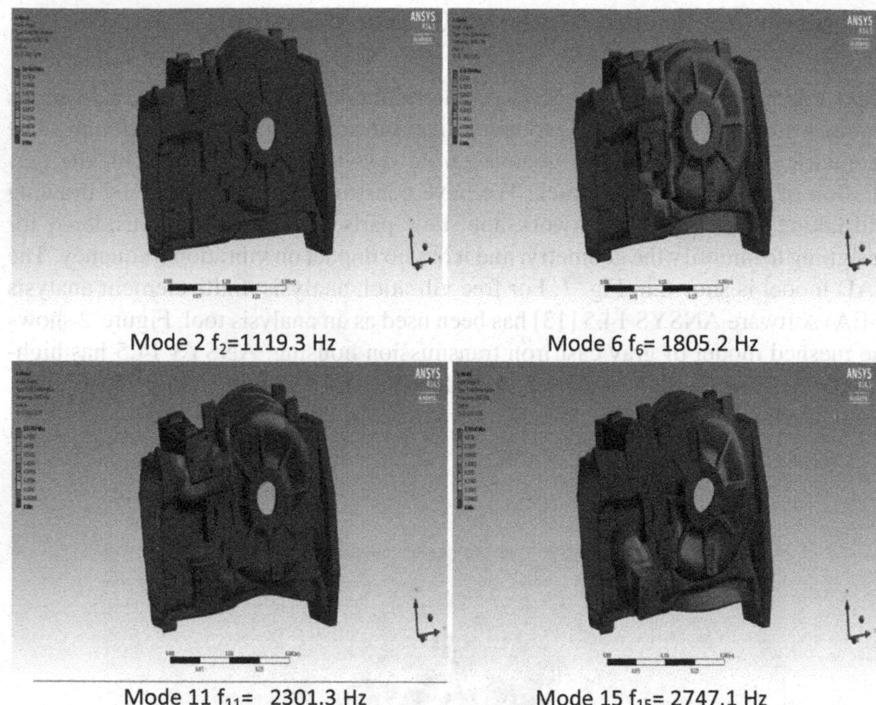

Fig. 3 Four different mode shapes (2, 6, 11, and 15) of gray cast iron transmission casing

3 Material Properties and Boundary Conditions

Transmission casing is manufactured by different process depending on size and various others factors. Casing is mounted on truck frame using fixtures. Loosing of fixtures may cause serious vibration and noise problem. Since past many years, cast iron is used for the truck transmission casing because of its damping properties. We have selected a series of four transmission casing materials gray cast iron grade FG 260, structural steel, Al alloy, and Mg alloy for the analysis of material influence on mode shape and natural frequency. In selection of materials, the main consideration is frequency analysis without considering manufacturing prospects. Elastic modulus, Poisson ratio, and material density are required for free vibration analysis.

Table 1 shows the material properties of different materials. The mechanical properties of gray cast iron grade FG 260 have been taken from the Metals Databook [14]. Structural steel, Al alloy, and Mg alloy are available as engineering material in the material library of ANSYS 14.5 [13]. Boundary conditions play important role in vibration analysis. There are two predefined boundary conditions in ANSYS for free vibration analysis. These are free–free and fixed–fixed boundary conditions.

Table 1 Material properties

S.No.	Material	Elastic modulus (Pa)	Poisson ratio	Material density (kg/m^3)
1	Gray cast iron grade FG 260	1.28e11	0.26	7,200
2	Structural steel	2.0e11	0.30	7,850
3	Al alloys	0.71e11	0.33	2,770
4	Mg alloys	0.45e11	0.35	1,800

Table 2 Variation of modal frequency of four different materials

Mode	Gray cast iron (modal frequency Hz)	Structural steel (modal frequency Hz)	Al alloy (modal frequency Hz)	Mg alloy (modal frequency Hz)
1	1,002.5	1,291.7	1,291.6	1,273.2
2	1,119.3	1,444.5	1,448	1,430
3	1,332.7	1,718.8	1,721.4	1,698.8
4	1,665.5	2,152.2	2,161.7	2,137.5
5	1,692.4	2,186.7	2,196	2,171.1
6	1,805.2	2,328.5	2,332.4	2,302.1
7	1,916.1	2,472.6	2,478.4	2,447.3
8	2,117.3	2,730.5	2,733.9	2,697.2
9	2,151.7	2,777.3	2,784.7	2,750.1
10	2,282	2,946.1	2,954.4	2,917.8
11	2,301.3	2,968.1	2,973.1	2,934.7
12	2,473.5	3,188.5	3,190.7	3,147.1
13	2,488.6	3,211.8	3,219.5	3,179
14	2,675.8	3,456.8	3,470.5	3,430.4
15	2,747.1	3,550.8	3,567.8	3,528.1
16	2,766.1	3,569.4	3,577.5	3,532.6
17	2,826.1	3,650.6	3,664.4	3,621.5
18	2,878.6	3,714.3	3,721.9	3,674
19	2,885.5	3,723.4	3,731.9	3,684.8
20	2,954.8	3,816	3,829.6	3,784.8

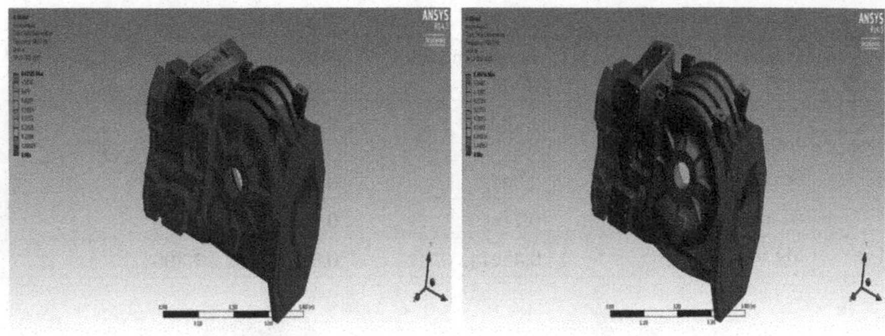

Mode 2 f_2=1463.7 Hz Mode 7 f_7=2495.7 Hz

Fig. 4 Two different mode shapes (2 and 7) of structural steel transmission casing

4 Modal Analysis Results and Discussion

The mode shapes and natural frequency of transmission housing have been evaluated using ANSYS 14.5 solver. Mathematical simulation is performed for fixed–fixed boundary condition. In free vibration analysis, load is selected automatically by solver. The first 20 mode shapes and corresponding natural or modal frequencies for four different materials are shown in Table 2. Different vibration mode shapes (bending, torsional, axial bending vibration, and combination of two vibrations) were obtained. Figure 4 shows the six different mode shapes and corresponding natural frequency of gray cast iron grade FG 260 transmission casing. The frequency range varies from 1,002.5 to 2,954.8 Hz. Modes 2, 4, and 6 are torsional vibration modes. This torsional vibration is performed at single left side on transmission housing. Modes 11, 15, and 19 are axial bending vibration. In axial bending vibration, the transmission body bends from the center line.

For structural steel, the frequency range varies from 1,291.7 to 3,816 Hz. Modes 2 and 6 are torsional vibration modes. Modes 7 and 13 are axial bending vibration, where transmission body is twisted about center line. Modes 15 and 19 are bending modes. The frequency range for all materials is 1,002.5–3,784 Hz (Table 2). This range of frequency variation is same 500–3,500 Hz as the experimental result obtained by Tuma [7]. The frequency range of gray cast iron 1,002.5–2,954.8 Hz is min (Table 2).

5 Conclusion

Analysis results show that transmission housing is subjected to torsional vibration, axial bending vibration, and axial bending with torsional vibration. The first 20 vibration mode shapes and corresponding natural frequencies have been calculated using ANSYS 14.5 FEA-based simulation software. The transmission housing motion is constrained by considering the fixed–fixed boundary conditions. The 3D solid model is generated using SOLIDEDGE software and is transferred to ANSYS 14.5. In this research work, we have considered the problem of influence of transmission casing material on mode shapes and natural frequency. The FEA result shows that on design and vibration index, all four materials can be used as a truck transmission casing without considering the manufacturing prospects. FEA offers satisfactory results. The simulation result is verified with the experimental result available in the literature.

References

1. Qin-man, F.: Modal analysis of a truck transmission based on ANSYS. In: Fourth International Conference on Information and Computing (2011)
2. Minfeng, H., Yingchun, J.: Fatigue life analysis of automobile component based on FEM. Mech. Res. Appl. **21**, 57–60 (2008)
3. Liu, D., Hou, W., Wang, F.: Fatigue life analysis of a component based on the finite elements technology. J. China Railway Soc. **26**, 47–51 (2004)
4. Saada, A., Velex, P.: An extended model for the analysis of the dynamic behavior of planetary trains. ASME J. Mech. Des. **117**, 241–247 (1995)
5. Velex, P., Flamand, L.: Dynamic response of planetary trains to mesh parametric excitations. J. Mech. Des. Trans. ASME **118**, 7–14 (1996)
6. Yu, L., Wu, G.: Analysis on fatigue life of rear suspension based on virtual test tig. Comput. Aided Eng. **15**, 128–130 (2006)
7. Tuma, J.: Gearbox noise and vibration prediction and control. Int. J. Acoust. Vibr. **14**, 1–11 (2009)
8. Åkerblom, M.: Gear noise and vibration—a literature survey. Report-volvo construction equipment components AB. Res. J. Appl. Sci. Eng. Technol. **5**, 1449–1453 (2013)
9. Nacib, L., Pekpe, K.M., Sakhara, S.: Detecting gear tooth cracks using cepstral analysis in gearbox of helicopters. Int. J. Adv. Eng. Technol. **5**, 139–145 (2013)
10. Gordon, T.J., Bareket, Z.: Vibration transmission from road surface features—vehicle measurement and detection. Technical Report for Nissan Technical Center North America, Inc., UMTRI-2007-4 (2007)
11. Kar, C., Mohanty, A.R.: Monitoring gear vibrations through motor current signature analysis and wavelet transform. Mech. Syst. Signal Process. **20**, 158–187 (2006)
12. SOLIDEDGE: Version 19.0 (2006)
13. ANSYS R 14.5: Academic, structural analysis guide (2013)
14. The Metals Databook, 4th edn. Tata McGraw-Hill (2008) ISBN-13: 978-0-07-462300-8

Ultra high-Speed InAlAs/InGaAs High Electron Mobility Transistor

Meryleen Mohapatra, Nutan Shukla and A.K. Panda

Abstract This work deals with the performance evaluation and characterization of an InAlAs/InGaAs-based high electron mobility transistor with different gate lengths viz. 50, 35, and 15 nm. A maximum drain current (I_{dss}) of 398 mA/mm is achieved for a 15-nm-gate-length device with a V_{ds} of 0.4 V as compared to 50- and 35-nm-gate-length HEMT with a current of 368 mA/mm and 384 mA/mm, respectively. A cutoff frequency (f_T) of 1.3 THz is reported for a 15 nm gate length, while a cutoff frequency of 625 GHz and 1.05 THz has been obtained for 50- and 35-nm-gate-length devices. The increase in cutoff frequency for a 15-nm-gate-length InAlAs/InGaAs-based HEMT results due to the decrease in transit time. A maximum oscillation frequency (f_{max}) of 1.8 THz has been obtained for a 15-nm-gate-length device whereas a f_{max} of 1.35 and 1.58 THz has been achieved for a 50- and 35-nm-gate-length HEMT. So, the device with ultrashort gate length performs better as compared to other two gate length devices.

Keywords Ultrashort gate length · Pseudomorphic HEMT · InAlAs · InGaAs

1 Introduction

The demand for high-frequency, high-bandwidth devices and the circuits with high-current capabilities is increasing day by day. Present generation's report says that the InP-based InAlAs/InGaAs pseudomorphic high electron mobility

M. Mohapatra (✉) · N. Shukla
ECE Department, Siksha 'O' Anusandhan University, Bhubaneswar, Odisha, India
e-mail: meryleenmohapatra@soauniversity.ac.in

N. Shukla
e-mail: nutan.shukla1992@gmail.com

A.K. Panda
ECE Department, NIST, Berhampur, Odisha, India
e-mail: akpanda@nist.edu

L.C. Jain et al. (eds.), *Intelligent Computing, Communication and Devices*,
Advances in Intelligent Systems and Computing 308,
DOI 10.1007/978-81-322-2012-1_57

transistors are currently the fastest and efficient transistors [1]. It has become very popular for high-frequency and low-noise applications [2]. To date, a great effort has been made on creating high-speed transistors for sub-100 nm gate lengths. Nguyen et al. [3] obtained a cutoff frequency of 340 GHz for AlInAs/GaInAs high electron mobility transistor having an Indium content of 80 % for a 50 nm gate length. Endoh et al. [4] reported a cutoff frequency of 362 GHz for a 50 nm gate. A cutoff frequency of 308 GHz with a gain of 4.4 dB for a 35 nm gate length has been reported by Mei et al. [5]. For a 30 nm gate, Shinohara et al. [6] reported a cutoff frequency of 472 GHz. Later, he used multilayer cap structure and found a cutoff frequency of 574 GHz for the 30 nm gate length [7]. Yamashita et al. [8] fabricated a 25-nm-gate-length InAlAs/InGaAs HEMT using two-step gate recess technology and found a cutoff frequency of 396 GHz. A cutoff frequency of 610 GHz is achieved by Yeon et al. [9] for a 15 nm gate length due to the enhanced electron velocity. Here, in this paper, to increase the device performance further, an InAlAs/InGaAs-based HEMT is proposed. Then, the gate length of the same device has been reduced from 50 to 35 nm and 15 nm. A comparison has been performed between them taking into account various device parameters and discussed for the different gate lengths viz. 50, 35, and 15 nm. The Sect. 2 describes the structure of proposed InAlAs/InGaAs-based HEMT. The mathematical modeling of device parameters is explained in Sect. 3 followed by Sect. 4, which describes the results and discussion. Conclusion is given in Sect. 5.

2 Device Descriptions

The proposed device structure is shown in Fig. 1 with different gate lengths, i.e., 50, 35, and 15 nm. Here, multi-cap layers consisting of 10 nm $In_{0.7}Ga_{0.3}As$, 20 nm InGaAs, and 3 nm InP have been taken. These cap layers prevent the epitaxial structure from oxidation and form low-resistance ohmic contacts. The barrier layer (InAlAs of thickness 6 nm) is a wider band gap material than the channel layer ($In_{0.7}Ga_{0.3}As$ of thickness 8 nm), which is a lower band gap material. A spacer layer (InAlAs of thickness 3 nm) plays major role in eliminating the possibility of reduced mobility in the device. A buffer layer (InAlAs of thickness 200 nm) creates well-isolated channel and eliminates any type of electrical influence from the substrate. A T-shaped gate is taken that helps in reducing the gate capacitance and gate resistance, which maximizes the current and transconductance, and the extrinsic noise is also reduced [10].

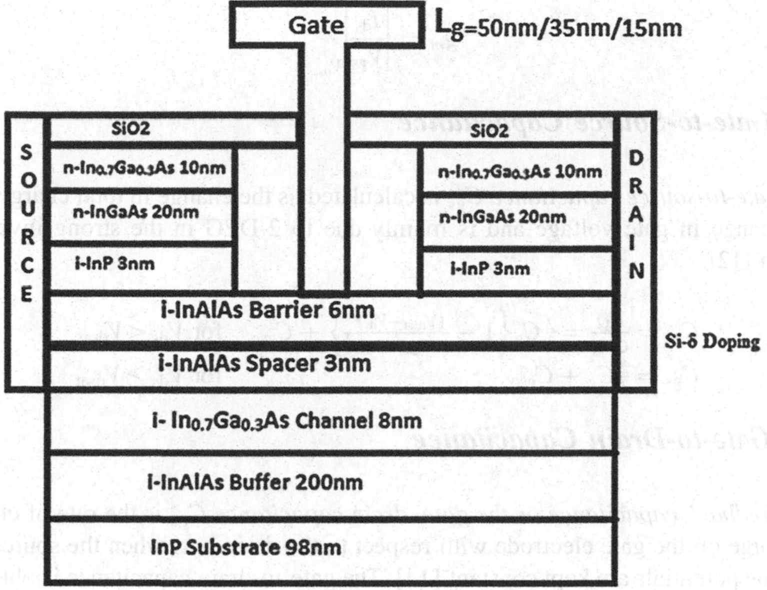

Fig. 1 Schematic cross section of an InAlAs/InGaAs-based HEMT

3 Mathematical Modeling

3.1 Current Model

The *drain current* (I_d) for InAlAs/InGaAs-based HEMT, along the channel at any point x can be obtained using drift diffusion model with an approximation of gradual channel effect and can be expressed as [11],

$$I_d = -\mu_{\text{eff}} W Q_{\text{ch}} \frac{\mathrm{d}\psi}{\mathrm{d}x} + \mu_{\text{eff}} W V_{\text{th}} \frac{\mathrm{d}Q_{\text{ch}}}{\mathrm{d}x} \tag{1}$$

where μ_{eff} is effective carrier mobility. W is the channel width and $Q_{\text{ch}} = qn_s = C_g(V_{go} - \psi)$ is the channel charge at x.

3.2 Transconductance

The *transconductance* (g_m) is a parameter used for estimating the microwave performance of the device. It governs the current driving capability. This is expressed as,

$$g_m = \left| \frac{I_{\text{ds}}}{V_{\text{gs}}} \right|_{V_{\text{ds}}} \tag{2}$$

3.3 Gate-to-Source Capacitance

The *gate-to-source capacitance* C_{gs} is calculated as the change in total charge with the change in gate voltage and is mainly due to 2-DEG in the strong inversion region [12].

$$\begin{aligned} C_{\text{gs}} &= \frac{\mathrm{d}Q}{\mathrm{d}V_{\text{gs}}} = \tfrac{2}{3}C_g\left\{1 - \frac{(V_{\text{dsat}} - V_{\text{ds}})^2}{(2V_{\text{dsat}} - V_{\text{ds}})^2}\right\} + C_{\text{gso}} \quad \text{for } V_{\text{ds}} < V_{\text{dsat}} \\ C_{\text{gs}} &= \tfrac{2}{3}C_g + C_{\text{gso}} \quad \text{for } V_{\text{ds}} \geq V_{\text{dsat}} \end{aligned} \tag{3}$$

3.4 Gate-to-Drain Capacitance

The *feedback capacitance or the gate–drain capacitance* C_{gd} is the rate of change of charge on the gate electrode with respect to the drain bias when the source and the gate potentials are kept constant [14]. The gate-to-drain capacitance is obtained as [13]

$$\begin{aligned} C_{\text{gd}} &= \tfrac{2}{3}C_g\left\{1 - \frac{(V_{\text{dsat}})^2}{(2V_{\text{dsat}} - V_{\text{ds}})^2}\right\} + C_{\text{gdo}} \quad \text{for } V_{\text{ds}} < V_{\text{dsat}} \\ C_{\text{gd}} &= C_{\text{gso}} \quad \text{for } V_{\text{ds}} \geq V_{\text{dsat}} \end{aligned} \tag{4}$$

3.5 Cutoff Frequency

The *unity gain cutoff frequency* f_T is an important figure of merit of HEMT performance at microwave frequency. The cutoff frequency is expressed as

$$f_T = \frac{g_m}{2\pi(C_{\text{gd}} + C_{\text{gs}})} \tag{5}$$

By substituting the value of g_m, C_{gs}, and C_{gd} from Eqs. (2), (3), and (4), cutoff frequency can be calculated.

3.6 Maximum Oscillation Frequency

The *maximum oscillation frequency* $f_{\max}$ can be described as [15]

$$f_{\max} = \frac{f_T}{2\sqrt{(R_i + R_s + R_g)R_{\text{ds}} + (2\pi f_T)R_g C_{\text{gd}}}} \tag{6}$$

where f_T is cutoff frequency, R_i, R_s, R_g represent internal resistance, source resistance, gate resistance, and output resistance, respectively.

4 Results and Discussion

To discuss the device performance, the DC and RF characteristics are observed. The drain characteristic is plotted in Fig. 2. The devices are well pinched off. There is an increase in current for 15 nm gate length compared to 35 and 50 nm. The maximum current density for 50, 35, and 15 nm is 368, 384, and 398 mA/mm, respectively. With the reduction of gate lengths, the potential barriers induced by the gate also become thinner. A substantial change in the electron density distribution is found to occur and the current become high for ultrashort gate lengths [16].

The transfer characteristics ($V_{gs} \sim I_{ds}$) and transconductance (g_m) are plotted in Fig. 2 for a V_{ds} of 0.6 V and for the gate lengths 50, 35, and 15 nm. The current is obtained using Eq. (1) and is high for 15 nm than 50- and 35-nm-gate-length InAlAs/InGaAs HEMT. The maximum transconductance for 50, 35, and 15 nm is 1,042, 973, and 760 mS/mm, respectively. The monotonic decrease in transconductance with the reduction of gate length is due to the prominent short gate geometry effects because the drain-induced barrier gets lower in short gate length devices [17] (Fig. 3).

To observe the RF characteristics of the device, the graph between the transcapacitance ($C_{gs} + C_{gd}$) and V_{gs} is plotted in Fig. 4 using the Eqs. (3) and (4). The transcapacitance is less for 15 nm gate length than for 50 and 35 nm. This reduction contributes to the ultrahigh-speed performance of the HEMTs with 15 nm gate length.

A current gain characteristic is shown in Fig. 5 and is obtained from the Eq. (5) to get the idea about the cutoff frequency. The cutoff frequency obtained for 15 nm gate length is 1.3 THz. For 50 and 35 nm gate length, f_T is found to be 620 GHz and 1.05 THz, respectively. The increase in cutoff frequency results from the decrease in transcapacitance in case of 15 nm gate length, which is as shown in Fig. 6.

A graph between unilateral power gain and frequency is plotted in Fig. 6 for 50-, 35-, and 15-nm-gate-length InAlAs/InGaAs HEMT. For 15 nm gate length, a maximum oscillation frequency (f_{max}) of 1.8 THz is achieved, whereas in case of 50 and 35 nm, it is 1.35 and 1.58 THz, respectively, which is as per Eq. (6). The decrease in gate capacitance and increase in transconductance are responsible for increase in maximum oscillation frequency for 15-nm-gate-length InAlAs/InGaAs-based HEMT [18].

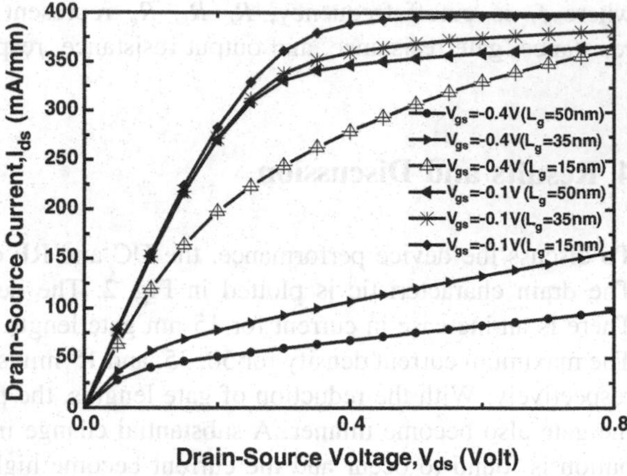

Fig. 2 Drain characteristics ($V_{ds} \sim I_{ds}$) of 50-, 35-, and 15-nm-gate-length HEMT

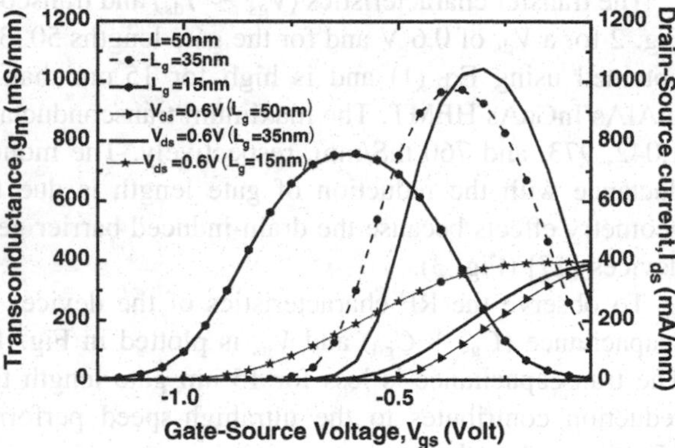

Fig. 3 Transfer characteristics and transconductance for a 50-, 35-, 15-nm-gate-length HEMT

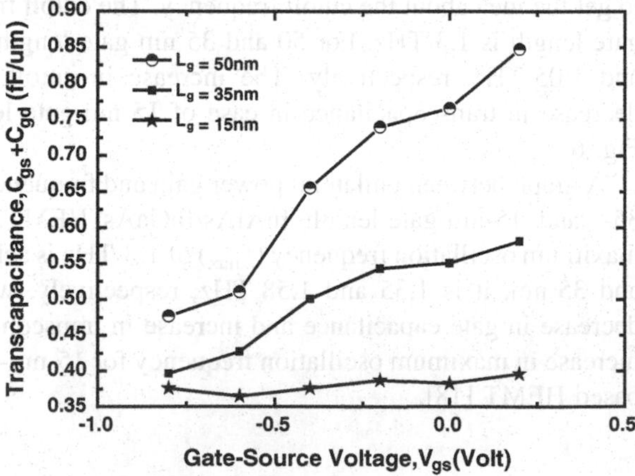

Fig. 4 Transcapacitance ($C_{gs} + C_{gd}$) characteristics of 50-, 35-, and 15-nm-gate-length HEMT

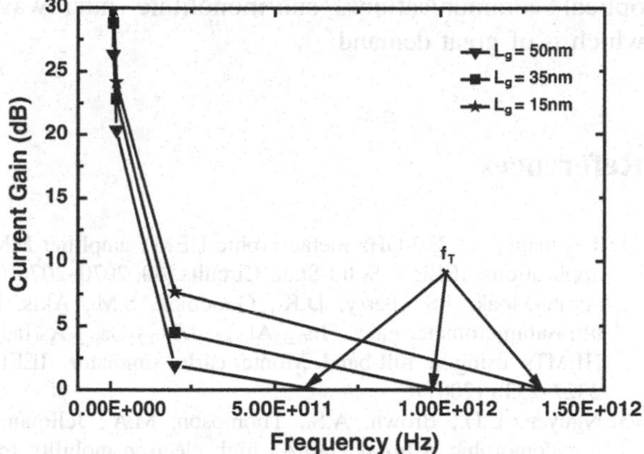

Fig. 5 Current gain (f_T) characteristics of 50-, 35-, and 15-nm-gate-length HEMT

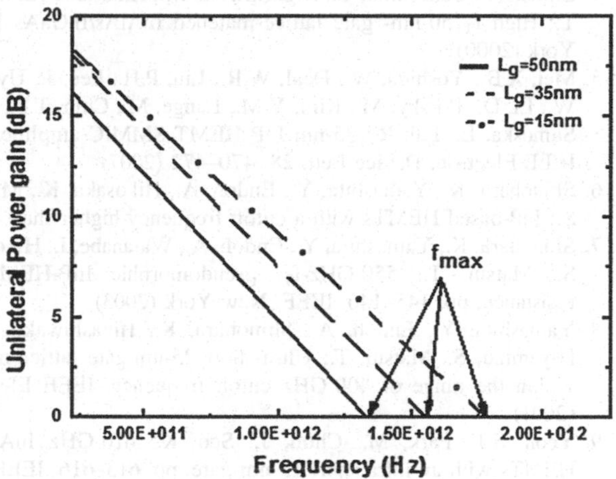

Fig. 6 Maximum oscillation frequency (f_{max}) of 50-, 35-, and 15-nm-gate-length HEMT

5 Conclusion

A systematic investigation of ultrasubmicrometer-gate-length InAlAs/InGaAs-based HEMT has been successfully conducted. In this paper, a major focus has been given upon observing the DC and RF characteristics of the devices. The most important aspect of this paper is to realize the performance of the device with reduced gate length. A cutoff frequency (f_T) of 625 GHz, 1.05, and 1.3 THz has been reported for a 50-, 35-, and 15-nm-gate-length device, respectively. A maximum oscillation frequency (f_{max}) of 1.8 THz has been obtained for a 15-nm-gate-length device where as a f_{max} of 1.35 and 1.58 THz has been achieved for a 50- and 35-nm-gate-length HEMT. So, these high-speed devices where frequencies obtained in THz range can be used in various applications including radar, optical communications, sub-monolithic microwave-integrated circuits, etc., which is of great demand.

References

1. Tessmann, A.: 220-GHz metamorphic HEMT amplifier MMICs for high resolution imaging applications. IEEE J. Solid-State Circuits. **40**, 2070–2076 (2005)
2. Ayubi-Moak, J.S., Ferry, D.K., Goodnick, S.M., Akis, R., Saraniti, M.: Simulation of ultrasubmicrometer-gate $In_{0.52}Al_{0.48}As/In_{0.75}Ga_{0.25}As/In_{0.52}Al_{0.48}As/InP$ pseudomorphic HEMTs using a full-band Monte carlo simulator. IEEE Trans. Electron. Devices **54**, 2327–2338 (2007)
3. Nguyen, L.D., Brown, A.S., Thompson, M.A., Jelloian, L.M.: 50-nm self-aligned-gate pseudomorphic AlInAs/ GaInAs high electron mobility transistors. IEEE Trans. Electron. Devices **39**, 2007–2014 (1992)
4. Endoh, A., Yamashita, Y., Higashiwaki, M., Hikosaka, K., Mimura, T., Hiyamizu, S., Matsui, T.: High f_T 50-nm- gate lattice-matched InAlAs/InGaAs HEMTs, pp. 87–90. IEEE, New York (2000)
5. Mei, X.B., Yoshida, W., Deal, W.R., Liu, P.H., Lee, J., Uyeda, J., Dang, L., Wang, J., Liu, W., Li, D., Barsky, M., Kim, Y.M., Lange, M., Chin, T.P., Radisic, V., Gaier, T., Fung, A., Samoska, L., Lai, R.: 35-nm InP HEMT SMMIC amplifier with 4.4-dB Gain at 308 GHz. IEEE Electron. Device Lett. **28**, 470–472 (2007)
6. Shinohara, K., Yamashita, Y., Endoh, A., Hikosaka, K., Matsui, T., Mimura, T., Hiyamizd, S.: InP-based HEMTs with a cutoff frequency higher than 450 GHz, pp. 166–169 (2002)
7. Shinohara, K., Yamashita, Y., Endoh, A., Watanabe, I., Hikosaka, K., Mimura, T., Hiyamizu, S., Matsui, T.: 550 GHz-f_T, pseudomorphic InP-HEMTs with reduced source-drain resistance, pp. 145–146. IEEE, New York (2003)
8. Yamashita, Y., Endoh, A., Shinohara, K., Higashiwaki, M., Hikosaka, K., Mimura, T., Hiyamizu, S., Matsui, T.: Ultra-short 25-nm-gate lattice-matched InAlAs/InGaAs HEMTs within the range of 400 GHz cutoff frequency. IEEE Electron Device Lett. **22**, 367–369 (2001)
9. Yeon, S.J., Park, M., Choil, J., Seo, K.: 610 GHz $InAlAs/In_{0.75}Ga_{0.25}As$ metamorphic HEMTs with an ultra-short 15-nm-gate, pp. 613–616. IEEE, New York (2007)
10. Waldron, N., Kim, D.H., Del Alamo, J.A.: 90 nm self-aligned enhancement-mode InGaAs HEMT for logic applications, pp. 633–636. IEEE, New York (2007)
11. Sourabh, K., Fjeldly, A.: Analytical modelling of surface-potential and drain current in AlGaAs/GaAs HEMT devices. In: IEEE International Symposium on Radio—Frequency Integration Technology, pp. 183–185 (2012)
12. Alam, M.K.: Gate capacitances of high electron mobility transistors. In: International Conference on Electrical and Computer Engineering, pp. 129–131 (2002)
13. Taguchi, T., Matsugatani, K., Hoshino, K., Yamada, H., Ueno, Y.: InAlAs/pseudomorphic-InGaAs MMICs for 76 GHz-band millimetre wave radar. In: International. Symposium on Compound Semiconductors, pp. 193–200 (1999)
14. Mohapatra,M., Mumtaz,A.,Panda,A.K.: Performance evaluation of GaSb/AlGaAs based high electron mobility transistors. In: Conf. on Advances in Recent Technologies in Communication and Computing, pp. 249–252 (2011)
15. Watanabe, I., Shinohara, K., Kitada, T., Shimomura, S., Yamashita, Y., Endoh, A., Mimura, T., Hiyamizu, S., Matsui, T.: Velocity enhancement in cryogenically cooled InP-based HEMTs on (411) a-oriented substrates. IEEE Trans. Electron. Devices **53**, 2842–2846 (2006)
16. Zhou, J.R., Ferry, D.K.: Modeling of quantum effects in ultrasmall HEMT devices. IEEE Trans. Electron. Devices **40**, 421–426 (1993)

17. Kizilyalli, I.C., Artaki, M., Shah, N.J., Chandra, A.: Scaling properties and short-channel effects in submicrometer AlGaAs.GaAs MODFET's: a Monte Carlo study. IEEE Trans. Electron. Devices **40**, 234–249 (1993)
18. Bhattacharya, M., Jogi, J., Gupta,R.S.,Gupta,M.: Temperature-dependent analytical model for microwave and noise performance characterization of $In_{0.52}Al_{0.48}As/In_mGa_{1-m}As$ ($0.53 \leq m \leq 0.8$) DG-HEMT. IEEE Trans. Device Mater. Reliab. **13**, 293–300 (2013)

Design and Simulation of CNTFET by Varying the Position of Vacancy Defect in Channel

Shreekant and Sudhanshu Choudhary

Abstract In this paper, we investigate the effect of vacancy defects in carbon nanotube field effect transistor (CNTFET) by applying the self-consistent solution of the Schrodinger and Poisson equations within the non-equilibrium Green's function (NEGF) formalism and calculated its characteristics by creating vacancies at different positions in the CNTFET channel. It has been observed that due to the creation of vacancy defects in the channel region of the CNTFET, the ON-current increases and OFF-current decreases in comparison to non-defective channel CNTFET. The results show that single, double, and triple vacancies have a very minor effect on the CNTFET conductivity. However, a huge reduction in CNTFET conductivity was observed when the density of vacancy defects in the channel was increased.

Keywords Carbon nanotube · CNTFET · Zigzag · Defects · Vacancy · Non-equilibrium Green's function (NEGF)

1 Introduction

In 1991, carbon nanotubes (CNTs) were invented by S. Ijima in Japan. CNTFET refers to a field effect transistor that utilizes a single carbon nanotube or an array of CNTs as the channel material instead of bulk silicon in the traditional MOSFET structure. Theoretically, a CNT can be considered as sheet of graphene rolled into a tube [1].

Shreekant (✉) · S. Choudhary
School of VLSI Design and Embedded System Design, NIT Kurukshetra,
Kurukshetra 136119, Haryana, India
e-mail: shreekant.sinha@gmail.com

S. Choudhary
e-mail: hellosudhanshubit@gmail.com

L.C. Jain et al. (eds.), *Intelligent Computing, Communication and Devices*,
Advances in Intelligent Systems and Computing 308,
DOI 10.1007/978-81-322-2012-1_58

CNT has metallic or semiconducting property on the basis of chirality vector [1, 2]. In 1998, CNT was first demonstrated, nowadays there have been major developments in CNTFETs [3–5]. In order to make CNTFETs competitive to conventional FETs, arrays of CNTs should be used [6]. CNTFETs have shown excellent electrical properties, including high transconductance, high ON/OFF-current ratio, and low-inverse sub-threshold swing [7]. With larger diameters, it is similar to that of MOSFETs because it has been proven that band gap of the CNT is proportional to inverse of its diameter [8]. Typical diameters are 1–2 nm, and the resulting band gaps are suitable for room temperature electronics. Since the introduction of CNTFETs, researches have started to solve numerical equations in order to obtain their current voltage relation [9, 10].

This paper shows a comparative study on $CNTFET_S$ with vacancy defects and CNTFET without defects. We use (13, 0) zigzag CNT in the CNTFET channel. The study is based on self-consistent solution of the Poisson and the Schrodinger equations, by means of the non-equilibrium Green's function (NEGF) formalism [11, 12]. We have created single, double, triple, and dense vacancies in the channel region to study their effect on the device performance.

2 Equilibrium and Non-equilibrium Transport Results and Analysis

All the simulations for CNTFET have been carried out on NanoTCAD ViDES [13, 14] which is an open-source tool. The software module has a set of predefined functions, which are able to compute transport in carbon nanotube FETs.

We have calculated the characteristics of CNTFETs by solving the Schrodinger equation using the NEGF formalism self-consistently with the 3-D Poisson equation [15], coherent transport is assumed. Double-gate [15, 16] geometry is used for the device. SiO_2-gate oxide ($\kappa = 3.9$) has a thickness of 2 nm, the lateral spacing S is equal to 1 nm. Source and drain are doped extension of the CNT's channel with a molar fraction equal to 5×10^{-3}. Source, drain, and channel extensions are 2 nm long and 5 nm long, respectively, accounting to a total length of 6 nm and 15 nm for nanotube transistor as seen in Fig. 1a, b. The transfer and

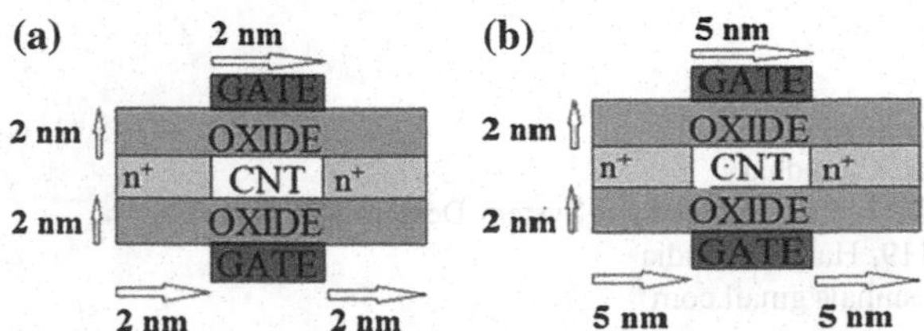

Fig. 1 **a** View along length of CNTFET with doped drain and source extension of (13, 0) CNT, 6 nm length. **b** View along length of CNTFET with doped drain and source extension of (13, 0) CNT, 15 nm length

output characteristics are calculated and drawn for normal case (with no vacancy or ideal case) and with single, double, and triple vacancies, respectively. The power supply voltage is $V_{DD} = 0.5$ V and room temperature $T = 300$ K. The atomistic vacancies are introduced by dislodging a carbon atom from the device channel.

Transfer characteristics are one of the most primary parameters for FETs. The effects of the defects on the transfer characteristics are analysed under different positions of vacancy defect in various locations in the channel. Figure 2 shows the transfer characteristics of the CNTFET with the single, double, triple, and dense vacancy defects at various positions and at different locations in channel with respect to the no defect (Ideal CNT channel) CNTFET under constant $V_{ds} = 0.1$ V. We observed a lower value of OFF-current (I_{OFF}) when triple vacancy defect is present in the channel in comparison to single and double vacancy defects.

Output characteristics are also one of the most primary parameters for FETs. The effects of the defects on the output characteristics are analysed under different positions of vacancy defect in various locations in the channel. Figure 3 shows the output characteristics of the CNTFET with the single, double, triple, and dense

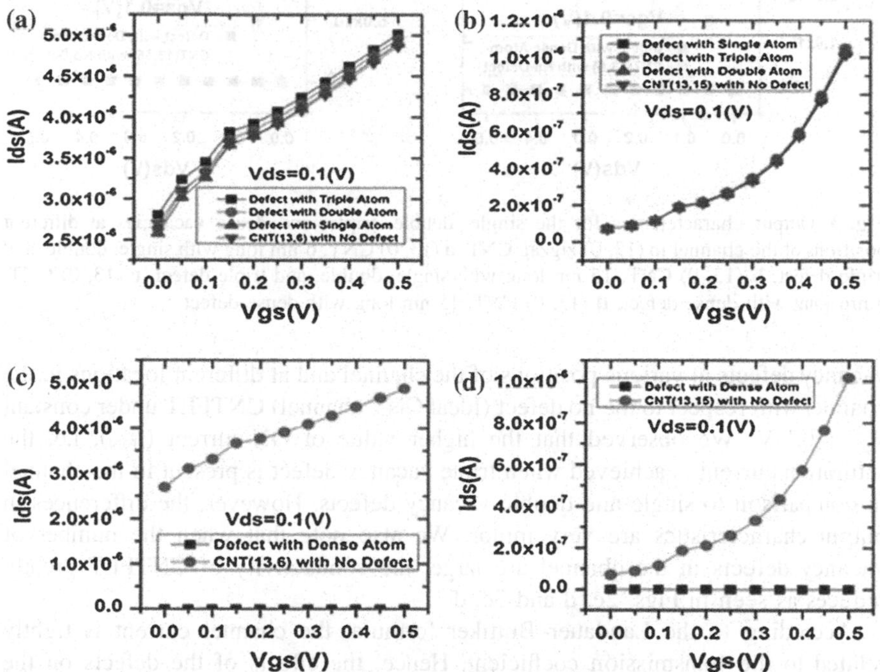

Fig. 2 Transfer characteristics for the single, double, triple, and dense vacancies at different positions of the channel in (13, 0) zigzag CNT: **a** (13, 0) CNT, 6 nm long with single, double, and triple defect, **b** (13, 0) CNT, 15 nm long with single, double, and triple defect, **c** (13, 0) CNT, 6 nm long with dense defect, **d** (13, 0) CNT, 15 nm long with dense defect

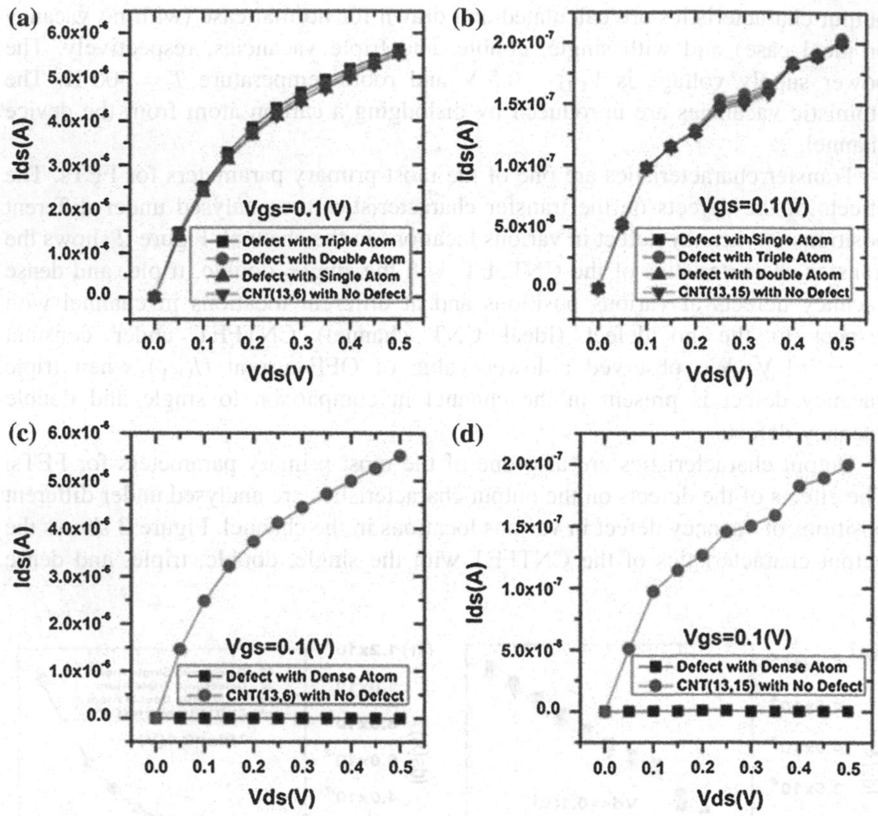

Fig. 3 Output characteristics for the single, double, triple, and dense vacancies at different positions of the channel in (13, 0) zigzag CNT: **a** (13, 0) CNT, 6 nm long with single, double, and triple defect, **b** (13, 0) CNT, 15 nm long with single, double, and triple defect, **c** (13, 0) CNT, 6 nm long with dense defect, **d** (13, 0) CNT, 15 nm long with dense defect

vacancy defects at various positions of the channel and at different locations in the channel with respect to the no defect (Ideal CNT channel) CNTFET under constant $V_{gs} = 0.1$ V. We observed that the higher value of ON-current (I_{ON}), i.e. the saturation current is achieved when triple vacancy defect is present in the channel in comparison to single and double vacancy defects. However, the differences in output characteristics are very minor. We also note that when the number of vacancy defects in the channel are large, the conductivity of CNTFET greatly reduces as seen in Figs. 2c, d and 3c, d.

According to the Landauer–Büttiker formula, the channel current is tightly related to the transmission coefficient. Hence, the effects of the defects on the transmission coefficient are important to study. Figure 4 shows the transmission coefficient as a function of energy of the CNTFET with the single, double, and triple vacancies defects at various positions of the channel. The transmission coefficient of the CNTFET without any defect is shown as a reference. The Fermi

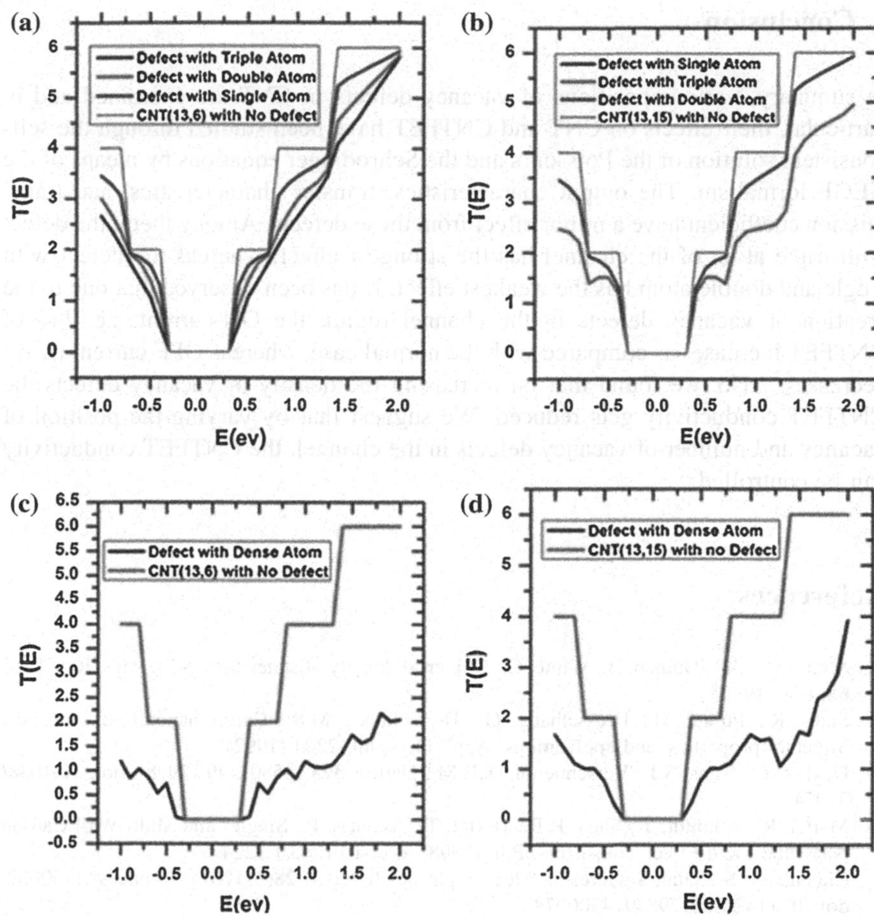

Fig. 4 Transmission coefficient for the single, double, triple, and dense vacancies at different positions of the channel in (13, 0) zigzag CNT: **a** (13, 0) CNT, 6 nm long with single, double, and triple defect, **b** (13, 0) CNT, 15 nm long with single, double, and triple defect, **c** (13, 0) CNT, 6 nm long with dense defect, **d** (13, 0) CNT, 15 nm long with dense defect

level is chosen as the reference energy level (i.e. energy level 0). Here, CNTFET channel is semiconducting, which is confirmed by transmission gap (transmission coefficient is zero) around Fermi level for semiconducting CNT. We find that the differences in transmission spectrum are not very significant unless the density of vacancy in the channel is large as seen in Fig. 4c, d. However, the current v_s voltage characteristics can be justified by comparing them with the transmission spectrum plots as reported in Choudhary et al. [5].

3 Conclusion

In summary, various positions of vacancy defects in CNT are examined and in particular, their effects on CNT and CNTFET have been studied through the self-consistent solution of the Poisson's and the Schrodinger equations by means of the NEGF formalism. The output characteristics, transfer characteristics, and transmission coefficient have a minor effect from these defects. Among them, the defect with triple atom of the channel has the strongest effect; whereas the defect with single and double atom has the weakest effect. It has been observed that due to the creation of vacancy defects in the channel region the ON-current, i.e. I_{ON} of CNTFET increases as compared with the normal case, whereas OFF-current (I_{OFF}) decreases. Also, we found that on increasing the density of vacancy defects the CNTFET conductivity gets reduced. We suggest that by varying the position of vacancy and number of vacancy defects in the channel, the CNTFET conductivity can be controlled.

References

1. Mintmire, J.W., Dunlap, B., White, C.: Universal density of states for CNTs. Phys. Rev. Lett. **68**, 631 (1992)
2. Saito, R., Fujita, M., Dresselhaus, G., Dresselhaus, M.S.: Caron nanotubes: synthesis, Structure properties, and applications. Appl. Phys. **60**, 2204 (1992)
3. Dekker, C., Tans, S.J., Verschueren, A.R.M.: Nature **393** (6680), 49 (1998). doi: 10.1038/29954
4. Martel, R., Schmidt, T., Shea, H.R., Hertel, T., Avouris, P.: Single- and Multi-Wall Carbon Nanotube Field-Effect Transistors. P.h. (1998). doi: 10.1063/1.122477
5. Choudhary, S., Saini, S. Qureshi.: Modern physics letters B **28**(2) (2014) (9 pages) 1450007. doi: 10.1142/S0217984914500079
6. Seidal, R., et al.: High-current nanotube transistors. Nano Lett. **4**(5), 831–834 (2004)
7. Knoch, J., Appenzellert, J.: A novel concept for field effect transistors—the tunneling Carbon nanotube FET. In: IEEE Device Research Conference USA, **1**, June (2005), pp. 153–156
8. Ijima, S.: Carbon Nanotubes. Pergamon Press (1996)
9. Avouris, P., Appenzeller, J., Martel, R., Wind, S.J.: Carbon nanotube electronics. In: Proceedings of the IEEE, **91**: 1772–1784 (2003)
10. McEuen, P.L., Fuhrer, M.S., Park, H.K.: Single-walled carbon nanotube electronics. IEEE Trans. Nanotechnol. **1**, 78–85 (2002)
11. Pourfath, M., Ungersboek, E., Gehring, A., Cheong, B.H., Park, W., et al.: Improving the ambipolar behavior of schottky barrier carbon nanotube field effect transistors. In: Proceedings ESSDERC pp. 429–432 (2004)
12. Datta, S.: Nanoscale device modeling: the Green's function method. Superlattices Microstruct. **5**, 253–278 (2000)
13. Fiori, G., Iannaccone, G., Klimeck, G.: A three-dimensional simulation study of the performance of carbon nanotube field-effect transistors with doped reservoirs and realistic geometry. IEEE Trans. Elec. Dev. **53**(8): 1782–1788 (2006)
14. Fiori, G., Iannaccone, G., Klimeck, G.: Coupled mode space approach for the simulation of realistic carbon nanotube field-effect transistors. IEEE Trans. Nanotechnol. **6**(4): 475–480 (2007)

15. Yoon, Y., Fiori, G., Hong, S., Iannaccone, G., Guo, J.: Performance comparison of graphene nanoribbon FETs with schottky contacts an dopped reservoirs. IEEE Trans. Elec. Dev. **55**(9):2314–2323 sept (2008)
16. Shim, M., Javey, A., Kam, N.W.S, Dai, H.: Polymer functionalization of air-stable n-type carbon nanotube field effect transistors. J. Amer. Chem. Soc. **123**: 11512–11513 (2001)

Performance Evaluation of InGaP/GaAs Solar Cell with Double Layer ARC

Praveen Priyaranjan Nayak, Jyoti Prakash Dutta and Guru Prasad Mishra

Abstract In order to obtain high-conversion percentage of the input available light, anti-reflection coating plays an important role in solar cell. In this work, the performance of InGaP/GaAs dual-junction solar cell has been investigated with single layer (Al_2O_3, TiO_2, and ITO) and double layers (Al_2O_3/TiO_2, and Al_2O_3/ITO) ARC. The work has been carried through computational numerical modeling TCAD tool ATLAS. The detailed photogeneration rates are determined, and the simulation results are validated with published experimental data. The model is implemented with optimized InGaP/GaAs dual-junction cell having Al_2O_3 and TiO_2 as double layer anti-reflection coating with effective 500 nm InAlGaP bottom BSF. A maximum conversion efficiency of 39.9724 % is obtained under AM1.5G illumination for 1,000 suns. The absorpvity and reflectance of different anti-reflection coatings are also studied.

Keywords Anti-reflection coating · ATLAS · Back surface field · Dual-junction solar cell · Short-circuit current · Tunnel diode

1 Introduction

One form of alternative energy that has shown useful space and terrestrial applications is the solar energy which is widely used in designing photovoltaic cells. In the beginning, developed photovoltaic cells were single-junction providing low

P.P. Nayak (✉) · J.P. Dutta · G.P. Mishra
Department of Electronics and Instrumentation Engineering, Institute of Technical Education and Research, Siksha 'O' Anusandhan University, Bhubaneswar, India
e-mail: praveennayak@soauniversity.ac.in

J.P. Dutta
e-mail: jpdutta.143@gmail.com

G.P. Mishra
e-mail: gurumishra@soauniversity.ac.in

L.C. Jain et al. (eds.), *Intelligent Computing, Communication and Devices*, Advances in Intelligent Systems and Computing 308,
DOI 10.1007/978-81-322-2012-1_59

efficiency, but design of tandem solar cells containing two or more junctions showed better performance as there is efficient absorption of available solar light by the different layers thereby reducing solar radiation loss to greater effect. But due to complexity such as high cost and time-consumption, it is difficult to optimize the solar cell structures. Thus, numerical modeling and simulation help to optimize the solar cell structure which helps us in understanding the main physical phenomena and behaviors of each layer. In this work, we report the design of InGaP/GaAs DJ solar cells with InGaP tunnel junction and compare with the previously reported results [4], through optimized V_{OC} and J_{SC}, using different combinations of anti-reflection coatings using the Silvaco ATLAS. The conversion efficiency, J_{SC} and V_{OC} of the present model, is investigated through simulation.

2 The Solar Cell Modeling

2.1 Device Structure

In order to model and build a multi-junction cell having many layers, it was necessary to start with individual simple cells of well-known characteristics. A detailed set of major material parameters shown in Table 1 is used in our design. A schematic of this model with detailed thickness and doping profile is given in Fig. 1.

3 Major Parameters of the Solar Cell

The major parameters of the solar cell are the total current (I), open-circuit voltage (V_{OC}), fill factor (FF), and conversion efficiency (η) which are given by [4]

$$I = I_0\left[\exp\left(\frac{qv}{nKT}\right) - 1\right] - I_L \tag{1}$$

where I_L is the light generated current

$$V_{OC} = \frac{nKT}{q}\ln\left(\frac{I_L}{I_0} + 1\right) \tag{2}$$

$$\mathrm{FF} = \frac{V_{OC} - L_n(V_{OC} + 0.72)}{V_{OC} + 1} \tag{3}$$

$$\eta = \frac{V_{OC} I_{SC} \mathrm{FF}}{P_{in}} \tag{4}$$

Table 1 Major parameters for the ternary $In_{0.49}$ $Ga_{0.51}P$ and quaternary $In_{0.5}$ $(Al_{0.7}Ga_{0.3})_{0.5}P$ lattice matched to GaAs materials used in this design [5]

Material	GaAs	InGaP	InAlGaP
Band gap E_g (eV) @300 K	1.42	1.9	2.3
Lattice constant (Å)	5.65	5.65	5.65
Permittivity	13.1	11.6	11.7
Affinity (eV)	4.07	4.16	4.2

Fig. 1 Schematic diagram of the cell model

Anode Contact
TiO_2 and Al_2O_3 as ARCs
0.030 μm Window $In_{0.5}(Al_{0.7}Ga_{0.3})_{0.5}P$ p = 2.0e18 cm^{-3}
0.050 μm Emitter $In_{0.49}Ga_{0.51}P$ p = 2.0e18 cm^{-3}
0.550 μm Base $In_{0.49}Ga_{0.51}P$ n = 7.0e16 cm^{-3}
0.030 μm BSF $In_{0.5}(Al_{0.7}Ga_{0.3})_{0.5}P$ n = 2.0e18 cm^{-3}
0.025 μm Tunnel $In_{0.49}Ga_{0.51}P$ n = 5.0e19 cm^{-3}
0.025 μm Tunnel $In_{0.49}Ga_{0.51}P$ p = 3.0e19 cm^{-3}
0.040 μm Window $In_{0.49}Ga_{0.51}P$ p = 2.0e18 cm^{-3}
0.500 μm Emitter GaAs p = 2.0e18 cm^{-3}
2.000 μm Base GaAs n = 2.0e17 cm^{-3}
0.500 μm BSF $In_{0.5}(Al_{0.7}Ga_{0.3})_{0.5}P$ n = 5.0e18 cm^{-3}
0.200 μm Substrate GaAs n = 1.0e18 cm^{-3}
Cathode Contact

For application of anti-reflection coating, absorpvity and reflectivity of different ARCs materials must be studied. Figures 2 and 3 show various comparisons of the reflectivity and absorpvity, respectively.

Here, use of the double layer anti-reflective coating (DLAR, 82 nm Al_2O_3/ 60 nm TiO_2) has greatly reduced the reflection losses thereby enhancing absorption and photocurrent of the cell.

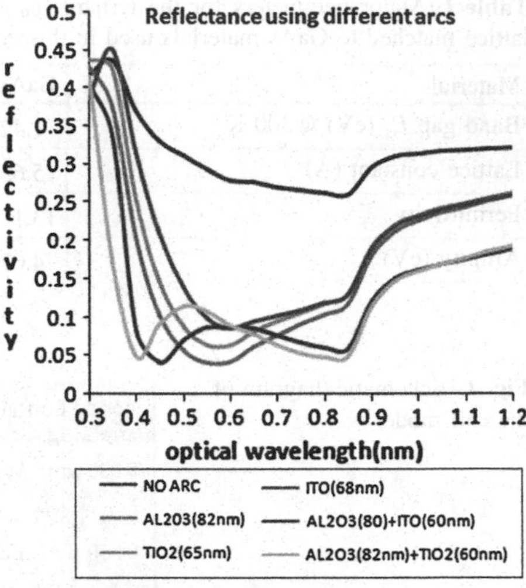

Fig. 2 Reflectivity of different arcs

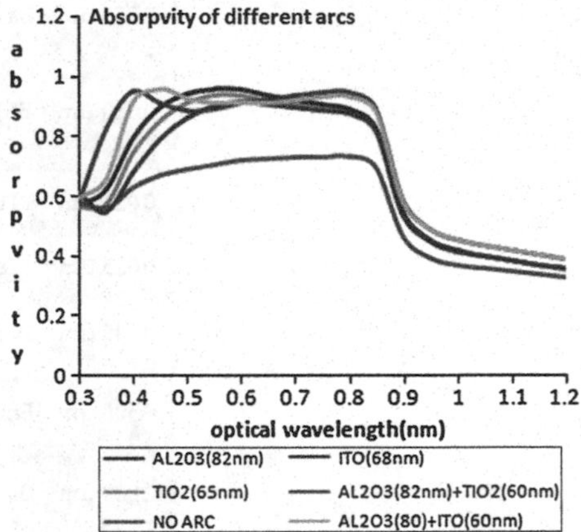

Fig. 3 Absorpvity of different arcs

4 Simulation Result

Figure 4 shows the design structure obtained along with different layers generated from ATLAS coding. The meshing on which the device will be constructed is shown in Fig. 5. Figure 6 illustrates the photogeneration rate of different layers. The I–V characteristics of the solar cell are shown in Fig. 7.

5 Comparison of Results with Experimental Data

The performance and the results observed from this model are similar to those published in [1, 4]. Minor differences are only due to the variations in the material

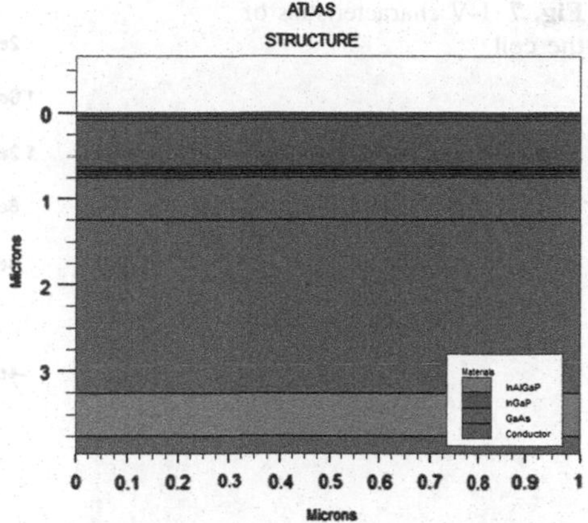

Fig. 4 ATLAS model of this InGaP/GaAs DJ cell

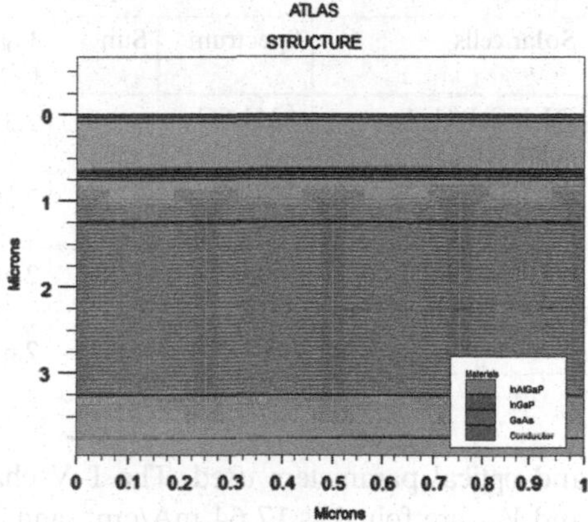

Fig. 5 Meshing of the cell

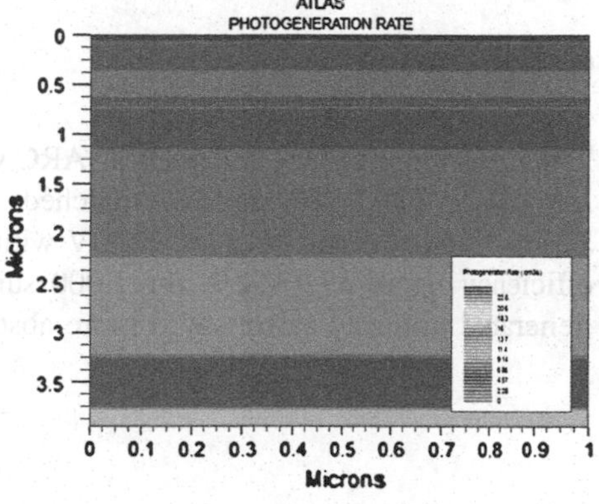

Fig. 6 Photogeneration rate

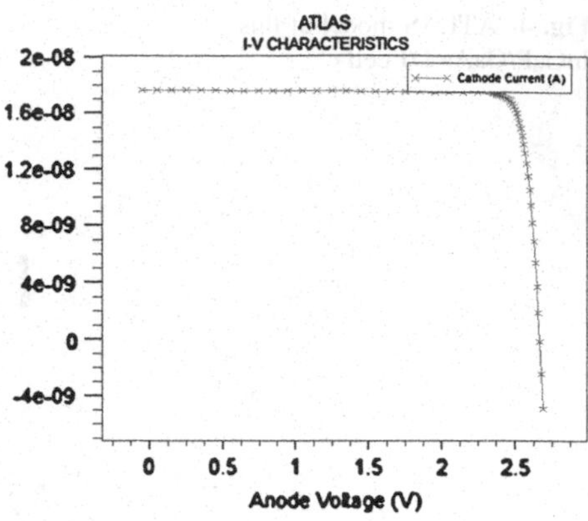

Fig. 7 I–V characteristics of the cell

Table 2 Comparison of the different optimized InGaP/GaAs DJ solar cell structures

Solar cells	Spectrum	Sun	V_{OC} (V)	J_{SC} (mA/cm^2)	FF (%)	Conversion efficiency (%)
DJ InGaP/GaAs solar cell [3]	AM1.5G	1	2.32	1.090E−10	79.00	23.8
DJ InGaP/GaAs solar cell [2]	AM1.5G	1	2.30	1.060E−10	87.55	25.14
DJ InGaP/GaAs solar cell [4]	AM1.5G	1,000	2.66	1.609E−08	89.50	36.678
This model	AM1.5G	1,000	2.66	1.764E−08	88.82	39.9724

and optical parameters used. The I–V characteristic is shown in Fig. 7. The J_{SC} and V_{OC} are found as 17.64 mA/cm^2, and 2.66 V, respectively. The V_{OC} is more or less same as [4] but J_{SC} is increased (Table 2).

6 Conclusion

The inclusion of Al_2O_3 and TiO_2 as ARC with the optimized 0.5-μm BSF material $In_{0.5}(Al_{0.7}\ Ga_{0.3})_{0.5}P$ lattice matched with InGaP and GaAs results in $J_{SC} = 17.64$ mA/cm^2, $V_{OC} = 2.66$ V with significant enhancement in conversion efficiency up to 39.9724 % for 1,000 suns due to a better collection of photo-generated minority carriers and more absorption of the solar spectrum.

References

1. King, R.R., Karam N.H., Ermer J.H., Haddad N., Colter P., Isshiki, T., Yoon, H., Cotal, H.L., Joslin D.E., Krut, D.D., Sudharsanan R., Edmondson K., Cavicchi B.T., Lillington, D.R.,: Next generation, high efficiency III–V multi-junction solar cells. In: Photovoltaic specialists conference. Conference record of the 28th IEEE, pp. 998–1001 (2000)
2. Leem, J.W., Lee, Y.T., Yu, J.S.: Optimum design of InGaP/GaAs dual-junction solar cells with different tunnel diodes. Opt. Quantum Electron. **41**(8), 605–612 (2010)
3. Lueck, M.R., Andre, C.L., Pitera, A.J., Lee, M.L., Fitzgerald, E.A., Ringel, S.A.: Dual junction GaInP/GaAs solar cells grown on metamorphic SiGe/Si substrates with high open circuit voltage. IEEE Electron Device Lett. **27**(3), 142–144 (2006)
4. Singh, K.J., Sarkar, S.K.: Highly efficient ARC less InGaP/GaAs DJ solar cell numerical modeling using optimized InAlGaP BSF layers. Opt. Quant. Electron. **43**, 1–21 (2012)
5. Vurgaftman, I., Meyer, J.R., Ram-Mohan, L.R.: Band parameters for III–V semiconductors and their alloys. J. Appl. Phys. **89**(11), 5815–5875 (2001)

Time-Delay Approximation: Its Influence on the Structure and Performance of the IMC-PI/PID Controller

P.V. Gopi Krishna Rao, M.V. Subramanyam and K. Satyaprasad

Abstract Proportional integral (PI) and proportional integral derivative (PID) controllers have been at the heart of control engineering practice for several decades. 95 % of the controllers employed in the industry are PI/PID. In process control, one often encounters systems described by transfer functions with time delays, which become transcendental functions. The design of the controller demands the rational transfer function approximation of the time-delay term. This paper focuses on the effect of time-delay approximation techniques, viz. Taylor series expansion and Padé approximation, on the structure and performance of PI/PID controllers designed with Internal Model Control (IMC). The performance of the PI/PID controllers was tested in simulation environment on various processes with time delay. For uniform comparison, the controllers were tuned to have a same robustness measure, in terms of maximum sensitivity (M_S). The results indicate, irrespective of time-delay approximation considered, the controllers provide good set point tracking and poor disturbance rejection.

Keywords Time delay · Padé approximation · Taylor series · Internal model control · PI/PID · Disturbance · Set point

1 Introduction

Proportional integral (PI) and proportional integral derivative (PID) controllers have been at the heart of control engineering practice for several decades [1]. The use of the PI or PID controller is ubiquitous in industry. In process control

P.V. Gopi Krishna Rao (✉) · K. Satyaprasad
Department of ECE, JNTUK, Kakinada 533003, India
e-mail: gopikrishnarao@gmail.com

M.V. Subramanyam
Shantiram Engineering College, Nandyal 518501, India

L.C. Jain et al. (eds.), *Intelligent Computing, Communication and Devices*,
Advances in Intelligent Systems and Computing 308,
DOI 10.1007/978-81-322-2012-1_60

applications, more than 95 % of the controllers are of PI or PID type [1–3]. PID controllers can assure satisfactory performances with a simple algorithm for a wide range of processes [3, 4]. In process control, one often encounters systems, described by transfer functions with time delays [2]. If a dynamic system with time delay modeled as a time invariant linear system, its transfer function (rational function) becomes a transcendental function because of time delay [5]. For design and analysis of controllers, these delays are usually approximated by rational transfer functions [6]. This is usually carried out using delay approximation methods, viz. Taylor series expansion and Padé approximation.

The internal model control provides a progressive, effective, natural, generic, unique, powerful, and simple framework for analysis and synthesis of control system performance [7–10]. The simplicity and enhanced performance of the IMC-based tuning rule, and the analytically derived IMC-PID tuning techniques have attracted the attention of the industrial users, in the past decade [9, 10]. The well-known IMC-PID tuning rule provides a clear compromise in the midst of closed-loop performance and robustness to model uncertainties, by only one user-defined tuning parameter λ, which is directly related to the closed-loop time constant [6, 7, 9–12], but the method used for time-delay approximation influences the nature of the IMC—PI/PID controller, controller parameters, and performance.

The organization of the paper is, Sect. 2 describes the basic design of IMC—PID controller and Sect. 3 discusses time-delay approximation techniques and the design of PI/PID controllers using IMC and time-delay approximation. Section 4 demonstrates the simulation results of the performance evaluation for set point tracking and disturbance rejection, and Sect. 5 draws the conclusions.

2 IMC: PID Design

The widely used approximate or predictive models of the chemical process are the first-order process with time delay (FOPTD) (1). Garcia and Morari [11, 13] introduced internal model control; it is characterized as a controller where the process model is explicitly an integral part. The design process of IMC involves factorizing the predictive plant model $G_M(s)$ as invertible$G_{M-}(s)$and non-invertible $G_{M+}(s)$parts depicted in (2) and (3) by simple factorization or all-pass factorization [7, 11, 12, 14–16]. The IMC in (4) is the inverse of invertible part of the plant model $G_M(s)$.

$$G_M(s) = \frac{Ke^{-\theta s}}{\tau s + 1} \tag{1}$$

$$G_M(s) = G_{M-}(s)G_{M+}(s) \tag{2}$$

$$G_{M-}(s) = \frac{K}{\tau s + 1}, G_{M+}(s) = e^{-\theta s} \tag{3}$$

The IMC controller

$$Q(s) = G_{M-}^{-1}(s) G_f(s) \tag{4}$$

where $G_f(s)$ is a low-pass filter of the form $G_f(s) = 1/(1 + \lambda s)$ used to physically realize the IMC controller. The IMC controller will take the form of ideal feedback controller of Fig. 2 with rearrangement of Fig. 1, expressed mathematically in terms of $Q(s)$ and $G_M(s)$as (5) and (6).

$$G_C(s) = \frac{Q(s)}{1 - Q(s) G_M(s)} \tag{5}$$

$$G_C(s) = \frac{(\tau s + 1)}{K[(\lambda s + 1) - e^{-\theta s}]} \tag{6}$$

The feedback controller of (5) and (6) lacks the standard PI/PID form and rearranges the equations with approximation of the time delay in the process model with Taylor series expansion or Padé approximate, to obtain PID controller form of (7).

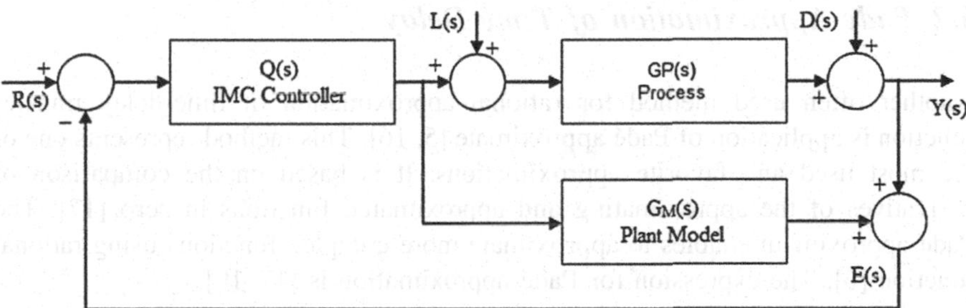

Fig. 1 Basic IMC structure

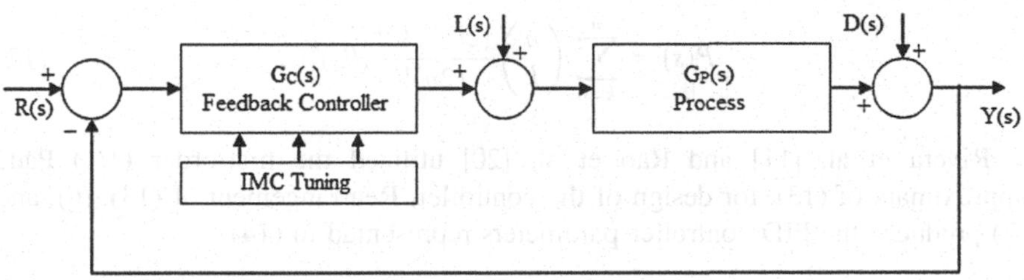

Fig. 2 Feedback control structure

$$G_C(s) = K_P\left(1 + \frac{1}{T_i s} + T_d s\right) \tag{7}$$

3 Time-Delay Approximation

3.1 Taylor Series

The Taylor series expansion of the time-delay term $e^{-\theta s}$ is [5]

$$e^{-\theta s} = 1 - \frac{\theta s}{1!} + \frac{(\theta s)^2}{2!} - \frac{(\theta s)^3}{3!} + \frac{(\theta s)^4}{4!} - \frac{(\theta s)^5}{5!} + \ldots + (-1)^n \frac{(\theta s)^n}{n!} \tag{8}$$

Considering the first two terms and truncation of other terms in (8) results in the first-order Taylor series (9).

$$e^{-\theta s} = (1 - \theta s) \tag{9}$$

Switching (6) and (9) and comparison with (7) produces a PI controller with proportional gain K_P and integral time T_i , represented in (10).

$$K_P = \frac{\tau}{K(\lambda s + 1)}, T_i = \tau \tag{10}$$

3.2 Padé Approximation of Time Delay

Another often used method for rational approximation of time-delay transfer function is application of Padé approximate [5, 16]. This method represents one of the most used and favorite approximations. It is based on the comparison of derivatives of the approximating and approximated functions in zero [17]. The Padé approximant enables to approximate more complex functions using rational function [6]. The expression for Padé approximation is [17–19].

$$e^{-\theta s} \approx \frac{P(-s)}{P(s)} \tag{11}$$

$$P(s) = \sum_{k=0}^{n} \binom{n}{k} \frac{(2n-k)!}{(2n)!} (\theta s)^k \tag{12}$$

Rivera et al. [11] and Rao et al. [20] utilized the first-order (1/1) Padé approximate of (13), for design of the controller. Rearrangement of (13), (6), and (7) produces the PID controller parameters represented in (14).

$$e^{-\theta s} \cong \left(\frac{1 - \frac{\theta}{2}s}{1 + \frac{\theta}{2}s}\right) \tag{13}$$

$$K_P = \frac{2\tau + \theta}{K(2\lambda + \theta)}, T_i = \tau + \frac{\theta}{2}, T_d = \frac{\theta\tau}{2\tau + \theta} \tag{14}$$

Lee et al. [21] utilized second-order (2/2) Padé approximate of (15) for design of PID controller cascaded with lead/lag filter, which is achieved from of (15), (6), and (16). The PID parameters and the filter coefficients are represented in (17).

$$e^{-\theta s} \cong \left(\frac{\frac{\theta^2}{12}s^2 - \frac{\theta}{2}s + 1}{\frac{\theta^2}{12}s^2 + \frac{\theta}{2}s + 1}\right) \tag{15}$$

$$G_c(s) = K_P\left(1 + \frac{1}{T_i s} + T_d s\right)\left(\frac{ds^2 + cs + 1}{bs^2 + as + 1}\right) \tag{16}$$

$$K_P(s) = \frac{\theta}{2K(\lambda + \theta)}, T_i = \frac{\theta}{2}, T_d = \frac{\theta}{6}, a = \frac{\theta\lambda}{2(\lambda + \theta)}, b = \frac{\theta^2\lambda}{12(\lambda + \theta)}, c = \tau, d = 0 \tag{17}$$

4 Simulation Results

Influence of time-delay approximation, viz. 1/1 Padé, 2/2 Padé approximations and Taylor series expansion, on structure and performance of PI/PID controllers for set point tracking and disturbance rejection on FOPTD systems with different θ/τratio, was tested in simulation environment. The process models used by other researchers were considered for study. For uniform comparison, the PI/PID controllers were tuned to have a same robustness measure, in terms of maximum sensitivity (M_S) using IMC technique. The closed-loop performance and robustness were evaluated using integral absolute error (IAE) criterion with step change in set point and load disturbance.

Example 1: The 500-MW superheated steam boiler system, with transfer function $G(s) = 0.7717e^{-56.278s}/(42.934s + 1)$ [20] and $\theta/\tau > 1$ was considered for study. The controllers were tuned to have the same robustness of $M_S = 1.59$ with the adjustment of single tuning parameter λ. The simulation results of Figs. 3, 4, and the Table 1 indicate that the IMC-PI/PID controllers provide good set point tracking but poor disturbance rejection. It is also inferred from the responses that the controllers with Padé approximation perform better than Taylor series for set point tracking and disturbance rejection.

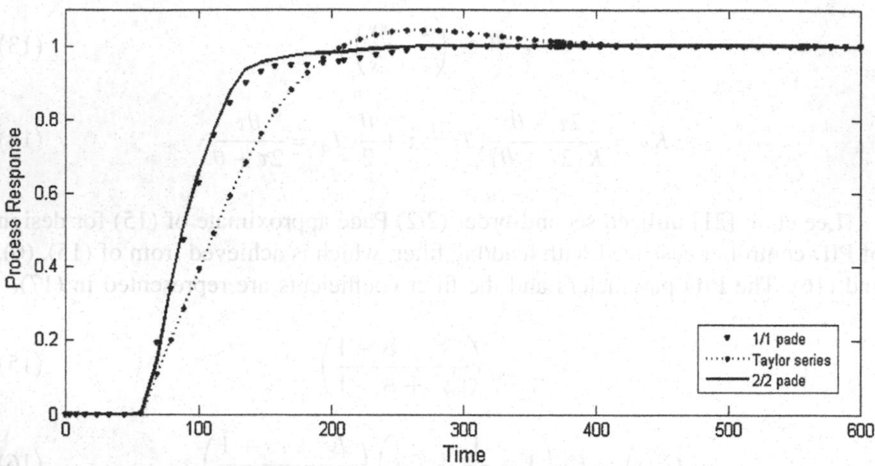

Fig. 3 Responses for step change in set point input of example 1

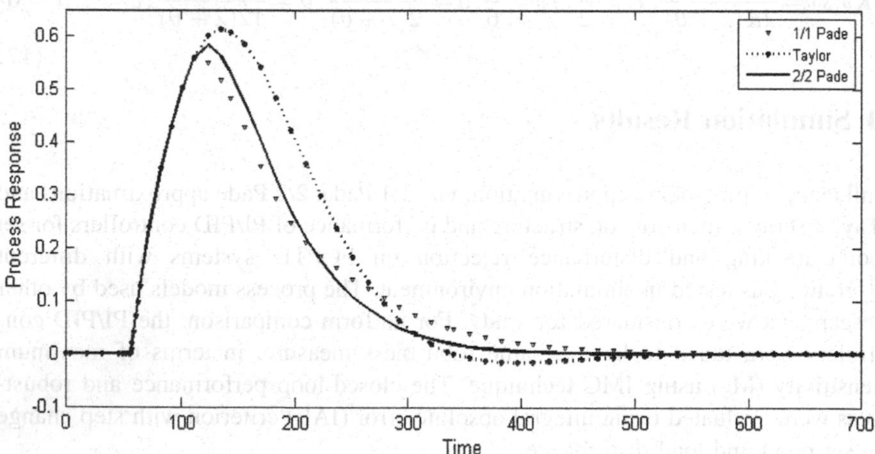

Fig. 4 Response for step-load disturbance input of example 1

Table 1 Performance of IMC-PI/PID controller for set point and disturbance of example 1

Delay approximation	K_P	T_i	T_d	λ	M_S	Set point		Disturbance	
						Peak	IAE	Peak	IAE
1/1 Padé	0.9290	71.073	16.998	71	1.59	0.998	99.14	0.559	76.48
[a]2/2 Padé	0.3839	28.139	9.379	38.7	1.59	1	94.99	0.584	73.3
Taylor	0.4955	42.934	–	56	1.59	1.042	122.1	0.613	90.38

[a] $\frac{ds^2+cs+1}{bs^2+as+1} = \frac{42.934s+1}{107.5435s^2+11.4656s+1}$

Example 2: The lag-time-dominant model $G(s) = 100e^{-1s}/(100s + 1)$ [22] with $\theta/\tau = 0.01$ was considered for study. The controllers were designed to have same robustness of $M_S = 1.4$ by adjusting single tuning parameter λ. The simulation results are represented in Figs. 5, 6, and the Table 2.

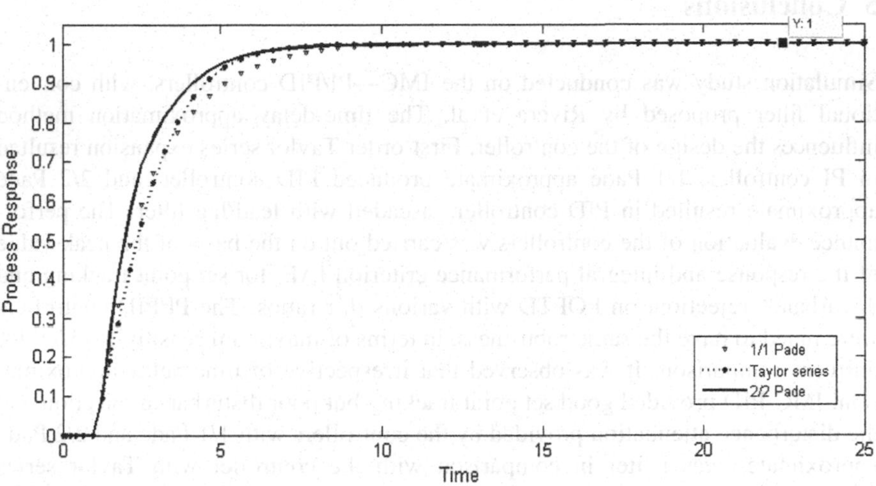

Fig. 5 Responses for step change in Set point input of example 2

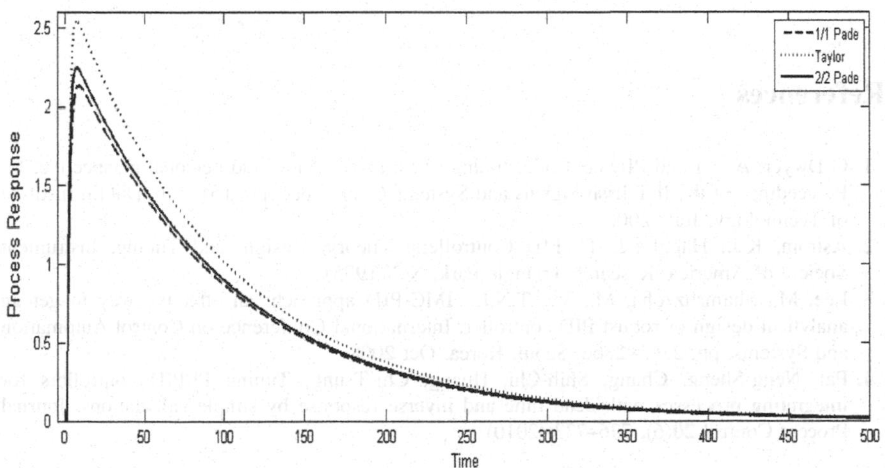

Fig. 6 Response for step-load disturbance input of example 2

Table 2 Performance of IMC-PI/PID controller for set point and disturbance of example 2

Delay approximation	K_P	T_i	T_d	λ	M_S	Set point		Disturbance	
						Peak	IAE	Peak	IAE
1/1 Padé	0.437	100.5	0.4975	1.8	1.4	1.004	2.812	2.131	229.4
[a]2/2 Padé	0.0021	0.5	0.1667	1.36	1.4	1	2.381	2.243	237.5
Taylor	0.3724	100	–	1.685	1.4	1	2.685	2.544	267.8

[a] $\frac{ds^2+cs+1}{bs^2+as+1} = \frac{100s+1}{0.0480s^2+0.2881s+1}$

5 Conclusions

Simulation study was conducted on the IMC—PI/PID controllers, with conventional filter proposed by Rivera et al. The time-delay approximation method influences the design of the controller. First-order Taylor series expansion resulted in PI controller, 1/1 Padé approximate produced PID controller, and 2/2 Padé approximate resulted in PID controller cascaded with lead/lag filter. The performance evaluation of the controllers was carried out on the basis of the peak value of the response and integral performance criterion IAE, for set point tracking and disturbance rejection, on FOPTD with various θ/τ ratios. The PI/PID controllers were tuned to have the same robustness, in terms of maximum sensitivity (M_S), for uniform comparison. It was observed that irrespective of time-delay approximation, IMC-PID provided good set point tracking but poor disturbance rejection and the disturbance attenuation provided by the controllers with 1/1 Padé and 2/2 Padé approximates was better in comparison with the controller with Taylor series expansion. It is suggested to use PID controller cascaded with lead/lag filter for disturbance rejection, obtained with 2/2 Padé approximant with conventional IMC filter or 1/1 Padé approximant with improved IMC filter.

References

1. O'Dwyer, A.: PI and PID controller tuning rules: an overview and personal perspective. In: Proceedings of the IET Irish Signals and Systems Conference, pp. 161–166, Dublin Institute of Technology, June 2006
2. Åström, K.J., Hägglund, T.: PID Controllers: Theory, Design, and Tuning, Instrument Society of America. Research Triangle Park, NC (1995)
3. Lee, M., Shamsuzzoha, M., Vu, T.N.L.: IMC-PID approach: an effective way to get an analytical design of robust PID controller. International Conference on Control Automation and Systems, pp. 2861–2866. Seoul, Korea, Oct 2008
4. Pai, Neng-Sheng, Chang, Shih-Chi, Huang, Chi-Tsung: Tuning PI/PID controllers for integrating processes with dead time and inverse response by simple calculations. Journal Process Control **20**(6), 726–733 (2010)

5. Hanta, V., Procházka, A.: Rational approximation of time delay. Institute of Chemical Technology in Prague. Department of computing and control engineering. Technická 5, 166 28 Praha 6 (2009)
6. Silva, G.J., Datta, A., Bhattacharyya, S.P.: Controller design via Padé approximation can lead to instability. In: Proceedings of the 40th IEEE Conference on Decision and Control. Orlando, Florida USA, Dec 2001
7. Saxena, Sahaj, Yogesh, V., Hote, : Advances in internal model control technique: a review and future prospects. IETE Tech. Rev. **29**, 461–472 (2012)
8. Morari, M., Zafiriou, E.: Robust Process Control. Prentice Hall, Englewood Cliffs, NJ (1989)
9. Shamsuzzoha, M., Lee, M.: Design of advanced PID controller for enhanced disturbance rejection of second-order processes with time delay. AIChE J. **54**(6), 1526–1536 (2008)
10. Shamsuzzoha, M., Lee, M.: IMC: PID Controller Design for Improved Disturbance Rejection of Time-Delayed Processes. Ind. Eng. Chem. Res. **46**(7), 2077–2091 (2007)
11. Rivera, D.E., Morari, M., Skogestad, S.: Internal model control. 4. PID controller design. Ind. Eng. Chem. Process Des. Dev. **25**, 252–265 (1986)
12. Horn, I.G., Arulandu, J.R., Christopher, J.G., VanAntwerp, J.G., Braatz, R.D.: Improved filter design in internal model control. Ind. Eng. Chem. Res. **35**, 3437–3441 (1996)
13. Garcia, C.E., Morari, M.: Internal model controls 1. A unifying review and some new results. Ind. Eng. Chem. Process Des. Dev. **21**, 308 (1982)
14. Lee, Y., Park, S., Lee, M., Brosilow, C.: PID controller tuning for desired closed-loop responses for SI/SO systems. AIChE J. **44**, 106–115 (1998)
15. Shamsuzzoha, M., Lee, M.: Analytical design of enhanced PID filter controller for integrating and first order unstable processes with time delay. Chem. Eng. Sci. **63**, 2717–2731 (2008)
16. Seborg, D.E., Edgar, T.F., Mellichamp, D.A.: Process Dynamics and Control, 2nd edn. Wiley, New York (2004)
17. Pekar, L., Kureckova, E.: Rational approximations for time-delay systems: case studies. Mathematical Methods and Techniques in Engineering and Environmental Science. pp. 217–222. ISBN: 978-1-61804-046-6
18. Partington, J.R.: Some frequency-domain approaches to the model reduction of delay systems. Ann. Rev. Control **28**, 65–73 (2004)
19. Battle, C., Miralles, A.: On the approximation of delay elements by feedback. Automatica **36**(5), 659–664 (2000)
20. Rao, P.V.G.K., Subramanyam, M.V., Satyaprasad, K.: Model based tuning of PID controller. J. Control Instrum. **4**(1), 16–22 (2013)
21. Shamsuzzoha, M., Lee, S., Lee, M.: Analytical design of PID controller cascaded with a lead-lag filter for time-delay processes. Korean J. Chem. Eng. **26**(3), 622–630 (2009)
22. Chen, D., Seborg, D.E.: PI/PID controller design based on direct synthesis and disturbance rejection. Ind. Eng. Chem. **41**, 4807–4822 (2002)

A Study on Nonlinear Classifier-Based Moving Object Tracking

Ajoy Mondal, Badri Narayan Subudhi, Moumita Roy, Susmita Ghosh and Ashish Ghosh

Abstract In the present article, we have provided a performance analysis of different classifier-based object tracking techniques. Here, object tracking has been considered as a binary classification problem. Different classifiers used in the present work are k-nearest neighbor (k-NN), fuzzy k-nearest neighbor (fuzzy k-NN), multilayer perceptron (MLP) neural network, and radial basis function (RBF) neural network. The object scale changes are controlled by considering an adaptive tracking technique. The present work is tested with different video sequences. However, for page constraints, we have provided results on two benchmark video sequences only. The performance of different object tracking techniques were evaluated by two evaluation measures: overlapped area and centroid distance.

Keywords Object tracking · k-nearest neighbor · Fuzzy k-nearest neighbor · Multilayer perceptron · Radial basis function

A. Mondal (✉) · B.N. Subudhi · A. Ghosh
Machine Intelligence Unit, Indian Statistical Institute, Kolkata 700108, India
e-mail: ajoy.mondal83@gmail.com

B.N. Subudhi
e-mail: subudhi.badri@gmail.com

A. Ghosh
e-mail: ash@isical.ac.in

M. Roy · S. Ghosh
Department of Computer Science and Engineering, Jadavpur University, Kolkata, India
e-mail: moumita2009.roy@gmail.com

S. Ghosh
e-mail: susmitaghoshju@gmail.com

L.C. Jain et al. (eds.), *Intelligent Computing, Communication and Devices*,
Advances in Intelligent Systems and Computing 308,
DOI 10.1007/978-81-322-2012-1_61

1 Introduction

Moving object tracking can be defined as the problem of estimating the trajectory of the object in the image plane as it moves around a scene. Tracking is widely used in automated surveillance, traffic monitoring, event detection, etc. [1, 2]. Object tracking is a difficult problem due to abrupt object motion, change in object and scene patterns, non-rigid or flexible object structure, and camera motion. In literature, there exists many object tracking techniques to handle these problems [2]. A commonly used method of object tracking is by using overlapping block-based similarity measure [1]. To provide better result in real-time environment, a hierarchical block-matching scheme is addressed by McErlean in [3]. Recently, tracking is being viewed as a classification problem [4] and a classifier is trained to distinguish the object from the background. Instead of trying to build a complex model to describe the object, classification-based trackers seek a decision boundary that can separate the object and the background. To adapt object appearance changes, tracking methods update the decision boundary instead of the object appearance model [2]. Avidan [4] proposed a support vector machine (SVM)-based object tracking scheme, where an SVM was learned off-line. Classification results were embedded into an optical-flow-based tracker. The performance of the object tracking result is also affected by selection of good set of features. Collins et al. [5] have proposed an online feature selection mechanism which selects the best set of features to improve tracking performance. Avidan [6] proposed an ensemble of weak classifiers which were trained online to distinguish the object and the background pixels. Weak classifiers were combined into a strong classifier using AdaBoost. The strong classifier was then used to classify the pixels of the target frame. Recently, Babenko et al. [7] proposed a novel online multiple instance learning (MIL) algorithm for tracking the moving objects. The MIL framework allows to update the appearance model with a set of image patches. In this article, a comparative study on different classifier-based object tracking techniques is made. Here, object tracking has been considered as a binary classification problem. Some of the popular classifiers used in the present work are k-NN [8], fuzzy k-NN [9], MLP [10], and RBF [10]. The HOG and R, G, and B features are used for classification task. The present work is tested with two different benchmark video sequences. The performance of different object tracking techniques was evaluated using overlapped area and centroid distance [11] measures. The organization of the present article is as follows. In Sect. 2, the HOG feature extraction and different steps of the classifier-based object tracking are discussed. Results and discussion are presented in Sect. 3. Section 4 draws the conclusion of the study.

2 Classifier-Based Object Tracking

The proposed object tracking technique follows two steps: feature extraction and object classification or tracking.

2.1 Histograms of Oriented Gradients Feature Extraction

Histograms of oriented gradients have been widely used as an important feature for different computer vision applications including hand gesture recognition, pedestrian detection, and object tracking [12]. It captures the local object appearance and shape perfectly. To extract these features, at each pixel location in the image frame, we consider a small spatial region called mask/window of size $m \times n$. In each window, a local 1D histogram of gradient directions or edge orientations is accumulated. For better illumination invariance, shadowing, etc., window histograms are locally normalized. The normalized descriptor windows are referred to as histograms of oriented gradients (HOG) descriptors. For gradient computation, initially, each image frame is filtered to obtain x and y derivatives (I_x and I_y) of pixel (x, y) using two 1D filters: $D_x = (-1\,0\,1)$ and $D_y = (-1\,0\,1)^T$. The magnitude and orientation of the gradient are computed as: $|G| = \sqrt{I_x^2 + I_y^2}$ and $\theta = \arctan\left(\frac{I_y}{I_x}\right)$. We have considered unsigned gradient (orientation value goes from 0° to 180°).

2.2 Classifier-Based Object Tracking

In the present work, we have considered the classifier (say L) to distinguish the required objects from the background in the target frame. To classify each pixel of the target frame, we have extracted 11-dimensional feature vector (as discussed in Sect. 2.1). Initially, the classifier L is trained with the training set generated from the candidate model (considered by bounding the required object with a rectangular tracker). Here, the pixels within the rectangular tracker region are considered to belong to the object (positive) class. In the candidate frame, a rectangle of double in size and the same centroid to the candidate model tracker is constructed. The pixel outside the candidate model rectangular tracker but within the larger rectangle is considered to belong to background (negative) class. The classifier L is trained using the considered label patterns and is further used to classify the pixels of the target frame. During the process of tracking, it is found that brute force tracking [13] is complex in nature. Hence, to reduce the complexity of search, we define a region in the target frame as double in size and the same centroid (coordinate) to the considered tracker in the candidate frame and called it as region of search. During the classification process, the trained classifier is used to classify the pixels within the region of search as positive or negative class. The classification results give two connected compact regions: object and background. New object location in the target frame is represented as the centroid of the object region by computing,

$$\mathrm{DC}_x = \frac{\sum_{i=1}^{N} x_i d_i}{\sum_{i=1}^{N} d_i} \quad \text{and} \quad \mathrm{DC}_y = \frac{\sum_{i=1}^{N} y_i d_i}{\sum_{i=1}^{N} d_i}, \tag{1}$$

where (x_i, y_i) is the coordinate of the pixel, N is total number of pixels, and

$$d_i = \begin{cases} 1 & \text{if it classified as object pixel} \\ 0 & \text{otherwise} \end{cases} . \qquad (2)$$

2.3 Object Scale Change

The change in object scale is very common in daily use video sequences. Object scale change occurs in a video due to enlargement or shrinking of the target in the consecutive frames. In a video, the target scale change is quite smooth rather than abrupt. Hence, in all object tracking algorithms, some prior assumption or adaptive tracking scheme is followed. In the present work, the tracker is adaptively modified or changed in scale. At any (frame) instant, we assume that N_c is the number of object pixels in the candidate frame and N_t is the number of object pixels classified in the target image frame. For each target image frame, we take a ratio (N_t/N_c), if it exceeds a predefined threshold value th1, we increase the object tracker size m times both in x and y directions, where m is a small positive constant and $m \propto (N_t/N_c)$. Similarly, if the ratio (N_t/N_c) is below a predefined threshold th2, we decrease the object tracker size m times both in x and y directions.

3 Experimental Results and Discussion

In this section, we show the experimental results of different classifiers-based object tracking techniques. All the considered algorithms were implemented in C/C++ and is run on Pentium D, 2.8 GHz PC with 2 GB RAM. The results are tested on several video sequences; however, for space constraint, we have provided results on two benchmark video (color) sequences. The considered video sequences are Coastguard and Tennis. The first video we have considered in our experiment is Coastguard video sequence. Here, the first frame is considered as the candidate frame and the eighth frame is considered as the target image frame. Features (HOG, R, G, and B) corresponding to each pixel of all the considered image frames were initially extracted. The test set is generated from region of search in the target frame. Thereafter, the classifier is trained with the training set and then, the trained classifier is used to classify all pixels in the test set. If we want to track the person in the fifteenth frame, then the eighth frame is considered as candidate frame and the fifteenth frame is considered as the target frame and the tracking process is repeated. The object (man) tracking results for Coastguard sequence obtained by the k-NN classifier is shown in Fig. 1a, where it is observed that in last two image frames, few portions of the moving objects were not properly tracked. Figure 1b shows the object (man) tracking results obtained by the fuzzy k-NN classifier. It is observed from these results that better object tracking result is obtained. However, some misclassifications are also there. The moving object tracking results obtained by the MLP classifier is shown in Fig. 1c.

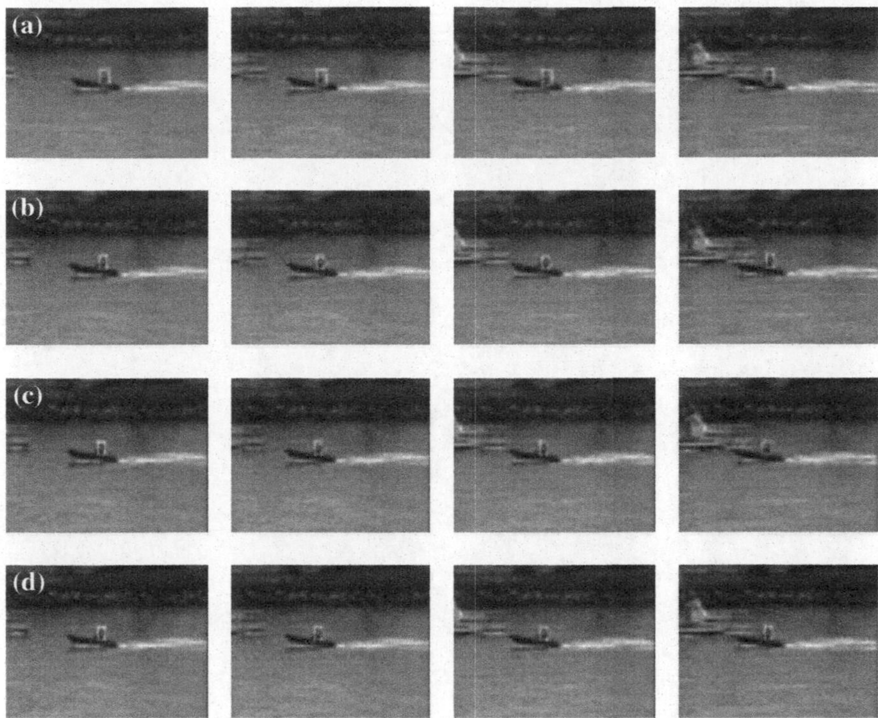

Fig. 1 Target tracking in Coastguard video sequence for frames: 8th, 15th, 22th, 29th, 36th, and 43th. **a** Target man racket tracking by k-NN classifier. **b** Target racket tracking by fuzzy k-NN classifier. **c** Target racket tracking by MLP. **d** Target racket tracking by RBF

Similarly, tracking results obtained by the RBF classifier is shown in Fig. 1d. A visual analysis of these results reveals that the moving object tracking results obtained by the RBF classifier is better than the k-NN, fuzzy k-NN, and MLP classifiers. It may also be noted that MLP classifier gives similar results as that of the fuzzy k-NN technique. However, RBF gives better results.

The last video sequence we have considered in our experiment is Tennis. Here, a person holds a tennis racket and playing with a ball. The aim is to track the racket in the scene. The object tracking results obtained by the k-NN, fuzzy k-NN, MLP, and RBF classifiers are shown in Fig. 2a–d, respectively. In this sequence also, better object tracking result is obtained by RBF classifier rather than k-NN, fuzzy k-NN, and MLP.

3.1 Evaluation Measures

In the present article, two approaches, area-based metric and distance-based metric [10], are used to measure the performance of the proposed tracking algorithm in a

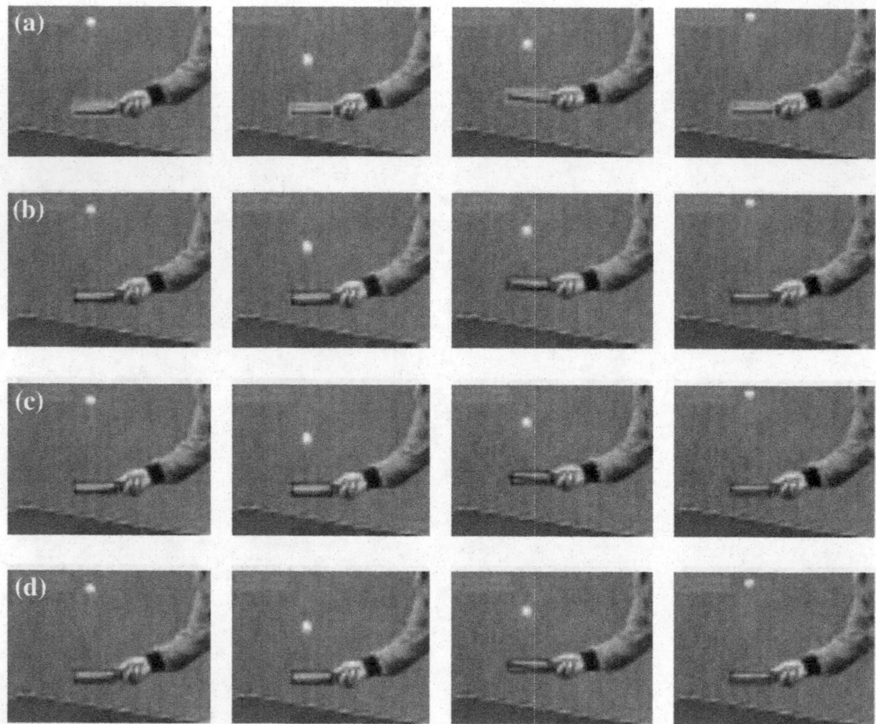

Fig. 2 Target tracking in Tennis video sequence for frames: 5th, 9th, 13th, 17th, 21th, and 25th. **a** Target racket tracking by k-NN classifier. **b** Target racket tracking by fuzzy k-NN classifier. **d** Target racket tracking by RBF

Table 1 Performance measures using area-based metric (ATA)

Considered video sequences	k-NN	fuzzy k-NN	MLP	RBF
Coastguard	0.650	0.673	0.677	0.679
Tennis	0.780	0.803	0.787	0.805

quantitative manner. For a good object tracking system, the ATA measure should be very close to one. Similarly, for a good object tracking system, D_m should be small. The performance measures for all the considered video sequences are given in Tables 1 and 2. It is found from these tables that the RBF classifier provides higher ATA value and lower D_m value. It is also obtained from these results that MLP classifier has provided better results than fuzzy k-NN classifier. However, for Tennis video sequence, fuzzy k-NN has provided better results than MLP classifier. In the above experiment, sometimes, the output of different classifier provides a disconnected region rather than a compact one. In such a scenario, a connected component analysis scheme [1] is followed to get connected/compact object and background regions.

Table 2 Performance measures using distance-based metric (D_m)

Considered video sequences	k-NN	Fuzzy k-NN	MLP	RBF
Coastguard	2.288	2.139	2.089	1.962
Tennis	2.288	2.126	2.139	1.962

4 Conclusions

The present article provides a performance analysis of different classifier-based object tracking techniques. HOG along with color (R, G, and B) features is used for object appearance modeling. Different classifiers used in the present work are k-NN, fuzzy k-NN, MLP, and RBF. Performance of different classifiers is tested on two different benchmark video sequences. It is observed from the result analysis that the performance of the RBF classifier is better than the k-NN, fuzzy k-NN, and MLP classifiers. In most of the cases, the performance of the MLP classifier is found to be better than the k-NN and fuzzy k-NN classifiers. However, both fuzzy k-NN and MLP classifiers are found to be better than the k-NN classifier. In future, we would like to develop some object tracking methods using combination of the above mention classifiers. It may be noted that the accuracy of combination of classifiers is expected to be higher than that of considering classifiers in isolation.

Acknowledgments Authors would like to thank Mr. Tuhin Dutta for his help during the coding.

References

1. Bovic, A.L.: Image and Video Processing. Academic, New York (2000)
2. Yilmaz, A., Javed, O., Shah, M.: Object tracking: a survey. ACM Comput. Surv. **38**(4), 1264–1291 (2006)
3. M. McErlean: Hierarchical motion estimation for embedded object tracking. In: Proceedings of 2006 IEEE international symposium on signal processing and information technology, pp. 997–802 (2006)
4. Avidan, S.: Support vector tracking. IEEE Trans. Pattern Anal. Mach. Intell. **26**(8), 1064–1072 (2004)
5. Collins, R.T., Liu, Y., Leordeanu, M.: Online selection of discriminative tracking features. IEEE Trans. Pattern Anal. Mach. Intell. **27**(10), 1631–1643 (2005)
6. Avidan, S.: Ensemble tracking. IEEE Trans. Pattern Anal. Mach. Intell. **29**(2), 261–271 (2007)
7. Babenko, B., Yang, M., Belongie, S.: Robust object tracking with online multiple instance learning. IEEE Trans. Pattern Anal. Mach. Intell. **33**(8), 1619–1632 (2011)
8. Bishop, C.M.: Pattern Recognition and Machine Learning. Springer, New Jersey (2006)
9. Keller, J.M., Gray, M.R., Givens, J.A.: A fuzzy K-Nearest neighbor algorithm. IEEE Trans. Syst. Man Cybern. **15**(4), 580–585 (1985)
10. Haykin, S.: Neural Networks: A Comprehensive Foundation. Prentice Hall, Singapore (1998)
11. Kasturi, R., Goldgof, D., Soundararajan, P., Manohar, V., Garofolo, J., Bowers, R., Boonstra, M., Korzhova, V., Zhang, J.: Framework for performance evaluation of face, text, and vehicle

detection and tracking in video: data, metrics and protocol. IEEE Trans. Pattern Anal. Mach. Intell. **31**(2), 319–336 (2009)
12. N. Dalal, B. Triggs: Histograms of oriented gradients for human detection. In: Proceedings of IEEE Computer Society Conference on Computer Vision and Pattern Recognition, vol 1, pp. 886–893 (2005)
13. Tekalp, A.M.: Digital Video Processing. Prentice Hall, New Jersey (1995)

Comparative Study of PCM, LPC, and CELP Speech Coders Used for VoIP Applications

Mahesh Chandra and Manas Ray

Abstract The quality of the speech signal in a voice over internet protocol (VoIP) is governed by the speech coding technique employed. Currently, various standard coders such as FS-1015 (LPC-10), ITU-G.711, and FS-1016 (ITU-G.728) are used to digitize the speech signal. This paper analyzes the performance of the above coders by comparing the coding capabilities of the coders on two Hindi and two English language sentences. The performance is then evaluated in terms of compression ratio (CR), peak signal-to-noise ratio (PSNR), and normalized root mean square error (NRMSE).

Keywords VoIP · LPC-10 · G.711 · G.728 · CR · PSNR · NRMSE

1 Introduction

The exponential growth and wide acceptance of voice over IP (VoIP) in various applications of communication technology is the driving force behind the evaluation of telephony technologies in the recent years and is fast becoming a dominant service for the overall telephony industry [1]. This had led to the rapid deployment of VoIP services over the recent years and had opened up various quality-of-service (QoS) challenges in the deployment, especially in the microwave domain [2]. The central component of the VoIP services delivery is the voice coder or the codec deployed for the application. There are different types of codec used, based on the application, complexity, bandwidth requirement, etc. Currently,

M. Chandra (✉) · M. Ray
Department of Electronics and Communication Engineering,
Birla Institute of Technology, Mesra, Ranchi 835215,
Jharkhand, India
e-mail: shrotriya69@rediffmail.com

M. Ray
e-mail: manas_ray.me@rediffmail.com

L.C. Jain et al. (eds.), *Intelligent Computing, Communication and Devices*,
Advances in Intelligent Systems and Computing 308,
DOI 10.1007/978-81-322-2012-1_62

various ITU/FS standard codecs are deployed for the VoIP applications. These codecs were initially designed and deployed in the legacy public switched telephone network (PSTN); however, due to the inherent qualities of the VoIP application to adapt to the existing infrastructure, these codecs have been seamlessly integrated to VoIP systems and have become the central theme of VoIP, dictating the QoS requirements. The most recent codec deployments in the VoIP setup are FS-1015/LPC-10, ITU-G.711, ITU-G.728, ITU-G.729.1, etc. The data rate generated by VoIP codecs differs based on the engineering trade-off between voice quality, available bandwidth, and complexity of the codec [3].

This paper attempts to analyze the working of the existing codecs by testing their characteristics on various sentences of a different speaker's speech in Hindi and English languages. The paper is organized into five sections. Section 1 introduces the motivation and nature of experimental work. The different types of codecs deployed in the VoIP applications along with a brief background on their signal processing techniques are presented in Sect. 2. The parameters measured for the performance evaluation of the codecs are defined in Sect. 3. The simulation results of the codecs and their response to the various Hindi and English sentences are presented in Sect. 4, and finally, the conclusions are drawn in Sect. 5.

2 Standard Codecs for VoIP Applications

There are various standard codecs that are used for speech coding applications. Some of them are pulse code modulation (PCM), differential pulse code modulation (DPCM), adaptive delta pulse code modulation (ADPCM), linear predictive coding (LPC), code excited linear predictive coding (CELP), etc. In this paper, three different speech coders are discussed for VoIP applications.

2.1 Pulse Code Modulation (ITU-G.711)

PCM is speech codec where the input speech is passed through a low-pass filter of bandwidth 4 kHz, and the resulting output is sampled at a rate of 8 kHz to generate a train of pulses as samples. These samples are then quantized, and each sample is encoded with 8-bit binary code. These binary codes are then transmitted from the transmitter to the receiver. At the receiving end, the binary codes are converted into the original speech with the help of digital-to-analog converter and a low-pass filter. It is a waveform-type speech coding technique where the waveform of the synthesized speech signal agrees with that of the original speech signal waveform. The operating bandwidth of the coder is 64 kbps. The reproduced speech signal has the highest quality, called as toll quality.

2.2 Linear Predictive Coding (LPC-10/FS-1015)

LPC is a digital method of speech coding in which the speech signal at a particular instant is represented by a linear sum of the 'p' previous samples [4]. This process is called as autoregressive process of speech coding. It is a lossy form of speech compression which operates at a data rate of 2.4 kbps. LPC is based on the concept of parametric coding, where the human vocal tract is mathematically approximated as a tube with varying diameter. Figure 1 provides an analogy of LPC voice coder with that of the human speech production. LPC is used to estimate the basic speech parameters such as voice/unvoiced decision, pitch, and formant. In the analysis stage of the LPC coding, the speech samples are broken down into segments or blocks, and each segment is then analyzed to determine the following:

(a) Nature of signal as voiced or unvoiced?
(b) Pitch of the signal
(c) The vocal tract filter is analyzed as per the equation

$$y_n = \sum a_i y_{i-1} + G\varepsilon_n \tag{1}$$

where y_n is the output and ε_n is the input and a_i and G are the parameters of the vocal tract filter which need to be estimated. These parameters are passed by the encoder to the decoder of the synthesis stage to reconstruct the signal. At the decoder end, the vocal tract filter is estimated from the parameters received from the analysis stage. The voice quality generated by the codec is of robotic nature, and the trade-off is the low bit rate employed for the coding purpose.

2.3 Code Excited Linear Prediction (ITU-G.728/FS-1016)

CELP is a hybrid coding technique which utilizes the features of the parametric coding as well as that of the waveform coding techniques in order to provide a robust low-bit speech coder [5]. In the analysis stage, the speech signal is passed through a cascade of formant predictor filter and pitch predictor filter. The formant predictor filter removes sample-to-sample correlation, and pitch predictor filter removes the long-term correlations. The residual signal *r*(*n*) noise like signal is compared with the entries of the code book and the index of the best-matched entry is selected. The parameters of the filters along with the index value of the codebook representing the residual filter are passed on to the synthesizer. Figure 2 provides the block diagram of the CELP analysis. In the synthesis stage, the excitation waveform is chosen from a dictionary of waveforms which drives a

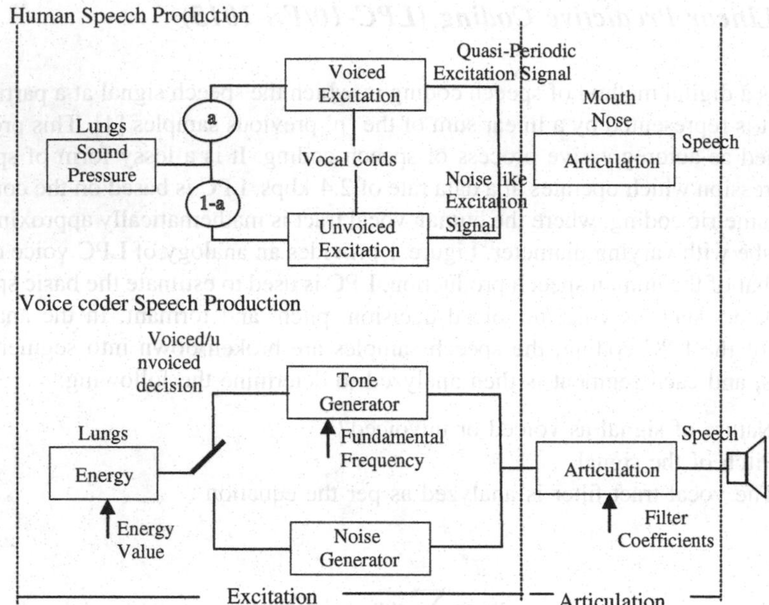

Fig. 1 Block diagram of LPC speech coder

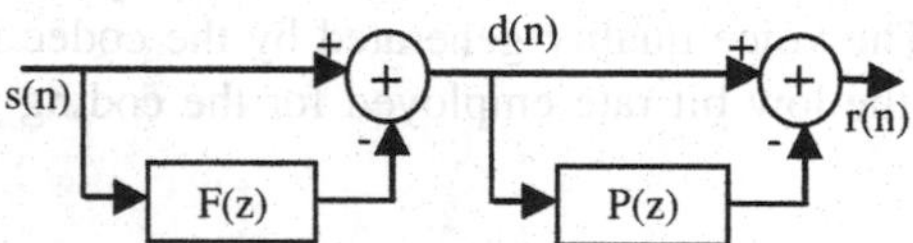

Fig. 2 CELP analysis stage

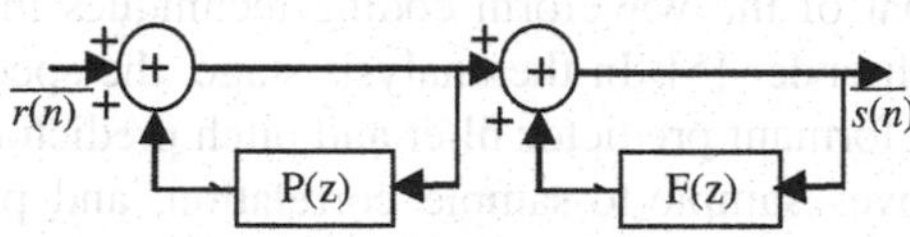

Fig. 3 CELP synthesis stage

cascade of filters synthesized from the parameters received from the analysis stage to approximate the input speech signal. Figure 3 provides the block diagram of the CELP synthesis stage.

3 Performance Evaluation of Speech Codecs

The performance evaluation of the codecs is carried out by objective testing for six speech samples containing both male and female voices of Hindi and English languages. The subjective test for each sample can also be carried out by mean opinion score, and the objective tests were carried out by evaluating the performance in terms of compression ratio (CR), SNR, PSNR, and NRMSE [6, 7]. The expressions of these parameters are given below:

$$CR = \frac{\text{Length of } (x(n))}{\text{Length of } (r(n))} \tag{2}$$

$$\text{SNR} = 10\log_{10}\left(\frac{\sigma_x^2}{\sigma_e^2}\right) \tag{3}$$

$x(n)$ and $r(n)$ are the original and reconstructed signals, respectively. σ_x^2 and σ_e^2 are mean square of the speech signal and the mean square difference between the original and reconstructed signals, respectively. Peak signal-to-noise ratio (PSNR) and normalized root mean square error (NRMSE) can be given by

$$\text{PSNR} = 10\log_{10}\frac{NX^2}{\|X - r\|} \tag{4}$$

$$\text{NRMSE} = \sqrt{\frac{(x(n) - r(n))^2}{(x(n) - \mu x(n))^2}} \tag{5}$$

In Eq. 4, N is the length of the reconstructed signal, X is the maximum absolute square value of the signal x, and $\|X-r\|^2$ is the energy difference between the original and reconstructed signals. Where $x(n)$ is the speech signal, $r(n)$ is the reconstructed signal and $\mu x(n)$ is the mean of the speech signal.

4 Experimental Setup

MATLAB simulation model is prepared for PCM [8], CELP [9], and LPC [10] as per the standard available. A total of four sentences were used for testing the performance of PCM, LPC, and CELP codecs as given in Table 1.

Table 1 Details of the sample sentences used in the experiment

Sample No.	Sentence	Language	Speaker
1	**धोबिन जब सोकर उठती तब देखती कि चौंका साफ़ पढ़ा हैं और बर्तन मजे हु ये हैं।**	Hindi	Male
2	**यहाँ से लगभग पाँच मील दक्षिण पश्चिम में कटघर गाँव हैं।**	Hindi	Male
3	Welcome to the Internet my friend, How can I help you?	English	Male
4	I don't want to go out with mad people	English	Female
5	**धोबिन जब सोकर उठती तब देखती कि चौंका साफ़ पढ़ा हैं और बर्तन मजे हु ये हैं।**	Hindi	Female
6	**यहाँ से लगभग पाँच मील दक्षिण पश्चिम में कटघर गाँव हैं।**	Hindi	Female

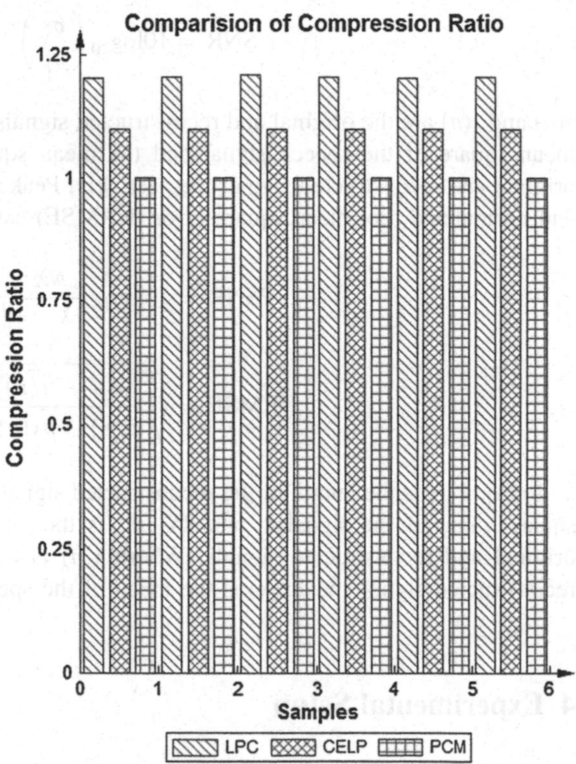

Fig. 4 Compression ratio comparison

Fig. 5 SNR comparison

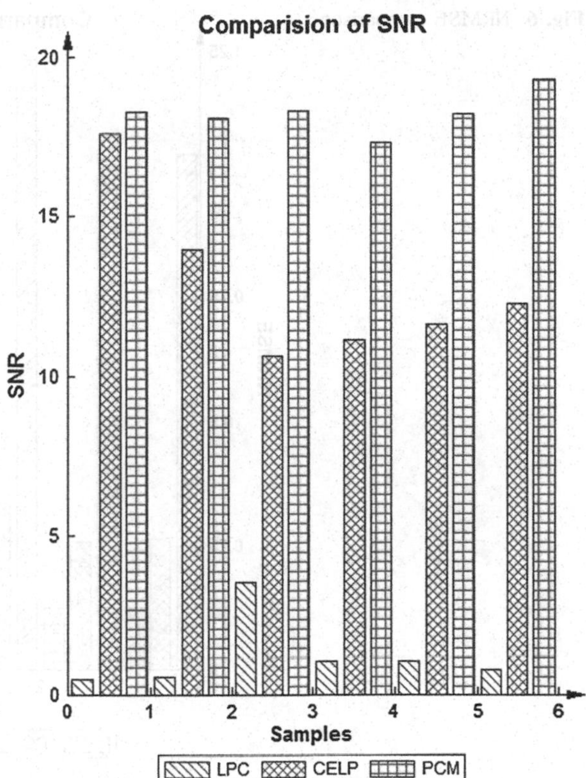

Comparative performance of LPC, CELP, and PCM is shown in Figs. 4, 5 and 6 in terms of compression ratio, SNR, and NRMSE. It is observed from results that LPC provides a greater degree of compression as compared to CELP and PCM for the entire sample sets. PCM provides excellent SNR, PSNR, and NRMSE measurements; however, it requires a huge bandwidth. CELP on the other hand provides an acceptable level of SNR, PSNR, and NRMSE with low bit rate requirements as compared to PCM. The quality of the reconstructed signal was tested and found to be in compliance with the MOS standard requirements [11, 12]. The MOS of PCM is best followed by CELP and LPC. Hence, it can be inferred from the above results that CELP provides a good alternative to PCM by conserving less bandwidth.

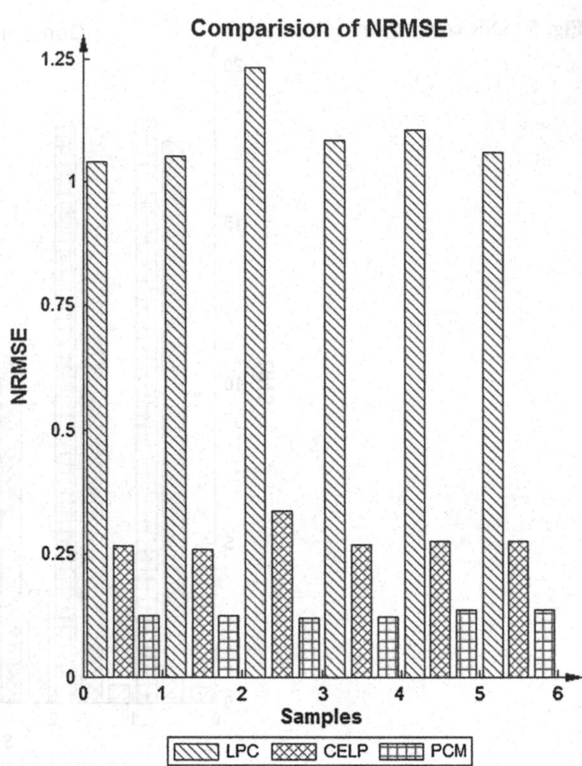

Fig. 6 NRMSE comparison

5 Conclusions

The performance of various standard codecs used for VoIP applications has been evaluated in this paper. It is observed that code excited linear prediction provides a comparable performance as compared to PCM with lower bandwidth requirements than that of PCM. LPC provides the good compression at lower bit rate; however, it lags in SNR as compared to the other two codecs. The results reveal that the performance of codecs under study remains unaffected by change in language or speakers.

References

1. Ogunfunmi, T., Narasimha, M.J.: Speech over VoIP networks: Advanced signal processing and system implementation. IEEE Circ. Syst. Mag. **2**(2), 35–55 (2012)
2. Ray, M., Chandra, M., Patil, B.P.: Evaluation of CDMA microwave links at different environments for VoIP applications. Int. J. Adv. Res. Comput. Commun. Eng. **1**(8), 508–512 (2012)

3. Ali, A.A., Vassilaras, S., Ntagkounakis, K.: A comparative study of bandwidth requirements of VoIP codecs over WiMax access networks. IEEE 3rd International Conference on Next Generation Mobile Applications, Services and Technologies, pp. 197–203 (2009)
4. Sunny, S., David, S.P., Jacob, K.P.: Recognition of speech signals: an experiment comparison of linear predictive coding and discrete wavelet transforms. Int. J. Eng. Sci. Technol. **4**, (2012)
5. Tandel, M., Shah, V., Patel, B.: Implementation of CELP CODER and to evaluate the performance in terms of bit rate, coding delay and quality of speech. In: 3rd IEEE International Conference on Electronics Computer Technology, pp. 86–89. Apr 2011
6. Ambika, D., Radha, V.: A comparative study between discrete wavelet transform and linear predictive coding. IEEE World Congress on Information and Communication Technologies, pp. 965–969 (2012)
7. Najih, A.M.M.A., Ramli, A.R., Ibrahim, A., Syed, A.R.: Comparing speech compression using wavelets with other speech compression schemes. In: IEEE Proceedings of Student Conference on Research and Development, pp. 55–58 (2003)
8. http://www.itu.int/ITU-T/recommendations/rec.aspx?rec=911
9. http://www.itu.int/ITU-T/recommendations/rec.aspx?rec=11674
10. Campbell, P.J., Tremain, T.E.: Voiced/unvoiced classification of speech with applications to the U.S. government LPC-10E algorithm. In: IEEE International Conference on Acoustics, Speech and Signal Processing, pp. 473–476 (1986)
11. http://www.cisco.com/c/en/us/support/docs/voice/h323/14069-codec-complexity.html#mos
12. Ray, A.K., Acharya, T.: Information Technology: Principles and Applications. Prentice-Hill of India Private Limited, New Delhi (2004)

Sound- and Touch-Based Smart Cane: Better Walking Experience for Visually Challenged

Rajesh Kannan Megalingam, Aparna Nambissan, Anu Thambi, Anjali Gopinath and K. Megha

Abstract Moving with the help of a white cane is an elusive task for the visually challenged unless they create a mental route map with recognizable reference elements. The Smart Cane is intended to provide the visually challenged a better walking experience. The design is incorporated with Bluetooth-enabled obstacle detection module, supported with heat detection and haptic modules. The ultrasonic range finders help in detecting obstacles. The calculated distance is send to an android device via Bluetooth. The user gets voice alerts about the distance through Bluetooth headset. Haptics module is designed to warn the user of moving obstacles with the help of vibratory motors. The goal of this project is to arm its wielder with the functional support of a walking cane without having to possess one physically.

Keywords Ultrasonic range finder · Haptics · Bluetooth

1 Introduction

According to WHO [1], there are about 39 million people who are blind. Out of these, majority of the people wish to get rid of the cane due to frustration. The Chairman of National Association for the Blind, Kottayam, suggested that it would be a great improvement in their society if the use of cane could be avoided.

To alleviate the issues of the visually challenged, Smart Cane is designed in such a way that it includes a Bluetooth-enabled obstacle detection module where the distance information from the Arduino board is send to the Bluetooth headset. The design also supports a temperature detection module and haptics module.

R.K. Megalingam (✉) · A. Nambissan · A. Thambi · A. Gopinath · K. Megha
Amrita School of Engineering, Amrita Vishwa Vidyapeetham University, Kollam 690525, India
e-mail: rajeshkannan@ieee.org

L.C. Jain et al. (eds.), *Intelligent Computing, Communication and Devices*, Advances in Intelligent Systems and Computing 308,
DOI 10.1007/978-81-322-2012-1_63

While the user gets voice feedback about the static obstacles, vibratory motors are used to inform about the moving obstacles. The intensity of vibration depends on the speed of the moving obstacles. Despite the simplicity, the integration module will emphatically be a solution to the visually challenged.

2 Related Works

A lot of study and research are being done to design a fine instrument that provides the user a better walking experience. One of them is Smart Vision [2]. It is an efficient design that can detect path borders using canny edge detector and an adapted version of Hough transform. The device can detect stationary as well as moving obstacles. The former is done through a camera attached on the user's chest, and the latter is achieved by multi-scale, annotated, and biologically inspired key points. HALO [3] is another device that can be mounted on the existing white cane and can detect low hanging obstacles such as branches of trees. It consists of ultrasonic range sensor with an eccentric-mass vibrating motor that vibrates distinctly for ground obstacle and low hanging obstacle. An intelligent guide stick [4] detects obstacles using ultrasonic sensors but it is unable to tell whether the obstacle is in motion or not. A wireless ultrasonic ranging system [5] detects obstacles using an ultrasonic sensors and the PIC16F877 microcontroller finds out the distance from the obstacle. The phone that is linked to the microcontroller converts the information to speech, and the data are sent to the Bluetooth earphone to alert the user.

The work done by Gallo et al. [6], 'Augmented White Cane with Multimodal Haptic Feedback' involves haptics feedback to imitate the behaviour of a longer cane. The feedback is given by a shock-generating module, which releases the kinetic energy stored in a spinning wheel in a controlled amount. In case of a moving obstacle, the spatiotemporal vibration pattern, stimulated on the user's hand, creates the sensation of an apparent movement.

3 System Architecture

Smart Cane consists of mainly three modules (Fig. 1) namely heat detection, obstacle detection and Bluetooth module. The presence of an obstacle in front of the user is identified by using an SRF05 ultrasonic sensor. The distance is measured in centimetres and corresponding to the distance; the user hears 'stop' and 'obstacle' for nearby and distant obstacle, respectively, in the Bluetooth headset. Arduino board that holds the sensor communicates to the Bluetooth headset via an android phone. If the obstacle is in motion, the vibration motor attached vibrates. The intensity of vibration would be high for fast moving obstacles. The presence of hot objects (above 70 °C) is informed to the user by the sound of a buzzer. The temperature is measured using an LM35 temperature sensor.

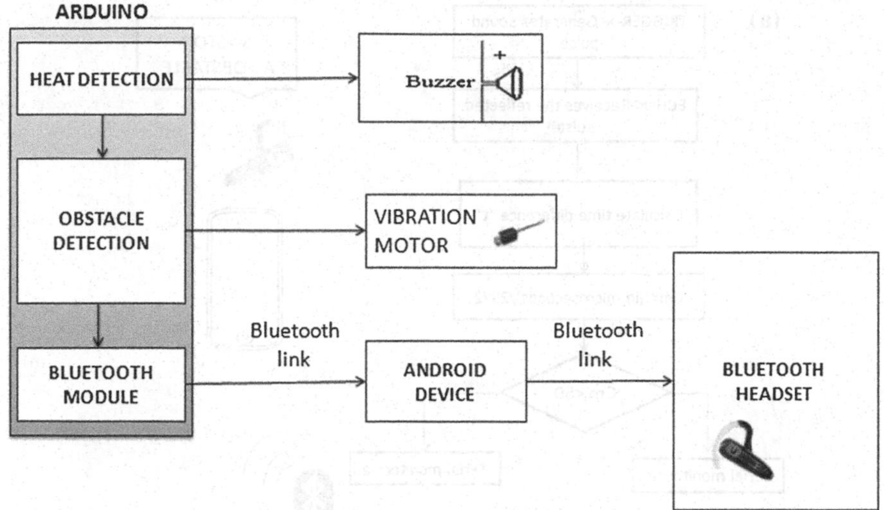

Fig. 1 Overall system

4 Implementation

See Fig. 2.

4.1 Algorithm for Obstacle Detection

A continuous stream of pulses is sent out through the trigger terminal of the ultrasonic sensor. The pulses reflected from the obstacle are received at the echo terminal. The time duration for which the echo pulse remains low gives the time taken by the ultrasonic pulse to travel twice the distance. Thus, relating the time taken (t) and distance between the obstacle (d), $2d = s \times t$; where $s = 340$ m/s (speed of sound). Converting speed to cm/us and time to microseconds, the distance d in cm can be written as

$$d = \text{microseconds}/29/2 \tag{1}$$

If the distance d is less than 50 cm, a message 's' is sent to the android through a Bluetooth link, and if the distance d is greater than 50 cm but less than 100 cm, a message 'a' is sent. On receiving's', android is programmed to speak out 'stop' and in case of 'a' it speaks out 'obstacle'.

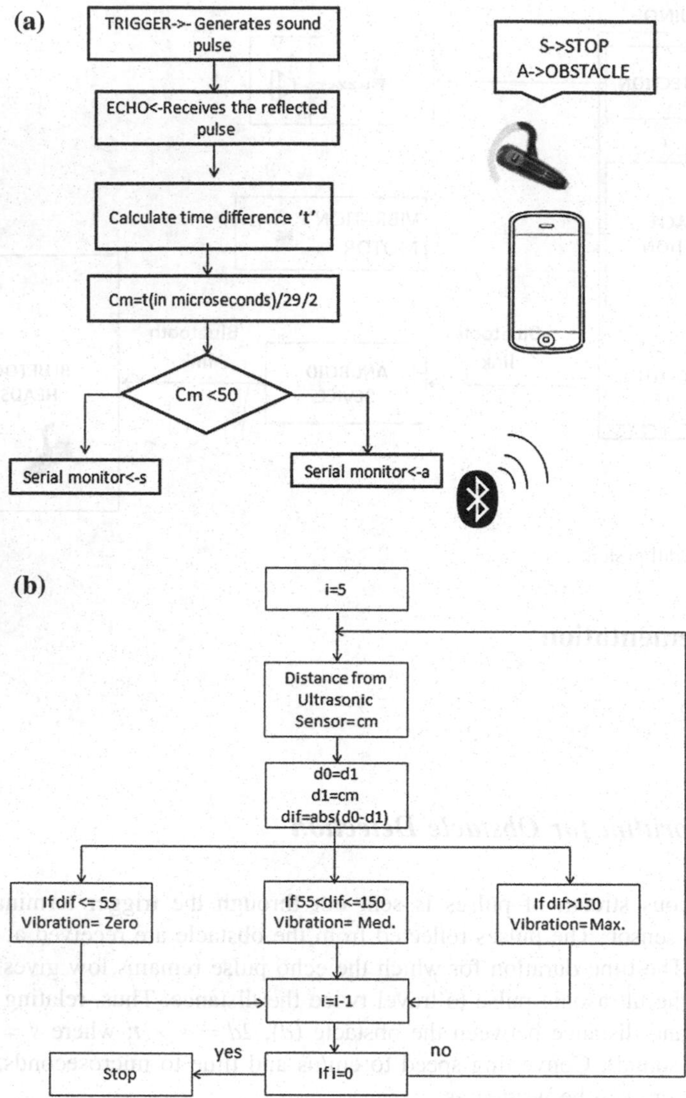

Fig. 2 **a** Flowchart for obstacle detection. **b** Flowchart for moving obstacle detection

4.2 Algorithm to Detect Moving Obstacle

The distance received from the ultrasonic sensor for that particular obstacle is measured 5 times. Each reading is subtracted from the previous reading, and the absolute value is taken. If the measured value is within 55 cm, which is equal to the average footstep of the user implies that the obstacle has not moved and the vibration given to the motor is zero. If the difference increases and lies between 55 and 150 cm, it means that the obstacle is moving at a faster rate and medium vibration is given. Finally, if the difference measured is greater than 150 cm, it gives an idea that the obstacle is moving at a greater rate and maximum vibration is given.

4.3 Algorithm to Detect Hot Objects

The temperature of the obstacle is measured using LM35 temperature sensor. The voltage corresponding to the temperature is received at the Arduino pin. To get the temperature reading from the board, the conversion,

$$\text{tempC} = (5 \times \text{Voltage} \times 100)/1{,}024 \quad (2)$$

is used. The transistor gives 10 mV for every degree rise in temperature, so the value measured by the analog pin needs to be multiplied by 100. To scale the voltage to 5 V, again the value needs to be multiplied by 5 and divided by 1,024 since the analog reading will be in the range of 1,024 (10 bit).

4.4 Arduino–Android Interfacing

Once the distance value is obtained, it is send to an android device via Bluetooth. Itead v2.2 Bluetooth shield is used for pairing the android device with Arduino board. Arduino is interfaced with the android phone to support the obstacle detection module. Distance information is sent to the android device using a Bluetooth shield mounted on the Arduino board (Fig. 3).

The device listener scans for all possible Bluetooth devices and pairs with that device with which the key is found to match. A socket connection is then established through the connection thread. The timer block is required to collect data in a timely manner, that is, data are collected after every 500 ms. The received data can be considered as a bundle and reaches the message handler. Message handler receives and processes the messages. Each instance in the message handler can be associated with a thread (process of speaking out ‘s’ or ‘a’ in this case). The received text is converted to speech. The UI gets updated automatically.

5 Experimental Results

The time delay in hearing each warning message was calculated using a stopwatch. The range for each messages is mentioned in Table 1.

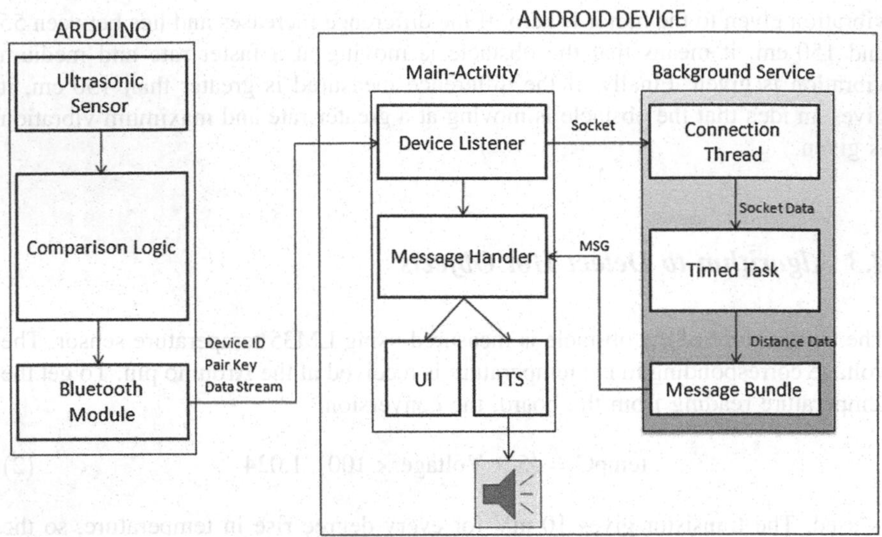

Fig. 3 Arduino–Android interfacing

Table 1 Distance-to-range mapping and speed to vibration mapping

Static obstacle		Moving obstacle	
Distance (cm)	Range	Speed (cm/s)	Vibration intensity
<50	1	4.9–49.01	Small
50–100	2	49.01–196.08	Medium
>100	3	196.08–490.01	Maximum

The count starts at the moment the obstacle comes in the range of the wielder, and the count stops when the warning message is heard. Values are taken for four scenarios (a) a faraway obstacle comes in range 2, (b) the obstacle comes in range 1, (c) the obstacle moves from range 2 to range 1 and (d) The obstacle moves from range 1 to range 2.

The average response time was found to be 1.96 s.

The distance measured by the ultrasonic sensor was verified using a measuring device. The detection of moving obstacle was simulated with the help of a small manually operated car. Three test cases were carried out and the following observations were verified.

6 Future Work

In this paper, we have presented the development and design of Smart Cane, which includes three inevitable modules. The temperature detection module is tested using lm35 sensor. Testing using a contactless sensor is in progress. Further, we are working on including a pit detection module and modifying the haptics module such that the person would get the feel of holding a cane without having to hold it physically.

7 Conclusion

The paper details the architecture and working algorithm of a device that scans the path of a visually challenged and alerts them in the event of any danger. An Arduino-based algorithm is constructed to detect hot objects and obstacles ahead of them. The Arduino algorithm combined with android interfacing warns the user of respective dangers through a Bluetooth headset. Bluetooth technology is exploited here to link android to the Arduino. In the event of an approaching obstacle, a tactile feedback is given on the hand. The vibratory motor attached to the hands vibrates with varying intensity depending on the speed of the approaching obstacle.

Acknowledgment We are extremely grateful to Electronics and Communication department of Amrita School of Engineering, Amritapuri Campus, Kollam, India, for providing us all the necessary laboratory facilities and support towards the successful completion of the project. We also thank the Amrita Research Laboratory and CAE Laboratory for their support and guidance.

References

1. World Health Organization, http://www.who.int/mediacentre/factsheets/fs282/en/
2. José, J., Miguel, F., Rodrigues, J.M.F., du Buf, J.M.H.: The smart vision local navigation aid for blind and visually impaired persons. Int. J. Digit. Content Technol. Appl. **5**(5) (2011)
3. Kuchenbecker, K.J., Wang, Y.: HALO: haptic alerts for low-hanging obstacles in white cane navigation. University of Pennsylvania (2012)
4. Kang, S.J., Kim, Y.H., Moon, I.H.: Development of an intelligent guide-stick for the blind. In: IEEE 2001 International Conference on Robotics and Automation, Seoul, Korea (2001)
5. Tahat, A.: A wireless ranging system for the blind long-cane utilizing a smart-phone. School of Electrical Engineering Princess Sumaya University for Technology Amman, Jordan (2009)
6. Gallo, S., Chapuis, D., Santos-Carreras, L., Kim, Y., Retornaz, P., Bleuler, H., Gassert, R.: Augmented white cane with multimodal haptic feedback. In: Proceedings of the 2010 3rd IEEE RAS and EMBS International Conference on Biomedical Robotics and Biomechatronics (2010)

Efficient VLSI Implementation of CORDIC-Based Direct Digital Synthesizer

N. Prasad, Manas Ranjan Tripathy, Ansuman DiptiSankar Das, Nihar Ranjan Behera and Ayaskanta Swain

Abstract This paper presents efficient VLSI implementation of a direct digital synthesizer (DDS). Coordinate rotation digital computer (CORDIC) architecture is used in realizing the phase-to-amplitude converter (PAC) block in the proposed design. The proposed synthesizer has a frequency control word (FCW) that can select up to three different values for the phase increment. The proposed design is realized in Xilinx Virtex II Pro FPGA development board and is tested for its functionality using ChipScope Pro. The proposed design is mapped on to several families of Xilinx FPGAs for comparing the performance. Proposed synthesizer is also implemented using ASIC design flow. In the reported design, quadrature outputs can be obtained simultaneously.

Keywords CORDIC · Direct digital synthesizer · FPGA · Xilinx · ASIC

N. Prasad (✉)
Department of E & ECE, Indian Institute of Technology, Kharagpur 721302, India
e-mail: nprasad@ece.iitkgp.ernet.in

M.R. Tripathy
Department of Electronics and Communication Engineering, ITER, SOA University, Bhubaneswar, India
e-mail: manastripathy@soauniversity.ac.in

A.D. Das
IBM India Private Ltd, Hyderabad, India
e-mail: ansuman.das.engg@gmail.com

N.R. Behera
Proxim Wireless, Hyderabad, India
e-mail: 1985nihar@gmail.com

A. Swain
Department of ECE, National Institute of Technology, Rourkela 769008, India
e-mail: swaina@nitrkl.ac.in

L.C. Jain et al. (eds.), *Intelligent Computing, Communication and Devices*,
Advances in Intelligent Systems and Computing 308,
DOI 10.1007/978-81-322-2012-1_64

1 Introduction

Coordinate rotation digital computer (CORDIC) is a special purpose computer to compute many transcendental and nonlinear functions. This was proposed by Volder in 1959 [1]. The functions that can be computed using a CORDIC computer include logarithmic, trigonometric, hyperbolic, etc. [2]. CORDIC has become a popular tool to implement several digital systems, especially in the areas of digital signal processing (DSP), communications, computer graphics, etc. The simplicity of CORDIC lies in the fact that it can compute any of the above-mentioned operations using shifts and additions. The operating mode and the coordinate system are two key factors to compute the desired functions in the CORDIC. Many signal processing and communication systems operate CORDIC in circular coordinate system.

Frequency synthesizers, sometimes also called oscillators, are essential units of many communication systems. Many communication systems employ digital subsystems in their units, thus making the usage of digital systems more ubiquitous. Direct digital synthesizers (DDS) are a class of frequency synthesizers in digital domain, which generate waveforms of desired frequencies [3]. These generate waveforms like sine, cosine, triangular, square or rectangular, saw tooth, etc. As mentioned earlier, these have wide applications in satellite communication systems, RF signal processing, etc. DDS offers many advantages over analog oscillators such as precise tuning resolution of output frequency, fast hopping of phase that reduces phase related errors, and many more.

Section 2 discusses the background, Sect. 3 discusses the proposed design, Sect. 4 discusses the implementation and results, and Sect. 5 concludes the paper.

2 Background

DDS or direct digital frequency synthesizer (DDFS) consists of the following blocks: phase accumulator (PA), phase-to-amplitude converter (PAC), digital-to-analog converter (DAC), and a low-pass filter. The block diagram of the system is shown in Fig. 1.

The PA accumulates the phase according to the frequency control word (FCW). The PAC converts the input phase into output amplitude. Several phases to amplitude mapping mechanisms are described in [4]. More details on how to implement a DDS are given in [5–7] that describes the implementation of DDFS by using analog interpolation method.

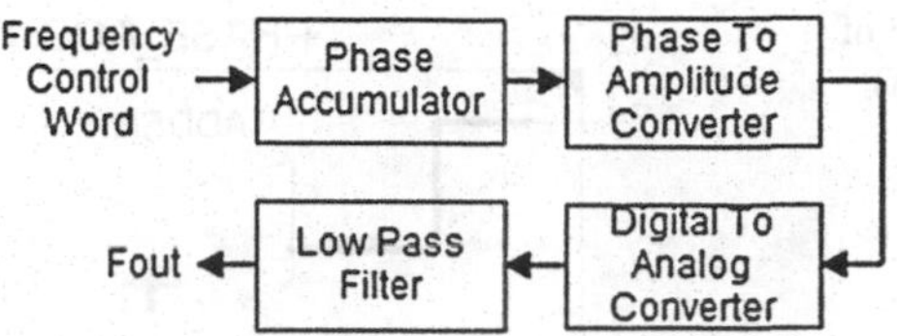

Fig. 1 Block diagram of direct digital synthesizer

3 Proposed Design

Figure 2 shows the RTL schematic of the proposed design, and Fig. 3 shows the RTL schematic of PA. Proposed design consists of implementing a DDS using CORDIC as a frequency–amplitude mapper block. Pipelined CORDIC is used in the proposed design to improve the throughput to one sample per one clock cycle. CORDIC block in the proposed design consists of 16 stages and the precision is 8 bits. To improve the accuracy, the difference between the precision and number of CORDIC stages is maintained. Mapped CORDIC is used to map the output of CORDIC block to the entire circular coordinate system. The phase input to CORDIC consists of 16 angle bits and 2 quadrant bits. The output consists of 8 amplitude bits. The input to the PA consists of 18 phase bits. These input values are controlled by a 2 bit 'EN' input. The PA consists of an 18-bit signed adder with a register following it. The 18-bit phase word is called FCW. The proposed design houses three values for FCW. As the CORDIC block gives both sine and cosine values simultaneously, the advantage of the proposed design lies in generating the quadrature outputs. The microarchitectures of mapped CORDIC block is same as the one mentioned in [11].

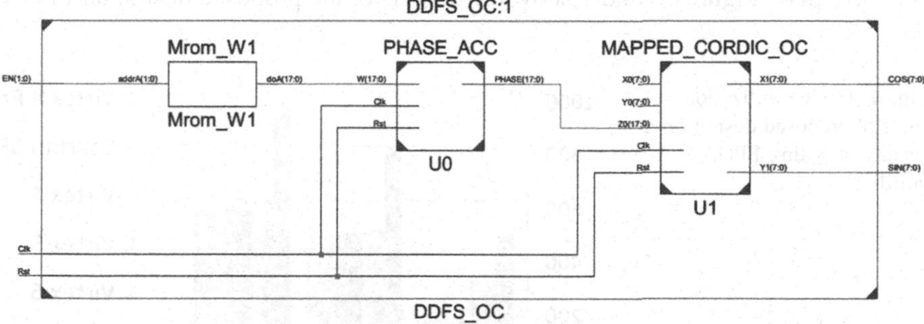

Fig. 2 RTL schematic of proposed DDFS

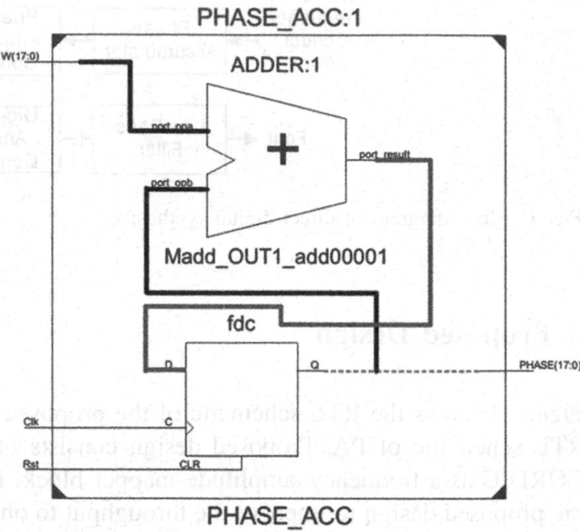

Fig. 3 RTL schematic of phase accumulator (PA)

4 Results and Discussion

The design is coded using VHDL, and the platforms considered for implementation are FPGA and ASIC. The output amplitude has a resolution of 8 bits. The phase difference between the quadrature outputs is maintained. The design is mapped on to a variety of Xilinx FPGA families to measure the performance. Figure 4 shows the chart of device utilization of the proposed design on a variety of FPGAs.

The design is physically mapped on the Xilinx Virtex II FPGA. Figure 5a shows the power distribution chart of the proposed design on the mapped FPGA. Figure 5b shows the total power consumed by the proposed design on the mapped FPGA. The maximum power consumed at the sampling frequency of 165.9 MHz is 658.15 mW. Figures 6 and 7 show the outputs of the proposed design on FPGA.

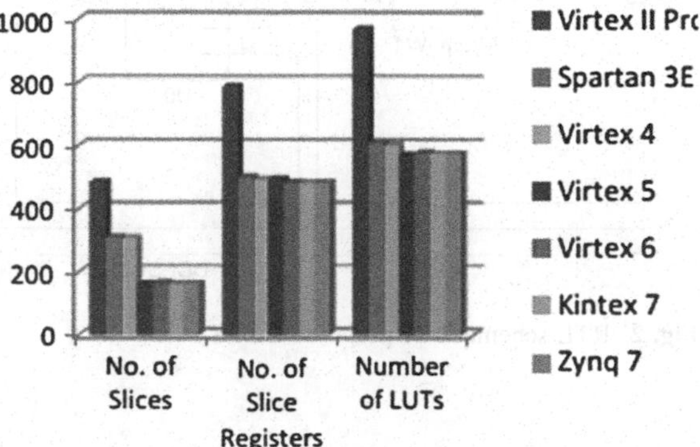

Fig. 4 Device utilization chart of proposed design on a variety of Xilinx FPGA families

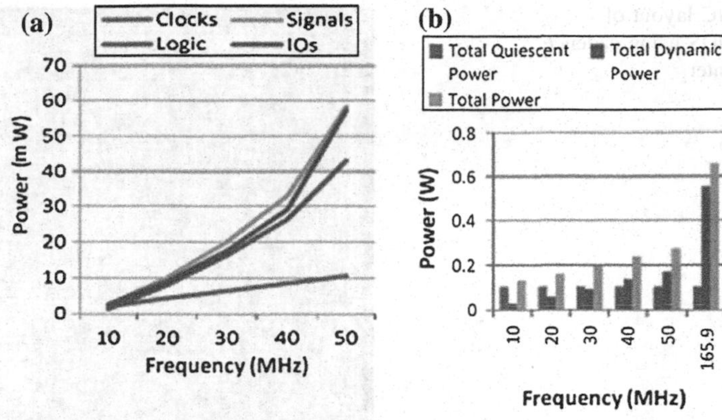

Fig. 5 **a** Power distribution chart of proposed design in Xilinx Virtex II Pro FPGA. **b** Power consumption of proposed design when mapped on Xilinx Virtex II Pro FPGA

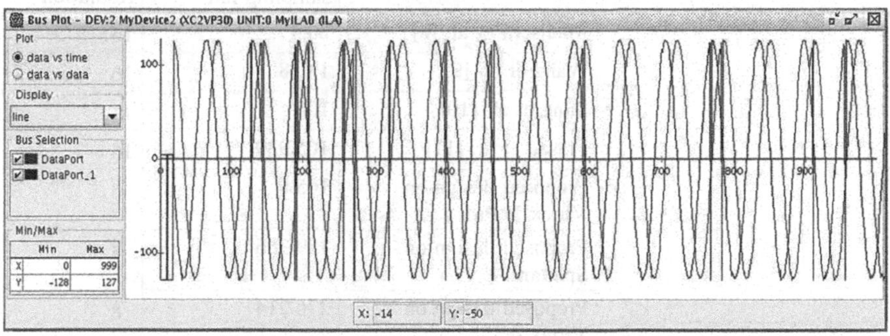

Fig. 6 DDS output in FPGA when EN = 01

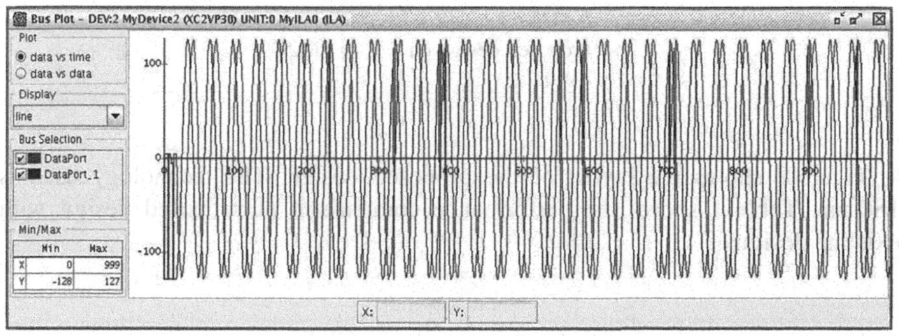

Fig. 7 DDS output in FPGA when EN = 11

Fig. 8 ASIC layout of proposed DDS using cadence SoC encounter

Table 1 Performance comparison of proposed design with existing designs

Device	Maximum sampling rate	Output bit resolution
Madisetti et al. [8]	80.4	16
Wang et al. [9]	1,018	16
Sung et al. [10]	100	16
Prasad et al. [11]	165.939	16
Proposed design on Virtex II Pro	165.9	8
Proposed design on Spartan 3E	129.366	8
Proposed design on Virtex 4	276.214	8
Proposed design on Virtex 5	346.141	8
Proposed design on Virtex 6	524.632	8
Proposed design on Kintex 7	511.509	8
Proposed design on Zynq 7	511.509	8

Figure 8 shows the layout of the proposed design. The technology used is 180 nm. Table 1 shows the performance comparison of proposed design with existing designs.

5 Conclusions

This paper reported efficient VLSI implementation of CORDIC-based DDS in both ASIC and FPGA design flows. CORDIC-based PAC has provided the advantage of giving simultaneous quadrature outputs. The amplitude resolution is 8 bits. The design is physically tested on Xilinx Virtex II Pro FPGA development board. The proposed design is mapped on to a number of Xilinx FPGA families to compare the performance in terms of operating frequency while observing the device utilization in each of them. The proposed design consumes a power of 658.15 mW when operated at 165.9 MHz on Xilinx Virtex II Pro FPGA. ASIC design of proposed architecture is also done using 180-nm process technology.

Acknowledgment This work is done as a part of master's thesis of the first author. Authors thank the VLSI Laboratory of NIT Rourkela for providing with the necessary tools and kits.

References

1. Volder, J.E.: The CORDIC trigonometric computing technique. IRE Trans. Electron. Comput. **EC–8**, 330–334 (1959)
2. Walther, J.S.: A unified algorithm for elementary functions. Joint Spring Comput. Conf. **38**, 379–385 (1971)
3. Bramble, A.L.: Direct digital frequency synthesis. In: 35th Annual Frequency Control Symposium, pp. 406–414 (1981)
4. Vankka, J.: Methods of mapping from phase to sine amplitude in direct digital synthesis. IEEE Trans. Ultrason. Ferroelectr. Freq. Control **44**(2), 526–534 (1997)
5. Cordesses, L.: Direct digital synthesis: a tool for periodic wave generation (Part 1). IEEE Signal Process. Mag. **21**(4), 50–54 (2004)
6. Cordesses, L.: Direct digital synthesis: a tool for periodic wave generation (Part 2). IEEE Signal Process. Mag. **21**(5), 110–112 (2004)
7. McEwan, A., Collins, S.: Direct digital-frequency synthesis by analog interpolation. IEEE Trans. Circuits Syst. II, Exp. Briefs **53**(11), 1294–1298 (2006)
8. Madisetti, A., Kwentus, A.Y., Willson Jr, A.N.: A 100-MHz, 16-b, direct digital frequency synthesizer with a 100-dBc spurious-free dynamic range. IEEE J. Solid State Circuits **34**(8), 1034–1043 (1999)
9. Wang, S., Piuri, V., Swartzlander, E.E.: Hybrid CORDIC algorithms. IEEE Trans. Comput. **46**(11), 1202–1207 (1997)
10. Sung, T.-Y., Kyo, L.-T., Hsin, H.-C.: Low-power and high-SFDR direct digital frequency synthesizer based on hybrid CORDIC algorithm. In: International Symposium on Circuits and Systems, pp. 249–252 (2009)
11. Prasad, N., Swain, A.K., Mahapatra, K.K.: FPGA implementation of pipelined CORDIC based quadrature direct digital synthesizer with improved SFDR. In: 2013 International Conference on Circuits, Power, and Computing Technologies, pp. 756–760 (2013)

A Novel Method for MP3 Steganalysis

Rinu Kuriakose and P. Premalatha

Abstract Modern communication through digital media is finding traction in our day-to-day life. Steganography plays its role in improving the security of communication by various means. However, all these technologies can also be used with malicious intent. Terrorist organizations have been using steganographic techniques to communicate for a while now. Thus, it is imperative that countermeasures are made to be efficient. Various forms of stego-media are available. Among the available media, digital audio is a very popular carrier for covert communication. Recently, many audio steganalysis methods have been proposed to detect existence of hidden data in stego-media of all formats. In this paper, we look into the analysis of MP3Stego an MP3 steganographic tool. We have considered features such as Markov transition and neighboring joint density. Experimental results show that our approach is successful in discriminating MP3 covers and the steganograms generated using MP3Stego.

Keywords MDCT · MP3Stego

1 Introduction

Any communication system that aims to conceal the very fact that the communication takes place can be classified as steganography. Steganography is the creation of a media embedded with secret content in such a way that no one apart from the sender and the intended recipients know the existence of the secret [1].

Digital steganography is based on the fact that artifacts such as bitmaps and audio files contain redundant information. That is why lossy compression

R. Kuriakose (✉) · P. Premalatha
TIFAC CORE in Cyber Security, Amrita Vishwa Vidyapeetham, Coimbatore, India
e-mail: rinukt90@gmail.com

P. Premalatha
e-mail: rppremalathaa@gmail.com

L.C. Jain et al. (eds.), *Intelligent Computing, Communication and Devices*,
Advances in Intelligent Systems and Computing 308,
DOI 10.1007/978-81-322-2012-1_65

techniques such as JPEG and MP3 work. Such techniques eliminate part of the redundancy, allowing the image or audio file to be compressed. The idea behind (digital) steganography is that instead of eliminating (all of) the redundant information, you replace (some of) it with other data. The human auditory system (HAS) perceives over a range of power greater than one billion to one and a range of frequencies greater than one thousand to one. HAS is fairly poor with respect to its differential range, e.g., loud sounds tend to mask out softer sounds easily. Psychoacoustic model can exploit this weakness [2]. There is a growing interest worldwide in MP3 because of near-CD quality at compression ratio of 11 to 1. This gives a good opportunity for information hiding in MP3. Several stego tools for MP3 audio have been arisen such as MP3Stego, UnderMP3Cover, and StegoLame. MP3Stego is the most typical one among them. It can achieve so good imperceptibility that it is hard to distinguish between background noise and steganographic distortion [3].

On the other hand, the objective of steganalysis is to identify the statistical differences between cover and steganogram. Although multiple steganalysis methods were designed for information hiding in uncompressed audios, the information hiding behavior in compressed audios such as MP3 has been rarely explored due to the complication and the variety of compression methods. Because of the difference in characteristics between compressed and uncompressed audio, most of the existing methods do not work for steganalysis of audios in compression domain, and the decompression attempt, which erases the hidden data through signal reconstruction, leads to a failure of those methods on decompressed audios [4].

In this article, we present an approach based on Markov transition features and neighboring joint density of the MDCT coefficients based on each specific frequency band on MPEG-l Audio Layer 3. We also extracted the features from second-order derivative of MDCT coefficients and combined the statistics with the targeted steganalysis feature called main_data_begin. Experimental results in SVM show that this new approach successfully detects the information hiding behavior of MP3Stego.

1.1 Preliminaries

There were some researchers to realize detecting MP3Stego in recent years. Westfeld [5] proposed a steganalytic method based on the variance of the block lengths. Qiao [4] introduced a detection method based on an inter-frame feature set which contains the moment statistical features on the second derivatives, as well as Markov transition features and neighboring joint density of the MDCT coefficients. In his other work [6], the statistical moments of generalized Gaussian distribution (GGD) shape parameters for MDCT coefficients are also taken as the steganalytic features. Ozer [7] presented a method for detecting audio steganography based on audio quality metrics. Although this method can expose the presence of MP3Stego, the detection performance is relatively poor. The dimensionality of the feature space is the main weakness of Qiao's and Ozer's methods. Later, Diqun Yan and Jie Zhu [8] introduced a parameter called main_data_begin, which is a parameter closely related to quantization. Detecting features directly from the domain where the steganography takes place is straightforward and effective. Since MP3Stego happens during quantization, more attentions should be paid to the parameters associated with the quantization. In their work, the main_data_begin in the side information which is closely related to quantization is introduced into steganalysis to detect MP3Stego.

The rest of this paper is organized as follows. Section 2 briefly covers the basic operations of the MP3Stego algorithm. Section 3 focuses on the essential idea of the proposed method. The effectiveness of the proposed method is verified in Sect. 4. Finally, conclusions and future work are drawn in Sect. 5.

2 MP3Stego Information Hiding Algorithm

MP3Stego exploits the degradation from lossy compression and embeds data by slightly expanding the distortion of the signal without attracting listener's notice. MP3Stego embeds compressed and encrypted data in an MP3 bit stream during the compression process. In the heart of layer III compression, two nested loops manipulate the trade-off between file size and audio quality. The hiding process takes place in the inner loop. The inner loop quantizes the input data and increases the quantizer step size until the quantized data can be coded with available number of bits. Another loop checks that the distortions introduced by quantization do not exceed the threshold defined by the psycho-acoustic model. The part2_3_length variable contains the number of main_data bits used for scalefactors and Huffman coded data in the MP3 bit stream. Encode the bits as its parity by changing the end loop condition of the inner loop. Only randomly chosen part2_3_length values are modified; the selection is done using a pseudo-random bit generator based on SHA-1. Once the inner loop is done, the outer loop will check the distortions introduced by the quantization operation. If the allowed distortion is exceeded, the inner loop will be called again [8]. The above process will be iterated until the bit rate and distortion requirements are both met. Thus, MP3Stego is a practical example of power of parity for information hiding.

3 Proposed System

The audio signal suitable [9] for MP3Stego application is given as input and produced the same amount of stego-audios by hiding random messages in these audios. Then, the features IM, INJ, and main_data_begin as discussed below are extracted. Here, IM and INJ denote inter-frame Markov and inter-frame neighboring joint density features. Here, second-order derivative-based audio

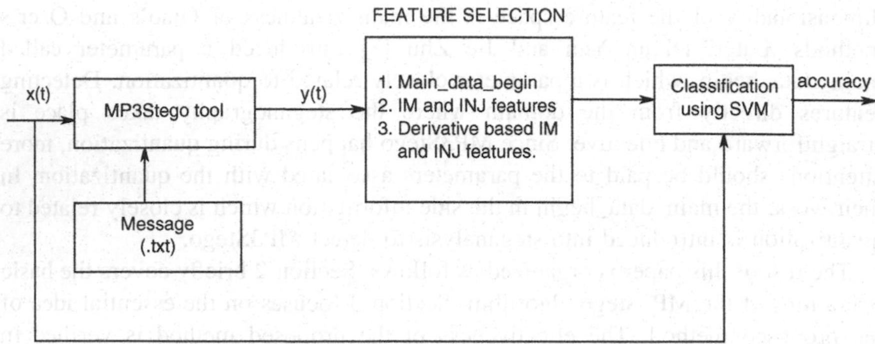

Fig. 1 Block diagram of proposed system

steganalysis is done. Previous works demonstrate that the second-order derivative-based audio steganalysis method gains a considerable advantage under all categories of signal complexity especially for audio streams with high signal complexity, which are generally the most challenging for steganalysis and thereby significantly improves the state of the art in audio steganalysis [10] (Fig. 1).

Since MP3Stego takes place in the quantization during MP3 encoding, the main_data_begin in the side information which is an important parameter closely related to quantization is considered to extract the steganalytic feature. This feature can be obtained directly from the MP3 bit stream without fully decoding.

3.1 Features Extraction for Detection

3.1.1 Side Information Feature

Since the hiding process of MP3Stego takes place during the quantization process, it is natural and reasonable to extract feature from the parameters related to quantization. Therefore, steganographic impact on main_data_begin is analyzed first. MP3 data streaming is encapsulated in the unit of frame after the process of quantization and Huffman coding. Each frame mainly consists of the following three parts: frame header, side information, and main data. Main data are the data streaming of the encoded original audio samples. The other two parts of the frame is mainly for decoding purpose. A bit reservoir has been adopted during MP3 coding. This mechanism provides the space to loan or deposit bits to control the audio quality under a bit rate constraint. If a frame does not need all the bits allocated to it, it can put the spare ones into this reservoir. A more complex frame that requires more bits than it is given can then take bits from this reservoir. Due to bit reservoir mechanism and the length of each frame are the same, one frame can contain other frames main data. In order to decode correctly, the side information will record each frames main_data_begin as illustrated in Fig. 2. The value of

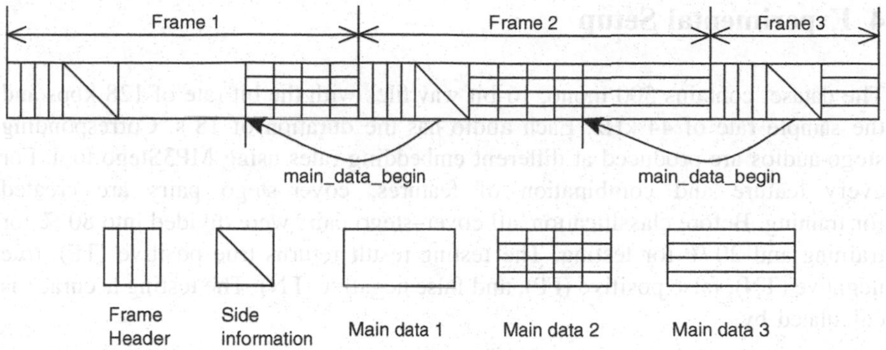

Fig. 2 Main_data_begin in the side information

main_data_begin in current frame is the negative offset to the frame header. As a result of embedding, more iterations are required to exit the inner loop in MP3Stego, due to which the quantization step size will be increased. The larger the quantization step is, the smaller the value of the quantized coefficients. Finally, the number of the actual allocation bits for quantization will be smaller. Hence in stego case, the length of frame's main data will be smaller and the frame's main_data_begin will be larger than in non-stego case [8].

3.1.2 Neighboring Joint Density and a Markov Approach on Inter-frame

The following equations describe IM and INJ features for MP3 audio steganalysis applied for MDCT_C (MDCT coefficients on MPEG-I Audio Layer 3) [4].

$$\mathrm{IM}(i,j) = \sum_{s=0}^{575} \frac{\sum_{t=0}^{N-2} \delta(\mathrm{MDCT\_}C(t*576+s) = i, \mathrm{MDCT\_}C((t+1)*576+s) = j)}{\sum_{t=0}^{N-2} \delta(\mathrm{MDCT\_}C(t*576+s) = i)} \tag{1}$$

$$\mathrm{INJ}(i,j) = \frac{\sum_{s=0}^{575} \sum_{t=0}^{N-2} \delta(\mathrm{MDCT\_}C(t*576+s) = i, \mathrm{MDCT\_}C((t+1)*576+s) = j)}{576*(N-1)} \tag{2}$$

Digital audio streams, especially speech audio clips, are normally band-limited. The difference between the spectrum of the cover- and the stego-signal at low and middle frequency is negligible; however, the stego-signal has higher magnitude than the cover-signal in the derivative spectrum for high-frequency components. So it is evident that the hidden data will be more exposed in derivative-based situations [10].

4 Experimental Setup

The dataset contains 500 mono, 16-bit wav files with the bit rate of 128 kbps and the sample rate of 44 kHz. Each audio has the duration of 18 s. Corresponding stego-audios are produced at different embedding rates using MP3Stego tool. For every feature and combination of features, cover–stego pairs are created for training. Before classification, all cover–stego pairs were divided into 80 % for training and 20 % for testing. The testing result returns true positive (TP), true negative (TN), false positive (FP), and false negative (FN). The testing accuracy is calculated by

$$\text{Testing accuracy} = \frac{\text{TP} + \text{TN}}{\text{TP} + \text{TN} + \text{FP} + \text{FN}} \tag{1}$$

Table 1 lists the testing accuracy values with the use of a support vector machine (SVM).

It shows that main_data_begin-based analysis produces better results than the Markov and neighboring joint density approaches. Derivative-based methods are more accurate in comparison with signal-based audio steganalysis. When we combine joint density statistics with side information feature, the detection performance is improved.

5 Conclusion and Future Works

In this paper, we proposed a scheme to detect the information hiding in MP3Stego based on merging of inter-frame Markov approach and neighboring joint density in MDCT domain and targeted steganalysis feature called main_data_begin. We also introduced the second-order derivative, which is widely used for isolation and

Table 1 Average testing accuracy with different feature sets

Features	Testing accuracy(%)
Main_data_begin	92.7
IM	66.3
INJ	60.7
2D-IM	73.2
2D-INJ	70.8
2D-IM + 2D-INJ	75.3
Main_data_begin + 2D-IM	94.3
Main_data_begin + 2D-INJ	93.5
Main_data_begin + 2D-IM + 2D-INJ	97.6

edge detection in image processing, to the steganalysis of MP3 audios. Experimental results show that our method successfully detects the MP3 audio steganograms created using MP3Stego.

Future work may include finding some more sensitive features and apply the concept of calibration in the assumption that, if we can obtain the estimation of the cover from the suspect one as effectively as possible, the steganalytic performance will be improved; extending the steganalysis performance evaluation framework to include analysis of computational complexity and building benchmark testing sets to facilitate cross-validation of new results.

References

1. Noto, M.: MP3Stego: Hiding Text in MP3 Files, p. 5. Sans Institute http://www.sans.org/readingroom/whitepapers/stenganography/mp3stego-hiding-textmp3-files-550 (2001)
2. Smith, S.W.: Digital Signal Processing : A Practical Guide for Engineers and Scientist. Massachusetts, Massachusetts (2003)
3. Petitcolas, F.: MP3Stego, 1998. http://www.petitcolas.net/fabien/steganography/mp3stego/ (2011)
4. Qiao, M., Sung, A.H., Liu, Q.: Steganalysis of MP3Stego. In: International Joint Conference on Neural Networks, 2009. IJCNN 2009. IEEE (2009)
5. Westfeld, A.: Detecting low embedding rates. In: Information Hiding. Springer, Berlin (2003)
6. Qiao, M., Sung, A.H., Liu, Q.: Feature mining and intelligent computing for MP3 steganalysis. In: International Joint Conference on Bioinformatics, Systems Biology and Intelligent Computing, 2009. IJCBS'09. IEEE (2009)
7. Özer, H., et al. Detection of audio covert channels using statistical footprints of hidden messages. Digital Signal Process. **16**(4), 389–401 (2006)
8. Yu, X., et al.: MP3 audio steganalysis using calibrated side information feature. J. Comput. Inf. Syst. **8**(10), 4241–4248 (2012)
9. Menon, R.J.: Mp3 steganography and steganalysis. http://www.dfcsc.uri.edu/docs/Menon_Thesis.pdf (2009)
10. Liu, Q., Sung, A.H., Qiao, M.: Derivative-based audio steganalysis. TOMCCAP **7**(3), 18 (2011)

Using Orthographic Transcripts for Stuttering Dysfluency Recognition and Severity Estimation

P. Mahesha and D.S. Vinod

Abstract Stuttering is a speech disorder characterized by a high frequency of disruption that obstructs the forward flow of speech. During the course of assessment, clinicians need to carefully measure client's speech fluency in order to obtain frequency of dysfluencies that can be used to determine the severity of stuttering and planning appropriate training program. Frequency of stuttering is also helpful as a snapshot measure of improvement during treatment. The purpose of this study is to introduce lexically driven algorithm to recognize and categorize dysfluencies in orthographic transcriptions. Further, frequencies of occurrence of these dysfluencies are used for assessment and severity rating.

Keywords Stuttering · Dysfluencies · Transcriptions and severity rating

1 Introduction

Stuttering is characterized by the occurrence of speech dysfluencies, which are used to describe and classify stuttering [1]. There are many types of dysfluencies that a speaker can produce. Different dysfluency classes are first defined in 1959 by Johnson et al. [2]. The dysfluency categories are as follows: interjections, sound or syllable repetitions, word and phrase repetitions, prolongations, revisions in which speaker try to correct something said previously, and incomplete phrases where a speaker abandons a topic without completing it.

P. Mahesha (✉)
Department of Computer Science and Engineering, S.J. College of Engineering,
Mysore, Karnataka, India
e-mail: maheshsjce@yahoo.com

D.S. Vinod
Department of Information Science and Engineering, S.J. College of Engineering Mysore,
Mysore, Karnataka, India
e-mail: ds.vinod@daad-alumni.de

L.C. Jain et al. (eds.), *Intelligent Computing, Communication and Devices*,
Advances in Intelligent Systems and Computing 308,
DOI 10.1007/978-81-322-2012-1_66

Assessment of stuttering is a widespread topic studied with different objectives. It typically looks at frequency of dysfluencies, types, duration, and its severity in spontaneous speech. The significance of each varies relatively based on the age and kind of treatment to be given for the client [3]. Frequency and type of dysfluencies are important and reliable measures of stuttering used for different purposes [4]. They are significant in initial assessment to distinguish a normal client from a client with borderline stuttering. In addition, it will be helpful as a snapshot measure of improvement during treatment. The type of assessment performed by clinicians depends on their theoretical perspectives.

Few clinicians perform speech dysfluency analysis based on detailed verbatim transcript for measuring stuttering [5, 6], whereas others give emphasis on real-time or online method that offers quick but less detailed evaluation of speech dysfluencies [7–9].

Transcript-based analysis is one of the assessment techniques which provides greater opportunities to assess qualitative aspects of speech dysfluencies [10]. In this method, clinician records a speech sample from the client and prepares detailed orthographic transcription in which occurrences of dysfluencies are marked. Further, each dysfluency is categorized into different types. This manual transcription-based analysis is time-consuming and not feasible to carry out detailed measurements on a regular basis [10].

To address the limitations of manual system, we propose lexically driven algorithm for automating the transcription analysis which makes the process of assessment accurate and faster. The proposed method also provides frequency of occurrences of each dysfluency and stuttering severity estimation for each transcription.

This paper is organized as follows. The ground truth data used for investigations are detailed in Sect. 2. The lexically driven transcript-based framework is discussed in Sect. 3. Further, Sect. 4 presents the experimental results and discussions. Subsequently, Sect. 5 concludes the paper with future work.

2 Experimental Data

Orthographic transcriptions used in this study were obtained from University College London Archive of Stuttered Speech (UCLASS) [11, 12]. This archive has been designed for research and clinical purposes to investigate language and speech behavior of speakers who stutter. Orthographic transcriptions use regular orthography based on dysfluencies that are spelled out as they are heard. The database consists of thirty transcription samples from 2 female and 28 male speakers age ranged between 8 years 1 month and 17 years 9 months. The entire transcription database is used for investigation.

3 Methodology

The basic idea of our method is to detect dysfluencies such as syllable repetition, word repetition, prolongation, and interjection in transcription using lexical procedures. Approximate severity rating for each transcription is then estimated using frequency of each dysfluency. The steps involved in this process are outlined in Fig. 1.

3.1 Dysfluency Structure

Speech dysfluencies are categorized into 'more typical' and 'less typical' dysfluencies [13–15]. Less typical dysfluencies are actual instances of stuttering such as repetitions of sounds or syllables (SR), word repetitions (WR), prolongations (P), and interjections (I). Therefore, majority of clinicians prefer to count these instances as dysfluencies [16]. The structure of dysfluencies is shown in Fig. 2. Our proposed system detects these instances in transcription.

3.2 Lexical Analysis Procedure

This section presents lexical procedures for the following tasks:

- Detecting and counting the occurrences of four types of dysfluencies such as syllable repetitions, WR, prolongations, and interjections.
- Computing the total dysfluent words and number of words spoken in the conversion.
- Calculating the percentage of syllables stuttered.

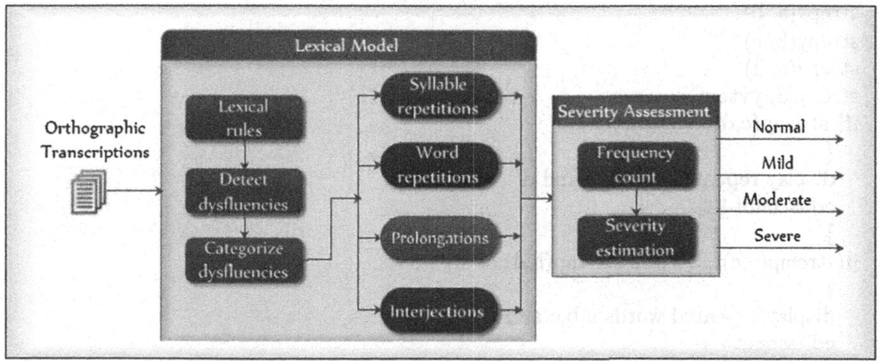

Fig. 1 Overall process involved in severity assessment

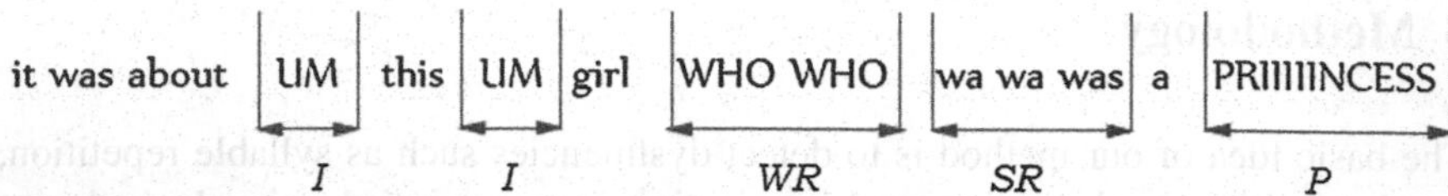

Fig. 2 Dysfluency structure

Syllable Repetition Lexical syllable repetitions' analysis will investigate the words with their starting letters repeated such as 'SSSSummer,' 'PPPark,' 'Fa Family,' and 'Pri Princess.' These are identified by using a defined substring function which returns 1 if the previous word is the substring of the current word as shown in algorithm 1.

Algorithm 1 : Procedure to Detect Syllable Repetitions

```
int  substring(char s1[ ], char s2[ ])
 {
 int i=0
   if( s1[0] =='\0')    return 0
   while( s1[i]!='\0')
   {
     if(s1[i]==s2[i])
       i++
     else
        return 0
   }
   return 1
 }
```

Word Repetitions Word repetitions such as 'who who,' 'own own,' and 'so what so what' are identified by using the strcmp function, and the logic is shown in algorithm 2.

Algorithm 2 : Procedure to Detect Word Repetitions

```
 char  a[10]='\0', b[10]='\0', c[10]='\0', d[10]='\0'
 strcpy(a, b)
 strcpy(b, c)
 strcpy(c, d)
 strcpy(d, yytext)
 if( strcmp(c,d)==0)
   {
   display repeated words c and d
   editscount++
   }
 if(strcmp(a,c)==0 && strcmp(b,d)==0)
   {
   display repeated words a,b,c and d
   editscount++
   }
```

Prolongations Words such as 'Livvvvved' and 'Offfff' are identified by using the logic defined in algorithm 3. The prolong function returns 2 when the successive three characters of a word (excluding the word at the beginning) are same.

Interjections The descriptions of interjections are specified as regular expressions to handle lower cases, upper cases, and its combinations. The logic to recognize different interjections is detailed in algorithm 4.

Algorithm 3 : Procedure to Detect Prolongations

```
int prolong(char s[])
 {
 char a=s[0], b=s[1], c=s[2]
 int i=3
   if( a==b ) return 1
   while( s[i] !='\0' )
   {
     if(a == b && b == c)
       return 2
     else
       {
        a=b
        b=c
        c=s[i]
        i++
       }
   }
   if( s[i]== '\0' ) return 0
}
```

Algorithm 4 : Procedure to Detect Interjections

```
int umcount=0, uhcount=0, ehcount=0, likecount=0
int imeancount=0, words=0, fillerscount=0, editscount=0
[uU][mM]    {umcount++; words++;}
[uU][hH]    {uhcount++; words++;}
[eE][hH]    {ehcount++; words++;}
[lL][iI][kK][eE]    {likecount++; words++;}
[iI][ ][mM][eE][aA][nN]    {imeancount++; words++;}
```

3.3 Severity Estimation

Severity estimation is useful for evaluating the disorder, monitoring the change, and assessing the outcome. It is most appropriate clinical assessment of stuttering behavior. Severity reflects an overall impression of the clinicians when they assess individual who stutter. Hence, it is an essential measure for judging the outcome of a treatment [1].

Table 1 Grading scheme for severity estimation [4]

Grade 0	Normal
Grade 1	Mild stutter communication unimpaired 0–5 % words stuttered
Grade 2	Moderate stutter communication slightly impaired 6–20 % words stuttered
Grade 3	Severe stutter communication definitely impaired above 20 % words stuttered

While evaluating stuttering outcome, percent syllables stuttered (%SS) and severity rating scales are the most typical measures to be used, either alone or in combination with additional measures [17]. The %SS stutter count measure is considered to be relatively objective.

Severity rating scales are simple where clinicians or listeners assign a numerical value indicating the perceived overall stuttering severity. This can be used as self-report measure enabling the clients to assess their speech eventually, in different situations, within or beyond the clinic [18].

Information gathered from lexical procedure consists of the total number of dysfluencies categorized by types of dysfluencies. The %SS is computed for each case using the following equation:

$$\%\text{Syllable Stuttered }(\%\text{SS}) = \frac{\text{No : of syllables stuttered}}{\text{No : of syllables spoken}} \times 100 \qquad (1)$$

Severity estimation is done after calculating %SS for each transcription. There are various ways of assessing severity. The grading system was proposed by Andrew et al. [4] and is one of the oldest, quickest, and most used methods. Table 1 shows this grading scheme employed for severity estimation.

4 Results and Discussions

As discussed in Sect. 2, ten transcriptions that are subset of UCLASS database are considered for the experimentation. Each transcription is analyzed to detect dysfluencies based on the lexical rules defined in Sect. 3.2.

Data gathered from the analysis contain the total number of words spoken and the total number of dysfluencies categorized by the type of dysfluency. Based on this information, %SS is calculated using Eq. 1. Further, severity estimation is done depending on the %SS, according to the grading scheme presented in Table 1. The details of frequency of occurrence of each dysfluency, total number of dysfluencies, total number of words spoken, %SS, and severity estimation for each transcription are tabulated in Table 2. Further, the distribution of the severity estimation of all thirty transcriptions is shown in Fig. 3. More samples fall in moderate category due to minimum number of prolongation dysfluencies.

Table 2 Frequency of dysfluencies and severity estimation

Transcription	SR	WR	P	I	TD	TW	%SS	Severity rating
F_0101_10y4m_1.orth	6	4	4	22	36	139	26	Severe
F_0101_13y1m_1.orth	4	6	1	15	26	202	13	Moderate
M_0028_15y11m_1.orth	1	3	1	11	16	244	7	Moderate
M_0030_17y9m_1.orth	6	1	2	22	31	338	9	Moderate
M_0052_16y4m_1.orth	12	10	2	34	58	333	17	Moderate
M_0061_14y8m_1.orth	9	13	2	27	51	168	30	Severe
M_0078_12y4m_1.orth	24	16	3	23	66	240	27	Severe
M_0078_14y4m_1.orth	29	31	1	14	75	148	50	Severe
M_0078_16y5m_1.orth	13	5	2	16	36	274	13	Moderate
M_0095_07y7m_1.orth	1	6	0	6	13	194	7	Moderate
M_0095_08y10m_1.orth	1	3	0	9	13	238	4	Mild
M_0098_07y8m_1.orth	10	41	2	9	62	216	5	Mild
M_0098_09y8m_1.orth	1	6	0	12	19	207	9	Moderate
M_0098_10y6m_1.orth	2	8	1	15	26	242	11	Moderate
M_0100_11y2m_1.orth	1	2	1	18	22	173	13	Moderate
M_0100_12y3m_1.orth	4	5	0	8	17	156	11	Moderate
M_0100_13y10m_1.orth	0	6	1	18	25	245	10	Moderate
M_0104_10y3m_1.orth	46	11	8	18	83	283	29	Severe
M_0138_12y2m_1.orth	7	6	1	18	32	246	12	Moderate
M_0138_13y3m_1.orth	1	2	0	46	49	361	14	Moderate
M_0210_11y3m_1.orth	13	45	0	12	70	236	30	Severe
M_0213_10y10m_1.orth	8	37	2	11	58	200	29	Severe
M_0219_11y2m_1.orth	2	4	0	8	14	240	5	Mild
M_0234_09y9m_1.orth	2	0	0	6	8	195	4	Mild
M_0394_08y10m_1.orth	18	3	0	3	24	220	11	Moderate
M_0394_09y2m_1.orth	10	3	1	14	28	221	13	Moderate
M_0394_09y5m_1.orth	19	3	2	19	43	238	18	Moderate
M_0399_12y4m_1.orth	2	12	2	25	41	344	12	Moderate
M_0556_07y8m_1.orth	14	16	2	13	45	160	28	Severe
M_0556_08y0m_1.orth	12	9	0	9	30	180	17	Moderate

SR syllable repetitions, **WR** word repetitions, **P** prolongations, **I** interjections, **TD** total dysfluencies, **TW** total words spoken, **%SS** % syllables stuttered

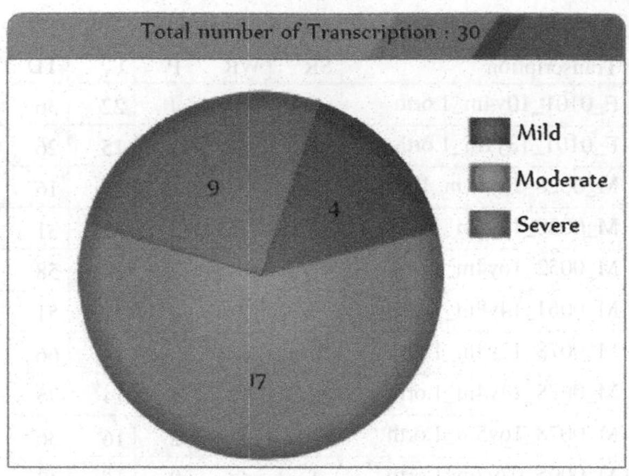

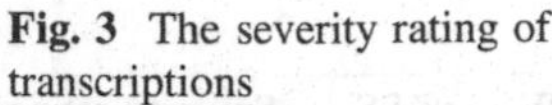
Fig. 3 The severity rating of transcriptions

5 Conclusions

We have presented lexical procedure to locate dysfluencies in orthographic transcription by defining rules. The procedure takes orthographic transcription as input, processes it, and identifies the different categories of dysfluent words along with frequency of occurrences.

Transcription analysis is a measurement technique used by clinicians to count dysfluencies. It provides more comprehensive analysis of dysfluencies compared to real-time analysis. However, real-time analysis is faster but gives less detailed analysis of dysfluencies. Real-time method was chosen by clinicians, because the transcription analysis is time-consuming though it provides better possibilities to assess qualitative aspects of dysfluencies. Our proposed method overcomes the limitation of transcription analysis by automating the process performed by clinicians during the evaluation.

Severity estimation is also presented based on the frequency of dysfluencies, which minimizes the manual effort made by clinicians to give final ratings. The strength of our approach lies in its simplicity and effectiveness of the methodology with respect to the detection of dysfluency, frequency analysis, and severity estimation.

References

1. Guiter, B.: Stuttering—An Integrated Approach to Its Nature and Treatment. Lippincot Williams and Wilkins (2006)
2. Johnson, W., Boehmler, R.M., Dahlstrom, W.G., Darley, F.L., Goodstein, L.D., Kools, J.A., Neeley, J.N., Prather, W.F., Sherman, D., Thurman, C.G., Trotter, W.D., Williams, D., Young, M.A.: The Onset of Stuttering; Research Findings and Implications. Minneapolis, University of Minnesota Press (1959)

3. Yaruss, J.S.: Real-time analysis of speech fluency: procedures and reliability training. Am. J. Speech-Lang. Pathol. **7**(2), 25–37 (1998)
4. Andrews, G., Ingham, R.J.: Stuttering: Considerations in the Evaluation of Treatment. Int. J. Lang. Commun. Disord. **6**, 129–138 (1971)
5. Campbell, J., Hill, D.: Systematic Disfluency Analysis: Accountability for Differential Evaluation and Treatment. Miniseminar presented to the Annual Convention of the American Speech-Language-Hearing Association. New Orleans, LA (1987)
6. Rustin, L., Botterill, W., Kelman, E.: Assessment and Therapy for Young Dysfluent Children: Family Interaction. Singular Publishing, San Diego (1996)
7. Conture, E.G.: Stuttering, 2nd edn. Prentice- Hall, Englewood Cliffs, NJ (1990)
8. Conture, E.G., Yaruss, J.S.: Handbook for Childhood Stuttering: A Training Manual. Bahill Intelligent Computer Systems, Tucson, AZ (1993)
9. Riley, G.: Stuttering Severity Instrument for Children and Adults, 3rd edn. Pro-Ed, Austin, TX (1994)
10. Yaruss, J.S., Max, M.S., Newman, R., Campbell, J.H.: Comparing real-time and transcript based techniques for measuring stuttering. J. Fluency Disord. **23**(2): 137–151, ISSN 0094-730X (1998)
11. Howell, P., Huckvale, M.: Facilities to assist people to research into stammered speech. Stammering Res., On-line J. Br. Stammering Assoc. pp. 130–242 (2004)
12. Howell, P., Devis, S., Batrip, J.: The UCLASS archive of stuttered speech. J. Speech, Lang. Hear. Res. (2009)
13. Gregory, H.: Stuttering: Differential Evaluation and Therapy. Pro-Ed, Austin, TX (1986)
14. Gregory, H.H., Hill, D.: Differential Evaluation—Differential Therapy for Stuttering Children. In: Curlee, R.F. (ed.) Stuttering and Related Disorders of Fluency. Thieme Medical Publishers, New York (1993)
15. Yaruss, J.S.: Clinical measurement of stuttering behaviors. Contemp. Issues Comm. Sci. Disord. **24**, 33–44 (1997)
16. O'Brian, S., Packman, A., Onslow, M.: Self-rating of stuttering severity as a clinical tool. Am. J. Speech Lang. Pathol. **13**, 219–226 (2004)
17. Andrews G., Harris M.: Clinics in Developmental Medicine. No. 17 Heinemann; London, The Syndrome of Stuttering, (1964)
18. O'Brian, S., Jones, M., Packman, A., Menzies, R., Onslow, M.: Stuttering Severity and Educational Attainment. Journal of Fluency Disorders **36**, 86–92 (2011)

Combining Cepstral and Prosodic Features for Classification of Disfluencies in Stuttered Speech

P. Mahesha and D.S. Vinod

Abstract The process of recognition and classification of dysfluencies are significant in objective assessment of stuttered speech. The main focus of this study is to combine prosodic features and cepstral features in order to improve the performance of dysfluency recognition. The term prosody represents several characteristics related to human speech such as speaking rate, loudness, duration, and pitch. In this study, pitch, energy, and duration are considered as prosodic features and Mel Frequency Cepstral Coefficient (MFCC), delta MFCC (DMFCC), and delta–delta MFCC (DDMFCC) are used as cepstral feature set. The efficacy of the considered features has been evaluated using support vector machine (SVM) classifier. Experimental results demonstrated considerable enhancement in the overall performance with respect to the conventional methods present in the literature of stuttering dysfluency recognition.

Keywords Dysfluency · MFCC · Prosodic features · Stuttering and SVM

1 Introduction

Speech is one of the effective ways of communication between people. The basic purpose of speech is to send and receive a message in the form of language communication. Sometimes, speakers who are normally fluent also experience dysfluencies due to emotional, physiological, or psychological factors. Speaker is dysfluent when there are disturbances or breaks in the smooth flow of speech.

P. Mahesha (✉) · D.S. Vinod
Department of Computer Science and Engineering, S.J. College of Engineering, Mysore, Karnataka, India
e-mail: maheshsjce@yahoo.com

D.S. Vinod
e-mail: ds.vinod@daad-alumni.de

L.C. Jain et al. (eds.), *Intelligent Computing, Communication and Devices*, Advances in Intelligent Systems and Computing 308,
DOI 10.1007/978-81-322-2012-1_67

Stuttering is a fluency disorder that affects the normal flow of speech that may yield involuntarily repetition of syllable, words or phrases, hesitation, interjection, and prolongations. Several studies revealed that stuttering affects about 1 % of the general population in the world and normally males are affected two to five times more often than females [1–4]. It is one of the oldest and serious speech problems noted in the history of speech and language pathology. Stuttering is the subject of interest to researchers from interdisciplinary domains such as speech physiology, pathology, psychology, acoustics, and signal analysis.

The repetition and prolongation dysfluencies are ubiquitous in stuttered speech. Hence, they are commonly used by speech pathologists for stuttering assessment. In conventional stuttering assessment, speech language pathologists (SLP) classify and count the occurrence of dysfluencies manually by transcribing the recorded speech. These types of assessments are based on the knowledge and experience of speech pathologists. However, making such assessments is time-consuming, subjective, inconsistent, and prone to error [5–8]. Hence, it would be sensible if stuttering assessment is often done through classification of dysfluencies using speech recognition technology and computational intelligence.

1.1 Related Work

In the last two decades, several studies have been carried out on the automatic detection and classification of dysfluencies in stuttered speech by means of acoustic analysis, parametric and nonparametric feature extraction, and statistical methods. In this section, we present a review of the recent papers published in this area.

In [9], author used artificial neural network (ANN) and rough set to detect stuttering events yielding accuracy of 73.25 % for ANN and about 91 % for rough set. The authors of [10, 11] proposed hidden markov model (HMM)-based classification for automatic dysfluency detection using MFCC features and achieved 80 % accuracy. In [5], automatic detection of syllable repetition was presented for objective assessment of stuttering dysfluencies based on MFCC and perceptron features. An accuracy of 83 % was achieved. Subsequently in [6], same author obtained 94.35 % accuracy using MFCC features and SVM classifier. Authors of [12] achieved 90 % accuracy with linear discriminant analysis (LDA), k-nearest neighbor (k-NN), and MFCC features. In [13], the same author used similar classifiers, LDA and k-NN for the recognition of repetitions and prolongations with linear predictive cepstral coefficient (LPCC) as feature-extraction method and obtained the best accuracy of 89.77 %. In [14], we proposed a procedure for classification of dysfluency using MFCC feature. Using k-NN classifier, an accuracy of 97.78 % is achieved.

In our previous work [15], comparison between three speech parameterization techniques for dysfluency recognition was proposed using multiclass SVM classifier and they achieved the average accuracy of 92.23 %.

In this study, we would like to explore the performance of cepstral and prosodic features combination for classification of repetition and prolongation dysfluencies using SVM classifier.

2 Experimental Data

The effectiveness of different features is evaluated with reference to dysfluency recognition using University College London Archive of Stuttered Speech (UCLASS) database [16, 17]. This database has been designed for research and clinical purposes to investigate language and speech behavior of speakers who stutter. It contains 107 reading recording contributed by 43 different speakers.

In this study, a subset of UCLASS database, which consists of 30 samples and includes 20 male and 10 female speakers with age ranging from 10 years 2 months to 20 years 1 month, is chosen. The samples were chosen to cover wide range of age and stuttering rate.

3 Features

This section describes MFCC, DMFCC, DDMFCC, and prosodic feature-extraction methods used in dysfluency recognition.

3.1 Mel Frequency Cepstral Coefficients

MFCC is one of the popular and most commonly used feature-extraction techniques to characterize the speech signal. It produces a multidimensional feature vector for every frame of speech. In this work, 36 MFCC features are considered, which consists of 12 MFCC, 12 DMFCC, and 12 DDMFCC. The strategy used to estimate MFCC features is discussed below. The block diagram of preprocessing and MFCC feature parameterization is shown in Fig. 1.

Preprocessing Due to the characteristics of human speech production, glottal airflow and lip radiations make the higher frequencies get dampened while the lower frequencies are boosted. To eliminate this effect and to avoid lower frequency components from dominating the signal, preprocessing is carried out before feature extraction. Normally, a first-order high-pass filter is used to emphasize the high frequency of the signal [18]. Next step is framing, where the signal is split into sequence of frames typically with the length of 10–30 ms and 25 to 70 % overlap between two adjacent frames. This is to ensure that each speech signal is approximately centered at some frame [19]. The steps involved preprocessing is shown in Fig. 1.

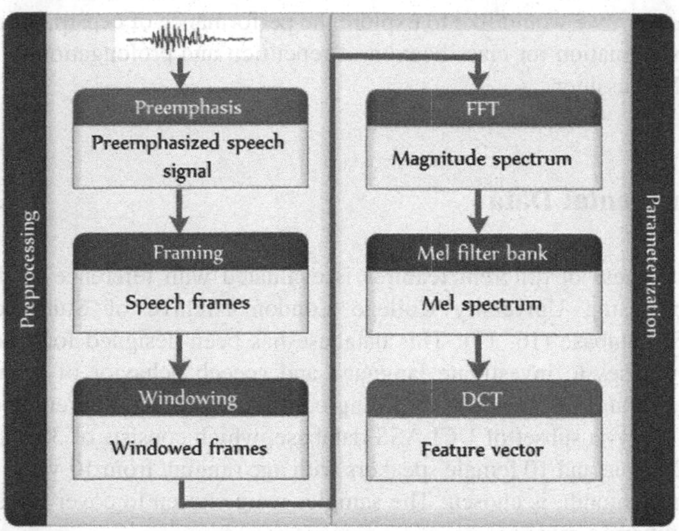

Fig. 1 Block diagram of speech signal preprocessing and feature extraction

Further, Hamming windowing is applied for each frame to reduce the signal discontinuities at the beginning and end of the frame. The Hamming window equation is given by:

$$w(n) = 5.4 - 0.46\ \cos\left[\frac{2\pi n}{n-1}\right], \quad 0 \leq n \geq N - 1 \tag{1}$$

MFCC Parameterization MFCC is based on known variation of the human perception of different frequencies. The Mel scale deals linearly with frequencies up to 1,000 Hz, and frequencies over 1,000 Hz are handled logarithmically; hence, it provides more emphasis to lower-frequency range [20]. Fast Fourier transform (FFT) is then applied to the windowed frame signal to obtain the magnitude frequency response. A magnitude spectrum (in human perception) is more important to model the magnitude spectra of speech than their phase [21].

The resulting spectrum is passed through a set of triangular band-pass filter to get the log energy of each triangular band-pass filter. The positions of these filters are equally spaced along the Mel frequency, which is related to the common linear frequency f by the following equation.

$$\text{Mel}\,(f) = 2{,}595\,\log_{10}\left(1 + \frac{f}{700}\right) \tag{2}$$

Mel frequency is proportional to logarithm of linear frequency, reflecting similar effects in the human's subjective aural perception. The relationship between the Mel and the linear frequencies is illustrated in Fig. 2.

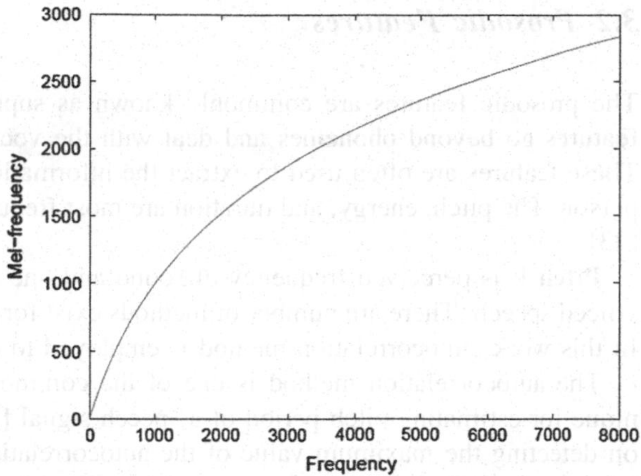

Fig. 2 Frequency to Mel frequency curve

Finally, discrete cosine transform (DCT) is applied on log energy obtained from triangular band-pass filter to have L Mel scale cepstral coefficients. The equation for DCT is given by

$$c_m = \sum_{k=1}^{N} \cos\left[m\left(k - \frac{1}{2}\right)\right] E_k, \quad m = 1,\ 2,\ 3 \ldots L \tag{3}$$

where N is the number of triangular band-pass filter and L is the number of Mel scale cepstral coefficients. In this work, we have used $N = 20, L = 12$ and E_k is log energy obtained from triangular band-pass filter.

Since FFT is applied in the beginning, DCT transforms the frequency domain into a time domain. The obtained features are similar to cepstrum; hence, it is referred as the Mel scale cepstral coefficients.

Delta and Delta–Delta Coefficients The first- and second-order derivatives of cepstral coefficients are known as Delta and Delta–Delta coefficients, respectively. Delta coefficients signify the speech rate and Delta–Delta coefficients are something similar to acceleration of speech [22]. Delta cepstral coefficients are computed using the following equation:

$$\Delta C_m = \frac{1}{2}\left[C_m(l+1) - C_m(l-1)\right] \tag{4}$$

$C_m(l)$ is delta cepstral coefficient at frame l. Delta–Delta coefficients are also computed using Eq. 4.

3.2 Prosodic Features

The prosodic features are commonly known as suprasegmental features. These features go beyond phonemes and deal with the vocal characteristics of speech. These features are often used to extract the information about speaking style of a person. The pitch, energy, and duration are most frequently used prosodic features [23].

Pitch It is perceived frequency of sound and one of the essential attributes of voiced speech. There are number of methods exist for determining the pitch value. In this work, autocorrelation method is employed to estimate the pitch.

The autocorrelation method is one of the commonly used time-domain technique for estimating pitch period of a speech signal [24]. This technique is based on detecting the maximum value of the autocorrelation function in the region of interest.

Consider a speech frame $(n), n = 0, 1, \ldots N - 1$, the autocorrelation function of the speech signal $s(n)$ is computed as follows:

$$R(m) = \frac{1}{N} \sum_{n=0}^{N-1-m} x(n) \cdot x(n+m), 0 \leq m < M_0 \tag{5}$$

where $R(m)$ 1 is autocorrelation function, N is total number of samples in window, m is the lag or delay index and M_0 is the number of autocorrelation points to be computed.

The first step in autocorrelation is to divide the given speech signal into 30–40 ms blocks of frames, and then, the auto correlation sequence of each frame is found.

The pitch period is computed by finding the time lag corresponding to the second largest peak from the central peak of autocorrelation sequence. The plot of speech segment and its autocorrelation is shown in Fig. 3.

When a segment of a signal is correlated with itself, the distance between the positions of the maximum and the second maximum correlation is considered as the fundamental period or pitch of the signal.

Energy and Duration The amplitude of speech signal varies considerably with time. The short-time energy is reflection of these amplitude variations [18]. Energy is computed as the sum of the values of samples over a period of time (in a window). The square of amplitude is summed up from 0 to, where N is the total number of samples in a given frame. This is expressed in equation given below:

$$\text{Energy} = \sum_{n=0}^{N} [x(n)]^2 \tag{6}$$

Duration refers to the length of speech sound. Some speech dysfluencies are inherently longer than others; for instance, prolongation dysfluency is longer than

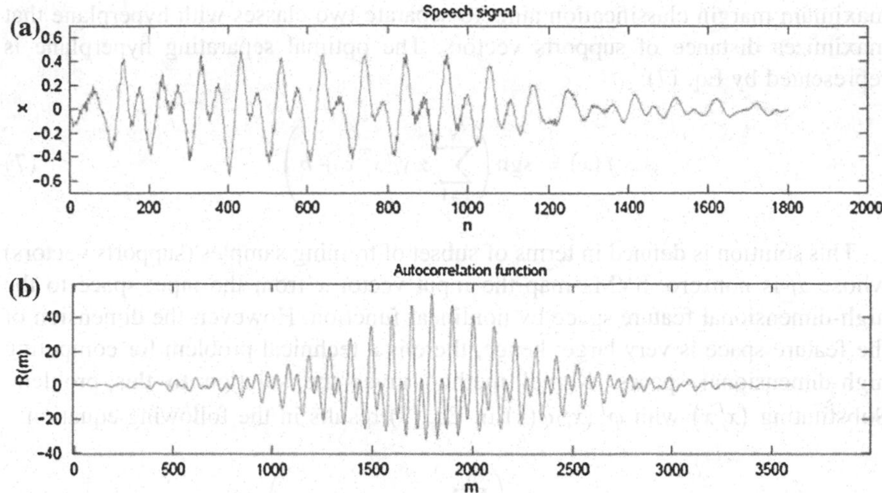

Fig. 3 Speech signal and its autocorrelation plot

Table 1 List of features used in the experiment

Abbreviation	Description
MFCC	12 MFCC
DMFCC	12 MFCC + 12 first-order delta coefficients
DDMFCC	12 MFCC + 12 first-order + 12 second-order delta coefficients
E, P and D	Energy, pitch, and duration

repetition dysfluencies. Hence, it is considered as one of the discriminative feature in dysfluency recognition. The features used in this work are listed in Table 1.

4 SVM Training and Classification

In this study, our main focus is to classify dysfluencies into repetition and prolongation categories. SVM is a classification technique based on the statistical learning theory [25].

SVM is a binary classification technique that offers the optimal linear decision surface based on the structural risk minimization. The decision surface is a weighted combination of elements of the training set. These elements are referred as support vectors, and they characterize the boundary between the two classes [26].

Consider a given training labeled data set $\{(x_i, y_i)\}_{i=1}^{N}$, where $x_i \in \Re^N$ is the input vector and $y_i \in \{-1, +1\}$ is its corresponding label. In case of linear separable data, maximum margin classification aims to separate two classes with hyperplane that maximizes distance of supports vectors. The optimal separating hyperplane is represented by Eq. (7).

$$f(x) = \text{sgn}\left(\sum_{i=1}^{N} \alpha_i y_i (x_i^T x + b\right) \tag{7}$$

This solution is defined in terms of subset of training samples (supports vectors) whose α_i is nonzero. SVMs map the input vector x from the input space to the high-dimensional feature space by nonlinear function. However, the dimension of the feature space is very large; hence, there is a technical problem for computing high-dimensional spaces. Kernel method gives the solution to this problem. Substituting $\left(x_i^T x\right)$ with $\varphi^T(x_i)\varphi(x)$ in Eq. (7) results in the following equation:

$$f(x) = \text{sgn}\left(\sum_{i=1}^{N} \alpha_i y_i k(x_i, x) + b\right) \tag{8}$$

In SVM, every mapping occurs in the form of inner product. Thus, kernel method is usually applied to SVM by replacing all occurrences of $\varphi^T(x_i)\varphi(x)$ by $k(x_i, x)$ instead of mapping φ explicitly. This kernel method is supported by Mercer's theorem [27]. Hence, the equation for nonlinear SVM with kernel is given by

$$f(x) = \text{sgn}\left(\sum_{i=1}^{N} \alpha_i y_i k(x_i, x) + b\right) \tag{9}$$

The requirement on the kernel is to satisfy Mercer's theorem. Within this requirement, there are few possible inner product kernels are available. In this paper, the most familiar learning machines' linear kernel is employed.

5 Results and Discussions

In this work, two types of dysfluencies, namely repetition and prolongation, are investigated. These dysfluencies were carefully identified and selected from the subset of UCLASS database. Hundred repetition and hundred prolongation dysfluency segments are prepared from the subset. The experiment was repeated three times, and each time 60 repetition and 60 prolongation dysfluency samples are randomly selected from prepared segments.

Experiments are conducted in two phases. Firstly, the SVM classifier was applied to MFCC, DMFCC, and DDMFCC coefficients. In the second-round prosodic features, such as energy and pitch, are combined with MFCC cepstral

Table 2 The performance results of different feature combination

Types of features	CA (%)	CR (%)	FR (%)	FA (%)	E (%)
MFCC	86.11	83.34	13.89	16.66	84.73
MFCC + E + P + D	88.89	86.11	11.11	13.89	87.50
DMFCC + MFCC	89.90	85.71	10.10	14.29	87.80
DMFCC + MFCC + E + P + D	91.67	87.90	8.33	12.10	89.80
DDMFCC + DMFCC + MFCC	95.15	93.16	4.85	6.84	94.16
DDMFCC + DMFCC + MFCC + E + P + D	**97.80**	**95.90**	**2.2**	**4.1**	**96.85**

features. To check the effectiveness of the extracted features, 60 % of the data were used for training and 40 % for testing.

The following measurements were used to evaluate the performance of the method [28]: correct acceptance (CA), correct rejection (CR), false acceptance (FA), and false rejection (FR), and efficiency is given by

$$E(\%) = 100 \times (CR + CA)/(CR + CA + FA + FR) \qquad (10)$$

Table 2 illustrates the results obtained for different feature combination. It can be noticed that DDMFCC with prosodic features gives the best efficiency and false-acceptance rate compares to MFCC and DMFCC methods.

6 Conclusion

In this study, two feature-extraction techniques are employed to classify dysfluency using SVM classifier. The experiments were carried out for different configurations of MFCC features and prosodic features. The combination of DDMFCC and prosodic features has produced the best efficiency rate of 96.85 %.

When prosodic features are combined with MFCC features, the performance is improved by 2–3 %. The results show that combination of DDMFCC and prosodic features yields significant performance improvement than individual ones. This indicates that feature combination considered is complementary and can be used to detect dysfluency in stuttered speech. In future work, different classification algorithms will be investigated to improve the classification results.

References

1. Young, M.: Predicting ratings of severity of stuttering [monograph]. pp. 31–54 (1961)
2. Sherman, D.: Clinical and experimental use of the Iowa scale of severity of stuttering. J. Speech Hear. Disord. **17**, 316–320 (1952)

3. Cullinan, W.L., Prathe, E.M., Williams, D.: Comparison of procedures for scaling severity of stuttering. J. Speech Hear. Res. **6**, 187–194 (1963)
4. Bloodstein, O.: A handbook on stuttering. Singular Publishing Group Inc., San-Diego, London (1995)
5. Ravikumar, K.M., Rajagopal, R., Nagaraj, H.C.: An approach for objective assessment of stuttered speech using MFCC features. ICGST Int. J. Digital Sig. Proc. DSP **9**, 19–24 (2009)
6. Ravikumar KM, Reddy B, Rajagopal R, Nagaraj H (2008) Automatic detection of syllable repetition in read speech for objective assessment of stuttered disfluencies. In: Proceedings of World Academy Science, Engineering and Technology, pp. 270–273 (2008)
7. Howell, P., Sackin, S., Glenn, K.: Development of a two stage procedure for the automatic recognition of dysfluencies in the speech of children who stutter: II. ANN recognition of repetitions and prolongations with supplied word segment markers. J. Speech Lang. Hear. Res. **40**, 1085 (1997)
8. Noth, E., Niemann, H., Haderlein, T., Decher, M., Eysholdt, U., Rosanowski, F., Wittenberg, T.: Automatic Stuttering Recognition Using Hidden Markov Models. Interspeech (2000)
9. Czyzewski, A., Kaczmarek, A., Kostek, B.: Intelligent processing of stuttered speech. J. Intell. Inf. Syst. **21**, 143–171 (2003)
10. Wisniewski, M., Kuniszyk-Jozkowiak, W., Smolka, E., Suszynsk, W.: Automatic Detection of Disorders in a Continuous Speech with the Hidden Markov Models Approach, vol. 45/ 2008. In: Computer Recognition Systems vol. 2. Springer, Berlin/Heidelberg, pp. 445–453 (2007)
11. Wisniewski, M., Kuniszyk-Jozkowiak, W., Smolka, E., Suszynski, W.: Automatic detection of prolonged fricative phonemes with the hidden Markov models approach. J. Med. Inf. Technol. **11**, 1–6 (2007)
12. Chee, L.S., Ai, O.C., Hariharan, M., Yaacob, S.: MFCC based recognition of repetition and prolongation in stuttered speech using K-NN and LDA. In: Proceedings of 2009 IEEE student conference on research and development (SCOReD), Malaysia (2009)
13. Chee, L.S., Ai, O.C., Hariharan, M., Yaacob, S.: Automatic detection of prolongations and repetitions using LPCC. In: Proceedings of International Conference for Technical Postgraduates (TECHPOS), pp. 1–4 (2009)
14. Mahesha, P., Vinod, D.S.: Automatic classification of dysfluencies in stuttered speech using MFCC. In: International Conference on Computing Communication and Information Technology, Chennai (2012)
15. Mahesha, P., Vinod, D.S.: Classification of speech dysfluencies using speech parameterization techniques and multiclass SVM. In: 9th International Conference, QShine 2013 vol. 115. Greader Noida, Springer Berlin, Heidelberg, pp. 298–308 (2013)
16. Howell, P., Huckvale, M.: Facilities to assist people to research into stammered speech. Stammering Research: An on-line journal published by the British Stammering Association, pp. 130–242 (2004)
17. Devis, S., Howell, P., Batrip, J.: The UCLASS archive of stuttered speech. J. Speech Lang. Hear. Res. (2009)
18. Rabiner, L., Juang, B.: Fundamentals of Speech Recognition. Prentice Hall, Englewood Cliffs (1993)
19. Proakis, J.G., Manolakis, D.G.: Digital Signal Processing, Principles, Algorithms and Applications. MacMillan, New York (2007)
20. Muda, L., Begam, K.M., Elamvazuthi, I.: Voice recognition algorithms using Mel frequency cepstral coefficient (MFCC) and dynamic time warping (DTW) techniques. J. Comput. **2**, 138–143 (2010)
21. O'Shaughnessy, D.: Linear predictive coding. Potentials. IEEE **7**, 29–32 (1988)
22. Feng, L.: Speaker recognition. Master's thesis, Institute of Informatics and Mathematical Modeling. Technical University of Denmark, DTU (2004)
23. Dehak, N., Dumouchel, P., Kenny, P.: Modeling prosodic features with joint factor analysis for speaker verification. IEEE Trans. Audio Speech Lang. Process. **15**, 2095–2103 (2007)
24. Hess, W.J.: Pitch Determination of Speech Signals. Springer, Berlin (1983)

25. Schoslkopf, B., Smola, A.: Learning with Kernals, Support Vector Machines. MIT Press, London (2002)
26. Wang, Q., Yang, J., Yang, W.: Face detection using rectangle features and SVM. Int. J. Intell. Technol. **1**(3), 228–232 (2006)
27. Mercer, J.: Functions of positive and negative type, and their connection with the theory of integral equations. Trans. London Philos. Soc. (A) **209**, 415–446 (1909)
28. Godino-Llorente, J., Gomez-Vilda, P., Blanco-Velasco, M.: Dimensionality reduction of a pathological voice quality assessment system based on gaussian mixture models and short-term cepstral parameters. IEEE Trans. Biomed. Eng. **53**, 1943–1953 (2006)

Retracted Chapter: Development of Magnetic Control System for Electric Wheel Chair Using Tongue

G. Hari Krishnan, R.J. Hemalatha, G. Umashankar, Nilofer Ahmed and Soumya Ranjan Nayak

Abstract The current status of disability in world shows that there are lots of people with some kind of disability, according to a census conducted in 2010 by the National Institute of Statistics and Geography (INEGI) an interesting fact is that the motor disability is the most common among the various disabilities of the population. People who have paralysis which is the motor disability cannot be able to move by themselves, their opportunities to socialize and get a job are reduced drastically, causing patients to fall into depression, isolation, and anxiety. A useful tool for those who can move their upper limbs are the mechanical or electric wheelchairs, since it is a way to recover part of their mobility, but to use it properly; users must acquire skills and knowledge through a program training, which leads patients to finally accept their disability. But in this project, we have to use one of the human organs which never get paralyzed so that the patients just have to move their tongue inside the mouth in order to move in specified direction. The design which implement this task consists of PIC microcontroller and accelerometer which produces 3 axes as *x*, *y*, and *z* which helps in identifying the direction in which the wheelchair has to be moved. The accelerometer is placed on both the sides of the cheeks and depending on the movement of the tongue, the wheel chair moves in four directions that are forward, backward, left, and right.

Keywords Magnetic control · Wheelchair · Tongue · H-bridge driver · MEMS · Microcontroller

---

The erratum of this chapter can be found under DOI 10.1007/978-81-322-2012-1_89

G.H. Krishnan (✉) · R.J. Hemalatha · G. Umashankar · N. Ahmed · S.R. Nayak
Department of Biomedical Engineering, Sathyabama University, Chennai, India
e-mail: haris_eee@yahoo.com

N. Ahmed
e-mail: nilofer.ahmed2@gmail.com

L.C. Jain et al. (eds.), *Intelligent Computing, Communication and Devices*,
Advances in Intelligent Systems and Computing 308,
DOI 10.1007/978-81-322-2012-1_68

1 Introduction

One of the most important problems for patients with severe disability is controlling electric wheelchairs, because the conventional methods as the joystick or keypads which are operated by hands are very difficult for motor disabled persons [1, 2]. The proposed system implements the development of a magnetic control system (MCS) to handle a power wheelchair as an alternative control system for patients with spinal cord injuries, as quadriplegics. The proposed system uses the movements of the patient's tongue to operate the power wheelchair, and also includes the development of new communication protocols for the wheelchair through a microcontroller, bridge H, and magnetic control. The movements of the hands, eyes, face, eyebrows, or the sound of the voice are few control methods have been used in conventional electric wheelchairs, but each of these methods still has efficiency problems. In voice-activated powered wheelchair for severely disable when user delivers voice command, a microphone in a throat detects the vibration of vocal cord but low intensity voice is not recognized [3, 4]. Also under noisy environment conditions, it is difficult to recognize the voice. Under comparative study on different adaptation approaches are concerning a sip and puff controller for a powered wheelchair which uses air pressure to control wheelchair by sipping and puffing, but this system cannot be efficiently applicable for people with week breathing [5].

Development of MCS for an electrical wheelchair using tongue is to develop a wheelchair that can work using tongue sensor. The total system divided into software and hardware section, in hardware section, accelerometer and its usage to control the wheel chair are focused more. In software section, the basic c language is used to program the process the data and to implement mathematic functions needed to design the control movements of the wheelchair. Besides that, apic microcontroller is programmed in the use of circuit.

Currently, there are many assistive devices for people with disabilities in their lower limbs, but people with a minimum capacity of movement or practically null have very limited options to care for themselves, so it is necessary to use specific skills of patients that can be used without problems, as the movement of the tongue. The brain is directly connected to the tongue by hypoglossal nerve, which usually is not damaged when people have spinal cord injuries, plus the muscle of the tongue is similar to the heart because of this muscle is not fatigued easily. Furthermore, the mouth cavity is accessible semi-invasively and is not influenced by the position of the head or body, for what it is possible to adapt a system that uses the movement of the tongue with ample comfort. Also the tongue and mouth have a lot of sensory cortex in the human brain, which is why the use of the tongue can be compared to the movements of the hand and fingers when it comes to accuracy. Therefore, these are capable of performing sophisticated motor tasks and manual activities with many degrees of freedom. It is noteworthy that the movement of the mouth on the tongue is intuitive and fast and that with proper training, if performed simple continuous motion can be induced neural plasticity. Among the different proposals to use the tongue as a method of controlling devices, most systems have direct contact with the patient, for example, the use of magnetic piercings, electrical contacts, a framework for mouse and keyboard emulation, or magnetic accessories. Generally, these systems use sensors to detect movement of the tongue, but there are two ways to place them. Some systems mount sensors and circuits in a dental retainer for use within the mouth and wirelessly sending the signals. But others use separate devices, placing sensors on cheeks, and then put on the tongue a magnet. Here MCS is to handle the wheelchair with the tongue and was designed to be used by people with severe disabilities to move again for themselves.

Block diagram with all phases of this system is as shown in Fig. 1. The main objective was to manage the electric wheelchair with simple commands to move seamlessly through an obstacle course with precision using an interface easy to use, minimally invasive, reliable, discreet, and inexpensive.

This project is to handle the wheelchair with the tongue and was designed to be used by people with severe disabilities to move again for themselves. It has the following advantages: it is very useful to the persons who were affected from spinal card injuries and other disabilities. It is useful to move inside the home or other places without any help. The accuracy was high because of using the magnetic controlled system. The cost is very less when compared to other method. It makes the patient convenient to use this device. It is less painful. It is helpful to control the movement by themselves. The efficiency is also high. The aim of this project was to handle the wheelchair with the tongue and was designed to be used by people with severe disabilities to move again for themselves. Two resistive sensors will be placed on the both side cheeks of the user. The sensors get the different axis from the movement of tongue inside the mouth. The data from the sensor is analog. So, we give the analog value to the ADC of the PIC microcontroller [6, 7]. Then the data converts into digital. Depend upon the data from the controller, we control the wheel chair by the help of H-Bridge motor driving circuit. Also the simulation of the operation of electric wheelchair is shown using temperature resistor sensor in the circuit. Since, we cannot use accelerometer sensor for simulation part in software instead of that we are using temperature register sensor. The language used for simulation is embedded C language.

2 Materials and Methods

2.1 System Overview

This project is to handle the electric wheelchair with the help of tongue and is designed for the people who are not able to move themselves. Two resistive type accelerometers have been placed on both sides of the cheeks of the user. The accelerometer generates the three different axes and detects the movement of tongue inside the mouth the sensors. The analog value from accelerometer is

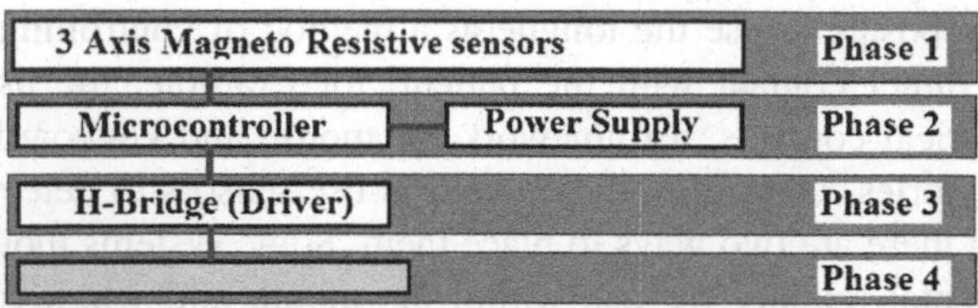

Fig. 1 Magnetic control systems

converted into digital using ADC of the PIC microcontroller. Depending upon the data from the controller wheel chair control can be implemented with the help of H-Bridge motor driving circuit. The PIC microcontroller and its interfacing circuits were placed in wheel chair circuit. The total system circuits were placed in the middle of the wheelchair.

In the first phase, the system uses a pair of 3-axis magneto resistive sensors (magnetometers); these are surface mount, of low-cost, compassing, and has magnetometer as shown in Fig. 1. They are anisotropic, directional sensors; they have precision in-axis sensitivity and linearity the magnetometer. The phase 2 uses +24 batteries to power the engines and the microcontroller. Nevertheless, microcontroller uses only +5 VDC through a voltage regulator. An H-bridge is an electronic circuit that enables a voltage to be applied across a load in either direction. These circuits are often used to allow DC motors to run forwards and backward. Depending upon the data from the PIC microcontroller, we can control the wheel chair with the help of H-bridge motor driving circuit.

MEMS accelerometers are micro-electromechanical systems that measure the static or dynamic force of acceleration. Accelerometer will generate 3 axes as *x*, *y*, and *z*. According to the axis, the movement of the wheel chair will be determined and it will be placed on both sides of the cheeks with the help of head set. In addition to the conventional accelerometer, the MEMS accelerometer contains a tiny cantilever beam with a seismic mass [8]. The sensor sends an analog signal as the proof mass deflects from its neutral position under the influence of external acceleration. MEMS accelerometer made by Analog Devices Inc., ADXL330 is particularly used. This model already deploys multi-axes sensor in three-dimensional space. This provides a great advantage for the current interest since one does not need to set up at least two small MEMS accelerometers perpendicularly to obtain two dimensional orientations. Using ADXL330, it is able to measure the inclinations of object relative to 2-dimensional plane simultaneously.

3 Results and Discussion

The component kit itself performs faultlessly and the completion of the prototype was constructed with the limited time-frame. There are four main directions of the wheelchair movement and the others are the mixture of four main directions. The

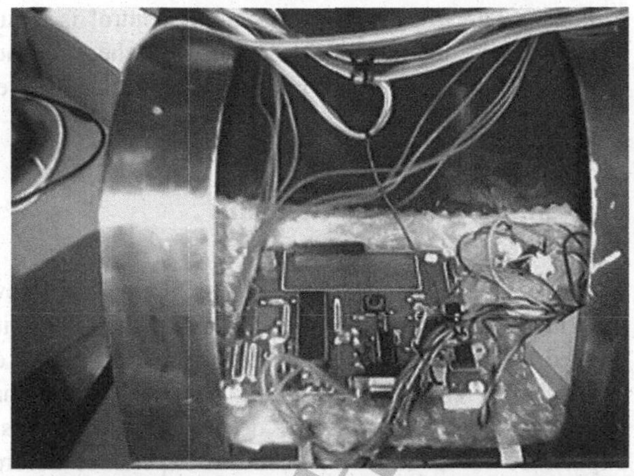

Fig. 2 Circuit placed in wheelchair

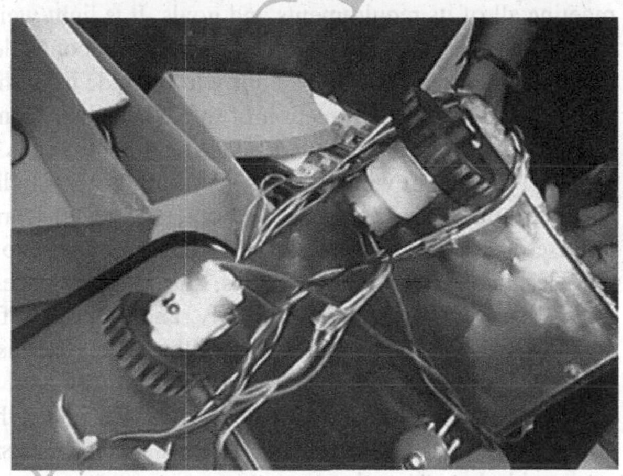

Fig. 3 Wheels placement in wheel chair

placement of overall hardware circuit in wheel chair is as shown in Fig. 2. The wheel position and its placement on wheel chair were as shown in Fig. 3. The overall project performance of constructing the electric wheelchair using tongue was success.

The final performance of designed wheel chair was achieved by making the system to move in four directions as forward movement, backward movement, right movement, and left movement. In the first case, the wheelchair is controlled to go forward and the experience is when the user moves the tongue to upward in right chick, the accelerometer sensor drive the wheelchair moving forward. In the second case, the wheelchair is controlled to be going backward. The tongue moves downwards in right cheek to operate the wheelchair moving backward. The third case is that the wheelchair rotates from right side. The tongue is moved to the upward side of the left cheek and then the wheelchair starts rotating from right direction. The last case of wheelchair control is to turn left side. The tongue is moved to the downward side of the left cheek. Hence by the movement of the tongue, the wheelchair can be moved and rotated in left direction.

4 Summary and Conclusion

The wheelchair designed is a successful electronic device that provides to disable individuals the ability to recover part of their mobility. It has been successfully created to adapt to wheelchair users by working in conjunction with automated wheelchairs. The component kit consists of two major features: it makes the patient convenient to use the device. These features allow wheelchair users to experience something they have not experience before while using an automated wheelchair. This product is less painful. The final prototype was successful in meeting all of its requirements and goals. It is light weight and cost effective. It has high efficiency and high accuracy. With microcontroller being use for this design, the component kit has a great ability to upgrade to fit the unique needs of its user in the near future. The completion of the Wheelchair Component Kit is a new step in improving people's lives and adapting to the needs of people with motor disabilities. The smart wheelchair is very helpful for handicapped people; the user can easily operate the wheelchair without needing to learn more skills. In particular, the user just needs to use the tongue and with the help of accelerometer sensor the wheelchair starts moving. Besides, the electric wheelchair controlled by tongue makes it easy to control, comfortably, and differently from other wheelchairs. This can improve the independent life of the user—the disabled people. However, by using the sensor to operate the electric wheelchair, sometimes there are some errors during the wheelchair movements. To make the electric wheelchair more convenient, the smart wheelchair system should be designed so that the wheelchair can avoid obstacles using a stereo camera system.

In this project, the tongue was used to control the electric wheelchair. The accelerometer sensor was embedded in the user's chick and the user needed to move his/her tongue to control the wheelchair. The signal is transmitted from a sensor to microcontroller and is processed by the C language. In the program, the mathematical functions were designed to calculate the input signal and the obtained results were match to the wheelchair controller to drive the wheelchair. Then the calculated result will be sent to the H-bridge in order to move the wheelchair. The experimental results show to illustrate the effectiveness of the proposed design.

References

1. Öztürk, A., Ucsular, F.D.: Effectiveness of wheelchair skills training programs for community-living users of manual wheelchairs in Turkey: a randomized controlled trial. Clin. Rehabil. **25**, 416–424 (2011)
2. Mattevi, B.S., Bredemeier, J., Fam, C., Fleck, M.P.: Quality of care, quality of life and attitudes toward disabilities: perspectives from a qualities focusgroup study in Porto Alegre, Brazil. Rev. Panam. Salud Publica **31**(3), 188–196 (2012)
3. Grøttlandand, H., Seide, P.M.: Assistive devices and distributed processes: reflections on activity systems and impairments. Scand. J. Disabil. Res. **12**(4), 305–319 (2010)
4. Sivanandan, K.S., Sirish, T.S.: Smart assisting device for partially paralyzed people during locomotion-design and control. Int. Rev. Modell. Simul. (I.RE.MO.S.) **4**(6), 3371–3375, 5p (2011)
5. Henao-Lema, C.P., Pérez-Parra, J.E.: Lesiones medulares discapacidad: revisión bibliográfica. AQUICHAN, Año 10 **10**(2), 157–172, (2010), Chia, Colombia
6. Nishimori, M., Saitoh, T., Konishi, R.: Voice controlled intelligent wheelchair. In: SICE Annual Conference 2007, International Conference on Instrumentation, Control and Information Technology, pp. 336–340 (2007)
7. Li, N., Xia, L., Shiming, D., Xu, X., Chan, M.Y.: Steady-state operating performance modelling and prediction for a direct expansion air conditioning system using artificial neural network. Build. Serv. Eng. Res. Technol. **33**(3) (2012)
8. Suk, S., Kojima, H.: Voice-activated powered wheelchair for severely disabled persons. IEICE Electron. Express **4**(18), 569–574 (2009)

An Autonomous Obstacle Avoiding and Target Recognition Robotic System Using Kinect

Anoop Velayudhan and T. Gireeshkumar

Abstract This article describes the development of an autonomous navigation system for mobile robots which can be used in unstructured and unknown indoor environment. Histogram of the range data captured using Kinect is utilized in developing the obstacle avoidance algorithm which avoids the static and dynamic obstacles. In addition to this, an object recognition task is also carried out by the robot using SURF algorithm. Once the object is recognized, it sends that information to the remote workstation through wireless communication.

Keywords Obstacle avoidance · Object recognition · SURF · Histogram of depth frame

1 Introduction

One of the inevitable features of an autonomous mobile robot is the ability to avoid both static and dynamic obstacles in its path. It is an essential safety measure and prerequisite for autonomous robot navigation. In the case of an unknown and unstructured environment, the information regarding the occupancy of obstacles as well as the structure of the surroundings has to be found by the robot in a reactive manner based on the real-time data acquired through its sensors during navigation [1, 2]. Here, the first objective of the system is to roam autonomously avoiding the static and dynamic obstacles with the help of depth data acquired using the Kinect sensor [3, 4]. The second objective is to recognize a target object (whose image and features are preloaded into the system) from the unknown environment where

A. Velayudhan (✉) · T. Gireeshkumar
Amrita Vishwa Vidyapeetham University, Coimbatore 641112, India
e-mail: anoop.t.velayudhan@gmail.com

T. Gireeshkumar
e-mail: gireeshkumart@gmail.com

L.C. Jain et al. (eds.), *Intelligent Computing, Communication and Devices*,
Advances in Intelligent Systems and Computing 308,
DOI 10.1007/978-81-322-2012-1_69

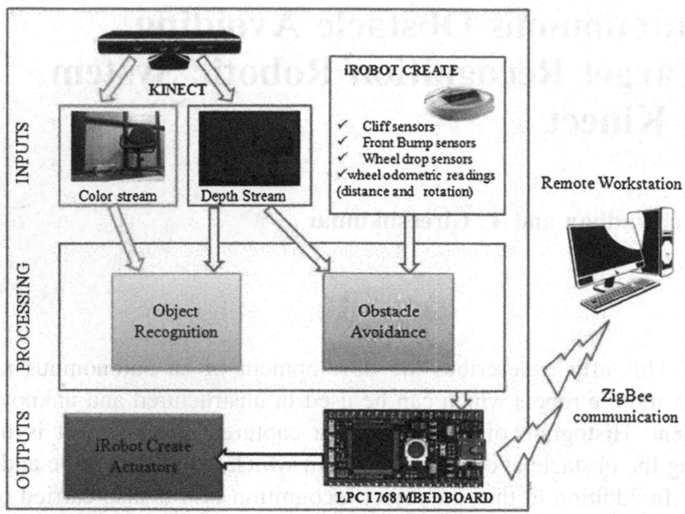

Fig. 1 Overview of the system

it is present. On finding the target, it reports to a remote workstation through wireless communication. This feature of the robot can be made use in search and rescue/find missions (e.g., retrieving survivors and valuables from earthquake wreckage where human intervention is difficult), intelligent pick and place of goods in industries, etc. This feature is also essential if the robot has to localize itself in the global map based on the landmarks in the environment which it identifies as target objects.

The rest of the paper is organized as follows: Section 2 describes the system architecture. Section 3 analyses the results obtained from experiments in the indoor environment, and Sect. 4 gives the conclusion and future works.

2 System Architecture

The whole robotic system is developed over the iRobot Create robotic development platform. The various modules of the system are depicted in Fig. 1. The main input devices to the system are the Kinect sensor and all the inbuilt sensor modules of the iRobot Create such as bump sensor, cliff sensors, shaft encoders, and wheel drop sensors. The main sensor, Kinect, acts as the eye of the robot giving visual perception, while all the other sensors ensure the appropriate and safe navigation. All the main computation tasks of the robot such as image processing and object recognition are done on a computer that is mounted over the robot, while all the low-level tasks of the robot such as motion control, reading onboard sensor values, communicating with remote workstation, and the onboard computer are taken care by the program running on an LPC 1768 MBED board [5]. Kinect is interfaced to the computer using OPENNI platform. OPENNI version 2.2 is used to capture the color and depth stream from the Kinect. OpenCV [6] libraries are used to implement the algorithms for obstacle avoidance and object recognition. The entire code development is done using Microsoft Visual Studio 2010 IDE.

2.1 Obstacle Avoidance

For obstacle avoidance algorithm, the depth frame captured by Kinect is utilized. To improve the efficiency, the original depth frame size of 640 (H) × 480 (V) pixels is divided into blocks of 40 (H) × 40 (V) pixels. Thus, each block consists of 1,600 pixels. There will be a total of 192 (16 × 12) blocks covering the entire depth frame. The depth frame is then grouped horizontally into three main segments, namely left, middle, and right. The left and right segments are of same size (200 × 480 pixels each) containing 60 blocks. The middle segment is of size 240 × 480 pixels containing 72 blocks. The average depth value of 1,600 pixels in each block is calculated and is assigned as the depth metric of that block. As a result, a down-sampled version of depth frame is obtained. Kinect gives reasonably correct readings at the range of 0.5–4 m. For every individual depth frame, three histograms are calculated for left, right, and middle segments separately which gives the number of blocks having different depth metric values in the reliable ranges [7]. Each histogram is divided into two main regions based on depth values, namely r_1: 500–800 mm and r_2: 800–1,100 mm. Let the variable '*i*' denote the depth value and *n*(*i*) denote the number of blocks having a depth metric value as '*i*'. Let N_{r_1} and N_{r_2} denote the total number of blocks in the range r_1 and r_2, respectively. Then,

$$N_{r_1} = \sum_{i=500}^{800} n(i) \tag{1}$$

$$N_{r_2} = \sum_{i=800}^{1100} n(i) \tag{2}$$

By observing how the number of blocks in each of these regions changes in consecutive depth frames, the movement of objects relative to the robot is found out. A transition of blocks from the farther region (r_2) to the closer region (r_1) in a short time suggests the approach of an obstacle and vice versa. A control action is to be taken when the transition of blocks exceeds a particular threshold value. Whenever it crosses the threshold, it means the obstacle is approaching and response action is triggered. It can be expressed formally as:

$$\text{Object detected} = \begin{cases} 1, & \text{if } \left(N_{r_{1\,\text{current}}} - N_{r_{1\,\text{previous}}}\right) > T \\ & \wedge\left(N_{r_{2\,\text{previous}}} - N_{r_{2\,\text{current}}}\right) > T \\ 0, & \text{otherwise} \end{cases} \tag{3}$$

where T is the threshold applied, $N_{r_{1\,\text{current}}}$ gives the current value of N_{r_1}, and $N_{r_{1\,\text{Previous}}}$ gives the value of N_{r_1} a fixed time length ago. Similarly, $N_{r_{2\,\text{current}}}$ gives the current value of N_{r_2} and $N_{r_{2\,\text{Previous}}}$ gives the value of N_{r_2} a fixed time length ago. By correctly setting the threshold, just sensitive enough for the robot to take control action only when needed, the false alarms can be avoided. Here, the threshold is set as 6,400 pixels, i.e., four times the block size. In this method only, objects which appear to move toward the robot's viewpoint when the robot moves forward will be detected as obstacles. So the added advantage is that floor will not be detected as an obstacle as there is no relative motion of the depth blocks related to the floor in the depth image, when the robot moves forward. The histogram information of the depth frames captured consecutively is stored in a cyclic buffer data structure, which follows the first-in first-out (FIFO) convention. The maximum possible frame rate of Kinect is 30 frames per second. A fixed buffer size which can store data of 10 frames is chosen. Each frame data has to retain itself in the buffer until data of nine more consecutive frames get entered to the buffer to make it full. Depending on this retention time, the length of the circular buffer is fixed as 400 ms. Whenever the buffer becomes full, the oldest data in the buffer are overwritten by the newest frame's data. In order to apply the obstacle avoidance algorithm, the data of the current frame and the oldest frame data stored in the buffer are considered. Traversal through the middle segment is always preferred, but when there are obstacles, the robot will check condition in the other two segments. It will eventually move toward that segment, which has more traversable space and lesser obstacles. In order to avoid moving obstacles, the robot should be able to calculate the current trajectory of the object in motion and see whether at any point of time its own trajectory will cross the former. When the camera viewpoint is kept static, the present and previous frames are subtracted to find out the relative displacement of the moving objects. And when the robot is in motion, the relative translation of the camera (obtained from the odometric readings) is negated from the depth data. Similarly, rotation effects of the camera can also be negated, but depth data captured just after the rotation of the robot have to be discarded since the present and previous data differ significantly due to change in view point of the robot.

2.2 Object Recognition

In this paper, speeded up robust features (SURF) [8] algorithm is used for object recognition task and is implemented with the help of OpenCV library functions. An image of the target object to be recognized is taken and saved in the robot's memory. From that image, interest points are extracted and descriptor matrix is computed. During navigation, the Kinect captures the real-time image frames from the surroundings and for each frame, features are extracted and descriptors are generated. Descriptors of the captured image and the target image are then compared to check whether the target appears in the robot's field of vision or not. Fast library for approximate nearest neighbors (FLANN) which is a library for performing fast approximate nearest neighbor searches in high-dimensional spaces is used for matching the reference image and the image procured in real time from the camera. Based on the number of matches and threshold value for the matching points, it determines whether the object is found or not. Both RGB and depth images are taken into consideration because depth image will be used to calculate physical distance and angle of the object from the camera in the real world. In order to find the perspective transformation between matched key points, a homography matrix (projection matrix) is found out. It is needed because the template image of object saved in the memory and that perceived by the robot from its current viewpoint may be scaled and rotated version of the former. There can be some possible errors while matching which affects the result. Homography transformation ensures that only good matches which provide correct estimation are selected and all other outliers are discarded.

3 Results

The algorithms were verified in various test environments. All those obstacles which Kinect cannot detect, such as glass elements (which do not reflect infrared waves) and nearby objects which fall within the error or blind range (0.5 m), were successfully avoided by using an ultrasonic sensor kept at the front end. The robot gave satisfactory results by crossing the path from start to stop. To test the object recognition algorithm, the box that came with Kinect served the purpose of target object. The image of the box was saved as a template image in the computer's memory and the robot successfully recognized it when the object was in its field of vision (see Fig. 2).

After matching, if the percentage of matching is greater than a set value (60 %), then the computer sends a signal to the microcontroller along with the time information at which the match was found (see Fig. 3). The microcontroller on receiving the signal sends out a message to a remote workstation that the object has been recognized, along with the temporal information using IEEE 802.15.4 standard ZigBee communication protocol. The time information helps in finding out how much delay the robot takes to recognize the object at different experimental runs. This helps in formulating the efficiency and reliability of the system. The system time of both the onboard computer and the workstation PC was synchronized before the deployment of the robot in the environment. The intensity reading corresponding to the object's centroid in the depth image will return the

Fig. 2 The image of the target object to be detected (*top left*); blob features identified on template image (*top right*); matching of key points of the object from the template image to the corresponding key points of the same object in the image captured by Kinect during navigation (*bottom*)

```
C:\Users\ANVE\Documents\Visual Studio 2010\Projects\robo_prjct\x64\Release\robo_prjct.exe
percentagetage of matching:   65.573770  matchedpts:40   objkeypts: 61
depth of object from kinect: 618
currentDateTime()=2014-03-12.08:58:55
percentagetage of matching:   65.573770  matchedpts:40   objkeypts: 61
depth of object from kinect: 618
currentDateTime()=2014-03-12.08:58:56
percentagetage of matching:   65.573770  matchedpts:40   objkeypts: 61
depth of object from kinect: 618
currentDateTime()=2014-03-12.08:58:56
percentagetage of matching:   65.573770  matchedpts:40   objkeypts: 61
depth of object from kinect: 618
currentDateTime()=2014-03-12.08:58:56
percentagetage of matching:   65.573770  matchedpts:40   objkeypts: 61
depth of object from kinect: 620
currentDateTime()=2014-03-12.08:58:56
percentagetage of matching:   65.573770  matchedpts:40   objkeypts: 61
depth of object from kinect: 618
currentDateTime()=2014-03-12.08:58:57
percentagetage of matching:   65.573770  matchedpts:40   objkeypts: 61
depth of object from kinect: 618
currentDateTime()=2014-03-12.08:58:57
percentagetage of matching:   65.573770  matchedpts:40   objkeypts: 61
depth of object from kinect: 618
currentDateTime()=2014-03-12.08:58:57
```

Fig. 3 The screenshot of the console window which displays the percentage of matching and distance of the object from the robot along with system time when the object is detected

distance of the object from the robot. The angle of the object is obtained by simply taking the arctangent of the ratio of distance between the template image center and the object center in the captured image to the focal length of the camera.

4 Conclusion and Future Works

The paper demonstrates an effective use of Kinect sensor for developing obstacle avoidance and target recognition modules for an autonomous navigation system. OPENNI 2.2 framework was made use in capturing the color and depth streams from Kinect. By analyzing the histogram formed from the depth image, the obstacles in the path were detected and were effectively avoided. Object recognition was done using the SURF algorithm and was integrated into the code using OpenCV libraries. Communication of the robot with a remote workstation was successfully implemented using ZigBee protocol. The work can be further extended by adding additional functionalities to the system such as 3D mapping of the environment, localizing the robot with respect to the global map, and developing path planning techniques to find out optimized paths among the many possible paths to reach at the target object.

References

1. Maaref, H., Barret, C.: Sensor-based navigation of a mobile robot in an indoor environment. Robot. Auton. Syst. **38**(1), 1–18 (2002)
2. Siegwart, R., Nourbakhsh, I.R., Scaramuzza, D.: Introduction to Autonomous Mobile Robots. MIT press, Cambridge (2011)
3. Oliver, A., Kang, S., Wünsche, B.C., MacDonald, B.: Using the kinect as a navigation sensor for mobile robotics. In: Proceedings of the 27th Conference on Image and Vision Computing New Zealand, pp. 509–514. ACM (2012)
4. Correa, D.S.O., Sciotti, D.F., Prado, M.G., Sales, D.O., Wolf, D.F., Osório, F.S.: Mobile robots navigation in indoor environments using kinect sensor. In: Critical Embedded Systems (CBSEC), 2012 Second Brazilian Conference, pp. 36–41. IEEE (2012)
5. LPC 1768 mbed board handbook, http://mbed.org/handbook/mbed-NXP-LPC1768
6. Bradski, G., Kaehler, A.: Learning Opencv: Computer Vision with the OpenCV library. O'Reilly Media, Inc., California (2008)
7. Simmonds, J.: Investigating low-power computer vision systems using the microsoft kinect to create autonomous litter bins. Doctoral dissertation, University of Bristol, (2013)
8. Bay, H., Ess, A., Tuytelaars, T., Van Gool, L.: Speeded-up robust features (SURF). Comput. Vis. Image Underst. **110**(3), 346–359 (2008)

Navigation Based on Adaptive Shuffled Frog-Leaping Algorithm for Underwater Mobile Robot

Kundu Shubhasri and Dayal R. Parhi

Abstract Navigational approach for underwater robot may be inferred as a numerical solution to the nonlinear optimal control problem. Adaptive shuffled frog-leaping algorithm has been chosen as a dynamic path planning scheme for underwater robot to track target position while avoiding obstacles. By introducing adaptation procedure in proposed algorithm, the optimization of path as well as time taken can be done through an iterative process by avoiding local minima situation. Objective function with adaptive parameter set as well as stopping criterion of iteration process has been chosen based on distance between robot and target as well as obstacles to regulate convergence rate toward optimal solution. The simulated as well as experimental analysis may validate the properties of the heuristic navigational approach such as faster decision-making, obstacle avoidance, and target seeking behavior during navigation of underwater robot in a messy environment.

Keywords Adaptation · Memplex · Navigation · Objective function · Obstacle avoidance · Target seeking behavior

---

Please note that the LNCS Editorial assumes that all authors have used the western naming convention, with given names preceding surnames. This determines the structure of the names in the running heads and the author index.

K. Shubhasri (✉) · D.R. Parhi
Robotics Lab, Department of Mechanical Engineering, National Institute of Technology Rourkela, Rourkela, India
e-mail: shubhasri_ee2006@yahoo.co.in

D.R. Parhi
e-mail: dayalparhi@yahoo.com

L.C. Jain et al. (eds.), *Intelligent Computing, Communication and Devices*,
Advances in Intelligent Systems and Computing 308,
DOI 10.1007/978-81-322-2012-1_70

1 Introduction

Motion control of AUV has been persisted to be challenging topic of research for decades as any rotational motion around any axis may initiate hydrodynamic translational forces and rotating moments [1]. Autonomous underwater vehicles are generally utilized for rescue operation, surveillance, inspection, recovery, maintenance, etc. [2]. The aim of the research was to plan a near about optimal path for underwater robot from source to target subjected to some optimization criteria in 3D environment.

To realize this purpose, researchers have offered evolutionary algorithms (EAs) for searching near optimal solutions [3] over the years. EAs are stochastic search methodologies that reproduce the metaphor of biological evolution and/or the social behavior of species. A performance index (fitness function) should be minimized based on an approximation of the weighted combination of energy and time consumption [4].

Path optimization method based on modified shuffled frog-leaping algorithm has been adopted here. Like particle swarm optimization (PSO) and genetic algorithm (GA), SFLA is also the optimization approach based on artificial population which can be applied for nonlinear, non-differential, and multimodal problems [3, 4]. In SFLA, after global search in population set, local search will be performed for all subgroups for replacing worst individual (frog) by the best one [5]. By commencing small modification in the fitness function as well as in algorithm, robot can follow comparatively smoother path by avoiding obstacles [6] than original algorithm.

This paper is coordinated as follows: Section 2 describes dynamic equations of underwater robot model. Section 3 presents the formulation of SFLA algorithm in a generalized manner. Section 4 states proposed adaptation in algorithm to solve the path planning problem. Sections 5 and 6 demonstrate simulation study and experimental view, respectively. Section 7 discusses outcome and future plan briefly.

2 Dynamics of the Underwater Vehicle

The robot has been modeled as a mass which is free to move in the 3D space. The body-fixed frame is assumed to be located at the center of gravity with neutral buoyancy. Only three actuators are here to guide the robot. Two thrusters for horizontal motion in forward, backward, and also for rotation about the z axis and another one for linear motion along z axis. The dynamic equations of motion are given by six coupled nonlinear differential equations according to previous researches [1]:

$$M(v)\dot{v} + C_D(v)v + g(\eta) + d = \tau \quad (1)$$

$$\dot{\eta} = J(\eta)v \quad (2)$$

where $\eta = [x \quad y \quad z \quad \phi \quad \theta \quad \psi]^T$; the position and orientation vector in reference with earth-fixed frame. Rotations about the x and y axes cannot exist, so the vector $\eta = [x \quad y \quad z \quad \psi]^T$ is in a reduced from. $v = [u \quad v \quad w \quad p \quad q \quad r]^T$ is the velocity and angular rate vector in body-fixed frame (Fig. 1), $M(v) \in R^{6\times6}$ is the inertia matrix (including added mass); $C_D(v) \in R^{6\times6}$ is the matrix of Coriolis, centripetal and damping term; d signifies the gravitational forces and moments vector along with uncertainty, and τ is the input torque vector. $J(\eta)$ is the transformation matrix defined as

$$J(\eta) = \begin{bmatrix} J(\eta_1) & 0_{6\times6} \\ 0_{6\times6} & J(\eta_2) \end{bmatrix} \quad (3)$$

where

$$J(\eta_1) = \begin{bmatrix} c\psi\theta & s\phi s\theta c\psi - c\phi s\psi & c\phi s\theta c\psi + s\phi s\psi \\ s\psi\theta & s\phi s\theta s\psi + c\phi c\psi & c\phi s\theta s\psi - s\phi c\psi \\ -s\theta & s\phi c\theta & c\phi c\theta \end{bmatrix} \text{ and}$$

$$J(\eta_2) = \begin{pmatrix} 1 & s\phi t\theta & c\phi t\theta \\ 0 & c\phi & -s\phi \\ 0 & s\phi/c\theta & c\phi/c\theta \end{pmatrix}$$

Here, $s\cdot = \sin(\cdot)$, $c\cdot = \cos(\cdot)$, and $t\cdot = \tan(\cdot)$

3 Architecture for Shuffled Frog-Leaping Algorithm

The shuffled frog-leaping algorithm as a memetic metaheuristic has already been applied in various research fields by merging the gains of GA and PSO [7]. In a swamp, group of frogs have been anticipated to find out the stone with maximum amount of food [8]. The SFL algorithm can be described as below:

In first step of SFLA, for P-dimensional problem, randomly initiated population is represented as $X_i = (x_{i1}, x_{i2},\ldots, x_{iP})$, $i = 1,2,\ldots, D$. Later, the frogs are ranked in a descending order based on their respective fitness value of triangular probability function [5]. The entire population is divided into m memeplexes, each containing f frogs (i.e., $D = m\ X\ f$) (Fig. 2). In this process, the first frog goes to the first memeplex, frog m goes to the mth memeplex, again frog $m + 1$ goes back to the first memeplex, etc. In each memplex, a local search (memetic evolution) will be done and also frogs with the best and the worst fitness (X_b and X_w) individually will be found. To improve X_w:

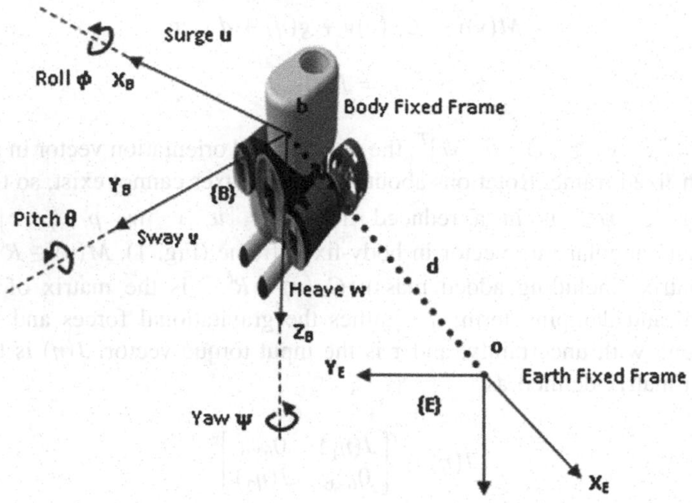

Fig. 1 Position and orientation of prototype underwater robot model

Change in frog position,

$$\delta X^n = \text{rand}() \cdot (X_b^n - X_w^n) \tag{4}$$

$$X_W^{n+1} = X_W^n + \delta X^n; \delta X_{\max} \geq \delta X^n \geq -\delta X_{\max} \tag{5}$$

where rand() $\in (0, 1)$; n: Iteration number and $\delta X_{\max}$: Maximum variation in a frog's position. If X_w^{n+1} is better than X_w, then it will be replaced by new one. Else, X_g will replace X_b in Eq. (4). After a defined number of memetic evolution steps, shuffling process will be done [6]. The whole process repeated until convergence criteria are contented.

4 Adaptation in SFLA for Path Planning

To attain the main objective of proposed method such as to reach the goal position (goalx^i, goaly^i) without collision with obstacles, at ith position, robot has to choose its next best position (robx_b^j, roby_b^j) based on obstacles' and target's position (Fig. 3). So a fitness function has been designed here for every probable position of the robot:

$$F(i) = C_L \frac{1}{\|\text{LOD}\|} + C_R \frac{1}{\|\text{ROD}\|} + C_F \frac{1}{\|\text{FOD}\|} + C_T \|\text{TD}\| \tag{6}$$

where

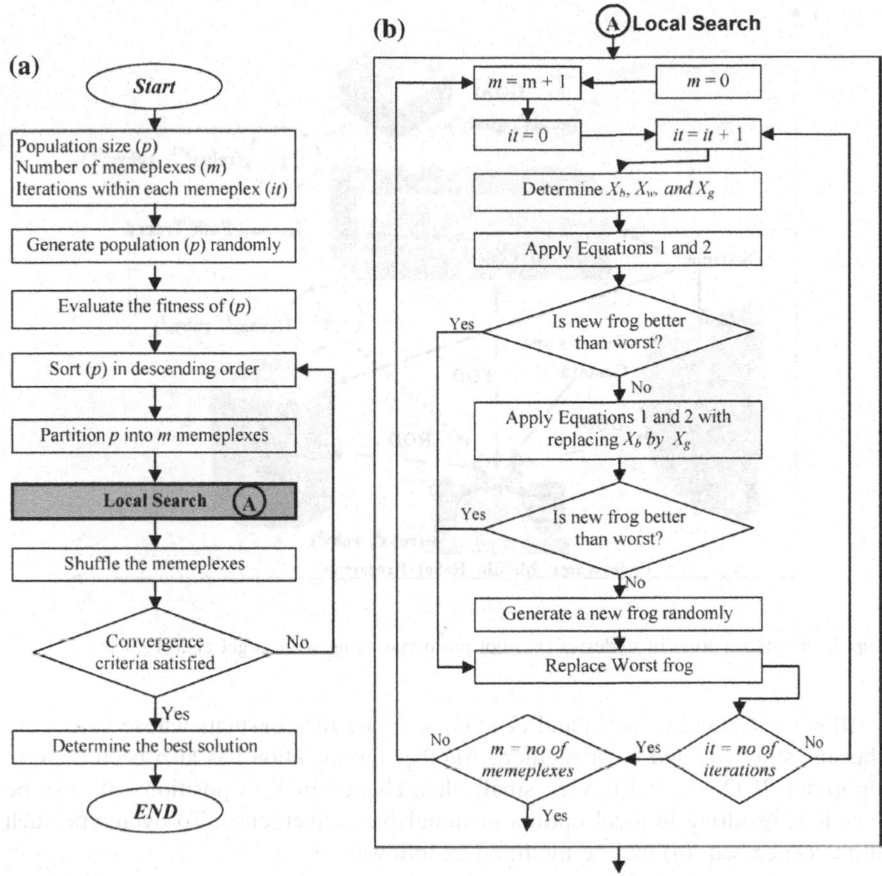

Fig. 2 Flowchart of the shuffled frog-leaping algorithm in a generalized manner

$$\|\text{LOD}\| = \left\| \sqrt{(\text{rob}x^i - \text{obl}x^i)^2 + (\text{rob}y^i - \text{obl}y^i)^2} \right\|$$

$$\|\text{ROD}\| = \left\| \sqrt{(\text{rob}x^i - \text{obr}x^i)^2 + (\text{rob}y^i - \text{obr}y^i)^2} \right\|$$

$$\|\text{FOD}\| = \left\| \sqrt{(\text{rob}x^i - \text{obf}x^i)^2 + (\text{rob}y^i - \text{obf}y^i)^2} \right\|$$

$$\|\text{TD}\| = \left\| \sqrt{(\text{goal}x^i - \text{rob}x^i)^2 + (\text{goal}y^i - \text{rob}y^i)^2} \right\|$$

Fitness will increase when obstacles are close to robot and will reduce as robot approaches target. So minimization of fitness function is highly desirable. The parameters of function can be adapted according to updated sensory information [4].

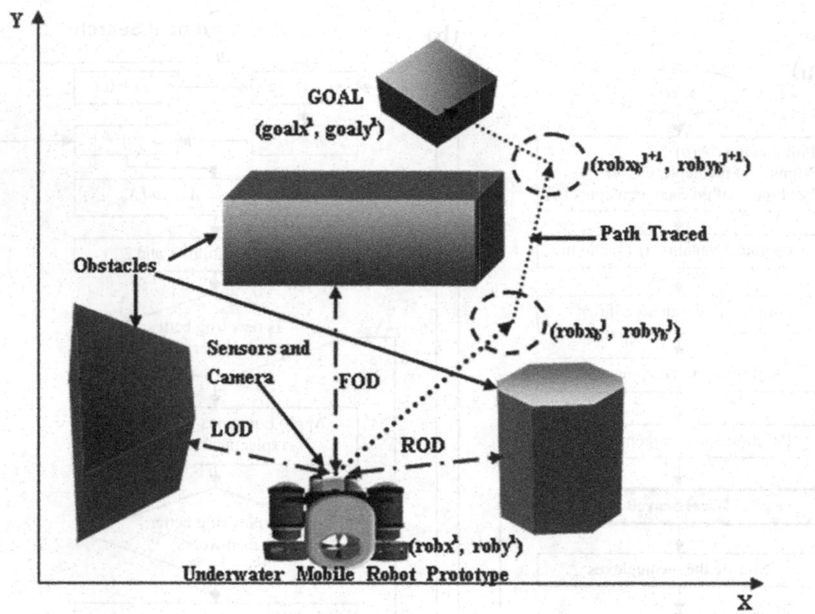

Fig. 3 2D presentation of underwater robot prototype along with target and obstacle

($0 < C_L$, C_R and $C_F < 1$) and ($1 < C_T < 5$) are the constants whose values are chosen using trial and error method. Another modification has also been done in algorithm. If ($X_b - X_w$) is very small, then change in X_w's position will also be very less, resulting in local optima or impulsive convergence. To overcome such an incidence, Eq. (4) can be modified as follows,

$$\delta X^n = \text{rand}() \cdot \frac{\lambda}{2}\left\{(X_g^n - X_w^n) + (X_b^n - X_w^n)\right\} \tag{7}$$

where λ denotes a multiplying factor termed as search acceleration factor taken as positive constant. The next part of algorithm will be same as given in previous section as well as in flowchart (Fig. 2). After getting globally best position, azimuth and elevation steering angle toward the selected position from the current position of the underwater robot has to be computed for each cycle through MATLAB code.

5 Simulation Results

In the MSFL algorithm, some conjectures are included such as the number of frogs and the number of memeplexes are 60 and 6, respectively, maximum step size $\delta X_{\max}$ is set to 0.0125, and number of evolution for each memeplex and number of

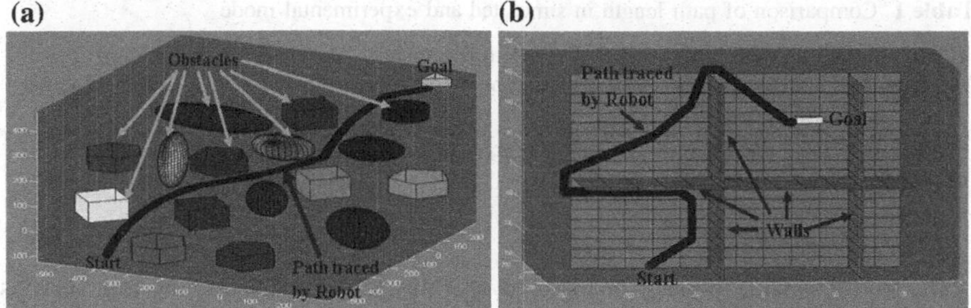

Fig. 4 **a** Implementation of SFLA toward obstacle avoidance and target seeking behavior. **b** Escaping from dead-end obstacle by wall following behavior

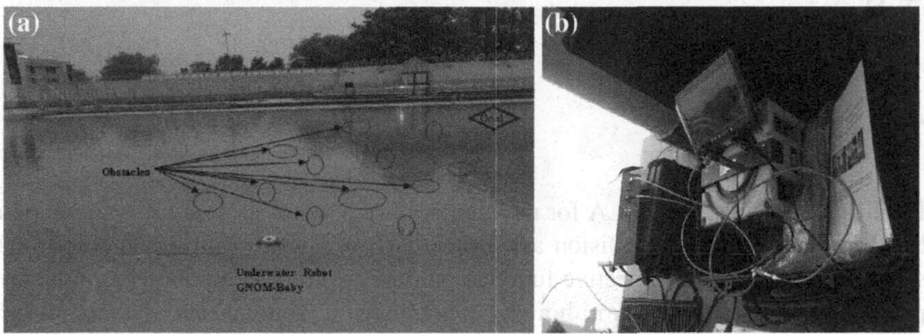

Fig. 5 **a** Navigational view in swimming pool of NIT Rourkela. **b** Robot's accessories

shuffling iterations are 30 and 300, respectively. The values for C_L, C_R, C_F, and C_T have chosen as 0.31, 0.57, 0.46, and 2.09, respectively. The obstacle avoidance behavior will be activated for sensor reading less than threshold (43 mm). When the distance between the center of robot and target is less than 3 cm, the path will be stopped there.

Collision avoidance must have the highest priority, as shown in Fig. 4a. When obstacle as well as target is in same way, wall following behavior should be executed to avoid local minima (Fig. 4b). For a particular obstacles arrangement, simulation has been executed a number of times using MATLAB code. The one simulated path which is shorter than others has been taken as nearly optimum and also shown here.

6 Experimental View

To validate simulated study based on proposed adaptive SFLA, experimental analysis has been performed. Underwater robots from GNOM family which has taken here for real-time navigation through underwater environment (swimming

Table 1 Comparison of path length in simulated and experimental mode

Environmental scenario	Path length in simulation mode (in pixel) (Fig. 4a)	Path length in experimental mode (in pixel) (Fig. 5a)	Percentage of error between simulation and experimental results
Figures 4a and 5a	237	279	15.05 %

pool) are most widely used for test-bed operations such as searching and surveys of wrecks. Pictorial views of underwater robot (GNOM baby) during navigation have been shown here (Fig. 5). The experiment has been repeated for 12 times to find out path length and speed of robot in average. The comparison of performance in both simulated and experimental environment has also been carried out (Table 1).

7 Conclusion

Realization of adaptive SFLA for navigation in real world as well as in simulation has been illustrated for collision avoidance with obstacles of different shape and sizes. The selection of objective function parameters have been done based on the performance of ASFL algorithm for a number of simulations. The proposed method may have been shown some optimal effect by using updated information about the target and obstacles concurrently, but the gained path may not be globally optimal. Further study with more experimental verification of simulated results as well as comparison with other navigation algorithms will be performed for justification of proposed technique.

References

1. Fossen, T.I.: Marine Control Systems: Guidance, Navigation and Control of Ships, Rigs and Underwater Vehicles. Marine Cybernetics, Trondheim (2002)
2. Aghababa, M.P.: 3D path planning for underwater vehicles using five evolutionary optimization algorithms avoiding static and energetic obstacles. Appl. Ocean Res. **38**, 48–62 (2012)
3. Lei, L., Wang, H., Wu, Q.: Improved genetic algorithms based path planning of mobile robot under dynamic and unknown environment. In: Proceedings of IEEE International Conference Mechatronics and Automation, 25–28 June 2006, Luoyang, China
4. Xin, C., Li, Y.M.: Smooth path Planning of a mobile robot using stochastic particle swarm optimization. In: Proceedings of IEEE International Conference Mechatronics and Automation Luoyang, China, pp. 1722–1727 (2006)
5. Eusuff, M.M., Lansey, K.E.: Optimization of water distribution network design using the shuffled frog leaping algorithm. J. Water Res. Plann. Manage. **129**(3), 210–225 (2003)

6. Hassanzadeh, I., Madani, K., Badamchizadeh, M.A.: Mobile robot path planning based on shuffled frog leaping optimization algorithm. In: 6th Annual IEEE Conference on Automation Science and Engineering, Canada, 21–24 Aug 2010, pp. 680–685
7. Elbeltagi, E., Hegazy, T., Grierson, D.: Comparison among five evolutionary-based optimization algorithms. Adv. Eng. Inf. **19**, 43–53 (2005)
8. Elbeltagiy, E., Hegazyz, T., Griersonz, D.: A modified shuffled frog-leaping optimization algorithm: applications to project management. Struct. Infrastruct. Eng. **3**(1), 53–60 (2007)

Power Efficiency with Localization for Tracking and Scrutinizing the Aquatic Sensory Nodes

Pushpendra R. Verma, D.P. Singh and R.H. Goudar

Abstract Underwater sensor network (UWSN's) localizations play a very crucial role in order to monitor or to track the underwater critters or any kind of underwater application. Many different methods for terrestrial sensor network localization have been developed, but for underwater sensor localization, it is still having many big challenges. The main reason is RF signals, as in underwater it will propagate very less. This paper deals with the design challenges that are faced by the UWSN localization, in which the locations of the unknown sensor node will be estimated by using the acoustic signals that is sent by the sensor nodes and after locating the UWSN. It will be then helpful for tracking and monitoring the underwater critters like fishes which is one of the major sources for countries economy. Hence, this UWSN localization will help the fisherman's to find the location of the aquatic creatures in well advanced and as a result will take a step forward toward in growing the economy of the country.

Keywords Aquatic critters · Acoustic signals · AUV · 3D localization

1 Introduction

Wireless sensor network made tremendous revolutionary changes in last few years. A large wireless sensor network is used in order to control or to monitor the different applications. As the concept of localization is a sizzling topic in present

P.R. Verma (✉) · D.P. Singh
Department of CSE, Graphic Era University, Dehradun, India
e-mail: pushplife80@gmail.com

D.P. Singh
e-mail: devesh.geu@gmail.com

R.H. Goudar
Department of CNE, Visvesvaraya Technological University, Belgaum, India
e-mail: rhgoudar@gmail.com

L.C. Jain et al. (eds.), *Intelligent Computing, Communication and Devices*,
Advances in Intelligent Systems and Computing 308,
DOI 10.1007/978-81-322-2012-1_71

days, it may be on the terrestrial plane or may be under water. Underwater sensor network (UWSN) is a very complicated task, because as the information or the data are normally gathered from the sea surface or from the coastline, but the problem arises at the time of collecting data from the sea bed [6]. There are many enabled applications for the underwater WSN: Few of them are underwater exploration, military applications such as oceanographic collection of data, and navigational assistance [9] are such enabled applications based on localization of the sensor network.

In this paper, we had proposed an efficient technique for underwater sensor node localization, so that this will help us in locating the direction as well as the habitats of the fishes. We had considered the following postulates for UWSN localizations:

1. All sensors are pressure sensors.
2. The anchor node should be lie on the above of all the sensor nodes which is placed on the sea bed.
3. The anchor node is having a higher acoustic signal strength.
4. All the sensor nodes that are placed should be homogeneous.

2 Literature Review

Underwater sensor localization is one of the tedious tasks, as RF signals get attenuated under water [3]. Therefore, underwater localization is a big challenge. Due to the signal attenuation, there will be a gradual loss in signal propagation. GPS is also not a practicable underwater approach [8]. Reference node or anchor node concept is being used. According to this, locations of the nodes are already known. Localization on the basis of reference node will be classified into two [1]:

Range-free scheme.
Range-based scheme.

Range-free scheme is the one in which it does not require any range or bearing information. DV-Hop, (DHL) density-aware hop-count localization is few that come under this scheme. Whereas, according to the range-based scheme it required range or bearing information, time of arrival (ToA), time difference of arrival (TDoA), and receive signal strength (RSSI) are few of them [9].

According to [5], autonomous underwater vehicle (AUV) coordinates with GPS, then dives into the sea up till the predefined depth, and then locate the sensor nodes. In order to get the updates from the deployed sensor node, AUV sends different messages to communicate with them [2] and deals with underwater localization by using a concept of Dive 'N' Rise (DNR) beacon. The DNR beacon is responsible for getting the GPS coordinates, and while floating down into the water, it will broadcast the information. A (3DUL) 3-D localization algorithm is proposed in [10] for UWSN. It is based on the 2-phase protocol, in which 1st phase is known for ranging and the other is known for projection and dynamic trilateration. The propagation delay is to be estimated by using 3DUL, and distance is calculated by using sound signals. A concept of threefold is used in [7], in which they had considered the range-based location scheme with a trilateration localization method. A concept of projection-based localization is proposed in [4] in which they are relating a projection method to transform the problem of (3D) localization to (2D) localization. And the coordinates of the nodes will find by using three reference nodes.

3 Proposed Scenario

Sensor node underwater localization is a very convoluted task. Therefore, in order to help the fishermen's to locate the habitats of the aquatic creatures, firstly we have to sort out the UWSN localization problem.

We had proposed a scenario that will help us in locating the aquatic sensory nodes. In order to monitor the underwater creatures, we are going to deploy sensors on the sea bed so that to track the regular movement of the fishes. An overview of the working system architecture is shown in Fig. 1. In the working of the system, once all the sensor nodes deployed, they will divide themselves into a small clusters. Afterward, the sensor node which is having highest acoustic signal range will become an anchor node of that cluster. Where the anchor node is responsible for locating each sensor node which is present in that cluster, the information of the sensor nodes from the cluster will then have to be forwarded to the AUV. AUV is well equipped with a navigation system. So that it will be able to correct its coordinates while floating into a high tide. AUV is then responsible for sending the location information to the surface buoys. As surface buoys are GPS-enabled, the coordinates of the surface buoys will be known, and it is then the responsibility of the surface buoys to forward the location information of the sensor nodes to the base station. As RF signals get attenuated under water, the deployed sensor nodes will communicate by using acoustic signals rather than RF signals. The detailed explanation for locating sensor nodes by using acoustic signals is shown in Algorithm A′.

3.1 Problem Formulation

Figure 2 shows the proposed structure for locating the unknown sensor network, in which the anchor node is responsible for locating the unknown sensor nodes and also the belonging cluster. The same information will be then sends to the AUV for forwarding to the surface buoys and then to base station. An algorithm is proposed, according to which the different possible ways are explained to achieve underwater sensor localization.

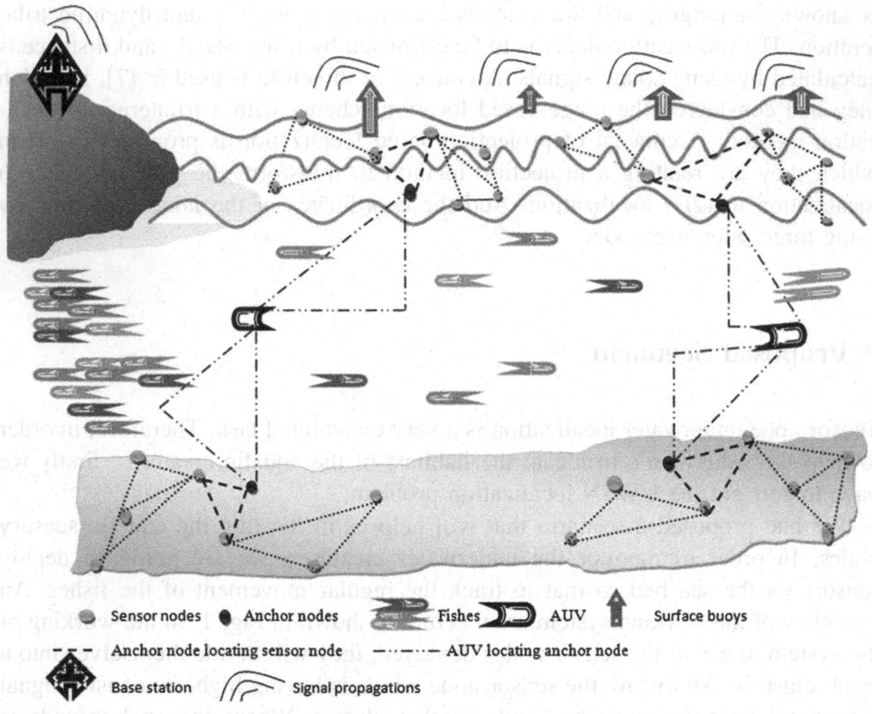

Fig. 1 An overview of the system architecture

Algorithm A:

1. S_n (sensor node) will send the calculated pressure Z^I to AUV.
2. Then AUV calculate the pressure exerted between AUV and point '*P*' and then simultaneously determine the distance between the AUV and *P*, i.e., $P^I = Z^I - Z$.
3. Now, P^I becomes the '*z*' coordinates of S_n in 3-D coordinates.
4. Afterward the *x*-coordinate and *y*-coordinate is then calculated by using the Algorithm A(1).

Algorithm A(1):

According to the Algorithm A(1), 4 case scenarios are discussed, and case 1 and case 2 are discussed by taking sea bed as a reference, whereas sea surface reference is taken in case 3 and case 4 scenarios for locating sensor nodes.

3.1.1 Sensor Node Localization by Taking Sea Bed as a Reference

Case 1 As AUV is GPS-enabled, then the coordinates of 'AUV' will be known, i.e., (*x*, *y*, *z*). In order to find the coordinate of 'S_n' suppose the pressure exists on

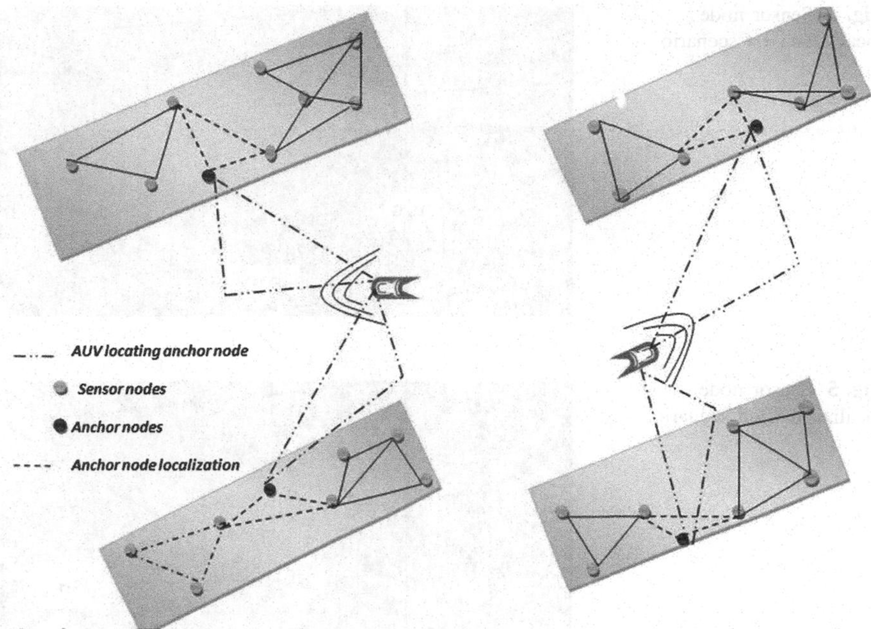

Fig. 2 Proposed localization structure

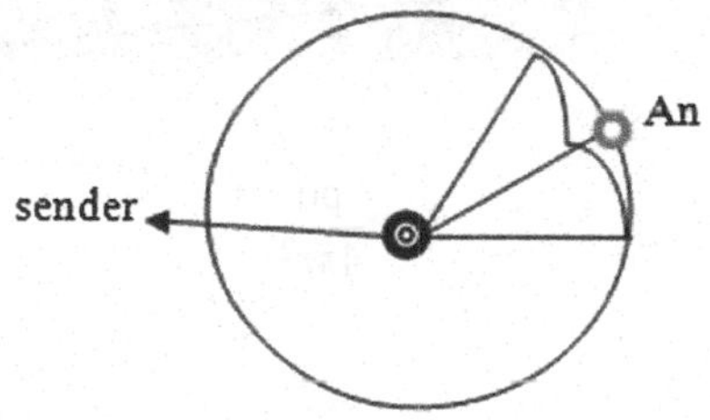

Fig. 3 Range determination for acoustic signal '*r*'

'*P*' with respect to sea surface is 'z^{I}', coordinates of '*P*' will become $[x, y, z^{I} - z]$. Therefore, by using the range of the acoustic signals, it is easier to locate S_n. Consider Fig. 3 for determining the range of acoustic signals send by the sensor nodes.
Where A_n = AUV and 'S_n' = sender (sensor node).
'S_n' sends the acoustic signal which is received by the 'AUV' to locate the sensor node distance.

Let us suppose po = power at which the acoustic signal sent, I = intensity at which acoustic signal received by AUV, r = range of the acoustic signal.

Then,

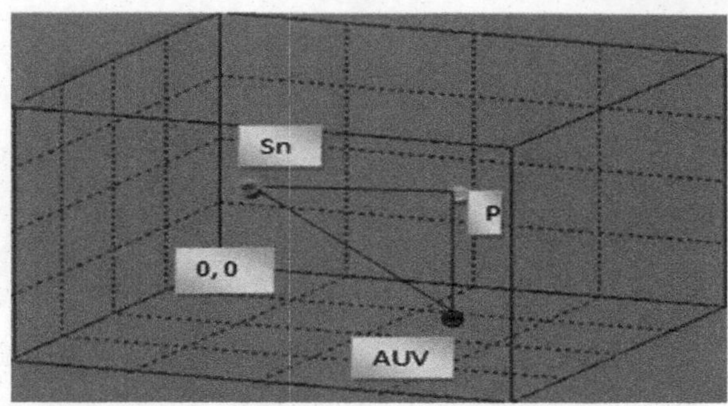

Fig. 4 Sensor node localization 1st scenario

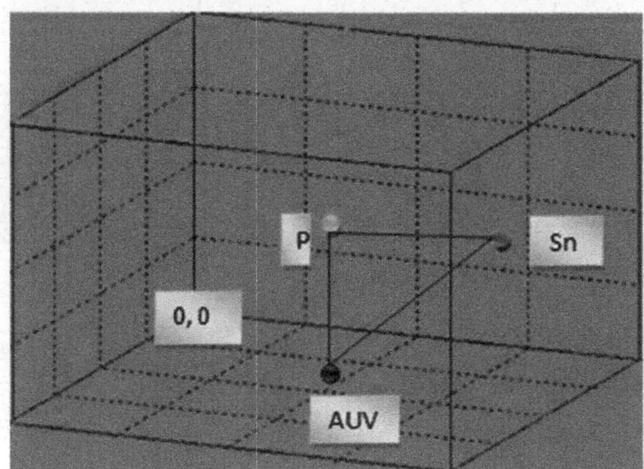

Fig. 5 Sensor node localization 2nd scenario

$$I = \frac{\text{po}}{4\pi r^2} \tag{1}$$

where

$$r = \sqrt{\frac{\text{po}}{4\pi I}} \tag{2}$$

Therefore, the calculated '*r*' is the distance between the AUV and S_n.'*d*' is the distance between '*P*' and AUV. Hence, by using '*r*' and '*d*', we can determine '*q*', the distance between S_n and '*p*'.

Then, $q^2 = r^2 - d^2$

$$q = \sqrt{r^2 - d^2} \tag{3}$$

Hence, once all the distances are known, then it is easier to locate S_n.

Therefore, the 2-D coordinates of the 'S_n' will be [*r*, *q*]. According to the Fig. 4, *x*-coordinate will be the same as it lies on *x*-axis, whereas '*q*' is in negative of

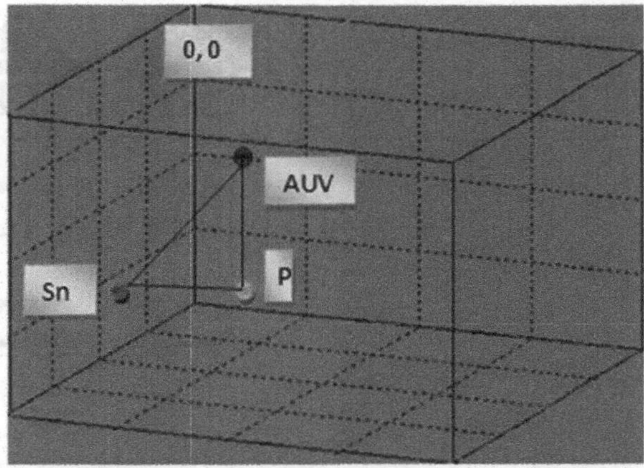

Fig. 6 Sensor node localization 3rd scenario

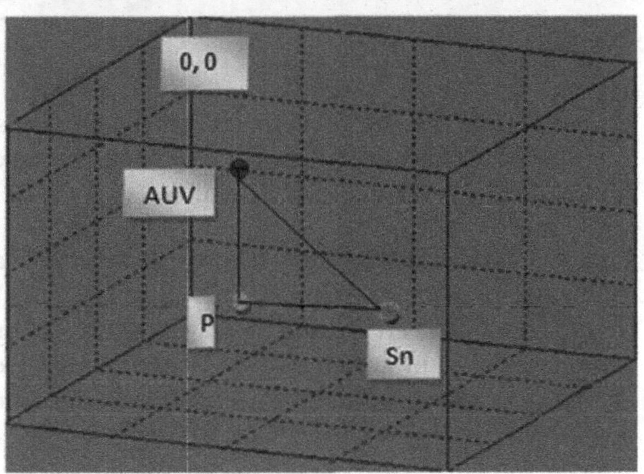

Fig. 7 Sensor node localization 4th scenario

y-coordinate; therefore, *y*-coordinate will become '$y - q$' and the *z*-coordinate will be '$z^{I} - z$'. Hence, the overall 3-D coordinate of S_n will be $(x, y - q, z^{I} - z)$.

Case 2 According to case-2, sea bed is taken to be as a reference to locate the sensor node.

2-D coordinates of S_n will be $[r, q]$, whereas for 3-D coordinates, *x*-coordinates will be same as it lies on *x*-axis, '*q*' is in positive of *y*-axis; therefore, the '*y*' coordinate is '$y + q$' and *z*-coordinate is '$z^{II} - z$'. Hence, the overall 3-D coordinate of S_n will be $[x, y + q, z^{II} - z]$ (Fig. 5).

3.1.2 Sensor Node Localization by Taking Sea Surface as a Reference

Case 3 According to case-3, As 'AUV' is GPS-enabled, then the coordinates will be known, i.e., (x, y, z). Suppose the pressure exists on '*P*' with respect to sea

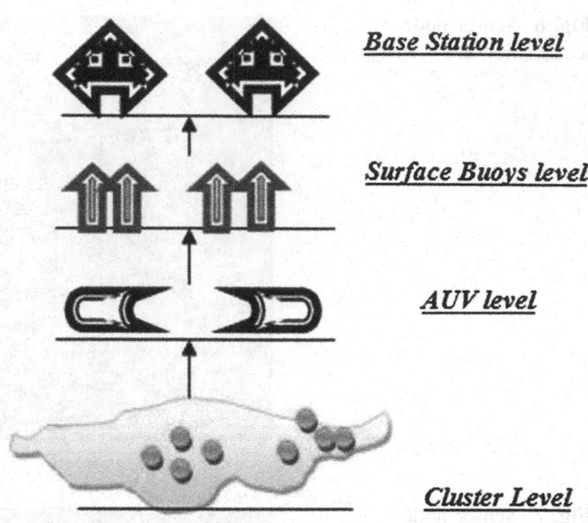

Fig. 8 Different level at which the power is required

surface will be (z^{I}). As x-coordinate will lie on x-axis and 'q' is in negative of y-axis, the y-coordinate will be '$-y-q$' and the z-coordinate will be '$z^{III} - z$'. Hence, the 3-D coordinate of S_n will be $[x, -y - q, z^{III} - z]$ (Fig. 6).

Case 4 Case-4 is discussed by taking the sea surface as a reference, and according to that, the pressure exerted on 'S_n' will be (z^{IV}). As 2-D coordinate of S_n will be $[r, q]$ and the 3-D coordinate of S_n will be as x-coordinate lies on x-axis, for y-coordinate 'q' is in negative of 'y'; therefore, the y-coordinate will be '$-y + q$' and the z-coordinate will be '$z^{IV} - z$'. Hence, the overall 3-D coordinate of S_n will be $[x, -y + q, z^{IV} - z]$ (Fig. 7).

All the different cases for locating the unknown sensor nodes are discussed in the above scenario. Once all the deployed sensor nodes are localized, then it will start scrutinizing the aquatic creatures and then send the collected data to the relevant base stations which will help the fishermen to locate the habitats of the aquatic creatures.

4 Power Consumption at Different Levels

This section deals with the power consumption for the proposed case scenario which shows the different level at which power is consumed for locating the sensor nodes (Fig. 8).

Suppose there is 'n' number of sensor nodes deployed where 'A_n' be the no. of anchor nodes present in each cluster and 'S' be the no. of messages send by the sensor nodes and receive by the anchor nodes in different clusters. Each node is sending and receiving a message of 'b' bits. The energy required for transmitting the data will be E_t and for receiving the data will be E_r. 'A_e' be the energy required for aggregating the data. Therefore,

$$E_t(b, d1) = E_t 1 * A_e * d2 \quad \text{and} \quad E_r(b, d1) = E_r 1 * d2 \tag{1}$$

4.1 Power Required at Different Levels

4.1.1 At Cluster Level

Consider the distance '$d1$' at which the sensor nodes send the data (location information) to anchor node. '$A_e 1$' is the energy for data aggregation. Energy required for receiving data from one S_n deployed on sea bed:

$$S * E_r(b, d1) \tag{2}$$

Energy for data transmission from anchor node to the AUV:

$$S * E_t(b, d1) * A_e 1 \tag{3}$$

Total power for sending and receiving the information from one cluster within the cluster level:

$$P_{CL} = S * E_t(b, d1) * A_e 1 + S * E_r(b, d1) \tag{4}$$

4.1.2 At AUV Level

Considered the distance '$d2$' between the anchor node and the AUV and '$A_e 2$' be the data aggregation energy. 'n' be the number of anchor nodes.

Similarly, the power required by the AUV to receive the data from the cluster level:

$$E_r = [n(S * E_r(b, d1) * A_e 1)] * d2 \tag{5}$$

The energy required for transmission of the same localized information:

$$E_t = [[n(S * E_t(b, d1) * A_e 1)] * A_e 2] * d3 \tag{6}$$

where $A_e 2$ = data aggregation by AUV, $d3$ = distance at which the aggregated data is send. Hence, the overall power required by one AUV at the AUV level for receiving and sending the information:

$$P_{\mathrm{AUVL}} = [(n(S * E_{\mathrm{r}}(b, d1) * A_e 1)) * d2 + [n * [(S * E_{\mathrm{t}}(b, d1) * A_e 1)A_e 2] * d3] \tag{7}$$

4.1.3 At Surface Buoys Level

Considered the distance between the AUV and the Surface buoys will be '$d3$', data aggregation energy at Surface buoys level will be '$A_e 3$', and '$n2$' will be the total number of AUV's. The energy required for receiving data from the AUV will be:

$$E_{\mathrm{r}} = [n2(n(S * E_{\mathrm{r}}(b, d1) * A_e 1)]A_e 2 * d3 \tag{8}$$

The energy required for data transmission from surface buoys to the base station will be:

$$E_{\mathrm{t}} = [n2(n(S * E_{\mathrm{t}}(b, d1) * A_e 1) * A_e 2 * d3)] * A_e 3 * d4 \tag{9}$$

where $d4$ = distance between a surface buoys and base station. Hence, the overall power required by the one surface buoys at surface buoys level will be:

$$P_{\mathrm{SBL}} = [n2(n(S * E_{\mathrm{r}}(b, d1)A_e 1) * A_e 2 * d3) + (n2(S * E_t(b, d1) * A_e 1)A_e 2 * d3)A_e 3 * d4 \tag{10}$$

4.1.4 At Base Station Level

Finally, the location information has to be received by the base station, and the power required by the base station to receive the location information will be:

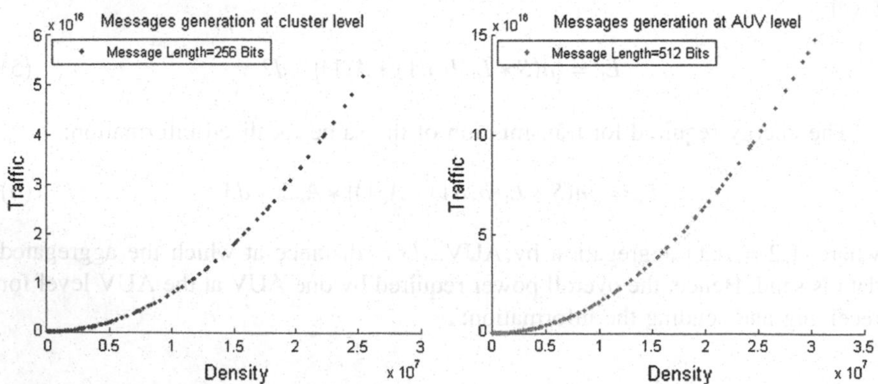

Fig. 9 Generation of messages at cluster level and at AUV level

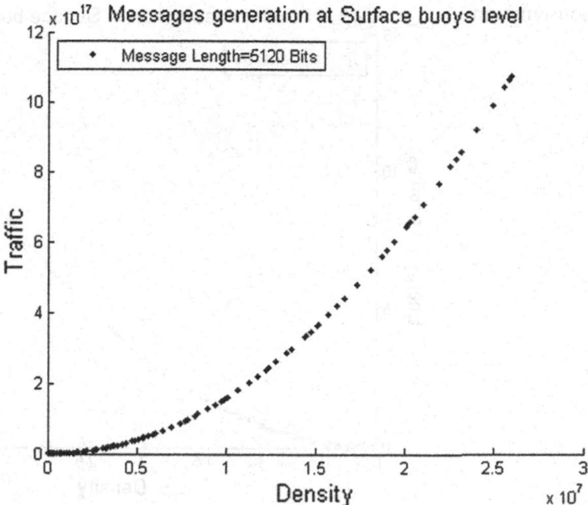

Fig. 10 Generation of messages at surface buoys level

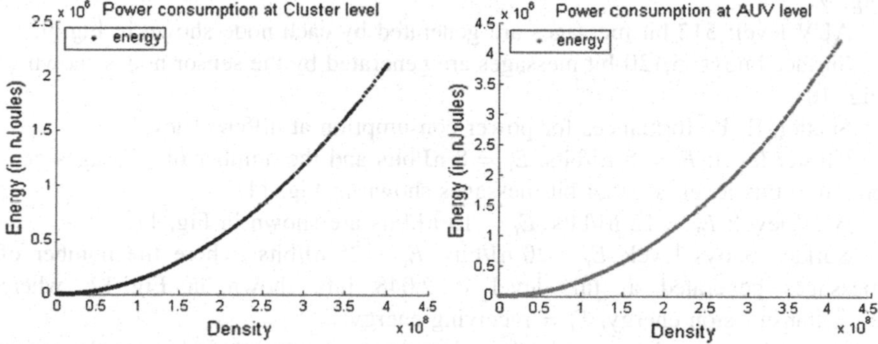

Fig. 11 Power consumption performance at cluster level and at AUV level

$$P_{\mathrm{BSL}} = n3[n2(n(S * E_{\mathrm{r}}(b, d1) * A_e1) * A_e2 * d3) * A_e3 * d4] \tag{11}$$

Hence, the Eqs. (4), (7), (10), and (11) show the overall power performance in order to localize the unknown sensor network.

5 Simulations

Power consumption and the traffic generated by the messages at each level from the sensor nodes are important aspects in our proposed system.

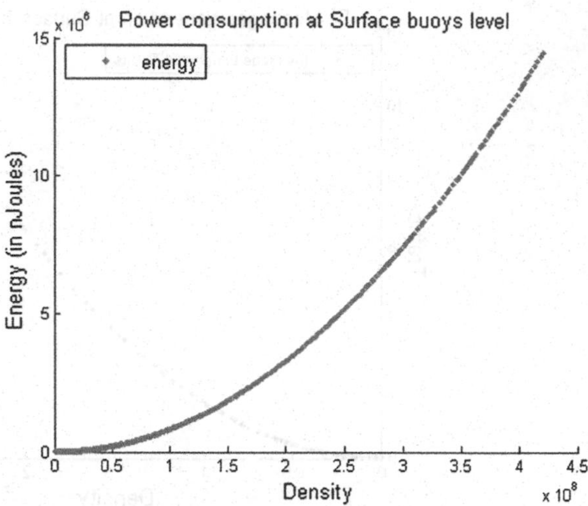

Fig. 12 Power consumption for surface level

Section-I: traffic generation by the messages

Cluster level: 256-bit messages are generated by the sensor nodes shown in Fig. 9.

AUV level: 512-bit messages are generated by each node shown in Fig. 9.

Surface buoys: 5,120-bit messages are generated by the sensor nodes shown in Fig. 10.

Section-II: Performances for power consumption at different levels.

Cluster level: $E_t = 8$ nJ/bits, $E_r = 8$ nJ/bits and the number of messages generated at this level is 1,024-bit messages shown in Fig. 11.

AUV level: $E_t = 12$ nJ/bits, $E_r = 15$ nJ/bits are shown in Fig. 11.

Surface buoys level: $E_t = 20$ nJ/bits, $E_r = 25$ nJ/bits, where the number of messages generated at this level is 2,048 bits shown in Fig. 12, where $E_t =$ transmission energy, $E_r =$ receiving energy.

Hence, from the above results, it is clear that as the density increases, the power consumption by the nodes is also increased as well it will lead to increase in the message trafficking. Therefore, a sleep and listen concept is used in our proposed system. So that instead of putting all the sensor nodes in active state, we can put some of the nodes in the sleep mode and the remaining nodes are in the listen mode and vice versa. Therefore, this will help the sensor nodes to work for a longer time and thus will lead to achieve a better cluster life.

6 Conclusion

In this paper, we had discussed the suitable structures which are more efficient for underwater sensor localization. An algorithm is proposed with different case studies for sensor node localization. And the simulated results show the performance of the power consumption and the message trafficking at each level. Moreover, our basic view to locate the underwater unknown sensor nodes is achieved, and thus, this study will definitely help those peoples who all are in commerce of the aquaculture in order to increase the country's production and the economy.

Future research direction: for future research work, we will try to monitor the temperature and the habitat for the underwater critters by using the seabed sensor localization in support of the concept of drift velocity.

References

1. Erol, M., Vieira, L.F.M., Gerla, M.: AUV-aided localization for underwater sensor networks. In: Wireless Algorithms, Systems and Applications 2007, pp. 44–54. IEEE, Chicago, IL (2007)
2. Erol, M., Vieira, L.F.M., Gerla, M.: Localization with Dive'N'Rise (DNR) beacons for underwater acoustic sensor networks. In: WuWNet'07 Proceedings of the Second Workshop on Underwater Networks, pp. 97–100. ACM, New York, USA (2007)
3. Han, G., Jiang, J., Shu, L., Xu, Y., Wang, F.: Localization algorithms of underwater wireless sensor networks: a survey. In: Sensors 2012, pp. 2026–2061 (2012)
4. Kurniawan, A., Ferng, H.-W.: Projection-based localization for underwater sensor networks with consideration of layers. In: TENCON Spring Conference, 2013 IEEE, pp. 425–429. IEEE, Sydney, NSW (2013)
5. Mohan, V., Mithun, T.P.: Virtual implementation of underwater wireless sensor networks and evaluation of its localization techniques. Int. J. Eng. Res. Appl. (IJERA) 1850–1856 (2012)
6. Naik, S.S., Nene, M.J.: Self organizing localization algorithm for large scale underwater sensor network. In: Recent Advances in Computing and Software Systems (RACSS), pp. 207–213. IEEE, Chennai (2012)
7. Ren, Y., Yu, N., Guo, X., Wan, J.: Cube-scan-based three dimensional localization for large-scale underwater wireless sensor networks. In: Systems Conference (SysCon), pp. 1–6. IEEE, Vancouver, BC (2012)
8. Waldmeyer, M., Tan, H.-P., Seah, W.K.G.: Multi-stage AUV-aided localization for underwater wireless sensor networks. In: Advanced Information Networking and Applications (WAINA), pp. 908–913. Biopolis (2011)
9. Yan, Y.-S., Wang, H.-Y., Shen, X.-H., Yang, F.-Z., Chen, Z.: Efficient convex optimization method for underwater passive source localization based on RSS with WSN. In: Signal Processing, Communication and Computing (ICSPCC), pp. 171–174. IEEE, Hong Kong (2012)
10. Zhou, Y., Chen, K., He, J., Chen, J., Liang, A.: A hierarchical localization scheme for large scale underwater wireless sensor networks. In: High Performance Computing and Communications, 2009, HPCC '09, pp. 470–475. Seoul (2009)

Off-line Handwritten Script Identification from Eastern Indian Document Images Using Logistic Model Tree

Sk Md Obaidullah, Nibaran Das and Kaushik Roy

Abstract Script identification from document images is a complex real-life problem for a multi-script country like India where 13 official scripts are present. To develop an optical character recognizer for a specific language, it is necessary to identify the script first by which the document is written. In this paper, scripts from the off-line handwritten document images written by any one of the four popular scripts in eastern India, namely Bangla, Roman, Devanagari, and Oriya, are identified. A document-level approach is followed for the same. Using some mathematical, structural, and script-dependent feature, a multi-dimensional feature set is constructed. Finally, logistic model tree (LMT) is applied for classification and an average accuracy rate of 95.5 % is obtained with a fivefold cross-validation.

Keywords Document image analysis · Handwritten script identification · Off-line documents · Classification · Optical character recognizer

S.M. Obaidullah (✉)
Department of Computer Science and Engineering, Aliah University, Kolkata, West Bengal, India
e-mail: sk.obaidullah@gmail.com

N. Das
Department of Computer Science and Engineering, Jadavpur University, Kolkata, West Bengal, India
e-mail: nibaran@gmail.com

K. Roy
Department of Computer Science, West Bengal State University, Barasat, West Bengal, India
e-mail: kaushik.mrg@gmail.com

L.C. Jain et al. (eds.), *Intelligent Computing, Communication and Devices*,
Advances in Intelligent Systems and Computing 308,
DOI 10.1007/978-81-322-2012-1_72

1 Introduction

Optical character recognition is an active area of research since many years. It is useful for converting the physical document into digital form for making a paperless world in future. Document digitization also helps for better indexing and retrieval of huge volume of data available in modern society. The work is more relevant for a multilingual and multi-script country like India where 13 different scripts including Roman and 23 different languages including English [4] are present. There are many languages which use same script for writing. As an example, Bangla is a popular script in the eastern part of India which is used to write Bangla, Assamese, and Manipuri languages, whereas Devnagari is a popular script which is used to write different languages such as Hindi, Marathi, Nepali, and Konkani. So, here, it is not possible to develop a general purpose optical character recognizer targeting a particular language. Before feeding the particular language to the optical character recognizer, script needs to be identified first. That is why development of a script identification system is an essential requirement. Another problem arises when a single document is written using multiple scripts. Postal documents, filled up preprinted application forms, commercial advertisement documents, etc., are example of such multi-script documents. In these cases, word-level, line-level, or block-level script identification is must before choosing language-specific optical character recognizer.

Script identification can be classified into two broad categories, namely printed script identification and handwritten script identification. Handwritten script identification can be classified into two categories, namely off-line script identification and online script identification. Few works are reported in literature on script identification based on Indic scripts and non-Indic scripts. Among the pieces of work, Zhou et al. [14] identified Bangla and English printed and handwritten scripts using connected component profile-based features. Singhal et al. [12] identified Roman, Devanagari, Bangla, and Telugu scripts from handwritten document images with the help of rotation invariant texture features using multichannel Gabor filter and graylevel co-occurrence matrix. Hochberg et al. [3] identified six scripts, namely Arabic, Chinese, Cyrillic, Devnagari, Japanese, and Latin, using some features such as horizontal and vertical centroid, sphericity, aspect ratio, and white holes. They performed the work at document level. In another work, Roy et al. [10] identified six popular Indian scripts, namely Bangla, Devnagari, Malayalam, Urdu, Oriya, and Roman, using features such as component-based features, fractal dimension-based features, and circularity-based features. This is a first kind of work involving six Indian scripts altogether. In a block-level script identification technique, Basu et al. [1] identified Latin, Devnagari, Bangla, and Urdu handwritten numeral scripts using similar-shaped digit pattern-based features. Using fractal-based features, Moussa et al. [8] identified Arabic and Latin scripts from line-level handwritten document.

Figure 1 shows block diagram of a multi-script document processing system. In Fig. 2, different multi-script documents are shown. The paper is organized as

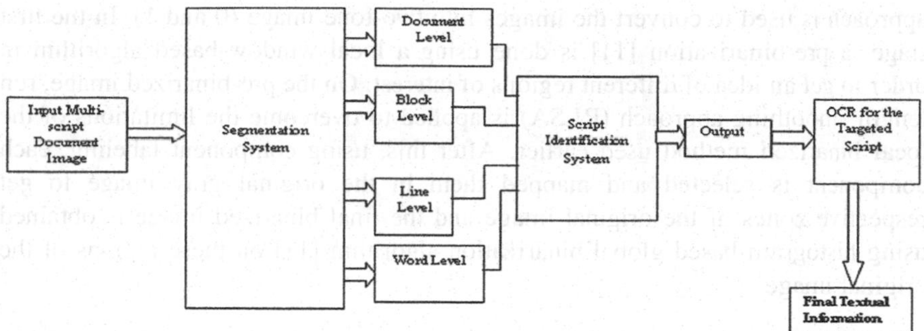

Fig. 1 Block diagram of multi-script document processing system

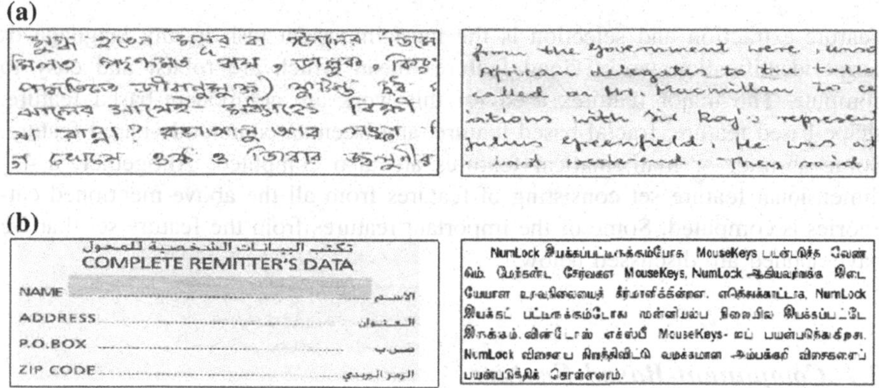

Fig. 2 Different multi-script documents. **a** Different document written by different scripts. **b** Same document written by different scripts

follows: In Sect. 2, data collection and preprocessing are described. In Sect. 3, feature extraction techniques are discussed, and the classification procedure with experimental result is described in Sect. 4. Finally, conclusion and scope of future works are described in Sect. 5. References are available in the last section.

2 Data Collection and Preprocessing

One of the major challenges in language and script identification work is absence of standard database. For this work, data are collected from different sources such as university and post office. From outside states, some data are collected through friends and different connections of the authors. Altogether, 32 Bangla, 32 Roman, 30 Devnagari, and 32 Oriya handwritten document pages are considered. Originally, the images are in gray tone and digitized at 300 dpi. A two-stage-based approach is used to convert the images into two-tone image (0 and 1). In the first stage, a pre-binarization [11] is done using a local window-based algorithm in order to get an idea of different regions of interest. On the pre-binarized image, run length smoothing approach (RLSA) is applied to overcome the limitations of the local binarized method used earlier. After this, using component labeling, each component is selected and mapped them in the original gray image to get respective zones of the original image and the final binarized image is obtained using histogram-based global binarization algorithm [11] on these regions of the original image.

3 Feature Extraction and Selection

Feature extraction and selection is the most important task in any language or script identification work. Good features mean which are robust and easy to compute. The major features used for this work are component-based feature, shape-based feature, fractal-based feature, and freeman chain-code-based feature. Some abstract or mathematical features are also computed. Altogether, a 41-dimensional feature set consisting of features from all the above-mentioned categories is computed. Some of the important features from the feature set that we have applied are discussed below.

3.1 Component-Based Feature

Component analysis is one of the most useful and widely used tools in image processing. Here, using component analysis, we have classified all the components into three categories, namely (i) large component, (ii) medium component, and (iii) small component. An experimental threshold value is assumed for categorizing the components. For example, to calculate small component, the threshold value is assumed to be five pixels. Dots and comma characters fall under this category.

3.2 Shape-Based Feature

Under shape-based feature, occurrence of circularity at component level is calculated in a particular script. Following are the steps followed:

- Minimum enclosing circle is drawn which will enclose the component minimally, and the radius (r_1) of the circle is being stored.

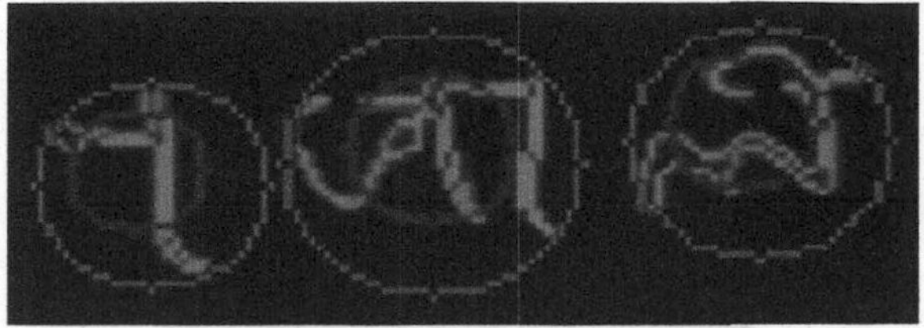

Fig. 3 Computation of circularity of component on Bangla script using fitted circles (*blue* minimum encapsulating and *red* best fitted)

- Circle fitting is done. Circle fitting refers to the fitting of a circle in the component in as minimum manner as possible. Its radius (r_2) is also stored.
- The difference of the two radii is stored to indicate the circularity of the component. The more the circularity of the component, the lesser will be the difference between the two radii (Fig. 3).

In fact, the circular components will have zero difference between the two radii or will have a difference tending to zero.

3.3 Fractal Dimension-Based Feature

Among the structural features, fractal dimension is one of the most important features. A fractal [7] is defined as a set for which the Hausdorff–Besikovich dimension is strictly larger than the topological dimension. The fractal dimension is a useful method to quantify the complexity of feature details present in an image. The fractal dimension is an important characteristic of the fractals because it contains information about their geometric structures. By employing fractal analysis, researchers typically estimate the dimension from an image. The fractal dimension of continuous object is an entity specified in terms of well-defined mathematical limiting processes. A fractal is an irregular geometric object with an infinite nesting of structure at all scales (self-similarity) (Fig. 4).

The upper part and the lower part play a significant role in feature extraction from the document image. In case of Devnagari script or Bangla script, the upper part will mainly contain matra or shirorekha pixels, whereas the lower part will contain the base pixels of the component. In case of Roman script or Urdu script, there will be no matra or shirorekha. So if pixel density is calculated, there will be difference in pixel density of upper part and lower part of the components of different scripts.

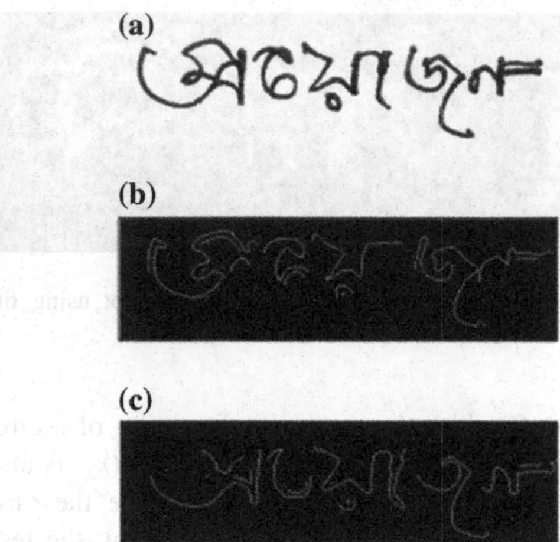

Fig. 4 Fractal dimension-based feature. **a** Original component. **b** *Upper* part of the contour. **c** *Lower* part of the contour

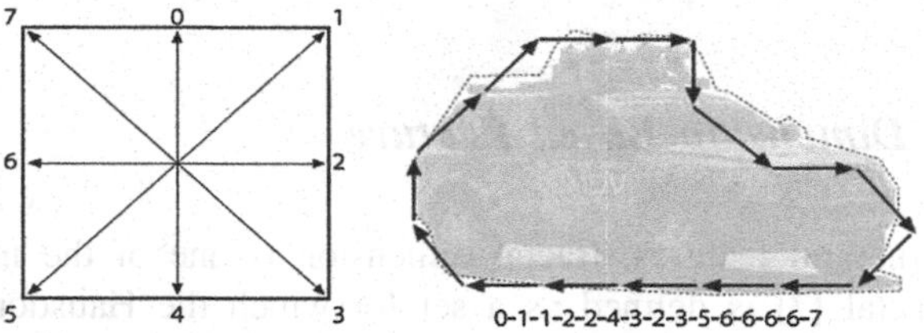

Fig. 5 Freeman chain code [2]

3.4 Feature Based on Freeman Chain Code

In Bangla and Devnagari scripts, horizontal lines present on the upper part of the writing are called 'Matra' or 'Shirorekha.' This is a unique distinguishing feature of these two scripts from the rest. We use cvFindContours() function in OpenCV [5] in CV_CHAIN_CODE mode for identifying these lines as a sequence of integers as shown in Fig. 5. Some slanting line presents in other scripts is also identified by the technique.

4 Classification Using Logistic Model Tree

Based on the above-normalized features, we employed logistic model tree (LMT) classifier under WEKA tool [2] for identification of handwritten Bangla, Roman, Devnagari, and Oriya scripts. WEKA is one of the widely used tools in the area of

Table 1 Confusion matrix

Script name	Bangla	Roman	Devnagari	Oriya	Average accuracy rate (%)
Bangla	96.8	0	3.2	0	95.5
Roman	4.2	95.8	0	0	
Devnagari	8.4	0	91.6	0	
Oriya	3.2	0	0	96.8	

machine learning. It contains tools for various applications such as data preprocessing, classification, clustering, regression, association rules, and visualization.

4.1 Logistic Model Tree Classifier

For present work LMT classifier is used. The model is build using a classification tree with logistic regression function at the leaves. The algorithm can deal with binary and multi-class target variables, numeric and nominal attributes, and missing values. For more detail, refer [6, 9, 13].

5 Result and Discussion

In the experiment a total of 126 document images are used consisting of 32 Bangla, 32 Roman, 30 Devnagari and 32 Oriya scripts. Table 1 shows confusion matrix where Bengali and Oriya obtain highest accuracy rate where as Devnagari obtained lowest among the four. Overall, 95.5 % average accuracy is obtained using LMT classifier with a fivefold cross-validation. In this result, observation is that Devnagari script gives lowest accuracy because of its similarity with Bengali script in some features such as presence of 'matra.' That is why 8.4 % Devnagari scripts are misclassified as Bengali script (Fig. 6).

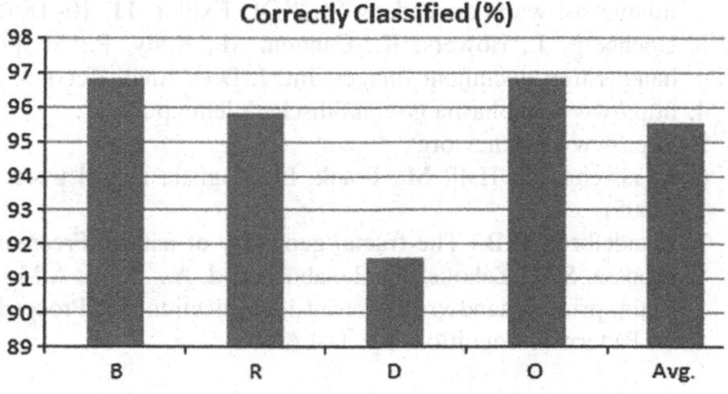

Fig. 6 Correctly classified percentage of all the scripts

Table 2 Comparative study

Name of algorithm	Scripts considered	Average accuracy rate (%)
Hochberg	Arabic, Chinese, Cyrillic, Devnagari, Roman, Japanese	88
M. Hangarge	Roman, Devnagari, Urdu	88.6
L. Zhou	Roman and Bangla	95
Proposed method	Bangla, Roman, Devnagari, Oriya	95.5

Table 2 provides a comparative study with other result available so far in handwritten script identification problems. The proposed method considering four scripts performs considerably well compared to other three available methods.

6 Conclusion

Script identification from four popular eastern Indian scripts in handwritten document images is proposed. Many works are available on printed script identification problem but attention is very less on handwritten script identification category. That is why emphasis needs to be given on the problem of handwritten script identification. So far, all the discussions were restricted to off-line script identification area. Future plan of the authors includes extending the work considering all 13 official Indian scripts and working in the online and video environment for real-life automatic script identification problem.

References

1. Basu, S., Das, N., Sarkar, R., Kundu, M., Nasipuri, M., Basu, D.K.: A novel framework for automatic sorting of postal documents with multi-script address blocks. Pattern Recogn. **43**(10), 3507–3521 (2010)
2. Hall, M., Frank, E., Holmes, G., Pfahringer, B., Reutemann, P., Witten, I.H.: The WEKA data mining software: an update. SIGKDD Explor. **11**, 10–18 (2009)
3. Hochberg, J., Bowers, K., Cannon, M., Kelly, P.: Script and language identification for handwritten document images. Int. J. Doc. Anal. Recogn. **2**(2/3), 45–52 (1999)
4. http://www.rajbhasha.gov.in/8thschedulehin.pdf
5. http://www.opencv.org
6. Landwehr, N., Hall, M., Frank, E.: Logistic model trees. Mach. Learn. **59**(1–2), 161–205 (2005)
7. Mandelbrot, B.B.: The fractal geometry of nature. Freeman, NY (1982)
8. Moussa, S.B., Zahour, A., Benabdelhafid, A., Alimi, A.M.: Fractal-based system for Arabic/Latin, printed/handwritten script identification. In: Proceedings of International Conference on Pattern Recognition, pp. 1–4 (2008)

9. Obaidullah, S.M., Roy, K., Das, N.: Comparison of different classifier for script identification from handwritten document. In: Proceedings of ISPCC 2013 at Shimla (2013)
10. Roy, K., Das, S.K., Obaidullah, S.M.: Script identification from handwritten document. In: Proceedings of the Third National Conference on Computer Vision Pattern Recognition, Image Processing and Graphics, pp. 66–69. Hubli, Karnataka, Dec 2011
11. Roy, K.: On the development of an optical character recognition system for Indian postal automation. PhD thesis, Jadavpur University (2008)
12. Singhal, V., Navin, N., Ghosh, D.: Script-based classification of hand-written text document in a multilingual environment. In: Research Issues in Data Engineering, p. 47 (2003)
13. Sumner, M., Frank, E., Hall, M.: Speeding up logistic model tree induction. In: 9th European Conference on Principles and Practice of Knowledge Discovery in Databases, pp. 675–683 (2005)
14. Zhou, L., Lu, Y., Tan, C.L.: Bangla/english script identification based on analysis of connected component profiles. In: Lecture Notes in Computer Science, 2006, vol. 3872/2006, 24354, doi:10.1007/11669487_22

Segmentation and Comparison of Water Resources in Satellite Images Using Fuzzy-Based Approach

P. Ganesan, V. Rajini, B.S. Sathish and Khamar Basha Shaik

Abstract It is necessary to monitor and control the changes to the river and other water bodies. The images received from satellite are useful for scientists and other officials to observe the changes in the water bodies, and it provides more information for decision making. This paper presents a simple and novel approach for the detection and segmentation of water resources in the satellite images and compares the water level in various years. The 26 years of changes in the Salmon River reservoir is detected and explained using possiblistic fuzzy c means (PFCM) clustering algorithm and threshold method. Experimental result illustrates the efficiency of the proposed approach.

Keywords Segmentation · Sharpening · Threshold · Histogram · PFCM

1 Introduction

Segmentation is the low level but important process which plays important role in the success of the image analysis [1, 2]. In the process, an image is segmented into number of cluster or sub-images based on color for color images or intensity for

P. Ganesan (✉) · B.S. Sathish · K.B. Shaik
Department of Electronics and Control Engineering,
Sathyabama University, Chennai, India
e-mail: gganeshnathan@gmail.com

B.S. Sathish
e-mail: subramanyamsathish@yahoo.co.in

K.B. Shaik
e-mail: rajiniv@ssn.edu.in

V. Rajini
Department of Electrical and Electronics Engineering,
SSN College of Engineering, Chennai, India
e-mail: khamars786@gmail.com

L.C. Jain et al. (eds.), *Intelligent Computing, Communication and Devices*,
Advances in Intelligent Systems and Computing 308,
DOI 10.1007/978-81-322-2012-1_73

grayscale images [3]. In the same segment or cluster, the entire pixels have similar characteristics as compared to the pixels in any other cluster. The monitoring and controlling of water resources is an important and difficult task. Water resources are backbone to the growth of economic development of any country. Water resources are mainly used for the safeguard of environment, production of electricity, irrigation, drinking and sanitary purpose, and transport and industrial applications [4]. So it is necessary to collect the information of level and purity of the water resources. The images gathered from satellite give more detailed information about water resources. Janahiraman and Kong [5] presented a segmentation algorithm based on self-organizing map (SOM) neural network with compression preprocessing by wavelet transform. A method based on hierarchical structure of spectrum and shape features for water extraction is explained in [6]. Kuang and He [7] proposed a novel geodesic active contour model based on an edge detector for rapid detection of water bodies from synthetic aperture radar (SAR) imagery with high speckle noise. Kalaivani [8] proposed support vector machine (SVM)- and morphological operation-based approach to identify the river spot in the satellite image. Zeki [9] described a method to represent water resources in satellite images using Voronoi diagrams (VD). Fuzzy-based segmentation and detection of water resources give more information than manual processing. Even necessary but tiny information is processed with greater accuracy with the help of fuzzy-based approach. In this paper, a simple method of detection and segmentation of water resources in satellite images based on possiblistic c means clustering algorithm is presented. The remainder of the paper is organized as follows. Section 2 explained the methodology used for the segmentation and detection of water resources in satellite images. The experimental results are discussed in detail in Sect. 3. Section 4 concluded the paper.

2 Methodology

In fuzzy-based approach, the degrees of membership are calculated from the distances of the data point to the cluster centers. The point which is very closer to the cluster center has higher degree of membership [10, 11]. Pal et al. proposed a method, called PFCM, for clustering object into number of class satisfying both the constraints of FCM and PCM [12]. The objective of this clustering algorithm is minimizing its objective function as given in (1)

$$PF_m(T, V, U; X, \gamma) = \sum_{i=1}^{n} \sum_{k=1}^{c} (a\mu_{ik}^{m} + bt_{ik}^{\eta}) d_{ki}^{2} + \sum_{i=1}^{c} \gamma_i \sum_{k=1}^{n} (1 - t_{ki})^{\eta} \quad (1)$$

where m >1 and a, b, $\gamma > 0$. All are user-defined constants. The following conditions are necessary for the objective function to reach the final optimum (minimum) value. The cluster centers and membership can be calculated using (2) and (4).

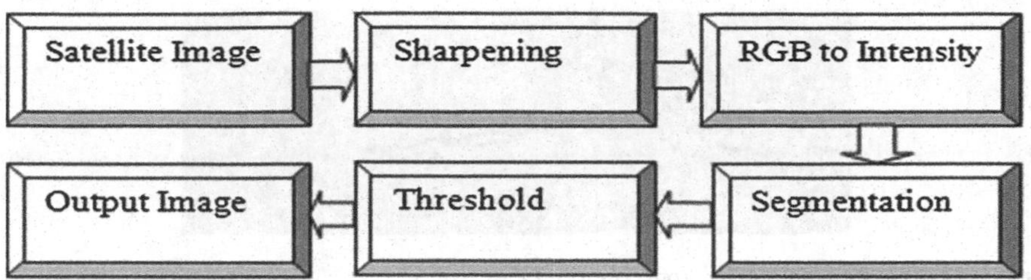

Fig. 1 Proposed approach

$$V_i = \frac{\sum_{k=1}^{n} \left(a\mu_{ik}^{m} + bt_{ik}^{\eta}\right)x_k}{\sum_{k=1}^{n} \left(a\mu_{ik}^{m} + bt_{ik}^{\eta}\right)} \tag{2}$$

$$t_{ik} = \left(1 + \left(\frac{d_{ik}^2}{\gamma_i}\right)^{1/(m-1)}\right)^{-1} \tag{3}$$

$$\mu_{ik} = \left(\sum_{j=1}^{c} \left(\frac{d_{ik}}{d_{jk}}\right)^{2/m-1}\right)^{-1} \tag{4}$$

Figure 1 shows the functional flow diagram for the proposed approach. The images received from satellite are stored in a database after initial processing. A database is a collection of 25 satellite image to test the efficiency of the proposed approach. Preprocessing operation sharpening is performed on the input image to enhance the visibility, i.e., the minute details of edges and lines are more sharpened. RGB to grayscale conversion is necessary to reduce the execution time. For example, RGB image has (255 × 255 × 255) variations, whereas grayscale has only 255 variations. The grayscale image is segmented using PFCM and then threshold by binary threshold method.

3 Experimental Result and Discussion

The image, Salmon River reservoir, shown in Fig. 2 is collected from the NASA's landsat imagery and taken by ETM+ multispectral sensor on July 21, 1985. This image is taken in bands 1, 2, 5, and 7 with spectral range of 0.450–2.35 μm and pixel resolution of 30 m. In 1914, the Salmon River reservoir was created in New York. Landsat satellite data illustrate the solid increase in the size of the reservoir due to heavy rainfall in the region.

Image is sharpened by using spatial domain sharpening algorithm with Laplacian mask. For this, the kernel size of sobel is 3. Low clip percentage and high clip percentage are maintained as 0.01 and 0.02, respectively. The image after

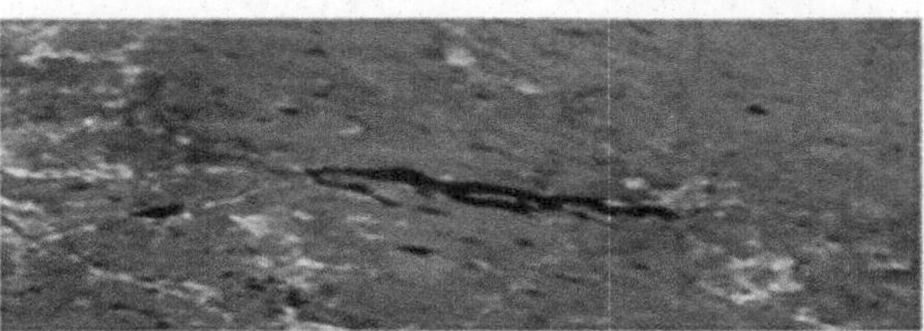

Fig. 2 Landsat satellite image-1985 (EOS GLCF)

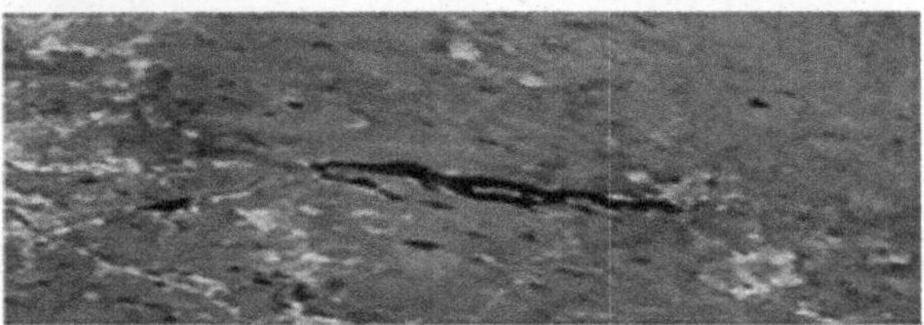

Fig. 3 Image after sharpening

sharpening is illustrated in Fig. 3. The grayscale image is segmented using PFCM and then threshold by binary threshold method. Figure 4 shows the intensity version of the RGB color image.

Possiblistic fuzzy C means (PFCM) clustering is performed on the grayscale image. Image is split into number of segment or cluster based on the gray level. This is shown in Fig. 5. Finally, binary threshold is applied on segmented image.

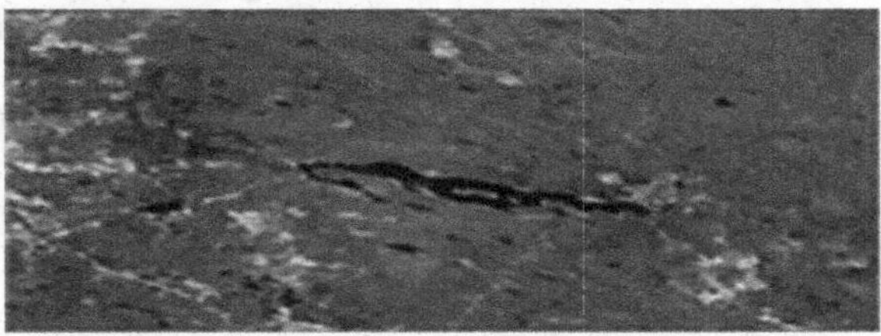

Fig. 4 Intensity version of input image

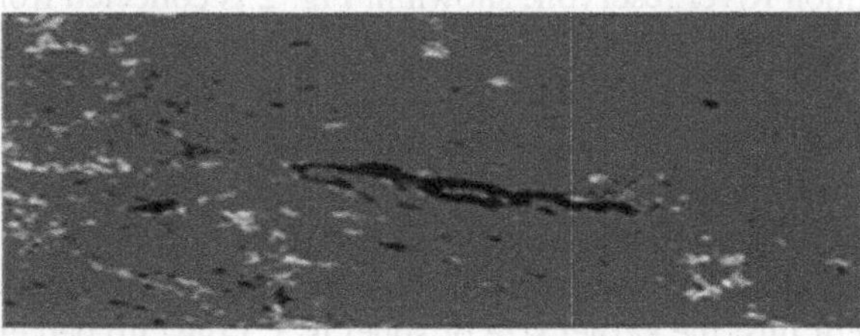

Fig. 5 Image after segmentation process

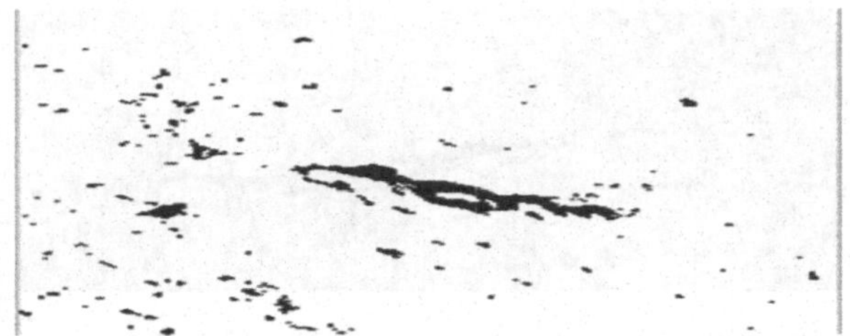

Fig. 6 The output image after threshold operation

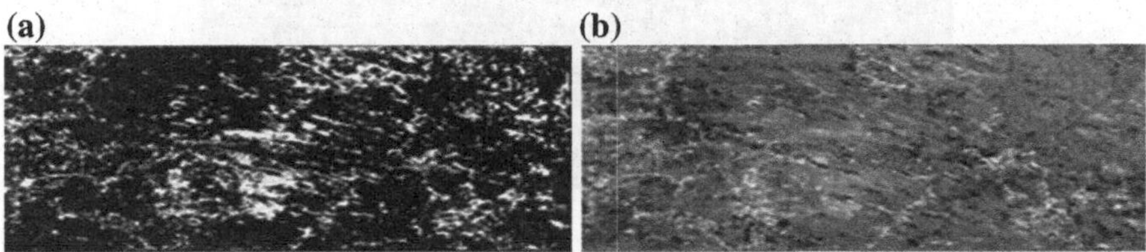

Fig. 7 Comparison of intensity and its segmented image by **a** exclusive OR and **b** subtraction operation

Threshold for this process is 80. The output image after threshold is shown in Fig. 6. The image contains only dark patches (water bodies).

The amount of segmentation performed is determined by comparison with intensity and segmented image by either exclusive OR operation or subtraction operation. This is shown in Fig. 7a and b, respectively.

The image shown in Fig. 8 is taken by landsat ETM+ multispectral sensor on July 05, 2011 to access the environmental and water-level changes in the reservoir. The same procedure is applied for this image to obtain the final output image (Figs. 9, 10, and 11).

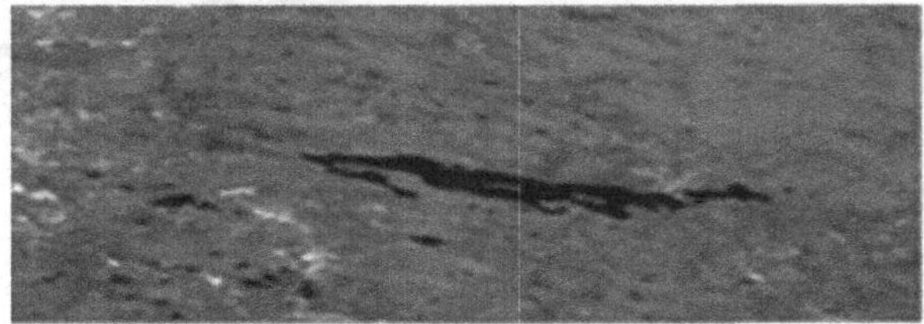

Fig. 8 Landsat satellite image-2011. (EOS GLCF)

Fig. 9 Image after sharpening

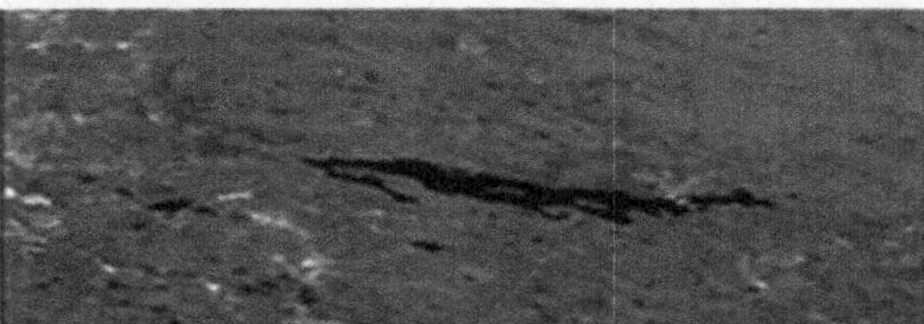

Fig. 10 Intensity version of input image

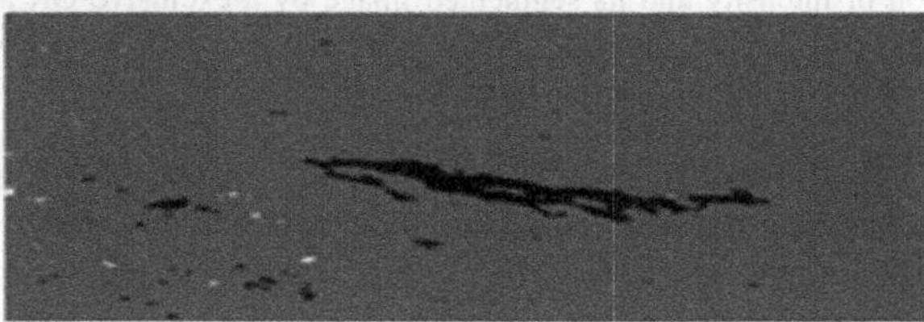

Fig. 11 Image after segmentation process

Fig. 12 The output image after threshold operation

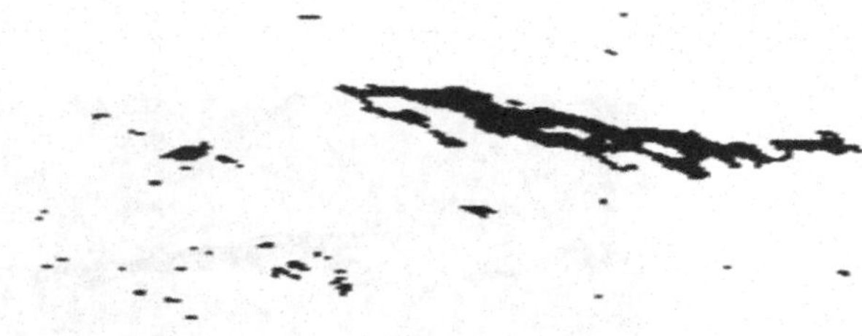

The changes in the water reservoir in this period (1985–2011) can be easily obtained by comparing two output images (Figs. 6 and 12) using exclusive OR and subtraction process. This is shown in Fig. 14. The white patches and lines indicate changes in the level of water (Fig. 13; Table 1).

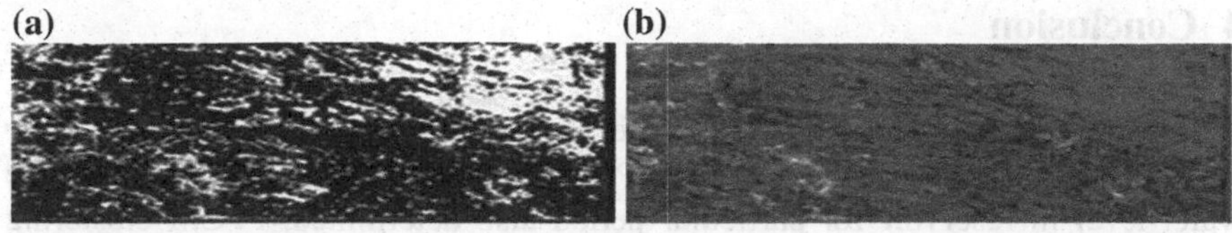

Fig. 13 Comparison of intensity and its segmented image by **a** exclusive OR and **b** subtraction operation

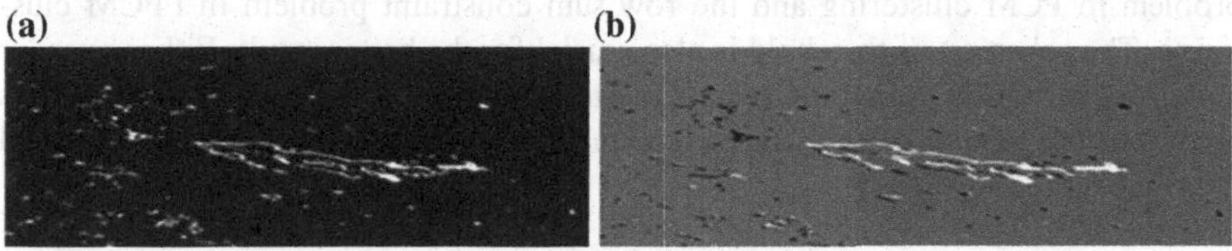

Fig. 14 Comparison of segmented images (Figs. 6 and 12) by **a** exclusive OR and **b** subtraction operation

Table 1 Comparison of segmentation and threshold result

Parameter	Intensity and segmented image (1985)	Intensity and segmented image (2011)	Threshold image 1 (1985) and image 2 (2011)
Centroid row	99	99	101
Centroid column	102	100	101
Orientation	−23	85	86
Perimeter	907	798	958
Euler number	−352	−56	−45
Thinness	0.575443	0.7682	0.550269
Aspect ratio	0.9950	1.0000	1.0000
Histogram mean	116.6980	120.0334	248.1980
Histogram STD	116.9156	17.6723	32.6013
Histogram skewness	0.002924	0.2888	−7.2817
Histogram entropy	6.067189	0.019908	0.1358
Texture energy	0.000492	0.02673	1.0000
Correlation average	0.7845	0.4312	0.0000

4 Conclusion

This paper presented a method for the segmentation and detection of water resources in a satellite image based on PFCM and threshold. The changes in the water level in reservoir for particular period also determined. PFCM clustering avoids various shortcomings of both FCM and PCM. The main disadvantage of FCM is that the less noise sensitivity. This means that FCM clustering algorithm considers noise pixels as image pixels. PFCM resolves this noise-sensitivity problem of FCM. Moreover, PFCM gives answer to the coincident clusters problem in PCM clustering and the row sum constraint problem in FPCM clustering. The selection of threshold is also crucial for the better result. Either over- or under-segmentation or incorrect value of threshold leads to undesirable result. This algorithm can be improved by the selection of best pixels by genetic algorithm and then segmented by PFCM.

References

1. Correa, C., et al.: A Comparison of Fuzzy Clustering Algorithms Applied to Feature Extraction on Vineyard. Lecture Notes in Computer Science. Springer, Berlin (2012)
2. Shamsi, H., Seyadarabi, H.: A modified fuzzy c-means clustering with spatial information for image segmentation. Int. J. Comput. Theory Eng. **4**(5), 762–766 (2012)
3. Ganesan, P., Rajini, V.: A method to segment color images based on modified fuzzy-possibilistic-c-means clustering algorithm. In: Recent Advances in Space Technology Services and Climate Change (RSTSCC), IEEE (2010)
4. Nath, R.K., Deb, S.K.: Water-body area extraction from high resolution satellite images-an introduction, review, and comparison. Int. J. Image Process. (IJIP) **3**(6), 353–372 (2009)
5. Janahiraman, T.V., Kong, W.: SOM based segmentation method to identify water region in LANDSAT images. IJECCT **2**(1), 13–18 (2011)
6. Li, B., Zhang, H., Xu, F.: Water extraction in high resolution remote sensing image based on hierarchical spectrum and shape features. In: 35th International Symposium on Remote Sensing of Environment (ISRSE35), IOP Conference Series: Earth and Environmental Science, vol. 17, p. 012123 (2014)
7. Kuang, G., He, Z.: Detecting water bodies on Radarsat imagery. GEOMATICA **65**(1), 15–25 (2011)
8. Kalaivani, R.: River water level prediction in satellite images using support vector machine. IJCA **43**(23) (2012)
9. Zeki, A.M.: Representation of water resources in satellite images using voronoi diagrams. In: The 2nd International Conference on Water Resources & Arid Environment (2006)
10. Krishnapuram, R., Keller, J.: The possibilistic c means algorithm: insights and recommendations. IEEE Trans. Fuzzy Syst. **1**, 385–393 (1996)
11. Zhang, J.S., Leung, Y.W.: Improved possibilistic c means clustering algorithm. IEEE Trans. Fuzzy Syst. **12**(2), 209–218 (2004)
12. Pal, N.R., Pal, K., Bezdek, J.C.: A possibilistic fuzzy c means clustering algorithm. IEEE Trans. Fuzzy Syst. **13**(4), 517–530 (2000)

Comparison between ANN-Based Heart Stroke Classifiers Using Varied Folds Data Set Cross-Validation

H.S. Niranjana Murthy and M. Meenakshi

Abstract This paper presents the design and development of an artificial neural network (ANN) model for the classification of heart stroke disease. The novelty of this work is training multilayer perceptron (MLP) neural network architectures with back-propagation algorithm for multivariate large data sets. Subsequently, the performances of the ANN models are evaluated using database of heart stroke obtained from Cleveland Clinic Foundation Database with all attributes are numeric-valued. The accuracy of the designed ANN models are cross-validated using varied folds of data set comprising 303 instances extracted from different age groups. This study exhibits ANN-based prognosis for early detection of level of heart stroke with testing classification accuracy of 85.55 %.

Keywords Artificial neural network · Multilayer perceptron · Coronary heart disease

1 Introduction

In general, the term heart disease comprises of different diseases that affect the heart. In 2007, heart disease was the main cause of deaths in the USA, England, Canada, and Wales. Heart disease kills one person for every 34 s in the USA [1]. It has been noticed from the estimate of World Health Organization that 12 million deaths occur worldwide, every year due to the coronary heart diseases (CHD). According to the statistics, the annual operational cost of coronary heart disease exceeded to $300 billion in 2010 in the USA [2]. The principal reason for heart stroke is contraction of the coronary artery due to CHD, resulting in the reduction

H.S. Niranjana Murthy (✉)
M.S. Ramaiah Institute of Technology, Bengaluru 560054, India
e-mail: hasnimurthy@rediffmail.com

M. Meenakshi
Dr. Ambedkar Institute of Technology, Bengaluru 560056, India
e-mail: meenakshi_mbhat@yahoo.com

L.C. Jain et al. (eds.), *Intelligent Computing, Communication and Devices*,
Advances in Intelligent Systems and Computing 308,
DOI 10.1007/978-81-322-2012-1_74

of blood and oxygen supply to the heart. Heart attack or heart stroke refers to an interruption of blood supply to the heart due to sudden blockage of a coronary artery which occurs due to a blood clot. Chest pains occur when the blood received by the heart muscles is insufficient [3].

The risk factors for CHD include blood pressure, cigarette smoking, cholesterol (total cholesterol), LDL-C, HDL-C, and diabetes. Several years of follow-up of the risk factors based on blood pressure, smoking history, and total cholesterol (TC), HDL-C levels, diabetes, and left ventricular hypertrophy on the ECG has facilitated the diagnosis of CHD [4]. To overcome the deaths due to CHD, there are many algorithms developed in the literature [5]. One among them is artificial neural network (ANN) which is widely used.

ANN-based decision support in medicine plays an important role in improving the reliability of health care of general population. This is because, it has the capability to detect unusual abnormal conditions which cannot be identified directly by clinician with available information bank. In some cases, ANN model-based diagnoses have been proved to be even more accurate than those by clinicians. This opens a significant platform for the research on development of an automated system for diagnosis of CHD.

The main drawback of earlier works is reduced testing accuracy of classification. Very little effort is directed toward the development of ANN model for predicting CHD using multivariate large data set. To overcome this drawback, this paper proposes different multilayer perceptron (MLP) architectures cross-validated with varied folds of data set for classification of level of heart stroke. The main contribution of this paper is training the different ANN architecture models using back-propagation learning algorithm for one and two hidden layers MLP. The accuracy of the designed ANN models are cross-validated using varied folds of a data set comprising 303 instances extracted from different age groups.

The organization of this paper is as follows. Section 2 presents the methodology adopted and a brief description of structure of MLP. Next, performance indices adopted for evaluation of performance and experimental results of ANN architectures in classification of heart strokes are highlighted in Sect. 3. Finally, conclusions are drawn at the end.

2 Methodology

2.1 Analysis of Patient Data

The data were collected from the open source Cleveland Clinic Foundation Data base which is in the instance format. The data for this study have been collected from 303 patients who have symptoms of angiographic CHD that leads to different levels of heart strokes. For each sample, there are 76 raw attributes, where the first four, age, sex, height, and weight, are the general descriptions of the patient and other attributes are extracted from the standard 12 lead ECG recordings. Only 13 significant attributes are actually used. All attributes are numeric-valued. The data sets have been analyzed using PC-based software package [6].

2.2 Varied Folds of Data Set

The input patient data set containing 303 instances are partitioned into different folds, and each fold is partitioned into two subsets, viz. training set and testing set as shown in Table 1.

2.3 General Structure of MLP Network

The general structure of a MLP network is shown in Fig. 1. MLPs are the simplest and most commonly used neural network architecture due to their structural

Table 1 Input patient data set partitioning

Data set name	% training set	% testing set	Training instances	Testing instances
Fold 1 (F1)	70	30	212	91
Fold 2 (F2)	75	25	227	76
Fold 3 (F3)	80	20	242	61
Fold 4 (F4)	85	15	258	45
Fold 5 (F5)	90	10	273	30

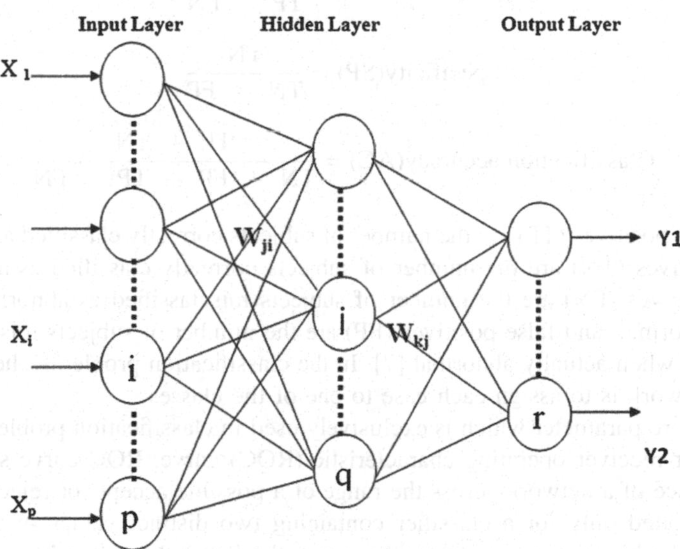

Fig. 1 General structure of a MLP network

flexibility, good representational capabilities, and large number of programmable algorithms.

In the training process of ANN model, the weight values are adjusted so that the actual output from the ANN model matches to the target values as closely as possible. The training of the ANN model was carried by using online back-propagation algorithm. The important step in training phase of the MLP network is to decide the number of neurons in the hidden layer. This is because, the MLP network may become incapable to model multivariate data, and the resulting fit will be poor, if an insufficient number of neurons are used in the network. Similarly, if too many neurons are used, network may over fit the data at the cost of training time. It is also observed experimentally that when overfitting occurs, the model fits the training data tremendously well, but its performance reduces to new and unseen data. The test set is used to test how well the ANN model will work on new and unseen data.

3 Performance Evaluation

3.1 Performance Indices

The performance of the proposed model is evaluated by computing the percentages of sensitivity (SE), specificity (SP) and correct classification, i.e., accuracy (AC). These validation parameters are defined as:

$$\text{Sensitivity(SE)} = \frac{\text{TP}}{\text{TP} + \text{FN}} \tag{1}$$

$$\text{Specificity(SP)} = \frac{\text{TN}}{\text{TN} + \text{FP}} \tag{2}$$

$$\text{Classification accuracy(AC)} = \frac{\text{TP} + \text{TN}}{\text{TN} + \text{FP} + \text{TP} + \text{FN}} \tag{3}$$

where true positives (TP) are the number of subjects correctly classified as healthy, true negatives (TN) are the number of subjects correctly classified as abnormal; false negatives (FN) are the number of subjects misclassified as abnormal when actually normal, and false positives (FP) are the number of subjects misclassified as normal when actually abnormal [7]. In the classification problems, the purpose of the network is to assign each case to one of the classes.

One more parameter which is exclusively used in classification problems is the area under receiver operating characteristic (ROC) curve. ROC curve shows the performance of a network across the range of a possible accepts or reject limit. It can be plotted only for a classifier containing two distinct values in the target column. The bigger the area under the curve, the better the network is. Since the area under curve is a portion of the area of the unit square, its value will always be between 0 and 1.

3.2 Experimental Results

Neural network model used is MLP network with back-propagation algorithm with learning rate and momentum coefficient. The optimized learning rate used is 0.5, and the momentum coefficient used is 0.25. The hidden layers are changed from one to two on each multifold data set. Performance indices, viz. sensitivity (SE), specificity (SP), and classification accuracy (AC), are calculated using the Eqs. 1, 2, and 3, respectively, from entries of confusion matrix. Both training and testing classification results for one and two hidden layers of proposed MLP network are given in Table 2.

Figure 2 shows the comparison between training data set specificity, sensitivity, and accuracy. The training data sets have showed good performance in terms of specificity, sensitivity, and accuracy. Figure 3 shows the testing data set accuracy on all data set folds of one- and two-layered MLP networks. From Fig. 3, it is clear that testing accuracy is 85.55 % for the network architecture with two hidden layers for the data set fold 1. This accuracy is considerably more in comparison with classification accuracy for heart stroke classification problem we have considered and existing classifiers designed by other researchers. Observing classification results in Table 1, it is clear that the data sets fold 1 gives better performance both in training and testing the network model. This is also supported by performance index, the area under ROC curve. From Table 2, it is proved that the two-layered MLP network has given the best classification accuracy in both

Table 2 Training and testing classification results for varied folds of data set

Data set folds	Hidden layers	Training			Testing			Area under ROC
		SE (%)	SP (%)	AC (%)	SE (%)	SP (%)	AC (%)	
Fold 1 (F1)	1	91.34	100	93.86	82.6	80.95	82.22	0.9432
	2	90.67	96.77	92.45	85.5	85.71	85.55	0.9569
Fold 2 (F2)	1	94.54	100	96.03	81.48	61.9	76	0.9437
	2	95.15	100	96.52	83.33	85.71	84	0.9584
Fold 3 (F3)	1	96.64	100	97.52	84.61	57.14	75	0.9375
	2	93.29	98.41	94.62	87.5	76.19	83.6	0.9486
Fold 4 (F4)	1	96.21	100	97.27	81.81	66.67	77.78	0.9585
	2	96.75	84.72	93.38	90.9	50	80	0.9429
Fold 5 (F5)	1	93.43	97.29	94.48	80.95	66.67	76.67	0.9584
	2	97.47	94.59	96.69	80.95	44.45	70	0.9737

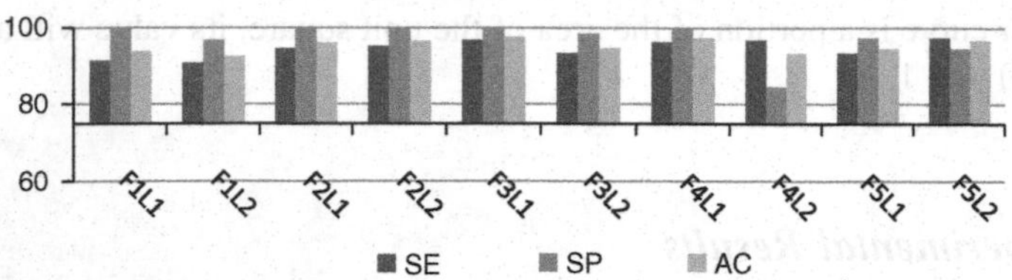

Fig. 2 Comparison between training set performance indices on all data set folds

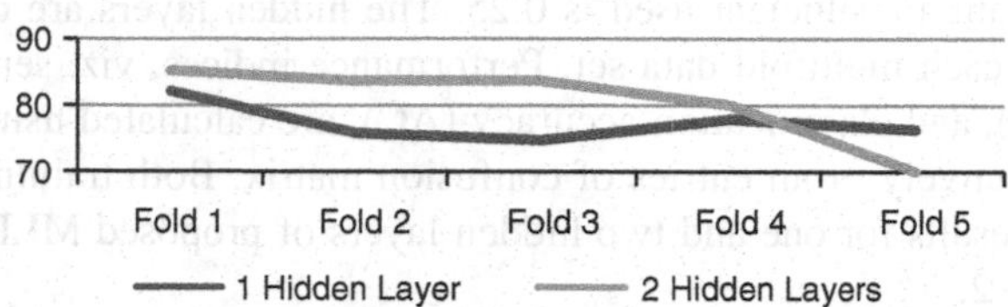

Fig. 3 Testing data set accuracy on all data set folds for one- and two-layered MLP

Table 3 Comparison between classification accuracies obtained by other methods in literature

A novel pruning method [8]	KDFW KNN [9]	SVM with Gaussian Kernel [10]	HLVQ [11]	Fuzzy weighted AIRS [12]	New FM [13]	Modular NN [14]	MLP with 2 HLs (our work)
68.47 %	70.66 %	76.1 %	76.92 %	80.71 %	81.32 %	82.22 %	85.55 %

training and testing phase for fold 1 and it is 92.45 % for training and 85.55 % for testing (Table 3).

4 Conclusion

This paper describes the ANN model for classification of levels of heart stroke based on risk factors. In this study, experiment is conducted by adopting one and two layers of hidden neurons ANN architectures for classification of heart disease. The ANN models are trained with back-propagation algorithm with momentum and variable learning rate. The results generated by this system have been verified with medical expert's data and are found correct. The experimental results proved that ANN technique provides adequate results for classification of heart stroke.

Acknowledgments Our thanks to the experts who have contributed in development of Cleveland Heart Disease database.

References

1. McGovern, P.G., Pankow, J.S., Shahar, E., Doliszny, K.M., Folsom, A.R., Blackburn, H., Luepker, R.V.: Recent trends in acute coronary heart disease: mortality, morbidity, medical care, and risk factors. New England J. Med. **334**(14), 884–890 (1996)
2. Mateny, M., McPheeters, M.L, Glasser, A., Mercaldo, N., Weaver, R.B, Jerome, R.N, et al.: Systematic review of cardiovascular disease risk assessment tools. Evidence Synthesis/ Technology Assessment No. 85. AHRQ Report No. 11-05155-EF-1. Rockville, Agency for Healthcare Research and Quality (2011)
3. Chen, J., Greiner, R.: Comparing Bayesian network classifiers In: Proceedings of UAI-99, pp. 101–108. Morgan Kaufmann, San Francisco (1999)
4. The Expert Panel: National cholesterol education program second report of the expert panel on detection, evaluation and treatment of high blood cholesterol in adults (adult treatment panel II). Circulation **89**(3), 1333–1445 (1994)
5. Niranjana Murthy, H.S., Meenakshi, M.: ANN model to predict coronary heart disease based on risk factors. Bonfring Int. J. Man Mach Interface **3**(2), 13–18 (2013)
6. Neuro Intelligence using Alyuda: Source available at www.alyuda.com. Last accessed 10 Jan 2014
7. Hagan, M.T., Menhaj, M.B.: Training feedforward networks with the Marquardt algorithm. IEEE Trans. Neural Netw. **5**(6), 989–993 (1994)
8. Mehmood, A.M., Kuppa, M.R.: A novel pruning approach using expert knowledge for data-specific pruning. Eng. Comput. **28**(1), 21–30 (2012)
9. Zuo, W.M., et.al: Diagnosis of cardiac arrhythmia using kernel difference weighted KNN classifier. In: Computers in Cardiology, pp. 253–256 (2008)
10. Uyar, A., Gurgen, F.: Arrhythmia classification using serial fusion of support vector machine and logistic regression. In: 4th IEEE workshop on intelligent data acquisition and advanced computing systems, pp. 560–565 (2007)
11. Elsayad, A.M.: Classification of ECG arrhythmia using Learning Vector Quantization Neural Networks. In: International Conference on Computer Engineering and Systems, pp. 139–144 (2009)
12. Polat, K., et al.: A new method to medical diagnosis; Artificial immune recognition system (AIRS) with fuzzy weighted pre-processing and application to ECG arrhythmia. J. Expert Syst. Appl. **31**(2), 264–269 (2006)
13. Lee, S.H., et al.: Extracting input features and fuzzy rules for detecting ECG arrhythmia based on NEWFM. In: International Conference on Intelligent and Advanced systems, pp. 22–25 (2007)
14. Jadhav, S., et al.: Modular neural network based arrhythmia classification system using ECG signal data. Int. J. Inf. Technol. Knowl. Manage. **4**(1), 351–356 (2011)

A Novel Approach for DDoS Mitigation with Router

S. Deepthi, K.S.S. Hemanth, Duvvuru Rajesh and M. Kalyani

Abstract Security is one of the critical attributes of any communication network. Various attacks have been reported over the last years but mainly denial of service effects entire network in a drastic way. So many mechanisms are developed but they are lagging in aspects like identifying IP spoofing (by session hijacking) and attack source. Existed Mechanisms identifies the attack after it's effect is being experienced by the victim. So we propose a new mechanism that is implemented on a router to identify the attack by monitoring the traffic flow, for that router uses a routing table with newly proposing attributes, i.e., timer, MAC address, and packet count. By using MAC address, there is possibility of finding actual attacker.

Keywords DDoS · Mitigation · Routing table · MAC address

1 Introduction

The distributed denial of service attack is performed mainly either by consuming the resources of the network with numerous data packets called as network centric attack or attacking the application which is used to communicate called as application layer attack.

S. Deepthi (✉) · K.S.S. Hemanth · M. Kalyani
Department of Computer Science and Engineering, Vignan's Lara Institute of Technology, Guntur, Andhra Pradesh, India
e-mail: deepthiklu@gmail.com

K.S.S. Hemanth
e-mail: karamsettyhemanth@gmail.com

M. Kalyani
e-mail: kalayanisrav@gmail.com

D. Rajesh
Department of Computer Science and Engineering, National Institute of Technology Jamshedpur, Jamshedpur, Jharkhand, India
e-mail: rajeshduvvuru.cse@nitjsr.ac.in

L.C. Jain et al. (eds.), *Intelligent Computing, Communication and Devices*,
Advances in Intelligent Systems and Computing 308,
DOI 10.1007/978-81-322-2012-1_75

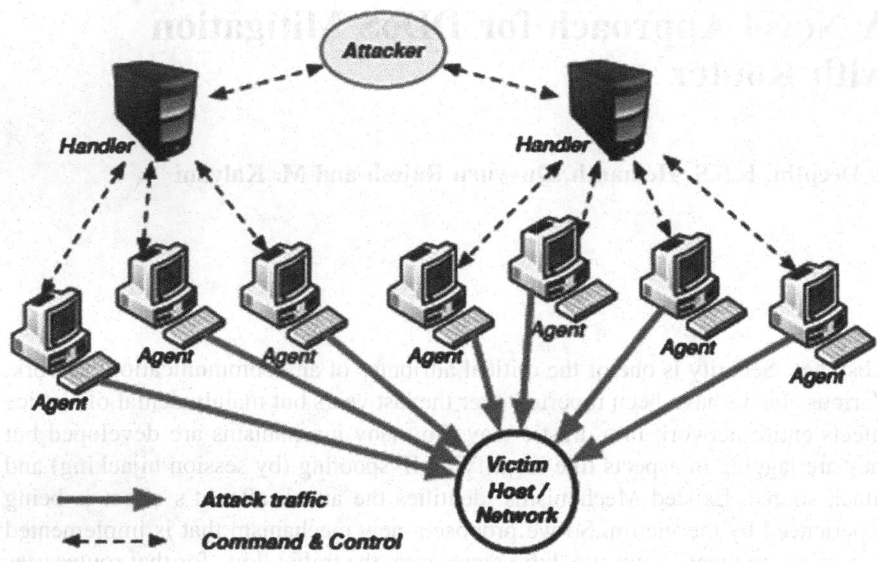

Fig. 1 Distributed denial of service

Most recent years, the DDoS attacks are carried out by using a botnet [1]. So that actual attacker hides himself from the scenario. Figure 1 explains the attack. A botnet is a large network of compromised agents (bots) controlled by attacker. The attacker can launch synchronized attacks, such as DDoS, by sending orders to the bots via a command and control channel. DoS attack may be caused by a single user or system, which can be easily identified unlike the DDoS which is performed by the multiple users or systems [2]. These attacks become as popular as they do not require identifying and using any vulnerabilities of the target system to perform the attack.

2 Related Works

There are so many methods proposed to overcome this DoS and DDoS attacks, but most of them applied on the network by using an external applications or devices like FireCol [3], IDS&IPS [4], IHoneycol [5], and active Internet traffic filtering mechanisms [6]. In all these cases, the attack can only be identified after entering the network which may cause the severe problems like congestion of packets and even the source of attack could not be identified easily by using these techniques. Consider the router itself which consists of so many filtering mechanisms to control various attacks: (1) ingress/egress filter, (2) route-based filtering (3) history-based IP address [7].

3 Proposed Method

Now proposed technique controls the DDoS and DoS attacks by using the already existed devices which needs a few enhancements. Router, which is the fastest performance hardware, that plays an important role during the communication in network.

Router mainly consists of the routing table [8] which specifies packet flow between source and destination addresses along with hop count. Consider Table 1.

Adding the two more fields called as **timer** and **counter** to the already existed table and making the MAC address [9] as visible as in the Table 2.

Using the TCAM [10] mechanism (fastest memory accessing) to trace the MAC address in the packet and compare with the routing table. So by matching the each packet address with stored MAC address in the routing table, the session hijacking can be identified. The routing table stores all the details up to a particular interval of time, later refreshes it. Between this refresh intervals time, the IP address of the system in the network will be maintained constant in the router. With the help of MAC address of the system, router checks whether it came from the IP address already stored or any new IP address with the same MAC address. Whenever a packet is spoofed, its IP address changes but the actual MAC address of the packet cannot be changed easily in case of using the proxies.

Counter is used to count the number of data packets flowing through the network for a particular interval of time and to limit the packet rate in the router whenever the data count reached the max limit, after that limit again the data count initiates from starting (zero). Here, the counter limit factor depends on the timer which is an increment timer or decrement timer. A particular time range is assigned to count the packets after it initiates from starting. Both the inflow and outflow should be considered separately.

Whenever data counter reaches max value within time range, then automatically the flow of the packets is delayed for particular amount of time and the data

Table 1 Normal routing table

Source address	Destination address	Hop count
172.2.2.141	178.3.3.165	5
157.5.78.19	172.3.6.9	12
171.2.2.224	172.3.6.9	9

Table 2 Proposed routing table

Source address	Destination address	Hop count	MAC address	Timer		Counter	
				In	Out	In	Out
172.6.0.11	172.68.0.11	3	00:13:A9:02:01:FC	–	10	–	640
172.6.0.11	172.68.6.11	3	00:13:A9:02:01:FC	–	–	–	–
172.6.0.11	172.68.6.11	3	00:13:A9:02:01:FC	–	5	–	280

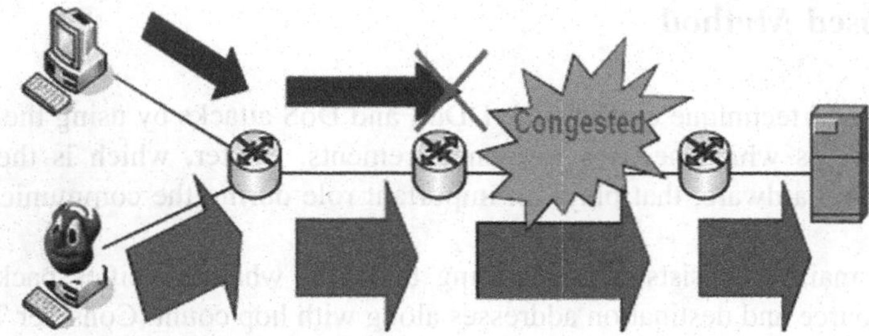

Fig. 2 Blocking paths through continuous data

packets are stored in the router buffer and then again they are sent to the receiver after time delay. DDoS attack actually happens if the network bandwidth is completely used by single host and not allowing the others to communicate, as shown in Fig. 2.

By making the delay in packets flow, the continuous usage can be avoided and by this others can communicate.

So, here Fig. 3 shows the architecture that is used to perform the analysis and perform the delay of packets. Dynamic cache (counter value dependent) and the static cache (predefined addresses) used for comparison of packets in the network and they are forwarded to the delay component for temporary storage. Delay time should be very less then only the packets can reach the destination without any loss, if the delay of the packets is more than the TTL of the packet, causing the self-destruct of the packet.

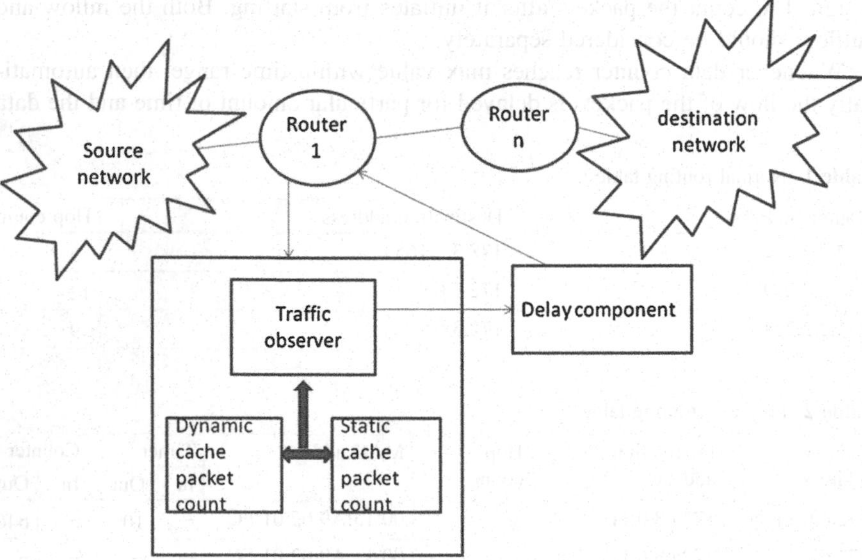

Fig. 3 Architecture of the proposed DDoS defense system

Here, static cache is used analyze the packets (for IP address, MAC address) and compare with previously identified and stored IP addresses, MAC addresses of attackers in the history database. No need to use the constant time range values to count the packets in the dynamic cache, it may vary continuously satisfying required characteristics. So, here attacker may not able to guess the actual limit value. Based on these, packets are forwarded to the network or delay component.

As proposed mechanism is applied in the router shown in Fig. 4, the attack can be identified on before it is entered into network. Here, each router stores the values independently, without the consideration of other routers. If all the routers present in the network use this mechanism, then the attack can be easily identified and controlled.

Rate limit: Consider network with a particular speed of "*n* bps," so it allows to flow only max of n bits of data within a second. So, if a host in a network sending *n* bps continuously, then it fully uses the bandwidth causing congestion of data. Based on the counter value only, the delay performed.

For example, consider a network of 64 kbps which means it send data of 64 kb in a second. Now node 1 is sending the data with the help of router 1 (by observing routing table).

172.6.0.11	**172.68.0.11**	**3**	**00:13:A9:02:01:FC**	**-**	**10**	**-**	**640**	**Delay of packets**
172.6.0.11	172.68.0.11	3	08:13:C9:02:03:FC	-		-		Not allowed
172.6.0.11	172.68.6.11	3	00:13:A9:02:01:FC	-	5	-	280	No delay

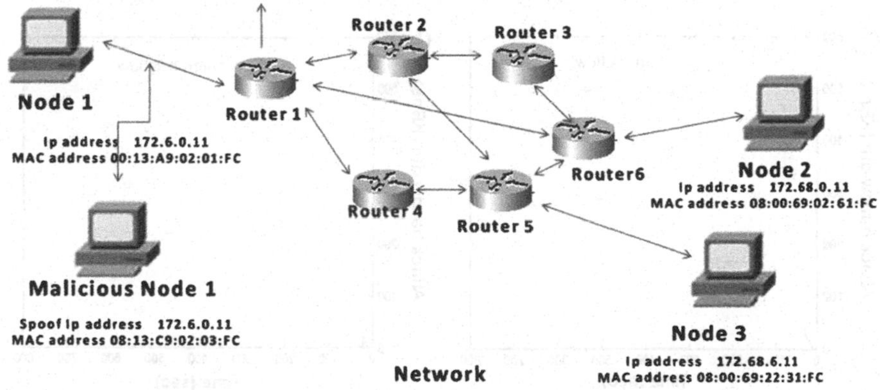

Fig. 4 Showing the routing table with transmission of data between nodes

First row in routing table indicates:
Here, the node 1 is sending the data to node 2 continuous.

For 1 s 64 k data

10 s 640 k data

So, delay of packets needs to be performed.

Second row in routing table indicates:
Here, the connection will not be allowed as already the IP address existed in routing table with diff MAC address.

Third row in routing table indicates:
Here, the transfer rate is normal so no need to delay the packets (for 5 s 280 < 320 kb).

Rate limit = min (actual rate, safe rate).

Safe rate < time × *n* bps.

Delay = time in microseconds when rate is more and in seconds when rate is normal.

4 Simulations

Mitigation of DDoS attack can be observed by following graphs as shown in Fig. 5. It is attack traffic which flows continuously for a particular amount of time and causing congestion. Instead of continuous flow ,delay the packet flow using interval gap by storing them in router buffer.

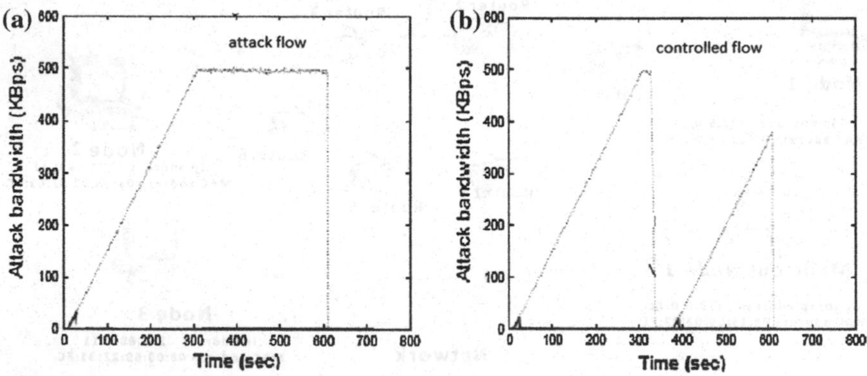

Fig. 5 **a** Graphs showing the continuous traffic flow and **b** controlled flow

5 Conclusion

Distributed denial of service is one of the major challenges a network is facing. So many mechanisms are proposed in order to mitigate it but those are failing in some issues .By considering that in mind, proposed a new mechanism to prevent DDoS attack at the router itself which could be possible by adding few attributes to the routing table, i.e., timer and counter which blocks the floods at router. From simulations can observe how proposed method mitigates the DDoS attack.

References

1. Alomari,E., Manickam, S., Gupta, B.B., Karuppayah, S., Alfaris, R.: Botnet-based distributed denial of service (DDoS) attacks on web servers: classification and art. Int. J. Comput. Appl. **49**(7), 0975–8887 (2012)
2. AL-Musawi, B.Q.M.: Mitigating DoS/DDoS attacks using iptables. Int. J. Eng. Technol. IJET-IJENS **12**(03) (2012)
3. François, J., Aib, I., Boutaba R.: FireCol: a collaborative protection network for the detection of flooding DDoS attacks. IEEE 2012 Transaction on Networking, vol. 99
4. Saritha, M., Chinta M.: Countering varying dos attacks using snort rules, Int. J. Adv. Res. Comput. Sci. Softw. Eng. **3**(10) (2013)
5. Buvaneswari, M., Subha, T.: Ihoneycol: a collaborative technique for mitigation of DDoS attack. Int. J. Emerg. Technol. Adv. Eng. **3**(1) (2013)
6. Deepthi, S., Prashanti, G., Sandhya, R.K.: Blocking of flooding attacks: using active internet traffic filtering mechanism. Int. J. Eng. Trends Technol. **4**(7) (2013)
7. Mahajan, D., Sachdeva M.: DDoS attack prevention and mitigation techniques–a review. Int. J. Comput. Appl. **67**(19), 0975–8887 (2013)
8. http://en.wikipedia.org/wiki/Routing_table
9. Garg, U., Verma, P., Moudgil, Y.S. Sharma, S.: MAC and logical addressing (A Review Study). Int. J. Eng. Res. Appl. (IJERA)
10. Naik, R.B., Shefali, R.R.: Low-area low-power and high-speed TCAMS. In: International Conference on VLSI, Communication and Instrumentation (ICVCI) 2011. Proceedings published by International Journal of Computer Applications (IJCA)

Enhancing Battery Lifetime of Wireless Heterogeneous and Ad hoc NW

Debashree Nayak, Saumendra Kumar Mohanty and Amiya B. Sahoo

Abstract The principal objective of next generation wireless access networks (NWs) is to provide quality of service at anywhere, anytime, and any type of environment. As the user mobile nodes (MNs) basically depend on their battery power, to provide best service, it is most important to minimize their energy consumption. Considering this factor, battery lifetime is maximized through cooperative game theory. Here, we have also applied a route selection algorithm to ad hoc MNs to get them attached with the appropriate attachment point with less power consumption. The experimental MATLAB simulation results show the cooperative act of MNs and NWs to maximize the battery lifetime, and the route selection algorithm can also balance the load among proxy MNs by which they will not be overloaded.

Keywords MANET · Cooperative game process · Battery lifetime · Ad hoc NW · Route selection

1 Introduction

In mobile ad hoc NW (MANET), the power consumption means the maximization of battery lifetime of the whole system. So, the ad hoc mobile nodes (MNs) should choose their appropriate attachment points accordingly.

In [1], we have applied cooperative game theory to maximize battery lifetime of the MNs, taking load balancing into account to avoid NW congestion. To solve the problem of NW selection, we have applied a multi-tenderee bidding model to the

---

D. Nayak · S.K. Mohanty (✉)
Department of EIE, ITER, SOA University, Bhubaneswar, Odisha, India
e-mail: saumendramohanty@soauniversity.ac.in

D. Nayak
e-mail: deb.shree09@gmail.com

A.B. Sahoo
Department of ECE, SIT, BPUT, Rourkela, Odisha, India
e-mail: amiyabhusana@gmail.com

L.C. Jain et al. (eds.), *Intelligent Computing, Communication and Devices*,
Advances in Intelligent Systems and Computing 308,
DOI 10.1007/978-81-322-2012-1_76

whole cellular coverage area. Other VHO decision algorithms such as SAW, TOPSIS, etc., are implemented in Ref. [2] with their standard deviations. In [3], the network selection algorithm is formulated as a cooperative bidding process, taking only the overlapping region of all the heterogeneous access networks. The idea of network selection with cooperative act of NWs is presented in Ref. [4]. In [5], a VHO decision algorithm is presented that enables a wireless access NW to not only balance the overall load among all attachment points but also maximize the collective battery lifetime of MNs. In addition, when an ad hoc mode is applied to 3G or 4G wireless data NWs, a route selection algorithm is proposed for forwarding data packets to the most appropriate attachment point to maximize collective battery lifetime and maintain load balancing. In [6], a series of experiments is described which gives detail measurements of the energy consumption of an IEEE 802.11 wireless NW interface operating in an ad hoc NW environment. In [7], minimum energy dynamic source routing (MEDSR) and hierarchical MEDSR protocols for MANET are proposed. The comprehensive analytic model is developed in [8] for the performance study of the route DSR protocol for MANET. This paper is organized as follows: In Sect. 2, a VHO scenario with ad hoc NW is introduced. In Sect. 3, the CGP is applied to enhance the battery lifetime of MNs in the heterogeneous NW and a route selection algorithm is applied to maximize the collective battery lifetime for the ad hoc NW. Section 4 shows the experimental results with discussion. In Sect. 5, conclusion and future works are provided.

2 Handoff Scenarios

NW selection is a great challenge for a heterogeneous environment in terms of its QoS, bandwidth, mobility, etc. VHO in urban areas may take place at anywhere and at anytime. Here, we have introduced a VHO scenario, taking ad hoc NW into consideration as shown in Fig. 1b. That means in addition to the vertical handoff region, as shown in Fig. 1a, if there exist some MNs that are not placed in the coverage areas of any access NWs, handoff process may take place to get connected with the nearest access NWs. In such a situation, the MNs may form a MANET and forward their data packets to the most appropriate APs or BS. In this system, the MNs act cooperatively to form a MANET using the IEEE 802.11 interface in an ad hoc mode as shown in Fig. 1b.

3 Battery Lifetime

3.1 Cooperative Game Process

Let p_{il} be the power consumption per unit of time needed in lth MN at ith NW. The value of p_{il} depends on the number of MNs attached to the ith NW and the data rate requested by lth MN, that is, larger the number of nodes attached to the same

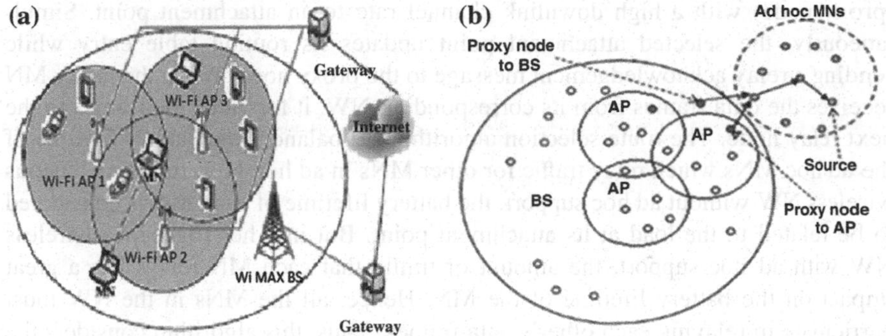

Fig. 1 **a** Handoff scenario of heterogenous NW. **b** Handoff scenario of ad hoc NW

AP, the more power consumed by each MN, as they each get lower data rates and hence need to connect longer [5]. Thus, the amount power consumed by MNs is as follows:

$$p_{il} \propto R_i \tag{1}$$

where R_i is the utility of ith NW and is given as

$$R_i = \frac{U\_B_i}{T\_B_i} \tag{2}$$

where U_B_i: Used bandwidth of ith NW and T_B_i: Total bandwidth of ith NW.

To maximize their battery lifetime, the MNs calculate their load on each NW to avoid NW congestion. If the load on a particular NW is higher than a threshold value, the MNs under that NW will be affected. So, the MN having largest battery power in that NW will go for the NW having second lowest power consumption rate for it. In this process, the game will be played in rounds till the load on each NW is balanced.

3.2 Route Selection Algorithm

For the NW of Fig. 1b, when an MN is actively receiving the data frames from a BS or AP of a cellular NW, experiences a low downlink channel rate and the VHDC cannot find an alternative direct access point for the MN, a route will be selected via the MANET to allow the MN to access an appropriate attachment point of its choice. The source MN in ad hoc NW sends out a route request message using its IEEE 802.11 interface which is broadcasted through the ad hoc NW to all the MNs according to the route discovery protocol [5]. The objective is to find an optimal relay route in terms of overall battery lifetime, to reach a MN (proxy node) with a high downlink channel rate to an attachment point. Simultaneously, the selected attachment point updates its routing table entry while sending a relay acknowledgement message to the proxy node. When the proxy MN receives the data frames from its corresponding NW, it forwards the frames to the next relay node. The route selection algorithm [5] balances the battery lifetime of the ad hoc MNs which relay traffic for other MNs in ad hoc NW. In heterogeneous wireless NW without ad hoc support, the battery lifetime of each MN is considered to be related to the load at its attachment point. But in a heterogeneous wireless NW with ad hoc support, the amount of traffic that each MN relays has a great impact on the battery lifetime of the MN. Hence, all the MNs in the NW must participate in relaying each other's data frames. Thus, this algorithm considers the amount of traffic load to be forwarded and maximizes the battery lifetime of the MNs in the available routes, which results in maximizing the overall battery lifetime of the system.

Let J be the amount of traffic that has to be routed via some MNs in the cellular coverage area and p_{kb} be the power consumption amount per byte of transmission at a given MN k. Then, the cost function is defined as

$$E_k = \frac{p_k}{p_{kb}J} \tag{3}$$

The maximum battery lifetime resulting from the selection of a given route r_s is determined by the minimum value of E_k over the path, i.e.,

$$L_s = \mathrm{Min}_{\forall k \in r_s} E_k \tag{4}$$

Let A be the set of all possible routes between kth MN that is experiencing a low downlink channel rate and candidate attachment points via proxy MNs in the ad hoc NW. Here, the attachment point has been already selected by the NW selection algorithm (bidding model and cooperative game process), with which the candidate proxy node is associated. Then, the route $r_{\max}$ with maximum battery lifetime value from the set A is selected as follows:

$$r_{\max} : \mathrm{Max}_{\forall r_s \in A} L_s = \mathrm{Max}_{\forall r_s \in R}(\mathrm{Min}_{\forall k \in r_s} E_k) \tag{5}$$

4 Experimental Results

Simulation Parameters:

Number of MNs in the VHO region: 22
Number of traffic classes: 4 (best effort, background, video, and VoIP)
Number of candidate NWs: 4 (AP-1, AP-2, AP-3, and WIMAX BS)
Number of parameters: 5 (bandwidth, delay, jitter, packet error rate, and price)

After the final round of bidding process, the MNs will act cooperatively to maximize their battery lifetime. From Fig. 2, the battery lifetime of 12 MNs is

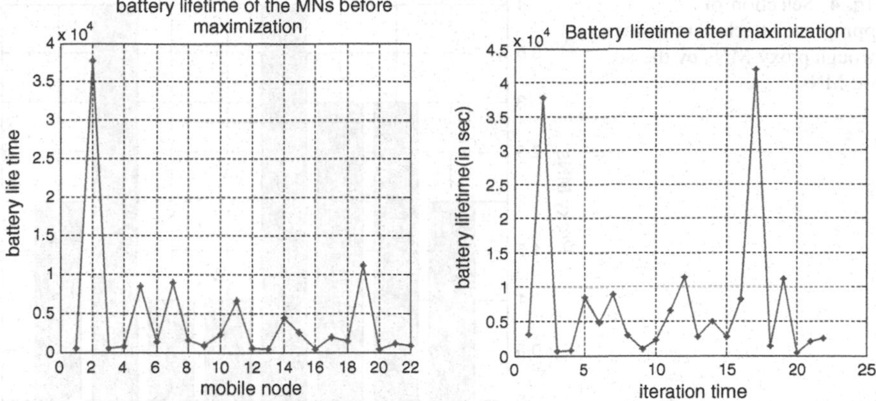

Fig. 2 Battery lifetime of each NW in the bidding process before and after maximization

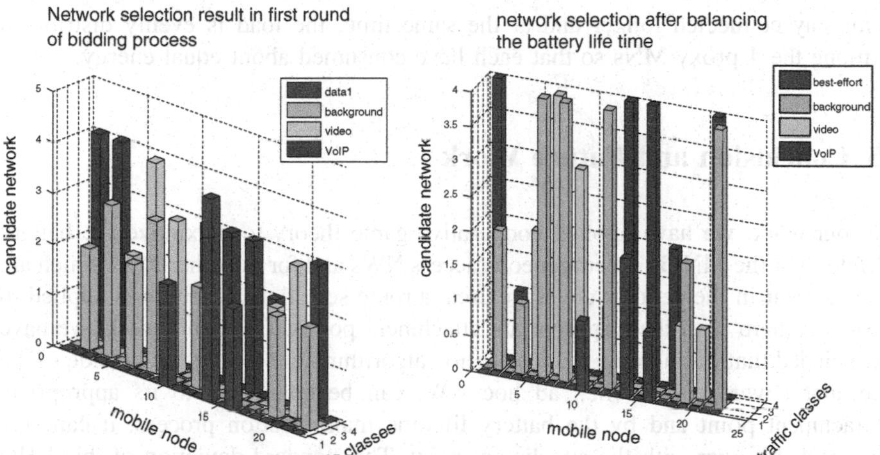

Fig. 3 Final NW selection result before and after maximization of the battery lifetime

increased by the CGP and the 10 MNs have same battery lifetime as before, that is, they will get the maximum battery lifetime in that NW, whose bid is the lowest one for them.

Figure 3 shows the final NW selection results after maximization where the NW utility of the four access NWs are 0.9649, 0.9604, 0.9617, and 0.9687, respectively.

By applying the route selection algorithm, the MNs in ad hoc NW select the attachment points which are shown in the above Fig. 4. In our experiment, the MN having maximum battery lifetime in each NW is acting as the proxy MN. So, the power consumption in the four proxy MNs (0.8 + 2.4 mj/kB × J) are 39.6226, 32.5013, 42.4925, and 45.6066, respectively. From the route selection algorithm described above, the ad hoc MNs have selected their attachment point, where they

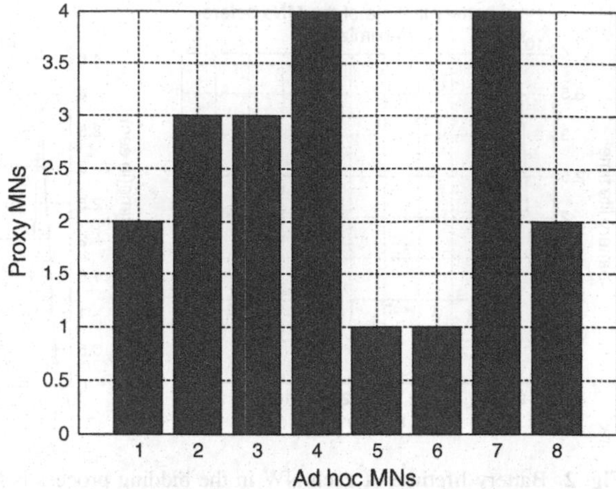

Fig. 4 Selection of appropriate attachment points through proxy MNs by the ad hoc MNs

can stay connected longer and at the same time, the load is evenly distributed among the 4 proxy MNs so that each have consumed about equal energy.

5 Conclusion and Future Work

In our work, we have applied cooperative game theory to maximize the battery lifetime of the MNs in heterogeneous access NWs and for the other MNs which are not present in the cellular coverage area, a route selection algorithm is applied to connect them with their appropriate attachment points. From this work, we have concluded that through the route selection algorithm, an MN present outside of the cellular coverage area (i.e., ad hoc NW) can be connected to its appropriate attachment point and by the battery lifetime maximization process, it can stay connected longer with the attachment point. The standard deviation of this VHO decision algorithm is obtained as 0.0032, whereas the standard deviation of some other algorithms such as SAW and TOPSIS are 0.0137 and 0.0117, respectively.

As the cooperative game process has successfully applied to maximize the battery lifetime of the MNs in cellular coverage area, it can also be applied to the multi-hop ad hoc NW. Apart from the MANET, the route selection algorithm can also be applied to vehicular ad hoc NW.

References

1. Mohanty, S.K., Nayak, D.: Analysis of VHO and battery lifetime using co-operative game process. In: Proceeding of IET-SEISCON, Chennai (2013)
2. Zhang, W.: Handover decision using fuzzy MADM in heterogeneous networks. IEEE Commun. Soc. **2**, 653–658 (2004)

3. Liu, X., Fang, X., Chen, X., Peng, X.: A bidding model and cooperative game-based vertical handoff decision algorithm. Network Comput. Appl. **34**, 1263–1271(2011)
4. Antoniou, J., Pitsillides, A.: 4G converged environment: modeling network selection as a game. In: Proceeding of MWCS, Budapest (2007)
5. Lee, S., Sriram, K., Kim, K., Kim, Y.H., Golmie, N.: Vertical handoff decision algorithms for providing optimized performance in heterogeneous wireless networks. IEEE Trans. Veh. Technol. **58**, 865–881 (2009)
6. Feeney, L., Nilsson, M.: Investigating the energy consumption of a wireless network interface in an ad-hoc networking environment. In: Proceedings of IEEE INFOCOM, pp. 1548–1557, Anchorage, AK. 3 (2001)
7. Tarique, M., Tepe, K.: Minimum energy hierarchical dynamic source routing for mobile Ad-hoc networks. Ad-hoc Networks **7**, 1125–1135 (2009)
8. Li, J., Pan, Y., Xiao, Y.: Performance study of multiple route dynamic source routing protocols for mobile ad-hoc networks. J. Parallel Distrib. Comp. **65**, 169–177 (2005)

Analyzing the Interaction Between TCP Variants and Routing Protocols in Static Multi-hop Ad hoc Network

Sukant Kishoro Bisoy and Prasant Kumar Pattnaik

Abstract In recent year, wireless Internet become more popular due to growth of mobile devices. TCP is designed to perform well in traditional wired networks, where packet loss is used as measure of congestion. However, TCP connections in ad hoc networks are plagued by problems such as high bit error rates, frequent route changes, multi-path routing, and temporary network partitions. The throughput of TCP over such connection is not satisfactory, because TCP misinterprets the packet loss or delay as congestion and invokes congestion control and avoidance algorithm. Hence, it is of utmost importance to identify the most suitable and efficient TCP variants that can perform well in MANET. Main objective of this paper is to find suitable routing protocols for TCP variants and analyze the performance differential variation in static multi-hop ad hoc network in terms of throughput, packet delivery ratio, and packet loss. Result using NS2 shows that AODV is best routing protocol with respect to throughput and packet delivery ratio irrespective of TCP variants. Vegas is the best protocol among TCP variants due to its higher throughput and higher PDR and lower packet loss in most situations, and DSR has lower packet loss irrespective of TCP variants.

Keywords AODV · DSDV · DSR · OLSR · TCP-NewReno · TCP-Sack1 · TCP-Vegas

S.K. Bisoy (✉)
SOA University, Bhubaneswar, India
e-mail: sukantabisoyi@yahoo.com

P.K. Pattnaik
School of Computer Engineering, KIIT University, Bhubaneswar, India
e-mail: patnaikprasantfcs@kiit.ac.in

L.C. Jain et al. (eds.), *Intelligent Computing, Communication and Devices*,
Advances in Intelligent Systems and Computing 308,
DOI 10.1007/978-81-322-2012-1_77

1 Introduction

MANET forms a random network by consisting of mobile nodes, which communicates over wireless path. This kind of network is more appropriate where networking infrastructure is not available, setup time is very less, and temporary network connectivity is required. Many routing protocols are available in MANET. Among them, ad hoc on demand distance vector (AODV) [1] and optimized link state routing (OLSR) [2] are standardized by IETF MANET working groups [3]. AODV and dynamic source routing (DSR) [4] are two reactive routing protocols. OLSR and destination sequenced distance vector (DSDV) [5] are proactive routing protocols.

TCP [6] is reliable protocol of transport layer, which provides in-order delivery of data to the TCP receiver. In fact, TCP has its variants of protocols like TCP-NewReno, TCP-Sack1, and TCP-Vegas. All these were proposed to improve the performance of TCP protocols.

The rest of the paper is structured as follows. Section 2 presents related works. Section 3 will present simulation environment, and Sect. 4 will explain result and analysis. Finally, we conclude our work in Sect. 5.

2 Related Work

Author in [7] studied the inter-layer interaction between MAC and physical layer and demonstrated the performance differences between DSR and AODV. The performance varies because they adopt different mechanism for routing.

Proactive routing protocol performs better than reactive at the cost of higher routing load [8]. In order to show which TCP variants works well, AODV, DSR and OLSR, and DSDV routing protocols in MANETs were considered and evaluated with various network conditions.

Author in [9] analyzed the interaction between internet-based TCP variants protocols and routing protocols in MANET. Result shows that OLSR is best routing protocol irrespective of TCP variants, and it provides a lower packet loss rate than others in most situations.

3 Simulation Environment

We study the performance of ad hoc routing protocol that may affect TCP performance with three metrics such as throughput, packet delivery ratio, and packet loss. We evaluate the performance of these protocols using NS2 simulator [10].

Fig. 1 Chain topology with 6 hops

Table 1 Parameter values

Parameter	Value	Parameter	Value
Channel type	Wireless channel	Propagation model	Two-ray ground
MAC	802.11	Number of nodes	2–7
Routing protocol	OLSR, DSDV, DSR, AODV	Transmission Range	250 m
Topology	Chain	Topography	1,000 m × 1,000 m
Packet type	FTP of 512 B	Time of simulation	100 s

3.1 Simulation Tool and Parameter

We create a linear string topology of 7 nodes as shown in Fig. 1. A single TCP connection is created to cover hop 1 to hop 7. Each connection stays for 100 s long. Each node is separated by 150 m distance whereas transmission range of each node is 250 m. Other parameters used for simulations are shown in Table 1.

4 Result and Analysis

In this section, we present the results gathered from the simulation experiments from various scenarios. To analyze the performances, those protocols' throughput (Kbps), packet delivery ratio, and packet loss are used as metrics.

4.1 Throughput Measurement

Initially, throughput of Vegas is measured over AODV, DSDV, DSR, and OLSR in static multi hop network. We found Vegas protocol over AODV achieves better throughput (see Fig. 2), because Vegas protocol able to maintain its window size in an appropriate range, which helps to reduce the unnecessary route discovery. However, when DSDV is used, Vegas achieves least throughput among all TCP variants. Next, we calculated the throughput of TCP-Sack1 and TCP-NewReno over AODV, DSDV, DSR, and OLSR with different number of hops. Figure 3 shows that AODV is better routing protocol with respect to throughput without

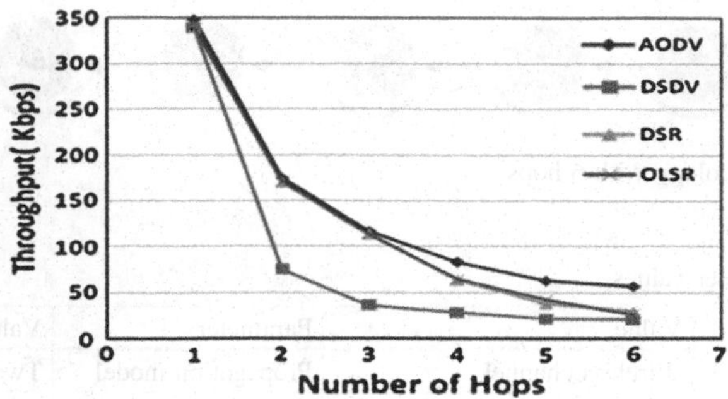

Fig. 2 Throughput of Vegas over AODV, DSDV, DSR, and OLSR

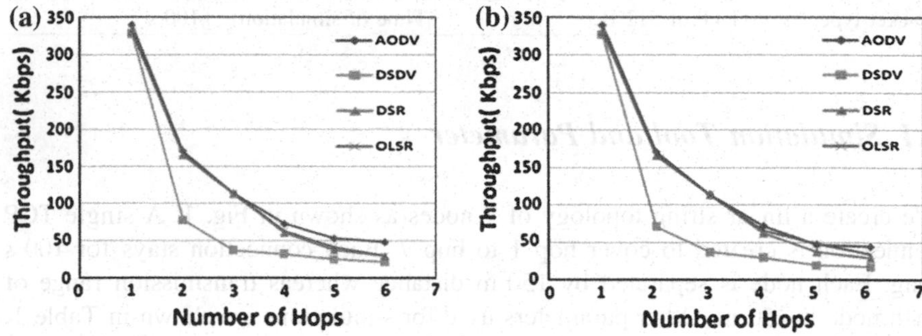

Fig. 3 Throughput of NewReno and Sack1 over AODV, DSDV, DSR, and OLSR. **a** NewReno, **b** Sack1

regards to TCP variants used in string topology. However, irrespective of ad hoc routing protocol, throughput decreases with increase of hop length in string topology. When hop length increases, the collision between data packet and ACK packet occurs more frequently and interference may cause more packet loss and decrease throughput. In a static environment, the maximum achievable throughput is limited by the interaction (at the MAC level) between neighboring nodes [11]. Throughput of NewReno protocol decreases due to increase of window size aggressively and hidden node problem [12].

Then, we measured the throughput TCP variants like Vegas, Sack1, and NewReno over selected ad hoc routing protocols. From Fig. 4, we realize that AODV has higher throughput over all TCP variants. As shown in Fig. 5, performances of Vegas are least over DSDV because it relies on accuracy of measurement of RTT value. This mechanism tries to avoid congestion by reducing its sending rate if RTT of a connection increases beyond certain threshold, thus resulting in lowest throughput. When DSDV is used, Vegas implements proactive mechanism to increase and decrease its congestion window (cwnd). In this, all variants of TCP perform better over AODV routing protocol in static multi hop, ad

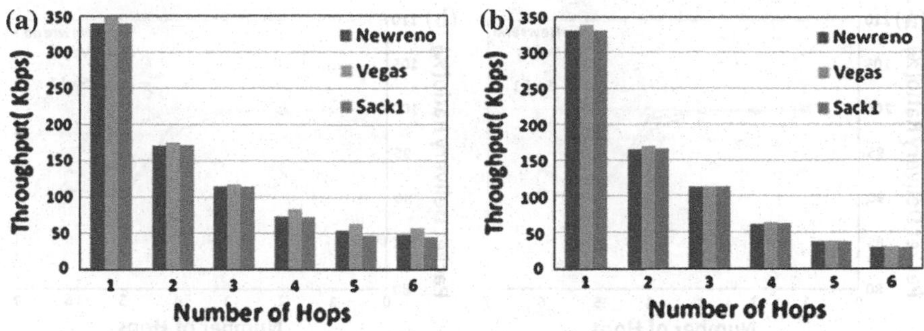

Fig. 4 Throughput of NewReno, Vegas and Sack1 over AODV and DSR. **a** AODV, **b** DSR

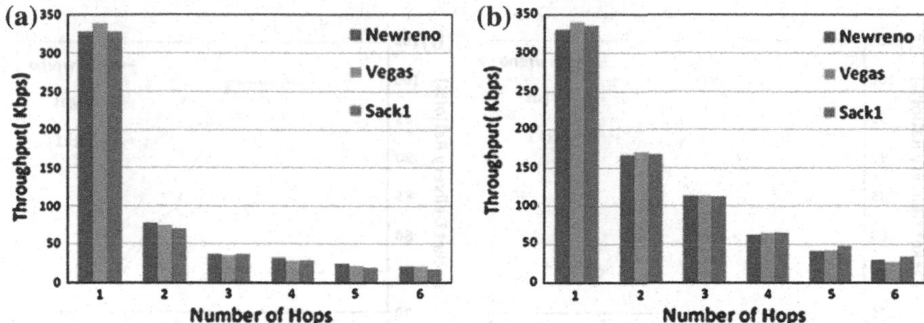

Fig. 5 Throughput of NewReno, Vegas, and Sack1 over DSDV and OLSR. **a** DSDV, **b** OLSR

hoc network. The main reason is AODV is reactive routing protocol and maintain same route until the path is broken.

4.2 Packet Delivery Ratio

Packet delivery ratio is the ratio between number of data packet received and number of data packet sent. Initially, packet delivery ratio (PDR) of NewReno, Vegas, and Sack1 is measured over selected routing protocol like AODV and DSDV. Vegas protocol achieves higher PDR than NewReno and Sack1 over AODV (see Fig. 6). However, there is a drastic change in PDR when hop length is more than $4(h > 4)$ for all TCP variants over AODV and DSDV. Then, we calculated the PDR of NewReno, Vegas, and Sack1 over DSR and OLSR. The PDR of DSR protocol is lower than other routing protocols (Fig. 7). There is sudden decrease of PDR for all TCP variants over DSR and OLSR when number of hops reaches more than $3(h > 3)$. When Vegas is used as transport protocol, DSR achieves somewhat better performance than NewReno and Sack1. However, AODV achieves above 90 % PDR for all number of hops, which is higher than

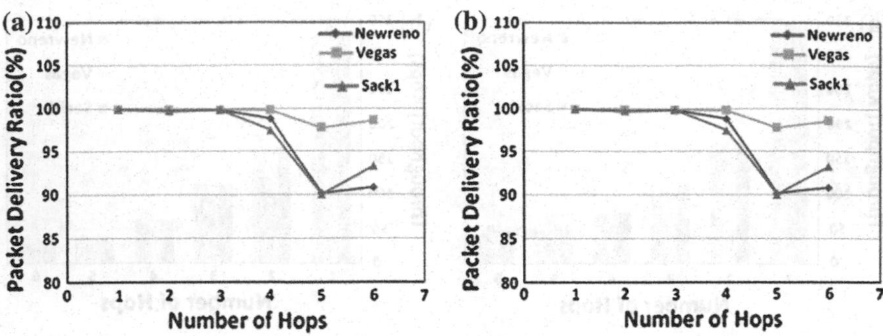

Fig. 6 PDR versus number of hops of Newreno, Vegas, and Sack1. **a** AODV, **b** DSDV

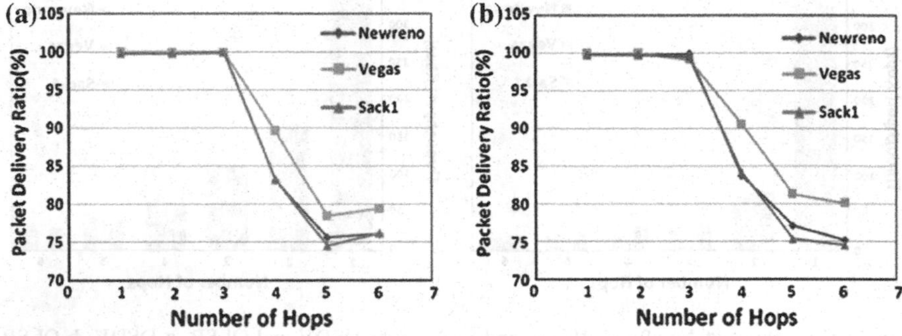

Fig. 7 PDR versus number of hops of NewReno, Vegas, and Sack1. **a** DSR, **b** OLSR

DSR, DSDV, and OLSR. This higher value of PDR is achieved when TCP-Vegas is used as transport layer protocol. The PDR of all TCP variants decreases with the number of hop increases due to hidden node problem in chain topology.

4.3 Packet Loss Measurement

Next, we measure the packet loss of TCP variants over OLSR, DSDV, DSR, and AODV. As Fig. 8 shows, packet loss range varies from 1.0 to 57.0 for all TCP variants we have considered. In case of lossy link, AODV shows worst performance than DSDV, OLSR, and DSR, which happens due to long path in static network. AODV creates more routing packets, which incurs more collision and high number of packet drop. In most situation, Vegas shows better performance among all TCP variants and provides low packet loss (Fig. 8), because Vegas accurately estimates the available bandwidth in the network by finding the difference between expected flow rates and actual flow rates. The packet loss of

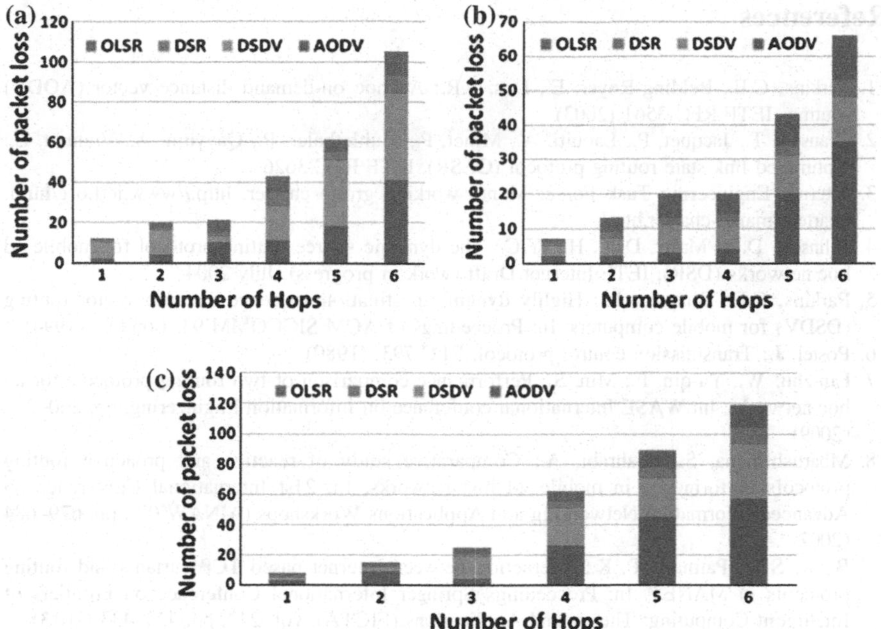

Fig. 8 Number of packet loss of OLSR, DSR, DSDV, and AODV. **a** Newreno, **b** Vegas, **c** Sack1

OLSR and DSR is lower than AODV and DSDV. With respect to packet loss, DSR is better routing protocol, while Vegas is used as transport protocol. In the other hand, TCP variants like Sack1 and NewReno are not good as Vegas with respect to packet loss for growing hop.

5 Conclusions

Aim of this paper was to find suitable routing protocol for particular TCP variants in static multi-hop wireless ad hoc network with respect to throughput and packet delivery ratio and packet loss. Our result shows that irrespective of TCP variants, AODV is better routing protocol than DSR, OLSR, and DSDV. Among TCP variants, Vegas performs better over AODV routing protocol. However, performance of all routing protocol degrades with the increase of number of hops due interaction between neighbor nodes, and interference may cause packet loss at sink node. Vegas is the best protocol among TCP variants due to its higher throughput and PDR and lower packet loss in most situations, and DSR achieves fewer packet loss irrespective of TCP variants.

References

1. Perkins, C.E., Belding-Royer, E., Das, S.R.: Ad hoc on-demand distance vector (AODV) routing, IETF RFC 3561 (2003)
2. Clausen, T., Jacquet, P., Laouiti, A., Minet, P., Muhlethaler, P., Qayyum, A., Viennot, L.: Optimized link state routing protocol (OLSR), IETF RFC 3626
3. Internet Engineering Task Force: Manet working group charter, http://www.ietf.org/html.charters/manet-charter.html
4. Johnson, D.B., Maltz, D.A., Hu, Y.C.: The dynamic source routing protocol for mobile ad hoc networks (DSR), IETF Internet Draft (work in progress), July 2004
5. Parkins, C.E., Bhagwat P..: Highly dynamic destination sequence distance vector routing (DSDV) for mobile computers. In: Proceedings of ACM SIGCOMM'94, London, (1994)
6. Postel, J.: Transmission control protocol. RFC 793, (1980)
7. Lin-zhu, W., Ya-qin, F., Min, S.: Performance comparison of two routing protocols for ad hoc networks. In: WASE International conference on Information Engineering, pp. 260–262 (2009)
8. Mbarushimana, S., Shahrabi, A.: Comparative study of reactive and proactive routing protocols performance in mobile ad hoc networks. In: 21st International Conference on Advanced Information Networking and Applications Workshops (AINAW'07), pp. 679–684 (2007)
9. Bisoy, S.K., Pattnaik, P. K.: Interaction between internet based TCP variants and routing protocols in MANET. In: Proceedings Springer International Conference on Frontiers of Intelligent Computing: Theory and Applications (FICTA), vol. 247, pp. 423–433 (2013)
10. Information Sciences Institute: The network simulator Ns-2, http://www.isi.edu/nanam/ns/, University of Southern California
11. Li, J, Blake, C., De Couto, D., Lee, H., Morris, R.: Capacity of ad hoc wireless networks. In: Proceedings of ACM/IEEE International Conference in Mobile Computing and Networking (MobiCom 2001), pp. 61–69 (2001)
12. Freeney, L.M.: An energy consumption model for performance analysis of routing protocols for mobile ad hoc networks. Mob. Networks Appl. **6**(3), 239–249 (2001)

Capacity Estimation for Cellular LTE Using AMR Codec with Semi-persistent Scheduling

Ajay Pratap and Hemanta Kumar Pati

Abstract Long-term evolution (LTE) network is a fully IP-based and does not include a circuit-switched domain for voice communication as known from GSM and UMTS networks. Continuous switching from active to inactive state is one of the challenging tasks for VoIP in LTE. Because of these switching and retransmissions, control channels come into function. Control channel is one of the major limitations for capacity in VoIP. In this paper, we analyzed different scheduling techniques to reduce the number of control channels in the network. We estimated the maximum number of LTE users can be supported over enhanced node B (eNodeB) using different bandwidth levels. From the numerical results, we observed that AMR codec with semi-persistent scheduling scheme is utilizing less number of control channels and accommodates more number of users.

Keywords LTE · 3GPP · TTI · VoIP · AMR

1 Introduction

The long-term evolution (LTE) is one of the most advance and growing mobile telecommunication system. The LTE was initiated in Third Generation Partnership project (3GPP) in 2004. The main aim of LTE was to support not only data services but also voice service with high efficiency [1]. LTE operates only in the packet-switched mode. So to transmit voice over IP-based network, circuit switch (CS) should be replaced by a real-time packet-switched voice service called voice

A. Pratap · H.K. Pati (✉)
Department of Computer Science and Engineering, IIIT Bhubaneswar, Bhubaneswar 751003, Odisha, India
e-mail: hemanta@iiit-bh.ac.in; h_pati_hindol@yahoo.com

A. Pratap
e-mail: a112002@iiit-bh.ac.in; jpratap6@gmail.com

L.C. Jain et al. (eds.), *Intelligent Computing, Communication and Devices*,
Advances in Intelligent Systems and Computing 308,
DOI 10.1007/978-81-322-2012-1_78

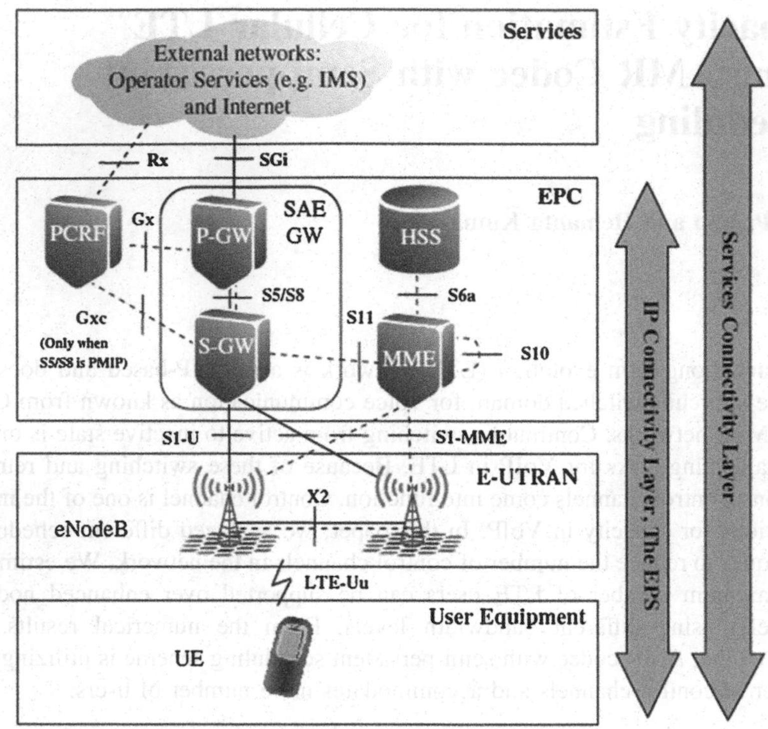

Fig. 1 EPS (LTE) architecture

over IP (i.e., VoIP). The architecture for LTE is presented in Fig. 1. The aim of LTE is to provide Internet protocol (IP) connectivity between user equipment (UE) to the packet data network (PDN). So to cope up with the entire IP network, new system architecture called evolved packet core system (EPS) has been proposed [2]. This system is a combination of two types of networks called as core network and access network. The core network called system architecture evolution (SAE) is based on evolved packet core (EPC). The other network (i.e., access network) is based on LTE, and this relies on evolved universal mobile telecommunication terrestrial radio access network (E-UTRAN). The radio network controller (RNC) functionality splits between enhanced node B (eNodeB) and mobility management entity (MME). eNodeB keeps track about all functionality that do not require from other cells. MME keeps track about the information which is required by other cells. EPS provides the user with IP connectivity to the packet data network (PDN) so that user can access VoIP as well as Internet. The quality of service (QoS) information is tracked by the EPS bearer.

The protocol stacks describing the user plane and the control plane for cellular LTE are presented in Figs. 2 and 3, respectively. In the user plane, an IP packet for a UE is encapsulated in an EPC-specific protocol and tunneled between packet data network gateway (P-GW) and eNodeB for transmission to the UE. GPRS-tunneling

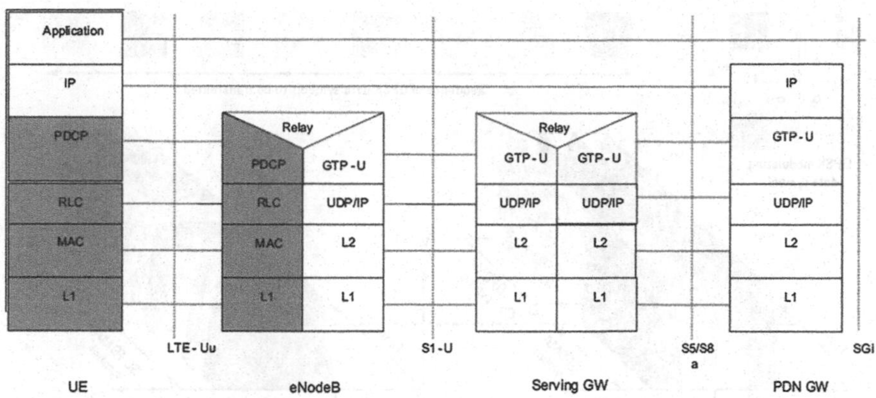

Fig. 2 E-UTRAN message exchange protocol

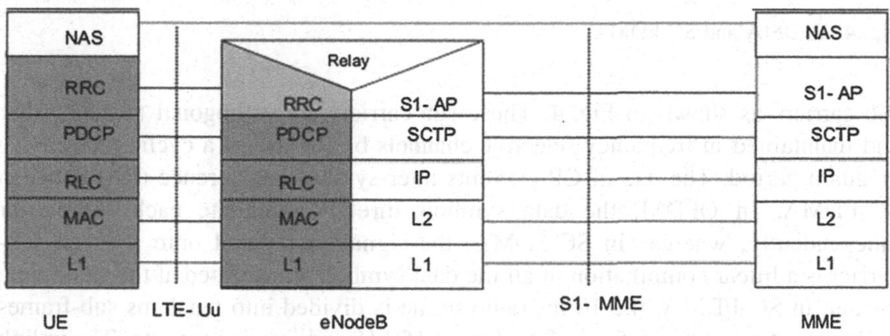

Fig. 3 Control plane exchange protocol

protocol (GTP) [3] is used over the core network interfaces, S1 and S5/S8. SAE also provides an option to use proxy mobile IP (PMIP) on S5/S8 [4]. The radio resource control (RRC) protocol is mainly responsible for establishing the radio bearers and configuring all the lower layers using RRC signaling between the eNodeB and the UE. The packet data convergence protocol (PDCP) layer is responsible for processing RRC messages in the control plane and IP packets in the user plane. The main functions of the PDCP layer are header compression, security, and support for reordering and retransmission during handover [5]. The main function of radio link control (RLC) layer is segmentation and assembling the packets. It is also responsible for retransmission of loss packets [6]. The medium access control (MAC) layer performs multiplexing of data from different radio bearers. MAC layer aims to achieve the negotiated QoS for each radio bearer [7].

LTE uses orthogonal frequency division multiple access (OFDMA) for downlink and single carrier frequency division multiple access (SC-FDMA) for uplink. SC-FDMA is valid for both FDD and TDD modes of operation. Both OFDMA and SC-FDMA divide the transmission bandwidth with multiple parallel

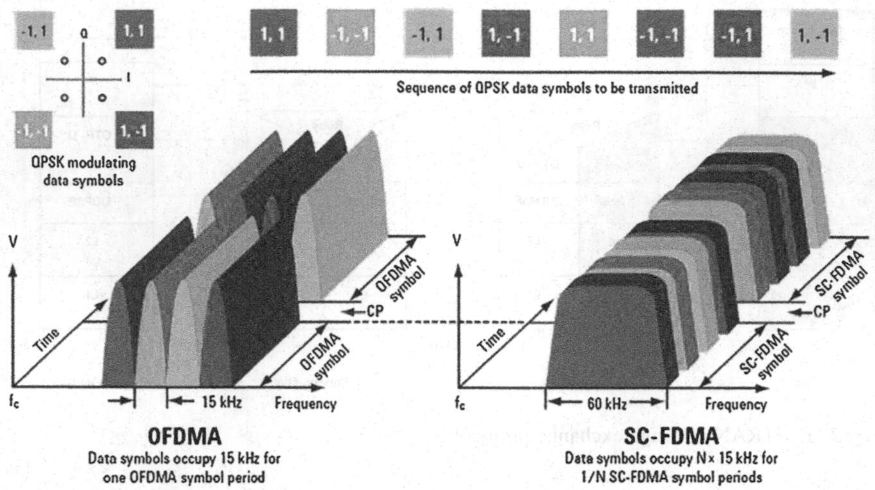

Fig. 4 OFDMA and SC-FDMA

sub-carriers as shown in Fig. 4. These sub-carriers are orthogonal to each other and maintained in frequency selective channels by the use of a cyclic prefix (CP) or guard period. The use of CP prevents inter-symbol interference (ISI) between SC-FDMA. In OFDM, the data symbols directly modulate each sub-carrier independently, whereas in SC-FDMA, the signal modulated onto a given sub-carrier is a linear combination of all the data symbols transmitted at the same time instant. In SC-FDMA, the 10 ms radio frame is divided into ten 1 ms sub-frames each consisting of two 0.5 ms slots. It uses 15 kHz sub-carrier spacing. The uplink transmission resources are also defined in the frequency domain, with the smallest unit of resource being a resource element (RE), consisting of one SC-FDMA data block length on one sub-carrier. The 1 ms sub-frame allows a 1 ms scheduling interval this is called as transmission time interval (TTI). There might be possibility that a single VoIP packet cannot be transmitted in a 1 ms sub-frame with an acceptable error rate. This may further require segmentation of the VoIP packet at higher layer. Improving uplink VoIP coverage at the cell edge, we use TTI bundling, where a single transport block from the MAC layer is transmitted repeatedly in multiple consecutive sub-frames, with only one set of signaling messages for the whole transmission. Uplink allows groups of 4 TTIs to be 'bundled.'

The modulation methods for user data are quadrature phase shift keying (QPSK), 16QAM and 64QAM. 16QAM and 64QAM provide lower maximum power for UE for control information transmission either BPSK or QPSK is used. For uplink, 16QAM or 64QAM is used, where control data and user data are multiplexed together. The adaptive multi-rate (AMR) codec is used for running voice service in GSM, WCDMA, and HSPA for its ability to codec rate adaptation to the radio conditions. AMR uses a sampling rate of 8 kHz, which provides 300–3400 Hz audio bandwidth. AMR can also be used for LTE. The extended

Fig. 5 AMR voice codec radio bandwidth

AMR-NB	AMR-WB
12.2 kbps	23.85 kbps
10.2 kbps	19.85 kbps
7.95 kbps	18.25 kbps
7.4 kbps	15.85 kbps
6.7 kbps	14.25 kbps
5.9 kbps	12.65 kbps
5.15 kbps	8.85 kbps
4.75 kbps	6.6 kbps
1.8 kbps	1.75 kbps

form of AMR-wideband(AMR-WB) codec was added to 3GPP Release 5. It uses 16-kHz sampling rate and provides 50–7000 Hz audio bandwidth. Figure 5 shows the different ranges of bandwidths supported in AMR (i.e., AMR-NB) and AMR-WB.

Rest of this paper is organized as follows. Section 2 summarizes related work for capacity estimation for LTE systems. Section 3 discusses about the scheduling techniques used in capacity analysis. Section 4 presents the numerical results. Finally, Sect. 5 concludes this paper.

2 Related Works

The capacity of cell is defined as the number of connections that can simultaneously exist with an acceptance level of mutual interference. The work reported in [8] described semi-persistent scheduling technique. It has proven better performance comparing the dynamic scheduling technique. In [9], 3GPP has published its VoIP capacity, proving LTE-advanced capable to outperform the IMT-A VoIP capacity requirement by factor of two to three depending on the scenario. The work reported in [10] discussed about the scheduling algorithm and link-to-system interface for their IMT-A compliant VoIP capacity evaluation. In this work, author used dynamic resource allocation technique for capacity estimation, considering the downlink capacity only. The work reported in [11] considered dynamic resource allocation procedure but did not consider the control channel limitation. The work reported in [12] considers the semi-persistent-based resource allocation for both uplink and downlink under realistic control channel constraints. In this work, authors mainly concentrate over channel quality variations caused by inter-cell interference. The work reported in [13] presents a novel modulation and coding scheme selection method to optimize the performance of VoIP semi-persistent scheduling in LTE downlink. They found their simulation results efficiently decreases the BLER, HARQ retransmission, and packet delay. The work done in [14] considered semi-persistent scheduling for VoIP by random access and proven better performance. This method does not work well in the case of large number of control channels. The work reported in [15] compared the semi-persistent and group scheduling technique and found that the semi-persistent has the larger capacity than group scheduling in the case of no group interaction. The work reported in [16] considered semi-persistent scheduling for the VoIP service in the LTE-advanced relaying networks. They found that VoIP capacity for LTE-advanced relaying networks is much higher than traditional networks without RNs. The main drawback of this scheme is the increase in delay when nodes are two hops away. In the following section, the theoretical capacity for VoIP users in LTE is found and it is further used to obtain the number of control channels used for different scheduling mechanisms.

3 Capacity Analysis for LTE Providing VoIP Service

If we consider VoIP system utilizes 100 % resources, then we can roughly formulate the maximum number of users using the work reported in [17]. The total number of available resources for VoIP traffic (i.e., T_{total}) can be found using the following expression.

$$T_{\text{total}} = (\text{Num of TTI per 20 ms}) \times (\text{Num of RB per TTI}) \tag{1}$$

Now, the number of used resources for VoIP traffic (i.e., T_{used}) can be found using the following expression.

$$T_{\text{used}} = \sum_{k=1}^{K} N_{\text{Ack},k}(1 + R_{\text{ave\_}k} \times \varepsilon_k) + \sum_{l=1}^{L} N_{\text{Non }A,l}(1 + R_{\text{ave\_}l} \times \varepsilon_l) \tag{2}$$

where K and L are the number of the active and non-active users, respectively, $N_{\text{Ack},k}$ and $N_{\text{Non }A,l}$ are number of needed TTI-RB grids to transmit one speech packet or SID packet of user k, l under active status and DTX (discontinuous transmission) status, respectively, $R_{\text{ave\_}k}$ and $R_{\text{ave\_}l}$ are the average retransmission number for the user under active and non-active state, respectively, and ε_k and ε_l are the ratio of resources for retransmissions to resources for initial transmissions of a specific user in different states. So, it is obvious that

$$\frac{T_{\text{used}}}{T_{\text{total}}} < 1 \tag{3}$$

Using Eqs. (1)–(3), we can calculate total number of users. Since the above equations need the basic information for all the users, so it is quite difficult. To simplify the calculation, we make the following assumptions: (1) v be the average activity factor for all the VoIP users, (2) the same average retransmission $R_{ave\,R}$ be used for both active and non-active state for all users, (3) $_{k}=_{l}=1$ means same number of frequency resources is used for new transmission and retransmission, (4) $N_{Ack,k} = N_{Non\,A,l} = N_{ave}$, (5) the proportion of packets arriving in active and non-active state be p with a default value of $p = 1/8$. Let the number of users supported per cell is N_{sup}. Now applying all these assumptions in Eq. (2), we obtain the following expression.

$$T_{used} = \left(1 + R_{avg\,R}\right) \times N_{ave} \times N_{sup} \times ((1-p) \times v + p) \tag{4}$$

Using Eqs. (3) and (4), we obtain the following expression.

$$N_{sup} \le \frac{T_{total}}{\left(1 + R_{avg\,R}\right) \times N_{ave} \times (0.875 \times v + 0.125)} \tag{5}$$

3.1 VoIP Scheduling Techniques

Several scheduling techniques available for VoIP in LTE are discussed in the following.

3.1.1 Dynamic Scheduling

In this procedure, UE sends resource request to eNB for each VoIP packet. For retransmission of packets, UE sends separate resource request to eNB. A lot of signaling is one of the major drawbacks of this scheme. The average number of control channels (i.e., No.CCH) can be formulated as given in the following expression [8].

$$\text{No.CCH} = m[\frac{nv}{I_1} + \frac{n(1-v)}{I_2}] \tag{6}$$

where n is the total number of VoIP users, m is the mean average transmission number, I_1 and I_2 denotes inter-arrival time for voice packet and SID, respectively.

3.1.2 Persistence Scheduling

To overcome with the signaling constraint, this scheme has been proposed. RRC signal is responsible for resource allocation and fixed modulation scheme for UE. This allocation also keeps track about the resource required for hybrid ARQ (HARQ) retransmission. If channel condition or AMR codec changes, then new persistence allocation will be done and previous will be overridden with newer one. Persistent allocation might be implemented in such a way that

HARQ ACK/NAK is not needed but instead each packet is sent a fixed number of times [18]. The main drawback of this technique is less resource utilization.

3.1.3 Semi-persistent Scheduling

This technique is combination of dynamic and persistence scheduling techniques. It uses persistence scheduling for initial transmission and dynamic scheduling for retransmissions. The initial allocation is done either by L1/L2 control channel or by MAC controls PDU, whereas retransmission is done by L1/L2 control channels dynamically. SID packets were allocated either by dynamically or by semi-persistently. If initial transmission is done in MAC control PDU and SID packet dynamically scheduled, then No.CCH per TTI can be obtained by using the following expression [8].

$$\text{No.CCH} = (m-1)\frac{nv}{I_1} + m\frac{n(1-v)}{I_2} \tag{7}$$

If SID packets are scheduled through semi-persistent, then No.CCH per TTI can be obtained by using the following expression [8].

$$\text{No.CCH} = (m-1)\frac{nv}{I_1} + (m-1)\frac{n(1-v)}{I_2} \tag{8}$$

3.2 Traffic Model for VoIP

The two-state Markov model is shown in Fig. 6. In this model, the probability of transitioning from state 1 (the active speech state) to state 0 (the inactive or silent state) while in state 1 is equal to a, while the probability of transitioning from state 0 to state 1 while in state 0 is c [19]. The model is assumed and updated at the speech encoder frame rate $R = 1/T$, where T is the encoder frame duration (typically, 20 ms). The steady state of this model requires that

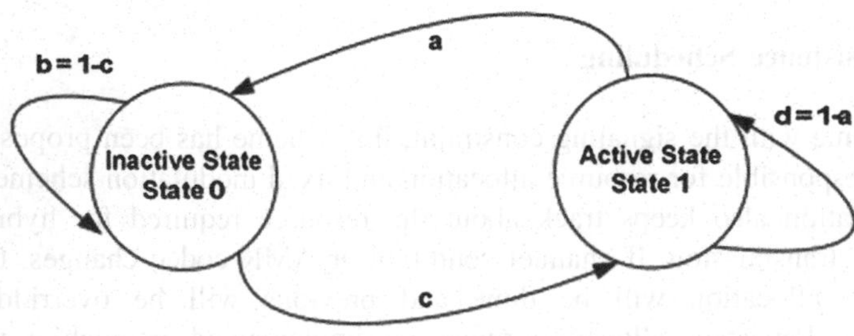

Fig. 6 Two-state voice activity model

$$P_0 = \frac{a}{a+c}, \quad P_1 = \frac{c}{a+c} \tag{9}$$

where P_0 and P_1 are respectively the probability of being in state 0 and state 1.

The voice activity factor (VAF) (i.e., v) is given by the following expression.

$$v = P_1 = \frac{c}{a+c} \tag{10}$$

Here, we assume the VAF = 50 % (i.e., half probability of active and inactive periods ($a = c = 0.01$)). Here, active and inactive duration periods are negative and exponentially distributed with an average of 2 s. Since probability for state transition is 0.01, so mean state time duration will be $\frac{1}{0.01} \times 20$ ms = 2 s. When user is in active state, the VoIP arrival rate is 20 ms, while in inactive state, a silence insertion descriptor (SID) packet arrival rate is 160 ms.

4 Numerical Results

We estimated the capacity of the cell on the basis of different scenarios. We considered different scheduling techniques and different codec schemes on different system bandwidths. The parameter values used to obtain numerical results are summarized in Table 1.

Table 1 Parameter values used

Parameters	Values
Encoder frame length	20 ms
Voice activity factor (VAF)	50 % (a = c = 0.01)
Total voice payload on air interface	40 bytes (AMR 12.2) and 30 bytes (AMR 7.95)
Bandwidth in MHz (corresponding resource blocks) [20]	1.4(6), 3(15), 5(25), 10(50), 15(75), 20(100)
TTI length	1 ms
VoIP Codec	AMR 12.2 kbps, AMR 7.95 kbps
Layout	1 cell
Mean average transmission number (m)	1.2
$R_{\text{avg}\,R}$	0.4

4.1 Capacity for Different Codecs and Bandwidths

In Fig. 7, we have plotted results obtained for the number of users on different bandwidths w.r.t. codec 12.2 and 7.95 kbps. We found from Fig. 7 that when the bandwidth increases, the total number of users also increases proportionally. Further, we observed that at different bandwidths the total number of users supported in 7.95 kbps is more than the users supported in 12.2 kbps.

4.2 Number of Control Channels for Different Scheduling Techniques

In this subsection, we estimated the number of control channels for different scheduling techniques with respect to number of users. Figures 8 and 9 show number of control channels using different scheduling techniques. From these results, we found that dynamic scheduling takes more numbers of control channels comparing semi-persistent scheduling. So, semi-persistent is one of the best scheduling techniques among the dynamic and persistent. Unlike the persistent scheduling scheme applying constraint on resource utilization, semi-persistent scheduling scheme gives better optimized resource utilization.

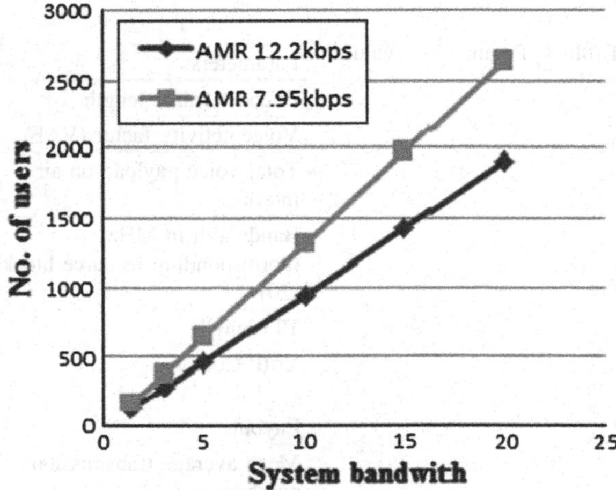

Fig. 7 Capacity comparison between AMR 12.2 and 7.95 kbps

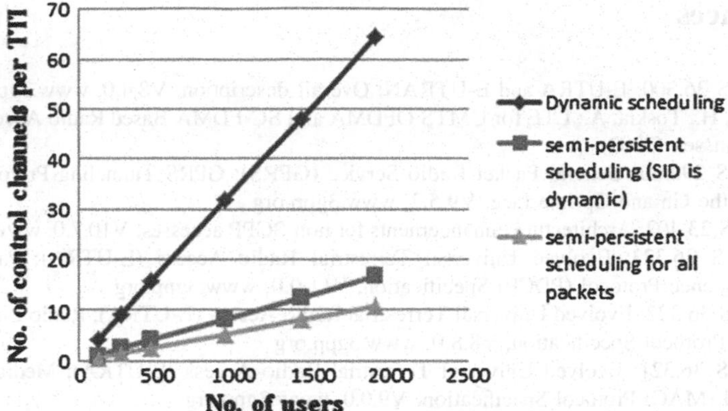

Fig. 8 Number of control channels comparison among scheduling techniques using AMR 12.2 kbps

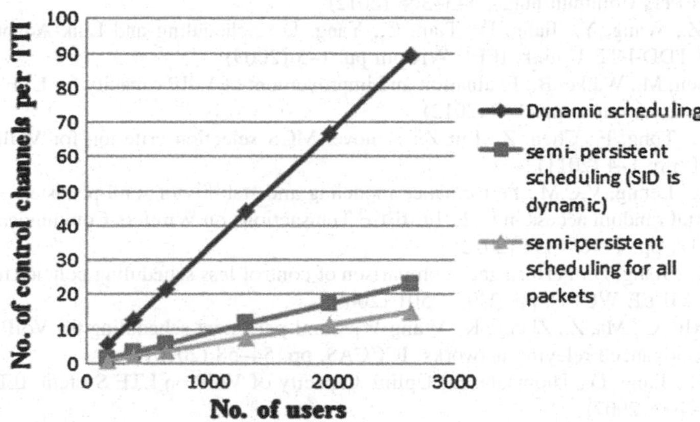

Fig. 9 Number of control channels comparison among scheduling techniques using AMR 7.95 kbps

5 Conclusion

In this paper, we discussed and analyzed different scheduling techniques and codec schemes used for improving the LTE UL VoIP cell capacity. We also found out the maximum number of users LTE can support in different bandwidths. From the results obtained, we found that carrier aggregation will further improve the cell capacity if we use the semi-persistent scheduling with TTI bundling.

References

1. GPP TS 36.300: E-UTRA and E-UTRAN: Overall description. V8.4.0, www.3gpp.org
2. Holma, H., Toskla, A.: LTE for UMTS OFDMA and SC-FDMA Based Radio Access. Wiley, West Sussex (2009)
3. GPP TS 29.060: General Packet Radio Service (GPRS); GPRS Tunneling Protocol (GTP) across the Gn and Gp interface, V9.5.2, www.3gpp.org
4. GPP TS 23.402: Architecture enhancements for non-3GPP accesses, V10.7.0, www.3gpp.org
5. GPP TS 36.323: Evolved Universal Terrestrial Radio Access (E-UTRA); Packet Data Convergence Protocol (PDCP) Specification, V11.0.0, www.3gpp.org
6. GPP TS 36.322: Evolved Universal Terrestrial Radio Access (E-UTRA); Radio Link Control (RLC) Protocol Specification, V8.8.0, www.3gpp.org
7. GPP TS 36.321: Evolved Universal Terrestrial Radio Access (E-UTRA); Medium Access Control (MAC) Protocol Specification, V9.0.0, www.3gpp.org
8. Jiang, D., Wang, H., Malkamaki, E., Tuomaala, E.: Principle and performance of semi-persistent scheduling for VoIP in LTE system. IEEE WiCom, pp. 2861–2864 (2007)
9. GPP TR 36.814: Evolved Universal Terrestrial Radio Access (E-UTRA); Further Advancements for E-UTRA Physical Layer Aspects, V9.0.0, March 2010, www.3gpp.org
10. Fan, Y., Valkama, M.: Enhanced VoIP support in OFDMA-based packet radio networks. Wireless Pers Commun **66**(2), 343–366 (2012)
11. Wang, Z., Wang, Y., Jiang, D., Tian, C., Yang, D.: Scheduling and Link Adaptations for VoIP in TDD-LTE Uplink. IEEE WiCom pp. 1–5 (2009)
12. Muhleisen, M., Walke, B.: Evaluation and improvement of VoIP capacity for LTE. European Wireless Conference, pp. 1–7 (2012)
13. Ding, L., Tong, F., Chen, Z., Liu Z.: A novel MCS selection criterion for VOIP in LTE. WiCOM, pp. 1–4 (2011)
14. Seo J.-B., Leung, V.C.M.: Performance modeling and stability of semi-persistent scheduling with initial random access in LTE. In: IEEE Transactions on Wireless Communications, vol. 11, no. 12, pp. 4446–4456 (2012)
15. Wang, H., Jiang, D.: Performance comparison of control-less scheduling policies for VoIP in LTE UL. IEEE WCNC, pp. 2497–2501 (2008)
16. Liu, J., Hu, C., Ma, Z., Zheng, K., Wang W.: Semi-persistent scheduling for VoIP service in the LTE-advanced relaying networks. ICCCAS, pp. 54–58 (2010)
17. Wang, H., Jiang, D., Tuomaala, E.: Uplink Capacity of VoIP on LTE System. IEEE APCC, pp. 397–400 (2007)
18. GPP TSG RAN WG2 #54, R2-062164 (2006) Uplink Resource Allocation Scheme, Tallinn, Estonia, 2006, www.3gpp.org
19. GPP TSG-RAN WG1 Meeting #48, R1-070674 LTE Physical Layer Framework for Performance Verification, St. Louis, USA, Feb 2007, www.3gpp.org
20. GPP TS 36.104, Evolved Universal Terrestrial Radio Access (E-UTRA) Base Station (BS) radio transmission and reception, V9.4.0, www.3gpp.org

Development of a Nonintrusive Driver Drowsiness Monitoring System

M. Harisanker and R. Shanmugha Sundaram

Abstract Driver errors and carelessness contribute most of the road accidents occurring nowadays. The major driver errors are caused by drowsiness, drunken, and reckless behavior of the driver. This paper focuses on a driver drowsiness detection system in Intelligent Transportation System, which focuses on abnormal behavior exhibited by the driver. In the proposed system, a nonintrusive driver drowsiness monitoring system has been developed using computer vision techniques. Based on the simulation results, it was found that the system has been able to detect drowsiness in spite of driver wearing spectacles as well as the darkness level inside the vehicle. Moreover, the system is capable of detecting drowsiness within time duration of about 2 s. The detected abnormal behavior is corrected through alarms in real time.

Keywords Drowsiness · Image processing · Eye tracking · Intelligent transportation system (ITS)

1 Introduction

Automotive population is increasing exponentially in the country. The biggest problem regarding the increased traffic is the rising number of road accidents. Road accidents are undoubtedly a global menace in our country. The Global Status Report on Road Safety published by the World Health Organization (WHO) identified the major causes of road accidents are due to driver errors and carelessness [1]. Driver sleepiness, alcoholism, and carelessness are the key players in

M. Harisanker (✉) · R. Shanmugha Sundaram
Amrita Vishwa Vidyapeetham, Coimbatore 641112, India
e-mail: hsmadai64@gmail.com

R. Shanmugha Sundaram
e-mail: rshanmugha@gmail.com

L.C. Jain et al. (eds.), *Intelligent Computing, Communication and Devices*,
Advances in Intelligent Systems and Computing 308,
DOI 10.1007/978-81-322-2012-1_79

the accident scenario. The fatalities and associated expenses as a result of road accidents are very serious problems. The related dangers resulted have been recognized as a serious threat to many families in every country.

All these factors led to the development of Intelligent Transportation Systems (ITS). Taking into account of these factors, the driver's behavioral state is a major challenge for designing advanced driver assistance systems. The major driver errors are caused by drowsiness, drunken, and reckless behavior of the driver. The real-time detection of these behaviors is a serious issue regarding the design of advanced safety systems in automobiles.

2 Background

Several works have been done in the field of driver abnormality monitoring and detection systems using a wide range of methods [2]. Among the possible methods, the best techniques are the ones based on human physiological phenomena [3]. These techniques can be implemented by measuring brain waves (EEG), heart rate (ECG), and open/closed state of the eyes [4]. The former two methods, though being more accurate are not realistic since sensing electrodes to be attached directly onto the driver's body and hence be annoying and distracting the driver. The latter technique based on eye closure is well suited for real world driving conditions, since it can detect the open/closed state of the eyes nonintrusively using a camera [5]. Eye tracking-based drowsiness detection systems have been done by analyzing the duration of eye closure and developing an algorithm to detect the driver's drowsiness in advance and to warn the driver by in-vehicle alarms.

3 System Architecture

The proposed system comprises of three phases.

1. Capturing: Eye camera mounted on the dashboard is used for capturing the facial image of the driver.
2. Detection: The analysis of the captured image is done to detect the open/closed state of the eyes. The driver's current driving behavior style is deduced using inbuilt haar classifier cascades in OpenCV.
3. Correction: This phase is responsible for doing the corrective actions required for that particular detected abnormal behavior. The corrective actions include in-vehicle alarms and displays [5]. The ARM7 microcontroller board connected serially to the PC performs the necessary corrective actions.

4 Procedure

In the proposed system shown in Fig. 1, the primary focus is given to the faster drowsiness detection and processing of data. The number of frames in which the eyes are kept closed is monitored and then counted. If the number of frames exceeds a threshold value, then a warning message is generated on the display showing that the drowsiness is detected. The system should be capable of detecting drowsiness in spite of the skin color and complexion of the driver, spectacles used by the driver and the darkness level inside the vehicle. All these objectives have been well satisfied by choosing the system using appropriate classifiers in OpenCV for eye closure detection.

In this algorithm, first a driver's image is acquired by the camera for processing. In OpenCV, the face detection of the driver's image is carried out first followed by eye detection [6]. The eye detection technique detects the open state of eyes only. Then the algorithm counts the number of open eyes in each frame and calculates the criteria for detection of drowsiness. If the criteria are satisfied, then the driver is said to be drowsy. The display and buzzer connected to the system perform actions to correct the driver abnormal behavior (Fig. 2).

For this system, the face and eye classifiers are required. The HaarClassifier-Cascade files inbuilt on OpenCV include different classifiers for the face detection and the eyes detection [7]. The inbuilt OpenCV xml "haarcascade_frontal-face_alt2.xml" is used to search and detect the face in individual frames. The classifier "haarcascade_eye_tree_eyeglasses.xml" is used to detect eyes in the open state from the detected face. The system does not detect in the closed state of the

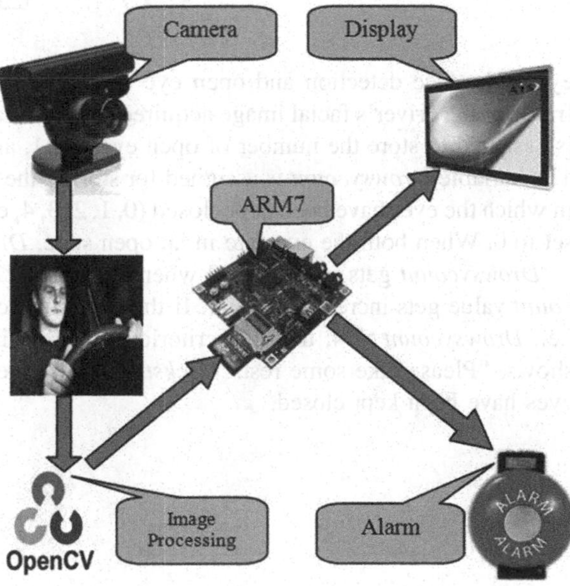

Fig. 1 Block diagram of the proposed system

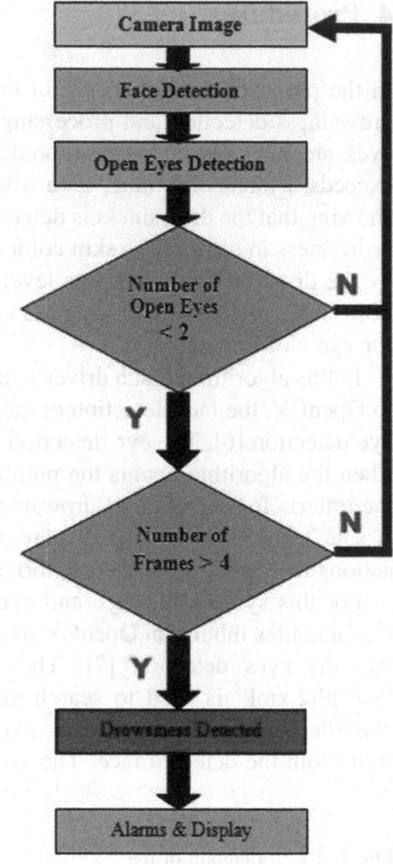

Fig. 2 Flow diagram of the proposed algorithm

eyes. The face detection and open eye detection have been carried out on each frame of the driver's facial image acquired from the camera. The variable *Eyestotal* is assigned to store the number of open eyes (0, 1, and 2) detected in each frame. The variable *Drowsycount* is assigned for storing the number of successive frames in which the eyes have been kept closed (0, 1, 2, 3, 4, etc.). Initially, *Drowsycount* is set to 0. When both the eyes are in an open state, *Drowsycount* is 0.

Drowsycount gets incremented when *Eyestotal* <2. For an eye blink, *Drowsycount* value gets incremented to 1. If the eyeblink occurs for more than 4 frames, i.e., *Drowsycount* >=4, then the criterion for drowsiness is satisfied. The display shows, "Please take some rest." *Tickstart* shows the real-time duration in which eyes have been kept closed.

5 Results

The proposed drowsiness detection system detects the drowsiness of the driver when the eyes are closed for 4 frames or more (i.e., more than 2 s). The detection system differentiates the normal eye blink and drowsiness. The system is nonintrusive and can be easily equipped with any vehicle.

The visual studio express simulation results of the drowsiness system are illustrated in the following figures. In Fig. 3, the normal state of the driver is shown in which the open eyes are detected. In Fig. 4, the driver eyes have been kept closed for successive 4 frames. The display shows "Please take some rest." The system is capable of detecting drowsiness in spite of the spectacles worn by the driver. The condition of the driver wearing spectacles is shown in Fig. 5. In normal driving conditions, the open eyes are detected. The drowsiness is detected when the eyes have been kept closed for successive 4 frames as shown in Fig. 6. The system developed here yields good results in the daytime as well as at night depending upon the camera performance.

Fig. 3 Normal state

Fig. 4 Drowsiness state

Fig. 5 Normal state with glass

Fig. 6 Drowsiness state with glass

Fig. 7 Execution window for drowsiness

```
eye count = 2  0
eye count = 2  0
eye count = 2  0
eye count = 2  0
eye count = 2  0
eye count = 0  0
eye count = 0  1
eye count = 0  2
eye count = 0  3
Drowsiness detected for 2.058237 secs
eye count = 0  4
Drowsiness detected for 2.592939 secs
eye count = 0  5
Drowsiness detected for 3.116029 secs
eye count = 2  6
eye count = 2  0
eye count = 2  0
```

In normal driving condition, both the eyes are open. Hence the system could detect the open eyes. Open eyes detection was illustrated by drawing rectangles around the eyes as shown in Figs. 3 and 5. In the drowsiness condition, the eyes get closed. Hence open eye detection could not be performed in that case shown in Figs. 4 and 6.

In Fig. 7, it is shown that for normal driving, the number of open eyes is 2 and the number of successive frames for closed eyes is 0. When the eyes are closed in one frame, the number of open eyes becomes 0 and number of closure frames increment by 1. If this condition persists for continuous 4 frames and more, the execution window shows drowsiness detected. Also, it calculates the duration for which the eyes have been kept closed. The system is capable to detect drowsiness in the case of closed eyes for more than 2 s.

6 Conclusion and Future Work

The drowsiness detection and correction system developed here is capable of detecting drowsiness in a rapid manner. The system which can differentiate normal eye blink and drowsiness can prevent the driver from entering the state of sleepiness while driving. The system works well even in case of drivers wearing spectacles and under low light conditions also. As a complete abnormality detection system, this system can be further developed by adding different sensors and lane detection camera with appropriate hardware units and controller, which can deliver highly accurate detection techniques. The system can be commercially generalized and well employed in today's vehicles with comparatively fewer expenses. As a whole, the system when equipped with the vehicles can reduce the traffic collisions occurring, related dangers and expenses in our country.

References

1. Global Status Report on Road Safety: Country profiles: India (Report). World Health Organization. Retrieved 3 May 2012
2. Williamson, A., Chamberlain, T.: Review of on-road driver fatigue monitoring devices, NSW Injury Risk Management Research Centre, University of New South Wales, April 2005
3. Rogado, E., García, J.L., Barea, R., Bergasa, L.M., López, E.: Driver fatigue detection system. In: Proceedings of the 2008 IEEE International Conference on Robotics and Biometics, Bangkok, Thailand, February 21–26, 2009
4. Lee, B.G., Chung, W.Y.: Driver alertness monitoring using fusion of facial features and bio-signals. IEEE Sens. J. **12**(7) (2012)
5. Singh, H., Bhatia, J.S., Kaur, J.: Eye tracking based driver fatigue monitoring and warning system. In: Proceedings of the IEEE in Power Electronics IICPE, New Delhi, India (2011)
6. Bajaj, P.R.Dr., Devi, M.S.: Driver fatigue detection based on eye tracking. In: Proceedings of the First International Conference on Emerging Trends in Engineering and Technology, IEEE Computer Society, 2008
7. Bradski, G., Kaehler, A.: Learning OpenCV, O'Reilly Media, Inc., 1005 Gravenstein Highway North, Sebastopol, CA 95472, September 2008

Wearable Medical Devices in Preventive Health Care: Cuffless Blood Pressure Measurement

Rajesh Kannan Megalingam, Ushma Unnikrishnan, Athira Subash, Goutham Pocklassery, Athul Asokan Thulasi, Galla Mourya and Vivek Jayakrishnan

Abstract The advent of technology has made great leaps from the conventional methods of disease treatment. Nowadays, there has been a great shift toward continual monitoring, early detection, and prevention of diseases. Complimenting to this is the innovation of wearable medical devices. Wearable medical devices are portable electronic devices embedded in the user's outfit. This not only facilitates faster treatment but also ensures comfortable diagnostic environment for the patient. ECG and PPG signals are of prime importance when the continual monitoring of the cardiac state of the patient is concerned. Hence, using a wearable medical device, the biomedical signals can be easily obtained and interpreted. This paper discusses a method to measure blood pressure from ECG and PPG signals acquired from the device. To realize the method, various algorithms for the implementation were coded in MATLAB. The algorithm was tested on records taken from PhysioNet ATM and a private database.

R.K. Megalingam (✉) · U. Unnikrishnan · A. Subash · G. Pocklassery · A.A. Thulasi · G. Mourya · V. Jayakrishnan
Amrita School of Engineering, Amrita Vishwa Vidyapeetham University, Kollam 690525, India
e-mail: megakannan@gmail.com

U. Unnikrishnan
e-mail: ushma.ushus@gmail.com

A. Subash
e-mail: athira3003@gmail.com

G. Pocklassery
e-mail: gouthamravindran@gmail.com

A.A. Thulasi
e-mail: athulat.002@gmail.com

G. Mourya
e-mail: mouryagalla@gmail.com

V. Jayakrishnan
e-mail: vivekjayakrishnan@gmail.com

L.C. Jain et al. (eds.), *Intelligent Computing, Communication and Devices*, Advances in Intelligent Systems and Computing 308,
DOI 10.1007/978-81-322-2012-1_80

Keywords Electrocardiogram · Photoplethysmography · Preventive health care

1 Introduction

The continuous monitoring of the patient's ECG and activity is of heightening demand as it has an increasing impact on the mortality and morbidity. This signal is easily accessible with the help of placing electrodes on the patient's body. Unlike the sophisticated and wired devices, wearable devices prove to be beneficial in easy acquisition of the signal for a long-term use. The electrocardiogram and photoplethysmography signals reflecting the cardiac state of a person provide ample scope for preventive health care. The inherent features of these biomedical signals are exploited to help predict abnormalities related to the heart. A simple and efficient algorithm was implemented to extract the vital features from these signals. This processing enables improved interpretation of the signals and facilitates the acquisition of clinically important information. The calculation of the pulse transit time from ECG and PPG signals can be used for the estimation of blood pressure. This method is noninvasive, and unlike the conventional methods, it ensures comfortability of the patient during examination.

2 Related Works

Various publications on the topic discuss about the digital filtering for the preprocessing, nonlinear transformations for signal enhancement, and various decision rule algorithms for peak detection of ECG and PPG signals. One of the very early publications, Pan and Tompkins algorithm in paper [1], represent a simple analytical method for QRS detection. It proposes a very sensitive method of detection based on continual tracking of two thresholds. Using parameters such as sensitivity and positive predictivity of beat detection, the performance of the proposed algorithm was evaluated in this paper [2]. They discuss algebraic derivative-based algorithm for R wave detection of noisy ECG signals. The results are then obtained by testing the algorithm on MIT-BIH arrhythmia database. Paper [3] not only describes the preprocessing and QRS detection methods but also details about the direct and transform-based methods for data compression. Authors of this paper [5] discuss the preprocessing required for PPG analysis and then propose a method for peak detection. This algorithm describes initial local maxima identification, false peak detection, and missed peak relocation.

3 Block Diagram

Figure 1.

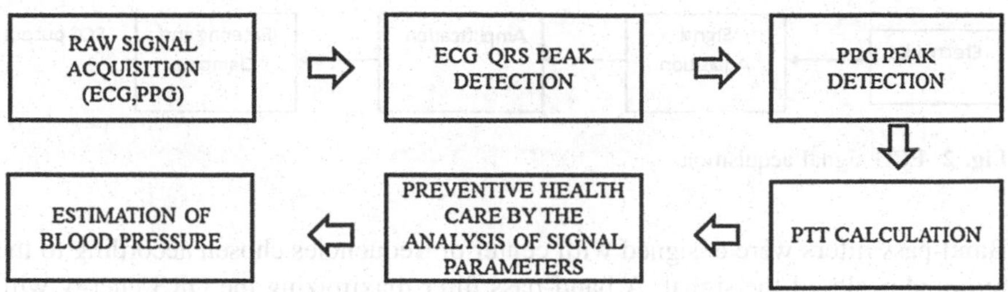

Fig. 1 Block diagram

3.1 Signal Acquisition

ECG: The first stage is differential-mode amplification by AD620 instrumentation amplifier using the right leg electrode as right leg drive and then comes the filtering stage which consists of three stages, namely high-pass, low-pass, and notch filters. High-pass filter removes the low-frequency noise components, whereas the low-pass filter removes the high-frequency noise components. The purpose of the notch filter is to eliminate the 50-Hz power supply interference. A clamper is kept at the output of the notch filter to restrict amplitude range of the ECG signal to positive axis (greater than zero) by adding a DC offset.

PPG: Photoplethysmography is a method in which the light absorption property of fluids is used to measure the volumetric flow of blood through the blood vessels in the finger. Two LEDs, namely red and infrared, are alternatively switched on and off. Red is absorbed by deoxygenated hemoglobin, whereas infrared is absorbed by oxygenated hemoglobin. A photodiode at the other end of the finger can be used to detect the intensity of light. These signals are passed through amplifying and filtering stages. Two different circuits are used to process the waveforms corresponding to each LED, and a sample-and-hold IC is used to switch between these circuits, depending on which LED is active. From the resulting two waveforms, the heart rate as well as the oxygen saturation level of the patient can be calculated.

3.2 Methodology

Figure 2 describes the methodology followed for peak detection. The raw ECG and PPG signals are mixed up with several noise components such as power line interference, breathing artifacts, and muscle movements. The acquired data need to be cleaned effectively before taken for analysis.

The inherent trend in the raw signals was removed by the method of piecewise polynomial fitting. The detrended signal is then ready for further analysis. Signals were then conditioned by removing the DC components prior to filtering.

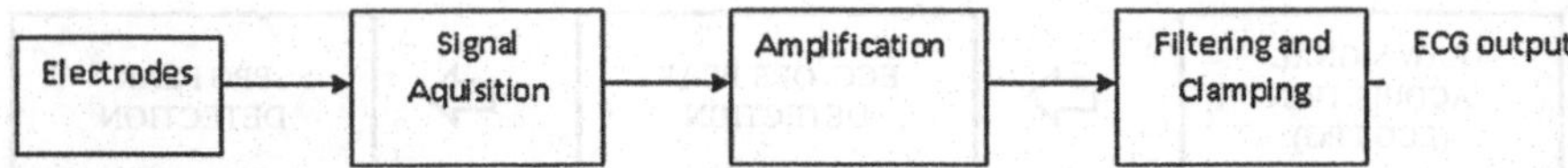

Fig. 2 ECG signal acquisition

Band-pass filters were designed with center of frequencies chosen according to the required peaks of the signal. A band-pass filter maximizing the QRS energy with lower and upper cutoff frequencies of 8 and 30 Hz, respectively, was used for the ECG signals. In the case for PPG signals, a band-pass filter with cutoff frequencies 1 and 15 Hz was used. The filtered signals were then conditioned to enhance the peaks. The peak detection is explained in Fig. 3. For the peak detection of ECG signal, a threshold value equal to the mean value of the integrated signal was set. This gives clusters of points crossing this threshold corresponding to the potential peaks of the QRS complex. The maxima of each cluster correspond to the R peak in each complex. With the location of R peak as reference, the minima on the left and right of the R peak give the Q and S peaks, respectively (Fig. 4).

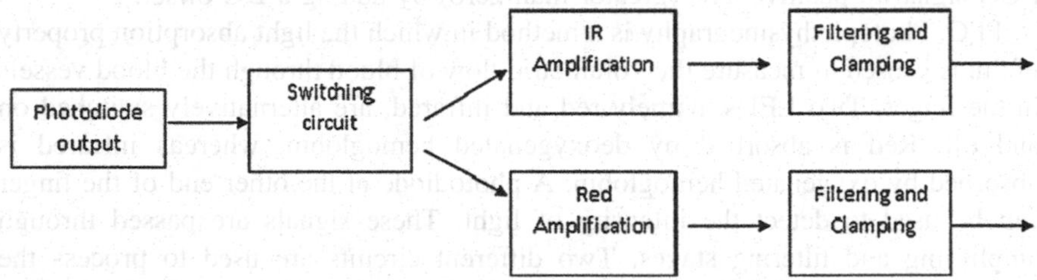

Fig. 3 PPG signal acquisition

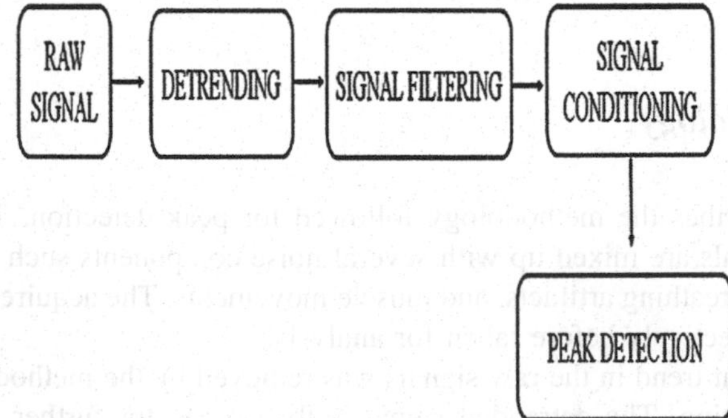

Fig. 4 Methodology

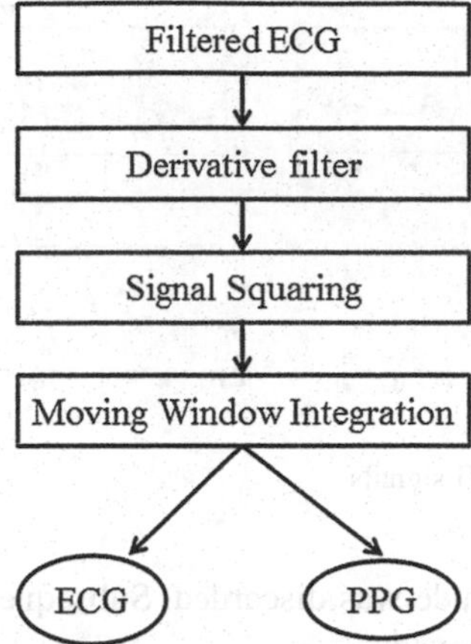

Fig. 5 Algorithm

In the case of PPG peak detection, thresholding was done similar to the previous case and the maxima in each cluster of points crossing the threshold give the required peak [4] (Figs. 5, 6 and 7).

After the implementation of the peak detection algorithm in cases where a false peak was detected, a corrective algorithm was implemented to remove the false peaks. R peak locations with R–R interval less than 0.3 s were taken, and the R

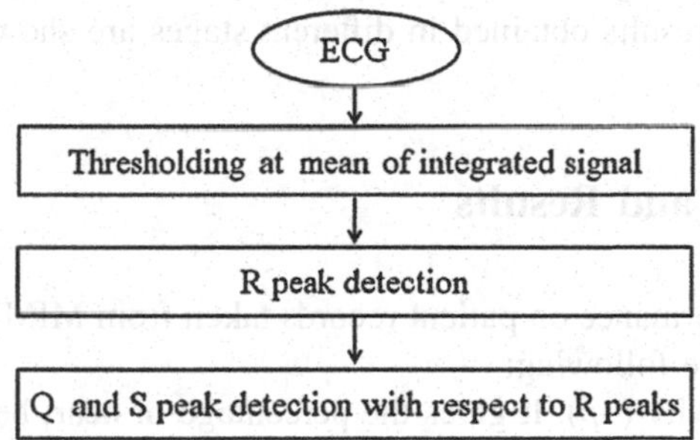

Fig. 6 Peak detection in ECG signals

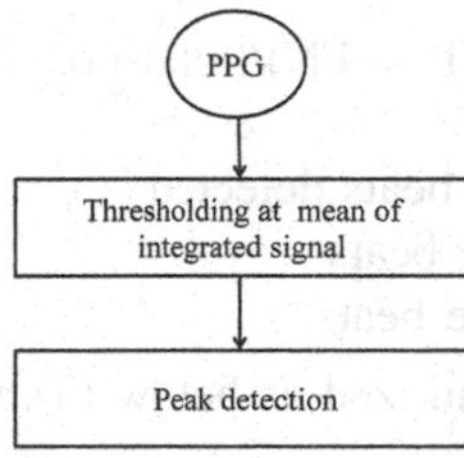

Fig. 7 Peak detection in PPG

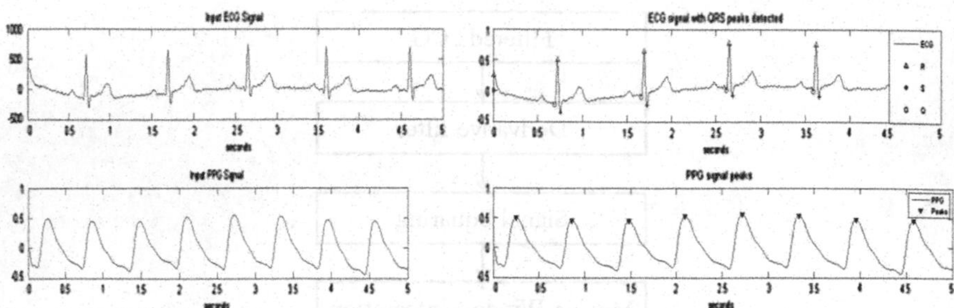

Fig. 8 Measured ECG and PPG signals

peak with the lower amplitude was discarded. Subsequently, the corresponding Q and S peaks were also removed.

4 Inference

The proposed algorithm was observed to be effective in detecting the Q, R, and S peaks in ECG and the peaks of the PPG signal. The false peaks detected were effectively removed using the proposed false peak detection algorithm. Further, the positive predictivity and the detection error rate of the algorithm were also calculated by testing the algorithm on patient records taken from MIMIC II Database of PhysioNet. The results obtained in different stages are shown in Fig. 8.

5 Experiments and Results

The algorithm performance on patient records taken from MIMIC II Database was used to calculate the following:

Positive predictivity (+*p*): It gives the percentage of heart beat detection which are really true beats.

$$+\mathrm{p} = \mathrm{TP}/(\mathrm{TP} + \mathrm{FP})$$

Detection error rate (%): (FP + FN)/(total no. of QRS complex)

where

TP Number of true-positive beats detected

FP Number of false-positive beats

FN Number of false-negative beats.

The results obtained are summarized as below (Table 1):

Table 1 Predictivity and detection error rate

S No.	Total no. of peaks	Number of false peaks	Number of true peaks	(+) p%	DER
1	4,921	12	4,909	99.756	0.244
2	4,727	1	4,726	99.978	0.021159
3	4,683	4	4,679	99.914	0.08547
4	5,422	0	5,422	100	0
5	6,082	0	6,082	100	0
6	4,321	1	4,320	99.976	0.023148
7	5,456	11	5,445	99.798	0.20202
8	4,264	15	4,249	99.648	0.35269
9	4,158	2	4,156	99.951	0.048123
10	5,125	3	5,122	99.941	0.0585708
11	4,931	0	4,931	100	0
12	6,403	23	6,380	99.641	0.360501
13	6,280	0	6,280	100	0
14	5,796	2	5,794	99.965	0.0345184
15	4,003	2	4,001	99.95	0.049987
16	4,928	0	4,928	100	0
17	4,778	2	4,776	99.958	0.041876
18	5,633	0	5,633	100	0

6 Future Works

In this paper, the R peaks were detected by setting a fixed value of threshold which is determined from the analysis of several data. But from the preventive health care point of view, the threshold needs to be adaptive, depending on the nature of the biomedical signal. The peak detection algorithm can be further improved by incorporating the missed peak detection algorithm so that the method succeeds in situations of all types of abnormal data.

There is scope for intensive research for the estimation of blood pressure from the calculation of pulse transit time. Further, the algorithm can be extended to predict the disease of the patient using the statistical and distribution studies.

7 Conclusion

Our objective is to develop an effective algorithm to analyze biomedical signals and extract vital information from them to aid preventive health care. Several records were collected from the database for testing the algorithm. The raw signals were initially preprocessed, and the signal was enhanced for further analysis. The detection algorithms were implemented, and the corresponding waveforms were observed. Using peak detection algorithm, the Q, R, and S peaks and the peaks of the PPG signal were detected. The time interval between the peaks of the ECG signal was also obtained. Comparison of this information with standard database aids the prediction of cardiac abnormalities.

Acknowledgments We would like to express our profound gratitude and deep regards to our guide Mr. Rajesh Kannan Megalingam for his exemplary guidance, cordial support, and valuable information. We also extend our thanks to Dr. G Jayachandran Nair, Distinguished Professor of ECE Department, for the technical support he provided in different stages of progress. His expertise has helped us simplify much of the complex concepts.

References

1. Pan, J., Tompkins, W.J.: A real time QRS detection algorithm. IEEE Trans. Biomed. Eng. **32**, 230–236 (1985)
2. Rezk, S., Join, C., El Asmi, S.: An algebraic derivative-based method for R wave detection. In: 19th European signal processing conference (2011)
3. Sornmo, L., Laguna, P.: Electrocardiogram signal processing. Wiley Encyclopedia of Biomedical Engineering (2006)
4. Smilyjeya Jothi, E., Preetha, S.: Detection of peak and ONSET of PPG signal. Int. J. Adv. Sci. Tech. Res., ISSN **1**(3)2249–9954, Jan–Feb 2013
5. Rabiner, Lawrence R., Gold, Bernard: Theory and Applications of Digital Signal Processing, 15th edn. Prentice Hall of India, New Delhi (2001)

Analysis of Voice for Parkinson's Disease Persons Using Dynamic Time Warping Technique

Vikas and R.K. Sharma

Abstract This paper presents a dynamic time warping (DTW) technique-based analysis of voice for distinguishing Parkinson's disease (PD) persons from healthy persons. Mel frequency cepstral coefficient (MFCC) algorithm with MATLAB coding has been used to process voice samples. MFCC is converted into vector using MATLAB. DTW is useful for matching of voice samples. DTW-based matching percentage between PD-affected persons is 80.2163 %, whereas it is 72.2588 % between healthy persons. First coefficient of MFCC shows large values in case of PD-affected persons.

Keywords Analysis · MFCC · Dynamic time warping · Parkinson's disease · Matching · Voice

1 Introduction

Parkinson's disease (PD) is a neurological disorder that affects motor action in human body. Tremors in hands, legs, neck, speech problems, muscle rigidity, etc., are the symptoms of PD. Age group of PD persons varies from 45 to 85 years. Due to speech problems in PD-affected persons, voice samples of PD-affected person and healthy person were analyzed using Mel frequency cepstral coefficients (MFCC) and dynamic time warping (DTW) technique in MATLAB and PRAAT. A very important key element in feature-matching process is feature extraction [1]. MFCC are most widely used for feature extraction process. According to Hossan et al. [2], MFCC is a popular technique that is based on the known variation of

Vikas (✉) · R.K. Sharma
School of VLSI Design and Embedded Systems, NIT Kurukshetra, Kurukshetra, India
e-mail: vikas.rohilla11@gmail.com

R.K. Sharma
e-mail: mail2drrks@gmail.com

L.C. Jain et al. (eds.), *Intelligent Computing, Communication and Devices*,
Advances in Intelligent Systems and Computing 308,
DOI 10.1007/978-81-322-2012-1_81

human ear's critical frequency bandwidth [2]. MFCC coefficients can be obtained by de-correlating the output log energies of a filter bank that consists of triangular filters spaced linearly on Mel scale. DTW technique is used basically for feature-matching process. DTW is a powerful technique used to find out the feature similarities between two time series which are extracted for two voice samples using MFCC and vectorization [3]. Vectorization can be done for converting MFCC coefficients in vector form using MATLAB.

2 Database

For feature matching, a voice database of 36 persons (especially five vowels a, e, i, o, and u) has been recorded using Sony IC Recorder (ICD-UX513F). Each person was asked to pronounce the vowels loudly twice or thrice. Thus, the total voice samples in the study were '180.' In the available database, '20' were of PD-affected persons and rest were of normal healthy persons. The male persons whose voices were recorded have an age between 50 and 74 years and for females it was between 45 and 85 years.

3 Feature Extraction and Feature Matching

3.1 Feature Extraction Using MFCC

MFCC are the most acoustic and robust in nature, therefore popularly used for feature extraction process [4]. MFCC coefficients are derived from a type of cepstral representation of voice signal. Frequency bands are spaced equally on Mel scale in Mel cepstral, which calculates human auditory system's response more closely than normal cepstral (in which frequency bands are linearly spaced).

Figure 1 shows the complete process to calculate MFCC coefficients. Conversion of voice sample to Mel frequency scale is done using (1), where '*f*' is frequency.

$$\mathrm{Mel}(f) = 2{,}595 \log_{10}(1 + f/100) \tag{1}$$

As shown in Fig. 1, an audio signal is passed through pre-emphasis block. Compensation of high-frequency part that was suppressed during recording of voice can be done using pre-emphasis block. Pre-emphasized signal splits into smaller frames of 15–30 ms duration. A voice signal shows large variations in its amplitude if it is of large duration, and therefore, smaller signals are preferred for frame blocking. After splitting voice signal into frames, hamming window function is multiplied with total number of frames. Then, signal is passed through fast Fourier transform (FFT) block to convert signal into frequency domain. Mel filter

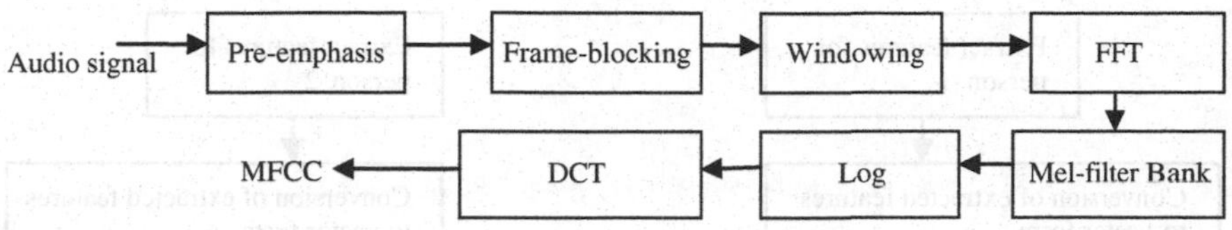

Fig. 1 Block diagram for MFCC

bank block converts signal frequency into Mel frequency using (1). Logarithm is used for channel normalization. Finally, discrete cosine transform (DCT) of logarithm block output is done that de-correlates overlapped energies of Mel filter bank. Output of DCT block provides MFCC values.

3.2 Feature Matching Using DTW

DTW is one of the algorithms for finding out similarities between two sequences. It is based on dynamic programming which proved to be a very reliable method [5]. Optimal path between two time series can also be founded out if one of the sequences may be 'wrapped' on other by stretching or shrinking on its time axis. Figure 2 shows how one sequence may be wrapped on other [6].

Let us assume two time series *X* and *Y* of length *m* and *n*, respectively, where

$$X = x_{1,} x_2, x_3, \ldots x_u, \ldots x_n \tag{2}$$

$$Y = y_1, y_2, y_3, \ldots y_v, \ldots y_m \tag{3}$$

To align these two sequences based on a common time axis, an *n*-by-*m* matrix is constructed. Distances are calculated between two sequence values by following equation

$$d(x_u, y_v) = (x_u - y_v)^2, \tag{4}$$

and accumulated distance '*D*' is measured by

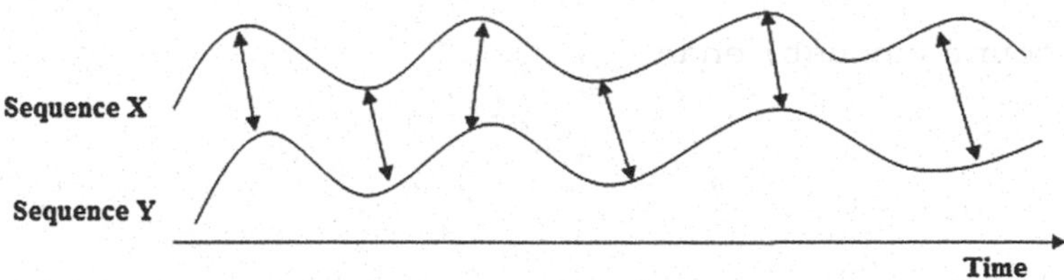

Fig. 2 Time alignment of two dependent sequences and aligned points are indicated by *arrows*

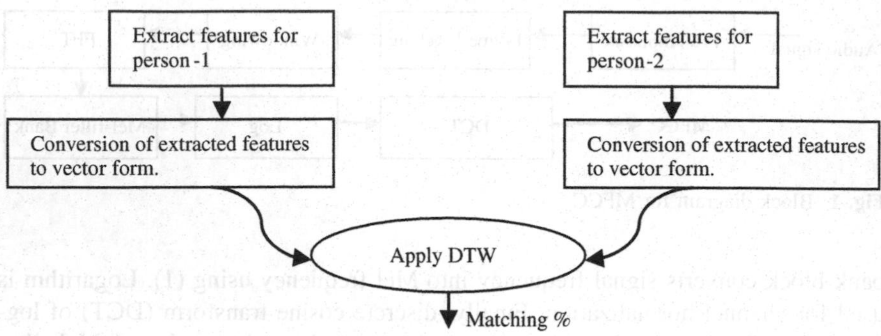

Fig. 3 Block diagram for feature matching

$$D(u, v) = \min[D(u-1, v-1), D(u-1, v), D(u, v-1)] + d(u, v) \quad (5)$$

Sakoe and Chiba made some conditions for good optimal path through grid [7]; these are monotonic condition, continuity condition, boundary condition, etc. Monotonic condition says that path through grid never turns back itself. The path proceeds with one single step at a time according to continuity condition [7]. Feature-matching process is shown in Fig. 3, and extract features for two voice samples were found using MFCC. Computed coefficients can be in the form of matrix, so convert these coefficients into vector form or time series before applying these inputs to DTW block and finally find out the percentage of matching between two sequences. A code written in MATLAB for finding out the percentage of matching between two sequences is shown below:

```
count = 0;
for i = 1:total values
   p(i) = (0.03)*x(i); p(i) = ((p(i))'; q(i) = x(i) + p(i);
   r(i) = x(i)-p(i); i = i+1;end

for i = 1:total values

   if(y(i) <=q(i) && r(i) <=y(i))
   count = count + 1;

  else

     count = count; end

end
```

4 Results

4.1 Analysis of MFCC

All the voice samples of database were analyzed using MFCC in MATLAB. In this analysis, '20' MFCC coefficients were founded out for all voice samples including PD-affected persons and healthy persons. First, MFCC coefficient in case of PD-affected persons shows more variation as compared with healthy persons as shown in Figs. 4 and 5.

4.2 Analysis of DTW

DTW technique was applied for all voice samples. Results of this analysis were done using MATLAB and PRAAT. Percentage of feature matching for PD-affected persons is more as compared with healthy persons in both cases.

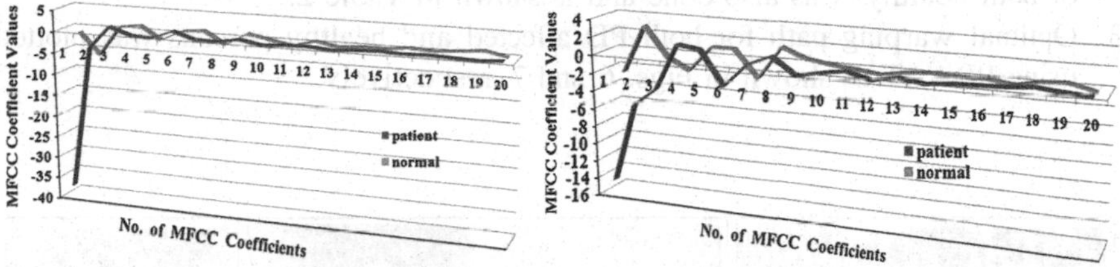

Fig. 4 Graph showing MFCC coefficients plot between (one PD-affected person and one healthy person for vowel 'a')

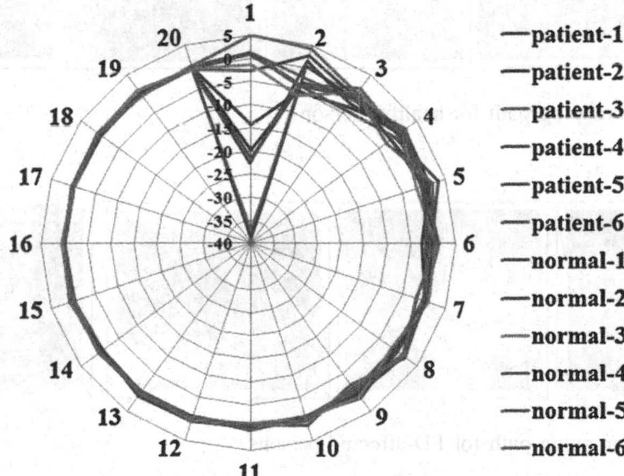

Fig. 5 Graph showing MFCC coefficients plot of all investigated PD-affected persons and healthy persons with '20' coefficients shown outside of the circle

Table 1 Feature matching percentage of same persons for vowel 'a'

S. no.	1.	2.	3.	4.	5.	6.	Avg. %
PD A.P	98.41	98.58	97.87	98.65	98.84	98.62	98.495
H.P	51.98	88.03	99.25	99.62	63.87	98.99	83.623

Table 2 Feature matching percentage of person to person matching for vowel 'a'

S. no.	1.	2.	3.	4.	5.	6.	7.	8.	Avg. %
PD. A.P	98.89	81.84	54.59	98.55	98.36	98.29	47.52	63.69	80.2163
H.P	35.80	86.84	73.93	31.27	98.86	98.96	99.53	52.88	72.2588

1. DTW-based matching percentage of the vector sequences (based on voice samples) of same subject (PD affected or healthy) is listed in Table 1 where 'PD. A. P' is PD-affected persons, 'H. P' is healthy persons, and 'Avg. %' is average percentage.
2. Feature matching between the voice samples of two persons (both PD affected or both healthy) was also done and is shown in Table 2.
3. Optimal warping path for both PD affected and healthy persons was plotted using PRAAT as shown in Figs. 6 and 7, respectively.

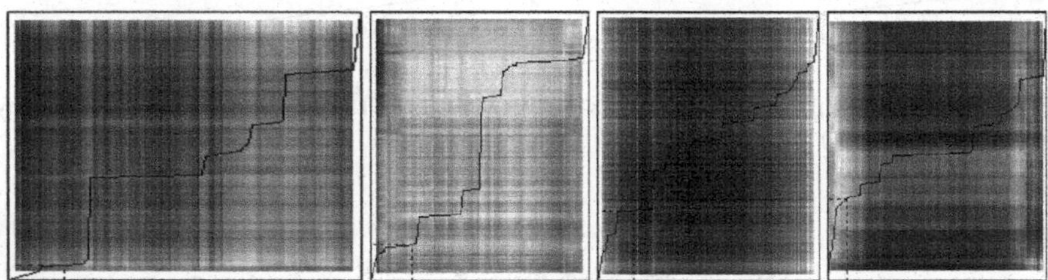

Fig. 6 Optimal wrapping path for healthy persons

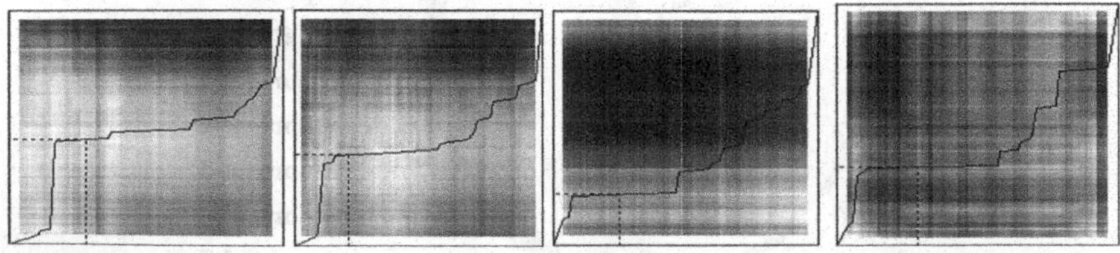

Fig. 7 Optimal wrapping path for PD-affected persons

5 Conclusion

MFCC and DTW have been applied for both Parkinson's-affected persons and healthy persons. First, coefficient of MFCC shows more variation between PD-affected persons and healthy persons for all voice samples for vowel 'a.' More features were matched in case of Parkinson's-affected persons as compared with healthy persons. Feature-matching percentage in case of PD-affected persons is 98.495 and 83.623 % in case of healthy persons.

References

1. Lee, S.M., Fang, S.H., Hung, J.W., Lee, L.S.: Improved MFCC feature extraction by PCA-optimized filter bank for speech recognition. In: IEEE Workshop on Automatic Speech Recognition and Understanding, pp. 49–52 (2001)
2. Hossan, MA., Memon, S., Gregory, M.A.: A novel approach for MFCC feature extraction. In: 4th international conference on signal processing and communication systems (ICSPCS), pp. 1–5 (2010)
3. Bhalke, D.G., Rama Rao, C.B., Bormane, D.S.: Dynamic time warping technique for musical instrument recognition for isolated notes. In: Proceedings of ICETECT, pp. 768–771 (2011)
4. Sahidullah, M., Saha, G.: Design, analysis and experimental evaluation of block based transformation in MFCC computation for speaker recognition. Sci. Dir. Speech Commun. **54**, 543–565 (2012)
5. Giorgino, T.: Computing and visualizing dynamic time warping alignments in R: the dtw package. J. Stat. Softw. **31**(7), 1–24 (2009)
6. Muller, M.: Information retrieval for music and motion, XVI, 318p. 136 illus. 39 in color., Hardcover. ISBN: 978-3-540-74047-6. http://www.springer.com/978-3-540-74047-6 (2007)
7. http://web.science.mq.edu.au/~cassidy/comp449/html/ch11s02.html

A Dynamic Model for FECG Synthesis and Preprocessing: A Signal Processing Approach

S. Ravindrakumar and K. Bommanna Raja

Abstract A dynamic model based on three-coupled ordinary different equations representing the electrical activity of the heart is presented for the fetal electrocardiogram (FECG) synthesis and preprocessing. The proposed method is based on the dipole theory of the heart and a previously exiting dynamic model for single channel adult electrocardiogram (ECG). The dynamic model is adopted and extended for the generation of maternal and FECG. An arbitrary number of synthetic ECG channel in single and multiple pregnancies for different fetal positions with variable maternal ECG interference and noises such as baseline wander, white noise, and Gaussian noise are analyzed, and the results are presented. For multiple fetal ECG, the applicability of the model is presented. Signal preprocessing algorithms were used with the noise modeling. The dynamic model is extended to generate abdominal ECG compressing of mixtures of maternal and fetal/multifetal ECG. Noise analysis and removal are done using different signal processing techniques. The results are been compared for various noises and signal-to-noise ratios (SNRs).

Keywords Fetal ECG · FECG extraction · ECG · Dynamic model · Noise · Gaussian model · Signal processing

S. Ravindrakumar (✉)
Department of Electronics and Communication Engineering,
Chettinad College of Engineering and Technology, Karur 639114, TN, India
e-mail: gsravindrakumar7@gmail.com

K. Bommanna Raja
Biomedical Engineering Department, PSNA College of Engineering
and Technology, Dindigul 624622, TN, India
e-mail: dr.k.bommannaraja@gmail.com

L.C. Jain et al. (eds.), *Intelligent Computing, Communication and Devices*,
Advances in Intelligent Systems and Computing 308,
DOI 10.1007/978-81-322-2012-1_82

1 Introduction

Heart defects are among the most common birth defects and leadmg cause of birth defects related deaths. Most cardiac defects have some manifestation in the morphology of electrocardiography and are believed to contain much more information as compared with convention sonographic methods. However, due to low signal-to-noise ratio (SNR) of fetal electrocardiogram (FECG) recorded from the maternal body surface, the application of fetal electrocardiography has been limited to heart beat analysis and invasive ECG recordings during labor. In this paper, the improvement in signal processing aspects of fetal cardiography and an improved modeling along with filtering of noises from FECG signal is done.

In recent years, research has been conducted toward the generations of synthetic ECG signals. Dynamic models have been developed which reproduce the morphology of PQRST complex and their relationship to beat timing in a single nonlinear dynamic model. The simple and flexibility of the model make it easy to adapt to broad class of normal and abnormal ECGs. Real ECG recordings are always contaminated with noise and artifacts. Hence, besides the modeling of cardiac sources and propagation media, it is very important to have realistic models for noise sources. Since common ECG contaminants are nonstationary and temporally correlated, time-varying dynamic models are required for generation of realistic noises. Here, a three-dimensional canonical model of the single dipole vector of heart is applied. This model that is inspired by the single channel ECG dynamic model is presented in [1]. A realistic synthetic ECG generator was first proposed by MCSharry et al., using a set of 3D state equations to generate a trajectory in the Cartesian coordinates. The dynamic equations were then transformed into polar form for a simpler compact set by Sameni et al. [2]. In this model, several Gaussians are utilized to approximate the feature waves (P, Q, R, S, and T waves) in one heart beat of ECG.

The other works which include the generation of synthetic ECG are by using Asymmetric Gaussians [3], knowledge based system using qualitative ECG simulation [4], Augmented mono domain model [5] , ECG simulations with realistic human membrane, heart, and torso models [6], lab view simulink [7], ECG SIM [8] , simulation of ECG under ischemic condition[9], a heart model with reaction diffusion action potentials [10], a dynamic model for phonocardiogram[11] and synthesizing the standard 12-lead ECG from three differential leads formed by pairs of proximal electrode on the body surface [12]. Boundary element approach, Tehron–Cairo formula, volume conductor model and time series for HRV by modified Zeeman model. However, the previous works are concentrated only on single channel ECG modeling, meaning that the parameter of the model should be re calculated for each of recording channels. For maternal and fetal mixtures analysis, only few works are done which considers the cardiac source and propagation media. The works also fail to include the noise model and filtering.

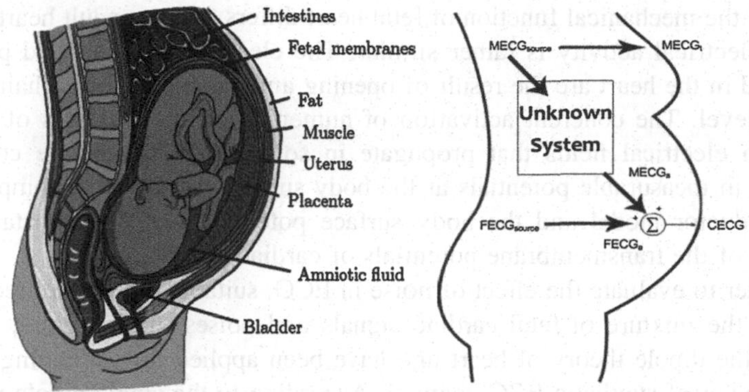

Fig. 1 The major fetomaternal compartments and components of fetal ECG system

2 Mathematical Model

The major components of the fetal monitoring system include the maternal ECG, FECG, and the noise (Fig. 1). They are represented by $\mathrm{MECG}_i(t)$, $\mathrm{FECG}_i(t)$, and $(N_{i\mathrm{lf}}(t) + N_{i\mathrm{hf}}(t))$. In general, the abdominal and the thoracic signal can be represented as

(i) abdominal signal(CECG)

$$\mathrm{Ab}_i(t) = N_{i\mathrm{lf}}(t)[F_{\mathrm{ECGi}}(t) + M_{\mathrm{ECGi}}(t) + N_{i\mathrm{hf}}(t)] \tag{1}$$

where

$N_{i\mathrm{lf}}(t)$ is the DC noise due to muscle and breathing movements.
$F_{\mathrm{ECG}i}(t)$ is the FECG signal. The amplitude of the FECG signal will be in μv.
$M_{\mathrm{ECG}i}(t)$ is the maternal ECG signal. The amplitude of the maternal ECG signal will be in few mv.
$N_{i\mathrm{hf}}(t)$ is the noise signal due to EMG, 50 Hz hum noises.

(ii) Thoracic signal

$$T_{\mathrm{hi}}(t) = N_{i\mathrm{lf}}(t)\,[M_{\mathrm{ECGi}}(t) + N_{i\mathrm{hf}}(t)] \tag{2}$$

3 Dynamic Model of FECG and Abdominal ECG

In the maternal abdomen compartments, the skin and the subcutaneous fat also have a poor conductivity about ten times smaller than muscle tissues. Accordingly, the fetus is surrounded by several different anatomical layers with different electrical conductivities (Fig. 1).

While the mechanical function of fetal heart differs from an adult heart, its beat to beat electrical activity is rather similar. The electrical currents and potentials generated in the heart are the result of opening and closure of ionic channels of a cellular level. The coherent activation of numerous cellular reactors of this sort results in electrical fields that propagate in so-called body volume conductor, resulting in measurable potentials at the body surface. However, a complete volume conductor model and the body surface potentials are linear instantaneous mixtures of the transmembrane potentials of cardiac myocytes.

In order to evaluate the effect of noise in ECG, suitable models are required to generate the mixture of fetal cardiac signals and noises. In this work, we have adopted the dipole theory of heart and have been applied for generating an arbitrary number of synthetic ECG channel. According to the single dipole model of the heart, the myocardium's electrical activity may be represented by a time-varying rotatory vector, the origin of which is assumed to be at the center of the heart as its end sweeps out a quasi-periodic path through the torso. This vector may be mathematically represented in the Cartesian coordinates as follows:

$$\mathrm{d}(t) = x(t)\hat{a}_x + y(t)\hat{a}_y + z(t)\hat{a}_z \tag{3}$$

ax, *ay*, *az*—unit vectors of three body axes.

The body volume conductor is assumed to be a passive resistive medium. The ECG signals recorded from the body surface would be a linear projection of the dipole vector d(t) onto the direction of recording electrode axes.

$$\begin{gathered} V = a\widehat{a}_x + b\widehat{a}_y + c\widehat{a}_z \\ \mathrm{ECG}(T) = <\mathrm{d}(t) \\ V > a \cdot x(t) + b \cdot y(t) + c \cdot z(t) \end{gathered} \tag{4}$$

The potential generated by dipole at a distance r (where $r = \mathbf{r_x}\bar{\mathbf{a}}_\mathbf{x} + \mathbf{r_y}\bar{\mathbf{a}}_\mathbf{y} + \mathbf{r_z}\bar{\mathbf{a}}_\mathbf{z}$ is the vector which connects the center of dipole to observation point, and conductivity of volume conductor [24] is

$$\phi(t) - \phi_0 = \frac{\mathrm{d}(t) \cdot r}{4\pi\sigma|r|^3} \tag{5}$$

$$\phi(t) - \phi_0 = \frac{1}{4\pi\sigma}\left[x(t)\frac{r_x}{|r|^3} + y(t)\frac{r_y}{|r|^3} + z(t)\frac{r_z}{|r|^3}\right] \tag{6}$$

From the single dipole model of heart adopted from MCSharry et al. model [24], it is well known that the different ECG leads can be assumed as projections of heart's dipole vector onto the recording electrode axes. All leads are time synchronized with each other and have quasi-periodic shape. The three-dimensional extension is given by

$$
\begin{aligned}
\dot{\theta} &= \omega \\
\dot{x} &= -\sum_{t} \frac{a_i^x \omega}{(b_i^x)^2} \Delta\theta_i^x \exp\left[-\frac{(\Delta\theta_i^x)^2}{2(b_i^x)^2}\right] \\
\dot{y} &= -\sum_{t} \frac{a_i^y \omega}{(b_i^y)^2} \Delta\theta_i^y \exp\left[-\frac{(\Delta\theta_i^y)^2}{2(b_i^y)^2}\right] \\
\dot{z} &= -\sum_{t} \frac{a_i^z \omega}{(b_i^z)^2} \Delta\theta_i^z \exp\left[-\frac{(\Delta\theta_i^z)^2}{2(b_i^z)^2}\right] \\
\Delta\theta_i^x &= (\theta - \theta_i^x) \bmod (2\pi) \\
\Delta\theta_i^y &= (\theta - \theta_i^y) \bmod (2\pi) \\
\Delta\theta_i^z &= (\theta - \theta_i^z) \bmod (2\pi)
\end{aligned}
\tag{7}
$$

where $\omega = 2\pi$

Each of the three coordinates of the dipole vector d(t) is modeled by a summation of Gaussian functions [25] with amplitudes α_i^x, α_i^y and α_i^z widths $b_i{}^x, b_i{}^y$ and $b_i{}^z$, located at rotational angle θ_i^x, θ_i^y and θ_i^z. In our work, the model for orthogonal lead VCG coordinates is altered using different scaling factors for attenuation of volume conductor. The model is extended to fetal signal by varying the rotational angle θ_i^x, θ_i^y and θ_i^z. The generated FECG signal for singleton, multiple fetal is shown in Figs. 2 and 3. The model is extended to generate the abdominal ECG signal which consists of 4 channels (Fig. 4).

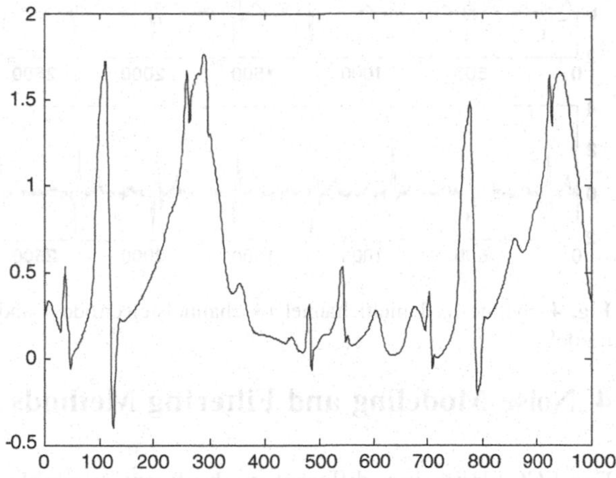

Fig. 2 Synthesized maternal and fetal ECG using the dynamic model

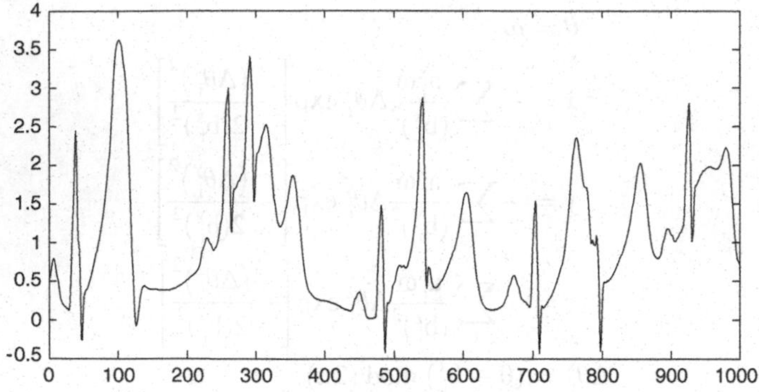

Fig. 3 Synthesized maternal and multiple fetal ECG using the dynamic model

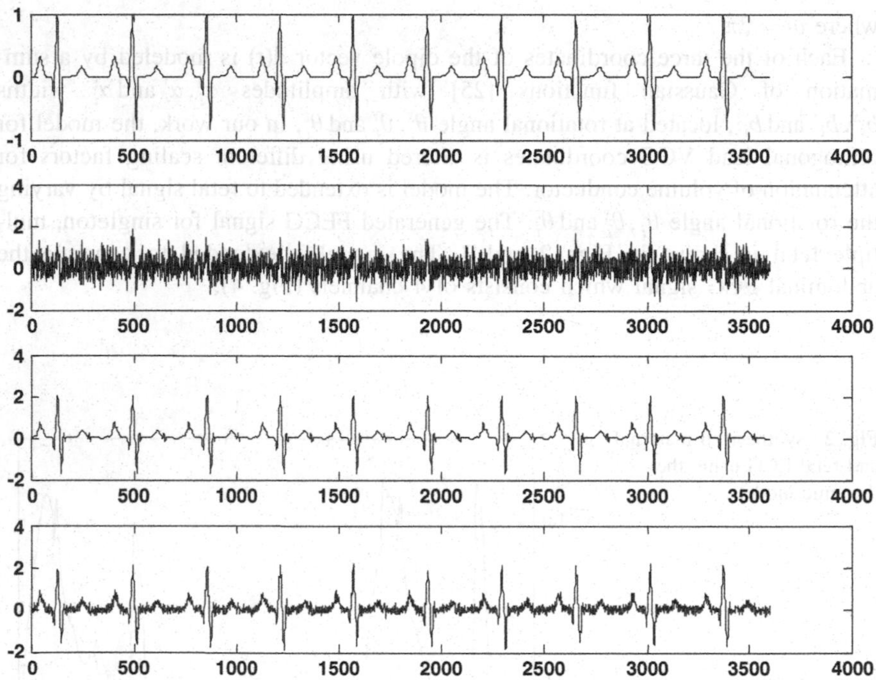

Fig. 4 Synthesized multichannel (4 channels/electrodes) abdominal ECG using the dynamic model

4 Noise Modeling and Filtering Methods

For ECG denoising, different methods are available in the literature. The FECG exibits a BW of 0.05–100 Hz. The amplitude of FECG QRS is 50 mv and that of MECG is 100–150 mv. Moreover, the spectrum of both signals overlap. The main

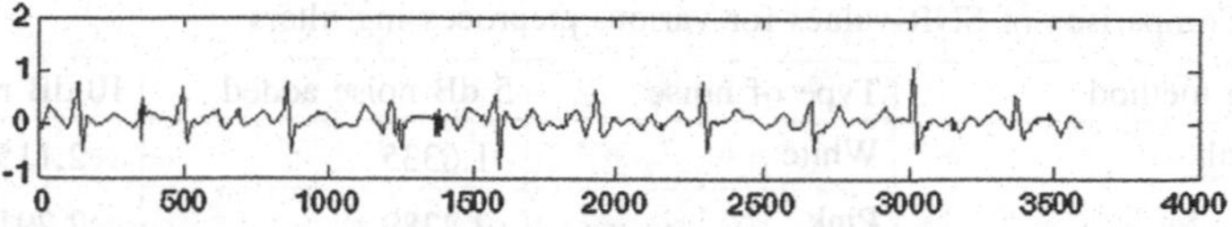

Fig. 5 Denoised abdominal ECG using wavelet transform (channel 2 of multichannel)

source of interference is the maternal electrical activity. Other common noise sources are power line interference, muscle contractions, respirations, in addition to EMG and electro-hysterogram. Various research areas have been carried out for removal of noise and interference from FECG, including Adaptive noise cancellation and blind source separation [14], auto and cross correlation [15] , adaptive real-time ICA [16], adaptive impulse correlated filter [17], Extended Barros's extraction algorithm [18], blind source separation [19, 20], blind source separation along with wavelet de-noising [21], blind source separation with adaptive noise cancellation techniques [22], Independent component analysis [23], matched filtering, linear regression, neural networks [24], IIR adaptive filtering combined with genetic algorithm, wavelet transform [25, 26] (Fig. 5)., subtraction of an averaged pattern [26], SVD method, moving average and Hilbert transform, and ICA [16, 19] (Fig. 5).

Due to the overlap of fetal signals and interference in different domains, the methods that use the information in only one of these domains do not usually succeed in extraction of FECG. Other methods used for the extraction of FECG from mixed signal are principal component analysis (PCA) or singular value decomposition and independent component analysis. These methods called blind source separation (BSS) consist of contracting unknown signals (called sources) which are statistically independent from known mixtures of signals. The PCA [14] uses second-order statistics which higher order statistics is performed by independent component analysis.

5 Methodology and Results

In this work, the ECG signals for fetal and maternal mixtures are generated and noises have been added. In addition to that real time data been used from Physionet MIT-BIH database [13]. Most of data are contaminated with noises, and the source separation methods are also susceptible to noises. Here, digital filters are used to remove the baseline wander and other noises. The filtering improves the SNR of the source separation procedure and reduces the susceptibility of noise (Table 1). The algorithm is tested for variable attenuation factor and SNR values. Various noise signals such as muscle artifact, electrode movements, baseline wander, white noise, and colored noise are analyzed and filtered. From the observation, the adaptive filtering techniques serve the best in preprocessing of the data or contaminated abdominal ECG. Here, LMS-based algorithm is used (MATLAB). To improve the SNR value, the adaptive filter stage will be extended with

Table 1 Comparison of SNR values for various preprocessing filters

Denoising method	Type of noise	5 dB noise added	10 dB noise added
Butterworth	White	1.6335	2.1152
	Pink	2.2289	2.2917
	Brown	2.480	2.4556
	Muscle artifact	2.3309	2.3899
	Electrode	2.4747	2.4398
	Baseline wander	2.4984	2.4561
Median	White	16.0899	24.6955
	Pink	27.5155	40.8177
	Brown	53.3405	50.6073
	Muscle artifact	44.3503	47.5741
	Electrode	52.384	50.8049
	Baseline wander	53.6322	51.5194
Adaptive LMS algorithm	White	0.0013	0.0016
	Pink	0.0023	0.0020
	Brown	0.0024	0.0021
	Muscle artifact	0.0024	0.0020
	Electrode	0.0024	0.0019
	Baseline wander	0.0027	0.0021
Savitzky Golay	White	14.6888	24.3182
	Pink	30.3015	38.6357
	Brown	91.1922	92.6525
	Muscle artifact	72.4127	81.4280
	electrode	84.6718	93.2652
	Baseline wander	97.6275	104.0710
Zero phase	White	13.6457	22.7116
	Pink	23.9571	36.5736
	Brown	46.9149	48.0604
	Muscle artifact	42.1658	45.6598
	Electrode	46.95	48.1145
	Baseline wander	47.1783	48.1517

multireference and multichannel. The SNR values are calculated for data, and so less value gives better performance. But if the noise level gets increased, the FECG will be hard to the extract. So additional improvements required in the data acquisition units which would provide high CMRR and rejection ratio. The work will be further extended to extract the FECG using hybrid ICA algorithms with wavelet transform and SVD with adaptive filtering.

6 Conclusion and Future Work

Depending upon the environment, gestational age, and physiological needs, different synthesis methods are reviewed. The correlation and subtraction methods are ineffective in removal of noise from FECG. In noisy environment, adaptive filtering technique is suitable for multichannel adaptive filtering. In future, the BSS method, independent component analysis, and singular value decomposition will be used to detect fetal heart R peaks. Finally, the interpretation problem will be solved by using fuzzy logic models. By comparison with literature survey done, the BSS method produces a clear fetal signal and also improves the SNR of the system but with more number of channels involved which is difficult during labor. A method is to be proposed for enhancement of FECG using signal processing technique. A new signal processing technique will be applied for MECG removal and FECG extraction.

References

1. McSharry, P.E., Clifford, G.D., Tarassenko, L., Smith, L.A.: A dynamic model for generating synthetic electrocardiogram signals. IEEE Trans. Biomed. Eng. **50**, 289–294 (2003)
2. Sameni, R., Jutten, C.: A Nonlinear Bayesian Filtering Framework for ECG Denoising. IEEE Trans. Biomed. Eng. **54**(12), 2172–2185 (2007)
3. Lu, Y., Yan, J., Yam, Y.:A generalized ECG dynamic model with asymmetric Gaussians and its application in model-based ECG denoising, IEEE (2009)
4. Wang', J.T., Sehmi3, A.S., Jones', N.B., de Bono', D.P.: A knowledge-based system for qualitative ECG simulation and ECG analysis, IBBB (1992)
5. Bishop, M.J., Plank, G.: Bidomain ECG simulations using an augmented monodomain model for the cardiac source. IEEE Trans. Biomed. Eng. **58**(8), 2297–2307 (2011)
6. Potse, M., Dubt, B., Gulrajani, R.M.: ECG simulations with realistic human membrane, heart, and torso models. In: Proceedings of the 25th Annual International Conference of the IEEE EMBS Cancun, Mexico, pp. 17–21 (2003)
7. Jóśko, A., Rak, R.J.: Effective simulation of signals for testing ECG analyzer. IEEE Trans. Instrum. Meas. **54**(3), 1019–1024 (2005)
8. van Dam, P.M., Oostendorp, T.F., van Oosterom, A.: ECGSIM: interactive simulation of the ECG for teaching and research purposes. In: Computing in Cardiology (2010)
9. Lu, W., Wang, K., Zhang, H., Zuo, W.: simulation of ECG under ischemic condition in human ventricular tissue. In: Computing in Cardiology (2010)
10. Berenfeld, O., Abboud, S.: Simulation of ECG using a heart model with reaction-diffusion action potentials. In: IEEE (1993)
11. Almasi, A., Shamsollahi, M.B., Senhadji, L.: A dynamical model for generating synthetic phonocardiogram signals. In: IEEE (2011)
12. Trobec, R., Tomasic, I.: Synthesis of the 12-lead electro-cardiogram from differential leads. In: IEEE transactions on information technology in biomedicine, vol. 15, no. 4 (2011)
13. www.Physionet.Org
14. Jafari, M.G., Chambers, J.A.: Adaptive noise cancellation and blind source separation. In: IEEE (2000)
15. Song, M.H., Cho, S.P., Park, H.D., Lee, K.J.: The novel method for the fetal electrocardiogram extraction from the abdominal signal. In: Conference of the IEEE EMBS (2007)

16. Waldert, S., et al.: Real-time fetal heart monitoring in biomagnetic measurements using adaptive real-time ICA. IEEE Trans. Biomed. Eng. **54**(10), 1867–1874 (2007)
17. Martinez, M., et al.: Application of the adaptive impulse correlated filter for recovering fetal electrocardiogram. In: IEEE (1997)
18. Zhang, Z.L., Ye, Y.: Extended Barros's extraction algorithm with its application in fetal ECG extraction. In: IEEE (2005)
19. Gupta, A., et al.: A Novel approach to fetal ECG extraction and enhancement using blind source separation (BSS-ICA) and adaptive fetal ECG enhancer (AFE). In: IEEE (2007)
20. Kam, A., Cohen, A.: Separation of twins fetal ECG by means of blind source separation (BSS). In: IEEE (2000)
21. Ming, M.A., Ning, W., San-Ya, L.: Extraction of FECG based on time frequency blind source separation and wavelet de-noising. In: IEEE (2009)
22. Zarzoso112, V., Millet-Roig', J., Nandi2, A.K.: Fetal ECG extraction from maternal skin electrodes using blind source separation and adaptive noise cancellation techniques. In: IEEE (2000)
23. Sato, M., et al.: A novel extraction method of fetal electrocardiogram from the composite abdominal signal. IEEE Trans. Biomed. Eng. **54**(1), 49–58 (2007)
24. Peters, C., Vullings, R., Bergmans, J., Oei, G., Wijn, P.: Heart rate detection in low amplitude non-invasive fetal ECG recordings. In: EMBS Proceedings, IEEE (2006)
25. Fazlul Haque, A.K.M., et al.: Detection of small variations of ECG features using wavelet. ARPN J. Eng. Appl. Sci. **4**(6), 27–30 (2009)
26. Echeverria, J.L., et al.: Fetal QRS extraction based on wavelet analysis and pattern matching. In: IEEE (1996)

Providing Mother and Child Care Telemedicine Through Interactive Voice Response

R. Subhashini, R. Sethuraman and V. Milani

Abstract In India, telemedicine services are provided through satellite communication, Web portals and mobile call centre services. These services enable people to interact with a specialized doctor who is in the other end of the country using videoconferencing and various channels as medium. This saves time and cost of the patient using the service. He/she gets all specialized medical instructions without having the need to see the doctor. This service may be ultimately useful when available to all the people in the country. But, it does not reach everyone in the country. Satellite communication is possible only in places where satellite link is established between the patient and a doctor and Web portals are accessible only by people who have complete knowledge about the existing system mechanism. These methods are mainly available only in urban areas. This is where the problem arises in India. Villages being the backbone of the country and farmers being its soul, they are not aware of these existing technologies and telemedicine application is completely new to them. Providing specialized medical services in cities is easy, but this is not the same for rural India. Therefore, a more efficient, convenient and user-friendly method of implementing telemedicine service is needed. In this paper, we extend this approach of telemedicine to the rural areas using interactive voice response (IVR) system. Since there has been a similar methods already established in urban areas, we are providing this newly proposed system for the rural areas for a better health care.

R. Subhashini (✉)
Faculty of Computing, Sathyabama University, Chennai, India
e-mail: subhaagopi@gmail.com

R. Sethuraman
Faculty of Computing, Department of Computer Science,
Sathyabama University, Chennai, India
e-mail: srssethuraman@gmail.com

V. Milani
Department of Information Technology,
Sathyabama University, Chennai, India
e-mail: milaniv345@gmail.com

L.C. Jain et al. (eds.), *Intelligent Computing, Communication and Devices*,
Advances in Intelligent Systems and Computing 308,
DOI 10.1007/978-81-322-2012-1_83

Keywords Telemedicine · IVR system · Web portals

1 Introduction

Telemedicine services [1] deal with providing medical services from a distance via telecommunications. Interactive voice response (IVR) is a technology that involves the interaction of a computer with a human through the use of voice and DTMF tones as input via keypad. In recent years, telemedicine has found its application globally and is now rooted in several fields of medicine. Telemedicine applications are provided for various services that include telecardiology, teleradiology, telepathology, teledermatology, teleophthalmology, teleoncology and telepsychiatry.

In India, only some of the above services are provided by certain hospitals over Web portals and satellite communication. Internationally, countries such as USA, Germany, Norway and other European union countries have implemented telemedicine through satellite communication, IVR, Web portals and have linked it to computers and mobile phones making it easy to access for all people. In general medicines, the government of India has taken all the necessary steps to help its people for the maximum medical facilities. The state governments along with the central government are playing a major role in providing medical facilities. Recent days, government-run hospitals are providing equal speciality services as compared to that of private speciality hospitals. This is indeed a great deed for the people as speciality services are available at cost-free and more reasonable price. These services reach to a maximum of almost all the people in urban cities and towns. Providing telemedicine [2] using the technology of IVR is a cost-effective way to make it to reach people at all levels of the society. In modern days, it is possible to implement numerous applications by using the above technologies. If a methodology of similar type is implemented in India, then it will be highly benefited for all the people including the people in rural India.

Child care is a broad spectrum that encompasses a varied range of issues that are concerned with children from the age of 0 and above. Child care covers a wide area concerning their general health after they are born. Some major common health problems in babies include colds, coughs, fevers and vomiting. Babies also commonly have skin problems, like diaper rash or cradle cap. Such situations tend to be highly disturbing to their mothers, who here basically are from the uneducated strata of the society. These issues may or may not be of a very frightening situation that may demand extra medical care from an expert.

The telemedicine service using IVR technology is a self-service which intends to provide immediate services to the concerned mothers in the rural and undeveloped areas by using the medium of a toll-free telephone service. This IVR service would first enlist out the problems based on which the customer would be directed to touch or choose the required option. Furthermore, there would be a step-by-step guidance that would ensure that the customer receives what he/she has been expecting.

IVR can be implemented in a number of ways when it comes to telemedicine in the form of self-services. Currently, it is being used by doctors to provide appointments to their patients. This way it helps the patients to save time by avoiding the waiting time for doctors. It is also implemented by hospitals to provide regular patient care once the patient gets discharged from their hospitals. A sequential order of questions is put into the IVR; the IVR calls the patient on regular intervals and questions the patients corresponding to their health problems. Furthermore, if the patient has some queries, then he/she can call the service and can get their doubts cleared.

2 Related Works

1. Pilot project initiated by ISRO [3] in 2001 involves linking a specialized hospital from an urban centre with a general hospital in rural areas through INSAT satellites. Currently, ISRO's telemedicine network stretches to around 100 hospitals all over the country with 78 rural/remote district hospitals/health centres connected to 22 speciality hospitals in the major cities.
2. States such as Karnataka, Chhattisgarh, Kerala and Jharkhand are some of the states that provide mobile telephone hospital services through satellite-based telemedicine.

   Apollo [4] through its ATNF(Apollo Telemedicine Networking Foundation) provides telemedicine services through Web portals.
3. Under the National Blindness Control Programme, teleophthalmology services are provided by Shankar Nethralaya at Chennai; Meenakshi Eye Mission and Aravind Eye Hospital at Madurai; and four other corporate hospitals with the support of ISRO.
4. Uninor and Handygo Technologies have launched "wellness world" through IVR and SMS. It provides services on heart care, women's health, diet and nutrition, pain management, etc. Subscribers are charged Rs. 30 for 7 days. Nnamdi Azikiwe University, Nigeria, has designed a medical first-aid self-service on IVR platform in Nigeria.
5. The United States Agency for International Development (USAID), the Government of Norway, the Bill and Melinda Gates Foundation, Grand Challenges Canada along with the World Bank have created a baby care monitoring system using Verboice.
6. Community home health services have implemented an IVR telehealth programme for their network to monitor patients' health care from Cantata Adult Life Services in Brookfield, USA.
7. University of Utah (USA) under the Utah telehealth network (UTN) links patients to health-care providers across state and country using user-interactive video to deliver patient care [5].

8. Care Bridge palliative care services, New Jersey, provides electronic house call (EHC) and interactive voice response(IVR) services to authenticate provided equipment are at patients reach.

3 Structure of the Work

Telemedicine services deal with providing medical services from a distance via telecommunications. interactive voice response (IVR) is a technology that involves the interaction of a computer with a human through the use of voice and DTMF tones as input via keypad. In recent years, telemedicine has found its application globally and is now rooted in several fields of medicine. Recent days, governments-run hospitals are providing equal speciality services when compared to that of private speciality hospitals. This is indeed a great deed for the people as speciality services are available at cost-free and more reasonable price. These services reach to a maximum of almost all the people in urban cities and towns. But these methods are very frequently used in the urban areas and are very helpful for those who are living in cities. For the people who live in rural areas, they find it really strenuous to approach doctors or any hospitals in case of any emergencies such as child care, heart attack, pregnancy, lung problems. We extend this approach of telemedicine to the rural areas using a new system called the IVR system. This is an interactive voice communication system. This system is through a "Verboice".

In this system, the approach for consulting a doctor during an emergency situation is done via mobile phones. The speciality of this system is that the process can be carried out even on basic mobiles. There is completely no requirement necessary for GPS, Internet, etc. There are toll-free numbers which are mainly used for contacting hospitals during a critical situation (Fig. 1). Initially, toll-free numbers are dialled for any urgent queries. The IVR system automatically links with the dialled toll-free number. Once the person has been connected with the IVR system, he/she is asked to choose their preferred language. After this process, the concern person is given two types of options to choose from. The two types of choices are general and emergency. The person can choose the option according to his convenience. The entire process takes place in matter of seconds. The general option is used when the person has queries related to normal health issues such as cold, fever. The emergency option is used for critical situations. If he/she has a problem which requires immediate attention, this option directly gives connection to the doctor who is experienced in the field relating to the problem reported. The IVR system sends a voicemail to the concern doctor. Once the voicemail is received by the doctor, he immediately gives the required information to the IVR system, and hence, the IVR system passes this on to the concern person having the health problem. Every call that is received by the IVR system is stored in their database. The IVR system creates a backup for every

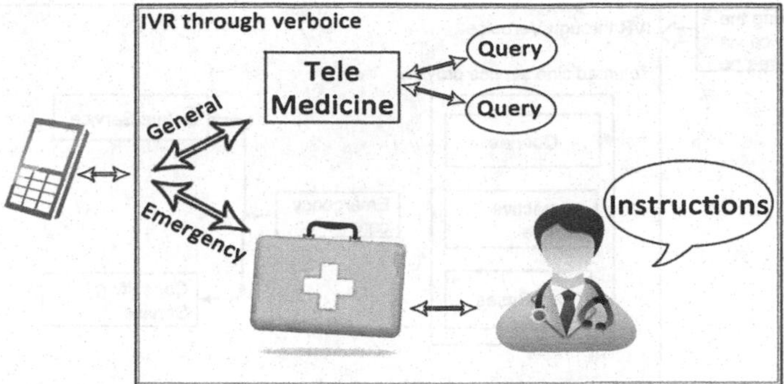

Fig. 1 Block diagram

caller and stores their issues under the concern person's specified profile information received by the IVR system.

4 Proposed System

In India, only some of the above services are provided by certain hospitals over Web portals and satellite communication. Internationally, countries such as USA, Germany, Norway and other European union countries have implemented telemedicine through satellite communication, IVR, Web portals and have linked it to computers and mobile phones making it easy to access for all people. In general medicines, the government of India has taken all the necessary steps to help its people for the maximum medical facilities. Telemedicine services are provided through satellite communication, Web portals and mobile call centre services. But, it does not reach everyone in the country. Mainly the people living in rural areas are finding it strenuous during critical health situations.

Our proposed approach to the problem for the people in rural areas to get a solution for an immediate response regarding health problems is the newly generated system named IVR system. Here, the person first makes a call regarding health service via toll-free number. The toll-free number is interlinked with the IVR system. So once the person makes the call, they are automatically connected to the IVR system. The IVR system is through a "Verboice". Once the person is connected with the system, he/she is made to choose their preferred language for communication. After this, the concern person can ask their queries and an interactive service takes place (Fig. 2). There are two kinds of options available for the callers: one is the "general" and the other is "emergency". The general option is used when the person has queries related to normal health issues such as cold and fever. The emergency option is used for critical situations. The general option does the querying to the caller. The person when asks a query related to his/her health, if the answer given is satisfied by the person, then further the process will not be continued. If the query asked by the concern person has not received

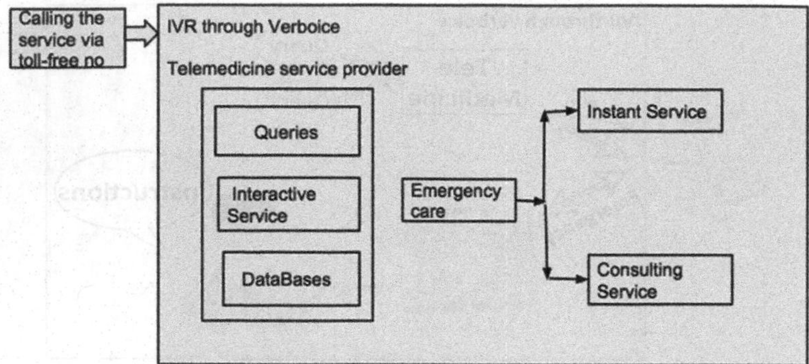

Fig. 2 Architecture for the system

the expected reply, then they can reply by saying "NO". This will intimate to the system and further the system will move on to more number of queries until the person has got a perfect solution to his problem.

In the Emergency care, there are two different ways of service (Fig. 3). They are known as the instant service and the other is known as the consulting service.

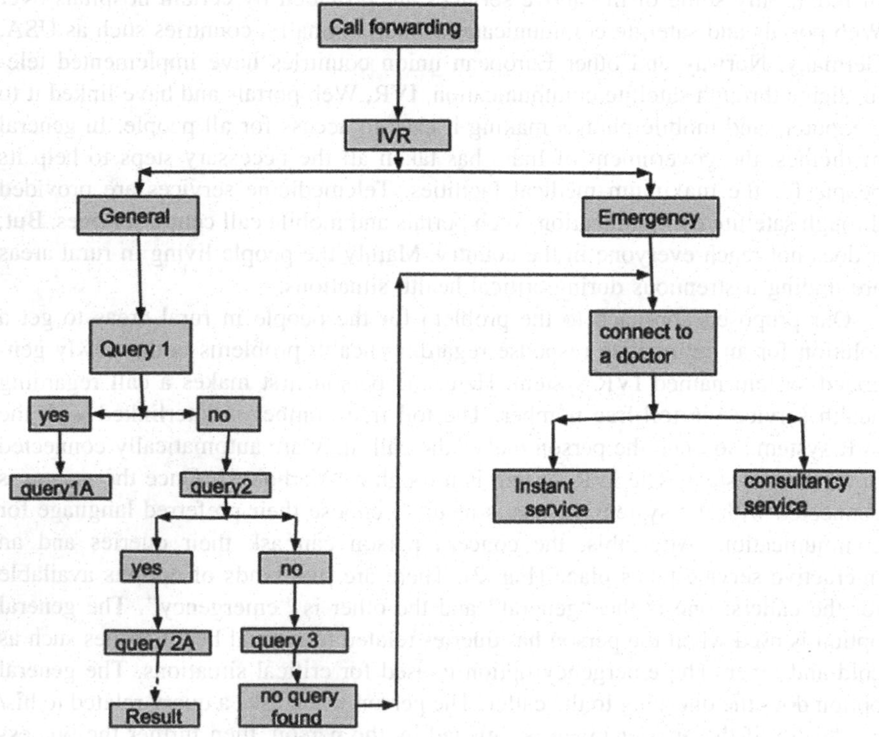

Fig. 3 Work flow of the IVR system

Both these services mainly collect keywords in order to analyse the problem faced by the caller. The instant service will immediately connect to the doctor via voicemail and notify the doctor about the situation and problem faced by the concern patient. Once the doctor receives this voicemail, he will immediately provide a solution to the problem stated. This problem will be received by the IVR system and, hence, will reach the concern person. During a consulting service call, if the doctor is unavailable or his mobile is not reachable, etc., then the concern person who has stated the problem will be given a temporary medication to ease their health and simultaneously a voice mail will be sent to the doctor. Once the doctor receives this, immediately a proper solution is issued to the concern caller. Every call that is received by the IVR system is stored in their database. So if the caller makes a call, the next time his health reports will automatically be diagnosed by the system and medication or solution to their health will be given.

5 Analysis Tool

The purposes of the required systems are reported below:

- IVR system
- Telemedicine

IVR System:

IVR is known as interactive voice response system. IVR can be implemented in a number of ways when it comes to telemedicine. Currently, it is being used by doctors to provide appointments to their patients. This way it helps the patients to save time by avoiding the waiting time for doctors. It is also implemented by hospitals to provide regular patient care once the patient gets discharged from their hospitals. A sequential order of questions is put into the IVR; the IVR calls the patient on regular intervals and questions the patients corresponding to their health problems. Furthermore, if the patient has some queries, then he/she can call the service and can get their doubts cleared. This system can be used for pregnancy or child care and any major or minor health emergencies.

Telemedicine:

Telemedicine services deal with providing medical services from a distance via telecommunications. Interactive voice response (IVR) is a technology that involves the interaction of a computer with a human through the use of voice and DTMF tones as input via keypad. In recent years, telemedicine has found its application globally and is now rooted in several fields of medicine. Telemedicine applications are provided for various services that includes telecardiology, teleradiology, telepathology, teledermatology, teleophthalmology, teleoncology and telepsychiatry. The telemedicine service using IVR technology intends to provide immediate services to the concerned mothers in the rural and undeveloped areas by using the medium of a toll-free telephone service. This IVR service would first enlist out the problems based on which the customer would be directed to touch or choose the required option. Furthermore, there would be a step-by-step guidance that would ensure that the customer receives what he/she has been expecting.

6 Conclusion

Telemedicine holds substantial potential to reduce the impact of illness on health and education of children, on time lost from work in parents and on absenteeism in the economy. Hence, the newly introduced system known as IVR system can be implemented for rural areas. Since villages being the backbone of the country and farmers being its soul, they are not aware of these existing technologies and telemedicine application is completely new to them. Therefore, a more efficient, convenient and user-friendly method of implementing telemedicine service has been introduced. Also providing of toll-free telemedicine service has been introduced along with a system of service that will reach all the mobile users in a user-friendly manner. Finally, we come to a conclusion that the IVR system provides a high-quality service, referral and patient management services.

References

1. Verulkar, S.M., Limkar, M.: Real time health monitoring using GPRS technology. Int. J. Comput. Sci. Netw. **1**(3), (2012). ISSN 2277-5420
2. Pattichis, C.S., Kyriacou, E., et al.: Wireless telemedicine systems: an overview. IEEE Antennas Propag. Mag. **44**(2), 143–153 (2002)
3. ISRO—Pilot project. http://www.isro.org/publications/pdf/Telemedicine.pdf
4. Telemedicine in India: the Apollo story. http://www.ncbi.nlm.nih.gov/pubmed/19659414
5. Utah Telehealth Network (UTN). http://www.authentidate.com/university-of-utah-selects-authentidate-telehealth-solutions-for-three-year-utah-telehealth-network-project

Wheeled Patient Monitoring System

Rajesh Kannan Megalingam, Pranav Sreedharan Veliyara, Raghavendra Murali Prabhu, Rithun Raj Krishna and Rocky Katoch

Abstract The most common and serious problem that the elderly suffers due to old age is their degrading health conditions that restrict their ability to perform life activities. Most of them require wheelchair to move even within their residence. According to WHO, there are about 1 % of world population that require wheelchair for mobility. People with debility in moving from one location to another are either bedridden or depend upon the wheelchair. An elder, dependent on wheelchair, often requires regular health check-up to keep track of their health status, and it is very difficult for them to have frequent visits to hospitals. Wheeled patient monitoring system (WPMS), in such a situation, is a system that continuously monitors the health condition of the patient by measuring body temperature, blood oxygen saturation in blood and heart rate. These three parameters are the crucial parameters of a body and help the doctor to diagnose easily. WPMS constantly monitors health parameters and identifies critical situations to inform physicians and healthcare centres so that the users are taken care at the earliest. The system can be easily attached to or detached from the wheel chair. WPMS can also be used by patients at hospitals.

Keywords Wheelchair · Temperature · SpO2 · ECG · Pulse oximetry

---

R.K. Megalingam (✉) · P.S. Veliyara · R.M. Prabhu · R.R. Krishna · R. Katoch
Amrita School of Engineering, Amrita Vishwa Vidyapeetham University,
Kollam 690525, India
e-mail: megakannan@gmail.com

P.S. Veliyara
e-mail: pranavsveliyara@gmail.com

R.M. Prabhu
e-mail: raghavendramprabhu@gmail.com

R.R. Krishna
e-mail: rithunrajkrishna@gmail.com

R. Katoch
e-mail: roc.katoch@gmail.com

L.C. Jain et al. (eds.), *Intelligent Computing, Communication and Devices*,
Advances in Intelligent Systems and Computing 308,
DOI 10.1007/978-81-322-2012-1_84

1 Introduction

Wheeled patient monitoring system (WPMS) is a system that constantly monitors the different health parameters such as body temperature, oxygen content in the blood and heart rate of the patient on the wheelchair. The microcontroller on the Arduino forms the control unit. It analyses the measured parameters based on the threshold values set. The parameter which exceeds the threshold value is then noted, and based on the levels of the measured value, either a message or a call is made to the caretaker/doctor.

2 Different Units in WPMS

The different units integrated with the wheelchair for the continuous monitoring of the health are sensor unit, control unit, communication unit and power supply unit. The following sections describe the different units, its functionality and working.

3 Sensor Unit

The sensor unit forms the basis of the health parameter measuring system which directly senses the different parameters mentioned from the patient's body.

3.1 Temperature Sensor

Body temperature is one of the basic health parameters usually measured. A negative temperature coefficient (NTC) thermistor is preferred over other temperature sensor because of its measuring accuracy and fast response. For a NTC thermistor, as the temperature increases, its resistance decreases. Since it is a passive device, it can also work for a wide range of supply voltage.

Equation involved. Steinhart–Hart equation is used here. The coefficients a, b and c in the equation are called the Steinhart–Hart Parameters. T is the temperature measured in Kelvin (K), and R is the resistance in ohms. This equation translates the resistance into temperature.

$$\frac{1}{T} = a + b\ln(R) + c\ln(R)^3$$

Working. The change in resistance with temperature change can be read through an Arduino [1]. A voltage divider circuit is used to measure the changes in the resistance. The circuit schematic is given in Fig. 1.

3.2 Electrocardiogram

Electrocardiogram (ECG) refers to the graphical representation of electrical activities in the heart and can be recorded fairly easily with surface electrodes such as Ag-AgCl electrodes, shown in Fig. 2, placed on the limbs or chest. A typical ECG signal will look like one shown in Fig. 3. At rest, each heart muscle has a negative charged called membrane potential across its cell membrane. Decreasing this charge towards zero is called depolarization, and this results in heart muscle to contract [2]. This is detected as tiny waves of voltages across the electrodes. A three lead system is used in our set-up.

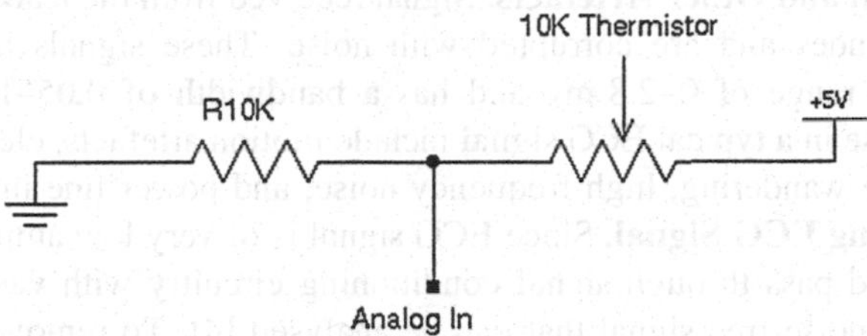

Fig. 1 Temperature sensor—schematic

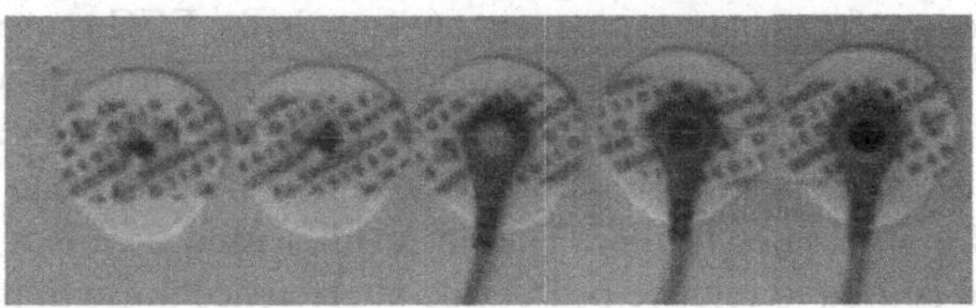

Fig. 2 Ag-AgCl electrodes

Fig. 3 Typical ECG signal

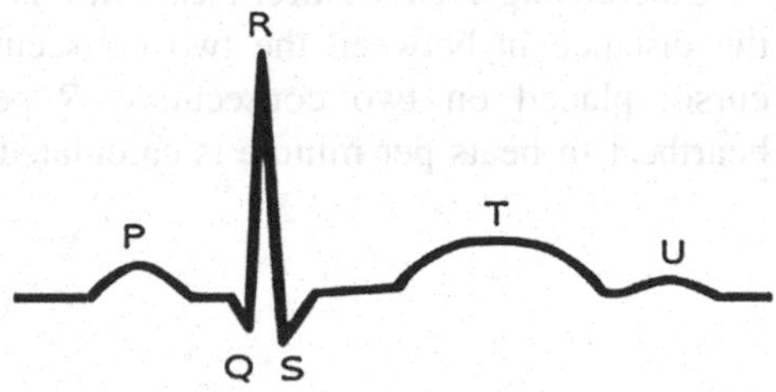

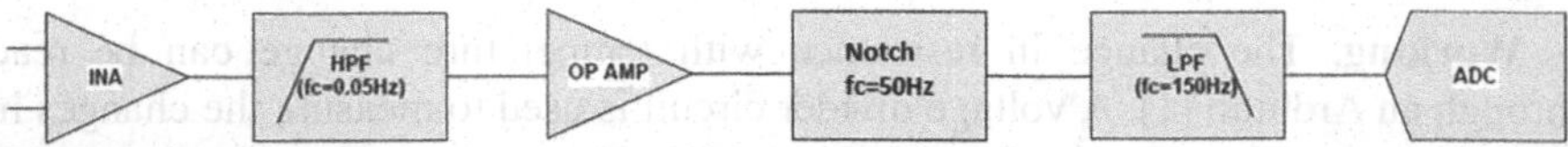

Fig. 4 Block diagram—ECG

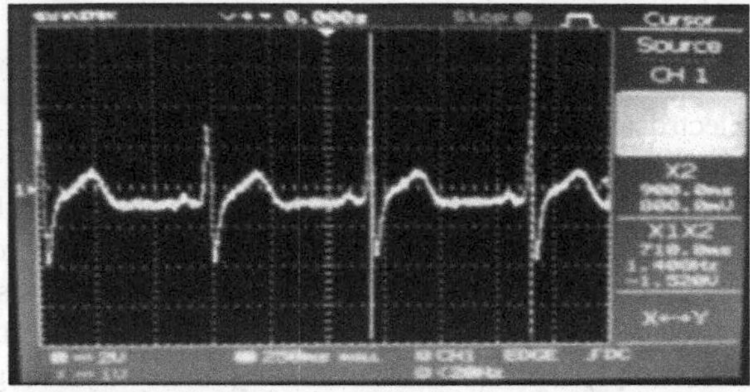

Fig. 5 Waveform obtained on oscilloscope

ECG Signal and Other Artefacts. Signal received from the leads are generally of low amplitudes and are corrupted with noise. These signals are differential signals in the range of 0–2.8 mv and has a bandwidth of 0.05–150 Hz. Major sources of noise in a typical ECG signal include motion artefacts, electrode contact noise, baseline wandering, high-frequency noise, and power line interference [3].

Conditioning ECG Signal. Since ECG signal is of very low amplitude and has noise, it should pass through signal conditioning circuitry with various stages to get clean and noise-free signal that can be analysed [4]. To remove the common mode noise and to provide the necessary differential amplification, an instrumentation amplifier having high input impedance and CMRR is used. The instrumentation amplifier used is AD620 from analogue devices. Baseline wandering refers to low-frequency component presents in the ECG caused due to offset voltage in the electrodes and is removed using a high-pass filter. The signal which is now devoid of low-frequency components is then passed through a second-stage amplifier to further amplify the signal for easy detection of peak and analysis. Next stage involves a 50-Hz notch filter. The notch filter used here is a TWIN-T-type notch filter and is used to remove power line interference noise in the ECG. Since our signal of interest is within the bandwidth of 0.05 and 150 Hz, the last stage in the circuit is a low-pass filter [5]. Figure 4 shows the various stages and its implementation.

Calculating Heart Rate. Heart rate is obtained from the signal by measuring the distance in between the two consecutive *R* peaks. Figure 5 also shows the cursor placed on two consecutive *R* peaks. After calculating *R*–*R* distance, heartbeat in beats per minute is calculated using the formula.

$$\text{Heartrate} = \frac{60{,}000}{R_{\text{PK1}} - R_{\text{PK2}}} \text{bpm}$$

3.3 Pulse Oximetry

Pulse oximetry is a non-invasive method of measuring the saturation of oxygen in the blood (SpO2). The principle of pulse oximetry is based on the light absorption characteristics of the blood. It uses infrared light and the red light for calculating the SpO2.

Circuit. A red LED, an infrared LED and phototransistor comprise the sensor part which harvests the signal off the pulsating arterial blood from the tip of the finger. The output of the phototransistor is converted to the voltage using a trans-impedance amplifier. The resulting signal is passed through a pre-amplifier which is then sent through a sample-and-hold circuit. Further, unwanted signals are removed using a band-pass filter of frequency range 0.5–2.5 Hz. The signal of interest is AC which corresponds to the pulsating arterial blood. The band-pass filter removes the DC part and the noise. The AC mains interference frequency of 50 Hz is also rejected, and the resulting signal is passed through a summing amplifier to make the voltage range to 0–5 V. The block diagram of the pulse oximetry module is shown in Fig. 6.

The resultant signal is then read by the Arduino where IR/R ratio can be calculated. R represents the red and IR represents the infrared. The equation relating the IR/R ratio and the SpO2 is given as [6]:

$$R = \frac{\frac{\text{DC}_{\text{Infrared}}}{\text{AC}_{\text{Infrared}}}}{\frac{\text{DC}_{\text{Red}}}{\text{AC}_{\text{Red}}}} = \frac{\frac{\min(\text{PPG})_{\text{Infrared}}}{\max(\text{PPG})_{\text{Infrared}}}}{\frac{\min(\text{PPG})_{\text{Red}}}{\max(\text{PPG})_{\text{Red}}}}$$

If the SpO2 value is greater than 95 %, it is considered to be normal.

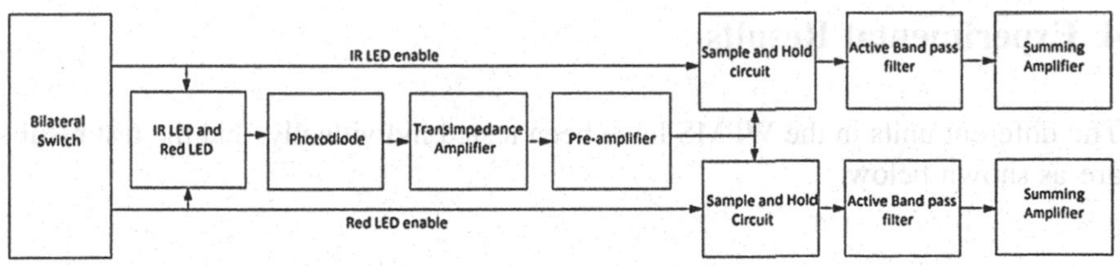

Fig. 6 Block diagram—pulse oximeter

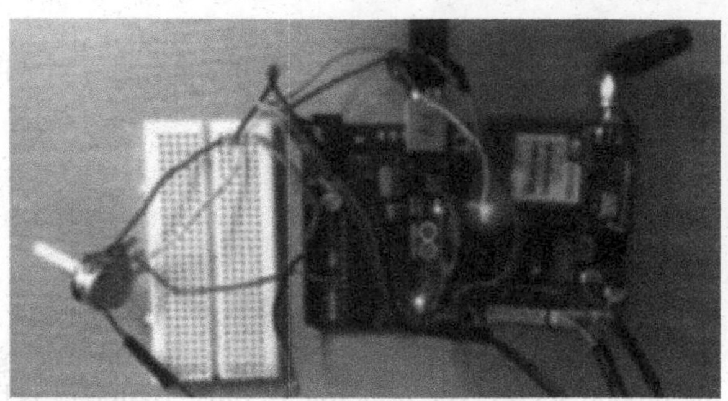

Fig. 7 Communication unit—circuit setup

3.4 Communication Unit

SIM900-TTL GSM modem [7] forms the heart of the communication unit. When there is an abnormality in the health parameter values measured, based on the threshold values set, the control unit will analyse the health parameter continuously and instructs the communication unit to take the necessary action. The unit can either make a call or send a message to the caretaker depending on the measured signal range. Various threshold values are set depending upon the severity of different parameters measured. GSM modem circuit set-up is shown in Fig. 7.

The GSM modem and the Arduino use AT commands, (ATD <space>caretaker_number) to make a call. Additional library functions, 'sim900.h' and 'sms.h' are included to send messages. A GSM supported SIM card is used by the modem, and the mobile number of the caretaker is fed into the program.

3.5 Power Supply Unit

The power supply unit mainly consists of two 12 V, 26Ah batteries used to power the wheelchair motor drivers, and the sensor unit uses rechargeable batteries.

4 Experimental Results

The different units in the WPMS have been tested individually, and the test results are as shown below.

Table 1 Temperature measurement of three subjects

Readings	1	2	3	4	5
Subject 1	99.63	99.44	99.44	99.63	99.26
Subject 2	97.07	96.89	97.25	97.07	97.07
Subject 3	97.61	97.61	97.80	97.80	97.80

4.1 Temperature Unit

The following tabular column shows the temperature in Fahrenheit measured for three different subjects (Table 1).

4.2 ECG Unit

The following tabular column shows the heartbeat in beats per minute obtained on the serial monitor for three different subjects (Table 2).

4.3 Pulse Oximetry Unit

The following tabular column shows the SpO2 value obtained on the serial monitor measured for three different subjects (Table 3).

Table 2 Heart rate in beats per minute for three subjects

Readings	1	2	3	4	5
Subject 1	75	77	71	70	74
Subject 2	83	83	89	87	87
Subject 3	73	73	74	79	77

Table 3 SpO2 measurements in percentage for three subjects

Readings	1	2	3	4	5
Subject 1	95.44	95.67	95.54	96.78	95.76
Subject 2	98.69	98.67	98.72	99.05	99.20
Subject 3	99.07	99.04	99.08	99.24	99.54

5 Conclusion

WPMS was successfully built with the patient monitoring unit integrated with the communication module. The various modules were tested individually, and the results are also included under the experimental results. Further optimization of the system can lead to its implementation in hospitals and other rehabilitation centres.

Acknowledgments We are really grateful to HuTLabs, Electronics and Communication department of Amrita School of Engineering, Amritapuri Campus, Kollam, India, for providing us all the necessary laboratory facilities and support towards the successful completion of the project. We also thank the IEEE and Amrita Vishwa Vidyapeetham University for funding this project.

References

1. Reading a Thermistor, http://playground.arduino.cc/ComponentLib/Thermistor2#.Uz6z1fm1bto
2. Rangaraj, M.R.: Biomedical signal analysis—a case study approach. Calgary (2002)
3. Yousuf, J.: Design of an infrared blood oxygen saturation and heart rate monitoring device. In: Project Report, McMaster University, Ontario (2009)
4. Leif, S., Pablo, L.: Electrocardiogram signal processing. wiley encyclopedia of biomedical engineering (2006)
5. Ajay, B., Umanath, K.: Accurate signal processing, cypress semiconductor corporation (2011)
6. Peter, S.S., Sorin, A.M., Florian, N., Breckling, J.: Signal conditioning techniques for health monitoring devices. In: 35th International Conference on Telecommunications and Signal Processing, Prague (2012)
7. Sim900-TTL GSM Modem, Datasheet, Rhydo Technologies Ltd, Cochin (2011)

Analysis of Medical X-ray Bone Images Using Image Segmentation

Shweta Jena, Barnali Sahu and Alok Kumar Jagadev

Abstract Image enhancement is the preprocessing stage of the image processing. The objective is to improve the visual effects and the perception of knowledge in images for viewers and to provide a better input for automated image processing technique. It gives emphasis on the whole or part features of the graphics in the designated image applications to enlarge the objects in the graphics. The enhanced image by use of threshold segmentation method identifies bone fracture in medical X-ray images. In this research work, the image is taken as input, and after noise removal, the X-ray image of right hand and hairline bone fracture image are being segmented using simple thresholding, multiple thresholding, and optimal thresholding method and they are compared with each other so as to choose the best technique for threshold image segmentation.

Keywords Medical image · X-ray · Segmentation · Thresholding

1 Introduction

In recent years, many efforts have been taken in developing automatize systems in the area of biomedical and bioinformatics applications. Medical image processing is one of the research areas with an interdisciplinary character that has dramatically grown in recent decades, and it has a large application domain for redundant

S. Jena (✉) · B. Sahu · A.K. Jagadev
Department of Computer Science and Engineering, Siksha 'O' Anusandhan University, Jagamohan Nagar, Khandagir, Bhubaneswar, Odisha, India
e-mail: shwetajena98@gmail.com

B. Sahu
e-mail: barnalisahu@soauniversity.ac.in

A.K. Jagadev
e-mail: alokjagadev@soauniversity.ac.in

L.C. Jain et al. (eds.), *Intelligent Computing, Communication and Devices*, Advances in Intelligent Systems and Computing 308,
DOI 10.1007/978-81-322-2012-1_85

clinical problems [1]. It sums up expertise from multiple disciplines such as computer sciences, engineering, mathematics, statistics, and medicine. For some applications such as image enhancement and compression, we cannot process the whole image at a time. Therefore, several image segmentation algorithms are being introduced. Image enhancement is the preprocessing stage for image segmentation, and the aim is to enhance the visual effect of the image by noise removal. Image segmentation is the process of partitioning a digital image into multiple clusters or super pixels according to the attributes of the image [2, 3]. Actually, the segments (clusters) are different objects in image which have the same color. Image segmentation results with a set of regions or sections that collectively cover the whole image. In a region (cluster), all of the pixels are similar with respect to some characteristic or computed property, such as intensity, color, or texture. The adjacent regions are significantly different with respect to the same characteristics. The main goal of medical image segmentation is to extract clinically relevant information or knowledge such as fractures present in bone X-ray images. Medical imaging focuses on the computational analysis of the images, not their acquisition [2]. Fracture detection based on image classification is an area of research which has proved to be challenging for the past several decades. This field has gained more importance due to the new challenges got by voluminous image databases. In medical imaging, the different segments often referred to different tissue classes, living organs, pathologies, or other biologically relevant forms. Medical image segmentation is made difficult due to low contrast, noise, and other imaging ambiguities [4]. Image segmentation is basically used to locate objects and boundaries of interest in images [5]. Basically, the segmentation method is divided into three categories as mentioned in Fig. 1 given below.

In our paper, we have used threshold-based image segmentation method for implementation. Thresholding is a simple approach used for image segmentation, where we need to fix a threshold value, and the output of thresholding is an image represented as groups of pixels with values greater or equal to the threshold or value less to threshold value [6]. The main objective of X-ray bone image segmentation is to subdivide the various portions of the image, so that it can help medical practitioners during the study of bone structure, identification of bone fracture, measurement of fracture treatment, and treatment planning prior to surgery. It has been considered as a challenging task because the bone X-ray images are complex in nature and the output of segmentation algorithm is affected due to various factors such as presence of noise and artifacts.

Goal of the paper: The goal is to analyze medical X-ray images through thresholding methods for medical imaging applications. For analysis of X-ray bone images, different thresholding methods are being used and the results are compared with each other. The general thresholding techniques used in our research are simple thresholding with different thresholding values, multiple thresholding, and optimal thresholding. The segmentation generates an enhanced image using intensity-based segmentation on the X-ray image.

Paper outline: The contents of the paper are organized as follows: In Sect. 2, the description of the thresholding-based segmentation method such as simple, multiple, and optimal thresholding are briefly discussed; in Sect. 3, the experimental results are presented and discussed; and conclusion is given in Sect. 4.

2 Thresholding-Based Segmentation

The threshold technique is the most widely used and the simplest technique to segment an image. Thresholding technique is basically used for segmenting image, having bright object on a dark background. It is useful in discriminating foreground from the background or object and background. According to Sezgin and Sankur, thresholding method is classified into the following categories:

1. Histogram-shape based
2. Clustering method
3. Entropy-based method
4. Object-attribute method
5. Spatial method
6. Local method

Generally, digital images are viewed as a 2D matrix or 2 variables function consisting of discrete points called pixels. There is a measure difference between a gray-level image and color images in the range of pixel counts. In image processing, it is simple to take a gray-level image in comparison with color image. In clustering method of thresholding, the gray-level samples are clustered into two parts such as foreground and background. For simplicity, the gray-level image is converted to a binary image. By selecting an adequate threshold value T, the binary image can be created by converting gray-level image [1, 7]. The advantage of obtaining first a binary image is that it reduces the complexity of the data and simplifies the process of recognition and classification. The most widely used way to convert a gray-level image to a binary image is to select a single threshold value (T). Then, all the gray-level values below T are treated as background and above T are considered to be the part of the object. The color code of the background is black (1), and the foreground color is white (0). The segmentation problem becomes one of selecting the proper value for the threshold T. Threshold technique can be expressed in the equation form given below.

$$T = T[x,\ y,\ m(x,\ y),\ n(x,\ y)] \tag{1}$$

T is the threshold value, where x and y are the coordinates of the threshold value point, $m(x, y)$, $n(x, y)$ are points of the gray-level image pixels. Threshold image $f(x, y)$ can be defined as:

$$f(x,y)=\begin{cases}1 & \text{if } f(x,y)>T\\ 0 & \text{if } f(x,y)\le T\end{cases} \quad (2)$$

In many types of images, the gray values of an object are different from the background value and thresholding is many a time a well-suited method to segment an image into objects and background. If the objects are without overlapping, then we can make a separate segment from each object on the threshold binary image, thus assigning a unique pixel value to each object. Many methods exist to select a suitable threshold value for a segmentation task. The simplest method is to set the threshold value interactively; the user manipulating the value and reviewing the thresholding result until a satisfying segmentation has been obtained. The histogram is generally a valuable tool in establishing a suitable threshold value.

2.1 Multiple Thresholding

Threshold-based algorithms are divided into single and multiple thresholding categories. Multiple thresholding approach focuses on finding multiple thresholds which aims to separate multiple objects. When several desired segments in an image are introduced, then threshold segmentation can be extended to use multiple thresholds to segment an image into more than two segments: All pixels with a value lesser than the first threshold are assigned to segment number 0, all pixels with values between the first and second threshold are assigned to segment number 1, all pixels with values between the second and third threshold are assigned to segment 2, etc. If n thresholds ($t1$, $t2$ … tn) are used,

$$f(v)=\begin{cases}0 & \text{if } \quad v<t1\\ 1 & \text{if } t1\le v<t2\\ 2 & \text{if } t2\le v<t3\\ \vdots & \vdots \quad \vdots\\ n & \text{if } \quad tn\le v\end{cases} . \quad (3)$$

After thresholding, the existing image has been segmented into $n+1$ distinct segments identified by the gray values 0 to n, respectively.

2.2 Optimal Thresholding

Optimal thresholding is applied where the histogram is the sum of two overlapping distributions. It is a technique that approximates the histogram using a weighted sum of distribution functions. The distribution function sets a threshold in such a way that the number of incorrectly segmented pixels (as predicted from the approximation) is minimal. In optimal thresholding, a criterion function is devised that yields some measure of separation between regions. A criterion function is applied for the intensity and that which maximizes this function is chosen as the threshold.

3 Results and Discussions

The threshold segmentation was implemented using (MATLAB R2010a, 7.4a), and the segmentation techniques are tested on the two images illustrated in the Fig. 2. Three techniques are applied on these images such as simple thresholding, multiple thresholding, and optimal thresholding techniques.

3.1 Simulation Results

Figure 1 showing the classification of Image Segmentation.

These are various results obtained by thresholding through implementation on an image of X-ray showing right hand (Figs. 3 and 4, Table 1).

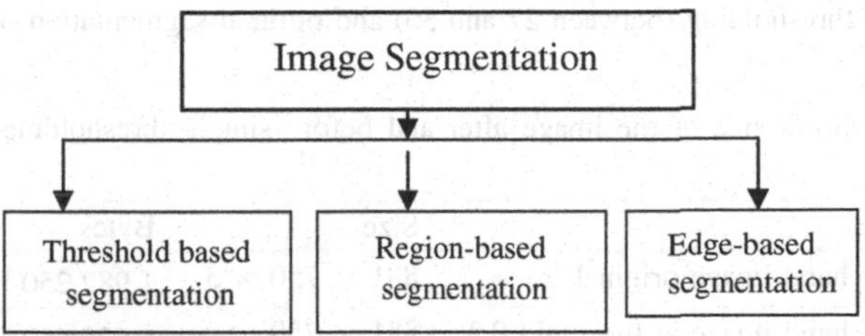

Fig. 1 Model for classification of image segmentation

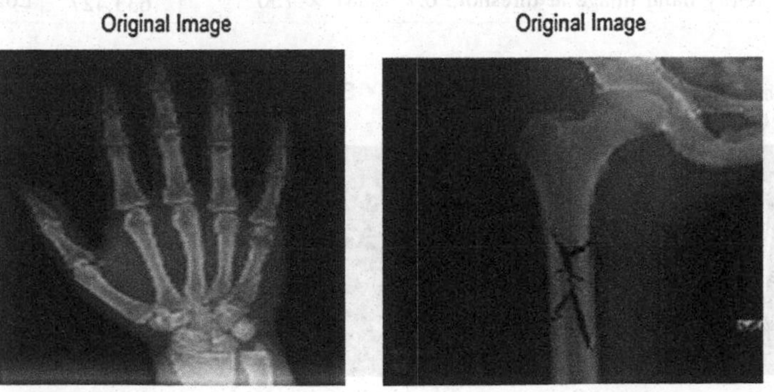

Fig. 2 The original images of *right hand* and hairline bone fracture

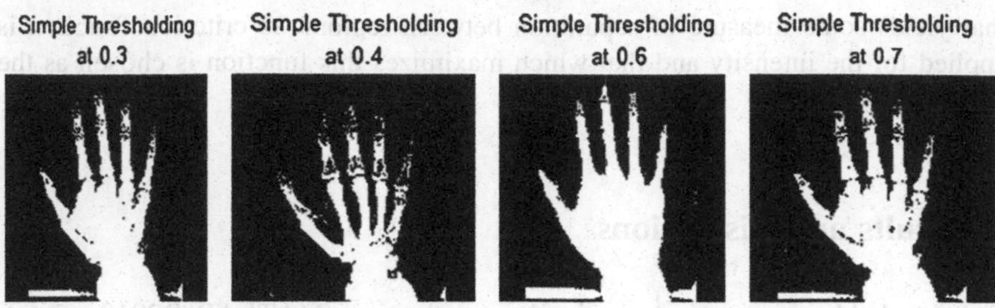

Fig. 3 Simple thresholding on original *right hand* image at 0.3, 0.4, 0.6, and 0.7 thresholding

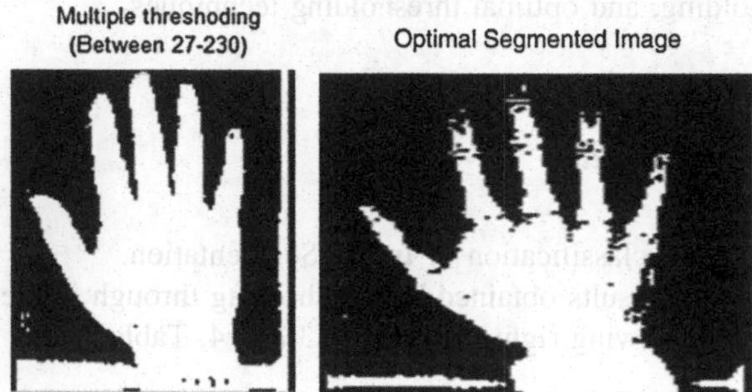

Fig. 4 Multiple thresholding (between 27 and 30) and optimal segmentation of original image

Table 1 Table shows size of the image after and before simple thresholding for X-ray hand image

Sl. no.	Name	Size	Bytes	Class attribute
1	X-ray hand image original	881 × 750 × 3	1,982,250	Unit 8
2	X-ray hand image at threshold 0.3	881 × 750	620,450	Logical
3	X-ray hand image at threshold 0.4	881 × 750	573,521	Logical
4	X-ray hand image at threshold 0.6	881 × 750	660,750	Logical
5	X-ray hand image at threshold 0.7	881 × 750	633,427	Logical

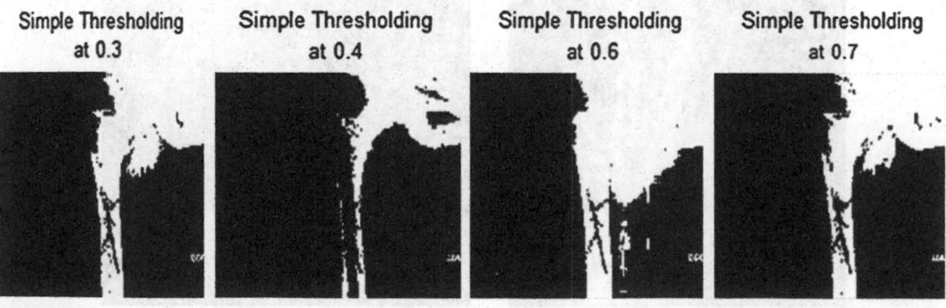

Fig. 5 Simple thresholding on original hairline bone fracture image at 0.3, 0.4, 0.6, and 0.7 thresholding

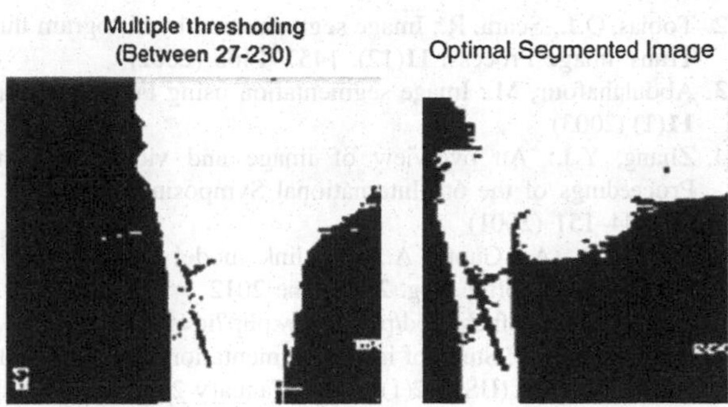

Fig. 6 Multiple thresholding (between 27 and 30) and optimal segmentation of original image

Table 2 Table shows size of the image after and before simple thresholding for X-ray hairline bone fracture image

Sl. no.	Name	Size	Bytes	Class attribute
1	X-ray hand image original	881 × 750 × 3	1,982,250	Unit 8
2	X-ray hand image at threshold 0.3	881 × 750	620,450	Logical
3	X-ray hand image at threshold 0.4	881 × 750	573,521	Logical
4	X-ray hand image at threshold 0.6	881 × 750	660,750	Logical
5	X-ray hand image at threshold 0.7	881 × 750	633,427	Logical

Various results obtained by thresholding through implementation on an image of X-ray showing hairline bone fracture (Figs. 5 and 6, Table 2).

4 Conclusion

In this paper, the effectiveness of the simple thresholding techniques at different levels, multiple thresholdings, and optimal thresholdings are compared to medical images. In our paper, we have taken 2 medical X-ray images one without fracture and another with fracture, after applying different segmentation techniques discussed above, in hand X-ray image with simple thresholding at 0.4 threshold value it is separating the object and background in a proper way but for hair line fracture image the fracture is clearly visible with optimal thresholding. From this, we can analyze that the efficiency of different segmentation algorithms are dependable on type of images.

References

1. Al-Amri, S.S., Kalyankar, N.V., Khamitkar, S.D.: Image segmentation by using threshold techniques. J. Comput. **2**(5), May 2010, ISSN: 2151-9617

2. Tobias, O.J., Seara, R.: Image segmentation by histogram thresholding using fuzzy sets. IEEE Trans. Image Process. **11**(12), 1457–1465 (2002)
3. Abdulghafour, M.: Image segmentation using Fuzzy logic and genetic algorithms. J. WSCG **11**(1) (2003)
4. Zhang, Y.J.: An overview of image and video segmentation in the last 40 years. In: Proceedings of the 6th International Symposium on Signal Processing and Its Applications, pp. 144–151 (2001)
5. Amandeep, A., Gupta, A.: Simulink model based image segmentation. Int. J. Adv. Res. Comput. Sci. Softw. Eng. **2**(6), June 2012
6. http://moodle.epfl.ch/mod/page/view.php?id=12593
7. Abubakar, F.M.: Study of image segmentation using thresholding technique on a noisy image. Int. J. Sci. Res. (IJSR) **2**(1), 49–51 January 2013, India Online ISSN: 2319-7064

Image Change Detection Using Particle Swarm Optimization

Santwana Sagnika, Saurabh Bilgaiyan and Bhabani Shankar Prasad Mishra

Abstract Image change detection can be expressed as a function of time period, whose main objective is to find the changes on the same area at different time intervals, which is a complex and intractable one. Due to large search space, general optimization algorithm fails to give the solution in a promising amount of time. So particle swarm optimization (PSO), one of the swarm-based approaches, can be used as an efficient tool, which the authors have explored in this paper. This mechanism aims to find a change mask that performs partitioning of image into changed and unchanged areas so that the weighted sum of mean square errors of both areas is minimized. This leads to accurate change detection with less noise in a feasible time period.

Keywords Image processing · Change detection · PSO · Difference image · Mean square error

1 Introduction

Currently, the data processing and computation scenario in the world are becoming exceedingly multimedia-based, where graphics and motion pictures are widely replacing traditional textual and less informative data. Change detection involves finding the differences that have occurred in a region or scene, by comparing images of that region taken at different times, especially in remote sensing

S. Sagnika (✉) · S. Bilgaiyan · B.S.P. Mishra
School of Computer Engineering, KIIT University, Bhubaneswar, Odisha, India
e-mail: santu.hmm@gmail.com

S. Bilgaiyan
e-mail: saurabhbilgaiyan01@gmail.com

B.S.P. Mishra
e-mail: mishra.bsp@gmail.com

L.C. Jain et al. (eds.), *Intelligent Computing, Communication and Devices*,
Advances in Intelligent Systems and Computing 308,
DOI 10.1007/978-81-322-2012-1_86

domains. This process finds use in a wide variety of applications, such as urban growth assessment, damage estimation after natural calamities, finding forest cover changes, and analyzing troops' placement during military operations, involving large images [1]. Besides, technical advancement in the equipments has also brought about availability of better quality and more detailed images, which require more intensive processing. Hence, the requirement arises for automation of the change detection process that can reduce the time and effort needed for analyzing the images [2, 3]. General change detection techniques take place in three steps—preprocessing of input images, comparing old and new images, and analyzing the comparison results [4].

Various methods for change detection broadly fall under supervised and unsupervised categories as per existing research. Supervised approaches learn from training datasets that can help estimate the types of changes occurred, subject to sufficient availability of suitable training data. Such methods include direct multidata classification (DMC) and post-classification comparison (PCC). On the other hand, unsupervised approaches consider only current input images that are preprocessed and then compared and analyzed. The popular methods in this category are change vector analysis (CVA), image differencing, normalized difference vegetation index (NDVI), etc. [2, 5].

Particle swarm optimization (PSO) is one of the efficient swarm-based optimization techniques, which is used to explore vast search spaces at multiple points simultaneously, so as to arrive at the required solution in a feasible time period. As the size of images involved in remote sensing tends to be quite large, the complexity of comparing and analyzing them for finding changes increases and needs more time. In such cases, swarm-based techniques are suitable to reduce computation time and handle complex problems efficiently. Thus, the authors have selected PSO to solve the change detection problem in images.

2 Related Work

Initial work on change detection focused on multispectral images captured using normal cameras, which include algebra, classification, transformation, visual analysis, GIS, and other advanced methods [6]. Recent research has developed alternative approaches to address for high-resolution and more detailed images, which is comprised of object-based change detection (OBCD) [7], hyperspectral image processing [8], and change detection in SAR images [9], among others. Owing to the popularity and generic nature of soft computing approaches to solve complex real-world problems, some work has also been done to exploit soft computing techniques for image change detection.

Kang et al. [10] used multiple histograms to perform scene change detection through Otsu thresholding to find out the number of changed blocks in a frame. This method is a fast and efficient technique, but the major drawback here is the manual selection of fixed threshold difference that can cause errors in change detection if not properly set.

Dai et al. [11] introduced a swarm-based approach that has hybridized ant colony optimization (ACO) and PSO to construct rules for a supervised change detection method based on Bayesian network model, which has proved to be an efficient mechanism for change detection.

Celik [12] carried out a genetic algorithm (GA)-based method that uses weighted sum of mean square errors of changed and unchanged regions as the fitness function to find out the change detection mask. This approach is advantageous as it overcomes fixed thresholding problem and reaches solution faster than completely randomized approach.

Celik [13] then refined the above-mentioned technique by integrating it with Gaussian mixture model (GMM), which exhibits improved performance and accuracy when applied to both optical and SAR images.

Celik and Yetgin [14] also extended the use of GA in change detection by specifying a multiobjective fitness function that considers correlation, spectral distortion, universal image quality index, and noise factors between changed and unchanged pixels as the multiple objectives. It does not require explicit computation of a difference image or any assumptions. This makes it useful for processing satellite images.

3 Mathematical Model

Consider a given area for which change detection is to be carried out. It is required to have two input images, which are taken at different times. The comparison between both images gives the required changes that have occurred in the interim time period in that particular area.

Let the first image be represented as follows:

$$I_1 = \{i_1(a,b) | a \in [1,R],\, b \in [1,C]\} \tag{1}$$

and the second image as

$$I_2 = \{i_2(a,b) | a \in [1,R],\, b \in [1,C]\} \tag{2}$$

where the images are of size $R \times C$. The final result of this process will yield a binary matrix of same dimensions $R \times C$, otherwise known as change mask, which will denote changed and unchanged regions in the form of 0 and 1. Mathematically, change mask is

$$M = \{m(a,b) | a \in [1,R],\, b \in [1,C]\}, \tag{3}$$

where $m(a,b) = \begin{cases} 0, & \text{for unchanged areas} \\ 1, & \text{for changed areas} \end{cases}$

To find this mask, the first requirement is to generate a dissimilarity matrix, which will contain the absolute difference of corresponding pixels in both images. This matrix can be represented by

$$DM = |I_2 - I_1| \tag{4}$$

Then, it is required to generate random masks and assess their fitness value. This fitness value defines whether the current mask is the actual solution or not. Let M be the current mask. Then, it can be partitioned into two sets, i.e.,

$$P_0 = \{(a,b)|\, m(a,b) = 0\} \text{ and } P_1 = \{(a,b)|\, m(a,b) = 1\} \tag{5}$$

Let $N_0 = |P_0|$ and $N_1 = |P_1|$, i.e., number of elements in sets P_0 and P_1, respectively. So, the mean of set P_0 will be

$$\text{Mean}\,(P_0) = (1/N_0) * \sum_{\forall (a,b)\, \epsilon\, P_0} DM(a,b) \tag{6}$$

and of set P_1 will be

$$\text{Mean}\,(P_1) = (1/N_1) * \sum_{\forall (a,b)\, \epsilon\, P_1} DM(a,b) \tag{7}$$

Thus, the weighted sum of mean square error can be represented as follows:

$$\begin{aligned} \text{WS} = {} & [(N_0 * \sum_{\forall (a,b)\, \epsilon\, P_0} DM(a,b) - \text{Mean}(P_0))^2) \\ & + (N_1 * \sum_{\forall (a,b)\, \epsilon\, P_1} DM(a,b) - \text{Mean}(P_1))^2)]/(R * C) \end{aligned} \tag{8}$$

Hence, the fitness function is

$$\text{Minimize (WS)} \tag{9}$$

By finding the minimum value of fitness function, the best mask can be found out, since the mean square errors' weighted sum will be minimum for a better partitioning.

4 Algorithm

4.1 General PSO Algorithm

PSO is one of the most efficient swarm-based optimization techniques developed by Kennedy and Eberhart in 2007, and this method can approximate solutions in polynomial for those problems for which conventional algorithms would take non-polynomial time to solve. Compared to other swarm-based and evolutionary approaches such as GA and ACO, PSO gives benefits of faster convergence and production of alternative solutions [15–17]. A large population formed by the collection of particles search and look around the solution space for the best solution which exists in that domain. The velocity and state of each particle get updated iteratively on the basis of self- and external experiences [18–20]. The general steps of PSO are as follows:

Step 1: Initialize the population of P agents with random positions and velocities in M-dimensional hyperspace.
Step 2: Assess the fitness value for each particle.
Step 3: Compare the current position of each particle with its local best position found so far.

a. If current position is better, then set it as the new local best position
b. else continue with the old local best position

Step 4: Choose the particle having best fitness value among all particles.
Step 5: Update the state and velocity of each particle iteratively on the basis of self-experience and global experience of all other particles in the same population.
Step 6: Check for the stopping criteria.

a. if stopping criteria is achieved then stop
b. else go to Step 2

where the state and velocity for each particle are calculated as follows:

$$V_i^{t+1} = V_i^t + a1 * r1 * (\text{lbest}_i^t - S_i^t) + a2 * r2 * (\text{gbest}_i^t - S_i^t) \tag{10}$$

$$S_i^{t+1} = S_i^t + V_i^{t+1} \tag{11}$$

Here, V_i^t = velocity of the ith particle in the tth iteration, V_i^{t+1} = velocity of the ith particle in the $t+1$th iteration, S_i^t = State of the ith particle in the tth iteration, S_i^{t+1} = State of the ith particle in the $t+1$th iteration, $a1$ and $a2$ are learning factors of local and global information, respectively, and $r1$ and $r2$ are random variables.

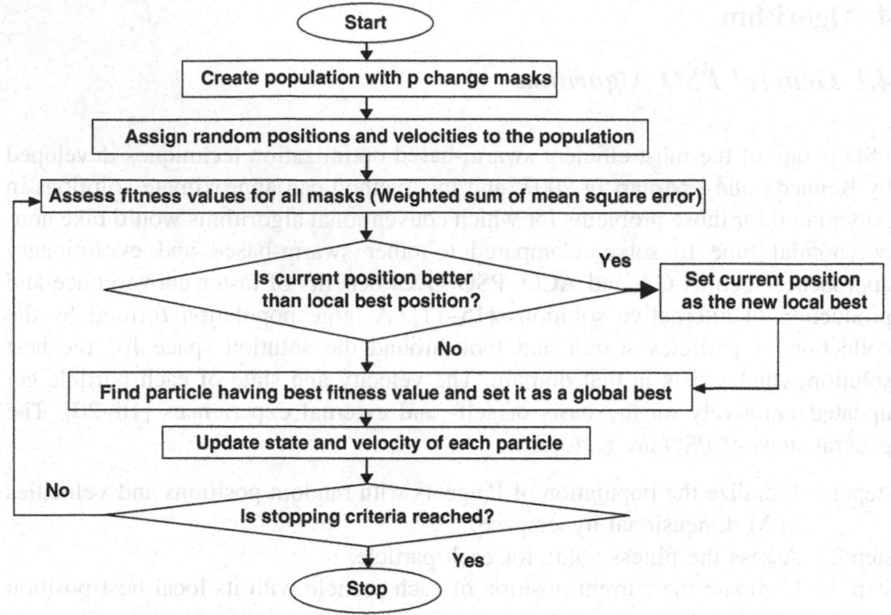

Fig. 1 Steps of the proposed technique

4.2 Proposed Technique

To solve the problem discussed in Sect. 3, the application of PSO serves as an effective technique to reach the solution faster. In this paper, the authors describe how PSO has been employed to solve the change detection problem. Each particle of the population represents a possible change mask for the input images, in the form of an $R \times C$ matrix. The fitness function is evaluated for each particle, and the best particle is found out. By following the PSO algorithm steps, the final change mask is attained. Figure 1 shows the procedure followed.

5 Implementation

5.1 Experimental Setup

The proposed method has been implemented using MATLAB as a tool, on a laptop having a 2-GB RAM and Intel i3 Core processor. The input optical images are taken from a dataset that contains images of a forest area before and after the occurrence of a fire as obtained from satellite [21]. The images are resized to size 50×50, for ease of experimentation on a laptop. Larger images can easily run on systems that will have higher processing ability or parallel architectures. For PSO technique, the population size is taken as 30, and the number of iterations is taken as 125,000.

5.2 Results and Discussion

Figure 2 shows the two input images, before and after the fire occurrence. Figure 3 shows the dissimilarity image as denoted by DM in matrix format. Figure 4 represents the results obtained at different stages of the experiment.

The results indicate that the proposed method exhibits high accuracy, as is visible when compared to the ground truth as shown in Fig. 5a. It is also seen that more detailed detection and less noise are obtained as regards results obtained by existing Markov random field (MRF) method (Fig. 5b), multiscale method (Fig. 5c), and principal Component Analysis(PCA) method (Fig. 5d), when each is compared with the ground truth. It takes up more resources and requires large processing time. But better detection is possible due to inherent quality of PSO technique, where searching of the complete search space takes place concurrently and particles converge toward the best position with varying velocities. For a 50×50 sized image having 2,500 pixels, a traditional algorithm would require approx. 2^{2500} computations in the worst case. The proposed technique achieves

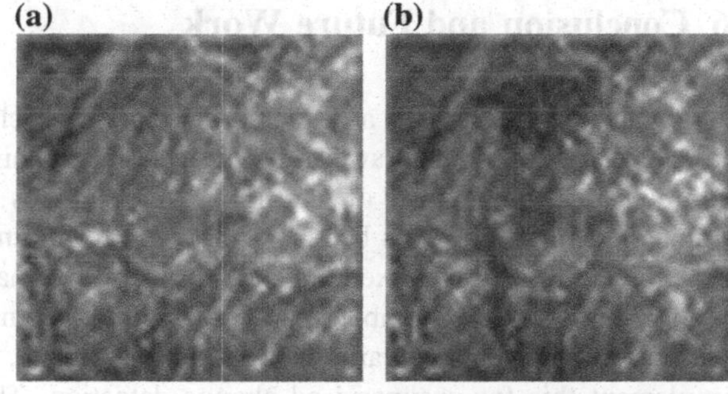

Fig. 2 Input images. **a** Image before fire. **b** Image after fire

Fig. 3 Dissimilarity image by intensity difference calculation

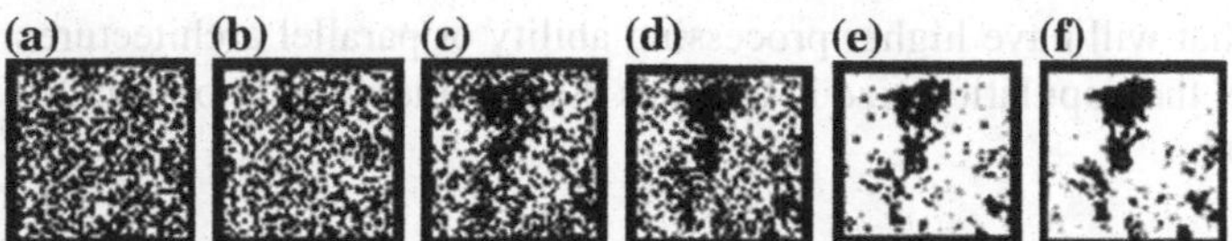

Fig. 4 Change masks at different stages. **a** Initial randomly generated mask. **b** 1,000 iterations. **c** 25,000 iterations. **d** 50,000 iterations. **e** 100,000 iterations. **f** 125,000 iterations (final mask)

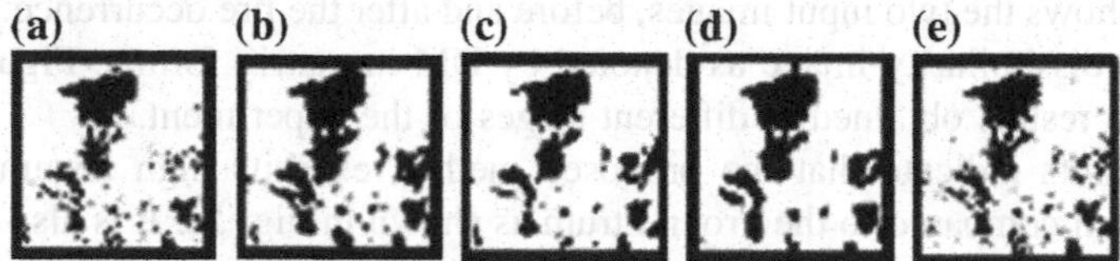

Fig. 5 **a** Ground truth as obtained by manual examination. **b** Result by MRF method. **c** Result by multiscale method. **d** Result by PCA method. **e** Result by proposed method

results in nearly 125,000 iterations, i.e., 30 × 125,000 = 37,50,000 computations. Hence, it proves to be a feasible mechanism. It also does not need fixing of any parameters, such as the threshold for determining change as is employed in a large number of existing techniques.

6 Conclusion and Future Work

This paper has described a new approach to solve change detection problem in images using an efficient swarm-based technique. This method is fast and attains solution in lesser number of iterations, which makes it more feasible than traditional algorithms. It gives better accuracy with minimal noise. It also overcomes the problem of setting a fixed threshold as is used in many existing methods, which is difficult to set in a suitable manner. The proposed mechanism does not use any parameters and handles variations efficiently. Hence, it is practically possible to implement this for unsupervised change detection. The disadvantage here is the high computational time which arises due to large size of images and can be overcome using parallel computing and high-capacity processor systems. Future work can be done in reducing the execution time of the algorithm and incorporating neighborhood information for pixels that will improve the accuracy of detection. Proper change detection of images can prove to be useful in various fields, particularly for remote sensing applications.

References

1. Ghosh, S., Patra, S., Ghosh, A.: An unsupervised context-sensitive change detection technique based on modified self-organizing feature map neural network. Int. J. Approximate Reasoning, Elsevier, 37–50 (2008)
2. Pacifici, F., Del Frate, F., Solimini, C., Emery, W.J.: An Innovative neural-net method to detect temporal changes in high-resolution optical satellite imagery. IEEE Trans. Geosci. Remote Sens. **45**, 2940–2952 (2007)
3. Patra, S., Ghosh, S., Ghosh, A.: Histogram thresholding for unsupervised change detection of remote sensing images. Int. J. Remote Sens. **32**(21), 6071–6089 (2011)
4. Ghosh, A., Subudhi, B.N., Bruzzone, L.: Integration of gibbs markov random field and hopfield-type neural networks for unsupervised change detection in remotely sensed multitemporal images. IEEE Trans. Image Process. **22**(8), 3087–3096 (2013)
5. Bruzzone, L., Pireto, D.F.: Automatic analysis of the difference image for unsupervised change detection. IEEE Trans. Geosci. Remote Sens. **38**(3), 1171–1182 (2000)
6. Lu, D., Mausel, P., Brondzio, E., Moran, E.: Change detection techniques. Int. J. Remote Sens. **25**(12), 2365–2401 (2004)
7. Tang, Y., Huang, X., Muramatsu, K., Zhang, L.: Object-oriented change detection for high-resolution imagery using a genetic algorithm. Int. Arch Photogrammetry Remote Sens. Spat. Inf. Sci. **38**(8), 769–774 (2007)
8. Eismann, M.T., Meola, J., Hardie, R.C.: Hyperspectral change detection in the presence of diurnal and seasonal variations. IEEE Trans. Geosci. Remote Sens. **46**, 237–249 (2008)
9. Huang, S.Q.: Change mechanism analysis and integration change detection method on SAR images. Int. Arch Photogrammetry Remote Sens. Spat. Inf. Sci. **37**, 1559–1568 (2008)
10. Kang, S.J., Cho, S.I., Yoo, S., Kim, Y.H.: Scene change detection using multiple histograms for motion-compensated frame rate up-conversion. IEEE J. Disp. Technol. **8**, 121–126 (2012)
11. Dai, Q., Liu, J., Liu, S.: Remote sensing image change detection based on swarm intelligent algorithm. In: IEEE International Conference on Multimedia Technology (ICMT), pp. 1–3 (2010)
12. Celik, T.: Change detection in satellite images using a genetic algorithm approach. IEEE Geosci. Remote Sens. Lett. **7**, 386–390 (2010)
13. Celik, T.: Image change detection using Gaussian mixture model and genetic algorithm. J. Vis. Commun. Image R., Elsevier, **21**(8), 965–974 (2010)
14. Celik, T., Yetgin, Z.: Change detection without difference image computation based on multiobjective cost function optimization. Turk. J. Elec. Eng. Comp. Sci. **19**, 941–956 (2011)
15. Kennedy, J., Eberhart, R.: Particle swarm optimization. In: Proceedings of IEEE International Conference on Neural Networks IV, pp. 1942–1948 (1995)
16. Yin, P.Y., Yu, S.S., Wang, P.P., Wang, Y.T.: A hybrid particle swarm optimization algorithm for optimal task assignment in distributed systems. Comput. Stand. Interfaces, Elsevier, **28**(4), 441–450 (2006)
17. Yang, X., Yuan, J., Yuan. J., Mao, H.: A modified particle swarm optimizer with dynamic adaptation. Appl. Math. Comput., Elsevier, **189**(2), 1205–1213 (2007)
18. Wu, Z., Ni, Z., Gu, L., Liu, X.: A revised discrete particle swarm optimization for cloud workflow scheduling. In: IEEE International Conference on Computational Intelligence and Security, pp. 184–188 (2010)
19. Parsopoulos, K.E.: Particle Swarm Optimization and Intelligence: Advances and Applications. Information science reference 24, pp. 2908–2914. Hershey, New York (2010)
20. Clerc, M.: discrete particle swarm optimization, illustrated by traveling salesman problem. In: Onwubolu, G.C., Babu, B.V. (eds.) New Optimization Techniques in Engineering, pp. 219–239. Springer, Berlin (2004)
21. Retrieved September 2009, http://geochange.er.usgs.gov/sw/changes/natural/reno-tahoe/burn.html

Efficient Microarray Data Classification with Three-Stage Dimensionality Reduction

Rasmita Dash, B.B. Misra, Satchidananda Dehuri and Sung-Bae Cho

Abstract High dimensionality and small sample size are the intrinsic nature of microarray data, which require effective computational methods to discover useful knowledge from it. Classification of microarray data is one of the important tasks in this field of work. Representation of the search space with thousands of genes makes this work much complex and difficult to classify efficiently. In this work, three different stages have been adopted to handle the crush of dimensionality and classify the microarray data. At the first stage, statistical measures are used to remove genes that do not contribute for classification. In the second stage, more noisy genes are removed by considering signal-to-noise ratio (SNR). In the third stage, principal component analysis (PCA) method is used to further reduce the dimension. Finally, these reduced datasets are presented to different classification techniques to evaluate their performance. Here, four different classification algorithms are used such as artificial neural network (ANN), naïve Bayesian classifier, multiple linear regression (MLR), and k-nearest neighbor (k-NN) to validate the

R. Dash (✉)
Department of Computer Science and Information Technology, SOA University, Bhubaneswar 751030, Odisha, India
e-mail: rasmitadash@soauniversity.ac.in

B.B. Misra
Department of Computer Science and Engineering, Silicon Institute of Technology, Bhubaneswar 751024, Odisha, India
e-mail: misrabijan@gmail.com

S. Dehuri
Department of Systems Engineering, Ajou University, San 5, Woncheon-dong, Yeongtong-Gu, Suwon 443-749, South Korea
e-mail: satchi@ajou.ac.kr

S.-B. Cho
Soft Computing Laboratory, Department of Computer Science, Yonsei University, 50 Yonsei-Ro, Sudaemoon-Gu, Seoul 120-749, South Korea
e-mail: sbcho@yonsei.ac.kr

L.C. Jain et al. (eds.), *Intelligent Computing, Communication and Devices*,
Advances in Intelligent Systems and Computing 308,
DOI 10.1007/978-81-322-2012-1_87

benefits of three-stage dimensionality reduction. The experimental results show that the use of statistical methods, SNR, and PCA improves the overall performance of the classifiers.

Keywords Classification · Microarray data · Feature selection

1 Introduction

Different microarray data analysis techniques are classification, clustering, identification of informative genes, and many more. Classification is one of the important works. Given a set of previously classified examples (e.g., different types of cancer classes such as AML and ALL), a classifier will find a rule that will allow to assign new samples to one of the above classes [1]. For classification task, a sufficient number of samples must be present to allow an algorithm to be trained known as training set and then to have it tested on an independent set of samples known as test set. A normalized gene expression data are used as input vectors, and classification rules can be built. There are a wide range of algorithms that can be used for classification. But in microarray data, the number of samples is too less (within the range of 100) as compared to the number of attributes (within the range of tens of 1,000) [2]. These are high-dimensional data, and analytical precision is influenced by a number of variables. So it is extremely useful to reduce the dataset to those genes that are best distinguished between the two cases or classes (e.g., normal vs. diseased). Hence, another issue in microarray data analysis is dimensionality reduction, and before classification, feature reduction is essential.

Several feature selection techniques are developed to address the problem of reducing genes, but identifying an appropriate feature reduction technique is challenging in microarray data analysis. This is due to the presence of enormous number of genes compared to the number of samples. Out of several genes, some are noisy and some are highly correlated and its presence degrades the classification performance. For that, it is significant to extract the informative genes from the original data. So before classification, identification of relevant or significant genes is important. Therefore, an ensemble feature reduction and classification model can be used for efficient processing of microarray data [3].

This paper proposes a three-stage dimensionality reduction technique for microarray data classification. Here, four different classifiers, namely artificial neural network (ANN) [4], multiple linear regression (MLR) [5], naïve Bayesian classifier [6], and k-nearest neighbor (k-NN), are used [7]. These classifiers are used to classify the original and reduced data separately, and a comparative study is presented.

This paper is organized as follows. Section 1 of the paper deals with the introductory concepts of microarray data, microarray data analysis and its challenges, the need of dimensionality reduction, and the goal of the paper. Our proposed method for dimensionality reduction is shown in Sect. 2. Section 3 describes the application of our proposed method for feature reduction and is classified as ANN, naïve Bayesian network, MLR, and k-NN algorithm, followed by conclusion in Sect. 4.

2 Proposed Three-Stage Dimensionality Reduction Scheme

In this proposed dimensionality reduction technique, the number of genes in microarray data is reduced using three different stages and the reduced data are represented as dimensionality-reduced microarray data (DRMD). The different levels of reductions are shown in Fig. 1:

DRMD1: In the first stage of analysis of the datasets, it is observed that the values of the genes in some of the attributes are highly similar, irrespective of the class labels. The dissimilarity in the feature value belonging to different classes contributes to design any classifier model. Therefore, less dissimilar genes can be considered less important for the task of classifier design and may be excluded. In this stage, the standard deviation of the attribute is evaluated. The attributes with high dissimilarity are allowed to remain in the database, and other attributes are removed. Comparing the standard deviation values of different attributes, a threshold value δ is fixed. All the genes having standard deviation value less than that of threshold value are eliminated from the microarray dataset. The remaining database is referred to as dimensionally reduced microarray database 1 (DRMD1).

DRMD2: After the removal of irrelevant genes at stage 1, it is observed that DRMD1 contains significant number of attributes which can be considered as noise for developing a robust classifier model. So in the second stage, signal-to-noise ratio (SNR) technique is used to measure the level of desired signal to the level of background noise. The SNR score identifies the expression patterns with a maximal difference in mean expression between two groups and minimal variation of expression within each group [9]. In this method, genes are first-ranked according to their expression levels using SNR test statistic. The SNR is defined as follows:

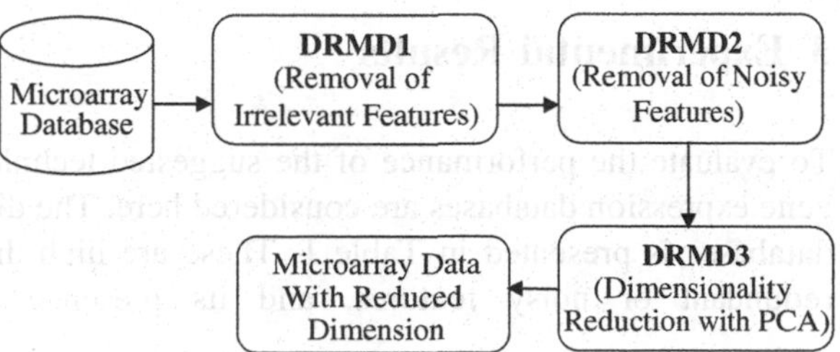

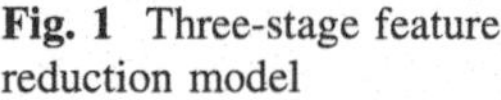
Fig. 1 Three-stage feature reduction model

$$\mathrm{SNR}(j) = \frac{(\mu_-^j - \mu_+^j)}{(\sigma_-^j - \sigma_+^j)} \quad (1)$$

where SNR(j) is the SNR of jth attribute, μ_-^j and μ_+^j are the mean values of jth attribute that belongs to negative class and positive class, respectively. σ_-^j and σ_+^j are the standard deviations of jth attribute for the respective classes. The higher SNR value indicates that the signal is more valuable than that of the noise. A predefined number of attributes with higher SNR value are selected, and the reduced data are termed as dimensionally reduced microarray database 2 (DRMD2).

DRMD3: Dimensionally reduced microarray database 3 represents another level of feature reduction that takes place at the third stage. After noise reduction, PCA is used for further reduction of the dimensionality. PCA by Yan et al. [8] and Valarmathie et al. [9] is an unsupervised feature reduction method for projecting high-dimensional data into a new lower-dimensional representation of the data that describe as much of the variance in the data as possible with minimum reconstruction error. Hence, PCA is a statistical technique for determining key variables in a high-dimensional dataset that explains the differences in the observations and can be used to simplify the analysis and visualization of high-dimensional dataset, without much loss of information. The transformation of the dataset to the new principal component axis produces the number of PCs equivalent to the number of original variables. But for many databases, the first several PCs explain the most of the variances, so the rest can be eliminated with minimal loss of information. The various criteria used to determine how many PCs should be retained for the interpretation are as follows:

- Scree diagram plots are used where PCs are considered as variance in percentage and dimension is reduced by eliminating the PCs with low variance.
- A threshold value is fixed for variance; PCs having variance below threshold value are rejected, and rests are considered.
- Drop PCs whose eigenvalues are smaller than a fraction of the mean eigenvalue.

3 Experimental Results

To evaluate the performance of the suggested technique, three publicly available gene expression databases are considered here. The dimension of these microarray databases is presented in Table 1. These are high-dimensional data that contain redundant or noisy features, and its presence degrades the classification

Table 1 Description of datasets used

Dataset	Sample	Genes
All/Aml [11]	38	5,000
Colon [12]	62	2,000
Adenoma [13]	8	7,086

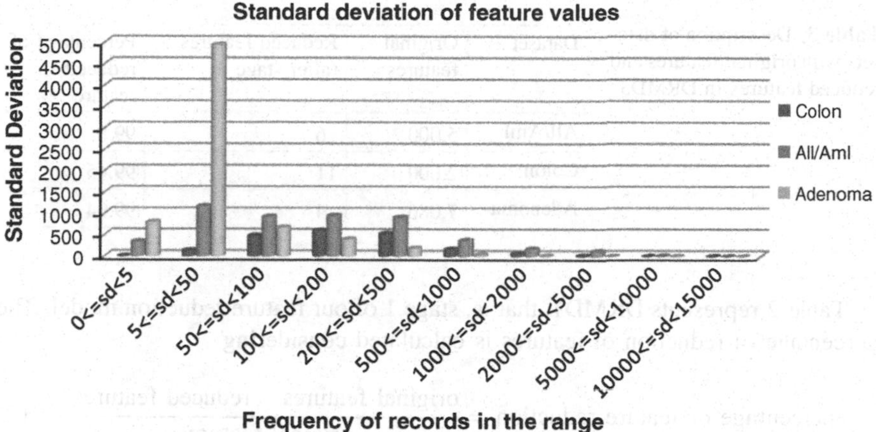

Fig. 2 Frequency distribution of records in different ranges of standard deviation of the databases

performance, requires huge memory, and consumes more computational time. So in this experiment, the primary task is to reduce the dimension without compromising the performance of such models. In general, feature reduction is considered as a one-step process in many research studies, but it may not be good enough to remove the noisy and insignificant features to a great extent with such effort [10]. Therefore, to improve the effectiveness of feature reduction, here a three-stage data reduction technique is performed before classification task is taken up.

In the initial stage, less important features are removed from the database. Standard deviation of different features in the microarray database is evaluated, and a threshold value of the standard deviation is considered. It is observed that these standard deviation values for each feature range from very small value to very large value. A large standard deviation in feature value may not be good enough to classify the records, but very small standard deviation value may not be much helpful in classifying the records. Figure 2 shows the frequency distribution of the standard deviation values obtained for different databases. For All/Aml and colon databases, threshold value is considered 500. In adenoma database, majority of the features lie in the range 5–50, so a lower threshold of 50 is kept for it.

Table 2 Description of datasets with original features and reduced features in DRMD1

Dataset	Threshold value (δ)	Original features	Reduced features (in stage 1)	Percentage of reduced feature
All/Aml	500	5,000	517	89.6
Colon	500	2,000	230	88.5
Adenoma	50	7,086	630	91.1

Table 3 Description of datasets with original features and reduced features in DRMD3

Dataset	Original features	Reduced features (after stage 3)	Percentage of reduced feature
All/Aml	5,000	6	99.8
Colon	2,000	11	99.45
Adenoma	7,086	4	99.94

Table 2 represents DRMD1, that is, stage 1 of our feature reduction model. The percentage of reduction of features is calculated considering

$$\text{percentage of feature reduction} = \frac{\text{original features} - \text{reduced features}}{\text{original features}} \tag{2}$$

In DRMD2, to eliminate the noisy data, SNR and top 100 genes are selected. To select the most significant features, PCA is applied in DRMD3. So at the end of stage 3, the reduced set of genes is presented in Table 3. It is observed that the high-dimensional gene expression data are reduced with a small number of features.

After the final reduction of the dimensionality of the databases, the DRMD3 is used for classification. Here, four different classification techniques such as ANN, naïve Bayesian, k-NN, and MLR are used to evaluate the performance of this data reduction scheme. To evaluate the classification, we have used different metrics such as sensibility, specificity, precision, and accuracy.

Table 4 shows the performance of colon database using four classifiers before and after feature reduction. Only in case of ANN, the sensitivity is reduced after reduction in dimensionality. But with the reduced dimension, sensitivity is highest in case of naïve Bayesian classifier, whereas k-NN yields the highest specificity, precision, and accuracy and achieves the second highest sensitivity with the reduced dimension data. Overall performance of k-NN can be considered better in comparison with all other models for colon database.

The performance metrics of adenoma database for four different classifiers is presented in Table 5. Naïve Bayesian and k-NN yield the best performance using reduced featured database.

The results obtained from the All/Aml database are presented in Table 6. Except naïve Bayesian, all the models yield best sensitivity. In case of MLR,

Table 4 Comparison of performance metrics for colon database

Performance metrics	ANN		MLR		Naïve Bayesian		*k*-NN	
	Before reduction	After reduction	Before reduction	After reduction	Before reduction	After reduction	Before reduction	After reduction
Sensitivity	85.37	80.5	10.95	34	–	**92.68**	77.4	85.7
Specificity	28.57	34	2	43	–	71.43	92.6	**100**
Precision	70	72	66.1	74	–	86.36	83.3	**100**
Accuracy	66.12	66.12	64.5	87.09	–	85.48	85.4	**95.16**

Table 5 Comparison of performance metrics for adenoma database

Performance metrics	ANN		MLR		Naïve Bayesian		*k*-NN	
	Before reduction	After reduction	Before reduction	After reduction	Before reduction	After reduction	Before reduction	After reduction
Sensitivity	50	100	–	6	25	**100**	50	**100**
Specificity	50	75	–	5	100	**100**	100	**100**
Precision	50	80	–	100	100	**100**	100	**100**
Accuracy	50	87.5	–	100	62.5	**100**	75	**100**

Table 6 Comparison of performance metrics for All/Aml database

Performance metrics	ANN		MLR		Naïve Bayesian		*k*-NN	
	Before reduction	After reduction	Before reduction	After reduction	Before reduction	After reduction	Before reduction	After reduction
Sensitivity	100	100	100	71	61.9	90.9	100	100
Specificity	0	63	00	10	82.93	100	0	91
Precision	71.05	87	100	75	65	100	100	96.43
Accuracy	71.05	89.4	71.05	81.57	75.8	97.3	71.05	97.3

sensitivity and precision reduce after dimensionality reduction. Also, in case of k-NN, the precision also reduces after dimensionality reduction.

4 Conclusion

This paper applies a three-stage dimensionality reduction to microarray databases and applies four different classifiers to observe improvement in performance. In the first stage, statistical measures are used to remove the irrelevant genes from the database. In this stage, about 90 % of the less important features are dropped from the database. In the second stage, another subset of highly noisy genes is removed by using SNR. Finally, PCA method is used to further reduce the dimension at the last stage. These reduced data are presented to ANN, MLR, naïve Bayesian, and k-NN classifier to evaluate their performance. Using different performance metrics, the performance of the classification is compared for the database without dimensionality reduction and with dimensionality reduction. It is observed that in majority of the cases, performance improves significantly after three-stage reduction of the dimension of the microarray databases.

Acknowledgment The authors gratefully acknowledge the support of the Original Technology Research Program for Brain Science through the National Research Foundation (NRF) of Korea (NRF:2010-0018948) funded by the Ministry of Education, Science, and Technology.

References

1. Quackenbush, J.: Computational analysis of microarray data. Nat. Rev. Genet. **2**(6), 418–427 (2001)
2. Zhou, X., Tuck, D.P.: MSVM-RFE extensions of SVM-REF for multiclass gene selection on DNA microarray data. Bioinformatics **23**(9), 1106–1114 (2007)
3. Mutch, D.M., Berger, A., Mansourian, R., Rytz, A., Roberts, M.A.: Microarray data analysis: a practical approach for selecting differentially expressed genes. Genome Biol. **2**(12) (2001)
4. Resul, D., Ibrahim, T., Abdulkadir, S.: Effective diagnosis of heart disease through neural networks ensembles. Expert Syst. Appl. **36**, 7675–7680 (2009)
5. Hsia, T.C.: System Identification: Least Squares Methods. D. C. Heath and Company (1997)
6. Lu, H., Setiono, R., Liu, H.: Effect data mining using neural networks. IEEE Trans. Knowl. Data Eng. **8**, 957–961 (1996)
7. Cover, T., Hart, P.: Nearest neighbor pattern classification. Proc. IEEE Trans. Inf. Theor. 21–27 (1967)
8. Yan, J., Zhang, B., Liu, N., Yan, S., Cheng, Q., Fan, W., Yang, Q., Xi, W., Chen, Z.: Effective and efficient dimensionality reduction for large-scale and streaming data preprocessing. IEEE Trans. Knowl. Data Eng. **18**(3), 320–333 (2006)
9. Valarmathie, P., Srinath, M., Dinakaran, K.: An increased performance of clustering high dimensional data through dimensionality reduction technique. J. Theor. Appl. Inf. Technol. **13**, 271–273 (2009)
10. Lee, C.-P., Leu, Y.: A novel hybrid feature selection method for microarray data analysis. Appl. Soft Comput. **11**, 208–213 (2011)
11. Mitchell, T.M.: Machine Learning. McGraw-Hill (1997)
12. Golub, T.R.: Molecular classification of cancer: class discovery and class prediction by gene expression monitoring. Science **286**, 531–537 (1999)
13. Notterman, D.A., Alon, U., Sierk, A.J., Levine, A.J.: Transcriptional gene expression profiles of colorectal adenoma, adenocarcinoma, and normal tissue examined by oligonucleotide arrays. Cancer Res. **61**(7), 3124–3130 (2001)

Trends in Citation Analysis

G. Parthasarathy and D.C. Tomar

Abstract Citation analysis is the process of examining the frequency, patterns of citations referred in articles, journals, and books. Citation refers a reference to a published or unpublished source of information. Though continuous research works endure for many years, different types of analysis have been conducted on citations over a period of time. Few important parameters rule the citation analysis research in a focused direction during all the time. These are the core requirements such as type of analysis technique and what kind of output the technique will give for various available citation databases. This paper surveys the important progresses of citation analysis and the important research works that facilitated this progress.

Keywords Citation analysis · Citation databases · Citation ranking · Citation index

1 Introduction

Citation analysis is a research area within the area of Web mining. It is the process of examining a citation which has been cited in many articles, journals, and books. Since the journals and articles are published in online, the research area of citation analysis emerged. An analysis refers to the combination of both the citation data such as the author, journal name, and institution and the technique using which the citation is analyzed. The goal of citation analysis depends on its application areas.

G. Parthasarathy (✉)
Department of CSE, Sathyabama University, Chennai, India
e-mail: amburgps@gmail.com

D.C. Tomar
Department of IT, Jerusalem College of Engineering, Chennai, India
e-mail: dctomar@gmail.com

L.C. Jain et al. (eds.), *Intelligent Computing, Communication and Devices*,
Advances in Intelligent Systems and Computing 308,
DOI 10.1007/978-81-322-2012-1_88

The research papers are having the highest number of citation counts which are considered as an important ingredient for performing citation analysis. Most of the earlier citation analysis techniques have applied the page rank algorithm for ranking the citations. The scope of citation analysis might differ from basic level of determining the referred citations to complex problem of determining the impact factor value of research journals. The citation analysis is an intermediary step in most of the applications that deal with determining the popularity of the journals, authors, and research papers. In recent years, citation analysis is mainly used collaboratively for finding out the reputation of author, journal, and institutions.

2 History of Citation Analysis

Advancements in citation analysis and its application areas such as citation classification, citation clustering, citation database analysis, and citation ranking have always be hand in hand. Original attempts on citation analysis involved citation ranking, largely based on frequencies of citations referred by various authors in different research journals and papers.

The initial experiments on citation analysis started with citation indexing using the measures of citation count from different publications for an article, author, or journal. The continuous development of Web technologies has made the information to spread among various places. But most of the information is not accessible to a different group of users. The information is available in the form of digital libraries, and the ways for accessing that information efficiently are not satisfactory. In the context of citations, if citations would become much easily accessible, indexing the citations automatically would evolve as a big analysis technique for evaluating the journals and publications of various authors. Online repositories incorporating indexes of citations help maintain scientific documents, and it can improve the efficiency of information access [1]. Most of the research articles, journals, and theses are published online from where other researchers can refer and incorporate the ideas or works discussed or experimented. The major problem for researchers is how to efficiently access those research papers. The traditional tree representation of citation indexes and query mechanism lead to complexities and problems in data access. Retrieving research papers topic-wise, identifying new research areas, retrieving non-semantic relative papers, and pinpointing a paper specifically in an area are all becoming a non-trivial problem. This is addressed by mapping the citation retrieval task onto a partitioning process. Every citation in the online library is mapped on graph through the reference links of papers. The citations are not evenly connected in the citation graph. Most of the times, the citation graph becomes highly connected with its links. Various subgraphs denote many topics, and the graph partition can disclose the topics at a very grain level. Every connection in the graph can help to find topics, related topic contents, and main citations. By doing all these tasks with automation, the search process is considerably reduced, but at the same time, the results are very accurate [2].

When the number of citations of a paper is high in number, then it is considered as a big indicator for paper quality, research quality with novel ideas, and journal quality. The quality of a researcher is measured by calculating the citations that have been received over a period of time. The number of citations or the comparative measurement of citation measures used this measurement for evaluating research scholars and the research journals. Various measurements are used for calculating the reputation of journals and authors. Mainly, those measurements are based on the citation counts. These are the key starting tasks for ISI Web of Knowledge, Citeseer citation index, and Google Scholar kind of citation analysis tools. The current citation mining techniques are not satisfied enough to discover all relevant citations. So the citation analysis technique must increase the efficiency of evaluating journal or author. Since we do not have any other better citation analysis technique, other than the techniques based on citation count, the current measurements of citations are considered as the best citation analysis [3].

Citation indices are not only the ultimate gateway source for researchers, but also the fundamental way for measuring the performance and the potential ideas present in the research. It is to be understood that the process of extracting citations and organizing those citation indices is ultimately important. The already existing citation extraction techniques are not so good in their accuracy. Many different techniques have been used for extracting the citations with high accuracy. The citations are extracted, the references are parsed, and the performance has been evaluated on various citation corpuses [4]. The Internet, citation networks, and scientific collaboration networks are nowadays in the center of attention. The study of the compactness of a general network, how the metric can be used in citation analysis, and the study of collaboration networks have become highly important one [5]. Citation analysis is mainly based on citation indexes from ISI citation databases. Currently, Scopus and Google Scholar are used as free citation tool. So the types of measurements have to be improved, and techniques have to be unleashed by comparing various tools. There are various measures to compute the similarity between rankings produced by various publications by sorting the citation count in the decreasing order using different citation tools. The experimented results have shown the great resemblances between rankings of Web of Science and Scopus databases [6]. One of the research studies on citation analysis has used the topics of 39 information papers. The statistical techniques between these citation links have allowed the discovery of most closed link among documents [7].

3 Components of Citation Analysis Techniques

Most of the citation analysis approaches have the similar components.

(1) Seed journals, (2) citation databases, (3) citation graph representation, (4) citation count computation, and (5) citation ranking.

3.1 Seed Journals

The selection of journals depends on the type of application for which the citation has to be analyzed. The citation analysis is mainly performed for finding out the popularity or the reputation of journals, articles, or authors. For example, one of the research works has picked out journal list from a recent perceptual study because it had included "pure" marketing journals. The work aimed to evaluate marketing journals. In many other analyses, conference publications have been taken as input seeds.

3.2 Citation Databases

The citation database is a vital component that highly influences the citation data for counting the citations based on its indices. Mainly, there are three citation tools commonly used. The first one is Web of Science which is the online version of the citation index available in ISI Web of Science. The second one is Scopus that provides citation databases from 1996 itself [8]. The third one is Google Scholar which is having number citations for different journals. Google Scholar is considered as the important citation analysis tools for analyzing the citation data. Ultimately, these citation analysis tools have taken into consideration highly cited papers of most popular researchers. The papers with more than 20 citations in the citation databases were retrieved. There are mainly two reasons for choosing highly cited data. One of the reasons is that retrieving the highly cited items is comparatively easy in a less span of time. And the second reason is that we need to find the publications that obtained the same number of citation counts in the citation databases.

3.3 Citation Graph Representation

The citation graph representation consists of two types of representation of citations as a graph: the text content-based graph and the citation-based graph. The data related to citation can be used for constructing a graph. Each journal can be considered as vertex, and the similar content between the journals can be considered as the edge. Even some times, the textual attributes between the journals can be used for modeling the citation graph. Many different graphs are being constructed from various citation data sources, and graphs are combined using coupling of subgraphs [9]. Table 1 summarizes the partial listing of citation graph representation algorithms.

Table 1 List of citation graph representation algorithms

Ref. No.	Algorithm	Reported performance	Comments on results
[12]	Modularity-based Louvain method	Integrate citation relation and textual attribute from a graph view. Weighting schemes AMIW to combine multiple graphs for partition	Clustering based on citation relation and textual relation
[13]	Scientific theme detection algorithm	Detects new topic timely with only a subset of data	Detection of new topic in a citation graph
[14]	A graph-based approach to author disambiguation on large-scale citation networks	Achieving precision of 0.997 and recall of 0.818 over a test group of eight surname clusters	Resolves name ambiguity in a citation graph
[15]	Detects temporal abnormalities in the number of citations and exploiting visibility of article	Relevance of the proposed ways and highlights their differences on real data extracted from Web	Detection of new topic in a citation graph
[5]	Merging/re-ranking method using characteristics of co-citation graph	Improvement against the basic method as well as some well-known meta-search engines	Ranking of citations based on characteristics of citations
[16]	Statistical hypothesis testing method	Generation of a directed graph from a matrix of non-symmetric interactions among elements of a system	Generation of graph from citations

3.4 Citation Count Computation

The productivity and impact of the research can be assessed using the most basic metric known as citation count. Most of the current ranking methods do employ the use of citation count for ranking the citation to reveal the popularity of the journals or authors. The citation count is the basic derivative measure for most of the recent measures. Based on the citation count, there are various counting methods: whole counting (university of the research paper), straight counting using first author collaborating with university, straight counting using corresponding author only, and fractional counting [10]. Table 2 summarizes the listing of citation count algorithms.

3.5 Citation Ranking

There are various reasons for effectively ranking the journals. First, the research scholars prefer to publish their works in the highly ranking journals. It is a belief in academic that papers that are published in popular journals are having good

Table 2 List of citation count algorithms

Ref. No.	Reported performance	Comments on results
[11]	Highly cited articles are found significantly more often in higher positions than articles that are cited less often	Reverse-engineered Google Scholar to rank citations
[4]	Gives an understanding of structures of a collection of documents that are related to each other by links	Understanding the structure of interlinked citation collections
[6]	Identifies the ingredients which affect the citation count	Helps to detect citation count
[2]	The use of citation-based indicators has the advantage of providing large-scale evaluations at relatively low cost	Citation count as an score indicator for large-scale citation evaluation

Table 3 List of citation ranking algorithms

Ref. No.	Reported performance	Comments on results
[17]	MeSH (MEDLINE Supervised HITS) outperforms text words in ranking citations	Ranking of citations using HITS and Text words
[12]	Reduce the bias against the recent papers which need less time for being studied and consequently cited by the researchers as compared to the older papers	Reduces bias of research papers based on ranking the citation using PageRank
[18]	A solution to address recent bias in the analysis of dynamic complex networks to improve the predictive performance	Analysis of citation networks to avoid recent bias
[19]	When two articles have an equal number of (direct) citations, the one that has triggered more research activity is assigned a higher impact factor rating and ranked to be better	Ranked the papers with similar citation counts
[20]	Exploit the potential power of bibliographical citations for information retrieval in digital libraries	Retrieval of books/journals using link-based citation ranking
[21]	Merit and shortcomings and the scope of application that the various algorithms are used to rank journal	Comparative analysis of citation ranking using different ranking methods

research ideas, so those papers are well cited. Second, researchers intend to know about popular theories and concepts with novel approaches and algorithms. Third, most of the funding organizations look at the top journal list while evaluating the applications to be granted. Table 3 summarizes the listing of citation ranking algorithms [11].

4 Complexity in Citation Analysis Techniques

While the performance depends largely on the components of the citation analysis techniques used, some features cause the performance to be considered as less optimal. Although citation analysis has grown rapidly in the Web, the citation analysis method is still an important factor for finding out the reputation of journal papers or authors. Some challenges encountered in the techniques are as follows:

4.1 Techniques to Extraction Citations

Citation databases and representation of citations as citation graph are important for the effectiveness of citation analysis. Recent developments have driven the availability of better citation databases such as Scopus, Web of Science, and Google Scholar. Though these citation databases are available, they are only suitable for checking the referred journals or articles. In addition, citations that are referred in many different research papers or articles have to be extracted and invoked to check its impact level.

4.2 Determination of Standard Technique

There is not exactly only one method for performing citation analysis to find impact factor or reputation of journal or author. Most of the citation databases have to be incorporated to distinguish the varying impact level of a particular journal or author. Then only, it would become a standard for measuring the impact factor by performing citation analysis.

4.3 Domain Dependence

Unlike other research areas, citation analysis is highly dependent on the area of journal or author. Each area is having different usage, and even authors might have published or referred journals in different areas. The problem arises, when we want to find out the impact level of particular paper which has been referred in various papers by different authors. Thus, it needs to be interlinked, and then, the impact level has to be measured.

5 Conclusion

The survey of citation analysis techniques and trends shows that

- There is no single algorithm or technique that is capable of analyzing the citations in all the aspects. As a best solution, this may be addressed by incorporating all the citation databases to perform citation analysis which may be citation count, citation ranking, or clustering of citations for finding out the reputation of journals.
- Almost all the citation analysis techniques such as citation classification, citation graph representation, citation clustering, and citation ranking are just focused to count the citations for analysis purposes. But it is important to perform citation analysis for finding out the reason that is on what basis the citations are highly referred or for what purpose the particular citation has been referenced many times.
- Citation ranking algorithms only work on large volume of citation databases such as Web version of ISI citation indices, Scopus database, and Google Scholar article citations. High-accuracy ranking methods to rank citations with a limited number of citations are not available.
- In spite of advances in citation count, citation databases and citation analysis techniques for all parts of a citation are yet to mature.

Based on the trends, it is predicted that there are citation analysis researches that are likely to progress in two directions. One main direction is the evolution of citation databases with citation analysis to all parts of citations and better citation graph representation for interlinking citations. Another direction is likely to be the area of improving citation ranking methods to rank citations from sources such as Web of Science, Scopus, and Google Scholar.

References

1. Ding, C., Chi, C.-H.: Citation retrieval in digital libraries. In: IEEE ICSMC (1999)
2. McAllister, P.R., Narin, F.: Programmatic evaluation and comparison based on standardized citation scores. IEEE Trans. Eng. Manage. **30**, 205–211 (1983)
3. Martino, J.P.: Citation indexing for research and development management. IEEE Trans. Eng. Manage. **18**, 146–151 (2013)
4. Noel, S., Chu, C.H.: Visualization of document co-citation counts. In: IEEE ICIV (2002)
5. Keyhanipour, A.H.: User-based meta-search with the co-citation graph. In: ICADIWT (2008)
6. Yi, H.S.: Ingredients for high citation index. In: IEEE ICEE, pp. 250–255 (2009)
7. Yang, C., Wu, S.-H., Lee, J.: A study of collaborative product commerce by co-citation analysis and social network analysis. In: IEEE ICIEEM, vol. 24, pp. 209–213 (2007)
8. Afzal, M.T., Balke, W.-T.: Improving citation mining. First ICNDT, pp. 116–121 (2009)
9. Lin, C.S., Huang, M.H., Chen, D.Z.: The influences of counting methods on university rankings based on paper count and citation count. J. Inf. 611–621 (2013)
10. Serenko, A.: The development of an AI journal ranking based on the revealed preference approach. J. Inf. **4**, 447–459 (2010)

11. Beel, J., Gipp, B.: Google Scholar's ranking algorithm: the impact of citation counts (an empirical study). In: IEEE ICRCIS, pp. 439–446 (2009)
12. Singh, A.P., Shubhankar, K., Pudi, V.: An efficient algorithm for ranking research papers based on citation network. In: IEEE ICDMO, pp. 88–95 (2011)
13. Moussa, S., Touzani, M.: Ranking marketing journals using the Google Scholar-based hg-index. J. Inf. **4**, 107–117 (2010)
14. Duncan, M.: A citation graph approaches to name disambiguation. In: ACM/IEEE (2006)
15. Egghe, L., Rousseau, R.: BRS-compactness in networks: theoretical considerations related to cohesion in citation graphs, collaboration networks and the internet. ScienceDirect Math Comput Model **37**, 879–899 (2003)
16. Miyamoto, S., Nakayama, K.: A directed graph representation based on a statistical hypothesis testing and application to citation and association structures, vol. –14. IEEE (1984)
17. Liu, Y., Yongjing, L.: Supervised HITS algorithm for MEDLINE citation ranking. In: IEEE ICBB, pp. 1323–1327 (2007)
18. Ghosh, R.: Time-aware ranking in dynamic citation networks. In: IEEE ICDMW (2011)
19. Dervos, D.A., Kalkanis, T.: A cascading citations impact factor framework for the automatic ranking of research publications. In: IEEE ICIDAACS: T&A, pp. 668–673 (2005)
20. Larsen, B.: Using citations for ranking in digital libraries. In: ACM, pp. 370 (2006)
21. Cheng, S., YunTao, P., JunPeng, Y.: PageRank, HITS and impact factor for journal ranking. In: JCSIE (2009)

Retraction Note to: Development of Magnetic Control System for Electric Wheel Chair Using Tongue

G. Hari Krishnan, R.J. Hemalatha, G. Umashankar, Nilofer Ahmed and Soumya Ranjan Nayak

Retraction Note to: 'Development of Magnetic Control System for Electric Wheel Chair Using Tongue' in: L.C. Jain et al. (eds.), *Intelligent Computing, Communication and Devices*, Advances in Intelligent Systems and Computing 308, DOI 10.1007/978-81-322-2012-1_68

The publisher regrets to announce that the following chapter entitled "Development of Magnetic Control System for Electric Wheel Chair Using Tongue" by G. Hari Krishnan, R.J. Hemalatha, G. Umashankar, Nilofer Ahmed, and Soumya Ranjan Nayak, pp. 635–641, published in *Intelligent Computing, Communication and Devices* has been retracted. This chapter contains reused and uncited material which has been previously published by a different group of authors.

The online version of the original chapter can be found under
DOI 10.1007/978-81-322-2012-1_68

G.H. Krishnan (✉) · R.J. Hemalatha · G. Umashankar · N. Ahmed · S.R. Nayak
Department of Biomedical Engineering, Sathyabama University, Chennai, India
e-mail: haris_eee@yahoo.com

N. Ahmed
e-mail: nilofer.ahmed2@gmail.com

L.C. Jain et al. (eds.), *Intelligent Computing, Communication and Devices*,
Advances in Intelligent Systems and Computing 308,
DOI 10.1007/978-81-322-2012-1_89

Author Index

L.C. Jain et al. (eds.), *Intelligent Computing, Communication and Devices*,
Advances in Intelligent Systems and Computing 308,
DOI 10.1007/978-81-322-2012-1

Printed in the United States
By Bookmasters